Springer Monographs in Mathematics

Springer
Berlin
Heidelberg
New York
Barcelona
Budapest
Hong Kong
London
Milan
Paris
Santa Clara
Singapore
Tokyo

J. Elstrodt F. Grunewald J. Mennicke

Groups Acting on Hyperbolic Space

Harmonic Analysis and Number Theory

Jürgen Elstrodt
Universität Münster
Mathematisches Institut
Einsteinstraße 62
D-48149 Münster

Fritz Grunewald
Universität Düsseldorf
Mathematisches Institut
Universitätsstraße 1
D-40225 Düsseldorf

Jens Mennicke
Universität Bielefeld
Fakultät für Mathematik
Universitätsstraße 25
D-33615 Bielefeld

CIP data applied for

Die Deutsche Bibliothek – CIP-Einheitsaufnahme

Elstrodt, Jürgen:
Groups acting on hyperbolic space: harmonic analysis and number theory / J. Elstrodt; F. Grunewald; J. Mennicke. - Berlin; Heidelberg; New York; Barcelona; Budapest; Hong Kong; London; Milan; Paris; Santa Clara; Singapore; Tokyo: Springer, 1997
(Springer monographs in mathematics)
ISBN 3-540-62745-6

Mathematics Subject Classification (1991): 11F72, 11F55, 11E39, 11E45, 11E45, 11M26, 20F55, 20F55, 20H05, 58C40

ISBN 3-540-62745-6 Springer-Verlag Berlin Heidelberg New York

Typesetting by the authors using a Springer TEX macro package
SPIN 10467814 41/3143-5 4 3 2 1 0 – Printed on acid-free paper

Dedicated to the Memory of

Hans Maaß

1911–1992

Preface

This book is concerned with discontinuous groups of motions of the unique connected and simply connected Riemannian 3-manifold of constant curvature -1, which is traditionally called hyperbolic 3-space. This space is the 3-dimensional instance of an analogous Riemannian manifold which exists uniquely in every dimension $n \geq 2$. The hyperbolic spaces appeared first in the work of Lobachevski in the first half of the 19th century. Very early in the last century the group of isometries of these spaces was studied by Steiner, when he looked at the group generated by the inversions in spheres. The geometries underlying the hyperbolic spaces were of fundamental importance since Lobachevski, Bolyai and Gauß had observed that they do not satisfy the axiom of parallels.

Already in the classical works several concrete coordinate models of hyperbolic 3-space have appeared. They make explicit computations possible and also give identifications of the full group of motions or isometries with well-known matrix groups. One such model, due to H. Poincaré, is the upper half-space $\mathbb{H}$ in $\mathbb{R}^3$. The group of isometries is then identified with an extension of index 2 of the group $\mathbf{PSL}(2, \mathbb{C})$. Another model, due to F. Klein, is a certain subset of 3-dimensional projective space $\mathbb{P}^3\mathbb{R}$, the group of isometries being identified with $\mathbf{PO}_4(1,3)$. Thirdly there is the hyperboloid model which identifies the group of isometries with a certain subgroup of index 2 in the 4-dimensional Lorentz group $\mathbf{O}(1,3)$. We treat these models in an introductory Chapter 1. The emphasis will lie on the upper half-space model $\mathbb{H}$.

The subject of discontinuous groups of isometries of hyperbolic 3-space started with works of H. Poincaré and F. Klein who searched for higher dimensional analogues of Fuchsian groups, these being the discontinuous groups of orientation preserving motions of hyperbolic 2-space. Although many important general facts on groups of isometries of hyperbolic 2-space were found in the last century, the 3-dimensional case was mainly studied through examples. Over the last twenty years the subject has risen to particular importance due to the discoveries of Thurston. It is nowadays expected that the 3-manifolds carrying a complete Riemannian metric with constant sectional curvature equal to -1 are the most important building blocks for all 3-manifolds. To give such a 3-manifold is of course the same thing as to give a subgroup Γ in the isometry group which acts discontinuously and fixed point freely on hyperbolic space. Chapter 2 collects some of the now known general facts about discontinuous groups Γ of isometries of hyperbolic 3-space. Of particular importance are those Γ which are not too small. This can mean that the quotient space $\Gamma\backslash\mathbb{H}$ is compact or more generally that it is of finite hyperbolic volume. The hyperbolic measure is obtained from the volume form dv given by the Riemannian structure on $\mathbb{H}$. We call these groups Γ cocompact and cofinite respectively.

Considering discontinuous groups Γ of isometries of the upper half-space model $\mathbb{H}$ it is natural to study functions on $\mathbb{H}$ which are invariant under Γ. Here one thinks of possibly generalizing parts of the theory of modular functions and forms from the 2-dimensional case. The theory of holomorphic modular forms has no direct counterpart since $\mathbb{H}$ carries unlike the upper half-plane no complex structure. So we are left with generalizations of the Maaß-Selberg theory of non-holomorphic modular functions. We start off with the more elementary parts of this theory in Chapter 3.

A basic ingredient in the theory of Maaß and Selberg is the Laplace-Beltrami operator Δ attached to the Riemannian structure on $\mathbb{H}$. This is a partial differential operator of degree two which is left invariant by all isometries of $\mathbb{H}$. Given a discontinuous group Γ of orientation preserving isometries of hyperbolic space Δ defines an operator (with suitable domain) on the Hilbert space $L^2(\Gamma\backslash\mathbb{H}, dv) =: L^2(\Gamma\backslash\mathbb{H})$ of Γ-invariant functions on $\mathbb{H}$ which are square-integrable over a measurable fundamental domain of Γ. It turns out that $-\Delta$ is essentially self-adjoint and positive on the subspace $\mathcal{D} \subset L^2(\Gamma\backslash\mathbb{H})$ consisting of all C^2-functions $f \in L^2(\Gamma\backslash\mathbb{H})$ such that $\Delta f \in L^2(\Gamma\backslash\mathbb{H})$. The operator $-\Delta$ has a unique self-adjoint extension denoted by $-\tilde{\Delta}$. The spectra of these self-adjoint operators are supposed to carry information on the discontinuous groups Γ. In Chapter 4 we discuss the general spectral theory of $-\Delta$ acting on its natural domain in the Hilbert space $L^2(\Gamma\backslash\mathbb{H})$.

For the understanding of the spectra of the above mentioned operators it is of importance to have some supply of interesting Γ-invariant functions on $\mathbb{H}$. One of the basic constructions of examples is an averaging process which leads to the Poincaré series. The Eisenstein series introduced in Chapter 3 are particularly important special cases of these.

Chapters 5 and 6 discuss the finer points of the spectral theory of the operator $-\tilde{\Delta}$ on the Hilbert space $L^2(\Gamma\backslash\mathbb{H})$ in case Γ is a discontinuous group of orientation preserving isometries which is cocompact (Chapter 5) or cofinite but not cocompact (Chapter 6). In both cases we aim at the development of the Selberg trace formula and some of its applications.

In the cocompact case we prove that the spectrum of $-\Delta$ on $L^2(\Gamma\backslash\mathbb{H})$ is purely discrete. In the way the theory is presented here this comes from an explicit construction of an integral kernel function for the resolvent of $-\Delta - (1 - s^2)$ ($s \in \mathbb{C}$ suitable) which is shown to give a Hilbert-Schmidt integral operator. If the resolvent kernel is applied twice, one gets a kernel of trace class. The evaluation of the trace of this kernel leads to Selberg's trace formula. It enables us to study the fascinating properties of the Selberg zeta function Z_Γ attached to Γ. This function is defined by some kind of Euler product. The trace formula yields a continuation of the zeta function as an entire function of s into the whole s-plane and also proves that it satisfies a simple functional equation. The zeros of Z_Γ are precisely the numbers $\pm s_n$ such that $\lambda_n = 1 - s_n^2$ is an eigenvalue of $-\tilde{\Delta}$ in $L^2(\Gamma\backslash\mathbb{H})$. This means that the analogue of the Riemann hypothesis is true for the Selberg zeta function except for the finitely many zeros of Z_Γ in $[-1, 1]$. As a first application of the Selberg zeta function we prove Weyl's asymptotic law on the distribution of the eigenvalues of $-\Delta$. We also produce the standard bound for the error term in this asymptotic law. A second application is the so-called prime geodesic theorem for the asymptotic distribution of the norms of the primitive hyperbolic or loxodromic elements of Γ.

In case Γ is cofinite but no longer cocompact the spectrum of the self-adjoint operator $-\tilde{\Delta}$ on $L^2(\Gamma\backslash\mathbb{H})$ is absolutely continuous but no longer purely discrete. In fact the Hilbert space $L^2(\Gamma\backslash\mathbb{H})$ decomposes naturally as an orthogonal sum $L^2(\Gamma\backslash\mathbb{H}) = L^2_{\text{disc}}(\Gamma\backslash\mathbb{H}) \oplus L^2_{\text{cont}}(\Gamma\backslash\mathbb{H})$ respected by the Laplace operator. The restriction of $-\Delta$ to $L^2_{\text{cont}}(\Gamma\backslash\mathbb{H})$ has purely continuous spectrum equal to the interval $[1, \infty[$ with multiplicity equal to the number of classes of cusps of Γ. This is proved by giving an explicit system of eigen-packets, or what is almost the same, a system of continuous spectral families for $-\Delta$ on this space. These are constructed with the help of the Eisenstein series $E(P, s)$, depending on the parameters $P \in \mathbb{H}$ and $s \in \mathbb{C}$. They are defined (in Chapter 3) as a generalized Dirichlet series in s converging in some right half-plane of $\mathbb{C}$. For the application here they have to be meromorphically continued to all of $\mathbb{C}$. The restriction of $-\Delta$ to $L^2_{\text{disc}}(\Gamma\backslash\mathbb{H})$ has purely discrete spectrum, the set of eigenvalues being discrete in $\mathbb{R}_{\geq 0}$ and each one occurring with finite multiplicity. After having established an explicit spectral expansion, analogous to the Fourier expansion, for elements of $L^2(\Gamma\backslash\mathbb{H})$ we proceed to derive Selberg's trace formula in a rather explicit form. One of the principal applications of the trace formula is to the counting of eigenvalues of $-\Delta$ below a bound T. Weyl's asymptotic law cannot be established unrestrictedly in the cofinite case due to an additional term which prevents us from getting any result on the number of eigenvalues between 0 and T. This term comes from the continuous spectrum, it is the integral of the logarithmic derivative of the determinant of the scattering matrix attached to $-\Delta$ (which is a meromorphic function on $\mathbb{C}$) over the imaginary axis from 0 to T. So in general the nature of the discrete spectrum of $-\Delta$ remains rather mysterious.

In the rest of the book (from Chapter 7 onwards) we discuss concrete examples of discontinuous groups of isometries of $\mathbb{H}$ and try to make the theory as explicit as possible. Thinking of the 2-dimensional hyperbolic space, its most important discontinuous group of motions seems to be $\mathbf{PSL}(2,\mathbb{Z})$. The theory of modular functions and forms of this group is intimately related to the arithmetic of the rational number field. The natural analogues of $\mathbf{PSL}(2,\mathbb{Z})$ are the groups $\mathbf{PSL}(2,\mathcal{O})$ where $\mathcal{O}$ is the ring of integers of an imaginary quadratic number field K. These groups are introduced in Chapter 7. They give rise to discontinuous cofinite but not cocompact groups of motions of $\mathbb{H}$. Conjecturally the modular functions and forms for $\mathbf{PSL}(2,\mathcal{O})$ should play the same role for the arithmetic of K as is expected for the pair $\mathbf{PSL}(2,\mathbb{Z})$ and $\mathbb{Q}$. For example the Artin L-functions of the 2-dimensional irreducible complex representations of the absolute Galois group of K should occur as Mellin transforms of certain eigenfunctions with eigenvalue 1 for $-\Delta$ acting on $L^2(\mathbf{PSL}(2,\mathcal{O})\backslash\mathbb{H})$. Examples are contained in Section 6 of Chapter 7. Another conjectured correspondence relates the elliptic curves defined over K and the Hecke eigenvectors in $\Gamma^{\mathrm{ab}} \otimes_{\mathbb{Z}} \mathbb{Q}$. Here Γ runs through all congruence subgroups of $\mathbf{PSL}(2,\mathcal{O})$ and Γ^{ab} are the corresponding commutator quotient groups. This is dicussed in Section 5 of Chapter 7.

In Chapter 8 we discuss the Eisenstein series attached to the groups $\mathbf{PSL}(2,\mathcal{O})$. This will result in number theoretic applications such as a Kronecker limit formula and non-vanishing results for certain Dirichlet L-functions. We also identify the determinant of the scattering matrix of the Laplace operator with the zeta function of the Hilbert class field of K. This makes it possible to deduce Weyl's asymptotic law for the eigenvalues of the Laplace operator from well known properties of the zeta function.

Chapter 9 shows that the groups $\mathbf{PSL}(2,\mathcal{O})$ play the same role in the theory of integral binary hermitian forms as $\mathbf{PSL}(2,\mathbb{Z})$ does for integral binary quadratic forms. Among other things we prove Siegel's mass-formula for binary hermitian forms.

The final chapter contains some more constructions of cofinite groups of isometries on $\mathbb{H}$. To describe them let L be a number field, $\sigma_1,\dots,\sigma_{r_1}$ its embeddings into the reals, τ_1, $\overline{\tau_1},\dots,\tau_{r_2}$, $\overline{\tau_{r_2}}$ its pairs of conjugate complex embeddings and $\mathcal{O}_L$ its ring of integers. Suppose $r_2 = 1$ and $\mathcal{H}$ is a quaternion algebra over L which is a skew field when extended to the real numbers through $\sigma_1,\dots,\sigma_{r_1}$. The norm 1 group $\mathbf{J}^1$ of any $\mathcal{O}_L$-order $\mathbf{J} \subset \mathcal{H}$ then gives rise via the complex embedding to a subgroup of $\mathbf{PSL}(2,\mathbb{C})$ which acts discontinuously on $\mathbb{H}$. This subgroup is always cofinite and cocompact if $\mathcal{H}$ is a skew field. For the second construction the Kleinian model is used. We start with a number field L having $r_2 = 0$ and a quadratic form Q in 4 variables with coefficients in L. We assume that Q has signature $(1,3)$ when extended to the reals through σ_1 and is definite when extended through $\sigma_2,\dots,\sigma_{r_1}$. The groups $\mathbf{PO}(\mathcal{O}_L, Q)$ act discontinuously on the Kleinian model of hyperbolic space. They are always cofinite and cocompact if and only if Q is L-anisotropic. We also discuss the relationship between these two constructions. The discontinuos groups described so far are (roughly speaking) the arithmetic groups of motions of hyperbolic space.

Another construction, which also gives non-arithmetic groups, is to start with a polyhedron $\mathcal{P}$ in hyperbolic space and consider the group $W_{\mathcal{P}}$ generated by the hyperbolic reflections in the faces of $\mathcal{P}$. The Poincaré combination theorem gives conditions (on the angles occurring in $\mathcal{P}$) when $W_{\mathcal{P}}$ will act discontinuously on hyperbolic space. There are for example 32 hyperbolic tetrahedra which satisfy these conditions. The corresponding groups are discussed in detail.

In the study of discontinuous isometry groups on $\mathbb{H}$ several highly developed mathematical theories meet such as operator theory on Hilbert spaces, automorphic theory, number theory and 3-manifold theory. We have tried to give an account of parts of the subject which are important in todays mathematics. Our treatment is by no means complete. A serious omission is for example Thurston's Dehn-surgery theory of hyperbolic 3-manifolds. The subject being so vast, we were also sometimes forced to refer the reader to the literature for proofs. We still hope that this book will be of help as an introduction to this interesting area of mathematics.

The authors express their sincere thanks to many colleages, especially to Hans Maaß and Walter Roelcke who helped over the years to complete this long manuscript. We also thank Stefan Kühnlein and Leonid Parnovski for proof-reading and many helpful discussions. Thanks go to Igor Lysionok who prepared some pictures for us. J.W.V. Channing helped to reduce the grammatical errors. Special thanks go to our wives Bärbel, Barbara and Gerda for their understanding and unselfish support over the years of preparation of this manuscript and others. Last but not least we express our sincere gratitude to Dr. J. Heinze and the editorial staff of Springer-Verlag for their continuous support and unending patience.

In recognition of the seminal contributions of Hans Maaß and bearing in mind our personal indeptedness to him we dedicate the present work to this outstanding mathematician.

Münster, Düsseldorf, Bielefeld, April 1997

Table of Contents

Chapter 1. Three-Dimensional Hyperbolic Space

Three-dimensional hyperbolic space is the unique 3-dimensional connected and simply connected Riemannian manifold with constant sectional curvature equal to -1. This space has certain concrete models which all have certain advantages. We discuss here the four most classical ones.

For each $n > 1$ there are similar global coordinates for n-dimensional hyperbolic space. A coherent treatment of them can be found in Wolf (1977) or Elstrodt, Grunewald, Mennicke (1987b).

Historically the theory of hyperbolic spaces started with work of Lobachevski in the first half of the 19th century. The group of isometries appeared first as the group of Möbius transformations that is the group generated by those inversions in spheres which leave the unit ball invariant. We do not discuss this realization here, see Beardon (1977), (1983) for an account.

Throughout the beginning chapters we assume a certain familiarity of the reader with the basic facts of hyperbolic geometry. An excellent treatment can be found in Alekseevskij, Vinberg, Solodovnikov (1991) and also in Beardon (1983).

1.1 The Upper Half-Space Model

The upper half-space $\mathbb{H}$ in Euclidean three-space gives a convenient model of 3-dimensional hyperbolic space which in its properties closely resembles the well-known upper half-plane as a model of plane hyperbolic geometry. We use the following coordinates:

$$\begin{aligned} \mathbb{H} :&= \mathbb{C}\times]0,\infty[\\ &= \{\ (z,r)\ \mid\ z\in\mathbb{C},\quad r>0\ \} \\ &= \{\ (x,y,r)\ \mid\ x,y\in\mathbb{R},\quad r>0\ \}. \end{aligned} \tag{1.1}$$

To facilitate computations we shall often think of $\mathbb{H}$ as a subset of Hamilton's quaternions $\mathcal{H} := \mathcal{H}(-1,-1)$. As usual we write $1, i, j, k$ for the standard $\mathbb{R}$-basis of $\mathcal{H}$. The notation for points in $\mathbb{H}$ is:

$$P = (z,r) = (x,y,r) = z + rj, \tag{1.2}$$

where

$$z = x + iy, \qquad j = (0, 0, 1). \tag{1.3}$$

We equip $\mathbb{H}$ with the hyperbolic metric coming from the line element:

$$ds^2 = \frac{dx^2 + dy^2 + dr^2}{r^2}. \tag{1.4}$$

We denote the hyperbolic distance between two points $P, Q \in \mathbb{H}$ by $d(P, Q)$. An explicit formula for d will be given in Proposition 1.6.

The geodesics with respect to the hyperbolic metric which are sometimes called hyperbolic lines are half-circles or half-lines in $\mathbb{H}$ which are orthogonal to the boundary plane $\mathbb{C}$ in the Euclidean sense. This can be proved in essentially the same way as in the case of plane hyperbolic geometry (cf. Maaß (1964)). The hyperbolic planes (also called geodesic hyperplanes), that is the isometrically embedded copies of 2-dimensional hyperbolic space, are Euclidean hemispheres or half-planes which are perpendicular to the boundary $\mathbb{C}$ of $\mathbb{H}$. The hyperbolic angles between hyperbolic lines or planes coincide with the corresponding Euclidean angles.

In the sense of Riemannian geometry the hyperbolic metric gives rise to the hyperbolic volume measure v with corresponding volume element

$$dv = \frac{dx\, dy\, dr}{r^3}. \tag{1.5}$$

Moreover, a general Riemannian metric

$$ds^2 = \sum_{\mu,\nu} g_{\mu\nu} dx^\mu dx^\nu$$

has an associated Laplace-Beltrami operator

$$\Delta = \frac{1}{\sqrt{g}} \sum_{\mu,\nu} \frac{\partial}{\partial x^\mu} \sqrt{g}\, g^{\mu\nu} \frac{\partial}{\partial x^\nu},$$

where $g = \det(g_{\mu\nu})$ and $(g^{\mu\nu}) = (g_{\mu\nu})^{-1}$. So the hyperbolic Laplace-Beltrami operator associated with (1.4) is given by

$$\Delta = r^2 \left(\frac{\partial^2}{\partial x^2} + \frac{\partial^2}{\partial y^2} + \frac{\partial^2}{\partial r^2} \right) - r \frac{\partial}{\partial r}. \tag{1.6}$$

The group $\mathbf{PSL}(2, \mathbb{C})$ of complex 2×2-matrices with determinant one modulo its center $\{\pm I\}$ has a natural action on $\mathbb{H}$ which may be described in geometric terms as follows: An element $M \in \mathbf{PSL}(2, \mathbb{C})$ induces a biholomorphic map of the Riemannian sphere $\mathbb{P}^1\mathbb{C} = \mathbb{C} \cup \{\infty\}$. This map may be represented as a composition of at most four inversions in certain circles in $\mathbb{P}^1\mathbb{C}$. Here Euclidean lines are thought of as circles passing through the point at infinity. Now consider $\mathbb{P}^1\mathbb{C}$ as the boundary of $\mathbb{H}$ and construct

Euclidean hemispheres in $\mathbb{H}$ which intersect $\mathbb{P}^1\mathbb{C}$ orthogonally along the circles mentioned above. If a circle under consideration passes through ∞, the corresponding hemisphere is an Euclidean half-plane which is orthogonal to $\mathbb{C}$. Now take the product, in appropriate order, of the inversions with respect to the hemispheres. Then it turns out that the action of this product on $\mathbb{H}$ does not depend on the particular choice of the circles (cf. Poincaré (1916), Vol. II, p. 261 or Greenberg (1977), p. 433) and thus defines a natural action of M on $\mathbb{H}$.

The action of $\mathbf{PSL}(2,\mathbb{C})$ on $\mathbb{H}$ leads to an action of $\mathbf{SL}(2,\mathbb{C})$. For the whole of this book we shall use the notation $M = \begin{pmatrix} a & b \\ c & d \end{pmatrix}$ both for elements of $\mathbf{SL}(2,\mathbb{C})$ and for their classes in $\mathbf{PSL}(2,\mathbb{C})$.

The actions of $\mathbf{PSL}(2,\mathbb{C})$ on $\mathbb{H}$ and on its boundary $\mathbb{P}^1\mathbb{C}$ may be described by simple formulas. We represent an element of $\mathbb{P}^1\mathbb{C}$ by $[x,y]$, where $x,y \in \mathbb{C}$ with $(x,y) \neq (0,0)$. Then the action of the matrix $M = \begin{pmatrix} a & b \\ c & d \end{pmatrix} \in \mathbf{PSL}(2,\mathbb{C})$ on $\mathbb{P}^1\mathbb{C}$ is given by:

$$[x,y] \mapsto M[x,y] := [ax+by, cx+dy]. \tag{1.7}$$

If we represent points $P \in \mathbb{H}$ as quaternions whose fourth component equals zero (cf. (1.2)), then the action of M on $\mathbb{H}$ is given by

$$P \mapsto MP := M(P) := (aP+b)(cP+d)^{-1}, \tag{1.8}$$

where the inverse is taken in the skew field of quaternions. An obvious computation in $\mathcal{H}(-1,-1)$ shows that (1.8) in fact defines an action of $\mathbf{PSL}(2,\mathbb{C})$ on $\mathbb{H}$. More explicitly, (1.8) may be written in the form $M(z+rj) = z^* + r^*j$, with coordinates

$$z^* = \frac{(az+b)(\bar{c}\bar{z}+\bar{d}) + a\bar{c}r^2}{|cz+d|^2 + |c|^2r^2}, \tag{1.9}$$

$$r^* = \frac{r}{|cz+d|^2 + |c|^2r^2} = \frac{r}{\|cP+d\|^2}. \tag{1.10}$$

Here we have set $P = z + rj$, and $\|cP+d\|$ denotes the Euclidean norm of the vector $cP + d \in \mathbb{R}^4$. This can also be understood as the square root of the norm of $cP + d$ in $\mathcal{H}(-1,-1)$.

Another simple proof of the fact that (1.9), (1.10) actually define an action of $\mathbf{SL}(2,\mathbb{C})$ on $\mathbb{H}$ (i.e. that $(\mathbf{SL}(2,\mathbb{C}),\mathbb{H})$ is a topological transformation group) can be given in another model of $\mathbb{H}$ which will be discussed later in this chapter.

Proposition 1.1. *The stabilizer of* $j = (0,0,1) \in \mathbb{H}$ *with respect to the action of* $\mathbf{SL}(2,\mathbb{C})$ *on* $\mathbb{H}$ *is equal to* $\mathbf{SU}(2) = \{\, V \mid V \in \mathbf{U}(2),\ \det(V) = 1 \,\}$.

Proof. An element $M = \begin{pmatrix} a & b \\ c & d \end{pmatrix} \in \mathbf{SL}(2,\mathbb{C})$ belongs to the stabilizer of j if and only if $|c|^2 + |d|^2 = 1$ and $a\bar{c} + b\bar{d} = 0$. Since we have $ad - bc = 1$, these conditions are equivalent to $M = \begin{pmatrix} \bar{d} & -\bar{c} \\ c & d \end{pmatrix} \in \mathbf{SL}(2,\mathbb{C})$, i.e. $M \in \mathbf{SU}(2)$. □

It is convenient to note that $\mathbf{SL}(2,\mathbb{C})$ has a simple set of generators, which we will describe in the next Proposition.

Proposition 1.2. *The group* $\mathbf{SL}(2,\mathbb{C})$ *is generated by the elements*

$$\begin{pmatrix} 1 & a \\ 0 & 1 \end{pmatrix}, \quad \begin{pmatrix} 0 & -1 \\ 1 & 0 \end{pmatrix} \qquad (a \in \mathbb{C}). \tag{1.11}$$

These generators operate on $\mathbb{H}$ *as follows:*

$$\begin{pmatrix} 1 & a \\ 0 & 1 \end{pmatrix}(z,r) = (z+a,r), \qquad \begin{pmatrix} 0 & -1 \\ 1 & 0 \end{pmatrix}(z,r) = \left(\tfrac{-\bar{z}}{|z|^2+r^2}, \tfrac{r}{|z|^2+r^2} \right).$$

Proof. Suppose $M = \begin{pmatrix} a & b \\ c & d \end{pmatrix} \in \mathbf{SL}(2,\mathbb{C})$. If $c \neq 0$, we have

$$\begin{pmatrix} a & b \\ c & d \end{pmatrix} = \begin{pmatrix} 1 & ac^{-1} \\ 0 & 1 \end{pmatrix}\begin{pmatrix} 0 & -1 \\ 1 & 0 \end{pmatrix}\begin{pmatrix} c & 0 \\ 0 & c^{-1} \end{pmatrix}\begin{pmatrix} 1 & dc^{-1} \\ 0 & 1 \end{pmatrix},$$

and for $c = 0$ we obtain the factorization

$$\begin{pmatrix} a & b \\ 0 & a^{-1} \end{pmatrix} = \begin{pmatrix} a & 0 \\ 0 & a^{-1} \end{pmatrix}\begin{pmatrix} 1 & a^{-1}b \\ 0 & 1 \end{pmatrix}.$$

We thus conclude that the matrices (1.11) together with the matrices

$$D_\beta := \begin{pmatrix} \beta & 0 \\ 0 & \beta^{-1} \end{pmatrix} \qquad 0 \neq \beta \in \mathbb{C}$$

generate $\mathbf{SL}(2,\mathbb{C})$. But the elements D_β may be represented as products of the matrices (1.11), as is evident from the following formulas:

$$\begin{pmatrix} \beta & 0 \\ 0 & \beta^{-1} \end{pmatrix} = \begin{pmatrix} 1 & \beta^2-\beta \\ 0 & 1 \end{pmatrix}\begin{pmatrix} 1 & 0 \\ \beta^{-1} & 1 \end{pmatrix}\begin{pmatrix} 1 & 1-\beta \\ 0 & 1 \end{pmatrix}\begin{pmatrix} 1 & 0 \\ -1 & 1 \end{pmatrix},$$

$$\begin{pmatrix} 1 & 0 \\ \alpha & 1 \end{pmatrix} = \begin{pmatrix} 0 & -1 \\ 1 & 0 \end{pmatrix}\begin{pmatrix} 1 & -\alpha \\ 0 & 1 \end{pmatrix}\begin{pmatrix} 0 & -1 \\ 1 & 0 \end{pmatrix}\begin{pmatrix} -1 & 0 \\ 0 & -1 \end{pmatrix},$$

$$\begin{pmatrix} -1 & 0 \\ 0 & -1 \end{pmatrix} = \begin{pmatrix} 0 & -1 \\ 1 & 0 \end{pmatrix}\begin{pmatrix} 0 & -1 \\ 1 & 0 \end{pmatrix}.$$

This proves the Proposition. □

In our particular case the notion of invariance may be introduced ad hoc as follows: Let $G = (g_{\mu\nu})$ be our Riemannian metric on $\mathbb{H}$ and let $T : \mathbb{H} \to \mathbb{H}$ be a C^∞- diffeomorphism of $\mathbb{H}$ with associated Jacobian matrix

$$J = \begin{pmatrix} u_x & u_y & u_r \\ v_x & v_y & v_r \\ w_x & w_y & w_r \end{pmatrix},$$

where $T(P) = (u, v, w)$. Then T is called an isometry for G and G is called *T-invariant* iff

$$G(P) = J(P)^t G(T(P)) J(P)$$

for all $P \in \mathbb{H}$. If T is an isometry the arc-length and the volume belonging to the metric are T-invariant. We use the notation

$$\mathbf{Iso}(\mathbb{H}) \quad \text{and} \quad \mathbf{Iso}^+(\mathbb{H}) \tag{1.12}$$

for the groups of all isometries and all orientation preserving isometries of $\mathbb{H}$.

Proposition 1.3. *The hyperbolic metric (1.4) is* $\mathbf{PSL}(2,\mathbb{C})$*-invariant. Thus the hyperbolic distance and the hyperbolic volume are also* $\mathbf{PSL}(2,\mathbb{C})$*-invariant. Moreover, the determinant of the Jacobian matrix of any element of* $\mathbf{PSL}(2,\mathbb{C})$ *is positive. Hence* $\mathbf{PSL}(2,\mathbb{C})$ *is a group of orientation preserving motions for the hyperbolic geometry in* $\mathbb{H}$*. In fact we may make the identification* $\mathbf{PSL}(2,\mathbb{C}) = \mathbf{Iso}^+(\mathbb{H})$*. Moreover* $\mathbf{PSL}(2,\mathbb{C})$ *is of index 2 in the full group of isometries. A representative system of the cosets of* $\mathbf{PSL}(2,\mathbb{C})$ *in the group of all isometries is given be the identity and by an arbitrary reflection* Σ *in a hemisphere perpendicular to* $\mathbb{C}$*. The group* $\mathbf{Iso}(\mathbb{H})$ *is isomorphic to the semidirect product* $\mathbf{PSL}(2,\mathbb{C}) \rtimes \mathbb{Z}/2\mathbb{Z}$ *where the nontrivial element of* $\mathbb{Z}/2\mathbb{Z}$ *acts by complex conjugation on* $\mathbf{PSL}(2,\mathbb{C})$*.*

All of this is easily verified, see for example Beardon (1977). In the proof it is advantageous to use the generators (1.11) and of course also (1.9), (1.10). To see the last statement use the hyperbolic reflection Σ_0 where $\Sigma_0(z+rj) = \bar{z} + rj$ to obtain the splitting. Note also that $\mathbf{SL}(2,\mathbb{C})$ is arcwise connected. This is clear from the generators (1.11). Moreover, it is easy to show that the hyperbolic metric (1.4) is uniquely determined up to a positive constant factor by its $\mathbf{PSL}(2,\mathbb{C})$-invariance (cf. Siegel (1966b), Teil II, p. 31).

Proposition 1.4. *The group* $\mathbf{PSL}(2,\mathbb{C})$ *acts in the following sense doubly transitively on* $\mathbb{H}$*: For all* $P, P', Q, Q' \in \mathbb{H}$ *such that* $d(P,P') = d(Q,Q')$ *there exists an* $M \in \mathbf{PSL}(2,\mathbb{C})$ *such that* $M(P) = Q$, $M(P') = Q'$.

Proof. Let $P = z + rj, z \in \mathbb{C}, r > 0$. Then the element

$$T_1 := \begin{pmatrix} \frac{1}{\sqrt{r}} & 0 \\ 0 & \sqrt{r} \end{pmatrix} \begin{pmatrix} 1 & -z \\ 0 & 1 \end{pmatrix} \in \mathbf{SL}(2,\mathbb{C})$$

maps P onto j. Applying a suitable element T_2 of the stabilizer $\mathbf{SU}(2)$ of j in $\mathbf{SL}(2,\mathbb{C})$ to $T_1(P')$ we arrive at a transformation $T = T_2 T_1$ such that $T(P) = j$, $T(P') = tj$, with $t \geq 1$. Obviously

$$d(P,P') = d(T(P),T(P')) = d(j,tj) = \int_1^t \frac{dr}{r} = \log t,$$

i.e. $t = \exp(d(P,P'))$. Similarly, there exists an element $S \in \mathbf{SL}(2,\mathbb{C})$ such that $S(Q) = j$, $S(Q') = tj$ with the same t, since $d(P,P') = d(Q,Q')$. Thus $M := S^{-1}T$ has the required properties. □

Applying the same idea of proof as in Proposition 1.4 one may arrive at the following sharper conclusion: The group $\mathbf{PSL}(2,\mathbb{C})$ acts triply transitively on $\mathbb{H}$ in the following sense: For any $P_1, P_2, P_3, Q_1, Q_2, Q_3 \in \mathbb{H}$ such that $d(P_k, P_\ell) = d(Q_k, Q_\ell)$ for all $k, \ell \in \{1,2,3\}$ there exists an $M \in \mathbf{PSL}(2,\mathbb{C})$ such that $M(P_k) = Q_k$ for $k = 1,2,3$.

Definition 1.5. A map $f : \mathbb{H} \times \mathbb{H} \to \mathbb{C}$ is called a point-pair invariant if

$$f(MP, MQ) = f(P,Q)$$

for all $P, Q \in \mathbb{H}$, $M \in \mathbf{PSL}(2,\mathbb{C})$.

The concept of a point-pair invariant was introduced by A. Selberg (1956) who made fascinating use of it. We shall often meet point-pair invariants later on. The proofs of Propositions 1.6, 1.7, 1.8 are intended as some sort of hors d'oeuvre to introduce the reader to the use of point-pair invariants.

For later purposes we need an explicit expression for the hyperbolic distance. Using the notion of point-pair invariant, it is now easy to obtain such an expression.

Proposition 1.6. *For* $P = z + rj$, $P' = z' + r'j$ $(z, z' \in \mathbb{C}, r, r' > 0)$ *the hyperbolic distance* $d(P,P')$ *is given by*

$$\cosh d(P,P') = \delta(P,P') \tag{1.13}$$

where δ *is defined by*

$$\delta(P,P') := \frac{|z-z'|^2 + r^2 + r'^2}{2rr'}. \tag{1.14}$$

Proof. If $P = j$ and $P' = tj$ $(t \geq 1)$ we have

$$d(j,tj) = \int_1^t \frac{dr}{r} = \log t, \qquad \delta(j,tj) = \frac{1}{2}\left(t + \frac{1}{t}\right) = \cosh d(j,tj).$$

Hence (1.13) is true in the special case $P = j, P' = tj$. Now a direct check by means of the generators (1.11) gives that δ is a point-pair invariant. Since $\mathbf{PSL}(2,\mathbb{C})$ acts doubly transitively on $\mathbb{H}$, there exists for all $P, P' \in \mathbb{H}$ an element $M \in \mathbf{PSL}(2,\mathbb{C})$ such that $M(P) = j,\ M(P') = tj,\ t = \exp(d(P,P'))$. Hence by point-pair invariance and the above we have

$$\cosh d(P,P') = \cosh d(j,tj) = \delta(j,tj) = \delta(MP,MP') = \delta(P,P'). \quad \square$$

Since there is the simple expression (1.14) for δ we prefer to work with δ instead of d in the sequel. The use of δ is facilitated by means of the following Propositions 1.7, 1.8.

Proposition 1.7. *For all* $P = z + rj,\ P' = z' + r'j \in \mathbb{H},\ M = \begin{pmatrix} a & b \\ c & d \end{pmatrix} \in \mathbf{SL}(2,\mathbb{C})$ *the formula*

$$\begin{aligned} \delta(P, M(P')) = \frac{1}{2rr'} \big(&|(az'+b) - z(cz'+d)|^2 + |-cz+a|^2 r'^2 \\ &+ |cz'+d|^2 r^2 + |c|^2 r^2 r'^2 \big) \end{aligned} \tag{1.15}$$

holds. Hence, for fixed P, P' *the expression* $\delta(P, M(P'))$ *is a positive definite hermitian form in* a, b, c, d. *In particular, we have*

$$\delta(j, M(j)) = \frac{1}{2}(|a|^2 + |b|^2 + |c|^2 + |d|^2). \tag{1.16}$$

Proof. An elementary computation using (1.9), (1.10), (1.14) yields (1.16). Observe that

$$P = \begin{pmatrix} \sqrt{r} & \frac{z}{\sqrt{r}} \\ 0 & \frac{1}{\sqrt{r}} \end{pmatrix} (j)\ .$$

A similar formula holds for P'. Hence by point-pair invariance we have

$$\delta(P, MP') = \delta(j, T(j)) \tag{1.17}$$

where T denotes the matrix

$$T = \begin{pmatrix} \frac{1}{\sqrt{r}} & -\frac{z}{\sqrt{r}} \\ 0 & \sqrt{r} \end{pmatrix} \begin{pmatrix} a & b \\ c & d \end{pmatrix} \begin{pmatrix} \sqrt{r'} & \frac{z'}{\sqrt{r'}} \\ 0 & \frac{1}{\sqrt{r'}} \end{pmatrix}.$$

The matrix T can be computed as:

$$T = \begin{pmatrix} (-cz+a)\sqrt{\frac{r'}{r}} & (az'+b) - z(cz'+d))\frac{1}{rr'} \\ c\sqrt{rr'} & (cz'+d)\sqrt{\frac{r}{r'}} \end{pmatrix}. \tag{1.18}$$

If we now evaluate the right-hand side of (1.17) by means of (1.18), (1.14), we obtain (1.15). Of course, a computational proof of (1.15) by means of (1.9), (1.10), is also possible but rather tedious. □

The next Proposition provides us with a substitute for the triangle inequality for our function δ.

Proposition 1.8. *For all $P, Q, R \in \mathbb{H}$ the inequality*

$$\frac{1}{4}\,\frac{\delta(Q,R)}{\delta(P,Q)} \le \delta(P,R) \le 4\;\delta(P,Q)\;\delta(Q,R) \tag{1.19}$$

holds.

Proof. Since δ is a point-pair invariant, we have to check the inequality

$$\delta(P,R) \le 4\;\delta(P,Q)\;\delta(Q,R) \tag{1.20}$$

only in the special case $Q = j$. But for $P = z + rj$, $Q = j$, $R = w + tj$ with $z, w \in \mathbb{C}, r, t > 0$ we have

$$\begin{aligned} 4\;\delta(P,j)\;\delta(j,R) &= \frac{|z|^2 + r^2 + 1}{r} \cdot \frac{|w|^2 + t^2 + 1}{t} \ge \frac{|z|^2 + |w|^2 + r^2 + t^2}{rt} \\ &\ge \frac{|z-w|^2 + r^2 + t^2}{2rt} = \delta(P,R). \end{aligned}$$

This proves (1.20), which is the right-hand side of (1.19). Interchanging P and Q in (1.20) we obtain the left-hand side of (1.19). □

Proposition 1.9. *A map $f : \mathbb{H} \times \mathbb{H} \to \mathbb{C}$ is a point-pair invariant if and only if there exists a map $\varphi : [1, \infty[\to \mathbb{C}$ such that $f = \varphi \circ \delta$.*

Proof. Obviously all functions f of the form $f = \varphi \circ \delta$ are point-pair invariants.To prove the converse, assume that f is a point-pair invariant. Let $P, Q \in \mathbb{H}$. By Proposition 1.4 there exists an $M \in \mathbf{PSL}(2, \mathbb{C})$ such that $M(P) = j,\ M(Q) = e^{d(P,Q)}j$. This implies

$$f(P,Q) = f(j, e^{d(P,Q)}j) = \psi(d(P,Q)),$$

where $\psi(x) = f(j, e^x j)$ for $x \ge 0$. Hence $\varphi(t) := \psi(\operatorname{arcosh} t)$ serves our purpose. □

Finally we add a few remarks concerning the Laplace-Beltrami-operator associated with our Riemannian metric. The invariance of the hyperbolic metric with respect to the group $\mathbf{PSL}(2, \mathbb{C})$ implies that the Laplace-Beltrami operator is $\mathbf{PSL}(2, \mathbb{C})$-invariant, i.e.

$$\Delta(f \circ M) = (\Delta f) \circ M \tag{1.21}$$

for all $f \in C^2(\mathbb{H}), M \in \mathbf{PSL}(2,\mathbb{C})$. This may be seen by a direct check using the generators (1.11). Of course, differential geometers will deduce (1.21) from the general theorem saying that a diffeomorphism of a Riemannian manifold leaves Δ invariant if and only if it is an isometry (cf. Helgason (1962), p. 387, Prop. 2.1).

The set of all $\mathbf{PSL}(2,\mathbb{C})$-invariant differential operators acting on $C^\infty(\mathbb{H})$ evidently forms a ring. What does this ring look like? Obviously $\mathbb{C}[\Delta]$, the set of polynomials in Δ with complex coefficients, is a subring. It is a remarkable fact that $\mathbb{C}[\Delta]$ is equal to the ring of all invariant differential operators. This is shown in Helgason (1962), p. 397–398, Prop. 2.10; the general idea of the proof is also outlined in Selberg (1956), pages 49–50. The proof uses an induction argument on the order of the differential operator.

1.2 The Unit Ball Model

The upper half-space model of hyperbolic geometry is particularly suited for the discussion of parabolic linear fractional transformations which simply become translations when the fixed point is transformed to infinity. However, when questions involving rotational symmetry are discussed, the unit ball model of hyperbolic geometry is appropriate. Consider the unit ball

$$\mathbb{B} = \{\ u = u_0 + u_1 i + u_2 j \in \mathcal{H} \ \mid\ \|u\|^2 < 1\ \}. \tag{2.1}$$

The line element

$$ds^2 = 4 \cdot \frac{du_0^2 + du_1^2 + du_2^2}{(1 - u_0^2 - u_1^2 - u_2^2)^2} \tag{2.2}$$

defines a Riemannian metric on $\mathbb{B}$. The space $\mathbb{B}$ together with this metric is the unit ball model of hyperbolic space. Later we shall give an explicit isometry to the upper half-space model.

The corresponding volume element is

$$dv = \frac{8}{(1 - u_0^2 - u_1^2 - u_2^2)^3}\, du_0\, du_1\, du_2. \tag{2.3}$$

Putting $\rho := (u_0^2 + u_1^2 + u_2^2)^{\frac{1}{2}}$ we have the following formula for the Laplace operator

$$\begin{aligned} \Delta = {} & \frac{(1-\rho^2)^2}{4}\left(\frac{\partial^2}{\partial u_0^2} + \frac{\partial^2}{\partial u_1^2} + \frac{\partial^2}{\partial u_2^2}\right) \\ & + \frac{(1-\rho^2)}{2}\left(u_0\frac{\partial}{\partial u_0} + u_1\frac{\partial}{\partial u_1} + u_2\frac{\partial}{\partial u_2}\right). \end{aligned} \tag{2.4}$$

As a sample of hyperbolic geometry we compute some lengths and volumes.

Let $d = d_{\mathbb{B}}$ denote the hyperbolic distance in $\mathbb{B}$ derived from the above line element. The distance between $0 \in \mathbb{B}$ and an arbitrary point $u \in \mathbb{B}$ may be computed as follows:

$$d(0,u) = 2\int_0^\rho \frac{dt}{1-t^2} = \log\frac{1+\rho}{1-\rho} \tag{2.5}$$

where $\rho = \|u\|$. The geodesics in $\mathbb{B}$ are the segments in $\mathbb{B}$ of circles or straight lines intersecting the boundary of $\mathbb{B}$ orthogonally.

Let v denote the hyperbolic volume measure derived from (2.3). We write

$$B(Q,r) := \{\ P \in \mathbb{B}\ \mid\ d(P,Q) < r\ \}$$

for the ball of center $Q \in \mathbb{B}$ and radius r. The hyperbolic volume measure of $B(0,r)$ and hence (anticipating the invariance of v under hyperbolic motions) of $B(Q,r)$ is readily computed by means of polar coordinates. Putting

$$r = \log\frac{1+R}{1-R}$$

we get

$$\begin{aligned} v(B(0,r)) &= 8\int_{0\le\rho\le R} \frac{du_0\, du_1\, du_2}{(1-u_0^2-u_1^2-u_2^2)^3} = 32\pi \int_0^R \frac{\rho^2}{(1-\rho^2)^3}\, d\rho \\ &= 4\pi\left(\frac{R(1+R^2)}{(1-R^2)^2} - \frac{1}{2}\log\frac{1+R}{1-R}\right) \\ &= 2\pi(\sinh(r)\cosh(r) - r) = \pi(\sinh(2r) - 2r). \end{aligned} \tag{2.6}$$

Observe that

$$v(B(Q,r)) \sim \frac{4}{3}\pi r^3 \quad \text{as} \quad r \to 0 \tag{2.7}$$

which means that the hyperbolic volume of a ball is asymptotically equal to the Euclidean volume if the radius tends to zero. On the other hand, (2.6) implies that

$$v(B(Q,r)) \sim \frac{\pi}{2}\, e^{2r} \quad \text{as} \quad r \to \infty. \tag{2.8}$$

So the hyperbolic volume of a ball grows exponentially if the radius tends to ∞. We note also the inequality

$$v(B(Q,r)) \le \pi\sinh(2r) < \frac{\pi}{2}\, e^{2r} \quad \text{for all} \quad r \ge 0. \tag{2.9}$$

To describe the group $\mathbf{Iso}^+(\mathbb{B})$ of orientation preserving isometries of $\mathbb{B}$ we introduce the following selfmaps of the quaternions $\mathcal{H}$.

Definition 2.1. For $h = a + bi + cj + dk \in \mathcal{H}$ with $a, b, c, d \in \mathbb{R}$ define

$$h' := a - bi - cj + dk, \quad h^* := a + bi + cj - dk, \quad \bar{h} := h'^*.$$

The map $'$ is a skew field automorphism and * is a skew field anti-automorphism of $\mathcal{H}$. The maps $'$, * are involutory, commute with each other and for $h \in \mathcal{H}$ we have $\bar{h}' = h^*$.

The following subset of the ring $M(2, \mathcal{H})$ is a subgroup of its group of units. This can be checked elementarily, it will also become clear from the next Proposition.

Definition 2.2. We define

$$\mathbf{SB}_2(\mathcal{H}) := \left\{ \begin{pmatrix} a & b \\ c & d \end{pmatrix} \in M(2, \mathcal{H}) \;\middle|\; d = a', b = c', a\bar{a} - c\bar{c} = 1 \right\}.$$

Note that the inverse of an element $f = \begin{pmatrix} a & c' \\ c & a' \end{pmatrix} \in \mathbf{SB}_2(\mathcal{H})$ can be computed as $f^{-1} = \begin{pmatrix} \bar{a} & -\bar{c} \\ -c^* & a^* \end{pmatrix}$. The following Proposition describes the relation between our models $\mathbb{H}$ and $\mathbb{B}$, it also gives the group of orientation preserving isometries $\mathbf{Iso}^+(\mathbb{B})$ as a group of matrices.

Proposition 2.3.

(1) For $P \in \mathbb{H}$ the quaternion $-jP+1$ is invertible and $(P-j)\cdot(-jP+1)^{-1}$ is contained in $\mathbb{B}$. The map $\eta_0 : \mathbb{H} \to \mathbb{B}$, $\eta_0(P) := (P-j)\cdot(-jP+1)^{-1}$ is an isometry.

(2) Define $g := \frac{1}{\sqrt{2}} \cdot \begin{pmatrix} 1 & j \\ j & 1 \end{pmatrix} \in M(2, \mathcal{H})$. Then the map $\eta : A \mapsto \bar{g} \cdot A \cdot g$ defines a group isomorphism $\eta : \mathbf{SL}(2, \mathbb{C}) \to \mathbf{SB}_2(\mathcal{H})$.

(3) For $u \in \mathbb{B}$ and $f = \begin{pmatrix} a & c' \\ c & a' \end{pmatrix} \in \mathbf{SB}_2(\mathcal{H})$ the quaternion $cu + a'$ is invertible. The transformations $f \cdot u := (au+c')\cdot(cu+a')^{-1}$ are isometries of $\mathbb{B}$ and define an action of $\mathbf{SB}_2(\mathcal{H})$ on $\mathbb{B}$.

(4) The action under (3) gives rise to an exact sequence

$$1 \to \{1, -1\} \to \mathbf{SB}_2(\mathcal{H}) \to \mathbf{Iso}^+(\mathbb{B}) \to 1.$$

(5) The map η_0 is equivariant with respect to η, that is $\eta_0(f \cdot P) = \eta(f) \cdot \eta_0(P)$ for $P \in \mathbb{H}$ and $f \in \mathbf{SL}(2, \mathbb{C})$.

Proof. (1): In our coordinates we find the following formulas for the map η_0. Let $P \in \mathbb{H}$ have coordinates $P = x + yi + rj$ and $\eta_0(P) = u_0 + u_1 i + u_2 j$, then

$$(2.10)\qquad \begin{cases} u_0 = \frac{2x}{x^2+y^2+(r+1)^2}, \\ u_1 = \frac{2y}{x^2+y^2+(r+1)^2}, \\ u_2 = \frac{x^2+y^2+r^2-1}{x^2+y^2+(r+1)^2}. \end{cases}$$

The inverse tranformation is given by

$$(2.11)\qquad \begin{cases} x = \frac{2u_0}{u_0^2+u_1^2+(1-u_2)^2}, \\ y = \frac{2u_1}{u_0^2+u_1^2+(1-u_2)^2}, \\ r = \frac{1-u_0^2-u_1^2-u_2^2}{u_0^2+u_1^2+(1-u_2)^2}. \end{cases}$$

In geometric terms, the transformation from $\mathbb{H}$ to $\mathbb{B}$ is obtained by an inversion of $\mathbb{H}$ with respect to the sphere of radius 1 and centre $-j$, followed by the map $Q \mapsto 2Q + j$, followed by the reflection $(u, v, w) \mapsto (u, v, -w)$. By an elementary computation it is then verified that the pullback of the line element on $\mathbb{B}$ is our line element on $\mathbb{H}$.

(2): By some obvious computations it is verified that the map η transforms the defining conditions for $\mathbf{SL}(2)$ into those of $\mathbf{SB}_2$.

(3): This can be established by straightforward considerations about the arithmetic in $\mathcal{H}$. We skip the details.

(4), (5): All of this is obvious or follows using (1), (2), (3) from our discussion of the upper half-space model. □

For $u \in \mathbb{B}$ the isometry inverse to η_0 is given by: $\eta_0^{-1}(u) = (u + j) \cdot (ju + 1)^{-1}$. The corresponding isomorphism $\mathbf{SB}_2(\mathcal{H}) \to \mathbf{SL}(2, \mathbb{C})$ is given by $\eta^{-1}(A) = g \cdot A \cdot \bar{g}$.

1.3 The Exceptional Isomorphism

We shall describe here the construction of an isomorphism

$$\Psi : \mathbf{PSL}(2, \mathbb{C}) \to \mathbf{PSO}_4(\mathbb{R}, q_1) = \mathbf{PSO}(1, 3),$$

where q_1 is the quadratic form $q_1(x_0, x_1, x_2, x_3) = x_0^2 - x_1^2 - x_2^2 - x_3^2$. Concerning orthogonal groups we use the standard notation of Dieudonné (1971), that is $\mathbf{O}_4(q_1, \mathbb{R}) := \{\, g \in \mathbf{GL}(4, \mathbb{R}) \;\mid\; q_1 \circ g = q_1 \,\}$. $\mathbf{S}$ stands for considering only elements of determinant 1, $\mathbf{P}$ stands for factoring out the center. The isomorphism Ψ is one of finitely many isomorphisms between low dimensional semisimple Lie groups, see Helgason (1962). One way to obtain such a map is to use our isometries between the various models of hyperbolic space and the fact that the group of orientation preserving isometries comes as $\mathbf{PSL}(2, \mathbb{C})$ in case of the models $\mathbb{H}$, $\mathbb{B}$ and as $\mathbf{PSO}_4(q_1, \mathbb{R})$ for $\mathbb{K}$, $\mathbb{S}$ which will be described in Sections 1.4 and 1.5. We shall give here a more algebraic construction which will produce algebraic formulas respecting all sorts of rational and integral structures. The material is taken in specialized form from Elstrodt, Grunewald, Mennicke (1987b). The more general method explained there leads to similar definitions for other exceptional isomorphisms. A more geometric way to understand our exceptional isomorphism is contained in the book of van der Waerden (1948). We shall sketch it at the end of this section.

Another purpose of this paragraph is to lay the foundations for the construction of two further models $\mathbb{K}$, $\mathbb{S}$ of hyperbolic space which will be carried out in the next two paragraphs.

We start off with certain facts about Clifford algebras. These can all be verified by direct computation or can be found in Chevalley (1954). Let K be a field of characteristic $\neq 2$ which will be considered fixed for this paragraph. Let E be an n-dimensional K-vector space. Suppose that $Q : E \to K$ is a nondegenerate quadratic form with associated symmetric bilinear form $\Phi_Q : E \times E \to K$, that is

(3.1) $$\begin{aligned}\Phi_Q(x,y) &= \frac{1}{2}(Q(x+y) - Q(x) - Q(y)),\\ Q(x) &= \Phi_Q(x,x).\end{aligned}$$

Denote by $T(E)$ the tensor algebra of E and by $\mathfrak{a}_Q$ its two-sided ideal generated by the elements $x \otimes y + y \otimes x - 2 \cdot \Phi(x,y)$ where x, y run through the elements of E. The Clifford algebra of Q is defined as the quotient $\mathcal{C}(Q) := T(E)/\mathfrak{a}_Q$. The field K and the vector space E inject into $\mathcal{C}(Q)$. We will always identify K and E with their canonical images in $\mathcal{C}(Q)$. Let $e_1, ..., e_n$ be a basis of E orthogonal with respect to Q. Then we have in $\mathcal{C}(Q)$:

(3.2) $$e_i^2 = Q(e_i), \quad e_i \cdot e_j = -e_j \cdot e_i \qquad \text{for all } i, j = 1, ..., n.$$

Let $\mathcal{P}_n$ be the set of subsets of $\{1, ..., n\}$. For $M \in \mathcal{P}_n$, $M = \{i_1, ..., i_r\}$ with $i_1 < ... < i_r$ we define

(3.3) $$e_M := e_{i_1} \cdot ... \cdot e_{i_r}$$

with the additional convention $e_\emptyset = 1$. The 2^n elements e_M are a vector space basis of $\mathcal{C}(Q)$. For $M, N \in \mathcal{P}_n$ the product $e_M \cdot e_N$ can be calculated explicitly as a scalar factor times an appropriate e_L. The Clifford algebra $\mathcal{C}(Q)$ has a main anti-involution * and a main involution $'$ commuting with *. These act on the basis elements e_M as follows

(3.4) $$e_M^* = (-1)^{\frac{r(r-1)}{2}} \cdot e_M, \quad e_M' = (-1)^r \cdot e_M.$$

Here r stands for the cardinality of M. The anti-involution defined by $\bar{x} := x'^*$ satisfies

(3.5) $$\bar{e}_M = (-1)^{\frac{r(r+1)}{2}} \cdot e_M.$$

The span of the elements e_M where M has even cardinality is a subalgebra called $\mathcal{C}^+(Q)$.

The following examples are of particular importance for us. For $\epsilon \in K$ with $\epsilon \neq 0$ let $E_\epsilon = K \cdot f_3$ be the one-dimensional K-vector space with basis f_3. The symbol f_3 is used in anticipation of later notation. On E_ϵ we fix the quadratic form Q_ϵ given by $Q_\epsilon(f_3) = -\epsilon$. Then the Clifford algebra $\mathcal{C}(Q_\epsilon)$ is two-dimensional and commutative. In case $-\epsilon \in K^{*2}$ the K-algebra $\mathcal{C}(Q_\epsilon)$ is isomorphic to $K \times K$, if not then $\mathcal{C}(Q_\epsilon)$ is a quadratic extension field of K. Here K^{*2} stands for the subgroup of squares in K^*. In case $K = \mathbb{R}$ and $\epsilon = 1$ we call f_3 also i, that is we make the identification $\mathcal{C}(Q_1) = \mathbb{C}$.

For $a, b \in K$ which are both not 0 we take a two-dimensional K-vector space $H_{a,b} = K \cdot i \oplus K \cdot j$ with quadratic form $Q_{a,b}(\lambda i + \mu j) = a\lambda^2 + b\mu^2$. The Clifford algebra $\mathcal{C}(Q_{a,b})$ is 4-dimensional. It is a quaternion algebra, the notation is

$$\mathcal{H}(a, b; K) = \mathcal{H}(a, b) = \mathcal{C}(Q_{a,b}). \tag{3.6}$$

The basis elements i, j satisfy the familiar relations $i^2 = a$, $j^2 = b$, $ij = -ji$. More on quaternion algebras will follow in Chapter 10.

We fix now a 3-dimensional vector space $E_0 = K \cdot f_0 + K \cdot f_1 + K \cdot f_2$ and on it the quadratic form

$$Q_0(y_0 f_0 + y_1 f_1 + y_2 f_2) = Q_0(y_0, y_1, y_2) = y_0^2 - y_1^2 - y_2^2. \tag{3.7}$$

It is well known that $\mathcal{C}^+(Q_0) = M(2, K)$, the algebra of 2×2-matrices over K. We need here a special isomorphism between these two algebras.

Definition 3.1. We define the following elements of $\mathcal{C}(Q_0)$:

$$\tau_0 = \frac{1}{2}(f_0 + f_1), \quad \tau_1 = \frac{1}{2}(f_0 - f_1),$$

$$u = \tau_1\tau_0 = \frac{1}{2}(1 + f_0 f_1), \quad w_1 = \tau_1 f_2 = \frac{1}{2}(f_0 f_2 - f_1 f_2),$$

$$w_0 = \tau_0 f_2 = \frac{1}{2}(f_0 f_2 + f_1 f_2), \quad v = \tau_0\tau_1 = \frac{1}{2}(1 - f_0 f_1).$$

Obvious relations between these lead to the following

Proposition 3.2. *The map*

$$\begin{pmatrix} 1 & 0 \\ 0 & 0 \end{pmatrix} \mapsto u, \quad \begin{pmatrix} 0 & 1 \\ 0 & 0 \end{pmatrix} \mapsto w_1, \quad \begin{pmatrix} 0 & 0 \\ 1 & 0 \end{pmatrix} \mapsto w_0, \quad \begin{pmatrix} 0 & 0 \\ 0 & 1 \end{pmatrix} \mapsto v$$

extends to an algebra isomorphism $\psi : M(2, K) \to \mathcal{C}^+(Q_0)$. *We have*

$$\psi\left(\begin{pmatrix} \alpha & \beta \\ \gamma & \delta \end{pmatrix}\right)^* = \psi\left(\begin{pmatrix} \delta & -\beta \\ -\gamma & \alpha \end{pmatrix}\right).$$

The following construction will be important for our definition of the exceptional isomorphism.

Definition 3.3. For the K-vector space E with quadratic form Q we define the K-vector space $\tilde{E} := E_0 \oplus E$ with quadratic form $\tilde{Q} := Q_0 \perp Q$.

We always identify $\mathcal{C}(Q_0)$ with the subalgebra of $\mathcal{C}(\tilde{Q})$ generated by f_0, f_1, f_2. We then have:

Proposition 3.4. *The map* $^{\cdot} : E \to \mathcal{C}^+(\tilde{Q})$ *with* $\dot{x} := f_0 f_1 f_2 \cdot x$ *extends to an injective K-algebra homomorphism* $^{\cdot} : \mathcal{C}(Q) \to \mathcal{C}^+(\tilde{Q})$. *The map* $^{\cdot}$ *commutes with the anti-automorphism* *.

Proof. The existence of the extension of the map $^{\cdot}$ can be deduced from the universal property of Clifford algebras or can be proved by straightforward elementary considerations. To prove the injectivity we note the formula:

$$\left(\sum_{M\in\mathcal{P}_n} \lambda_M e_M\right)^{\bullet} = \sum_{M\in\mathcal{P}_n} \lambda_M \cdot (f_0 f_1 f_2)^{\epsilon_M} \cdot e_M, \tag{3.8}$$

where ϵ_M is 0 or 1 if the cardinality of M is even or odd. The e_M constitute a basis of $\mathcal{C}(Q)$ constructed from an orthogonal basis of the n dimensional vector space E. From this formula it is clear that $^{\cdot}$ commutes with *. □

Proposition 3.5. *The map* $\psi : M(2, \mathcal{C}(Q)) \to \mathcal{C}^+(\tilde{Q})$,

$$\psi\left(\begin{pmatrix} \alpha & \beta \\ \gamma & \delta \end{pmatrix}\right) := \dot{\alpha}u + \dot{\beta}w_1 + \dot{\gamma}w_0 + \dot{\delta}v$$

is a K-algebra isomorphism and satisfies

$$\psi\left(\begin{pmatrix} \alpha & \beta \\ \gamma & \delta \end{pmatrix}\right)^* = \psi\left(\begin{pmatrix} \delta^* & -\beta^* \\ -\gamma^* & \alpha^* \end{pmatrix}\right).$$

Proof. Since the elements u, w_1, w_0, v commute with $\mathcal{C}(Q)$ it follows from Proposition 3.2 that ψ is a homomorphism. Formula (3.8) then shows that ψ is injective. Since the dimensions of $M(2, \mathcal{C}(Q))$ and $\mathcal{C}(\tilde{Q})$ coincide ψ is also surjective. □

We are now ready to describe an isomorphism between $\mathbf{SL}(2, \mathcal{C}(Q_\epsilon))$ and the spin-group of a suitable quadratic form. The following is the standard definition of spin-groups, see Chevalley (1954).

Definition 3.6. Let U be an n-dimensional K-vector space with nondegenerate quadratic form q. Then the spin-group of q is defined as

$$\mathbf{Spin}_n(K,q) := \{\ s \in \mathcal{C}^+(q)\ \mid\ s \cdot U \cdot s^* \subset U,\ s \cdot s^* = 1\ \}.$$

The following is the first step in our construction of the exceptional isomorphism.

Proposition 3.7. *The K-algebra isomorphism $\psi : M(2, \mathcal{C}(Q_\epsilon)) \to \mathcal{C}^+(\tilde{Q}_\epsilon)$ defined in Proposition 3.5 restricts to a group isomorphism $\psi : \mathbf{SL}(2, \mathcal{C}(Q_\epsilon)) \to \mathbf{Spin}_4(\tilde{Q}_\epsilon)$.*

Proof. Let

$$A := \begin{pmatrix} a_0 + a_1 f_3 & b_0 + b_1 f_3 \\ c_0 + c_1 f_3 & d_0 + d_1 f_3 \end{pmatrix}$$

be the general element of $M(2, \mathcal{C}(Q_\epsilon))$ with $a_0, \dots, d_1 \in K$. Recalling the definition of ψ we find:

$$\begin{aligned} 2\psi(A) = a_0 + d_0 &+ (a_0 - d_0) f_0 f_1 + (b_0 + c_0) f_0 f_2 \\ &+ (b_1 - c_1) f_0 f_3 + (c_0 - b_0) f_1 f_2 - (b_1 + c_1) f_1 f_3 \\ &+ (a_1 - d_1) f_2 f_3 + (a_1 + d_1) f_0 f_1 f_2 f_3. \end{aligned} \tag{3.9}$$

The statement of the Proposition follows then from Proposition 3.2 together with some elementary considerations. □

The spin-group of a quadratic form has a canonical homomorphism to the corresponding orthogonal group which is constructed as follows. Start with a vector space U of dimension n with nondegenerate quadratic form Q. The space U is identified with a subspace of $\mathcal{C}(Q)$ and we have for $x \in U$: $Q(x) = x \cdot x^*$. Associate to $s \in \mathbf{Spin}_n(K, Q)$ the linear map: $\Lambda(s) \in \mathbf{GL}(U)$, with $\Lambda(s)(x) := s \cdot x \cdot s^*$. Then the computation

$$sxs^* \cdot (sxs^*)^* = sxs^* sx^* s^* = sxx^* s^* = xx^*$$

shows that $\Lambda(s)$ is in $\mathbf{O}_n(K, Q)$. Note that an element $s \in \mathbf{Spin}_n(K, Q)$ also satisfies $s^* s = 1$. We have constructed now a homomorphism

$$\Lambda : \mathbf{Spin}_n(K, Q) \to \mathbf{O}_n(K, Q). \tag{3.10}$$

We shall later need a description of the image and cokernel of Λ. In the following we report on this well-known theory. Let $\Omega_n(K, Q)$ be the commutator subgroup of $\mathbf{O}_n(K, Q)$ and $\Gamma(Q)$ the subgroup of K^*/K^{*2} generated by the expressions $Q(x)Q(y)$ where $x, y \in U$ satisfy $Q(x) \neq 0 \neq Q(y)$. Let furthermore $x \in U$ be a vector with $Q(x) \neq 0$ (such vectors are called anisotropic). The linear map $\sigma_x : U \to U$,

$$\sigma_x(v) := v - 2\,\frac{\Phi_Q(v, x)}{Q(x)} \cdot x \tag{3.11}$$

is called the reflection associated to x or the reflection in the hyperplane perpendicular to x. For the definition of Φ_Q see (3.1). Every element $g \in \mathbf{O}_n(K,Q)$ can be expressed as a product of reflections $g = \sigma_{x_1} \cdot \ldots \cdot \sigma_{x_r}$. Associating to g the product $Q(x_1) \cdot \ldots \cdot Q(x_r)$ in K^*/K^{*2} we get a well-defined homomorphism

$$\Sigma : \mathbf{SO}_n(K,Q) \to K^*/K^{*2} \tag{3.12}$$

which is called the spinorial norm homomorphism. The following is contained in Dieudonné (1971), Artin (1957).

Proposition 3.8. *Let U be an n-dimensional K-vector space with nonsingular quadratic form Q. Then the following hold.*

(1) We have $\Lambda(\mathbf{Spin}_n(K,Q)) \subset \mathbf{SO}_n(K,Q)$, and $\Sigma(\mathbf{SO}_n(K,Q)) \subset \Gamma(Q)$, the resulting sequence $1 \to \{1,-1\} \to \mathbf{Spin}_n(K,Q) \to \mathbf{SO}_n(K,Q) \to \Gamma(Q) \to 1$ is exact.
(2) We define $\mathbf{SO}_n^+(K,Q) := \mathrm{Im}(\Lambda)$ and get $\Omega_n(K,Q) \subset \mathbf{SO}_n^+(K,Q)$.
(3) If $n \geq 3$ then $\Omega_n(K,Q)$ is also the commutator subgroup of $\mathbf{SO}_n(K,Q)$.
(4) Suppose that $n \geq 3$ and that U contains an isotropic vector x, that is $x \neq 0$ satisfies $Q(x) = 0$, then $\Omega_n(K,Q) = \mathrm{Ker}(\Sigma) = \mathrm{Im}(\Lambda)$.

For the special quadratic forms chosen above we define now:

Definition 3.9. Let E_ϵ be the one-dimensional vector space with basis f_3 and quadratic form Q_ϵ, then $\Psi : \mathbf{SL}(2, C(Q_\epsilon)) \to \mathbf{O}_4(K, \tilde{Q}_\epsilon)$ is defined as $\Psi := \Lambda \circ \psi$.

The isomorphisms ψ, Ψ are usually called exceptional isomorphisms. We get from Proposition 3.8:

Proposition 3.10. *The map $\Psi : \mathbf{SL}(2, C(Q_\epsilon)) \to \mathbf{O}_4(K, \tilde{Q}_\epsilon)$ has the property $\Psi(\mathbf{SL}(2, C(Q_\epsilon))) = \mathbf{SO}_4^+(K, \tilde{Q}_\epsilon)$ and the resulting sequence*

$$1 \to \{1,-1\} \to \mathbf{SL}(2, C(Q_\epsilon)) \to \mathbf{SO}_4(K, \tilde{Q}_\epsilon) \to K^*/K^{*2} \to 1$$

is exact.

We give now a concrete matrix-expression for our map from $\mathbf{SL}(2)$ over the two-dimensional Clifford algebra to the corresponding orthogonal group. We shall later have to make use of it.

Proposition 3.11. *Let E_ϵ be the one-dimensional vector space with basis f_3 and quadratic form Q_ϵ, that is $Q_\epsilon(f_3) = -\epsilon$. For*

$$A = \begin{pmatrix} a_0 + a_1 f_3 & b_0 + b_1 f_3 \\ c_0 + c_1 f_3 & d_0 + d_1 f_3 \end{pmatrix} \in \mathbf{SL}(2, C(\tilde{Q}_\epsilon))$$

define

$$N_1 := a_0^2 + b_0^2 + c_0^2 + d_0^2 + \epsilon(a_1^2 + b_1^2 + c_1^2 + d_1^2),$$
$$N_2 := -a_0^2 - b_0^2 + c_0^2 + d_0^2 + \epsilon(-a_1^2 - b_1^2 + c_1^2 + d_1^2),$$
$$N_3 := -a_0^2 + b_0^2 - c_0^2 + d_0^2 + \epsilon(-a_1^2 + b_1^2 - c_1^2 + d_1^2),$$
$$N_4 := a_0^2 - b_0^2 - c_0^2 + d_0^2 + \epsilon(a_1^2 - b_1^2 - c_1^2 + d_1^2),$$

$$T_1 := -a_0c_0 - d_0b_0 - \epsilon(a_1c_1 + b_1d_1), \quad T_2 := a_0c_1 - d_0b_1 - a_1c_0 + b_0d_1,$$

$$T_3 := a_0c_0 - d_0b_0 + \epsilon(a_1c_1 - b_1d_1), \quad T_4 := -a_0c_1 - d_0b_1 + a_1c_0 + b_0d_1.$$

Then

$$\Psi(A) =$$

$$\begin{pmatrix} \frac{N_1}{2} & \frac{N_3}{2} & a_0c_0 + d_0b_0 - \epsilon(a_1b_1 + c_1d_1) & \epsilon(-a_0c_1 + d_0b_1 + a_1b_0 - d_1c_0) \\ \frac{N_2}{2} & \frac{N_4}{2} & a_0b_0 - d_0c_0 + \epsilon(a_1b_1 - c_1d_1) & \epsilon(a_0b_1 + d_0c_1 - a_1b_0 - d_1c_0) \\ T_1 & T_3 & a_0d_0 + c_0b_0 + \epsilon(b_1c_1 + a_1d_1) & \epsilon(a_0d_1 - d_0a_1 - c_1b_0 + b_1c_0) \\ T_2 & T_4 & -a_0d_1 + a_1d_0 - b_0c_1 + b_1c_0 & a_0d_0 - b_0c_0 + \epsilon(a_1d_1 - b_1c_1) \end{pmatrix}.$$

Proof. These formulas follow by straightforward computations from the definitions. □

We now discuss the situation for $K = \mathbb{R}$. We take the one-dimensional $\mathbb{R}$-vector space $E_1 = \mathbb{R} \cdot f_3$ with quadratic form $Q_1(\lambda f_3) = -\lambda^2$. We remind the reader of our identification $\mathcal{C}(Q_1) = \mathbb{C}$. We define $q_1 := \tilde{Q}_1$ which is a quadratic form on the 4-dimensional vector space $\tilde{E}_1$ defined in Definition 3.3. We have the following characterizations.

Proposition 3.12. *Let* $\mathbf{SO}_4^+(\mathbb{R}, q_1)$ *be defined following Proposition 3.8 as the image of our homomorphism* $\Psi : \mathbf{SL}(2, \mathbb{C}) \to \mathbf{SO}_4(\mathbb{R}, q_1)$. *Then the following hold.*

(1) $\mathbf{SO}_4^+(\mathbb{R}, q_1)$ *has index* 2 *in* $\mathbf{SO}_4(\mathbb{R}, q_1)$.
(2) $\mathbf{SO}_4^+(\mathbb{R}, q_1)$ *is the connected component of the identity in* $\mathbf{SO}_4(\mathbb{R}, q_1)$.
(3) $\mathbf{SO}_4^+(\mathbb{R}, q_1)$ *consists of those elements in* $\mathbf{SO}_4(\mathbb{R}, q_1)$ *which have a positive entry in the left upper corner.*
(4) $\mathbf{SO}_4^+(\mathbb{R}, q_1)$ *is equal to the commutator subgroup of* $\mathbf{SO}_4(\mathbb{R}, q_1)$ *and also equal to the commutator subgroup of* $\mathbf{O}_4(\mathbb{R}, q_1)$.
(5) The quotient map $\mathbf{SO}_4(\mathbb{R}, q_1) \to \mathbf{PSO}_4(\mathbb{R}, q_1)$ *maps* $\mathbf{SO}_4^+(\mathbb{R}, q_1)$ *isomorphically onto* $\mathbf{PSO}_4(\mathbb{R}, q_1)$.

Proof. The only statements which are not clear from Proposition 3.8 are (2) and (5). Statement (2) follows from the formulas in Proposition 3.11, while (5) is obvious. □

In the following we discuss a subgroup of $\mathbf{O}_4(\mathbb{R}, q_1)$ which comes up naturally in our study of two further models in the next two sections.

Proposition 3.13. *Define $\mathbf{O}_4^+(\mathbb{R}, q_1)$ to be the set of elements in $\mathbf{O}_4(\mathbb{R}, q_1)$ which have a positive entry in the left upper corner. Then the following hold:*

(1) Let $x \in \tilde{E}_1$ be an anisotropic vector and let $\sigma_x \in \mathbf{O}_4(\mathbb{R}, q_1)$ be the corresponding reflection (3.11). Then $-\sigma_x \in \mathbf{O}_4^+(\mathbb{R}, q_1)$.
(2) $\mathbf{O}_4^+(\mathbb{R}, q_1)$ is a subgroup of index 2 in $\mathbf{O}_4(\mathbb{R}, q_1)$.
(3) $\mathbf{O}_4^+(\mathbb{R}, q_1)$ maps isomorphically onto $\mathbf{PO_4}(\mathbb{R}, q_1)$.

Proof. We infer (2) from the obvious (1) and Proposition 3.12. Statement (3) is clear. □

In Chapter 10 we will have to use the following version of Proposition 3.7.

Proposition 3.14. *Let K be a field with* $\mathrm{char}(K) \neq 2$. *Let U be a 4-dimensional K-vector space with basis e_1, e_2, e_3, e_4 and quadratic form*

$$q(x_1e_1 + x_2e_2 + x_3e_3 + x_4e_4) = x_1^2 + ax_2^2 + bx_3^2 + cx_4^2.$$

Assume further that $d = abc$ is not a square in K. Then $L = K + Ke_1e_2e_3e_4$ is a quadratic extension field of K contained in the center of $\mathcal{C}^+(q)$. L is isomorphic to $K(\sqrt{d})$. The elements $i = e_1e_3,\ j = e_1e_2,\ k = e_2e_3$ are an L-basis of $\mathcal{C}^+(q)$. They give rise to a K-algebra isomorphism $\psi : \mathcal{H}(-c, -b; L) \to \mathcal{C}^+(q)$, where $\mathcal{H}(-c, -b; L)$ is the quaternion algebra defined in (3.6). The map ψ restricts to a group isomorphism $\psi : \mathcal{H}(-c, -b; L)^1 \to \mathbf{Spin_4}(q)$, where

$$\mathcal{H}(-c, -b; L)^1 = \{\ x \in \mathcal{H}(-c, -b; L) \ \mid \ x\bar{x} = 1\ \}$$

is the group of elements of norm 1 in $\mathcal{H}(-c, -b; L)$.

All of this follows by elementary considerations along the same lines as above.

Note that as in Proposition 3.10 we get a homomorphism

$$\Psi : \mathcal{H}(-c, -b; L)^1 \to \mathbf{SO}_4(K, q)$$

but we can in general not use Proposition 3.8, part (4) to determine the image of Ψ.

We shall now describe a more geometric way to see the exceptional isomorphism. This method is taken from the book of van der Waerden (1948). We start off with a field K of characteristic different from 2. For an $\epsilon \in K$ we consider on K^4 the quadratic form

$$f(x_1, x_2, x_3, x_4) := x_1x_2 - x_3^2 - \epsilon x_4^2.$$

If, as we shall assume now, $-\epsilon$ is not a square in K, the form f has Witt-index 1. If $K = \mathbb{R}$ then f has signature (1,3). We write $L := K(\sqrt{-\epsilon})$ for the quadratic extension field generated by $\sqrt{-\epsilon}$. We put $^{-}: L \to L$, $N : L \to K$ for the nontrivial element of the Galois group and the norm on L. Putting

$$x_1 = y_1, \quad x_2 = y_2, \quad x_3 = y_3 + \sqrt{-\epsilon} y_4, \quad x_4 = y_3 - \sqrt{-\epsilon} y_4$$

we obtain an equivalence A defined over L of the form f and the quadratic form

$$F(y_1, y_2, y_3, y_4) = y_1 y_2 - y_3 y_4$$

on L^4. We consider the quadratic map

$$T : L^4 \to L^4, \qquad T : (\lambda_1, \lambda_2, \mu_1, \mu_2) \mapsto (\lambda_1\mu_1, \lambda_2\mu_2, \lambda_1\mu_2, \lambda_2\mu_1).$$

The map T is a rational parametrization of the affine quadric $\{ \, y \in L^4 \mid F(y) = 0 \, \}$. The projective quadric $\{ \, [y] \in \mathbb{P}(L) \mid F([y]) = 0 \, \}$ is, using T, written in two different ways as the union of a family of projective lines. We put

$$\mathbf{GL}(2, L)^1 := \{ \, g \in \mathbf{GL}(2, L) \mid N(\det g) = 1 \, \}$$

and for $g = \begin{pmatrix} a & b \\ c & d \end{pmatrix} \in \mathbf{GL}(2, L)^1$

$$g(\lambda_1, \lambda_2, \mu_1, \mu_2) := (a\lambda_1 + b\lambda_2, c\lambda_1 + d\lambda_2, \bar{a}\mu_1 + \bar{b}\mu_2, \bar{c}\mu_1 + \bar{d}\mu_2).$$

Writing everything back in terms of our parameters x_1, x_2, x_3, x_4 we find that there is a uniquely determined matrix $\Psi_W(g) \in \mathbf{O_4}(K, f)$ with

$$\Psi_W(g)(A \circ T(\lambda_1, \lambda_2, \mu_1, \mu_2)) = A \circ T(g(\lambda_1, \lambda_2, \mu_1, \mu_2))$$

for all $(\lambda_1, \lambda_2, \mu_1, \mu_2) \in L^4$. Hence the map Ψ_W defines a group homomorphism

$$\Psi_W : \mathbf{GL}(2, L)^1 \to \mathbf{O_4}(K, f).$$

It is easy to see that the kernel of Ψ_W consists of the scalar matrices in $\mathbf{GL}(2, L)^1$. The map Ψ_W has the same image as our map Ψ from Definition 3.9.

1.4 The Hyperboloid Model

The isometry group of the hyperboloid model $\mathbb{S}$ comes naturally as a four-dimensional orthogonal group. This will later be used to construct discrete subgroups of $\mathbf{Iso}(\mathbb{S})$ by orthogonal groups satisfying integrality conditions.

We keep the notation introduced in the preceding paragraph, that is $E_1 = \mathbb{R} \cdot f_3$ stands for a one-dimensional $\mathbb{R}$-vector space with quadratic form $Q_1(\lambda f_3) = -\lambda^2$. The four-dimensional vector space $\tilde{E}_1 = \mathbb{R} f_0 + \mathbb{R} f_1 + \mathbb{R} f_2 + \mathbb{R} f_3$ is equipped with the quadratic form $q_1 := \tilde{Q}_1$ which is given by

$$q_1(y_0f_0 + y_1f_1 + y_2f_2 + y_3f_3) = y_0^2 - y_1^2 - y_2^2 - y_3^2.$$

We consider

$$\text{(4.1)} \quad \mathbb{S} := \{\ y = y_0f_0 + y_1f_1 + y_2f_2 + y_3f_3 \in \tilde{E}_1 \ \mid\ y_0 > 0,\ q_1(y) = 1\ \}.$$

On $\mathbb{S}$ we get a Riemannian metric from the line element

$$\text{(4.2)} \qquad ds^2 = -dy_0^2 + dy_1^2 + dy_2^2 + dy_3^2.$$

The set $\mathbb{S}$ together with this metric is the hyperboloid model of hyperbolic space.

To get the isometry group of $\mathbb{S}$ notice first that the groups $\mathbf{SO}_4^+(\mathbb{R}, q_1)$ and $\mathbf{O}_4^+(\mathbb{R}, q_1)$ defined in Propositions 3.12, 3.13 map $\mathbb{S}$ into itself. After this we shall define now what will be the isometry from $\mathbb{H}$ to $\mathbb{S}$.

Definition 4.1. For an element $P = x + yi + rj \in \mathbb{H}$ we define

$$\pi_0(P) = \frac{1}{2r}((1 + P\bar{P})f_0 + (1 - P\bar{P})f_1 - 2xf_2 - 2yf_3).$$

Notice that $\pi_0(P)$ actually lies in $\mathbb{S}$.

Proposition 4.2. *The map π_0 has the following properties:*

(1) π_0 is an isometry from $\mathbb{H}$ to $\mathbb{S}$.
(2) π_0 is equivariant with respect to the homomorphism $\Psi : \mathbf{SL}(2, \mathbb{C}) \to \mathbf{SO}_4^+(\mathbb{R}, q_1)$, that is $\pi_0(g \cdot P) = \Psi(g) \cdot \pi_0(P)$, for $g \in \mathbf{SL}(2, \mathbb{C})$ and $P \in \mathbb{H}$.

Proof. (1): Using our usual coordinates on $\mathbb{H}$ and $\mathbb{S}$ a simple computation shows that the map π_0 respects the Riemannian metrics.

(2): This needs only to be checked on our generators (1.11) of $\mathbf{SL}(2, \mathbb{C})$. On the basis f_0, f_1, f_2, f_3 of $\tilde{E}_1$ we find:

$$\Psi\left(\begin{pmatrix} 0 & 1 \\ -1 & 0 \end{pmatrix}\right) = \begin{pmatrix} 1 & 0 & 0 & 0 \\ 0 & -1 & 0 & 0 \\ 0 & 0 & -1 & 0 \\ 0 & 0 & 0 & 1 \end{pmatrix},$$

$$\Psi\left(\begin{pmatrix} 1 & b_0 + b_1 i \\ 0 & 1 \end{pmatrix}\right) = \begin{pmatrix} 1 + \frac{b_0^2+b_1^2}{2} & \frac{b_0^2+b_1^2}{2} & -b_0 & -b_1 \\ -\frac{b_0^2+b_1^2}{2} & 1 - \frac{b_0^2+b_1^2}{2} & b_0 & b_1 \\ -b_0 & -b_0 & 1 & 0 \\ -b_1 & -b_1 & 0 & 1 \end{pmatrix}.$$

From this everything easily follows. □

From the preceding Propositions we infer now the description of the isometry group of $\mathbb{S}$.

Proposition 4.3. *The linear action of* $\mathbf{O}_4^+(\mathbb{R}, q_1)$ *on* $\mathbb{S}$ *gives isomorphisms*

$$\mathbf{O}_4^+(\mathbb{R}, q_1) \to \mathbf{Iso}(\mathbb{S}), \qquad \mathbf{SO}_4^+(\mathbb{R}, q_1) \to \mathbf{Iso}^+(\mathbb{S}).$$

Proof. The second isomorphism is immediate from Propositions 1.3, 4.2. For the first we let $x \in \tilde{E}_1$ be an anisotropic vector then the linear map $-\sigma_x$ defines an orientation reversing isometry of $\mathbb{S}$. This is easily checked and Propositions 3.12, 3.13 complete the proof. □

To describe the group $\mathbf{Iso}^+(\mathbb{S})$ we can also use our isomorphism $\psi : \mathbf{SL}(2, \mathbb{C}) \to \mathbf{Spin_4}(\mathbb{R}, q_1)$ from Proposition 3.7. The equivariance of π_0 is then expressed as $\pi_0(g \cdot P) = \psi(g) \cdot \pi_0(P) \cdot \psi(g)^*$ for all $g \in \mathbf{SL}(2, \mathbb{C})$ and $P \in \mathbb{H}$.

1.5 The Kleinian Model

For this model the group of isometries is again naturally identified with an orthogonal group. We shall use the Kleinian model in the chapter on unit groups of quadratic forms and in our study of Coxeter groups.

We keep the notation introduced in paragraph 1.3 of this chapter and in the preceding paragraph, that is $E_1 = \mathbb{R} \cdot f_3$ stands for a one-dimensional $\mathbb{R}$-vector space with quadratic form $Q_1(\lambda f_3) = -\lambda^2$. Notice that we have identified the Clifford algebra $\mathcal{C}(Q_1)$ with $\mathbb{C}$. We denote the basis element f_3 then by i.

The four-dimensional vector space $\tilde{E}_1 = \mathbb{R}f_0 + \mathbb{R}f_1 + \mathbb{R}f_2 + \mathbb{R}f_3$ is equipped with the quadratic form $q_1 := \tilde{Q}_1$ which is given by

$$q_1(y_0 f_0 + y_1 f_1 + y_2 f_2 + y_3 f_3) = y_0^2 - y_1^2 - y_2^2 - y_3^2.$$

Consider the real three-dimensional projective space

$$\mathbb{P}\tilde{E}_1 = (\tilde{E}_1 \backslash \{0\})/\mathbb{R}^*, \tag{5.1}$$

where $\mathbb{R}^*$ stands for the multiplicative group $\mathbb{R}\backslash\{0\}$. We represent the equivalence class of $y = y_0 f_0 + y_1 f_1 + y_2 f_2 + y_3 f_3 \in \tilde{E}_1$ in $\mathbb{P}\tilde{E}_1$ by the homogeneous vector

$$[y] = [y_0 f_0 + y_1 f_1 + y_2 f_2 + y_3 f_3] = [y_0, y_1, y_2, y_3].$$

Here $y_0, \ldots, y_3 \in \mathbb{R}$ are not all equal to zero.

The set underlying the Kleinian model is:

$$\mathbb{K} := \{\ [y] \in \mathbb{P}\tilde{E}_1 \ \mid \ q_1(y) > 0\ \}. \tag{5.2}$$

Notice that the condition $q_1(y) > 0$ is well-defined on real projective space. To introduce the line element we will give a universal chart on $\mathbb{K}$. This space is homeomorphic to a three-dimensional ball, so we put

$$B^3 = \{ (z_1, z_2, z_3)^t \in \mathbb{R}^3 \mid z_1^2 + z_2^2 + z_3^2 < 1 \}. \tag{5.3}$$

The map

$$\mathbb{K} \to B^3, \qquad [y_0, y_1, y_2, y_3] \mapsto \left(\frac{y_1}{y_0}, \frac{y_2}{y_0}, \frac{y_2}{y_0} \right)$$

is a homeomorphism. Notice that $y_0 \neq 0$ for every $[y_0, y_1, y_2, y_3] \in \mathbb{K}$. On this chart the line element giving the Riemannian metric is

$$ds^2 = \frac{dz_1^2 + dz_2^2 + dz_3^2}{1 - z_1^2 - z_2^2 - z_3^2} + \frac{(z_1 dz_1 + z_2 dz_2 + z_3 dz_3)^2}{(1 - z_1^2 - z_2^2 - z_3^2)^2}. \tag{5.4}$$

The volume element in $\mathbb{K}$ is

$$dv = \frac{dz_1 dz_2 dz_3}{(1 - z_1^2 - z_2^2 - z_3^2)^3}. \tag{5.5}$$

The set $\mathbb{K}$ together with this line element is the Kleinian model of three-dimensional hyperbolic space.

Let $d(x, y)$ be the distance of $x, y \in \mathbb{K}$ derived from our line element. There is a very simple formula for the hyperbolic distance in the Kleinian model.

Proposition 5.1. *Suppose that $x, y \in \mathbb{K}$ are represented as vectors $x = [x_1, x_2, x_3, x_4]$, $y = [y_1, y_2, y_3, y_4]$ in homogeneous coordinates. Let q_1 also denote the symmetric bilinear form Φ_{q_1} corresponding to the quadratic form q_1 . Assume that the representatives of x, y are chosen so that $q_1(x, y) \geq 0$ then*

$$\delta(x, y) := \cosh d(x, y) = \frac{q_1(x, y)}{\sqrt{q_1(x)}\sqrt{q_1(y)}}, \tag{5.6}$$

where the positive values of the square roots are taken.

We shall prove Proposition 5.1 later after having discussed the isometries of $\mathbb{K}$ to the various other models.

The hyperbolic lines, that is the geodesics in $\mathbb{K}$, are Euclidean line segments in $\mathbb{K}$. This can be seen from the isometry between $\mathbb{K}$ and the unit ball model $\mathbb{B}$ on the next but one page and is one of the main advantages of the Kleinian model. It must be observed, however, that the hyperbolic angle between two hyperbolic lines in $\mathbb{K}$ does not agree with the corresponding Euclidean angle. This is the main drawback of the Kleinian model. Nevertheless, there exists an elegant formula for the angle between two hyperbolic planes in $\mathbb{K}$ (see (5.7)).

We add a few remarks on hyperbolic geometry in the Kleinian model $\mathbb{K}$. The boundary of $\mathbb{K}$ is given by $\partial\mathbb{K} = \{ b \in \mathbb{P}\tilde{E}_1 \mid q_1(b) = 0 \}$. Assume now that $b = [b_1, b_2, b_3, b_4] \in \mathbb{P}\tilde{E}_1$ is a homogeneous vector such that $q_1(b) < 0$. Then b is a point in $\mathbb{P}\tilde{E}_1$ outside $\mathbb{K}$. There is a unique tangent cone from b to the ellipsoid bounding $\mathbb{K}$. This tangent cone meets the ellipsoid in an ellipse, and this ellipse uniquely defines a hyperplane in $\mathbb{P}\tilde{E}_1$. It is customary in projective geometry to identify this hyperplane (also called the polar of b) with the point $b \in \mathbb{P}\tilde{E}_1$ (the pole of the hyperplane). This identification enables us to say: *A hyperplane in* $\mathbb{K}$ *is a homogeneous vector* b *such that* $q_1(b) < 0$. We have a simple criterion for the incidence of a point $x \in \mathbb{K}$ with a hyperplane b: The point $x \in \mathbb{K}$ lies on the hyperplane b if and only if $q_1(x, b) = 0$. There is also an analogue of formula (5.6) for hyperplanes in $\mathbb{K}$. Let $a, b \in \mathbb{P}\tilde{E}_1$ be two hyperplanes. Then the angle α between a and b satisfies

$$\cos\alpha = \pm\frac{q_1(a,b)}{\sqrt{q_1(a)}\sqrt{q_1(b)}}. \tag{5.7}$$

Note that the square roots are purely imaginary. This formula determines $\cos\alpha$ only up to a sign factor ± 1. This is inherent to the present situation since there is no way to distinguish, in homogeneous coordinates, between α and $\pi - \alpha$. Formula (5.7) should be compared with (10.2.40) which does distinguish between the two angles in question. If the absolute value of the right-hand side of (5.7) is greater than 1, then this absolute value is the distance function δ for the planes a, b measured along the unique common perpendicular.

We shall now give an isometry between $\mathbb{H}$ and $\mathbb{K}$.

Definition 5.2. For an element $P = x + yi + rj \in \mathbb{H}$ we define

$$\psi_0(P) = [(1 + P\bar{P})f_0 + (1 - P\bar{P})f_1 - 2xf_2 - 2yf_3].$$

Notice that $\psi_0(P)$ actually lies in $\mathbb{K}$.

The linear action of $\mathbf{O}_4(\mathbb{R}, q_1)$ on $\tilde{E}_1$ induces an action of $\mathbf{O}_4(\mathbb{R}, q_1)$ on $\mathbb{P}\tilde{E}_1$ which maps $\mathbb{K}$ to itself. Our homomorphism $\Psi : \mathbf{SL}(2, \mathbb{C}) \to \mathbf{O}_4(\mathbb{R}, q_1)$ defined in Definition 3.9, induces a homomorphism $\Psi : \mathbf{SL}(2, \mathbb{C}) \to \mathbf{PO}_4(\mathbb{R}, q_1)$ which we also call Ψ. We shall formulate now some properties of ψ_0 which will result in a description of the isometry group of $\mathbb{K}$.

Proposition 5.3. *The map* ψ_0 *has the following properties:*

(1) ψ_0 *is an isometry from* $\mathbb{H}$ *to* $\mathbb{K}$.
(2) ψ_0 *is equivariant with respect to the homomorphism* $\Psi : \mathbf{SL}(2, \mathbb{C}) \to \mathbf{PO}_4(\mathbb{R}, q_1)$, *that is* $\psi_0(g \cdot P) = \Psi(g) \cdot \psi_0(P)$.

The proof goes along the same lines as in Proposition 4.2. We also get the following description of the isometry group.

Proposition 5.4. *The linear action of* $\mathbf{O}_4(\mathbb{R}, q_1)$ *on* $\tilde{E}_1$ *induces isomorphisms*

$$\mathbf{PO_4}(\mathbb{R}, q_1) \to \mathbf{Iso}(\mathbb{K}), \qquad \mathbf{PSO_4}(\mathbb{R}, q_1) \to \mathbf{Iso}^+(\mathbb{K}).$$

Identifying the group $\mathbf{O}(3) = \{ A \in \mathbf{GL}(3, \mathbb{R}) \mid A^t A = I \}$ *with the subgroup*

$$\left\{ \begin{pmatrix} 1 & 0 \\ 0 & A \end{pmatrix} \;\middle|\; A \in \mathbf{O}(3) \right\} < \mathbf{PO_4}(\mathbb{R}, q_1),$$

the above maps induce the identification of the stabilizer $\mathbf{Iso}(\mathbb{K})_{[1,0,0,0]} = \mathbf{O}(3)$.

The action of $\mathbf{Spin}_4(\mathbb{R}, q_1)$ on $\tilde{E}_1$ induces an action on $\mathbb{P}\tilde{E}_1$ which maps $\mathbb{K}$ to itself. The equivariance of ψ_0 can then also be formulated using our isomorphism $\psi : \mathbf{SL}(2, \mathbb{C}) \to \mathbf{Spin}_4(\mathbb{R}, q_1)$. We have $\psi_0(g \cdot P) = \psi(g) \cdot \psi_0(P) \cdot \psi(g)^*$. The natural map $\tilde{E}_1 \setminus \{0\} \to \mathbb{P}\tilde{E}_1$ induces an isometry from $\mathbb{S}$ to $\mathbb{K}$. The resulting identifications of isometry groups are the isomorphisms $\mathbf{O}_4^+(\mathbb{R}, q_1) \to \mathbf{PO}_4(\mathbb{R}, q_1)$, $\mathbf{SO}_4^+(\mathbb{R}, q_1) \to \mathbf{PSO}_4(\mathbb{R}, q_1)$ from Propositions 3.12, 3.13.

The result in Proposition 5.4 can be used to describe the hyperbolic spheres of constant distance to the point $[1, 0, 0, 0] \in \mathbb{K}$. They are in fact euclidean spheres given by $\{ [1, x_1, x_2, x_3] \mid x_1^2 + x_2^2 + x_3^2 = r \}$. Here the radius r has to satisfy $r < 1$.

There is a nice classical and geometric way to define an isometry from $\mathbb{K}$ onto the unit ball model $\mathbb{B}$ which we will explain now. Many of the geometric properties of $\mathbb{K}$ can be seen from the geometry of this map. We construct such a map by means of a stereographic projection followed by an orthogonal projection. Start with a point (u_0, u_1, u_2) of the closed unit ball $\bar{B}^3$ in $\mathbb{R}^3$, then the u_0, u_1, u_2 satisfy: $u_0^2 + u_1^2 + u_2^2 \le 1$. Consider the 3-sphere of radius 1 and centre 0 in $\mathbb{R}^4$, that is

$$S^3 = \{ (z_1, z_2, z_3, z_4) \in \mathbb{R}^4 \mid z_1^2 + z_2^2 + z_3^2 + z_4^2 = 1 \}. \tag{5.8}$$

Use stereographic projection from the north pole $(0, 0, 0, 1)$ of the sphere S^3 to map the point $(u_0, u_1, u_2, 0)$ from the hyperplane $z_4 = 0$ in $\mathbb{R}^4$ to the southern hemisphere $z_4 \le 0$ of (5.8). This yields a point $(z_1, z_2, z_3, z_4) \in S^3$. Project this point orthogonally to the hyperplane $z_4 = 0$ to obtain the point $(z_1, z_2, z_3) \in \bar{B}^3$. The map

$$\bar{B}^3 \ni (u_0, u_1, u_2) \mapsto (z_1, z_2, z_3) \in \bar{B}^3$$

obviously is a bijection which carries 2-spheres in $\mathbb{R}^3$ which are orthogonal to the boundary S^2 of $\bar{B}^3$ into Euclidean planes which meet S^2 in the same circle. Remember that hemispheres or parts of Euclidean planes in $\mathbb{B}$ which are perpendicular to the boundary S^2 of $\mathbb{B}$ are the hyperbolic planes in the unit ball model. An informative picture of our map in slightly different, but equivalent notation in the corresponding two-dimensional case is in Hilbert and Cohn-Vossen (1932), fig. 246. Analytically, our map $(u_0, u_1, u_2) \mapsto (z_1, z_2, z_3)$ is described by

$$\begin{pmatrix} z_1 \\ z_2 \\ z_3 \end{pmatrix} = \frac{2}{1+\rho^2} \begin{pmatrix} u_0 \\ u_1 \\ u_2 \end{pmatrix}, \qquad \rho^2 := u_0^2 + u_1^2 + u_2^2. \tag{5.9}$$

From this formula it can easily be checked that the resulting map from $\mathbb{B}$ to $\mathbb{K}$ is an isometry. This map obviously leaves the boundary S^2 of $\bar{B}^3$ fixed pointwise. The origin $(0,0,0) \in \bar{B}^3$ is also left fixed. These properties are also evident from the geometry of our construction.

Note that the inverse map of (5.9) is

$$\begin{pmatrix} u_0 \\ u_1 \\ u_2 \end{pmatrix} = (1 + \sqrt{1-r^2})^{-1} \begin{pmatrix} z_1 \\ z_2 \\ z_3 \end{pmatrix}, \qquad r^2 := z_1^2 + z_2^2 + z_3^3. \tag{5.10}$$

Proof of Proposition 5.1. Let x, y be chosen as required and put $z_\nu := \frac{x_\nu}{x_4}$, $w_\nu := \frac{y_\nu}{y_4}$ for $\nu = 1, 2, 3$. Then we have

$$\frac{q_1(x,y)}{\sqrt{q_1(x)}\sqrt{q_1(y)}} = \frac{1 - z_1w_1 - z_2w_2 - z_3w_3}{\sqrt{1-z_1^2-z_2^2-z_3^2}\sqrt{1-w_1^2-w_2^2-w_3^2}}.$$

Let (ξ_1, ξ_2, ξ_3), $(\eta_1, \eta_2, \eta_4) \in \mathbb{B}$ be the image points of (z_1, z_2, z_3), (w_1, w_2, w_3) under the mapping (5.9), respectively. Then

$$\frac{q_1(x,y)}{\sqrt{q_1(x)}\sqrt{q_1(y)}} = \frac{(1+\xi_1^2+\xi_2^2+\xi_3^2)(1+\eta_1^2+\eta_2^2+\eta_3^2) - 4(\xi_1\eta_1+\xi_2\eta_2+\xi_3\eta_3)}{(1-\xi_1^2-\xi_2^2-\xi_3^2)(1-\eta_1^2-\eta_2^2-\eta_3^2)}.$$

Mapping to $\mathbb{H}$ by means of ${\psi_0}^{-1}$, we obtain two points $P, Q \in \mathbb{H}$ corresponding to $x, y \in \mathbb{K}$, respectively, and computing the right-hand side of this formula, we obtain

$$\frac{q_1(x,y)}{\sqrt{q_1(x)}\sqrt{q_1(y)}} = \delta(P,Q),$$

which implies our assertion. □

We now eliminate the restriction to the special form q_1. Assume that q is an arbitrary quadratic form on $\mathbb{R}^4$ which over $\mathbb{R}$ is equivalent to q_1. That is we require q to have signature (1,3). Use the letter q also to denote the symmetric bilinear form corresponding to q. Remember that a nonsingular linear map T from $\tilde{E}_1$ to $\mathbb{R}^4$ defines an equivalence of q and q_1 whenever $q = q_1 \circ T^{-1}$.

The set

$$\mathbb{K}(q) := \{\ y \in \mathbb{P}^3\mathbb{R} \ \mid \ q(y) > 0\ \} \tag{5.11}$$

will serve as the set of points for the Kleinian model of hyperbolic geometry associated with q. Notice that the condition $q(y) > 0$ is well defined on projective space. If T defines an equivalence of q and q_1, T also defines a linear homeomorphism $T : \mathbb{K} \to \mathbb{K}(q)$. We transfer the hyperbolic metric from $\mathbb{K}$ to $\mathbb{K}(q)$ by means of T^{-1} and the space $\mathbb{K}(q)$ becomes the Kleinian model of hyperbolic geometry attached to q. If we choose another linear isomorphism S from $\tilde{E}_1$ to $\mathbb{R}^4$ such that $q = q_1 \circ S^{-1}$, then $T^{-1}S$ is in $\mathbf{O}_4(\mathbb{R}, q_1)$ and hence by Proposition 5.4 defines an isometry of $\mathbb{K}$. This proves that the hyperbolic geometry on $\mathbb{K}(q)$ does not depend on the choice of T.

Directly from Proposition 5.4 we also find the groups of isometries and orientation preserving isometries of $\mathbb{K}(q)$: $\mathbf{Iso}(\mathbb{K}(q)) = \mathbf{PO_4}(\mathbb{R}, q)$, $\mathbf{Iso}^+(\mathbb{K}(q)) = \mathbf{SO_4}(\mathbb{R}, q)$.

These groups have obvious discrete subgroups. Assume for example that q has rational coefficients, then the group of integral matrices $\mathbf{PO}_4(q, \mathbb{Z}) := \mathbf{PO}_4(\mathbb{R}, q) \cap \mathbf{PSL}(4, \mathbb{Z})$ is discrete in $\mathbf{PO}_4(\mathbb{R}, q)$. Hence the Kleinian model is particularly well-suited for the discussion of certain arithmetically defined discrete transformation groups of hyperbolic space.

It is now an easy reformulation of Proposition 5.1 to write down the formula for the hyperbolic distance between two points in $\mathbb{K}(q)$. Let d_q denote the hyperbolic distance on $\mathbb{K}(q)$. Suppose that $x, y \in \mathbb{K}(q)$ are represented by vectors $x = [x_1, x_2, x_3, x_4]$, $y = [y_1, y_2, y_3, y_4]$ in homogeneous coordinates. Assume that the representatives of x, y are chosen so that $q(x, y) > 0$. Then

$$\delta_q(x, y) := \cosh d_q(x, y) = \frac{q(x, y)}{\sqrt{q(x)}\sqrt{q(y)}}. \tag{5.12}$$

As in the model $\mathbb{K}$ we identify the geodesic hyperplanes in $\mathbb{K}(q)$ with the homogeneous vectors $b \in \mathbb{P}^3\mathbb{R}$ such that $q(b) < 0$. This again yields the criterion for the incidence of a point $x \in \mathbb{K}(q)$ with a hyperplane b: The point $x \in \mathbb{K}(q)$ lies on the hyperplane b if and only if $q(x, b) = 0$. There is also an analogue of the formula for the angle between two hyperbolic planes in $\mathbb{K}(q)$. Let $a, b \in \mathbb{P}^3$ be two hyperplanes. Then the angle α between a and b satisfies

$$\cos \alpha = \pm \frac{q(a, b)}{\sqrt{q(a)}\sqrt{q(b)}}. \tag{5.13}$$

The remarks following (5.7) also apply here.

1.6 Upper Half-Space as a Symmetric Space

The action of $\mathbf{SL}(2,\mathbb{C})$ on the upper half-space model $\mathbb{H}$ may also be looked at in the spirit of the theory of topological transformation groups and of Lie theory. Considered as a Lie group over $\mathbb{R}$, $\mathbf{SL}(2,\mathbb{C})$ has dimension 6. The special unitary group

$$\mathbf{K} := \mathbf{SU}(2) := \{\ V \mid V \in \mathbf{U}(2),\ \det(V) = 1\ \}$$

is one of its maximal compact subgroups. The symmetric space associated to $\mathbf{SL}(2,\mathbb{C})$ is $\mathbf{SL}(2,\mathbb{C})/\mathbf{K}$ with $\mathbf{SL}(2,\mathbb{C})$ acting by multiplication from the left. The book of Helgason (1962) contains the general theory of this situation. Here we have the map

$$\text{(6.1)} \qquad \pi : \mathbf{SL}(2,\mathbb{C}) \to \mathbb{H}, \qquad \pi(g) = g \cdot j,$$

which gives rise to an $\mathbf{SL}(2,\mathbb{C})$-equivariant bijection between the symmetric space of $\mathbf{SL}(2,\mathbb{C})$ and $\mathbb{H}$. Note that π is infinitely often differentiable with respect to the natural manifold structures on $\mathbf{SL}(2,\mathbb{C})$ and $\mathbb{H}$.

The identification $\mathbb{H} = \mathbf{SL}(2,\mathbb{C})/\mathbf{K}$ allows us to derive the Riemannian structure given by (1.4) from the Killing form of the Lie algebra $\mathcal{SL}(2,\mathbb{C})$ of the Lie group $\mathbf{SL}(2,\mathbb{C})$. We have $\mathcal{SL}(2,\mathbb{C}) = \{\ x \in M(2,\mathbb{C}) \mid \operatorname{tr}(x) = 0\ \}$ with the Lie bracket defined by $[x,y] := xy - yx$. The Lie algebra $\mathcal{SL}(2,\mathbb{C})$ is identified as usually with the tangent space $T(\mathbf{SL}(2,\mathbb{C}))_1$ to $\mathbf{SL}(2,\mathbb{C})$ at the identity. The Killing form is the non-degenerate bilinear form on $\mathcal{SL}(2,\mathbb{C})$ given by $B(x,y) := \operatorname{tr}(\operatorname{ad}(x)\circ\operatorname{ad}(y))$ where $\operatorname{ad}(x)(z) := [x,z]$. The Lie algebra $\mathcal{SL}(2,\mathbb{C})$ has an involutory automorphism θ called the Cartan involution of $\mathbf{K}$ defined as $\theta(x) := -\bar{x}^t$. The fixed point set $\mathcal{SU}(2)$ of θ is the Lie algebra of the Lie group $\mathbf{K} = \mathbf{SU}(2)$, it consists of the skew hermitian 2×2 matrices. Let $\mathcal{P}$ be the orthogonal complement of $\mathcal{SU}(2)$ with respect to the Killing form. The space $\mathcal{P}$ consists of hermitian matrices, that is $\mathcal{P} = \{\ x \in \mathcal{SL}(2,\mathbb{C}) \mid x = \bar{x}^t\ \}$. The Killing form B restricts to a positive definite bilinear form B_0 on the 3-dimensional real vector space $\mathcal{P}$, an orthonormal basis is given by the matrices $u/2$, $v/2$, $h/2$ where

$$\text{(6.2)} \qquad u := \begin{pmatrix} 0 & 1 \\ 1 & 0 \end{pmatrix}, \qquad v := \begin{pmatrix} 0 & i \\ -i & 0 \end{pmatrix}, \qquad h := \begin{pmatrix} 1 & 0 \\ 0 & -1 \end{pmatrix}.$$

This scalar product on $\mathcal{P}$ is invariant under the adjoint action of $\mathbf{K} = \mathbf{SU}(2)$.

All of this being simple linear algebra we have

Proposition 6.1. *The differential $d\pi : T(\mathbf{SL}(2,\mathbb{C}))_1 = \mathcal{SL}(2,\mathbb{C}) \to T(\mathbb{H})_j$ is zero on $\mathcal{SU}(2)$ and an isomorphism on $\mathcal{P}$. Let B_0 also denote the scalar product induced by (6.1) on the tangent space $T(\mathbb{H})_j$. Using the $\mathbf{SL}(2,\mathbb{C})$ action, B_0 may be extended to a C^∞-Riemannian structure B_0 on $\mathbb{H}$. This structure coincides with (1.4), that is the corresponding quadratic form Q is given in local coordinates by*

$$Q_{(x,y,r)} = \frac{dx^2 + dy^2 + dr^2}{r^2}.$$

Proof. What remains to be shown are the formulas

$$(6.3) \qquad d\pi\left(\frac{u}{2}\right) = \frac{\partial}{\partial x}\bigg|_j, \qquad d\pi\left(\frac{v}{2}\right) = \frac{\partial}{\partial y}\bigg|_j, \qquad d\pi\left(\frac{h}{2}\right) = \frac{\partial}{\partial r}\bigg|_j.$$

Note first that we have $d\pi(A) = \frac{d}{dt}\exp(tA)\cdot j\,|_0$ for every $A \in \mathcal{SL}(2,\mathbb{C})$. We infer (6.3) then from

$$(6.4) \quad \exp(tu) = \begin{pmatrix} \cosh t & \sinh t \\ \sinh t & \cosh t \end{pmatrix}, \qquad \exp(tv) = \begin{pmatrix} \cosh t & i\sinh t \\ -i\sinh t & \cosh t \end{pmatrix},$$

$$(6.5) \qquad \exp(th) = \begin{pmatrix} \exp(t) & 0 \\ 0 & \exp(-t) \end{pmatrix}.$$

□

The coset space $\mathbf{SL}(2,\mathbb{C})/\mathbf{K}$ can also be thought of as a space of hermitian forms. This will play an important rôle in Chapter 9. It can be done as follows.

Associate with the coset $M\mathbf{K}$ ($M \in \mathbf{SL}(2,\mathbb{C})$) the positive definite hermitian matrix $Y := M\bar{M}^t$, where the bar denotes entry-wise complex conjugation. This gives an injection of $\mathbf{SL}(2,\mathbb{C})/\mathbf{K}$ into the set $\mathcal{H}^+$ of positive definite Hermitean matrices. If we choose by Iwasawa decomposition in each coset $M\mathbf{K}$ the uniquely determined representative

$$\begin{pmatrix} 1 & z \\ 0 & 1 \end{pmatrix}\begin{pmatrix} \sqrt{r} & 0 \\ 0 & \frac{1}{\sqrt{r}} \end{pmatrix} = \begin{pmatrix} \sqrt{r} & \frac{z}{\sqrt{r}} \\ 0 & \frac{1}{\sqrt{r}} \end{pmatrix}$$

($z \in \mathbb{C}, r > 0$), we can parametrize the image of $\mathbf{SL}(2,\mathbb{C})/\mathbf{K}$ in $\mathcal{H}^+$ by means of

$$Y = M\bar{M}^t = \begin{pmatrix} \frac{|z|^2+r^2}{r} & \frac{z}{r} \\ \frac{\bar{z}}{r} & \frac{1}{r} \end{pmatrix}.$$

This leads us to introduce the quaternion $P = z + rj \in \mathbb{H}$ as a coordinate for Y (hence for $M\mathbf{K}$). Now a computation yields that the natural action of $\mathbf{SL}(2,\mathbb{C})$ on $\mathbf{SL}(2,\mathbb{C})/\mathbf{K}$ (multiplication of cosets from the left) is expressed in terms of the coordinate P exactly by (1.9), (1.10). The details of this computation are contained in Maaß (1964), p. 6–7. Kubota (1968a) chooses the familiar representation of the quaternions by complex 2×2-matrices which makes the computations very transparent.

We finally recall that there are natural homeomorphisms $\mathbf{SU}(2) \cong S^3$, $\mathbf{SO}_3 \cong \mathbb{P}^3$. The group $\mathbf{SU}(2)$ is the simply connected universal covering group of $\mathbf{SO}_3$. For these and related facts, see Chevalley (1946).

1.7 Notes and Remarks

The first mathematical works on non-Euclidean geometry were published by N. Lobachevski (1829) and J. Bolyai (1832). E. Beltrami (1868) introduced the upper half-space model, the unit ball model, and the projective disc model (i.e. the Kleinian model) of n-dimensional hyperbolic geometry; see e.g. Milnor (1982) for a brief and lucid exposition of the history of the subject. These models were well-known to the classical writers who used them in their investigations on discontinuous groups; see e.g. Bianchi (1952), Vol. I, 1, Fricke (1914), Fricke and Klein (1897), Vol. 1, Humbert (1919a), Klein (1872), (1884), (1893), Klein and Fricke (1890), Vol. 1, Picard (1978), Vol. 1, Poincaré (1916), Vol. 2. The classical literature is extensively quoted in Fricke (1914).

The upper half-space model of three-dimensional hyperbolic geometry received less attention outside the field of differential geometry in subsequent decades, notable exceptions being e.g. the work of Fueter (1927), (1931), (1936), Got (1933), (1934) and Maaß (1953). Its importance was emphasized in the recent past in particular in the course of the modern development of the theory of Kleinian groups. The reader will find the accounts of three-dimensional hyperbolic geometry in Beardon (1983) and Magnus (1974) well-suited for our purposes. The upper half-space model also arises quite naturally in connection with the theory of binary Hermitian forms, see Chapter 9 of this book for an exposition of this topic.

Hyperbolic spaces are typical examples of spaces of constant curvature. Wolf (1977) is the standard reference for this subject; Section 2.4 of this book contains the various classical models of hyperbolic n-space and describes the isometries between them.

Proposition 1.6 is Fricker's (1968) Satz 4 for which we have given a simple proof using Selberg's elegant method of point-pair invariants. Equation (1.15) is also in Beardon and Nicholls (1972) where a more complicated proof is given. Proposition 1.8 is adapted from Elstrodt (1973a), Lemma 1.1.

The Kleinian model of hyperbolic geometry is used extensively in Got (1933), (1934). A slightly more general approach is in Chen (1974).

The exceptional isomorphism described in section 1.3 is a particular example of isomorphism of orthogonal groups in dimensions 3, 4, 5, 6 with lower dimensional linear groups. They seem to have appeared first in Klein's Erlanger program (1872), (1884), (1893).

The formulas of elementary hyperbolic geometry become particularly nice if we represent the points of hyperbolic 3-space by matrices in $\mathbf{SL}(2,\mathbb{C})$. The following formalism is due to H. Helling. We define

$$\mathbb{W}_0 := \left\{ \begin{pmatrix} a & b \\ c & -\bar{a} \end{pmatrix} \in \mathbf{SL}(2,\mathbb{C}) \;\middle|\; a \in \mathbb{C},\ b, c \in \mathbb{R},\ c > 0 \right\}. \tag{7.1}$$

The group $\mathbf{SL}(2,\mathbb{C})$ acts on $\mathbb{W}_0$ by $w \mapsto w' = Xw\overline{X}^{-1}$ for $w \in \mathbb{W}_0$, $X \in \mathbf{SL}(2,\mathbb{C})$. Consider two points $v, w \in \mathbb{W}_0$, and let the functions δ, d be defined by

$$\delta(v,w) = \cosh d(v,w), \qquad \delta(v,w) := -\frac{1}{2}\,\mathrm{tr}(v \cdot \overline{w})\,. \tag{7.2}$$

Then d is a metric on $\mathbb{W}_0$. For $P = (z,r) \in \mathbb{H}$ we put

$$\omega(P) := \begin{pmatrix} zr^{-1} & b \\ r^{-1} & -\bar{z}r^{-1} \end{pmatrix} \in \mathbb{W}_0 \tag{7.3}$$

with suitable $b \in \mathbb{R}$. The map $\omega : \mathbb{H} \to \mathbb{W}_0$ is then a $\mathbf{SL}(2,\mathbb{C})$-equivariant isometry. We furthermore put

$$\mathbb{W}_2 := \left\{ \begin{pmatrix} a & bi \\ ci & -\bar{a} \end{pmatrix} \in \mathbf{PSL}(2,\mathbb{C}) \;\middle|\; a \in \mathbb{C},\ b, c \in \mathbb{R} \right\}. \tag{7.4}$$

The group $\mathbf{SL}(2,\mathbb{C})$ acts on $\mathbb{W}_2$ by $E \mapsto E' = XE\overline{X}^{-1}$ for $E \in \mathbb{W}_2$, $X \in \mathbf{SL}(2,\mathbb{C})$. For $E \in \mathbb{W}_2$ we define

$$\tilde{E} := \{\, v \in \mathbb{W}_0 \;\mid\; \mathrm{tr}(v\overline{E}) = 0 \,\}. \tag{7.5}$$

Then $\hat{E} := \omega^{-1}(\tilde{E})$ is a hyperbolic plane in $\mathbb{H}$. The map $E \mapsto \hat{E}$ is a $\mathbf{SL}(2,\mathbb{C})$-equivariant bijection from $\mathbb{W}_2$ to the set of hyperbolic planes in $\mathbb{H}$. The additional sign which may be gained by looking at the preimage of $\mathbb{W}_2$ in $\mathbf{SL}(2,\mathbb{C})$ can be used to determine the orientations of the planes in $\mathbb{H}$. For $E, F \in \mathbb{W}_2$, consider the quantity

$$\delta(E,F) = \frac{1}{2}\,\mathrm{tr}(E\overline{F})\,. \tag{7.6}$$

The planes $\hat{E}, \hat{F}$ intersect iff $-1 \le \delta(E,F) \le +1$, and the angle α of intersection is given by $\cos\alpha = \delta(E,F)$. If $|\delta(E,F)| > 1$, then $\hat{E}, \hat{F}$ do not intersect, they then have a unique common perpendicular. The distance between $\hat{E}, \hat{F}$, measured along the common perpendicular, is given by $\cosh d(E,F) = |\delta(E,F)|$. For $v \in \mathbb{W}_0$ and $E \in \mathbb{W}_2$ the shortest distance between $\omega(v)$ and $\hat{E}$ is given by

$$\sinh \delta(v,E) = \frac{1}{2}\,|\mathrm{tr}(v\overline{E})|\,.$$

Let $X_1,\ X_2,\ X_3 \in \mathbb{W}_0$ be three pairwise distinct points. Define $x_i = \delta(X_j, X_k)$ where i, k, j run through the distinct triples of the numbers 1,2,3. Put

$$Y = \frac{i}{2}\,\frac{1}{\sqrt{1 + 2x_1x_2x_3 - x_1^2 - x_2^2 - x_3^2}}\,\left(X_1\overline{X}_2X_3 - X_3\overline{X}_2X_1\right)$$

Then $\hat{Y}$ is the plane through the points $\omega(X_1)$, $\omega(X_2)$, $\omega(X_3)$. Let X_1, X_2, $X_3 \in \mathbb{W}_2$ be three pairwise distinct elements. Define $x_i = |\delta(X_j, X_k)|$ where i, k, j run through the distinct triples of the numbers 1,2,3. Put

$$Y = \frac{1}{2} \frac{1}{\sqrt{1 + 2x_1x_2x_3 - x_1^2 - x_2^2 - x_3^2}} \left(X_1\overline{X}_2X_3 - X_3\overline{X}_2X_1\right).$$

If $1+2x_1x_2x_3-x_1^2-x_2^2-x_3^2 > 0$, then $\omega(Y)$ is the point in which $\hat{X}_1$, $\hat{X}_2$, $\hat{X}_3$ meet. If $1 + 2x_1x_2x_3 - x_1^2 - x_2^2 - x_3^2 < 0$, then $\hat{Y}$ is the unique plane which is perpendicular to $\hat{X}_1$, $\hat{X}_2$, $\hat{X}_3$.

Chapter 2. Groups Acting Discontinuously on Three-Dimensional Hyperbolic Space

In this part we shall describe fundamental facts from the theory of transformation groups on 3-dimensional hyperbolic space. The more elementary theory is quoted from Ford (1951) or Beardon (1977), (1983). We set up the theory as far as is necessary for the chapters to come. We can only quote the more difficult theorems on the subject. To include the proofs of all of them would have blown up the size of this book.

This chapter does not contain any particular examples. We have postponed their discussion until Chapters 7 and 10.

We use the upper half-space $\mathbb{H}$ as reference model for hyperbolic 3-space. A lot of our discussion will be correct in any other model.

2.1 Discontinuity

As in Chapter 1 the symbol $\mathbb{H}$ stands for the upper half-space model of 3-dimensional hyperbolic space and $\mathbf{Iso}(\mathbb{H})$ is its group of isometries. We write $\mathbf{Iso}^+(\mathbb{H})$ for the group of orientation preserving isometries of $\mathbb{H}$. The group $\mathbf{Iso}^+(\mathbb{H})$ is isomorphic with $\mathbf{PSL}(2,\mathbb{C})$. We think of these two groups as being identified.

Definition 1.1. A subgroup $\Gamma < \mathbf{Iso}(\mathbb{H})$ is called a discontinuous group if and only if for every $P \in \mathbb{H}$ and for every sequence $(T_n)_{n\geq 1}$ of distinct elements of Γ the sequence $(T_n P)_{n\geq 1}$ has no accumulation point in $\mathbb{H}$. In this case Γ is also said to act discontinuously on $\mathbb{H}$.

Remember that a subgroup $\Gamma < \mathbf{PSL}(2,\mathbb{C})$ is called discrete if and only if its inverse image in $\mathbf{SL}(2,\mathbb{C}) \subset \mathbb{C}^4$ is discrete in the vector space topology. The topology on $\mathbf{SL}(2,\mathbb{C})$ can be defined by the following norm of matrices:

$$\|A\| = +\sqrt{|a|^2+|b|^2+|c|^2+|d|^2} \tag{1.1}$$

for $A = \begin{pmatrix} a & b \\ c & d \end{pmatrix}$. See Beardon (1983), Chapter 1 for a discussion. Note that $\|\ \|$ defines also a function on $\mathbf{PSL}(2,\mathbb{C})$ and for $A \in \mathbf{PSL}(2,\mathbb{C})$ we have by Proposition 1.1.7

$$\delta(j, A(j)) = \frac{1}{2}\,\|A\|^2 = \cosh(d(j, A(j))).$$

See Chapter 1 for the definition of the quantities δ, d.

Discontinuous groups $\Gamma < \mathbf{PSL}(2, \mathbb{C}) = \mathbf{Iso}^+(\mathbb{H}) \leq \mathbf{Iso}(\mathbb{H})$ can now be characterized as follows.

Theorem 1.2. *A subgroup $\Gamma < \mathbf{PSL}(2, \mathbb{C})$ is a discontinuous group if and only if Γ is discrete in $\mathbf{PSL}(2, \mathbb{C})$.*

This is a classical theorem of Poincaré (1916), Tome II, page 268. See for example Beardon (1977), Section 4, Magnus (1974), page 58. We remark that the following third characterisation of discontinuous groups can easily be proved. A subgroup $\Gamma < \mathbf{Iso}(\mathbb{H})$ is discontinuous if and only if the set $\{\, \gamma \in \Gamma \mid \gamma K \cap K \neq \emptyset \,\}$ is finite for every compact set $K \subset \mathbb{H}$.

A very clear treatment of the more elementary properties of discontinuous groups is contained in Beardon (1983). We shall take these results for granted here.

We introduce now the standard notations for parabolic, hyperbolic, elliptic and loxodromic elements. Our conventions are the same as in Ford (1951), chapter I.

Definition 1.3. An element $\gamma \in \mathbf{SL}(2, \mathbb{C}), \gamma \neq \pm I$ is called

$$\begin{array}{rlcll} \text{parabolic} & \text{iff} & |\mathrm{tr}\,(\gamma)| = 2 & \text{and} & \mathrm{tr}\,\gamma \in \mathbb{R}, \\ \text{hyperbolic} & \text{iff} & |\mathrm{tr}\,(\gamma)| > 2 & \text{and} & \mathrm{tr}\,\gamma \in \mathbb{R}, \\ \text{elliptic} & \text{iff} & 0 \leq |\mathrm{tr}(\gamma)| < 2 & \text{and} & \mathrm{tr}\,\gamma \in \mathbb{R}, \end{array}$$

and loxodromic in all other cases. An element of $\mathbf{PSL}(2, \mathbb{C})$ is called parabolic, elliptic, hyperbolic, loxodromic if its preimages in $\mathbf{SL}(2, \mathbb{C})$ have this property.

An element $\gamma \in \mathbf{PSL}(2, \mathbb{C})$ is parabolic if and only if it is conjugate in $\mathbf{PSL}(2, \mathbb{C})$ to $\begin{pmatrix} 1 & 1 \\ 0 & 1 \end{pmatrix}$. In all other cases γ is diagonalizable, the eigenvalues being real in the hyperbolic case and of absolute value 1 in the elliptic case.

There is also a geometric characterisation of these concepts, which we are going to describe in the next Proposition. Let γ be an orientation preserving isometry of $\mathbb{H}$, then γ extends by continuity to a uniquely defined map $\gamma : \mathbb{P}^1\mathbb{C} \to \mathbb{P}^1\mathbb{C}$ of the boundary $\mathbb{P}^1\mathbb{C}$ of $\mathbb{H}$. Our characterisation will refer to properties of the map given by a $\gamma \in \mathbf{PSL}(2, \mathbb{C})$ both on $\mathbb{H}$ and $\mathbb{P}^1\mathbb{C}$.

Proposition 1.4. *Let $\gamma \neq I$ be an element of* $\mathbf{PSL}(2, \mathbb{C})$*. Then the following hold:*

γ is parabolic iff γ has exactly one fixed point in $\mathbb{P}^1\mathbb{C}$.

γ is elliptic iff it has two fixed points in $\mathbb{P}^1\mathbb{C}$ and if the points on the geodesic line in $\mathbb{H}$ joining these two points are also left fixed. γ is then a rotation around this line.

γ *is hyperbolic iff it has two fixed points in* $\mathbb{P}^1\mathbb{C}$ *and if any circle in* $\mathbb{P}^1\mathbb{C}$ *through these two points together with its interior is left invariant. The line in* $\mathbb{H}$ *joining these two fixed points is then left invariant, but* γ *has no fixed point in* $\mathbb{H}$.

γ *is loxodromic in all other cases.* γ *has then two fixed points in* $\mathbb{P}^1\mathbb{C}$ *and no fixed point in* $\mathbb{H}$. *The geodesic joining the 2 fixed points is the only geodesic in* $\mathbb{H}$ *which is left invariant.* γ *may leave the circles joining the two fixed points invariant, but then it interchanges exterior and interior.*

Note that the Brouwer fixed point theorem implies that every $\gamma \in \mathbf{PSL}(2,\mathbb{C})$ has at least one fixed point in $\mathbb{H} \cup \mathbb{P}^1\mathbb{C}$. Notice also that $\gamma \in \mathbf{PSL}(2,\mathbb{C})$, $\gamma \neq I$ is elliptic iff it has a fixed point in $\mathbb{H}$. We introduce some more notation.

Definition 1.5. Let $\Gamma \leq \mathbf{PSL}(2,\mathbb{C})$ be a subgroup. For a $P \in \mathbb{H} \cup \mathbb{P}^1\mathbb{C}$ define $\Gamma_P := \{\, \sigma \in \Gamma \mid \sigma P = P \,\}$ to be the stabilizer of P in Γ.

It is then clear that for a discontinuous group Γ and $P \in \mathbb{H}$, the stabilizer Γ_P is a finite group. It is advantageous to know which finite groups are possible. We quote from Ford (1951):

Theorem 1.6. *The finite subgroups of* $\mathbf{PSL}(2,\mathbb{C})$ *are exactly the groups from the following list:*

(1) the cyclic groups $\mathbb{Z}/m\mathbb{Z}$ *of order* m,
(2) the dihedral groups $\mathbf{D_m}$ *of order* $2m$,
(3) the tetrahedral group, it is of order 12 and isomorphic to the alternating group $\mathbf{A_4}$ *on 4 symbols,*
(4) the octahedral group, it is of order 24 and isomorphic to the symmetric group $\mathbf{S_4}$ *on 4 symbols,*
(5) the icosahedral group, it is of order 60 and isomorphic to the alternating group $\mathbf{A_5}$.

The above groups are just the groups of orientation preserving symmetries of the Euclidean regular solids and their subgroups.

We shall now give a description of the stabilizer Γ_ζ for $\zeta \in \mathbb{P}^1\mathbb{C} = \mathbb{C} \cup \{\infty\}$. We will do this by transforming ζ to ∞ and assuming without loss of generality that $\zeta = \infty$. Let us first introduce the following two groups:

$$B(\mathbb{C}) = \left\{ \begin{pmatrix} a & b \\ 0 & a^{-1} \end{pmatrix} \;\middle|\; 0 \neq a \in \mathbb{C},\ b \in \mathbb{C} \right\} / \{\pm I\} < \mathbf{PSL}(2,\mathbb{C}), \tag{1.2}$$

$$N(\mathbb{C}) := \left\{ \begin{pmatrix} 1 & b \\ 0 & 1 \end{pmatrix} \;\middle|\; b \in \mathbb{C} \right\} / \{\pm I\} < \mathbf{PSL}(2,\mathbb{C}). \tag{1.3}$$

The group $B(\mathbb{C})$ is a Borel-group of $\mathbf{PSL}(2,\mathbb{C})$ and $N(\mathbb{C})$ is the unipotent radical of $B(\mathbb{C})$. The group $N(\mathbb{C})$ is isomorphic to the additive group $\mathbb{C}^+$. The Borel-group $B(\mathbb{C})$ is isomorphic to the semidirect product $\mathbb{C}^+ \rtimes (\mathbb{C}^*/\{\pm 1\})$, the action of $\mathbb{C}^*/\{\pm 1\}$ on $\mathbb{C}^+$ being defined by $\sigma_a(b) = a^2 b$. The bijective map

$$\varphi: B(\mathbb{C}) \to \mathbb{C}^+ \rtimes (\mathbb{C}^*/\{\pm 1\}), \qquad \varphi: \begin{pmatrix} a & b \\ 0 & a^{-1} \end{pmatrix} \to (ab, a) \tag{1.4}$$

is in fact an isomorphism. Note the relation $D_a U^b D_a^{-1} = U^{a^2 b}$, if

$$D_a = \begin{pmatrix} a & 0 \\ 0 & a^{-1} \end{pmatrix}, \quad U^b = \begin{pmatrix} 1 & b \\ 0 & 1 \end{pmatrix}.$$

The Borel-group $B(\mathbb{C})$ also appears as the stabilizer of ∞ in the action of $\mathbf{PSL}(2,\mathbb{C})$ on $\mathbb{P}^1\mathbb{C}$ that is $\mathbf{PSL}(2,\mathbb{C})_\infty = B(\mathbb{C})$. If $\zeta \in \mathbb{P}^1\mathbb{C}$ then there is an $A \in \mathbf{PSL}(2,\mathbb{C})$ with $A\zeta = \infty$ and we have the formula $\Gamma_\zeta = \Gamma \cap A^{-1}B(\mathbb{C})A$. We furthermore define:

Definition 1.7. If $\zeta \in \mathbb{P}^1\mathbb{C}$ and $A \in \mathbf{PSL}(2,\mathbb{C})$ with $A\zeta = \infty$ we define

$$\Gamma'_\zeta = \Gamma \cap A^{-1}N(\mathbb{C})A = \Gamma_\zeta \cap A^{-1}N(\mathbb{C})A.$$

Note that Γ'_ζ consists of the parabolic elements in Γ_ζ together with I. The possibilities for Γ_∞ and Γ'_∞ are then summarized in the following theorem.

Theorem 1.8. *Let $\Gamma < \mathbf{PSL}(2,\mathbb{C})$ be a discrete group. Then there are the following possibilities for Γ_∞:*

(1) $\Gamma'_\infty = \{I\}$, then Γ_∞ is equal to one of the following two.

(a) A cyclic group of order $m \geq 1$, which for $m > 1$ is generated by an elliptic element γ of order m which is a rotation around the hyperbolic line joining the fixed point ∞ of γ with the second fixed point of γ.

(b) $< \sigma > \times < \gamma >$, where σ is hyperbolic or loxodromic, γ is elliptic or equal to the identity, and where the set of fixed points $\{a, \infty\}$ of σ is contained in the set of fixed points of γ.

(2) $\Gamma'_\infty = \left\{ \begin{pmatrix} 1 & k\omega \\ 0 & 1 \end{pmatrix} \;\middle|\; k \in \mathbb{Z} \right\}$ with $\omega \neq 0$, then one of the following two holds.

(a) $\Gamma_\infty = \Gamma'_\infty$.

(b) Γ_∞ is conjugate in $B(\mathbb{C})$ to the group $< X, U >$ where

$$X = \begin{pmatrix} i & 0 \\ 0 & -i \end{pmatrix}, \quad U = \begin{pmatrix} 1 & 1 \\ 0 & 1 \end{pmatrix}.$$

Here Γ_∞ is isomorphic to the semidirect product $\mathbb{Z} \rtimes \mathbb{Z}/2\mathbb{Z}$ where the nontrivial element of $\mathbb{Z}/2\mathbb{Z}$ acts on $\mathbb{Z}$ by inversion.

(3) Γ'_∞ is a lattice in $N(\mathbb{C}) \cong \mathbb{C}$, then one of the following three holds.
(a) $\Gamma_\infty = \Gamma'_\infty$.
(b) Γ_∞ is conjugate in $B(\mathbb{C})$ to a group of the form

$$\left\{ \begin{pmatrix} \epsilon & \epsilon b \\ 0 & \epsilon^{-1} \end{pmatrix} \;\middle|\; b \in \Lambda,\ \epsilon \in \{1, i\} \right\} / \{\pm I\}$$

where $\Lambda < \mathbb{C}$ is an arbitrary lattice. The abstract group Γ_∞ is isomorphic to $\mathbb{Z}^2 \rtimes \mathbb{Z}/2\mathbb{Z}$ where the nontrivial element of $\mathbb{Z}/2\mathbb{Z}$ acts by multiplication by -1.
(c) Γ_∞ is conjugate in $B(\mathbb{C})$ to a group of the form

$$\Gamma(n,t) = \left\{ \begin{pmatrix} \epsilon & \epsilon b \\ 0 & \epsilon^{-1} \end{pmatrix} \;\middle|\; \begin{array}{l} b \in \mathcal{O}_n, \\ \epsilon = \exp\left(\dfrac{\pi i \nu t}{n}\right) \text{ for } 1 \le \nu \le 2n \end{array} \right\} / \{\pm I\}$$

where $n = 4$ or $n = 6$ and $t|n$ and where $\mathcal{O}_n$ is the ring of integers in the quadratic number field $\mathbb{Q}(\exp(\frac{2\pi i}{n}))$. Hence, as an abstract group Γ_∞ is isomorphic to the group $\mathbb{Z}^2 \rtimes \mathbb{Z}/m\mathbb{Z}$ for some $m \in \{1,2,3,4,6\}$. An element

$$\epsilon \in \mathbb{Z}/m\mathbb{Z} \cong \left\{ \exp\left(\frac{\pi i \nu}{m}\right) \;\middle|\; \nu \in \mathbb{Z} \right\} / \{\pm I\}$$

acts on $\mathbb{Z}^2 \cong \mathcal{O}^+_{m'}$ by multiplication with ϵ^2, where $m' = 4$ in case $m \in \{1,2,4\}$ and $m' = 6$ otherwise.

Proof. To treat the case (1), we determine the discrete subgroups of $\mathbb{C}^*/\{\pm 1\}$. The discrete subgroups of $\mathbb{C}^*/\{\pm 1\} \simeq S^1 \times]0,\infty[$ are cyclic groups of order $m \ge 1$ generated by the class of a root of unity and direct products of such a finite cyclic group with the cyclic group generated by the class of some $a \in \mathbb{C}^*, |a| \ne 1$. In case (1), the map (1.4) followed by the projection on the second component, yields an injection $\Gamma_\infty \to \mathbb{C}^*/\{\pm 1\}$. Hence Γ_∞ is abelian and consists of diagonalizable elements. Note that two commuting diagonalizable elements $\ne I$ of $\mathbf{PSL}(2,\mathbb{C})$ with one common fixed point in $\mathbb{P}^1\mathbb{C}$ also have their second fixed point in common. This implies (1).

In case (2), only part (b) requires a proof. Assume that $\Gamma_\infty \cap N(\mathbb{C}) \ne \Gamma_\infty$. After a conjugation by an element of $B(\mathbb{C})$, we can assume that $\Gamma_\infty \cap N(\mathbb{C}) = < U >$, where $U = \begin{pmatrix} 1 & 1 \\ 0 & 1 \end{pmatrix}$. Suppose that

$$X = \begin{pmatrix} \epsilon & b \\ 0 & \epsilon^{-1} \end{pmatrix} \in \Gamma_\infty, \qquad X \notin N(\mathbb{C}). \tag{1.5}$$

Then

$$XUX^{-1} = \begin{pmatrix} 1 & \epsilon^2 \\ 0 & 1 \end{pmatrix} \in \Gamma_\infty \cap N(\mathbb{C})$$

is also a generator of $\Gamma_\infty \cap N(\mathbb{C}) \triangleleft \Gamma_\infty$. Hence $\epsilon^2 = \pm 1$, i.e., $\epsilon = \pm i$ since $X \notin N(\mathbb{C})$. Performing a suitable conjugation in $N(\mathbb{C})$

$$\begin{pmatrix} 1 & x \\ 0 & 1 \end{pmatrix}\begin{pmatrix} \epsilon & b \\ 0 & \epsilon^{-1} \end{pmatrix}\begin{pmatrix} 1 & -x \\ 0 & 1 \end{pmatrix} = \begin{pmatrix} \epsilon & b+(\epsilon^{-1}-\epsilon)x \\ 0 & \epsilon^{-1} \end{pmatrix},$$

we can achieve that $b = 0$. Then we have $X = \begin{pmatrix} i & 0 \\ 0 & -i \end{pmatrix}$, $U = \begin{pmatrix} 1 & 1 \\ 0 & 1 \end{pmatrix}$, and our discussion of Definition 1.7 implies that for all $T \in \Gamma_\infty$, $T \notin N(\mathbb{C})$, we have $XT \in N(\mathbb{C})$. This yields (b).

In case (3), we treat the case $\Gamma_\infty \cap N(\mathbb{C}) \neq \Gamma_\infty$. A conjugation in $B(\mathbb{C})$ shows that we may assume from the outset that $\Gamma_\infty \cap N(\mathbb{C})$ is generated by

$$U = \begin{pmatrix} 1 & 1 \\ 0 & 1 \end{pmatrix}, \qquad U^\tau = \begin{pmatrix} 1 & \tau \\ 0 & 1 \end{pmatrix}$$

for some $\tau \in \mathbb{C} \backslash \mathbb{R}$. Consider the homomorphism

$$\Gamma_\infty \to \mathbb{C}^*/\{\pm I\}, \qquad \begin{pmatrix} \epsilon & b \\ 0 & \epsilon^{-1} \end{pmatrix} \mapsto [\epsilon], \tag{1.6}$$

where $[\epsilon] = \{\epsilon, -\epsilon\}$. We first prove that the image of (1.6) is a finite cyclic group of order 2,3,4 or 6. Note that the conjugation of $\Gamma_\infty \cap N(\mathbb{C}) \simeq \mathbb{Z} \oplus \mathbb{Z}\tau$ by elements of Γ_∞ defines a homomorphism $\Gamma_\infty/(\Gamma_\infty \cap N(\mathbb{C})) \to \mathbf{GL}(2, \mathbb{Z})$, such that the element

$$X = \begin{pmatrix} \epsilon & b \\ 0 & \epsilon^{-1} \end{pmatrix} \in \Gamma_\infty$$

is mapped to a 2×2 matrix with double eigenvalue ϵ^2. Assume $X \notin N(\mathbb{C})$, then ϵ^2 is either -1 or not a real number. The first case leads to (b). In the second ϵ^2 is an algebraic integer of degree 2, hence $\epsilon^2 = \exp(\frac{2\pi i k}{m}), m \in \{3, 4, 6\}, k = \pm 1$, hence the image of the homomorphism (1.6) is a finite cyclic group of order 2,3,4 or 6. Choose X such that $[\epsilon]$ generates the image of (1.6), and conjugate X in $N(\mathbb{C})$ such that $X = \begin{pmatrix} \epsilon & 0 \\ 0 & \epsilon^{-1} \end{pmatrix} = D_\epsilon$. Now $\Gamma_\infty =< D_\epsilon,\ U,\ U^\tau >$. Note that

$$D_\epsilon U D_\epsilon^{-1} = \begin{pmatrix} 1 & \epsilon^2 \\ 0 & 1 \end{pmatrix} \in N(\mathbb{C}), \qquad D_\epsilon U^\tau D_\epsilon^{-1} = \begin{pmatrix} 1 & \epsilon^2\tau \\ 0 & 1 \end{pmatrix} \in N(\mathbb{C}).$$

Hence the lattice $\mathbb{Z} \oplus \mathbb{Z}\tau$ is an ideal in $\mathbb{Q}(\epsilon^2)$. Note that the $\mathbb{Q}(\epsilon^2)$ has class number one here. This implies that there exists an element $\lambda \in \mathbb{Q}(\epsilon^2)$ such that $\mathbb{Z} \oplus \mathbb{Z}\tau = \mathbb{Z}[\epsilon^2]\lambda$ as a $\mathbb{Z}[\epsilon^2]$-module. Conjugating with $\begin{pmatrix} \lambda^{1/2} & 0 \\ 0 & \lambda^{-1/2} \end{pmatrix}$, we find that Γ_∞ is conjugate in $B(\mathbb{C})$ to $< D_\epsilon,\ U,\ U^{\epsilon^2} >$. This implies the assertion. □

Theorem 1.8 can also be drawn from Ford (1951), section 60. Ford also gives a complete list of pictures of fundamental domains (cf. Definition 2.1) for the action of Γ_∞ on $\mathbb{P}^1\mathbb{C}$. Using Theorem 1.8, it is easy to classify the abelian discrete subgroups of $\mathbf{PSL}(2,\mathbb{C})$. We also can state:

Corollary 1.9. *A fixed point of a parabolic element of a discrete group* $\Gamma < \mathbf{PSL}(2,\mathbb{C})$ *never is a fixed point of a hyperbolic or loxodromic element of* Γ.

This follows from our classification given in Theorem 1.8. See Beardon (1977), pages 63–64 for another proof. We shall often use the following concept.

Definition 1.10. An element $\zeta \in \mathbb{P}^1\mathbb{C}$ is called a cusp of a discrete group $\Gamma < \mathbf{PSL}(2,\mathbb{C})$ if Γ_ζ contains a free abelian group of rank 2. We write C_Γ for the set of cusps of Γ.

Transforming ζ to ∞ we are in the case (3) of our classification in Theorem 1.8. This also shows that the free abelian group in Γ_∞ consists of I and parabolic elements. Clearly, the group Γ leaves its set of cusps C_Γ invariant.

Theorem 1.8 is part of the classification of discrete elementary groups. We recall the definition

Definition 1.11. A subgroup $G < \mathbf{PSL}(2,\mathbb{C})$ is called elementary if $\mathbb{H}\cup\mathbb{P}^1\mathbb{C}$ contains a finite G-orbit. A G-orbit is a set of type $GP = \{\ \gamma P \ \mid\ \gamma \in G\ \}$ for some $P \in \mathbb{H} \cup \mathbb{P}^1\mathbb{C}$.

Elementary groups can also be characterized by the two following properties. A subgroup $G < \mathbf{PSL}(2,\mathbb{C})$ is elementary iff any two elements of infinite order in G have a common fixed point or iff the limit set of G in $\mathbb{P}^1\mathbb{C}$ contains at most two elements. The limit set of a group $G \leq \mathbf{PSL}(2,\mathbb{C})$ consists of the accumulation points of the orbits from $\mathbb{H}$ in $\mathbb{P}^1\mathbb{C}$. The discrete elementary groups can be completely classified:

Theorem 1.12. *Let* $\Gamma < \mathbf{PSL}(2,\mathbb{C})$ *be a discrete elementary group. Then one of the following cases holds:*

(1) Γ *has a fixed point in* $\mathbb{H}$ *and is finite,*
(2) $\Gamma = \Gamma_\zeta$ *for some* $\zeta \in \mathbb{P}^1\mathbb{C}$,
(3) there is a Γ*-orbit* $S \subset \mathbb{P}^1\mathbb{C}$ *consisting of 2 elements. Then there is a subgroup* $\tilde{\Gamma} < \Gamma$ *of index 2 with* $\tilde{\Gamma} = \tilde{\Gamma}_\zeta$ *for some* $\zeta \in \mathbb{P}^1\mathbb{C}$.

This classification result follows easily from Theorem 1.8. For all of this see e.g. Ford (1951), pages 117–147 or Beardon (1983), Chapter 5. The classification of the finite subgroups in $\mathbf{PSL}(2,\mathbb{C})$ in Theorem 1.6 gives more precise information about case (1).

Finally we shall give some information on the conjugacy classes in $\mathbf{O}^+(1,3)$. This will describe the classification of motions from $\mathbf{Iso}^+(\mathbb{H})$ given in Definition 1.3 and Proposition 1.4 in geometric terms related to the Kleinian model $\mathbb{K}$. As in chapter 1, section 5 we let

$$q_1(x_0, x_1, x_2, x_3) = x_0^2 - x_1^2 - x_2^2 - x_3^2$$

be our standard quadratic form. The group $\mathbf{O}_4(q_1, \mathbb{R})$ is also denoted by $\mathbf{O}(1,3)$. Taking the Kleinian model $\mathbb{K}$ for three-dimensional hyperbolic space, the group of orientation preserving isometries is identified with $\mathbf{O}^+(1,3)$.

Proposition 1.13. *Let $\gamma \in \mathbf{O}^+(1,3)$ be an elliptic element. Then γ is conjugate in $\mathbf{O}(1,3)$ to*

$$\begin{pmatrix} 1 & 0 & 0 & 0 \\ 0 & 1 & 0 & 0 \\ 0 & 0 & \cos\varphi & \sin\varphi \\ 0 & 0 & -\sin\varphi & \cos\varphi \end{pmatrix} \quad \textit{for some } \varphi \in \mathbb{R},\ \varphi \notin 2\pi\mathbb{Z}.$$

φ is the rotation angle around the invariant line ℓ joining the two fixed points of γ in the boundary of $\mathbb{K}$. Two elliptic elements $\gamma, \gamma' \in \mathbf{O}^+(1,3)$ are conjugate in $\mathbf{O}(1,3)$ iff they have the same set of eigenvalues.

Proof. We only give a (geometric) argument for the last statement, the first being obvious. Let ℓ, ℓ' be the lines in $\mathbb{K}$ which join the fixed points of γ, γ' in $\partial\mathbb{K}$. We shall first show that there is a $\sigma \in \mathbf{O}(1,3)$ with $\sigma\ell = \ell'$. If ℓ and ℓ' do not yet intersect take $P \in \ell$ and $P' \in \ell'$, let L be the line segment between P, P' and S the plane through the midpoint of L perpendicular to L. Apply the reflection in S to ℓ. If ℓ and ℓ' intersect, they lie on a common plane. Rotate now ℓ on this plane into ℓ' by means of a hyperbolic rotation. We have now shown there is a $\sigma \in \mathbf{O}(1,3)$ so that γ' and $\sigma\gamma\sigma^{-1}$ have the same fixed lines. They have the same eigenvalues if and only if their angles of rotation around this line are the same. □

Proposition 1.14. *Let $\gamma \in \mathbf{O}^+(1,3)$ be a hyperbolic element. Then γ is conjugate in $\mathbf{O}(1,3)$ to*

$$\begin{pmatrix} a & b & 0 & 0 \\ b & a & 0 & 0 \\ 0 & 0 & 1 & 0 \\ 0 & 0 & 0 & 1 \end{pmatrix} \quad \textit{with } a, b \in \mathbb{R},\ a^2 - b^2 = 1,\ b \neq 0.$$

If t is the translation length of γ on the invariant line joining the two fixed points of γ in $\partial\mathbb{K}$, then $\cosh(t) = |a|$. Two hyperbolic elements of $\mathbf{O}^+(1,3)$ are conjugate in $\mathbf{O}(1,3)$ iff they have the same set of eigenvalues.

Proposition 1.15. *Let $\gamma \in \mathbf{O}^+(1,3)$ be a loxodromic element. Then γ is conjugate in $\mathbf{O}(1,3)$ to*

$$\begin{pmatrix} a & b & 0 & 0 \\ b & a & 0 & 0 \\ 0 & 0 & \cos\varphi & \sin\varphi \\ 0 & 0 & -\sin\varphi & \cos\varphi \end{pmatrix} \qquad \begin{array}{l} \textit{with } \varphi \in \mathbb{R},\ \varphi \notin 2\pi\mathbb{Z}, \\ a, b \in \mathbb{R},\ a^2 - b^2 = 1,\ b \neq 0. \end{array}$$

φ is the rotation angle around the invariant line ℓ joining the two fixed points of γ in $\partial\mathbb{K}$. If t is the translation length of γ on ℓ then $\cosh(t) = |a|$. *Two loxodromic elements of* $\mathbf{O}^+(1,3)$ *are conjugate in* $\mathbf{O}(1,3)$ *iff they have the same set of eigenvalues.*

Proposition 1.16. *Let $\gamma \in \mathbf{O}^+(1,3)$ be a parabolic element. Then γ is conjugate in $\mathbf{O}(1,3)$ to*

$$\begin{pmatrix} 3 & 2 & -2 & 0 \\ -2 & -1 & 2 & 0 \\ -2 & -2 & 1 & 0 \\ 0 & 0 & 0 & 1 \end{pmatrix}.$$

The Jordan normal form of the above matrix is

$$\begin{pmatrix} 1 & 1 & 0 & 0 \\ 0 & 1 & 1 & 0 \\ 0 & 0 & 1 & 0 \\ 0 & 0 & 0 & 1 \end{pmatrix}.$$

Propositions 1.14, 1.15, 1.16 are proved analogously to Proposition 1.13. They can also easily be deduced from the formulas for the isomorphism $\mathbf{SO}^+(1,3) \cong \mathbf{PSL}(2,\mathbb{C})$ given in Chapter 1, section 3.

2.2 Fundamental Domains and Polyhedra

We now turn to the discussion of fundamental domains of discrete subgroups of $\mathbf{Iso}(\mathbb{H})$ and $\mathbf{Iso}^+(\mathbb{H}) = \mathbf{PSL}(2,\mathbb{C})$.

Definition 2.1. A closed subset $\mathcal{F} \subset \mathbb{H}$ is called a fundamental domain of the discontinuous group $\Gamma < \mathbf{Iso}(\mathbb{H})$ if the following conditions are satisfied:

(1) $\mathcal{F}$ meets each Γ-orbit at least once,
(2) the interior $\mathcal{F}^o$ meets each Γ-orbit at most once,
(3) the boundary of $\mathcal{F}$ has Lebesgue measure zero.

Each discontinuous group $\Gamma < \mathbf{Iso}(\mathbb{H})$ has a fundamental domain. A construction is given by the Poincaré normal polyhedron or Dirichlet fundamental domain which is defined as follows:

Definition 2.2. Let $\Gamma < \mathbf{Iso}(\mathbb{H})$ be a discontinuous group. Let $Q \in \mathbb{H}$ such that $\gamma Q \neq Q$ for all $\gamma \in \Gamma \backslash \{I\}$. Then the Poincaré normal polyhedron for Γ with center Q is

$$\mathcal{P}_Q(\Gamma) = \{ P \in \mathbb{H} \mid d(P,Q) \leq d(\gamma P, Q) \quad \text{for all } \gamma \in \Gamma \}.$$

Since Γ is a countable set, there is a $Q \in \mathbb{H}$ such that $\gamma Q \neq Q$ for all $\gamma \in \Gamma \backslash \{I\}$. The set $\mathcal{P}_Q(\Gamma)$ is a fundamental domain for Γ. The boundary of $\mathcal{P}_Q(\Gamma)$ consists of faces which are parts of the hyperbolic planes bounding the hyperbolic half-spaces whose intersection is $\mathcal{P}_Q(\Gamma)$. Any compact subset of $\mathbb{H}$ is met by only finitely many faces of $\mathcal{P}_Q(\Gamma)$. See Beardon (1983) for the properties of the Poincaré normal polyhedron.

If $\mathcal{F}$ is a fundamental domain of a discontinuous group $\Gamma < \mathbf{Iso}(\mathbb{H})$ then $\mathcal{F}$ is closed and hence a measurable set for the measure defined by the hyperbolic volume element

$$dv = \frac{dx\, dy\, dr}{r^3}.$$

If $M \subset \mathbb{H}$ is a Borel-measurable set we also use the notation

$$v(M) = \int_M dv. \tag{2.1}$$

Although the differential form dv is only up to sign invariant under the elements of $\mathbf{Iso}(\mathbb{H})$, the measure on the Borel-measurable subsets of $\mathbb{H}$ given by (2.1) is $\mathbf{Iso}(\mathbb{H})$ invariant.

Definition 2.3. Let $\Gamma < \mathbf{Iso}(\mathbb{H})$ be a discontinuous subgroup and $\mathcal{F} \subset \mathbb{H}$ a fundamental domain for Γ. We say that Γ is of finite covolume or a cofinite group if

$$\operatorname{vol}(\Gamma) = \int_{\mathcal{F}} dv < \infty.$$

We call $\operatorname{vol}(\Gamma)$ the covolume of Γ. We say that Γ is cocompact if Γ has a compact fundamental domain. A subgroup $\Gamma < \mathbf{SL}(2, \mathbb{C})$ is called cofinite or cocompact if its image in $\mathbf{PSL}(2, \mathbb{C}) = \mathbf{Iso}^+(\mathbb{H})$ has this property.

The above definition is justified by

Proposition 2.4. *Let $\Gamma < \operatorname{Iso}(\mathbb{H})$ be a discontinuous group and let $\mathcal{F}_1, \mathcal{F}_2$ be two fundamental domains for Γ. If $\int_{\mathcal{F}_1} dv < \infty$, then $\int_{\mathcal{F}_2} dv < \infty$ and $\int_{\mathcal{F}_1} dv = \int_{\mathcal{F}_2} dv$.*

Proof. By the usual properties of Lebesgue measure and using that Γ is countable we have

$$v(\mathcal{F}_1) = v(\mathcal{F}_1 \cap (\cup_{\gamma \in \Gamma} \gamma\mathcal{F}_2)) = \sum_{\gamma \in \Gamma} v(\mathcal{F}_1 \cap \gamma\mathcal{F}_2) = \sum_{\gamma \in \Gamma} v(\mathcal{F}_2 \cap \gamma^{-1}\mathcal{F}_1)$$
$$= v(\mathcal{F}_2 \cap (\cup_{\gamma \in \Gamma} \gamma\mathcal{F}_1)) = v(\mathcal{F}_2).$$

This computation holds since $(\mathcal{F}_1 \cap \gamma_1\mathcal{F}_2) \cap (\mathcal{F}_1 \cap \gamma_2\mathcal{F}_2)$ has by definition Lebesgue measure 0 for any pair $\gamma_1, \gamma_2 \in \Gamma$ with $\gamma_1 \neq \gamma_2$. □

From the definition it is also clear that for any cofinite subgroup $\Gamma <$ Iso($\mathbb{H}$) and any $\gamma \in$ Iso($\mathbb{H}$), the group $\gamma\Gamma\gamma^{-1}$ has the same property and $\mathrm{vol}\,(\Gamma) = \mathrm{vol}\,(\gamma\Gamma\gamma^{-1})$.

If $\Gamma < \mathbf{PSL}(2, \mathbb{C})$ is a discrete group, acting fixed-point freely on $\mathbb{H}$ then the quotient space $\Gamma\backslash\mathbb{H}$ inherits from $\mathbb{H}$ the structure of an orientable Riemannian manifold. The projection $\pi_\Gamma : \mathbb{H} \to \Gamma\backslash\mathbb{H}$ is a local isometry. It can easily be seen that Γ is cocompact iff $\Gamma\backslash\mathbb{H}$ is compact and that Γ is cofinite iff $\Gamma\backslash\mathbb{H}$ is of finite volume. In this case we have $\mathrm{vol}(\Gamma) = \int_{\Gamma\backslash\mathbb{H}} dv$, where dv is the volume form coming from the Riemannian structure. See Bourbaki (1964b), Chap. VII, §2. We also need the following result:

Lemma 2.5. *Let $\Gamma <$ Iso($\mathbb{H}$) be a discontinuous group of finite covolume. Let furthermore $M \subset \mathbb{H}$ be a Lebesgue measurable set with $v(M) > \mathrm{vol}\,(\Gamma)$. Then $M \cap \gamma M \neq \emptyset$ for some $\gamma \in \Gamma$ with $\gamma \neq I$.*

Proof. Let $\mathcal{F}$ be a fundamental domain for Γ. If $M \cap \gamma M = \emptyset$ for every $\gamma \in \Gamma\backslash\{I\}$ then $(\gamma_1 M \cap \mathcal{F}) \cap (\gamma_2 M \cap \mathcal{F}) = \emptyset$ for every pair $\gamma_1 \neq \gamma_2$ of elements of Γ. By the usual properties of Lebesgue measure we have

$$v(M) = v(M \cap \cup_{\gamma \in \Gamma} \gamma\mathcal{F}) = v(\cup_{\gamma \in \Gamma}(M \cap \gamma\mathcal{F}))$$
$$\leq \sum_{\gamma \in \Gamma} v(M \cap \gamma\mathcal{F}) = \sum_{\gamma \in \Gamma} v(\gamma M \cap \mathcal{F})$$
$$= v(\cup_{\gamma \in \Gamma}(\gamma M \cap \mathcal{F})) \leq v(\mathcal{F}).$$

This is a contradiction. □

In later chapters we shall describe nice fundamental domains $\mathcal{F}$ for certain groups Γ. These fundamental domains will be polyhedra.

A polyhedron $\mathcal{P}$ is a closed connected subset of $\mathbb{H}$ so that the boundary $\partial\mathcal{P}$ is the union of countably many faces $\mathcal{S}_i$, each $\mathcal{S}_i$ being a hyperbolic polygon in a geodesic plane X_i.

A polygon is a closed connected subset of X_i so that the boundary $\partial\mathcal{S}_i$ is the union of countably many faces s_{ij}, called edges of $\mathcal{P}$ which are closed connected intervals on geodesic lines. We require that for $j \neq k$ the intersection $s_{ij} \cap s_{ik}$ is either empty or a point, called a vertex of $\mathcal{S}_i$ and of $\mathcal{P}$. Furthermore if an edge s_{ij} has a boundary point $x \in \mathbb{H}$ then a unique edge s_{ik} with $j \neq k$ and with x as boundary point is supposed to exist.

If an edge s has a point $y \in \partial\mathbb{H} = \mathbb{P}^1\mathbb{C}$ in its closure, we call y a vertex of $\mathcal{P}$ at infinity. For $\mathcal{P}$ to be a polyhedron we further require that each edge s is the intersection of two faces, and that two faces are either disjoint, meet in a common edge or in a common vertex. An edge is supposed to be either disjoint from a side, lie on the side or meet it in a vertex. Two distinct edges are required to be either disjoint or meet in a common vertex. Finally we assume that a compact set of $\mathbb{H}$ meets $\mathcal{P}$ in only finitely many edges and that the intersection of $\mathcal{P}$ with any sufficiently small hyperbolic ball is connected.

It can be proved, see Beardon (1983), that the $\mathcal{F} = \mathcal{P}_Q(\Gamma)$ for discrete groups $\Gamma < \mathbf{Iso}(\mathbb{H})$ are polyhedra. We think of the translations of $\mathcal{F}$ by elements of Γ as giving a polyhedral tesselation of $\mathbb{H}$. See Chapters 7 and 10 for examples. In most cases these fundamental polyhedra will have finitely many faces. Since groups having such particular fundamental domains play a special role we make the following definition.

Definition 2.6. A discrete group $\Gamma < \mathbf{PSL}(2,\mathbb{C})$ is called geometrically finite if there is a $Q \in \mathbb{H}$ such that $\mathcal{P}_Q(\Gamma)$ has only finitely many faces.

Greenberg (1977) shows that if Γ has one Poincaré normal polyhedron with finitely many faces then every Poincaré normal polyhedron has this property. The following result is due to Garland and Raghunathan (1970).

Theorem 2.7. *Let $\Gamma < \mathbf{PSL}(2,\mathbb{C})$ be a discrete group of finite covolume. Then there is a $Q \in \mathbb{H}$ such that $\mathcal{P}_Q(\Gamma)$ has only finitely many faces, that is, Γ is geometrically finite.*

The paper of Wielenberg (1977) also contains a proof of this result. We shall only add a few comments on Wielenberg's proof. Since the result is clear for cocompact groups we assume that Γ does not have this property. By the Kazhdan, Margulis theorem (Theorem 3.6 below) we may also assume that Γ contains a parabolic element fixing ∞. Arguments analogous to those in section 3 show that Γ_∞ contains a full lattice in $N(\mathbb{C})$. The Poincaré fundamental polyhedron $\mathcal{P}_{rj}(\Gamma)$ for suitable $r \geq 1$ is then contained in a chimney the closure of which intersects $\mathbb{P}^1\mathbb{C} \setminus \{\infty\}$ in a compact set. The argument is finished by appealing to the discreteness of Γ.

See Greenberg (1977) for a discussion of the properties of geometrically finite groups. In general a discontinuous group Γ need not be geometrically finite. Every geometrically finite group is finitely generated. However a finitely generated group need not be geometrically finite, see Greenberg (1977).

Poincaré studied in his papers (1882), (1883) the question which polyhedra arise as fundamental polyhedra of discrete groups. He stated a complete solution. A convincing proof of his statement is given by Maskit in (1971). A proof of Poincaré's theorem which applies to all dimensions and all geometries of constant curvature is contained in Epstein, Petronio (1994). The general solution being difficult to formulate we only give the following very useful partial result.

Theorem 2.8. *Let $\mathcal{P} \subset \mathbb{H}$ be a (closed) polyhedron with faces $\mathcal{S}_i$ contained in geodesic planes L_i. Assume that for every pair $\mathcal{S}_i$, $\mathcal{S}_j$ of distinct faces of $\mathcal{P}$ the containing geodesic planes L_i, L_j are either disjoint or meet at an angle π/k for some $k \in \mathbb{N}$. Then the group $\Gamma_{\mathcal{P}} \leq \mathbf{Iso}(\mathbb{H})$ generated by the reflections in the geodesic planes L_i is discrete and has $\mathcal{P}$ as fundamental domain. The polyhedron $\mathcal{P}$ is said to tesselate $\mathbb{H}$. The group $\Gamma_{\mathcal{P}} < \mathbf{Iso}(\mathbb{H})$ is called the tesselation group associated to $\mathcal{P}$.*

Furthermore the following two conditions concerning the stabilzers of the vertices are satisfied.

(1) Let $V \in \mathbb{H}$ be a vertex of $\mathcal{P}$ and let Γ_V be the group generated by the reflections in the geodesic planes containing the faces of $\mathcal{P}$ which contain V. Assume that $\gamma \in \Gamma_V$. Then either $\gamma\mathcal{P} \cap \mathcal{P} = \mathcal{P}$ or $\gamma\mathcal{P} \cap \mathcal{P}$ is contained in the union of the faces of $\mathcal{P}$. Moreover there is an open neighbourhood B of V with

$$B \subset \bigcup_{\gamma \in \Gamma_V} \gamma\mathcal{P}.$$

(2) Let $V \in \partial\mathbb{H}$ be a vertex of $\mathcal{P}$ at infinity and let Γ_V be the group generated by the reflections in the geodesic planes containing the faces of $\mathcal{P}$ which contain V in their closure. Then either $\gamma\mathcal{P} \cap \mathcal{P} = \mathcal{P}$ or $\gamma\mathcal{P} \cap \mathcal{P}$ is contained in the union of the faces of $\mathcal{P}$. Moreover there is an $A \in \mathbf{PSL}(2, \mathbb{C})$ with $A\infty = V$ and $r_1, r_2 > 0$ such that

$$A \{ \, z + rj \in \mathbb{H} \mid |z| < r_1, \; r > r_2 \, \} \subset \bigcup_{\gamma \in \Gamma_V} \gamma\mathcal{P}.$$

Proof. We shall only sketch a proof here, details can easily be extracted from Maskit (1971). We write σ_i for the reflection in the geodesic plane L_i containing the side $\mathcal{S}_i$. We put $X := \Gamma_{\mathcal{P}} \times \mathcal{P}$, and define $\tilde{X}$ to be the quotient of X by the equivalence relation

$$(\gamma, P) \sim (\gamma\sigma_i, P) \qquad \text{for } P \in \mathcal{S}_i.$$

The topological space $\tilde{X}$ comes with a 3-dimensional combinatorial structure. The map $\varphi : X \to \mathbb{H}$, where $\varphi : (\gamma, P) \mapsto \gamma P$ descends to a map $\tilde{\varphi} : \tilde{X} \to \mathbb{H}$. The conditions of our theorem imply that $\tilde{\varphi}$ is a local homeomorphism. It is also clear that every path can be lifted from $\mathbb{H}$ to $\tilde{X}$. Hence $\tilde{\varphi}$ is a covering map. Since $\mathbb{H}$ is simply connected $\tilde{\varphi}$ is a homeomorphism. A moment of meditation shows that the theorem is proved. □

All our conditions are local around edges and are easy to verify in examples. Note that the conditions under (1) imply that Γ_V is finite. To analyze the vertex groups Γ_V we need the following facts. For the formulations it is advantageous to bring also the Kleinian model $\mathbb{K}$ into the picture. We use the terminology of Chapter 1, section 5, in particular we use the identifications of Proposition 5.4.

Proposition 2.9. *Let $\mathcal{P} \subset \mathbb{H} \cong \mathbb{K}$ be a polyhedron.*
(1) Let $V \in \mathbb{K}$ be a finite vertex of $\mathcal{P}$. The group Γ_V generated by the reflections in the geodesic planes containing the faces of $\mathcal{P}$ that contain V is called the vertex group at V. Choose an $A \in \mathbf{Iso}(\mathbb{K})$ so that $A[1,0,0,0] = V$. Then $\tilde{\Gamma}_V := A^{-1}\Gamma_V A$ is contained in $\mathbf{O}(3)$. Moreover it is a group generated by Euclidean reflections in planes from 3-dimensional Euclidean space.
(2) Let $V \in \partial\mathbb{H} \cong \partial\mathbb{K}$ be a vertex of $\mathcal{P}$ at infinity and let Γ_V be the group generated by the reflections in the geodesic planes containing the faces of $\mathcal{P}$ which contain V in their closure. The group Γ_V is again called the vertex group at V. Choose an $A \in \mathbf{Iso}(\mathbb{H})$ so that $A\infty = V$. Put $\tilde{\Gamma}_V := A^{-1}\Gamma_V A$. The group $\tilde{\Gamma}_V$ is a group of affine motions on $\partial\mathbb{H} \setminus \{\infty\} = \mathbb{C} = \mathbb{R}^2$. Moreover it is generated by Euclidean reflections.

This result is proved by using the explicit descriptions of geodesics and reflections in the models $\mathbb{H}$, $\mathbb{K}$ from Chapter 1.

We need the following reformulation of Theorem 2.8.

Theorem 2.10. *Let $\mathcal{P} \subset \mathbb{H}$ be a polyhedron with faces $\mathcal{S}_i$ contained in geodesic planes L_i. Assume that for every pair $\mathcal{S}_i$, $\mathcal{S}_j$ of distinct faces of $\mathcal{P}$ the containing geodesic planes L_i, L_j are either disjoint or meet at an angle π/k for some $k \in \mathbb{N}$. Then the group $\Gamma_{\mathcal{P}} \leq \mathbf{Iso}(\mathbb{H})$ generated by the reflections in the geodesic planes L_i is discrete and has $\mathcal{P}$ as a fundamental domain. The following conditions concerning the stabilizer groups of the vertices are satisfied.*
(1) Let $V \in \mathbb{H}$ be a vertex of $\mathcal{P}$ and let $\tilde{\Gamma}_V < \mathbf{O}(3)$ be the group constructed in Proposition 2.9. Then $\tilde{\Gamma}_V$ acts discontinuously on the unit sphere $S^2 \subset \mathbb{R}^3$.
(2) Let $V \in \partial\mathbb{H}$ be a vertex of $\mathcal{P}$ at infinity and let $\tilde{\Gamma}_V$ be the group of affine Euclidean motions of $\mathbb{C} = \mathbb{R}^2$ constructed in Proposition 2.9. Then $\tilde{\Gamma}_V$ acts discontinuously on $\mathbb{R}^2$.

Thinking of our explicit models $\mathbb{H}$, $\mathbb{K}$ it is clear that the hypotheses of Theorem 2.10 imply those of Theorem 2.8. Notice that $\mathcal{P}$ cuts out a spherical polygon on $S^2 \subset \mathbb{R}^3$ in case of a finite vertex and an affine polygon from $\mathbb{R}^2$ in case of a vertex at infinity. Conditions (1) and (2) can be formulated as conditions about the angles between the lines of reflection that belong to the generators of the $\tilde{\Gamma}_V$. Hypothesis (2) from the above implies that Γ_V contains an abelian subgroup of finite index.

Interesting examples of tesselation polyhedra are the ideal solids. A polyhedron is called ideal if every side is a regular hyperbolic p-gon, and each vertex has q of these p-gons containing it. It is also supposed that the angle beetween the faces is always $2\pi/r$ for some fixed $r \in \mathbb{N}$. Furthermore it is assumed that the vertex figure is a regular hyperbolic polygon for every vertex V. The vertex figure is the (hyperbolic) convex hull of the points of intersection of a small hyperbolic sphere with center V with the edges containing V. Such an object is in classical notation, see Coxeter (1956), denoted by

$$\{p, q, r\}.$$

Note that we have not supposed that $\mathcal{P}$ is compact. A non-compact ideal polyhedron will have all its vertices at infinity. We find the following complete list of tesselating ideal polyhedra, see Coxeter (1956).

Theorem 2.11. *There are the following compact ideal tesselating polyhedra.*

(1) $\{4, 3, 5\}$ *: a cube with angle of* 72° *between faces,*
(2) $\{5, 3, 5\}$ *: a dodecahedron with angle of* 72° *between faces,*
(3) $\{5, 3, 4\}$ *: a dodecahedron with angle of* 90° *between faces,*
(4) $\{3, 5, 3\}$ *: an icosahedron with angle of* 120° *between faces.*

There are the following non-compact ideal tesselating polyhedra.

(1) $\{3, 3, 6\}$ *: a tetrahedron with angle of* 60° *between faces,*
(2) $\{4, 3, 6\}$ *: a cube with angle of* 60° *between faces,*
(3) $\{3, 4, 4\}$ *: an octahedron with angle of* 90° *between faces,*
(4) $\{5, 3, 6\}$ *: a dodecahedron with angle of* 60° *between faces.*

Proof. A proof of the above is obtained by first using Euler's theorem on polyhedra to find that a hyperbolic ideal polyhedron $\mathcal{P}$ must have the same combinatorial type as one of the Euclidean ideal solids, that is $\mathcal{P}$ has to be a tetrahedron, cube, ... Then we embed such a solid with all points at infinity into the Kleinian model $\mathbb{K}$. While shrinking we compute angles and try to verify the conditions of Theorem 2.8. This leeds to the above list. □

The tesselation $\{5, 3, 5\}$ appears together with a spherical dodecahedral tesselation in Seifert, Weber (1933). The tesselation group corresponding to $\{5, 3, 5\}$ contains two torsion free subgroups of index 5, giving rise to interesting 3-manifolds, see Best (1971). One of them was discovered by Seifert, Weber (1933). That these two 3-manifolds are not homeomorphic was proved by Mennicke (unpublished) and Zimmermann (1995). The polyhedron $\{5, 3, 5\}$ is the first of an infinite series of tesselating polyhedra described by Helling, Kim, Mennicke (1995). The homeomorphism problem for these was solved by Zimmermann (1995). Also the polyhedron $\{3, 5, 3\}$ is the member of an infinite series of tessalating polyhedra. The associated manifolds are called Fibonacci manifolds. They are described in an unpublished manuscript by Helling, Kim and Mennicke and also in Hilden, Lozano, Montesinos-Amilibia (1992a), (1992b).

Sometimes polyhedra which are nearly ideal appear as fundamental domains for interesting discrete groups, see Thurston (1978), Hatcher (1983), Grunewald, Gushoff, Mennicke (1982).

2.3 Shimizu's Lemma

Fundamental for the study of parabolic elements in discrete subgroups of $\mathbf{PSL}(2,\mathbb{C})$ is Shimizu's Lemma (see Shimizu (1963)):

Theorem 3.1. *Let* $M = \begin{pmatrix} a & b \\ c & d \end{pmatrix} \in \mathbf{SL}(2,\mathbb{C})$, *and* $U^\lambda = \begin{pmatrix} 1 & \lambda \\ 0 & 1 \end{pmatrix}$ *where* $c \neq 0,\ 0 \neq \lambda \in \mathbb{C}$. *If the group* $\Gamma := < M, U^\lambda >$ *generated by* M *and* U^λ *is discrete, then* $|c\lambda| \geq 1$.

Proof. Since

$$\begin{pmatrix} \alpha & 0 \\ 0 & \alpha^{-1} \end{pmatrix}\begin{pmatrix} a & b \\ c & d \end{pmatrix}\begin{pmatrix} \alpha^{-1} & 0 \\ 0 & \alpha \end{pmatrix} = \begin{pmatrix} a & \alpha^2 b \\ \alpha^{-2} c & d \end{pmatrix},$$

$$\begin{pmatrix} \alpha & 0 \\ 0 & \alpha^{-1} \end{pmatrix}\begin{pmatrix} 1 & \lambda \\ 0 & 1 \end{pmatrix}\begin{pmatrix} \alpha^{-1} & 0 \\ 0 & \alpha \end{pmatrix} = \begin{pmatrix} 1 & \alpha^2 \lambda \\ 0 & 1 \end{pmatrix},$$

we may assume that $\lambda = 1$. We put $U := U^1$. Then we have to show that $|c| \geq 1$. Assume to the contrary that $0 < |c| < 1$, and define inductively $M_0 := M$, $M_{n+1} := M_n U M_n^{-1}$. Letting $M_n = \begin{pmatrix} a_n & b_n \\ c_n & d_n \end{pmatrix}$, we have the recursion formula

$$M_{n+1} = \begin{pmatrix} 1 - a_n c_n & a_n^2 \\ -c_n^2 & 1 + a_n c_n \end{pmatrix}$$

and hence $c_n = -c^{2^n} \quad (n \geq 1)$, in particular

$$\text{(3.1)} \qquad M_k \neq M_\ell \quad \text{if} \quad k \neq \ell.$$

Now we have $|a_{n+1} - 1| = |d_{n+1} - 1| = |a_n c_n|$, $b_{n+1} = a_n^2$ and we know that $c_n \to 0$. Hence, if we show that $(a_n)_{n \geq 0}$ is bounded, we can conclude that

$$\text{(3.2)} \qquad M_n \longrightarrow \begin{pmatrix} 1 & 1 \\ 0 & 1 \end{pmatrix}$$

as $n \to \infty$ which is a contradiction since Γ is discrete and the M_n are distinct. We show now that $(a_n)_{n \geq 0}$ is bounded. Choose $\gamma > 0$ such that $1/(1 - |c|) < \gamma$ and $|a| < \gamma$. Then we have $|a_0| < \gamma$, and the induction hypothesis $|a_n| < \gamma$ implies

$$|a_{n+1}| = |1 - a_n c_n| \leq 1 + |c||a_n| < 1 + |c|\gamma \leq \gamma.$$

Hence $(a_n)_{n\geq 0}$ is bounded, and we arrive at the contradiction (3.1), (3.2). □

Theorem 3.2. *Let $\Gamma < \mathbf{PSL}(2,\mathbb{C})$ be a discrete group and suppose that*

$$U^\lambda = \begin{pmatrix} 1 & \lambda \\ 0 & 1 \end{pmatrix} \in \Gamma \quad (0 \neq \lambda \in \mathbb{C}), \qquad M = \begin{pmatrix} a & b \\ c & d \end{pmatrix} \in \Gamma, \quad c \neq 0.$$

Define $\mathbb{H}_\lambda := \{\ z + rj \ \mid \ z \in \mathbb{C},\ r > |\lambda|\ \}$. *Then* $\mathbb{H}_\lambda \cap M\mathbb{H}_\lambda = \emptyset$.

Proof. $M\mathbb{H}_\lambda$ is a Euclidean ball in $\mathbb{H}$ touching $\mathbb{C}$ in $M\infty = a/c$. Now an elementary computation based on our formulas for the action of $\mathbf{PSL}(2,\mathbb{C})$ on $\mathbb{H}$ from Chapter 1 yields

$$\left\| MP - \frac{a}{c} \right\| = \frac{1}{|c| \|cP + d\|}.$$

Hence we have for $P = z + rj \in \mathbb{H}_\lambda$ in view of Shimizu's Lemma

$$\left\| MP - \frac{a}{c} \right\| \leq \frac{1}{|c|^2 r} < \frac{1}{|c|^2 |\lambda|} \leq |\lambda|.$$

This implies the assertion. □

A slight reformulation of Theorem 3.2 is given in Corollary 3.3.

Corollary 3.3. *Let $\Gamma \in \mathbf{PSL}(2,\mathbb{C})$ be a discrete group such that ∞ is a fixed point of a parabolic element of Γ, and suppose that $0 \neq \lambda \in \mathbb{C}$ is such that $\begin{pmatrix} 1 & \lambda \\ 0 & 1 \end{pmatrix} \in \Gamma$, with $|\lambda|$ minimal. Then two points contained in $\mathbb{H}_\lambda := \{\ z + rj \ \mid \ z \in \mathbb{C},\ r > |\lambda|\ \}$ are Γ-equivalent if and only if they are Γ_∞-equivalent.*

We shall also need the following theorem which describes the relation of the $\mathbb{H}_\lambda$ corresponding to Γ-inequivalent elements of $\mathbb{P}^1\mathbb{C}$.

Theorem 3.4. *Let $\Gamma < \mathbf{PSL}(2,\mathbb{C})$ be a discrete group and let $\zeta_1, \zeta_2 \in \mathbb{P}^1\mathbb{C}$ be Γ-inequivalent. Let furthermore $A_1, A_2 \in \mathbf{PSL}(2,\mathbb{C})$ such that $\zeta_1 = A_1\infty$ and $\zeta_2 = A_2\infty$. Assume that there are $\lambda_1, \lambda_2 \in \mathbb{C}\backslash\{0\}$ with*

$$A_1 \begin{pmatrix} 1 & \lambda_1 \\ 0 & 1 \end{pmatrix} A_1^{-1}, \qquad A_2 \begin{pmatrix} 1 & \lambda_2 \\ 0 & 1 \end{pmatrix} A_2^{-1} \in \Gamma.$$

Then $A_1\mathbb{H}_{\lambda_1} \cap \gamma A_2 \mathbb{H}_{\lambda_2} = \emptyset$ for all $\gamma \in \Gamma$.

Proof. We fix a $\gamma \in \Gamma$ and put $C = A_1^{-1}\gamma A_2 = \begin{pmatrix} a & b \\ c & d \end{pmatrix}$. We infer that $c \neq 0$ since otherwise $C\infty = \infty$ which implies $\gamma\zeta_2 = \zeta_1$ contrary to our assumption. It is also clear that

$$\begin{pmatrix} 1 & \lambda_1 \\ 0 & 1 \end{pmatrix}, \; C\begin{pmatrix} 1 & \lambda_2 \\ 0 & 1 \end{pmatrix}C^{-1} = \begin{pmatrix} * & * \\ -c^2\lambda_2 & * \end{pmatrix} \in A_1^{-1}\Gamma A_1.$$

Since $A_1^{-1}\Gamma A_1$ is discrete we get from Theorem 3.1 $|\lambda_1| \geq 1/|c|^2|\lambda_2|$. We have to show that $\mathbb{H}_{\lambda_1} \cap C\mathbb{H}_{\lambda_2} = \emptyset$. If $z_1 + r_1 j \in C\mathbb{H}_{\lambda_2}$ then there is a $z + rj \in \mathbb{H}_{\lambda_2}$ with

$$r_1 = \frac{r}{|cz+d|^2 + r^2|c|^2}.$$

Hence $r_1 \leq 1/r|c|^2 < 1/|\lambda_2||c|^2 \leq |\lambda_1|$. This means that $z_1 + r_1 j \notin \mathbb{H}_{\lambda_1}$. □

The set $A\mathbb{H}_\lambda$ for $A \in \mathbf{PSL}(2,\mathbb{C})$ and $\lambda \in \mathbb{C}\setminus\{0\}$ is either an open upper half-space or an open ball in $\mathbb{H}$ touching $\mathbb{P}^1\mathbb{C}$. The $A\mathbb{H}_\lambda$ are called horoballs. Taking the usual topology on $\mathbb{H}$ and the horoballs touching $\mathbb{P}^1\mathbb{C}$ at λ as a basis for the neighbourhoods of λ we get a topology on $\mathbb{H} \cup \mathbb{P}^1\mathbb{C}$. The group $\mathbf{PSL}(2,\mathbb{C})$ acts continuously on this space.

Another Corollary of Theorem 3.1 is:

Corollary 3.5. *Let $\Gamma < \mathbf{PSL}(2,\mathbb{C})$ be a discrete group. If Γ contains a parabolic element then Γ is not cocompact.*

Proof. Without loss of generality there is an element $U^\lambda = \begin{pmatrix} 1 & \lambda \\ 0 & 1 \end{pmatrix} \in \Gamma$ with $\lambda \neq 0$. Let U^λ be chosen so that $|\lambda|$ is minimal. Then we know from Corollary 3.3 that two points contained in $\mathbb{H}_\lambda$ are Γ-equivalent if and only if they are Γ_∞-equivalent. Hence the space $\Gamma\setminus\mathbb{H}$ cannot be compact. □

Much deeper than this is a Theorem of Kazhdan and Margulis which gives a converse to Corollary 3.5 for groups of finite covolume.

Theorem 3.6. *Let $\Gamma < \mathbf{PSL}(2,\mathbb{C})$ be a discrete group of finite covolume. If Γ is not cocompact then Γ contains a parabolic element.*

We will not go into the proof here, see Borel (1971) and Kazhdan, Margulis (1968). The Kazhdan-Margulis Theorem applies even in the more general situation where $\mathbf{SL}(2,\mathbb{C})$ is replaced by any semi-simple Lie group.

Another interesting consequence of Shimizu's Lemma is

Proposition 3.7. *Let $\Gamma < \mathbf{PSL}(2,\mathbb{C})$ be a discrete group of finite covolume. Assume that the stabilizer Γ_ζ of $\zeta \in \mathbb{P}^1\mathbb{C}$ contains a parabolic element. Then ζ is a cusp of Γ.*

Proof. We may, without loss of generality, assume that $\zeta = \infty$. By this assumption there is an element $U^\lambda = \begin{pmatrix} 1 & \lambda \\ 0 & 1 \end{pmatrix} \in \Gamma$ with $\lambda \neq 0$. Let U^λ be chosen so that $|\lambda|$ is minimal. Then we know from Corollary 3.3 that two points contained in $\mathbb{H}_\lambda$ are Γ-equivalent if and only if they are Γ_∞-equivalent. If there is no lattice $\Lambda \subset \mathbb{C}$ so that

$$\left\{ \begin{pmatrix} 1 & \omega \\ 0 & 1 \end{pmatrix} \;\middle|\; \omega \in \Lambda \right\} \subset \Gamma_\infty$$

we conclude from Theorem 1.8 that $< U^\lambda >$ has finite index in Γ_∞. Then $\mathbb{H}_\lambda$ clearly contains a Lebesgue measurable set M with $M \cap \gamma M = \emptyset$ for all $\gamma \in \Gamma_\infty$ and with $\int_M dv = \infty$. By Lemma 2.5 we are finished. □

Finally for this section we discuss fundamental domains of a particularly nice kind.

Proposition 3.8. *Let $\Gamma < \mathbf{PSL}(2, \mathbb{C})$ be a discrete group of finite covolume, then Γ has only finitely many Γ-classes of cusps.*

This follows from Proposition 2.7 together with Theorem 3.4. See Theorem 5.1 for another proof.

Given a discrete group Γ of finite covolume choose $A_1, \cdots, A_h \in \mathbf{PSL}(2, \mathbb{C})$ so that the

$$\eta_1 = A_1\infty, \cdots, \eta_h = A_h\infty \;\in \mathbb{P}^1\mathbb{C}$$

are representatives for the Γ-classes of cusps. Choose further closed fundamental sets $\mathcal{P}_i$ (which may be parallelograms) for the action of the stabilizers $A_i^{-1}\Gamma_{\eta_i}A_i$ on $\mathbb{P}^1\mathbb{C} \setminus \{\infty\} = \mathbb{C}$. Define for $Y > 0$

$$\tilde{\mathcal{F}}_i(Y) := \{\ z + rj \ \mid \ z \in \mathcal{P}_i,\ r \geq Y\ \}$$

Let further $Y_1, \cdots, Y_h \in \mathbb{R}$ be large enough that the $\mathcal{F}_i(Y_i) := A_i\tilde{\mathcal{F}}_i(Y_i)$ are contained in the horospheres $A_i\mathbb{H}_{\lambda_i}$ mentioned in Theorem 3.4. The $\mathcal{F}_i(Y_i)$ are called cusp sectors.

Proposition 3.9. *Let $\Gamma < \mathbf{PSL}(2, \mathbb{C})$ be a discrete group of finite covolume, and let $Y_1, \cdots, Y_h > 0$ and $\mathcal{F}_1(Y_1), \cdots, \mathcal{F}_h(Y_h)$ be chosen as just explained. Then there is a compact set $\mathcal{F}_0 \subset \mathbb{H}$ so that*

$$\mathcal{F} := \mathcal{F}_0 \cup \mathcal{F}_1(Y_1) \cup \cdots \cup \mathcal{F}_h(Y_h)$$

is a fundamental domain for Γ. The compact set $\mathcal{F}_0$ can be chosen so that the intersections $\mathcal{F}_0 \cap \mathcal{F}_i(Y_i)$ all are all contained in the boundary of $\mathcal{F}_0$ and hence have Lebesgue measure 0. The intersections $\mathcal{F}_i(Y_i) \cap \mathcal{F}_j(Y_j)$ will be empty if $i \neq j$.

Proposition 3.9 is proved by using Proposition 2.7 and Theorem 3.4.

A useful criterion for a point in the boundary of $\mathbb{H}$ to be a cusp of a cofinite group is contained in the following.

Proposition 3.10. *Let $\Gamma < \mathbf{PSL}(2,\mathbb{C})$ be a cofinite group with its classes of cusps represented by $\eta_1 = A_1\infty, \cdots, \eta_h = A_h\infty \in \mathbb{P}^1\mathbb{C}$. Assume that $A_1 = I$, the unit matrix. Define for $\eta \in \mathbb{C} \subset \mathbb{P}^1\mathbb{C}$ and $\kappa > 0$:*

$$\mathbf{S}(\eta, \kappa) := \{ (z, r) \in \mathbb{H} \mid |z - \eta|^2 \leq \kappa r^2 \}.$$

Then η is a cusp of Γ if and only if the sets

$$\mathbf{S}(\eta, \kappa) \setminus \left(\bigcup_{i=1}^{h} \bigcup_{\gamma \in \Gamma} \gamma A_i \mathbb{H}_A \right) \tag{3.3}$$

have finite hyperbolic volume for all κ, $A > 0$.

Proof. We shall only give the geometric intuition for the proof, the details are left as an exercise.

Assume that (3.3) holds. Let G be the vertical geodesic which has η in its closure. The set $\mathbf{S}(\eta, \kappa)$ is a tubular neighbourhood of G with a crossection of bounded diameter. Let G' be half of the geodesic G starting from some point on G and leading towards η. Let $\mathbf{S}'(\eta, \kappa)$ the part of $\mathbf{S}(\eta, \kappa)$ which encloses G'. Note that $\mathbf{S}'(\eta, \kappa)$ has infinite hyperbolic volume.

The quotient $\Gamma\backslash\mathbb{H}$ is the union of a compact set and the images of finitely many cusp sectors. The images of the $A_i\mathbb{H}_A$ in $\Gamma\backslash\mathbb{H}$ lie further and further out in the images of the cusp sectors while A grows. Condition (3.3) ensures that the image of every $\mathbf{S}'(\eta, \kappa)$ in $\Gamma\backslash\mathbb{H}$ meets the image of a $A_i\mathbb{H}_A$. Hence the image of G' has to tend towards a cusp and this has to be η.

The reverse conclusion is obvious since then there is a $\gamma \in \Gamma$ and an $i \in \{1, \ldots, h\}$ so that the horoballs $\gamma A_i \mathbb{H}_A$ $(A > 0)$ touch the boundary of $\mathbb{H}$ in the point η. □

For later use we include the following corollary of the above proposition.

Corollary 3.11. *Let $\Gamma < \mathbf{PSL}(2,\mathbb{C})$ be a cofinite group and let $\eta \in \mathbb{P}^1\mathbb{C}$ be a cusp of Γ. Assume that τ is an elliptic element in the stabilizer Γ_η and let $\{\eta, \eta_1\}$ be the set of fixed points of τ in $\mathbb{P}^1\mathbb{C}$. Then η_1 is a cusp of Γ.*

Proof. We may assume that $\eta = \infty$. Then τ is of the form

$$\tau = \begin{pmatrix} \epsilon & \epsilon\omega \\ 0 & \epsilon^{-1} \end{pmatrix}.$$

Where ϵ is a root of unity not equal ± 1 and $\omega \in \mathbb{C}$. Consider a smooth function $k : [1, \infty[\to \mathbb{R}$ of compact support and let $K := k \circ \delta$ where δ is the function from (1.1.14). Let $\mathcal{F}$ be a fundamental domain for Γ as described in Proposition 3.9 and let the cusp sectors $\mathcal{F}_1, \ldots, \mathcal{F}_h$ be defined accordingly. Define for large A

$$\mathcal{F}_A := \mathcal{F} \setminus (\mathcal{F}_1(A) \cup \ldots \cup \mathcal{F}_h(A)), \qquad \mathfrak{M}_A := \bigcup_{\gamma \in \Gamma} \gamma \mathcal{F}_A.$$

Then the integral

$$\int_{\mathfrak{M}_A} K(P, \tau P)\, dv(P) = \int_{\mathfrak{M}_A} k\left(\frac{|(1-\epsilon^{-2})z_P + \omega|^2}{2r_P^2} + 1\right) dv(P)$$

exists and is finite. Notcice that the function $P \mapsto K(P, \tau P)$ $(P \in \mathbb{H})$ has its support in a set of the form $\mathbf{S}(\eta_1, \kappa)$ for suitable κ. The hypothesis of Proposition 3.10 is now verified by choice of appropriate functions k. □

2.4 Jørgensen's Inequality

Another very important technical tool in the study of discrete groups is the following result of Jørgensen.

Theorem 4.1 (Jørgensen's inequality). *Let $A, B \in \mathbf{SL}(2, \mathbb{C})$ generate a discrete subgroup. Then*

$$|\mathrm{tr}(A)^2 - 4| + |\mathrm{tr}(ABA^{-1}B^{-1}) - 2| \geq 1 \tag{4.1}$$

unless one of the following holds.

(1) A is parabolic and B fixes the unique fixed point of A in $\mathbb{P}^1\mathbb{C}$,
(2) $BAB^{-1} \in \{A, A^{-1}\}$.

Note that if (1) or (2) applies to $< A, B >$ then the group $< A, B >$ is elementary. The group can then be further analyzed using Theorem 3.4. We reproduce here the proof of Jørgensen (1976). A crucial role is played by the following map on matrices.

Definition 4.2. For $A \in \mathbf{SL}(2, \mathbb{C})$ we define $\theta_A : \mathbf{SL}(2, \mathbb{C}) \to \mathbf{SL}(2, \mathbb{C})$, by $\theta_A : B \mapsto BAB^{-1}$.

In the proof of Theorem 4.1 the following result is used.

Proposition 4.3. *Let $A, B \in \mathbf{SL}(2, \mathbb{C})$ be matrices such that A does not have order exactly equal to 4. Assume further that there is an $n \in \mathbb{N}$ with $\theta_A^n(B) = A$. Then one of the following holds.*
(1) A is parabolic and B fixes the unique fixed point of A in $\mathbb{P}^1\mathbb{C}$,
(2) $BAB^{-1} \in \{A, A^{-1}\}$.

Proof. We define $\gamma_0 = B$ and $\gamma_m = \theta_A^m(B)$ for $m \in \mathbb{N}$. Note that $\gamma_{m+1} = \gamma_m A \gamma_m^{-1}$ for all $m \in \mathbb{N}$.

Case 1. A is parabolic with unique fixed point $\zeta = \infty \in \mathbb{P}^1\mathbb{C}$. The elements $\gamma_1, \gamma_2, \ldots$ are also parabolic with unique fixed points $\zeta_1, \zeta_2, \ldots \in \mathbb{P}^1\mathbb{C}$ say. We have $\zeta_m = \gamma_{m-1}(\infty)$ for $m \in \mathbb{N}$. Since $\zeta_n = \infty$ we may conclude by induction that $\zeta_1 = \zeta_2 = \ldots = \zeta_n = \infty$. Since A and BAB^{-1} have the same fixed point ∞ we conclude $B\infty = \infty$ and (1) applies.

Case 2. A is not parabolic. Conjugating A, B we may further assume without loss of generality that $A = \begin{pmatrix} \epsilon & 0 \\ 0 & \epsilon^{-1} \end{pmatrix}$ with $\epsilon \neq \pm 1, \pm i$. Otherwise (2) applies or A has order 4. For $m \geq 1$ we write S_m for the set of fixed points of γ_m. We have $S_m = \gamma_{m-1}(\{0, \infty\})$. Since $S_n = \{0, \infty\}$ we conclude that $\gamma_{n-1}\{0, \infty\} = \{0, \infty\}$. γ_{n-1} cannot permute $0, \infty$ since γ_{n-1} would then have order exactly equal to 4. This would then imply that A also has this property. Hence we find that $S_{n-1} = \{0, \infty\}$. By induction we get $S_1 = S_2 = \ldots = S_n = \{0, \infty\}$. But then A and BAB^{-1} are in diagonal form, this is only possible if (2) applies. □

Note that Proposition 4.3 implies that $\theta_A^2(B) = A$ once the hypotheses are satisfied.

Proof of Theorem 4.1. First of all we assume that $A \neq I, -I$ since otherwise (2) applies. We distinguish two cases.

Case 1. A is unipotent, that is its image in $\mathbf{PSL}(2, \mathbb{C})$ is parabolic. Conjugating A, B we may assume that $A = \begin{pmatrix} 1 & 1 \\ 0 & 1 \end{pmatrix}$, $B = \begin{pmatrix} a & b \\ c & d \end{pmatrix}$. The inequality (4.1) is then equivalent to $|c| \geq 1$. If $c = 0$, case (1) applies. If $c \neq 0$, then $|c| \geq 1$ is implied by Shimizu's Lemma, that is Theorem 3.1.

Case 2. A is diagonalizable, that is, its image in $\mathbf{PSL}(2, \mathbb{C})$ is hyperbolic, loxodromic or elliptic. After conjugation we have $A = \begin{pmatrix} \rho & 0 \\ 0 & \rho^{-1} \end{pmatrix}$, $B = \begin{pmatrix} a & b \\ c & d \end{pmatrix}$ with $\rho \neq \pm 1$. We put

$$\alpha = |\mathrm{tr}(A)^2 - 4| + |\mathrm{tr}(ABA^{-1}B^{-1}) - 2|.$$

A computation shows $\alpha = (1 + |b||c|)|\rho - \rho^{-1}|^2$. We also assume that (4.1) is false, that is $0 < \alpha < 1$.

Fist we consider the case $b \cdot c = 0$. If $b = c = 0$ then (2) applies. In case $b \neq 0$ and $c = 0$ the group $< A, B >$ is not discrete. This can be seen as follows. We define $C = ABA^{-1}B^{-1} = \begin{pmatrix} 1 & h \\ 0 & 1 \end{pmatrix}$. Then $h \in \mathbb{C}$, $h \neq 0$. Define sequences

$$D_0 = C, \quad D_{m+1} = AD_mA^{-1}D_m^{-1}, \qquad E_0 = C, \quad E_{m+1} = A^{-1}E_mAE_m^{-1}.$$

We find

$$D_m = \begin{pmatrix} 1 & \rho^m(\rho - \rho^{-1})^m h \\ 0 & 1 \end{pmatrix}, \qquad E_m = \begin{pmatrix} 1 & \rho^{-m}(\rho^{-1} - \rho)^m h \\ 0 & 1 \end{pmatrix}.$$

One of these sequences converges to I without becoming stationary. This follows from $|\alpha| = |\rho - \rho^{-1}|^2 < 1$. The case $b = 0$ and $c \neq 0$ is treated by transposition.

Next assume that $bc \neq 0$. We define for $m \in \mathbb{N}$: $\gamma_m = \begin{pmatrix} a_m & b_m \\ c_m & d_m \end{pmatrix} = \theta_A^m(B)$. Clearly we have

$$\gamma_{m+1} = \begin{pmatrix} a_m d_m \rho - b_m c_m \rho^{-1} & a_m b_m(\rho^{-1} - \rho) \\ c_m d_m(\rho - \rho^{-1}) & a_m d_m \rho^{-1} - b_m c_m \rho \end{pmatrix}.$$

From this formula we find

$$b_{m+1}c_{m+1} = -b_m c_m(1 + b_m c_m)(\rho - \rho^{-1})^2. \tag{4.2}$$

By formula (4.2) we find that $b_m c_m \neq 0$ for all $m \in \mathbb{N}$ or $b_{m_0}c_{m_0} = -1$ for some m_0. The second case is treated analogously to the case $bc = 0$. So we may assume that $b_m c_m \neq 0$ for all m. From formula (4.2) we then find by induction that $|b_m c_m| \leq \alpha^m |b \cdot c|$. This implies $\lim_{m\to\infty} b_m c_m = 0$. By the determinant relation we get $\lim_{m\to\infty} a_m d_m = 1$. Because of $a_{m+1} = a_m d_m \rho - b_m c_m \rho^{-1}$ we infer

$$\lim_{m\to\infty} a_m = \rho, \qquad \lim_{m\to\infty} d_m = \rho^{-1}. \tag{4.3}$$

Next we have $|b_{m+1}/b_m| = |a_m(\rho^{-1} - \rho)|$. This implies that

$$\lim_{m\to\infty} \left| \frac{b_{m+1}}{b_m} \right| = |\rho(\rho^{-1} - \rho)| \leq \sqrt{\alpha} \cdot |\rho|.$$

For large m we get $|b_{m+1}/\rho^{m+1}| < (1 + \sqrt{\alpha})/2\, |b_m/\rho^m|$. Hence

$$\lim_{m\to\infty} \frac{b_m}{\rho^m} = 0, \qquad \lim_{m\to\infty} c_m \rho^m = 0. \tag{4.4}$$

the second equality being found by a symmetric argument. Clearly we have

$$A^{-m}\gamma_{2m}A^m = \begin{pmatrix} a_{2m} & b_{2m}\rho^{-2m} \\ c_{2m}\rho^{2m} & d_{2m} \end{pmatrix}.$$

This sequence of matrices converges to A by (4.3), (4.4). Since $< A, B >$ is discrete we must have $A^{-n}\gamma_{2n}A^n = A$ for some $n \in \mathbb{N}$. But then $\theta_A^{2m}(B) = A$ and we may use Proposition 4.3, or A has exactly order 4. This implies that $\mathrm{tr}(A) = 0$ and (4.1) is satisfied anyway. □

We mention two interesting corollaries of Theorem 4.1.

Corollary 4.4. *Let $A, B \in$ **SL**$(2, \mathbb{C})$ generate a non-elementary discrete group, then* $\max\{\ \|A - I\|,\ \|B - I\|\ \} > 0,146$.

Corollary 4.5. *Let $\Gamma <$ **PSL**$(2, \mathbb{C})$ be a non-elementary subgroup. Then the following two statements are equivalent.*

(1) Γ is discrete.
(2) Every two generator subgroup of Γ is discrete.

We leave the straightforward proofs as exercises. See for example Jørgensen (1976) and Beardon (1983). We shall also make use of the following fact.

Corollary 4.6. *Let $A \in \mathbf{SL}(2,\mathbb{C})$ be a matrix. Then there is an $\epsilon \in]0,\infty[$ such that for any $B \in \mathbf{SL}(2,\mathbb{C})$ with $\|B - I\| \leq \epsilon$ the group $< A, B >$ is either not discrete or one of the following cases holds.*

(1) A is parabolic and B fixes the unique fixed point of A in $\mathbb{P}^1\mathbb{C}$,
(2) $BAB^{-1} \in \{A, A^{-1}\}$.

Proof. By making ϵ small enough we may enforce that

$$|\mathrm{tr}(B)^2 - 4| + |\mathrm{tr}(BAB^{-1}A^{-1}) - 2| < 1.$$

Then Theorem 4.1 applies. □

We shall need Corollary 4.6 in a more explicit form.

Corollary 4.7. *Let* $A = \begin{pmatrix} t & 1 \\ 0 & t^{-1} \end{pmatrix}, \quad B = \begin{pmatrix} a & b \\ c & d \end{pmatrix} \in \mathbf{SL}(2,\mathbb{C})$. *Assume that*

(1) $|t - 1| \leq \delta, \qquad |t^{-1} - 1| \leq \delta$,
(2) $\|B - I\| \leq \delta$ for $\delta = 0,432889$.
Then $< A, B >$ is either not discrete or one of the following holds:

(a) A is parabolic and $c = 0$,
(b) $BAB^{-1} \in \{ A, A^{-1} \}$.

Proof. A computation shows:

$$\mathrm{tr}(A)^2 - 4 = (t - t^{-1})^2,$$
$$\mathrm{tr}(ABA^{-1}B^{-1}) - 2 = c\{(a - d)(t - t^{-1}) + c + b(t - t^{-1})^2\}.$$

The function $\varphi(t) = |t - t^{-1}|$ attains its maximum on the set defined by (1) for

$$t = 1 - \frac{1}{2}\delta^2 \pm i\delta\sqrt{1 - \frac{1}{4}\delta^2},$$

and this maximum is

$$\varphi_{\max} = 2\delta\sqrt{1 - \frac{1}{4}\delta^2} =: \alpha.$$

Hence we have the estimate

$$|\mathrm{tr}(A)^2 - 4| + |\mathrm{tr}(ABA^{-1}B^{-1}) - 2| \\ \leq \alpha^2 + \alpha|a-d||c| + |c|^2 + \alpha^2|b||c| =: \phi(a,b,c,d).$$

Write

$$B = \begin{pmatrix} a & b \\ c & d \end{pmatrix} = \begin{pmatrix} 1+x+iy & r+is \\ p+iq & 1+u+iv \end{pmatrix}$$

with $x, y, p, q, r, s, u, v \in \mathbb{R}$. We shall determine the maximum of ϕ on the ball

$$\mathcal{B} = \{ \, x^2+y^2+p^2+q^2+r^2+s^2+u^2+v^2 \leq \delta^2 \, \}.$$

For $c = 0$, the function ϕ attains a minimum. Hence we may assume $c \neq 0$. Consider the domain

$$\mathcal{B}_1 = \{ \, B \in \mathcal{B} \mid a-d \neq 0, \; b \neq 0, \; c \neq 0 \, \}.$$

We determine the maximum of the function ϕ on $\mathcal{B}_1$, using Lagrange's method. It is obvious that the maximum is attained on the sphere bounding $\mathcal{B}$. We consider the function

$$\psi = \phi + \lambda(x^2+y^2+p^2+q^2+r^2+s^2+u^2+v^2-\delta^2).$$

For a maximum of ϕ on $\mathcal{B}$, the partial derivatives of first order of ψ all vanish:

$$\alpha\frac{|c|}{|a-d|}(x-u) + 2\lambda x = 0, \qquad \alpha\frac{|c|}{|a-d|}(y-v) + 2\lambda y = 0,$$

$$-\alpha\frac{|c|}{|a-d|}(x-u) + 2\lambda u = 0, \qquad -\alpha\frac{|c|}{|a-d|}(y-v) + 2\lambda v = 0,$$

$$\alpha p\frac{|a-d|}{|c|} + \alpha^2 p\frac{|b|}{|c|} + 2p + 2\lambda p = 0, \quad \alpha q\frac{|a-d|}{|c|} + \alpha^2 q\frac{|b|}{|c|} + 2q + 2\lambda q = 0,$$

$$\alpha r\frac{|c|}{|b|} + 2\lambda r = 0, \qquad \alpha s\frac{|c|}{|b|} + 2\lambda s = 0.$$

The first four equations yield $x+u=0$, $y+v=0$. These imply $a-1 = 1-d$ and also $|a-d| = 2|a-1|$. The above equations then reduce to

$$\alpha\frac{|c|}{|a-1|} + 2\lambda = 0, \qquad 2\alpha\frac{|a-1|}{|c|} + \alpha^2\frac{|b|}{|c|} + 2 + 2\lambda = 0, \qquad \alpha^2\frac{|c|}{|b|} + 2\lambda = 0.$$

Solve these equations for $|b|$, $|c|$, obtaining $|b| = \alpha|a-1|$ and also $|c| = (2+\alpha^2)|a-1|/\alpha$. Hence we obtain on the 7-sphere bounding $\mathcal{B}$:

$$2|a-1|^2 + |b|^2 + |c|^2 = \delta^2, \qquad |a-1|^2 = \frac{\alpha^2\delta^2}{2(\alpha^2+1)(\alpha^2+2)}.$$

The maximum of ϕ is

$$\phi_{\max} = \frac{\delta^2(\alpha^2+2)}{2} + \alpha^2.$$

Hence we obtain

$$|\mathrm{tr}(A)^2 - 4| + |\mathrm{tr}(ABA^{-1}B^{-1}) - 2| \leq 5\delta^2 + \delta^4 - \frac{1}{2}\delta^6.$$

For $\delta = 0,432889$, we obtain

$$|\mathrm{tr}(A)^2 - 4| + |\mathrm{tr}(ABA^{-1}B^{-1}) - 2| \leq 0,99997724 < 1.$$

Hence Jørgensen's Theorem 4.1 applies, and yields the conclusion. For $a = d$, the function ϕ reduces to $\phi = \alpha^2 + |c|^2 + \alpha^2|b||c|$. We consider ϕ on the domain

$$\mathcal{B}_2 = \left\{ \begin{pmatrix} p \\ q \\ r \\ s \end{pmatrix} \in \mathbb{R}^4 \;\middle|\; p^2 + q^2 + r^2 + s^2 \leq \delta^2,\ r^2 + s^2 \neq 0 \right\}.$$

The function ϕ attains its maximum on $\mathcal{B}_2$:

$$\phi_{\max} = \alpha^2 + \frac{1}{2}\delta^2(1 + \sqrt{1 + \alpha^4}) < \alpha^2 + \frac{1}{2}\delta^2(2 + \alpha^2).$$

Hence Theorem 4.1 applies for $\delta = 0,432889$.

In the case $b = 0,\ ad = 1,\ a - d = 0$, the condition (b) with $\delta < 1$ implies $a = d = 1, |c| \leq \delta$. Hence we obtain the inequality

$$|\mathrm{tr}(A)^2 - 4| + |\mathrm{tr}(ABA^{-1}B^{-1}) - 2| \leq 5\delta^2 - \delta^4 < 1$$

for $\delta = 0,432889$. Hence Theorem 4.1 applies. It remains to consider the domain

$$\mathcal{B}_3 = \{\ B \in \mathcal{B} \mid b = 0,\ ad = 1,\ a - d \neq 0\ \}.$$

We replace $\mathcal{B}_3$ with the larger domain $\mathcal{B}_4 = \{\ B \in \mathcal{B} \mid b = 0,\ a - d \neq 0\ \}$. The function ϕ restricted to $\mathcal{B}_4$ is $\phi = \alpha^2 + \alpha|c||a-d| + |c|^2$. The computation of the maximum is as before. The result is

$$\phi_{\max} = \alpha^2 + \frac{\delta^2(1 + \sqrt{1 + 2\alpha^2})}{2} < \alpha^2 + \frac{\delta^2(2 + \alpha^2)}{2}.$$

As before we can apply Theorem 4.1 to finish the proof. □

The constant δ in the Corollary 4.7 is crucial in later applications. One can slightly improve on δ by working out the maximum of $|\mathrm{tr}(A)^2 - 4| + |\mathrm{tr}(ABA^{-1}B^{-1}) - 2|$ on the domain defined by (1), (2). However, the computations involved are prohibitive. We shall show that Jørgensen's inequality does not apply for $\delta = 0,451$ by working out an example.

Let $t = 0,8982995 + 0,4393836687\, i$, $a = 1,028782750$, $b = 0,1364569505$, $c = 0,427929442$, $d = 1,028782750$. Then all assumptions of the Corollary hold, for $\delta = 0,451$, and we have

$$|\mathrm{tr}(A)^2 - 4| + |\mathrm{tr}(ABA^{-1}B^{-1}) - 2| = 1,000449317.$$

Of course, A, B in this example do not generate a discrete group, since the pair (B, A) instead of (A, B) does not satisfy Jørgensen's trace condition.

We do not know the maximal value of δ such that the Corollary holds. It is not clear whether for any $\epsilon > 0$, $\delta = 0,5 - \epsilon$ is a candidate.

There exist quite a number of papers motivated by Jørgensen's inequality and dealing with inequalities for Möbius transformations contained in discrete groups, see Brooks and Matelski (1981), Cao (1995), Tan (1989), Gehring and Martin (1989), (1991).

2.5 Covolumes

Here we shall analyse the covolumes of discrete groups $\Gamma < \mathbf{PSL}(2, \mathbb{C})$ using the Lemma of Shimizu and Jørgensen's inequality. If $\Gamma < \mathbf{PSL}(2, \mathbb{C})$ is a discrete group, we have defined C_Γ to be the set of cusps of Γ. C_Γ is a Γ-invariant subset of $\mathbb{P}^1\mathbb{C}$, see Definition 1.10.

Theorem 5.1. *Let $\Gamma < \mathbf{PSL}(2, \mathbb{C})$ be a discrete group of finite covolume. Then*

(1) $\mathrm{vol}(\Gamma) \geq \sqrt{3}/24 \cdot |\Gamma \backslash C_\Gamma|$,
(2) $|\Gamma \backslash C_\Gamma| < \infty$.

This theorem was already noted in Jørgensen, Marden (1988). In case Γ happens to be torsion free it can be improved, see Adams (1988). The above result gives no information if C_Γ is empty. In general we have

Theorem 5.2. *Let $\Gamma < \mathbf{PSL}(2, \mathbb{C})$ be a discrete group, then*

$$\mathrm{vol}(\Gamma) > 1,3751 \cdot 10^{-15}. \tag{5.1}$$

Theorem 5.2 is, with a non-explicit bound, due to Kazhdan and Margulis (1968). See also Borel (1971) and Wang (1969). We give here a variation of a proof of Wielenberg (1977). Our approach is elementary and makes it possible to give at least the rough lower bound (5.1). By using more hyperbolic geometry Meyerhoff (1988a), (1988b) arives at the following sharper conclusions:

$$\mathrm{vol}(\Gamma) > 1,7 \cdot 10^{-6} \qquad \text{if } \Gamma \text{ is discrete in } \mathbf{PSL}(2, \mathbb{C}),$$

$$\mathrm{vol}(\Gamma) > 8,2 \cdot 10^{-4} \qquad \text{if } \Gamma \text{ is discrete and torsion free in } \mathbf{PSL}(2, \mathbb{C}).$$

If Γ happens to contain nontrivial torsion elements there are even better bounds for $\mathrm{vol}(\Gamma)$ by Gehring, Martin (1993).

Theorem 5.2 has a classical counterpart for discontinuous groups of motions on 2-dimensional hyperbolic space due to Petersson (1938a), (1938b) and Siegel (1945). This says that the covolume of any such group is at least $\frac{\pi}{21}$. Even more is known on the set of covolumes of discrete groups $\Gamma < \mathbf{PSL}(2,\mathbb{C})$. We report here on the theorems of Thurston (1978) before we enter into the proofs of Theorems 5.1, 5.2. This theorem determines the structure of the set of covolumes and also shows that the covolume is reasonable invariant of the discrete group. See Meyerhoff (1992) for interesting perspectives of this result.

Theorem 5.3. *Let $\mathcal{V} \subset \mathbb{R}$ be the set of all covolumes of discrete, torsion free subgroups $\Gamma < \mathbf{PSL}(2,\mathbb{C})$ of finite covolume. Then the following hold.*

(1) For any $v \in \mathcal{V}$ the set

$$\{\, \Gamma < \mathbf{PSL}(2,\mathbb{C}) \mid \Gamma \text{ is discrete, torsion free and } \mathrm{vol}(\Gamma) = v \,\}$$

splits into finitely many $\mathbf{PSL}(2,\mathbb{C})$ conjugacy classes.

(2) $\mathcal{V}$ is a closed non-discrete subset of $\mathbb{R}$. $\mathcal{V}$ is well-ordered with respect to $\leq$ on $\mathbb{R}$. Its order type is ω^ω.

Part 2 of Thurston's result says that there is a discrete subgroup $\Gamma < \mathbf{PSL}(2,\mathbb{C})$ with smallest covolume v_1. Then there is the next smallest covolume v_2 and so on. The groups corresponding to these first ω covolumes are all cocompact. The above strictly ascending series of covolumes has a limit point v_ω which is the smallest covolume of a group with exactly one class of cusps. The next smallest covolume after v_ω is $v_{\omega+1}$ and again belongs to a cocompact group. The then following covolumes converge to $v_{2\omega}$ which is the covolume of a group with one class of cusps. This continues and gives an ascending sequence of covolumes $v_\omega,\ v_{2\omega},\ \ldots$ which converges to v_{ω^2} which is the smallest covolume of a group with exactly 2 classes of cusps. This picture then goes on forever.

A similar picture arises if one omits the condition of torsion freeness for the groups Γ. The difference being that the set

$$\{\, \mathrm{vol}(\Gamma) \mid \Gamma < \mathbf{PSL}(2,\mathbb{C}) \text{ is discrete } \}$$

contains some isolated points. This happens if a group Γ has a cusp ζ with $\Gamma_\zeta \neq \Gamma'_\zeta$.

It is an interesting question for which groups Γ the covolume is minimal or at the first limit point. There are at the moment no final answers to these questions. Let $\mathbf{CT}(22)_2$ be the twofold extension of the tetrahedral Coxeter group $\mathbf{CT}(22)$ discussed in Chapter 10. The discrete subgroup of $\mathbf{Iso}(\mathbb{H})$ of smallest covolume known at the time of this writing is $\mathbf{CT}(22)_2$. The discrete subgroup of $\mathbf{Iso}^+(\mathbb{H})$ of smallest covolume known is $\mathbf{CT}(22)_2^+ = \mathbf{CT}(22)_2 \cap \mathbf{Iso}^+(\mathbb{H})$. We have $\mathrm{vol}(\mathbf{CT}(22)_2^+) \sim 0.039$. The covolume of this group is proved to be smallest amongst the covolumes of the arithmetic subgroups of $\mathbf{Iso}^+(\mathbb{H})$ by Chinburg, Friedman (1986). Let $\mathcal{O}$ be the ring of integers in the field of sixth roots of unity $\mathbb{Q}(\sqrt{-3})$, and let Γ be the maximal discrete extensiongroup of $\mathbf{PSL}(2,\mathcal{O})$ in $\mathbf{PSL}(2,\mathbb{C}) = \mathbf{Iso}^+(\mathbb{H})$. This group is dicussed in Chapters 7, 10. We have $\mathrm{vol}(\Gamma) \sim 0.0846$. Using sphere packing arguments Meyerhoff (1985) proves that the covolume of Γ is smallest among all discrete subgroups of $\mathbf{Iso}^+(\mathbb{H})$ having one class of cusps. The discrete torsion free non-cocompact subgroup of $\mathbf{Iso}(\mathbb{H})$ having minimal covolume amongst groups of this class is known, see Adams (1987). This group is a subgroup of index 12 in the tetrahedral group $\mathbf{CT}(8)$ of Chapter 10. Its covolume is about 1.0149. The discrete torsion free non-cocompact subgroup of $\mathbf{Iso}^+(\mathbb{H})$ having the minimal covolume of groups of this class is conjectured to be the group Γ_8 discussed in Section 7.7.

In dimensions n different from 3 the structure of the set of covolumes of torsion free, discontinuous groups of motions of n-dimensional hyperbolic space is much less dramatic. In dimension 2 the Gauß-Bonnet theorem implies that the finite covolumes are 2π, 4π, 6π, $8\pi, \dots$. In dimensions $n > 3$ a result of Wang (1972) says that there are only finitely many covolumes below a given bound.

For arithmetic discrete groups $\Gamma < \mathbf{PSL}(2,\mathbb{C})$ the covolumes can be identified as an elementary multiple of the value at 2 of certain Dirichlet series. This is shown by the covolume formulas in Chapters 7 and 10. Apart from this the covolume of a given group can only be computed by describing a fundamental polyhedron, and then using formulas of Lobachevski, see Coxeter (1956) or Milnor (1982), for its volume. The work of Kellerhals (1995) contains useful information on this subject. Chapter 10 discusses the special case of tetrahedra.

It is not very difficult to modify our proof of Theorem 5.2 to give the following result.

Theorem 5.4. *There is a $\kappa \in]0,\infty[$ such that for any discrete subgroup $\Gamma < \mathbf{PSL}(2,\mathbb{C})$ there is an $A \in \mathbf{PSL}(2,\mathbb{C})$ with $B(j,\kappa) \subset \mathcal{P}_j(A\Gamma A^{-1})$.*

Here $B(j,\kappa)$ is the hyperbolic ball of radius κ and center j. $\mathcal{P}_j(A\Gamma A^{-1})$ is the Poincaré normal polyhedron as defined in Definition 2.2.

To prepare for the proof of Theorem 5.1 we need a lemma on lattices in $\mathbb{C}$.

Lemma 5.5. *Let $\Lambda \subset \mathbb{C}$ be a lattice, and suppose that $0 \neq \omega \in \Lambda$ is a vector of minimal length. Then the Euclidean area $|\Lambda|$ of a period parallelogram for Λ satisfies $|\Lambda| \geq \sqrt{3}\,|\omega|^2/2$. Equality holds if and only if $\Lambda = \mathbb{Z}\omega \oplus \mathbb{Z}\omega \cdot e^{\frac{\pi i}{3}}$.*

Proof. Define $\omega_1 := \omega$, and choose $\omega_2 \in \Lambda \backslash \mathbb{Z}\omega_1$ such that $|\omega_2|$ is minimal. Then it is known that $\{\omega_1, \omega_2\}$ is a $\mathbb{Z}$-basis for Λ (see e.g. Ahlfors (1979), pages 265–266). Let

$$\frac{\omega_2}{\omega_1} = \left|\frac{\omega_2}{\omega_1}\right| \cdot e^{i\alpha},$$

then $\alpha \notin \mathbb{Z}\pi$. The inequality $|\omega_1 \pm \omega_2|^2 \geq |\omega_2|^2$ implies

$$|\omega_1|^2 - 2|\omega_1|\,|\omega_2|\,|\cos\alpha| \geq 0,$$

and hence

$$|\cos\alpha| \leq \frac{1}{2}\left|\frac{\omega_1}{\omega_2}\right| \leq \frac{1}{2}.$$

This implies that $|\Lambda| = |\omega_1|\,|\omega_2|\,|\sin\alpha| \geq \frac{\sqrt{3}}{2}|\omega|^2$. Equality holds if and only if $\Lambda = \mathbb{Z}\omega \oplus \mathbb{Z}\omega e^{\frac{\pi i}{3}}$. □

We shall draw Theorem 5.1 from the following Proposition.

Proposition 5.6. *Let $\Gamma < \mathbf{PSL}(2,\mathbb{C})$ be a discrete group of finite covolume, then*

$$\mathrm{vol}\,(\Gamma) \geq \frac{\sqrt{3}}{4} \sum_{[\zeta]\in\Gamma\backslash C_\Gamma} \frac{1}{|\Gamma_\zeta : \Gamma'_\zeta|}. \tag{5.2}$$

Here $[\zeta]$ denotes the class of a cusp $\zeta \in C_\Gamma$ in $\Gamma\backslash C_\Gamma$. The groups Γ'_ζ are defined in Definition 1.7. Note that the sum on the right-hand side of (5.2) does not depend on the choices of the respresentatives $\zeta \in [\zeta]$.

Proof. Let ζ_t with $t \in T$ be a set of representatives for $\Gamma\backslash C_\Gamma$. For every $t \in T$ choose $A_t \in \mathbf{PSL}(2,\mathbb{C})$ and lattices $\Lambda_t \subset \mathbb{C}$ such that $\zeta_t = A_t\infty$ and

$$A_t^{-1}\Gamma'_{\zeta_t}A_t = \left\{ \begin{pmatrix} 1 & \omega \\ 0 & 1 \end{pmatrix} \;\middle|\; \omega \in \Lambda_t \right\}.$$

Furthermore let $0 \neq \omega_t \in \Lambda_t$ be an element of minimal length. We also choose a period parallelogram $\mathcal{P}_t$ for Λ_t. The sets $\mathcal{F}_t = A_t \cdot \{\, z + rj \;\mid\; z \in \mathcal{P}_t,\, r > |\omega_t| \,\}$ are fundamental domains for the action of Γ'_{ζ_t} on $A_t\mathbb{H}_{\omega_t}$. We compute the hyperbolic volume of $\mathcal{F}_t$:

$$v(\mathcal{F}_t) = |\Lambda_t| \int_{|\omega_t|}^{\infty} \frac{dr}{r^3} = \frac{1}{2}\,\frac{|\Lambda_t|}{|\omega_t|^2}.$$

By Lemma 5.5 we get $v(\mathcal{F}_t) \geq \sqrt{3}/4$. Replacing $\mathcal{P}_t$ by a fundamental domain $\mathcal{P}_t^1$ for the action of $A_t^{-1}\Gamma_{\zeta_t}A_t$ and defining $\mathcal{F}_t^1$ analogously to $\mathcal{F}_t$ we find $v(\mathcal{F}_t^1) \geq v(\mathcal{F}_t)/[\Gamma_{\zeta_t} : \Gamma'_{\zeta_t}]$. We put $\mathcal{F}_0 = \bigcup_{t\in T} \mathcal{F}_t^1$. By Theorem 3.4 and Corollary 3.3 there is a set $M_0 \subset \mathcal{F}_0$ of Lebesgue measure zero, such that $(\mathcal{F}_0\backslash M_0) \cap \gamma(\mathcal{F}_0\backslash M_0) = \emptyset$ for all $\gamma \in \Gamma\backslash\{I\}$. M_0 can, for example, be taken as the union of the boundaries of the $\mathcal{F}_t^1$. We then have

$$\mathrm{vol}\,(\Gamma) \geq v(\mathcal{F}_0 \backslash M_0) = v(\mathcal{F}_0) = \sum_{t \in T} v(\mathcal{F}_t^1) \geq \frac{\sqrt{3}}{4} \cdot \sum_{t \in T} \frac{1}{|\Gamma_{\zeta_t} : \Gamma'_{\zeta_t}|}.$$

The Proposition is proved. □

Proof of Theorem 5.1. (1) and (2) follow from Proposition 5.6 by noting that $[\Gamma_\zeta : \Gamma'_\zeta] \in \{1, 2, 3, 4, 6\}$. For this fact see Theorem 1.8. □

We proceed now with the proof of Theorem 5.2. We need some elementary Lemmas.

Lemma 5.7. *Let $A \in \mathbf{SL}(2, \mathbb{C})$ and $\epsilon \geq 0$. Assume that $\|A\|^2 \leq 2 + \epsilon$. Then there is a matrix $U \in \mathbf{SU}(2)$ such that $\|A - U\|^2 \leq \epsilon$.*

Proof. We put $A = \begin{pmatrix} a & b \\ c & d \end{pmatrix}$. We have $|a - \bar{d}|^2 + |b + \bar{c}|^2 = \|A\|^2 - 2$. Hence $\|A\|^2 \geq 2$ and we may assume $\|A\|^2 = 2 + \epsilon$. Define

$$U_0 = \begin{pmatrix} \frac{\bar{d}+a}{2} & \frac{b-\bar{c}}{2} \\ \frac{c-\bar{b}}{2} & \frac{\bar{a}+d}{2} \end{pmatrix}.$$

A computation shows that $\det U_0 = 1 + \frac{\epsilon}{4}$. Define $\epsilon_0^{-1} = \sqrt{1 + \epsilon/4}$, and put $U := \epsilon_0 U_0$. Then $U \in \mathbf{SU}(2, \mathbb{C})$. Another computation shows

$$\|A - U\|^2 = \|A\|^2 \left(1 - \epsilon_0 + \frac{\epsilon_0^2}{2}\right) - 2\epsilon_0 + \epsilon_0^2.$$

Insert $\|A\|^2 = 2 + \epsilon$ and ϵ_0, obtaining

$$\|A - U\|^2 = 4\sqrt{1 + \frac{\epsilon}{4}} \left(\sqrt{1 + \frac{\epsilon}{4}} - 1\right).$$

The function on the right-hand side is $\leq \epsilon$, for $\epsilon \geq 0$. Hence we obtain $\|A - U\|^2 \leq \epsilon$. □

The above lemma implies in case $\epsilon = 0$ that $\|A\|^2 = 2$ is equivalent to $A \in \mathbf{SU}(2)$. The following is an easy reformulation.

Proposition 5.8. *Let $A \in \mathbf{SL}(2, \mathbb{C})$ and $\epsilon \in [0, \mathrm{arcosh}(2)]$. Assume that $d(j, A(j)) \leq \epsilon$. Then there is a $U \in \mathbf{SU}(2)$ with $\|A - U\|^2 \leq 2(\cosh(\epsilon) - 1)$.*

Proof. We have $\|A\|^2 = 2 \cdot \cosh(d(j, A(j))) \leq 2 + 2(\cosh(\epsilon) - 1)$. □

Lemma 5.9. *Parametrize the matrices $A \in \mathbf{SU}(2)$ as follows:*

$$(5.3)\qquad A = \begin{pmatrix} a & b \\ -\bar{b} & \bar{a} \end{pmatrix}, \quad a = x + iy,\ b = u + iv$$
$$x = \cos\vartheta\,\cos\varphi\,\cos\psi,\ y = \cos\vartheta\,\cos\varphi\,\sin\psi,$$
$$u = \cos\vartheta\,\sin\varphi,\ v = \sin\vartheta,$$

where $0 \le \psi < 2\pi$, $-\pi/2 \le \varphi \le \pi/2$, $-\pi/2 \le \vartheta \le \pi/2$, *and let* $A' \in \mathbf{SU}(2)$ *be similarly parametrized with angles* $\varphi', \psi', \vartheta'$. *Then*

$$(5.4)\qquad \|A - A'\| \le \sqrt{6}\left\| \begin{pmatrix} \varphi \\ \psi \\ \vartheta \end{pmatrix} - \begin{pmatrix} \varphi' \\ \psi' \\ \vartheta' \end{pmatrix} \right\|,$$

where the norm on the left-hand side is defined as in (1.1) *and the norm on the right-hand side is the usual Euclidean norm on* $\mathbb{R}^3$.

Proof. Consider the map $F : \mathbb{R}^3 \to \mathbb{R}^4$ where $F(\varphi, \psi, \vartheta)^t = (x, y, u, v)^t$ with x, y, u, v given by the above formulae. Then the mean value theorem implies

$$(5.5)\qquad \left\| F\begin{pmatrix} \varphi \\ \psi \\ \vartheta \end{pmatrix} - F\begin{pmatrix} \varphi' \\ \psi' \\ \vartheta' \end{pmatrix} \right\| \le (\sup \|DF\|) \left\| \begin{pmatrix} \varphi \\ \psi \\ \vartheta \end{pmatrix} - \begin{pmatrix} \varphi' \\ \psi' \\ \vartheta' \end{pmatrix} \right\|,$$

where the norm of the Jacobian DF is the matrix norm associated with the Euclidean norm. An easy computation yields $\|DF\| \le \sqrt{3}$, and the assertion follows by comparing $\|A - A'\|$ with the right-hand side of (5.5). □

Proposition 5.10. *Let* $U_1, \dots, U_N \in \mathbf{SU}(2)$ *with* $N \ge 16$. *Then there exist* $k, \ell \in \{1, \dots, N\}$ *with* $k \ne \ell$ *such that*

$$(5.6)\qquad \|U_k - U_\ell\| \le \frac{3\sqrt{2}\,\pi}{\left[\sqrt[3]{\frac{N}{2}}\right] - 1}.$$

Proof. We parametrize $\mathbf{SU}(2)$ as in Lemma 5.9 and divide the parameter interval $[-\pi/2, \pi/2]$ into n subintervals of length π/n and $[0, 2\pi]$ into $2n$ subintervals of length π/n. Then we obtain $2n^3$ little cubes in the parameter space whose images under the parameter map cover $\mathbf{SU}(2)$. Now if $N > 2n^3$, then there exist $k, \ell \in \{1, \dots, N\}$ with $k \ne \ell$ such that U_k and U_ℓ belong to the image (in $\mathbf{SU}(2)$) of the same little cube. Hence we obtain from Lemma 5.9: $\|U_k - U_\ell\| \le \sqrt{6} \cdot \sqrt{3} \cdot \pi/n$. Choosing $n = [\sqrt[3]{N/2}] - 1$ (≥ 1) we obtain the assertion. □

We use the notation of Chapter 1: $B(j, r) = \{\ P \in \mathbb{H} \ \mid\ d(P, j) < r\ \}$ stands for the hyperbolic ball of radius r and center j. Its hyperbolic volume is $v(B(j, r)) = 2\pi(\sinh(r)\ \cosh(r) - r)$.

Theorem 5.11. *Let $\Gamma < \mathbf{SL}(2,\mathbb{C})$ be a discrete subgroup of finite covolume. Let $\epsilon, \delta \in]0,\infty[$ and $N \in \mathbb{N}$, $N \geq 16$, be given such that*

(1) $\delta \leq 8$,
(2) $\cosh(2\epsilon) \leq 1 + \delta^2/72$,
(3) $3\sqrt{2}\pi/([\sqrt[3]{N/2}] - 1) \leq \delta/6$,
(4) $\mathrm{vol}(\Gamma) \leq v(B(j,\epsilon))/N$.

Then there is an $A \in \Gamma \backslash \{I\}$ with $\|A - I\| \leq \delta$.

Proof. Let $\mathcal{F} \subset \mathbb{H}$ be a fundamental domain for Γ and $\mathring{\mathcal{F}}$ its open kernel. We have by definition $\mathrm{vol}(\Gamma) = v(\mathcal{F}) = v(\mathring{\mathcal{F}})$. Since Γ is a countable set, we may choose an enumeration $I = A_1, A_2, A_3, \ldots$ of Γ, such that from each possible pair $\{A, -A\} \subset \Gamma$ only one member is enumerated. We define inductively

$$
\begin{aligned}
L_1 &= B(j,\epsilon) \cap \mathring{\mathcal{F}}, \\
L_2 &= (A_2 \cdot B(j,\epsilon)) \cap (\mathring{\mathcal{F}} \backslash L_1) \\
&\vdots \\
L_k &= (A_k \cdot B(j,\epsilon)) \cap (\mathring{\mathcal{F}} \backslash \bigcup_{\ell=1}^{k-1} L_\ell).
\end{aligned}
$$

Put furthermore $T_k = A_k^{-1}(L_k)$ for $k = 1, 2, \ldots$ and $T = \cup_{k=1}^{\infty} T_k$. Using this notation we get

(a) $AT \cap T = \emptyset$ for every $A \in \Gamma \backslash \{\pm I\}$. If not there are k, ℓ with $AT_k \cap T_\ell \neq \emptyset$. This implies $\emptyset \neq (A_\ell A A_k^{-1} \cdot L_k) \cap L_\ell \subset (A_\ell A A_k^{-1} \cdot \mathring{\mathcal{F}}) \cap \mathring{\mathcal{F}}$, which is impossible.
(b) $B(j,\epsilon) \backslash (\cup_{A \in \Gamma} A \cdot T)$ has measure zero. This follows since $\mathbb{H} \backslash (\cup_{A \in \Gamma} A\mathring{\mathcal{F}})$ is a set of measure zero.
(c) $v(T) \leq \mathrm{vol}(\Gamma) \leq v(B(j,\epsilon))/N$. This follows from (a) and (4).

(b) and (c) now imply that there are N pairwise distinct elements $C_1, \ldots, C_N \in \Gamma$ with $B(j,\epsilon) \cap C_k T \neq \emptyset$ for $k = 1, \ldots N$. Since $T \subset B(j,\epsilon)$ we infer $B(j,\epsilon) \cap C_k B(j,\epsilon) \neq \emptyset$. This implies $d(j, C_k(j)) \leq 2\epsilon$. By Proposition 5.8 we may choose $U_1, \ldots, U_N \in \mathbf{SU}(2)$ with

$$
\|C_k - U_k\| \leq \sqrt{2} \cdot \sqrt{\cosh(2\epsilon) - 1} \leq \frac{\delta}{6}.
$$

By Proposition 5.10 there are $1 \leq k,\ \ell \leq N$ with $k \neq \ell$ such that

$$
\|U_k - U_\ell\| \leq \frac{3\sqrt{2}\pi}{\left[\sqrt[3]{\frac{N}{2}}\right] - 1} \leq \frac{\delta}{6}.
$$

From $d(j, C_k^{-1}(j)) \leq 2\epsilon$ we infer that $\|C_k^{-1}\| \leq \sqrt{2} \cdot \sqrt{\cosh(2\epsilon)} \leq 2$. By the triangle inequality we get

$$\|C_k - C_\ell\| \le \|C_k - U_k\| + \|C_\ell - U_\ell\| + \|U_k - U_\ell\| \le \frac{\delta}{6} + \frac{\delta}{6} + \frac{\delta}{6} = \frac{\delta}{2}.$$

For $A = C_k^{-1} C_\ell$ we have

$$\|A - I\| = \|C_k^{-1}(C_\ell - C_k)\| \le \|C_k^{-1}\| \|C_\ell - C_k\| \le 2 \cdot \frac{\delta}{2} = \delta.$$

□

The following is elementary, we leave the proof as an exercise.

Lemma 5.12. *Let t_1, t_2 be the roots of the characteristic polynomial of a matrix $A \in \mathbf{SL}(2, \mathbb{C})$. Assume that $\|A - I\| < \epsilon$ for some $\epsilon \in]0, \infty]$. Then $|t_1 - 1| < \epsilon$ and $|t_2 - 1| < \epsilon$.*

Now we have collected all prerequisites for the proof of Theorem 5.2.

Proof of Theorem 5.2. Take

$$\delta_1 = 0,09144, \ \delta = 0,432889, \ \epsilon = 0,007619, \ N = 1,346 \cdot 10^9.$$

Then $v(B(j,\epsilon)) = 1,851 \cdot 10^{-6}$ and $v(B(j,\epsilon))/N \ge 1,3751 \cdot 10^{-15}$. The requirements (1), (2), (3) of Theorem 5.11 are now satisfied (for δ_1, ϵ, N). Take $\Gamma < \mathbf{SL}(2,\mathbb{C})$ any discrete group of finite covolume with $\mathrm{vol}(\Gamma) \le 1,3751 \cdot 10^{-15}$. By Theorem 5.11 there is a matrix $A \in \Gamma \backslash \{I\}$ with $\|A - I\| \le \delta_1 = 0,09144$. If the zeroes of the characteristic polynomial of A are ± 1 we use Proposition 3.7 to conclude that Γ has a cusp. Hence Theorem 5.1 applies and gives $\mathrm{vol}(\Gamma) \ge \sqrt{3}/24$, contrary to our hypothesis on $\mathrm{vol}(\Gamma)$. Choose a $\gamma_1 \in \mathbf{SL}(2,\mathbb{C})$ such that $\gamma_1 A \gamma_1^{-1} = \begin{pmatrix} t & 1 \\ 0 & t^{-1} \end{pmatrix}$. A simple estimate shows that $|t^{\pm 1} - 1| \le \delta$. Put $\Gamma_1 = \gamma_1 \Gamma \gamma_1^{-1}$. Define the following set

$$S = \left\{ \begin{pmatrix} a & b \\ 0 & a^{-1} \end{pmatrix} \in \Gamma_1 \;\middle|\; |a - 1| \le \delta, \ |a^{-1} - 1| \le \delta, \ b \ne 0 \right\}.$$

The set S is not empty since $\gamma_1 A \gamma_1^{-1} \in S$. Choose $C_1 \in S$ such that $C_1 = \begin{pmatrix} a_1 & b_1 \\ 0 & a_1^{-1} \end{pmatrix}$ and such that $|b_1| \le |b|$ for all $C = \begin{pmatrix} a & b \\ 0 & a^{-1} \end{pmatrix} \in S$. We define $\gamma_2 = \begin{pmatrix} \sqrt{b_1}^{-1} & 0 \\ 0 & \sqrt{b_1} \end{pmatrix}$ and $\Gamma_2 = \gamma_2 \Gamma_1 \gamma_2^{-1}$ and furthermore $C_2 = \gamma_2 C_1 \gamma_2^{-1} \in \Gamma_2$. We have $C_2 = \begin{pmatrix} a_1 & 1 \\ 0 & a_1^{-1} \end{pmatrix}$. We infer that every $C = \begin{pmatrix} a & b \\ 0 & a^{-1} \end{pmatrix} \in \Gamma_2$ with $|a - 1| \le \delta$, $|a^{-1} - 1| \le \delta$, $b \ne 0$ satisfies $|b| \ge 1$. By Theorem 5.11 now applied to δ, ϵ, N there is a matrix $A_2' = \begin{pmatrix} a' & b' \\ c' & d' \end{pmatrix} \in \Gamma_2 \backslash \{I\}$ with $\|A_2' - I\| \le \delta$. If $c' \ne 0$ we apply Corollary 4.7 to the pair of matrices $C_2, A_2' \in \Gamma_2$, obtaining $C_2 A_2' C_2^{-1} \in \{A_2', A_2'^{-1}\}$. Note that

$$C_2 A_2' C_2^{-1} = \begin{pmatrix} a' + c' a_1^{-1} & * \\ c' a_1^{-2} & d' - c' a_1^{-1} \end{pmatrix}.$$

The case $C_2 A_2' C_2^{-1} = A_2'$ is clearly impossible, since $c' \neq 0$. If $C_2 A_2' C_2^{-1} = A_2'^{-1}$, then $a_1^2 = -1$, $a_1 = \pm i$, and hence $|a_1 - 1| = \sqrt{2} > \delta$. We conclude that $c' = 0$. If $b' \neq 0$, then $|b'| \geq 1$, by $|b| \geq 1$. Then

$$\|A_2' - I\|^2 = |a' - 1|^2 + |b'|^2 + |d' - 1|^2 \geq 1,$$

and hence $\|A_2' - I\| \leq \delta$ is impossible. We conclude that $b' = 0,\ c' = 0$.

We have

$$A_2' C_2 A_2'^{-1} C_2^{-1} = \begin{pmatrix} 1 & a_2^2 - 1 \\ 0 & 1 \end{pmatrix},$$

and $a'^2 - 1 \neq 0$. Again we use Proposition 3.7 and Theorem 5.1 and the result is proved. □

2.6 Hyperbolic Lattice Point Problems

Let $\Gamma < \mathbf{PSL}(2, \mathbb{C})$ be a discrete group. For $P, Q \in \mathbb{H}$ and $x \in \mathbb{R},\ x > 0$ we consider the counting function

$$\pi(P, Q, x) := |\{\ M \in \Gamma \mid \delta(P, MQ) \leq x\ \}|. \tag{6.1}$$

Here we, as before, use the function $\delta(P, Q) = \cosh d(P, Q)$, d being the hyperbolic distance on upper half-space $\mathbb{H}$. Obviously $\pi(P, Q, x) = 0$ for $0 < x < 1$. For $x \geq 1$ the number $\pi(P, Q, x)$ has the following geometric meaning:

$\pi(P, Q, x)$ is equal to $|\Gamma_Q|$ times the number of lattice points of the hyperbolic lattice ΓQ which are contained in the closed hyperbolic ball of center P and radius $\operatorname{arcosh} x$.

We shall be interested in the asymptotic behaviour of $\pi(P, Q, x)$ as x tends to ∞. These questions are reminiscent of the lattice point problems in Euclidean spaces. We shall discuss the so-called hyperbolic lattice point problems in Chapter 5, Section 1 where the hyperbolic lattice point theorem for cocompact groups is proved. There are two elementary a priori estimates of $\pi(P, Q, x)$ which are of crucial importance for our work in Chapter 5. We shall give these estimates in Lemmas 6.1 and 6.3.

Lemma 6.1. *Let $\Gamma < \mathbf{PSL}(2, \mathbb{C})$ be a discrete group. Then there exists a constant $C_1(P) > 0$ depending only on P and Γ such that for all $Q \in \mathbb{H}$ and $x > 0$ the estimate*

$$\pi(P, Q, x) \leq C_1(P)\ x^2 \tag{6.2}$$

holds. $C_1(P)$ may be chosen as a function which is bounded on compact subsets of $\mathbb{H}$.

Before we enter into the details of the proof of this lemma we add some remarks on the literature. If P is not a fixed point of any transformation from $\Gamma \setminus \{I\}$, Lemma 6.1 is identical with the first statement of Nicholls (1974) Theorem 1, page 194. Nicholls explicitly states that $C_1(P)$ is independent of Q, a fact which will be of importance to us later in Chapter 5. He arrives at his result by means of a geometrical argument employed by Tsuji (1959), page 516 in the case of Fuchsian groups. Our proof uses the same method.

Proof of Lemma 6.1. We may assume that $x \geq 1$. Writing $x = \cosh t$ $(t \geq 0)$, we obtain

$$\begin{aligned}\pi(P,Q,x) &= |\{ M \in \Gamma \mid d(P,MQ) \leq t \}| \\ &= |\{ M \in \Gamma \mid d(MP,Q) \leq t \}|.\end{aligned}$$

Let $\epsilon > 0$ be so small that the intersection of the hyperbolic balls $B(P,\epsilon)$ and $B(MP,\epsilon)$ for $M \in \Gamma$ is non-empty only if $M \in \Gamma_P$. Then we may estimate

$$\begin{aligned}\pi(P,Q,x) &\leq |\{ M \in \Gamma \mid M(B(P,\epsilon)) \subset B(Q,t+\epsilon) \}| \\ &\leq |\Gamma_P| \frac{v(B(Q,t+\epsilon))}{v(B(P,\epsilon))} \leq \frac{\pi}{2v(B(P,\epsilon))} |\Gamma_P|\, e^{2(t+\epsilon)} \\ &< \frac{2\pi e^{2\epsilon}}{v(B(P,\epsilon))} |\Gamma_P| \cosh^2(t) = C_1(P)\, x^2.\end{aligned}$$

Note that our computation uses the estimate of the volume of a hyperbolic ball from Section 1.2, equation (2.9). Obviously $C_1(P)$ can be chosen as a function which is bounded on compact subsets of $\mathbb{H}$. □

In general, the constant $C_1(P)$ in (6.2) will not be uniformly bounded in P. This is true whenever Γ contains a parabolic transformation. In this case, one only has to choose some fixed $x > 0$ and let P tend to the fixed point of a parabolic element from Γ. Then $\pi(P,P,x) \to \infty$.

If Γ has a fundamental domain of finite hyperbolic volume, estimate (6.2) will also hold from below but only with a constant depending possibly on Q. This is our next lemma.

Lemma 6.2. *Let $\Gamma < \mathbf{PSL}(2,\mathbb{C})$ be a discrete group of finite covolume. Then there exist constants $C_2(P,Q) > 0$, $x_0(P,Q) \geq 1$ depending only on P,Q and Γ, such that for all $P,Q \in \mathbb{H}$ and for all $x \geq x_0(P,Q)$*

$$\pi(P,Q,x) \geq C_2(P,Q)\, x^2. \tag{6.3}$$

$C_2(P,Q)$ and $x_0(P,Q)$ may be chosen as functions which are bounded on compact subsets of $\mathbb{H} \times \mathbb{H}$.

This lemma was proved by Nicholls (1974) if P is not a fixed point of Γ. His method of proof is adapted from Tsuji (1959), page 518. Chen (1979), page 80 uses the same idea. Our proof follows the same approach.

Proof of Lemma 6.2. The function $\pi(P, Q, x)$ is Γ-invariant with respect to P and Q. Hence we may assume from the outset that $P, Q \in K$, where K is a compact subset of some Poincaré normal polyhedron $\mathcal{F}$ for Γ. Choose $C_1(K) > 0$ according to Lemma 6.1 such that

$$\pi(P, Q, x) \leq C_1(K)\, x^2 \qquad \text{for all } x > 0,\ P \in K,\ Q \in \mathbb{H}. \tag{6.4}$$

Let $\mathcal{F}_0 \subset \mathcal{F}$ be a compact set which is so large that $K \subset \mathcal{F}_0$, and $v(\mathcal{F} \setminus \mathcal{F}_0) < \pi/2C_1(K)$. Then we have for all $P \in K,\ x > 1,\ t = \operatorname{arcosh} x$

$$\begin{aligned} v\left(B(P,t) \cap \bigcup_{M \in \Gamma} M(\mathcal{F} \setminus \mathcal{F}_0)\right) &= \sum_{M \in \Gamma} v(B(P,t) \cap M(\mathcal{F} \setminus \mathcal{F}_0)) \\ &= \sum_{M \in \Gamma} v((\mathcal{F} \setminus \mathcal{F}_0) \cap B(MP,t)) = \int_{\mathcal{F} \setminus \mathcal{F}_0} \pi(P, Q, x)\, dv(Q) \\ &\leq C_1(K)\, v(\mathcal{F} \setminus \mathcal{F}_0)\, x^2 < \frac{\pi}{2}\, x^2. \end{aligned} \tag{6.5}$$

Now let $\rho := \sup\{\, d(X,Y) \mid X, Y \in \mathcal{F}_0 \,\}$ denote the hyperbolic diameter of $\mathcal{F}_0$. Suppose that $M \in \Gamma$, $x \geq \cosh \rho$, $t = \operatorname{arcosh} x$, and $Z \in B(P, t-\rho) \cap M\mathcal{F}_0$, then we have $M\mathcal{F}_0 \subset B(Z, \rho)$, and hence $M\mathcal{F}_0 \subset B(P,t)$. Since for $M \in \Gamma$ the sets $M\mathcal{F}_0$ are disjoint up to sets of measure zero, we conclude that

$$\begin{aligned} v(B(P, t-\rho) \cap \bigcup_{M \in \Gamma} M\mathcal{F}_0) &\leq v(\mathcal{F}_0)\ |\{\, M \in \Gamma \mid M\mathcal{F}_0 \subset B(P,t) \,\}| \\ &\leq v(\mathcal{F}_0)\ \pi(P, Q, x) \end{aligned}$$

for all $Q \in \mathcal{F}_0$. Using (6.5) we conclude for all $P, Q \in K$, $x \geq \cosh(\rho + 2)$, $t = \operatorname{arcosh} x$:

$$\begin{aligned} \pi(P, Q, x) &\geq \frac{1}{v(\mathcal{F}_0)} v(B(P, t-\rho) \cap \bigcup_{M \in \Gamma} M\mathcal{F}_0) \\ &= \frac{1}{v(\mathcal{F}_0)} \left(v(B(P, t-\rho)) - v\left(B(P, t-\rho) \cap \bigcup_{M \in \Gamma} M(\mathcal{F} \setminus \mathcal{F}_0) \right) \right) \\ &\geq \frac{1}{v(\mathcal{F}_0)} \left(\pi - \frac{\pi}{2} \right) \cosh^2(t-\rho) > \frac{\pi}{8v(\mathcal{F}_0)} e^{-2\rho}\, e^{2t} \\ &\geq \frac{\pi}{8v(\mathcal{F}_0)} e^{-2\rho} \cosh^2 t = C_2(K, \Gamma)\, x^2. \end{aligned}$$

This finishes the proof. □

We recall the normal form of hyperbolic or loxodromic elements of $\mathbf{PSL}(2, \mathbb{C})$. Let $T \in \mathbf{PSL}(2, \mathbb{C})$ be a hyperbolic or loxodromic transformation. Then T is conjugate within $\mathbf{PSL}(2, \mathbb{C})$ to an element

(6.6) $$D(T) = \begin{pmatrix} a(T) & 0 \\ 0 & a(T)^{-1} \end{pmatrix}$$

uniquely determined by the condition $|a(T)| > 1$. Note that

(6.7) $$D(T)\,(z + rj) = K(T)z + N(T)rj \qquad (z \in \mathbb{C},\ r > 0),$$

where $K(T) := a(T)^2$ is called the *multiplier* of T and where $N(T) := |K(T)| = |a(T)|^2 > 1$ is called the *norm* of T (see Selberg (1956)). The hyperbolic line connecting the fixed points of T in $\mathbb{P}^1\mathbb{C}$ is called the *axis* of T. The axis of T is a T-invariant set.

Now let $\Gamma < \mathbf{PSL}(2,\mathbb{C})$ be a discrete group. Following Selberg (1956), we denote the Γ-conjugacy class of an element $T \in \Gamma$ by $\{T\}_\Gamma$ or briefly by $\{T\}$, if no confusion with the usual set theoretic notation may arise. Note that all elements in a Γ-conjugacy class of a hyperbolic or loxodromic element $T \in \Gamma$ have the same multiplier and the same norm which we also call the multiplier or the norm of $\{T\}$ respectively.

Lemma 6.3. *Suppose that $\Gamma < \mathbf{PSL}(2,\mathbb{C})$ is a discrete group such that $\Gamma\backslash\mathbb{H}$ is compact. Then the counting function*

(6.8) $$\pi_0(x) := |\{\, \{T\}_\Gamma \mid T \in \Gamma,\ T \text{ hyperbolic or loxodromic},\ N(T) \le x \,\}|$$

satisfies the growth restriction $\pi_0(x) = O(x^2)$ as $x \to \infty$.

Proof. It follows from Chapter 1, Section 1 that for all hyperbolic or loxodromic elements $T \in \mathbf{PSL}(2,\mathbb{C})$, $\log N(T) = \inf\{\, d(P,TP) \mid P \in \mathbb{H} \,\}$ and that $\log N(T) = d(P,TP)$ if and only if P is on the axis of T.

Since $\Gamma\backslash\mathbb{H}$ is compact, Γ has a compact fundamental domain $\mathcal{F}$. Let $T \in \Gamma$ be a hyperbolic or loxodromic element, and suppose that $P \in \mathbb{H}$ belongs to the axis of T. Then there exists an element $S \in \Gamma$ such that $Q := SP \in \mathcal{F}$ and hence

$$\log N(T) = d(P,TP) = d(SP, STS^{-1}(SP)) = d(Q, STS^{-1}Q),$$

with $V := STS^{-1} \in \{T\}_\Gamma$. Let $d(A,B) := \inf\{\, d(P,Q) \mid P \in A,\ Q \in B \,\}$ denote the hyperbolic distance between the non-empty sets A, $B \subset \mathbb{H}$, and let $d_0 := \sup\{\, d(P,Q) \mid P,\ Q \in \mathcal{F} \,\}$ be the hyperbolic diameter of $\mathcal{F}$. Choose a point $P_0 \in \mathcal{F}$. Then equation (2.9) from Chapter 1 implies

$$\begin{aligned}
\pi_0(x) &= |\{\, \{T\}_\Gamma \mid T \text{ hyperbolic or loxodromic},\ N(T) \le x \,\}| \\
&\le |\{\, V \in \Gamma \mid d(\mathcal{F}, V\mathcal{F}) \le \log x \,\} \\
&\le |\{\, V \in \Gamma \mid d(P_0, VP_0) \le \log x + 2d_0 \,\}| \\
&\le |\{\, V \in \Gamma \mid V\mathcal{F} \subset B(P_0, \log x + 3d_0)\}| \le \frac{v(B(P_0, \log x + 3d_0))}{v(\mathcal{F})} \\
&< \frac{\pi}{2v(\mathcal{F})} e^{6d_0}\, x^2.
\end{aligned}$$

This proves the lemma even with an explicit O-constant. $\square$

2.7 Generators and Relations

Consider a group $\Gamma < \mathbf{PSL}_2(\mathbb{C})$ which acts discontinuously on 3-dimensional hyperbolic space. Sometimes it is possible to construct a not too complicated fundamental domain $\mathcal{F}$. Examples can be found in Chapters 7 and 10. The following theorem can then be used to find generators or even a presentation for Γ. The explicit knowledge of a presentation of a group does of course not help too much in the understanding of the group. But sometimes presentations are helpful for heuristic investigations. At least the next theorem will imply the strong finiteness result in Theorem 7.3.

Theorem 7.1. *Let $\Gamma < \mathbf{Iso}(\mathbb{H})$ be a discontinuous group. Suppose there is a connected open set $V \subset \mathbb{H}$ such that*

(1) $\cup_{\sigma \in \Gamma} \sigma V = \mathbb{H}$,
(2) $S = \{ \gamma \in \Gamma \mid V \cap \gamma V \neq \emptyset \}$ is finite.
Then Γ is finitely presented and has the following presentation:

$$\begin{aligned} &Generators: \ [\gamma] \ \ for\ every\ \gamma \in S, \\ &Relations: \ [\gamma\sigma] = [\gamma][\sigma] \quad whenever\ V \cap \gamma V \cap \gamma\sigma V \neq \emptyset. \end{aligned}$$

Proof. Let Δ be the group generated by the symbols $[\gamma]$, $\gamma \in S$, defined by the relations $[\gamma\sigma] = [\gamma][\sigma]$ if $V \cap \gamma V \cap \gamma\sigma V \neq \emptyset$. Δ is a finitely presented group. The map $[\gamma] \mapsto \gamma$ induces a group homomorphism $\varphi : \Delta \to \Gamma$. We shall show that φ is an isomorphism.

(1) φ is surjective: Let $g \in \Gamma$ and $P \in V$. Take a path C in $\mathbb{H}$ joining P and gP, for example a geodesic joining P and gP. There are finitely many $\gamma_1, \ldots, \gamma_n \in \Gamma$ such that $C \subset \cup_{i=1}^n \gamma_i V$. This is proved using the compactness of C, the openness of V and $\cup_{\sigma \in \Gamma} \sigma V = \mathbb{H}$.

We construct now a sequence of elements $I = \sigma_1, \sigma_2, \ldots, \sigma_k = g \in \Gamma$ such that $(a): \ C \subset \cup_{i=1}^k \sigma_i V$, and $(b): \ \sigma_i V \cap \sigma_{i+1} V \neq \emptyset$. This is done starting from the elements $\gamma_1, \ldots, \gamma_n$ and then using the connectedness of C. Then $\sigma_i^{-1}\sigma_{i+1} \in S$ and

$$g = \sigma_1^{-1}\sigma_2\sigma_2^{-1}\sigma_3 \ldots \sigma_{k-1}^{-1}\sigma_k = \varphi([\sigma_1^{-1}\sigma_2][\sigma_2^{-1}\sigma_3] \ldots [\sigma_{k-1}^{-1}\sigma_k]).$$

So φ is surjective.

(2) φ is injective: We have $I \in S$, and $\sigma \in S$ iff $\sigma^{-1} \in S$. For every $\sigma \in S$ we have the intersections

$$V \cap \sigma V \cap \sigma \cdot 1 V \neq \emptyset, \qquad V \cap \sigma V \cap \sigma \cdot \sigma^{-1} V \neq \emptyset.$$

Hence the relations $[1] = 1$ and $[\sigma^{-1}] = [\sigma]^{-1}$. Using these relations it is enough to show that $\sigma_1, \ldots, \sigma_n \in S$ and $[\sigma_1] \ldots [\sigma_n] \in \ker\varphi$ implies $[\sigma_1] \ldots [\sigma_n] = 1$. We proceed now in several steps. Choose a point $P \in V$. Consider a loop $\omega : [0,1] \to \mathbb{H}$ based at P, that is $\omega(0) = \omega(1) = P$. A sequence $1 = \gamma_1, \gamma_2, \ldots, \gamma_k = 1$ of elements from Γ is called ω-adapted if

$$\omega([0,1]) \subset \bigcup_{i=1}^{k} \gamma_i V \qquad \text{and if} \qquad \gamma_i V \cap \gamma_{i-1} V \neq \emptyset.$$

For any loop ω based at P and any ω-adapted sequence $L = \{1 = \gamma_1, \gamma_2, \ldots, \gamma_k = 1\}$ define

$$\theta(\omega, L) := [\gamma_1^{-1}\gamma_2] \ldots [\gamma_{k-1}^{-1}\gamma_k] \in \Delta.$$

We first show that $\theta(\omega, L_1) = \theta(\omega, L_2)$ for any two ω-adapted sequences L_1 and L_2. If $L_1 = \{1 = \gamma_1, \gamma_2, \ldots, \gamma_k = 1\}$ and $\gamma \in \Gamma$ is such that

$$\gamma V \cap \gamma_\ell V \cap \gamma_{\ell+1} V \neq \emptyset, \qquad \text{for any } \ell, \text{ with } 1 \leq \ell < k,$$

then the sequence $L_1' = \{1 = \gamma_1, \ldots, \gamma_\ell, \gamma, \gamma_{\ell+1}, \ldots \gamma_k = 1\}$ is ω-adapted and we find $\theta(\omega, L_1) = \theta(\omega, L_1')$ because of $[\gamma_\ell^{-1}\gamma_{\ell+1}] = [\gamma_\ell^{-1}\gamma] \cdot [\gamma^{-1}\gamma_{\ell+1}]$. Choosing a common refinement of L_1, L_2 we have proved $\theta(\omega, L_1) = \theta(\omega, L_2)$.

Assume now that ω_1 and ω_2 are two loops based at P which are homotopic. We prove that $\theta(\omega_1, L_1) = \theta(\omega_2, L_2)$ for ω_i-adapted sequences L_i. Choose a homotopy

$$\Phi : [0,1] \times [0,1] \to \mathbb{H}$$

from ω_1 to ω_2. That is $\Phi(0,t) = \omega_1(t)$, $\Phi(1,t) = \omega_2(t)$. Since $\Phi([0,1] \times [0,1])$ is compact it can be covered by finitely many translates under Γ of V. By choosing $d+1$ equidistant points, with d an odd natural number, on each edge of the square we divide $[0,1] \times [0,1]$ into d^2 closed subsquares $Q_1, \ldots Q_{d^2}$ such that for every subsquare Q there is a $\sigma_Q \in \Gamma$ with $\Phi(Q) \subset \sigma_Q V$. The numbering of the Q_i starts in the left upper corner of $[0,1] \times [0,1]$ and proceeds in a snake to the right lower corner. Hence we may choose $\sigma_{Q_1} = \sigma_{Q_{d^2}} = 1$. Note that the openness of V guarantees the existence of such a subdivision. Let the subsquares be ordered as indicated above. Then the sequence

$$L = \{1 = \sigma_{Q_1}, \sigma_{Q_2}, \ldots, \sigma_{Q_{d^2}} = 1\}$$

has the property $\sigma_{Q_i} V \cap \sigma_{Q_{i+1}} V \neq \emptyset$ for $i = 1, \ldots, d^2 - 1$. Furthermore we have

$$\omega_1([0,1]),\ \omega_2([0,1]) \subset \bigcup_{i=1}^{d^2} \sigma_{Q_i} V.$$

Hence L is ω_1 and ω_2 adapted. Hence $\theta(\omega_1, L_1) = \theta(\omega_1, L) = \theta(\omega_2, L) = \theta(\omega_2, L_2)$. Note that θ does only depend on the sequence L.

Going back to elements $\sigma_1, \ldots, \sigma_n \in S$ with $\varphi([\sigma_1] \ldots [\sigma_n]) = \sigma_1 \ldots \sigma_n = 1$ we define

$$\gamma_0 = 1, \; \gamma_i = \sigma_1 \ldots \sigma_i, \; \ldots \; \gamma_n = \sigma_1 \ldots \sigma_n = 1.$$

Note that $\gamma_i V \cap \gamma_{i+1} V \neq \emptyset$. Choose a point $P_i \in \gamma_i V \cap \gamma_{i+1} V$ such that $P_0 = P_{n-1} \in V$. Connect these points in such a way that the arising loop ω from P_0 to P_{n-1} has the property $\omega([0,1)] \subset \cup_{i=1}^n \gamma_i V$. The sequence $L = \{1 = \gamma_0, \gamma_1, \ldots, \gamma_n = 1\}$ is then ω-adapted. Let $\omega_0 : [0,1] \to \mathbb{H}$ be the trivial loop mapping everything to P_0. Since $\mathbb{H}$ is simply connected there is a homotopy from ω_0 to ω. Hence

$$[\sigma_1] \ldots [\sigma_n] = \theta(\omega, L) = \theta(\omega_0, L_0) = 1.$$

Here $L_0 = \{1 = \gamma_0 = 1\}$ is an ω_0-adapted sequence. □

The following is an obvious consequence of Theorem 7.1.

Corollary 7.2. *Let $\Gamma < \mathbf{Iso}(\mathbb{H})$ be a discontinuous group. Let $\mathcal{F} \subset \mathbb{H}$ be a fundamental domain for Γ. Assume that there is an open connected set $V \supset \mathcal{F}$ such that for all $\sigma, \tau \in \Gamma$*

$$\begin{aligned} \mathcal{F} \cap \sigma\mathcal{F} \neq \emptyset, &\quad \textit{iff} \quad V \cap \sigma V \neq \emptyset \\ \mathcal{F} \cap \sigma\mathcal{F} \cap \tau\mathcal{F} \neq \emptyset &\quad \textit{iff} \quad V \cap \sigma V \cap \tau V \neq \emptyset. \end{aligned}$$

Let $S = \{ \gamma \mid \mathcal{F} \cap \gamma\mathcal{F} \neq \emptyset \}$ be finite. Then Γ is finitely presented and has the following presentation:

$$\begin{aligned} &\textit{Generators}: \; [\gamma] \; \textit{for every} \; \gamma \in S, \\ &\textit{Relations}: \; [\gamma\sigma] = [\gamma][\sigma] \qquad \textit{if} \; \mathcal{F} \cap \gamma\mathcal{F} \cap \gamma\sigma\mathcal{F} \neq \emptyset. \end{aligned}$$

In the examples in Chapter 7 or 10 we shall always use Corollary 7.2 to construct presentations of certain groups Γ. It will be clear from the fundamental domains that they can be enlarged to open sets V having the above property. The presentations obtained by Theorem 7.1 or Corollary 7.2 are usually highly redundant. It is sometimes useful to obtain a cut-down on the number of generators by taking

$$S = \{ \gamma \mid \mathcal{F} \cap \gamma\mathcal{F} \text{ has dimension } 2 \}.$$

We leave the justification of this fact as an exercise.

The above theorems are classical results originating from the work of Poincaré. The present formulation is adapted from Swan (1971). Corollary 7.2 implies the following result.

Theorem 7.3. *Let $\Gamma < \mathbf{PSL}(2, \mathbb{C})$ be a discrete group of finite covolume. Then Γ is finitely presented.*

Proof. By Theorem 2.7 Γ is geometrically finite, that is there is a $Q \in \mathbb{H}$ such that the polyhedron $\mathcal{F} = \mathcal{P}_Q(\Gamma)$ has only finitely many faces, and hence only finitely many edges and vertices. Note that the vertices may lie in $\mathbb{H}$ or also on its boundary. To construct an open set V satisfying the requirements of Corollary 7.2 we start by choosing small neighbourhoods of vertices and edges which only meet those $\gamma\mathcal{F}$ that intersect $\mathcal{F}$. This is possible since the covering of $\mathbb{H}$ by the $\gamma\mathcal{F}$, $\gamma \in \Gamma$, is locally finite. Finally we connect these neighbourhoods along the faces of $\mathcal{F}$ to a suitable neighbourhood of $\mathcal{F}$. Since the covering of $\mathbb{H}$ by the $\gamma\mathcal{F}$, $\gamma \in \Gamma$, is locally finite, there can only be finitely many $\gamma \in \Gamma$ so that $\gamma\mathcal{F} \cap \mathcal{F} \neq \emptyset$. □

For the fundamental concepts of combinatorial group theory we refer to the book of Magnus, Karrass and Solitar (1966). We shall use here the following notation.

Definition 7.4. Let G be a group and let $g_1, \ldots, g_n \in G$ be elements of G. Let $R_1, \ldots, R_m$ be words in the $g_1, \ldots, g_n, g_1^{-1}, \ldots, g_n^{-1}$. Then

$$G \cong < g_1, \ldots, g_n \quad | \quad R_1 = \ldots = R_m = 1 > \tag{7.1}$$

means that the canonical map

$$\varphi : \mathbf{F}(\{[g_1], \ldots, [g_n]\}) \to G \qquad \varphi : [g_k] \mapsto g_k, \qquad k = 1, \ldots, n$$

is surjective and has $<< [R_1], \ldots, [R_m] >>_{\mathbf{F}(\{[g_1],\ldots,[g_n]\})}$ as kernel. Here $\mathbf{F}(T)$ is the free group on the set T. The $[R_k]$ are the words in the $[g_k]$ corresponding to the words R_k. Finally $<< T >>_G$ denotes the normal subgroup generated by a subset T of a group G. We call (7.1) a presentation of G.

The following easy consequence of Corollary 7.2 will be useful in the sequel. We leave the proof to the reader, see the remarks just after Corollary 7.2.

Theorem 7.5. *Let $\Gamma < \mathbf{Iso}(\mathbb{H})$ be a discontinuous group. Suppose that Γ has a fundamental domain which is a polyhedron bounded by finitely many faces $S_1, \ldots, S_n$. Put*

$$T := \left\{ (k, \ell) \;\middle|\; \begin{array}{l} k \neq \ell \textit{ and } S_k \cap S_\ell \textit{ contains a proper} \\ \textit{segment of a hyperbolic line} \end{array} \right\}.$$

Assume that for any $k = 1, \ldots, n$ the reflection σ_k in the hyperbolic plane $\hat{S}_k$ containing the side S_k is contained in Γ. Then

$$\Gamma = < \sigma_1, \ldots, \sigma_n \quad | \quad \sigma_1^2 = \ldots = \sigma_n^2 = (\sigma_k \cdot \sigma_\ell)^{m_{k\ell}} = 1, \quad (k, \ell) \in T >$$

is a presentation of Γ. Here $m_{k,\ell}$ is the order of the rotation $\sigma_k \cdot \sigma_\ell$.

Note that $\frac{m_{k\ell}}{2} \cdot \pi$ is the angle between the two hyperbolic planes $\hat{S}_k$ and $\hat{S}_\ell$. Theorem 7.5 shows that Γ is a Coxeter group once the hypotheses are satisfied. See Chapter 10 for a further treatment of these groups. The hypotheses of Theorem 7.5 are satisfied for the reflection group $\Gamma_{\mathcal{P}}$ where $\mathcal{P}$ is any polyhedron satisfying the hypotheses of Theorem 2.8.

Theorem 7.5 is a special case of a result of Koszul (1965) concerning discontinuous groups of isometries on Riemannian manifolds generated by reflections.

2.8 Conjugacy and Commensurability

An important and deep theorem concerning discontinuous subgroups of $\mathbf{Iso}\,(\mathbb{H})$ is the Mostow Rigidity Theorem:

Theorem 8.1. *Let $\Gamma_1, \Gamma_2 < \mathbf{Iso}(\mathbb{H})$ be two discontinuous groups of finite covolume. Let $\varphi : \Gamma_1 \to \Gamma_2$ be a group isomorphism. Then there is a $g \in \mathbf{Iso}(\mathbb{H})$ with $\varphi(\gamma) = g\,\gamma\,g^{-1}$ for all $\gamma \in \Gamma_1$.*

See Mostow (1973) for the cocompact case and Prasad (1973) for the general case. Another proof due to Gromov is described in Thurston (1978). Theorem 8.1 says that there is up to conjugacy at most one embedding of a group Γ onto a discontinuous subgroup of finite covolume of $\mathbf{Iso}(\mathbb{H})$. This is in sharp contrast to groups acting on the hyperbolic plane, where a theory of deformations of embeddings into the isometry group exists, Harvey (1977).

Definition 8.2. Let G be a group. Two subgroups U_1, $U_2 < G$ are called commensurable if the indices

$$[U_1 : U_1 \cap U_2] \quad \text{and} \quad [U_2 : U_1 \cap U_2] \tag{8.1}$$

are both finite. If U_1, U_2 are commensurable subgroups of G we call

$$[U_1 : U_2] = \frac{[U_1 : U_1 \cap U_2]}{[U_2 : U_1 \cap U_2]} \tag{8.2}$$

the index of U_1 over U_2. For a subgroup $U < G$ the set

$$\mathbf{Co}\,(U) = \{\ g \in G \ \mid \ gUg^{-1} \ \text{ is commensurable with } \ U\ \}$$

is called the commensurator of U.

Clearly commensurability is an equivalence relation. We also have

Proposition 8.3. *Let G be a group with subgroup U. Then $\mathbf{Co}(U)$ is a subgroup of G with $U < \mathbf{Co}(U)$.*

Proof. For $\sigma, \gamma \in \mathbf{Co}(U)$ the following two pairs consist of commensurable groups

$$(\sigma\gamma)U(\sigma\gamma)^{-1} = \sigma(\gamma U\gamma^{-1})\sigma^{-1}, \ \sigma U\sigma^{-1} \qquad \text{and} \qquad \sigma U\sigma^{-1}, \ \ U.$$

Hence $\sigma\gamma \in \mathbf{Co}(U)$. Everything else is clear. □

Sometimes the following obvious consequence of the definitions is used.

Proposition 8.4. *Let Γ_1, $\Gamma_2 < \mathbf{PSL}(2, \mathbb{C})$ be two commensurable groups. Then*

(1) if Γ_1 is discrete then also Γ_2,
(2) if Γ_1 is of finite covolume then also Γ_2 and $\mathrm{vol}(\Gamma_2) = [\Gamma_1 : \Gamma_2] \cdot \mathrm{vol}(\Gamma_1)$.

We shall also need the following result.

Proposition 8.5. *Let K be a number field with ring of integers $\mathcal{O}_K$. Furthermore let $\gamma \in \mathbf{GL}(n, K)$. Then the groups $\mathbf{GL}(n, \mathcal{O}_K)$ and $\gamma^{-1}\mathbf{GL}(n, \mathcal{O}_K)\gamma$ are commensurable.*

Proof. Choose a $\lambda \in \mathcal{O}_K, \lambda \neq 0$ which has the property that $\lambda \cdot a \cdot b \in \mathcal{O}_K$ for any pair a, b of entries from γ and γ^{-1}. Let $\Gamma < \mathbf{GL}(n, \mathcal{O}_K)$ be the full congruence group with respect to the ideal $\mathfrak{a} = \lambda \cdot \mathcal{O}_K$,

$$\Gamma = \{\ \omega \in \mathbf{GL}(n, \mathcal{O}_K) \ \ | \ \ \omega \equiv I \bmod \mathfrak{a}\ \}.$$

Γ has finite index in $\mathbf{GL}(n, \mathcal{O}_K)$ and $\Gamma < \gamma^{-1}\mathbf{GL}(n, \mathcal{O}_K)\gamma$. This is so because by the choice of λ we have for any $\sigma \in \Gamma$: $\gamma\sigma\gamma^{-1} = \gamma(I+A)\gamma^{-1} = I+\gamma A\gamma^{-1} \in \mathbf{GL}(n, \mathcal{O}_K)$. Note that A has all its entries in $\mathfrak{a} = \lambda \cdot \mathcal{O}_K$. □

Proposition 8.5 implies that for any of the arithmetic groups Γ defined in Chapter 10 the index of its commensurator in $\mathbf{Iso}(\mathbb{H})$ over Γ is infinite. The following theorem of Margulis (1969) is a converse.

Theorem 8.6. *Let $\Gamma < \mathbf{Iso}(\mathbb{H})$ be a cofinite subgroup. Then Γ is arithmetic if and only if $[\mathbf{Co}(\Gamma) : \Gamma] = \infty$.*

For the notion of arithmeticity see Chapter 10. It follows from the above theorem that a non-arithmetic group of finite covolume has exactly one maximal group in its commensurability class. This maximal group is again discrete. For arithmetic groups there are, even up to conjugacy, infinitely many maximal discrete subgroups in the commensurability class. For explicit constructions see Helling (1966), Rohlfs (1978).

If $\Gamma < \mathbf{SL}(2, \mathbb{C})$ is a subgroup define $\mathrm{Tr}(\Gamma)$ to be the subfield of $\mathbb{C}$ generated by the traces of the elements of Γ. If Γ_1, $\Gamma_2 < \mathbf{SL}(2, \mathbb{C})$ are $\mathbf{SL}(2, \mathbb{C})$-conjugate subgroups then $\mathrm{Tr}(\Gamma_1) = \mathrm{Tr}(\Gamma_2)$. The following useful criterion for commensurability is contained in Reid (1990).

Proposition 8.7. *Let Γ_1, $\Gamma_2 < \mathbf{SL}(2,\mathbb{C})$ be two cofinite, commensurable subgroups then* $\mathrm{Tr}(\Gamma_1^{(2)}) = \mathrm{Tr}(\Gamma_2^{(2)})$, *where $\Gamma_1^{(2)}$ is the subgroup generated by the squares of all the elements of Γ_1.*

Note that if Γ is a finitely generated group, then $\Gamma^{(2)}$ is a normal subgroup of finite index in Γ.

For later use we include the following Proposition.

Proposition 8.8. *Let $\Gamma < \mathbf{Iso}(\mathbb{H})$ be a cofinite subgroup. Then the centralizer of Γ in $\mathbf{Iso}(\mathbb{H})$ is trivial.*

The triviality of the centralizer of Γ in $\mathbf{PSL}(2,\mathbb{C})$ follows from Borel's density theorem, Borel (1960). We give here a somewhat simpler proof. In our proof we use the following Lemma.

Lemma 8.9. *Let $\Gamma < \mathbf{Iso}(\mathbb{H})$ be a cofinite subgroup. Then Γ leaves no geodesic hyperplane invariant.*

Proof. By Selberg's Lemma proved in the next section we may assume without loss of generality that Γ is torsion free and $\Gamma < \mathbf{Iso}^+(\mathbb{H})$. Let S be a geodesic hyperplane stabilized by Γ. Let $P \in \mathbb{H}$ be a point outside S. It is easy to see that there is a unique point $P_0 \in S$ so that

$$d(P_0, P) = \inf\{\, d(Q,P) \;\mid\; Q \in S\,\}.$$

The hyperbolic line connecting P_0 and P meets S orthogonally. Let $B \subset S$ be a disc so that $\gamma B \cap B = \emptyset$ whenever $\gamma \neq 1$. Define $\tilde{B} = \{\, P \in \mathbb{H} \setminus S \;\mid\; P_0 \in B\,\}$. It is easy to see that $\gamma\tilde{B} \cap \tilde{B} = \emptyset$ whenever $\gamma \neq 1$. It is also clear that $\tilde{B}$ has infinite hyperbolic volume. By application of Lemma 2.5 we get a contradiction. □

Proof of Proposition 8.8. Let $g \in \mathbf{Iso}(\mathbb{H})$ be a nontrivial element centralizing Γ. The Brouwer fixed point theorem implies that the set M of fixed points of g in $\mathbb{H} \cup \mathbb{P}^1\mathbb{C}$ is non-empty. A simple computation using formulas (1.1.9), (1.1.10) and Proposition 1.1.3 shows that M is contained in the closure in $\mathbb{H} \cup \mathbb{P}^1\mathbb{C}$ of a geodesic hyperplane. Since Γ centralizes g the set M is left invariant by Γ. If $M \cap \mathbb{H}$ contains three non-collinear points we are finished by Lemma 8.9. The further cases to consider are easily treated directly. □

2.9 A Lemma of Selberg

Here we prove a result originally due to Selberg, see Selberg (1960), Lemma 8. It will allow us to replace groups with torsion by torsion free groups in some arguments.

Theorem 9.1. *Let $\Gamma < \mathbf{GL}(n, \mathbb{C})$ $(n \in \mathbb{N})$ be a finitely generated subgroup, then Γ has a torsion free subgroup of finite index.*

For the proof we need the following results from commutative algebra.

Lemma 9.2. *Let R be an integral domain which is finitely generated as a ring and let $a \in R$ with $a \neq 0$. Then there is an ideal $\mathfrak{a} \subset R$ with*

(1) $a \notin \mathfrak{a}$,
(2) $R/\mathfrak{a}$ is a finite ring.

Proof. Let R_a be the ring of fractions of R with respect to the set $\{ a^n \mid n \in \mathbb{N} \cup \{0\} \}$. The rings $R \subset R_a$ are Jacobson rings and any of their maximal ideals has finite index, see Bourbaki (1964a), Chap. 5, §3.4. Let $\mathfrak{b} \subset R_a$ be a maximal ideal and put $\mathfrak{a} = \mathfrak{b} \cap R$. Then $\mathfrak{a}$ satisfies (1) and (2). □

The following Lemma is used in the proof of Theorem 9.1.

Lemma 9.3. *Let L be an infinite field and let L_1 be a purely transcendental extension of L. Then every irreducible polynomial $f \in L[x]$ stays irreducible in $L_1[x]$.*

Proof. Taking transcendence bases we may assume that $L_1 = L(y)$ is of transcendence degree 1. By the lemma of Gauß we may also assume that f factors over $L[x, y]$ if it factors in $L_1[x]$. So if f is reducible in $L_1[x]$ we may write

$$f(x) = f_1(x, y) \cdot f_2(x, y) \tag{9.1}$$

with $f_1(x, y) = \sum_j b_j(y) \cdot x^j$ and $f_2(x, y) = \sum_j c_j(y) \cdot x^j$ and $b_j, c_j \in L[y]$. We may insert infinitely many values of $y \in L$ into equation (9.1). In order not to get a factorization of f in $L[x]$ we must have $b_j = 0$ for $j \geq 1$ or $c_j = 0$ for $j \geq 1$. In either case we see that the factorization is trivial. Note that b_0, c_0 have to be in L since their product is contained in L. □

Proof of Theorem 9.1. Let $\gamma_1, \dots, \gamma_m$ be a set of generators of Γ. R_0 is defined as the ring generated by all the entries of all the γ_i together with the $\det\gamma_i^{-1}$. Then R_0 is a finitely generated integral domain. We write K for the quotient field of R_0. We have $\Gamma < \mathbf{GL}(n, R_0)$. We shall construct a torsion free subgroup $U_0 < \mathbf{GL}(n, R_0)$ which has finite index. Let $\gamma \in \mathbf{GL}(n, R_0)$ be an element of finite order N. Let F be the splitting field of the characteristic polynomial of γ. F is a finite extension of K and γ is diagonalizable over F. Since the characteristic polynomial of γ has degree n we find $|F:K| \leq n!$. Here $|F:K|$ stands for the degree of the extension $K \subset F$. The eigenvalues of γ are all N-th roots of unity and since γ has order exactly N the group generated by them contains a primitive N-th root of unity. Writing $\psi_K(N)$ for the degree over K of the extension generated by a primitive N-th root of unity, we find $\psi_K(N) \leq n!$. K is a finitely generated field, hence there is a subfield $T \subset K$ such that T is purely transcendental over $\mathbb{Q}$ and K is a finite extension of T. We obviously have

$$\psi_T(N) \leq |K:T| \, \psi_K(N) \leq |K:T| \, n! \; .$$

Let $f \in \mathbb{Q}[x]$ be the minimal polynomial of a primitive N-th root of unity. f is irreducible in $T[x]$ by Lemma 9.3 and hence $\psi_T(N) = \psi_{\mathbb{Q}}(N)$. By the usual formulas for the degree of f in terms of Euler's φ-function we infer that there is an $N_0 \in \mathbb{N}$ so that the order of any element of finite order in $\mathbf{GL}(n, R_0)$ divides N_0. Put $R = R_0[\zeta]$ where ζ is a primitive N_0-th root of unity. R is still a finitely generated ring and contains the eigenvalues of all elements of finite order in $\mathbf{GL}(n, R)$. We put

$$S = \{ \, \zeta_1 + \ldots + \zeta_n - n \, \} \subset R \subset \mathbb{C}$$

where the ζ_i range over all N_0-th roots of unity excluding the case $\zeta_1 = \ldots = \zeta_n = 1$. We claim that $0 \notin S$. Assume by contradiction that $\zeta_1 + \ldots + \zeta_n = n$. Then the triangle inequality implies that $\zeta_1 = \ldots \zeta_n = 1$. By Lemma 9.2 there is an ideal $\mathfrak{a} \subset R$ with finite quotient $R/\mathfrak{a}$ and $S \cap \mathfrak{a} = \emptyset$. Put $\mathfrak{a}_0 = \mathfrak{a} \cap R_0$ and

$$U_0 = \{ \, \gamma \in \mathbf{GL}(n, R_0) \;\; | \;\; \gamma \equiv I \mod \mathfrak{a}_0 \, \}.$$

Since $R_0/\mathfrak{a}_0$ is also finite, U_0 is of finite index in $\mathbf{GL}(n, R_0)$. Note that U_0 is the kernel of the restriction map $\mathbf{GL}(n, R_0) \to \mathbf{GL}(n, R_0/\mathfrak{a}_0)$. Every element $\gamma \in U_0$ satisfies $\mathrm{tr}(\gamma) - n \in \mathfrak{a}_0$. By our construction no element $\gamma \in \mathbf{GL}(n, R_0)$ of finite order except I can satisfy this. The group $U = \Gamma \cap U_0$ satisfies the requirements of the theorem. □

Our proof of Theorem 9.1 can easily be adjusted to work for any finitely generated subgroup of $\mathbf{GL}(n, L)$ where L is any field of characteristic 0. Examples like $\mathbf{GL}(3, \mathbb{Z}/p\mathbb{Z}[x])$ show that the theorem is false in characteristic p. For another proof of Theorem 9.1 see for example Borel (1969).

Part (1) of the following is an immediate consequence of Lemma 9.2. Groups satisfying property (1) are traditionally called residually finite.

Theorem 9.4. *Let $\Gamma < \mathbf{GL}(n, \mathbb{C})$ ($n \in \mathbb{N}$) be a finitely generated subgroup, then the following are true.*

(1) Γ has a sequence $\Gamma = \Gamma_1 > \Gamma_2 > \ldots$ of normal subgroups of finite index which intersect in $\{1\}$.

(2) Γ has a sequence $\Gamma = \Gamma_1 > \Gamma_2 > \ldots$ of torsion free normal subgroups of finite index which intersect in $\{1\}$.

Proof. Let $\gamma_1, \ldots, \gamma_m$ be a set of generators of Γ. R_0 is defined as the ring generated by all the entries of all the γ_i together with the $\det\gamma_i^{-1}$. Then R_0 is a finitely generated integral domain. Choose an ordering $\Gamma\backslash\{1\} = \{g_1, g_2, \ldots\}$ of the nontrivial elements of Γ. For each of the elements g_i we pick a non zero matrix entry a_i and an ideal $\mathfrak{a}_i$ of finite index in R_0 with $a_i \notin \mathfrak{a}_i$. Define $\mathfrak{b}_k = \mathfrak{a}_1 \cap \ldots \cap \mathfrak{a}_k$. Then the congruence subgroups

$$\Gamma_i = \{\ \gamma \in \mathbf{GL}(n, R_0)\ \mid\ \gamma \equiv 1 \bmod \mathfrak{b}_i\ \}.$$

satisfy the requirements of (1). To prove part (2) let Γ_0 be one of the subgroups from Theorem 9.1. Put $\tilde{\Gamma} = \bigcap_{\gamma\in\Gamma} \gamma\Gamma_0\gamma^{-1}$. Then $\tilde{\Gamma}$ is a torsion free normal subgroup of finite index in Γ. Intersecting the sequence from (1) with $\tilde{\Gamma}$ we obtain (2). □

2.10 Notes and Remarks

One topic not treated here concerns the quotients $\pi : \mathbb{H} \to \Gamma\backslash\mathbb{H}$, for discontinuous subgroups $\Gamma < \mathbf{PSL}(2,\mathbb{C})$. This quotient always inherits from $\mathbb{H}$ a Hausdorff topology so that the quotient map π is continuous and open. If Γ acts fixed point freely on $\mathbb{H}$ then $\Gamma\backslash\mathbb{H}$ gets a 3-manifold structure and also a Riemannian metric of constant curvature -1 from $\mathbb{H}$. If Γ happens to have finite covolume, but is not cocompact then $\Gamma\backslash\mathbb{H}$ can be compactified by the addition of finitely many tori. These manifolds are studied extensively in Thurston (1978), (1982). He has a precise conjecture (proved in many instances) which 3-manifolds are of the form $\Gamma\backslash\mathbb{H}$.

Even if Γ is not fixed point free, the space $\Gamma\backslash\mathbb{H}$ inherits the structure of a topological 3-manifold. It is then called a V-manifold or an orbifold. See Grunewald, Mennicke (1980) for some interesting examples.

A serious omission in this chapter is the process of convergence of hyperbolic manifolds or Dehn surgery of discrete groups. This has been invented by Jørgensen and was developed by Thurston (1978) for his proof of Theorem 5.3.

Gromov (1987) introduced a class of so-called word hyperbolic groups. His approach is based on a suitable formulation of a property of a metric space to have a hyperbolic-like geometry in terms of the metric. Let X be a metric space with distance function $d(x,y)$. We denote

$$(x,y)_z = \frac{1}{2}(d(x,z) + d(y,z) - d(x,y))$$

for $x, y, z \in X$. According to Gromov, X is called a hyperbolic metric space if there exists $\delta > 0$ such that

$$(x, y)_z \geq \min\{ (x, w)_z, (y, w)_z \} - \delta$$

for any $x, y, z, w \in X$. A group G is called word hyperbolic if it is hyperbolic as a metric space, with the word metric respective to some finite set of generators of G. Here, given a set of generators A of G, the distance $d(x, y)$ is the length of a shortest word over A representing the element $x^{-1}y$. The word hyperbolic groups admit several definitions and form a rather general class of groups, containing all discrete cocompact groups of isometries of n-dimensional hyperbolic space $\mathbb{H}^n$, and, in particular, in our context, all discrete cocompact groups of isometries of $\mathbb{H} = \mathbb{H}^3$. Many properties of word hyperbolic groups are similar to those of discrete cocompact hyperbolic groups. However, some results are easier to prove in the more general context. As an example, we mention that any word hyperbolic group has solvable word and conjugacy problems. For more information on word hyperbolic groups and hyperbolic metric spaces, we refer the reader to Ghys, de la Harpe (1990), Alonso, Brady, Cooper, Ferlini, Lustig, Mihalik, Shapiro, Short (1991), Bowditch (1991) and Lysionok (1989). In particular Lysionok (1989) proves that the class of word hyperbolic groups is the same as the class satisfying Dehn's cancellation condition. We note also that there is a larger class of groups described in a combinatorial way, called automatic groups, which contains all geometrically finite discrete hyperbolic groups. As a consequence one gets that any geometrically finite discrete hyperbolic group has solvable word and conjugacy probles as well. The book Epstein, Holt, Levy, Paterson, Thurston (1992) contains a detailed exposition of the subject.

Chapter 3. Automorphic Functions

Considering subgroups $\Gamma < \mathbf{PSL}(2,\mathbb{C})$ which act discontinuously on 3-dimensional hyperbolic space $\mathbb{H}$ it is natural to study functions $f : \mathbb{H} \to \mathbb{C}$ which are invariant under Γ. This means that $f(\gamma P) = f(P)$ for all $\gamma \in \Gamma$ and $P \in \mathbb{H}$.

In this chapter we introduce certain fundamental tools for this task. We also introduce certain families of functions which will be of importance later.

A suitable way to define invariant functions is to consider the Poincaré summation process. By definition, a Poincaré series is a series of the type

$$f(P) = \sum_{M \in \Gamma} h(MP)$$

where h is a function on $\mathbb{H}$ which is subject to a suitable growth condition that ensures the (absolute) convergence of the series for all $P \in \mathbb{H}$. Of particular interest to us are Poincaré series where h is a point-pair invariant. The simplest point-pair invariant is $h(P,Q) = \delta(P,Q)^{-1-s}$, where s is a complex variable and δ is the function from Chapter 1, Section 1 computing the hyperbolic distance. If the Poincaré series converges it gives a Γ-invariant function in both the variables P and Q. This point of view leads to the functions studied in Section 1 of this chapter.

Sometimes the function h might already have invariance properties with respect to a subgroup of Γ. Then a full summation $\sum_{M \in \Gamma} h(MP)$ makes no sense in general and some restriction on the summation index has to be made. Consider $h : \mathbb{H} \to \mathbb{C}$, $h(z + rj) = r^s$ for some $s \in \mathbb{C}$. This function is invariant under all translations $\begin{pmatrix} 1 & \lambda \\ 0 & 1 \end{pmatrix}$ $(\lambda \in \mathbb{C})$. Writing Γ'_∞ for the subgroup of translations in Γ, it is natural to consider

$$f(P) = \sum_{g \in \Gamma'_\infty \backslash \Gamma} h(gP).$$

These functions are called Eisenstein series, their treatment is started in Sections 2 and 4.

3.1 Definition and Elementary Properties of some Poincaré Series

Throughout this section Γ denotes a discrete subgroup of $\mathbf{PSL}(2,\mathbb{C})$. We introduce several types of Poincaré series which will play an essential rôle in the sequel.

Our first Poincaré series is the function $H(P,Q,s)$ which is obtained by averaging the simplest point pair invariant.

Definition 1.1. For $P,Q \in \mathbb{H}$ put

$$H(P,Q,s) := \sum_{M\in\Gamma} \delta(P,MQ)^{-1-s}, \tag{1.1}$$

provided the series converges absolutely.

Here δ is the function introduced in Proposition 1.1.6, that is $\delta(P,Q) = \cosh d(P,Q)$ where d stands for the hyperbolic distance between P,Q. Since δ is a point-pair invariant, the function H is Γ-invariant in both variables P and Q. Looking at the particular example $\Gamma = \mathbf{PSL}(2,\mathbb{Z}[i])$ we see, using Proposition 1.1.7 the following analogue of Epsteins zeta-function, see Epstein (1903),

$$H(j,j,s) = 2^{1+s} \sum_{\left(\begin{smallmatrix} a & b \\ c & d\end{smallmatrix}\right)\in \mathrm{PSL}(2,\mathbb{Z}[i])} (|a|^2+|b|^2+|c|^2+|d|^2)^{-1-s}$$

as a value of H. Note furthermore that (1.1) is a generalized Dirichlet series, see Apostol (1976). Hence the series has an abscissa of absolute convergence σ_0, and defines a holomorphic function of the variable s in the half-plane $\operatorname{Re} s > \sigma_0$. It follows from the proof of Proposition 1.3 below that σ_0 does not depend on P,Q and that (1.1) converges uniformly on compact sets with respect to

$$(P,Q,s) \in \mathbb{H}\times\mathbb{H}\times\{\, s \mid \operatorname{Re} s > \sigma_0 \,\}.$$

Hence we may define

Definition 1.2. Let $\Gamma < \mathbf{PSL}(2,\mathbb{C})$ be a discrete group. Then the common abscissa of convergence σ_0 of the series (1.1) is called the abscissa of convergence of Γ.

It will turn out later that $H(P,Q,\cdot)$ has a meromorphic continuation into the complex plane if Γ has a fundamental domain of finite hyperbolic volume. For the proof which is contained in Chapter 6, Section 4 we use the spectral theory of the Laplace operator.

Proposition 1.3. *The series (1.1) converges absolutely and uniformly on compact subsets of* $\mathbb{H}\times\mathbb{H}\times\{\, s \mid \operatorname{Re} s > 1 \,\}$.

Proof. Let $K \subset \mathbb{H} \times \mathbb{H} \times \{ s \mid \operatorname{Re} s > 1 \}$ be a compact set and let $\sigma > 1$ be the minimum of the real parts of the third components of the elements of K. We have by Proposition 1.1.8 for $(P, Q, s) \in K,\ M \in \Gamma$

$$\begin{aligned} \delta(P, MQ)^{-1-\operatorname{Re} s} &\leq 4^{1+\sigma}\ \delta(P, P_0)^{1+\sigma}\ \delta(P_0, MQ)^{-1-\sigma} \\ &= 4^{1+\sigma}\ \delta(P, P_0)^{1+\sigma}\ \delta(M^{-1}P_0, Q)^{-1-\sigma} \\ &\leq 4^{2+2\sigma}\ \delta(P, P_0)^{1+\sigma}\ \delta(Q, Q_0)^{1+\sigma}\ \delta(P_0, MQ_0)^{-1-\sigma}. \end{aligned} \tag{1.2}$$

Hence it suffices to prove the convergence of $H(P_0, Q_0, \sigma)$ for some $P_0, Q_0 \in \mathbb{H}$.

Now let $\mathcal{F}$ be a measurable fundamental domain for Γ. Then we have by Levi's theorem on monotone convergence

$$\begin{aligned} \int_{\mathcal{F}} H(P, j, \sigma)\ dv(P) &= \int_{\mathbb{H}} \delta(P, j)^{-1-\sigma}\ dv(P) \\ &= \int_{\mathbb{H}} \left(\frac{2r}{x^2 + y^2 + r^2 + 1} \right)^{1+\sigma} \frac{dx\ dy\ dr}{r^3} \\ &= 2^{1+\sigma} \int_0^\infty \left(\int_{-\infty}^{+\infty} \int_{-\infty}^{+\infty} \frac{dx\ dy}{(x^2 + y^2 + r^2 + 1)^{1+\sigma}} \right) \frac{dr}{r^{2-\sigma}} \\ &= 2^{2+\sigma}\ \pi \int_0^\infty \int_0^\infty \frac{\rho\ d\rho}{(\rho^2 + r^2 + 1)^{1+\sigma}}\ \frac{dr}{r^{2-\sigma}} \\ &= \frac{2^{1+\sigma}}{\sigma}\ \pi \int_0^\infty \frac{dr}{(1 + r^2)^\sigma\ r^{2-\sigma}}, \end{aligned} \tag{1.3}$$

which is finite, since $\sigma > 1$. Hence $H(P, j, \sigma)$ is finite for almost all P. In particular, there exists some $P_0 \in \mathbb{H}$ such that $H(P_0, j, \sigma)$ converges. This finishes the proof of our proposition. □

By definition, a series $\sum_{n=1}^\infty f_n(x)$ of functions $f_n : X \to \mathbb{C}$ on a topological space X is called normally convergent if

$$\sum_{n=1}^\infty \sup \{ |f_n(x)| \mid x \in K \}$$

converges for all compact subsets $K \subset X$. It is evident from the above proof that $H(P, Q, s)$ is even normally convergent on $\mathbb{H} \times \mathbb{H} \times \{ s \mid \operatorname{Re} s > 1 \}$.

The proof of Proposition 1.3 given above does not use any a priori information on the behaviour of the orbits of Γ. If we use the estimates for

$$\pi(P, Q, x) := |\{ M \in \Gamma \mid \delta(P, MQ) \leq x \}| \tag{1.4}$$

contained in Lemmas 2.6.1, 2.6.2 we obtain growth estimates for the partial sums of the series (1.1) which give another proof of Proposition 1.3, see also Beardon, Nicholls (1972), Chen (1979).

Proposition 1.4. *If Γ is an arbitrary discrete subgroup of* $\mathbf{PSL}(2,\mathbb{C})$ *and* $K \subset \mathbb{H}$ *is a compact set, there exists a constant* $C_1(K)$ *depending only on* K *and* Γ, *such that for all* $P \in K,\ Q \in \mathbb{H},\ T \geq 1$ *the following estimates hold:*

$$\sum_{\delta(P,MQ)\leq T} \delta(P,MQ)^{-1-s} \leq C_1(K)\,\frac{s+1}{s-1} \qquad \text{if } s>1, \tag{1.5}$$

$$\sum_{\delta(P,MQ)\leq T} \delta(P,MQ)^{-2} \leq C_1(K)\,(1+2\log T), \tag{1.6}$$

$$\sum_{\delta(P,MQ)\leq T} \delta(P,MQ)^{-1-s} \leq C_1(K)\,\frac{2}{1-s}\,T^{1-s} \qquad \text{if } -1\leq s<1, \tag{1.7}$$

$$\sum_{\delta(P,MQ)\leq T} \delta(P,MQ)^{-1-s} \leq C_1(K)\,T^{1-s} \qquad \text{if } s<-1. \tag{1.8}$$

Proof. Let $K \subset \mathbb{H}$ be a compact set and assume $P \in K,\ Q \in \mathbb{H},\ s \in \mathbb{R},\ 0 < t < 1 \leq T$. We know from Lemma 2.6.1 that for all $P \in K,\ Q \in \mathbb{H},\ x > 0$

$$\pi(P,Q,x) \leq C_1(K)\,x^2, \tag{1.9}$$

where $C_1(K)$ depends only on K and Γ. We also have

$$\begin{aligned}\sum_{\delta(P,MQ)\leq T} \delta(P,MQ)^{-1-s} &= \int_t^T x^{-s-1}\,d\pi(P,Q,x)\\ &= [\,x^{-s-1}\,\pi(P,Q,x)\,]_t^T + (s+1)\int_t^T \frac{\pi(P,Q,x)}{x^{s+2}}\,dx.\end{aligned} \tag{1.10}$$

For $s \geq -1$ we deduce (1.5)–(1.7) from (1.9), (1.10), letting $t \to 1$.

$$\sum_{\delta(P,MQ)\leq T} \delta(P,MQ)^{-1-s} \leq C_1(K)\,(\,T^{1-s} + (s+1)\int_1^T x^{-s}\,dx\,). \tag{1.11}$$

This implies for $s \geq -1,\ s \neq 1$

$$\sum_{\delta(P,MQ)\leq T} \delta(P,MQ)^{-1-s} \leq C_1(K)\left(T^{1-s} + \frac{1+s}{1-s}\,(T^{1-s}-1)\right), \tag{1.12}$$

and for $s = 1$ we obtain

$$\sum_{\delta(P,MQ)\leq T} \delta(P,MQ)^{-2} \leq C_1(K)(1+2\log T). \tag{1.13}$$

Now (1.12) implies (1.5) and (1.7), and (1.13) is the same as (1.6). In the case $s < -1$, we simply omit the integral on the right-hand side of (1.10) and conclude

$$\sum_{\delta(P,MQ)\leq T} \delta(P,MQ)^{-1-s} \leq C_1(K)\, T^{1-s}$$

which is (1.8). □

If Γ has finite covolume we may apply Lemma 2.6.2 in order to estimate the right-hand side of (1.10) from below.

Proposition 1.5. *Let Γ be a cofinite subgroup of* $\mathbf{PSL}(2,\mathbb{C})$*, and let $K \subset \mathbb{H}$ be a compact set. Then there exist constants $C_1 > C_2 > 0$, $T_0 \geq 1$ depending only on K and Γ such that for all $P, Q \in K$, $T \geq T_0$ the following is true:*

$$\text{(1.14)} \quad \sum_{\delta(P,MQ)\leq T} \delta(P,MQ)^{-1-s} \geq C_2 \,\frac{s+1}{s-1}\, (T_0^{1-s} - T^{1-s}) + C_2\, T^{1-s} - C_1\, T_0^{1-s} \qquad \text{if } s > 1,$$

$$\text{(1.15)} \quad \sum_{\delta(P,MQ)\leq T} \delta(P,MQ)^{-2} \geq 2\, C_2\, (\log T - \log T_0) + C_2 - C_1,$$

$$\text{(1.16)} \quad \sum_{\delta(P,MQ)\leq T} \delta(P,MQ)^{-1-s} \geq C_2\, \frac{2}{1-s}\, T^{1-s} - \left(C_1 + C_2\, \frac{1+s}{1-s}\right) T_0^{1-s} \qquad \text{if } -1 \leq s < 1,$$

$$\text{(1.17)} \quad \sum_{\delta(P,MQ)\leq T} \delta(P,MQ)^{-1-s} \geq \left(C_2 + \frac{1+s}{1-s}\, C_1\right) T^{1-s} - C_1\, \frac{2}{1-s}\, T_0^{1-s} \qquad \text{if } s < -1.$$

Proof. We know from Lemma 2.6.2 that for all $P, Q \in K$, $x \geq T_0 \geq 1$

$$\text{(1.18)} \quad \pi(P,Q,x) \geq C_2\, x^2,$$

where C_2, T_0 depend only on K, Γ. Choosing $t = T_0$, $T \geq T_0$ on the right hand side in (1.10) and writing $\geq$ we obtain (1.14)–(1.17) from (1.10), (1.18). □

Observe that (1.17) suffers from the disadvantage that the right-hand side may be negative. This is due to the fact that the integral on the right-hand side of (1.10) must be estimated from below by means of (1.9) if $s < -1$. The corresponding statements in Beardon, Nicholls (1972) and in Chen (1979) are not proved completely. Similarly, the right-hand side in (1.14) may be negative.

Corollary 1.6. *If $\Gamma < \mathbf{PSL}(2,\mathbb{C})$ is a cofinite group, the series (1.1) diverges for $s \leq 1$.*

The proof is obvious from (1.15). We wish to point out that (1.6) and (1.15) will be sharpened to a statement on the asymptotic behaviour of the sum

$$\sum_{\delta(P,MQ)\leq T} \delta(P,MQ)^{-2}$$

in Corollary 6.5 of this chapter.

Corollary 1.7. *Let Γ be a discrete subgroup of $\mathbf{PSL}(2,\mathbb{C})$ and $K \subset \mathbb{H}$ be a compact set. Then there exists a constant $C_1(K) > 0$ depending only on K and Γ such that*

$$H(P,Q,s) \leq C_1(K)\,\frac{s+1}{s-1} \qquad \textit{for all } P \in K,\ Q \in \mathbb{H},\ s > 1. \tag{1.19}$$

If in addition Γ has finite covolume, there exists a constant $C_3(K)$ depending only on K and Γ such that

$$H(P,Q,s) \geq C_3(K)\,\frac{s+1}{s-1} \qquad \textit{for all } P,Q \in K,\ 1 < s \leq 2. \tag{1.20}$$

Proof. (1.19) is obvious from (1.5), and (1.20) follows from (1.14), letting $T \to \infty$. □

In Theorem 6.3 of this chapter, we shall sharpen (1.19), (1.20) to a theorem yielding the asymptotic behaviour of $H(P,Q,s)$ as $s \downarrow 1$.

It follows from (1.19) that $H(P,\cdot,s)$ is bounded on $\mathbb{H}$ for fixed $P \in \mathbb{H}$, $s > 1$. This remarkable fact can also be deduced by a transformation of $\mathbb{H}$ to the unit ball $\mathbb{B}$. Transform $\mathbb{H}$ to $\mathbb{B}$ as described in Chapter 1, Section 2 such that $P \mapsto 0$, $Q \mapsto x \in \mathbb{B}$. Let Γ' denote the corresponding group in $\mathbf{Iso}(\mathbb{B})$ and let $M \in \Gamma$ correspond to $g \in \Gamma'$. Let furthermore d' denote the hyperbolic distance in $\mathbb{B}$. Then we have by our formula for the distance in $\mathbb{B}$

$$\begin{aligned}\delta(P,MQ) &= \cosh d(P,MQ) = \cosh d'(0,g(x))\\ &= \cosh\left(\log\frac{1+\|g(x)\|}{1-\|g(x)\|}\right) = \frac{1+\|g(x)\|^2}{1-\|g(x)\|^2}.\end{aligned} \tag{1.21}$$

Hence the series corresponding to $H(P,\cdot,s)$ on $\mathbb{B}$ is

(1.22) $$F(x,s) := \sum_{g\in\Gamma'} \left(\frac{1-\|g(x)\|^2}{1+\|g(x)\|^2}\right)^{1+s} \qquad (x \in \mathbb{B}).$$

It follows from our previous considerations that this series converges for $\operatorname{Re} s > 1$ and defines a bounded function on $\mathbb{B}$. This is essentially known in the theory of automorphic functions as Godement's theorem (see Godement (1958), Earle (1969)). Using a beautiful idea of Ahlfors (1964) we shall give now another proof of this boundedness property which leads to an even sharper conclusion.

Theorem 1.8. *Let Γ be a discrete subgroup of* $\mathbf{PSL}(2,\mathbb{C})$ *and* $P \in \mathbb{H}$, $s \in \mathbb{R}$, $s > 1$. *Then the function* $Q \mapsto H(P,Q,s)$ *attains its maximum on* $\mathbb{H}$ *in the closed hyperbolic ball*

(1.23) $$\left\{ Q \in \mathbb{H} \mid \delta(P,Q) \le \sqrt{\frac{s+2}{s-1}} \right\}.$$

Before we enter into the details of the proof let us remark that the radius of the ball (1.23) depends only on s and not on Γ. Hence Theorem 1.8 contains a universal property of discrete groups in $\mathbf{PSL}(2,\mathbb{C})$.

Proof. We transform $\mathbb{H}$ onto $\mathbb{B}$ such that $P \mapsto 0$, $Q \mapsto x \in \mathbb{B}$. Then we have to show that the series (1.22) attains its maximum in the ball

(1.24) $$B_s := \left\{ x \in \mathbb{B} \;\middle|\; \frac{1+\|x\|^2}{1-\|x\|^2} \le \sqrt{\frac{s+2}{s-1}} \right\}.$$

Let $E \subset \Gamma'$ be a finite non-empty subset of Γ' and put

$$F_E(x) := \sum_{g\in E} \left(\frac{1-\|g(x)\|^2}{1+\|g(x)\|^2}\right)^{1+s}.$$

Obviously F_E is a continuous function on the closed unit ball $\overline{\mathbb{B}}$ in $\mathbb{R}^3$ which vanishes on the boundary S^2 of $\overline{\mathbb{B}}$ and is positive on the open unit ball $\mathbb{B}$. Hence F_E attains a maximum at some point $x_0 \in \mathbb{B}$. From our formula in Chapter 1 for the Laplace operator Δ on $\mathbb{B}$ and since the first order derivatives of F_E vanish at x_0 we get $\Delta F_E(x_0) \le 0$. So there exists an element $g_0 \in E$ such that

(1.25) $$\Delta \left(\frac{1-\|g_0(x)\|^2}{1+\|g_0(x)\|^2}\right)^{1+s} \Bigg|_{x=x_0} \le 0.$$

Since Δ commutes with the action of $\mathbf{Iso}^+(\mathbb{B})$, the left-hand side of (1.25) is readily computed by means of

$$\Delta\left(\frac{1-\|g(x)\|^2}{1+\|g(x)\|^2}\right)^{1+s} = \left(\Delta\left(\frac{1-\|x\|^2}{1+\|x\|^2}\right)^{1+s}\right)\Big|_{x\mapsto g(x)}. \tag{1.26}$$

We put $\|x\| = \rho$ and observe that the radial part of Δ is given by

$$\frac{1}{4}(1-\rho^2)^2\left(\frac{1}{\rho^2}\frac{\partial}{\partial\rho}\rho^2\frac{\partial}{\partial\rho} + \frac{2\rho}{1-\rho^2}\frac{\partial}{\partial\rho}\right).$$

Then an elementary computation yields

$$\begin{aligned}&\Delta\left(\frac{1-\rho^2}{1+\rho^2}\right)^{1+s}\\ &= (s+1)\left(\frac{1-\rho^2}{1+\rho^2}\right)^{3+s}\left((s-1)\left(\frac{1-\rho^2}{1+\rho^2}\right)^{-2} - (s+2)\right).\end{aligned} \tag{1.27}$$

Now (1.25)–(1.27) imply that there exists an element $g_0 \in E$ such that

$$\frac{1+\|g_0(x_0)\|^2}{1-\|g_0(x_0)\|^2} \le \left(\frac{s+2}{s-1}\right)^{\frac{1}{2}},$$

that is $g_0(x_0) \in B_s$. Observe that B_s is a Euclidean ball with centre 0 and Euclidean radius R, and

$$R^2 = \frac{\sqrt{\frac{s+2}{s-1}} - 1}{\sqrt{\frac{s+2}{s-1}} + 1}. \tag{1.28}$$

Note that R is strictly less than 1. Hence we have for all $x \in \mathbb{B}$

$$\begin{aligned}F_E(x) \le F_E(x_0) &= \max\ \{\ F_E(y)\ \mid\ y \in g_0^{-1}(B_s)\ \}\\ &\le \max\ \{\ F(y,s)\ \mid\ y \in g_0^{-1}(B_s)\ \} = \max\ \{\ F(y,s)\ \mid\ y \in B_s\ \}\end{aligned}$$

since $F(\cdot, s)$ is Γ'-invariant. But now the last bound on the right-hand side is finite and no longer depends on E. Hence we may take the sup of the left-hand side with respect to E to obtain $F(x,s) \le \max\ \{\ F(y,s)\ \mid\ y \in B_s\ \}$. □

We remark in passing that (1.21), (1.26)–(1.27) yield the following differential equation for our function $H(\cdot, Q, s)$:

$$(-\Delta - (1-s^2))\ H(P,Q,s) = (s+1)\ (s+2)\ H(P,Q,s+2). \tag{1.29}$$

This differential equation holds for all $s \in \mathbb{C}$ such that $\operatorname{Re} s > \sigma_0$, where σ_0 denotes the abscissa of convergence of the group Γ. Note that the right-hand side of (1.29) is defined for $\operatorname{Re} s > \sigma_0 - 2$. The differential equation (1.29) will be used in the next chapter for the analytic continuation of $H(P,Q,s)$ into the s-plane.

The method of proof for Theorem 1.8 can also be applied to the series

$$G(x,s) := \sum_{g\in\Gamma'} (1 - \|g(x)\|^2)^{1+s} \qquad (s > 1)$$

instead of (1.22). The result is that for fixed $s > 1$ the function $G(\cdot, s)$ attains its maximum in $\mathbb{B}$ in the closed Euclidean ball with centre 0 and radius

$$R_1 := \left(\frac{3}{2s+1}\right)^{1/2}.$$

This result also is a universal property of discrete groups. It is interesting to compare the radii R (cf. (1.28)) and R_1. Introducing the new parameter $t > 0$ by

$$\frac{s+2}{3} = \cosh^2 t,$$

we obtain, after an elementary computation, $R = e^{-t}$, $R_1 = (\cosh 2t)^{-\frac{1}{2}}$, and hence $R_1 > R$.

Definition 1.9. The discrete group Γ is said to be of convergence type if the series (1.22) converges for $s = 1$, otherwise Γ is said to be of divergence type.

It is a remarkable fact that the series $F(\cdot, 1)$ is bounded on $\mathbb{B}$ for every group of convergence type. We prove this in Theorem 1.10. All cofinite groups are of divergence type, see Corollary 1.6.

Theorem 1.10. *If Γ is of convergence type, the function $P \mapsto H(P,Q,1)$ is bounded on $\mathbb{H}$ by a bound depending on Q.*

Proof. We transform $\mathbb{H}$ to $\mathbb{B}$ such that $Q \mapsto 0$, $P \mapsto x \in \mathbb{B}$. Let Γ' denote the corresponding group of transformations on $\mathbb{B}$. Then we have to show that

$$F(x) := \sum_{g\in\Gamma'} (1 - \|gx\|)^2 \tag{1.30}$$

is bounded on $\mathbb{B}$. If h is an arbitrary element in $\mathbf{Iso}^+(\mathbb{B})$, the series F and the corresponding series for $h\Gamma'h^{-1}$ are both either bounded or unbounded on $\mathbb{B}$. Hence it is sufficient to prove the boundedness of F in the special case that 0 is not a fixed point of Γ'.

Suppose now that 0 is not a fixed point of Γ' and let B be a closed Euclidean ball of centre 0 contained in $\mathbb{B}$ whose radius is so small that $B \cap gB = \emptyset$ for all $g \in \Gamma'$, $g \neq id$. Let S denote the boundary of B.

We compare F with a function on $\mathbb{B}$ which is harmonic in the hyperbolic sense. The function

$$\psi(x) := \frac{(1 - \|x\|)^2}{\|x\|} \qquad (x \in \mathbb{B},\ x \neq 0)$$

is harmonic in the hyperbolic sense, which means it is a C^2-function satisfying the differential equation $\Delta\psi = 0$ in $\mathbb{B} \setminus \{0\}$. This is easily verified by computation. The following argument will show that the series

$$H(x) := \sum_{g\in\Gamma'} \frac{(1-\|g(x)\|)^2}{\|g(x)\|}$$

converges on $\mathbb{B} \setminus \{\, g(0) \mid g \in \Gamma' \,\}$. All terms of this series are nonnegative, harmonic in the hyperbolic sense on $\mathbb{B} \setminus \Gamma' B$ and vanish on the unit sphere. Observe that the maximum principle also holds for functions which are harmonic in the hyperbolic sense. This follows from the mean value formula contained in Theorem 5.4.

Consider an arbitrary finite partial sum $T(x)$ for $H(x)$. Then it follows from the maximum principle and the Γ'-invariance of H that for all $x \in \mathbb{B} \setminus \Gamma' B$

$$\begin{aligned} T(x) &\le \sup \{\, T(w) \mid w \in \cup_{g\in\Gamma'} gS \,\} \le \sup \{\, H(w) \mid w \in \cup_{g\in\Gamma'} gS \,\} \\ &= \max \{\, H(w) \mid w \in S \,\}. \end{aligned}$$

This is now true for all finite partial sums T for H. Hence we obtain

$$H(x) \le \max \{\, H(w) \mid w \in S \,\}$$

for all $x \in \mathbb{B} \setminus \Gamma' B$. But trivially we have $F \le H$, hence F is bounded on $\mathbb{B} \setminus \Gamma' B$. Since F is also bounded on $\Gamma' B$, we conclude that F is bounded on $\mathbb{B}$. □

We shall now introduce another function denoted by $\Theta(P,Q,t)$ which is an analogue of Jacobi's theta function.

Definition 1.11. For $P, Q \in \mathbb{H},\ t > 0$ put

$$\Theta(P,Q,t) := \sum_{M\in\Gamma} e^{-t\,\delta(P,MQ)}. \tag{1.31}$$

Proposition 1.12. *The series (1.31) is normally convergent in the whole of* $\mathbb{H} \times \mathbb{H} \times]0,\infty[$. *The function* $\Theta(P,Q,t)$ *is* Γ*-invariant with respect to* P *and with respect to* Q.

Proof. Since the exponential function e^{-x} decreases more rapidly than any polynomial as $x \to +\infty$, the proposition is clear from Proposition 1.4. □

There is a simple relation between $H(P,Q,s)$ and $\Theta(P,Q,t)$:

Proposition 1.13. *For* $\operatorname{Re} s > 1$ *the formula*

$$H(P,Q,s) = \frac{1}{\Gamma(s+1)} \int_0^\infty \Theta(P,Q,t)\ t^s\ dt \tag{1.32}$$

holds, the integral being absolutely convergent.

Up to normalization, Proposition 1.13 says that $H(P,Q,s)$ is essentially the Mellin transform of $\Theta(P,Q,t)$, see Magnus, Oberhettinger, Soni (1966), page 397.

Proof. It suffices to prove our proposition for real $s > 1$, since then the appropriate integrable majorants for the application of the dominated convergence theorem are obvious and imply the statement in full generality. But for $s > 0,\ a > 0$ the equation

$$a^{-s-1} = \frac{1}{\Gamma(s+1)} \int_0^\infty e^{-at}\ t^s\ dt \tag{1.33}$$

yields by the monotone convergence theorem

$$\frac{1}{\Gamma(s+1)} \int_0^\infty \Theta(P,Q,t)\ t^s\ dt = \sum_{M \in \Gamma} \delta(P, MQ)^{-1-s}.$$

Since the right-hand side is finite for $s > 1$, we are done. Actually our proposition holds for $\operatorname{Re} s > \sigma_0$, where σ_0 denotes the abscissa of convergence of the series (1.1). This is evident from the above proof. □

The following proposition is rather plausible from the definition of Θ.

Proposition 1.14. *If $K \subset \mathbb{H}$ is a compact set, there exist constants $c_1, c_2, \epsilon_1, \epsilon_2 > 0$ such that*

$$c_1\ e^{-\epsilon_1 t} \le \Theta(P,Q,t) \le c_2\ e^{-\epsilon_2 t} \tag{1.34}$$

for all $t \ge 1$ uniformly with respect to $(P,Q) \in K \times K$.

Proof. If we put

$$\mu(P,Q) := \min \{\ \delta(P, MQ)\ \mid\ M \in \Gamma\ \}, \tag{1.35}$$

we have

$$e^{-t\mu(P,Q)} \le \Theta(P,Q,t), \tag{1.36}$$

so that the lower estimate is obvious with $\epsilon_1 = \max\{\ \mu(P,Q)\ \mid\ (P,Q) \in K \times K\ \}$, $c_1 = 1$. To prove the upper estimate, we proceed as in the proof of Proposition 1.4 and conclude

$$
\begin{aligned}
\Theta(P,Q,t) &= \int_1^\infty e^{-tx}\, d\pi(P,Q,x) = t\int_1^\infty \pi(P,Q,x)\, e^{-tx}\, dx \\
&\le C_1(P)\, t\int_1^\infty x^2\, e^{-tx}\, dx = C_1(P)\left(1+\frac{2}{t}+\frac{2}{t^2}\right) e^{-t},
\end{aligned} \tag{1.37}
$$

where $C_1(P)$ is bounded on compact sets by Lemma 2.6.1. A slight refinement of this estimate leads to an upper estimate of the form

$$\Theta(P,Q,t) \le c\, e^{-t\mu(P,Q)}$$

with $\mu(P,Q)$ as in (1.35). □

Observe that (1.37) holds for all $t>0,\ P\in K,\ Q\in \mathbb{H}$. This implies the first part of our next Proposition.

Proposition 1.15. *If $K\subset \mathbb{H}$ is a compact set, there exists a constant $C_1>0$ such that for all $P\in K,\ Q\in\mathbb{H}$ and $0<t\le 1$*

$$\Theta(P,Q,t) \le \frac{C_1}{t^2}. \tag{1.38}$$

If Γ has finite covolume and $K\subset \mathbb{H}\times\mathbb{H}$ is a compact set, there exists a constant $C_2>0$ such that

$$\frac{C_2}{t^2} \le \Theta(P,Q,t) \tag{1.39}$$

for all $(P,Q)\in K$ and $0<t\le 1$.

Proof. The first part follows from (1.37). The second is done in the same way using the lower estimate (1.18). Compare the proof of Proposition 1.5. □

Later on we shall even show that for all groups with a fundamental domain $\mathcal{F}$ of finite hyperbolic volume

$$\Theta(P,Q,t) \sim \frac{4\pi}{v(\mathcal{F})}\cdot\frac{1}{t^2} \qquad \text{as} \quad t\to 0. \tag{1.40}$$

It is rather easy to see that this result also holds for arbitrary groups in the mean over the fundamental domain. This vague formulation is made more precise in the following proposition.

Proposition 1.16. *If $\mathcal{F}$ is a fundamental domain for Γ, we have for all $t>0$*

$$\int_{\mathcal{F}} \Theta(P,Q,t)\, dv(Q) = \frac{4\pi}{t} K_1(t), \tag{1.41}$$

where K_1 is the usual modified Bessel function.

Proof. For $t > 0$,

$$\int_{\mathcal{F}} \Theta(P,Q,t)\, dv(Q) = \int_{\mathbb{H}} e^{-t\delta(P,Q)}\, dv(Q) = \int_{\mathbb{H}} e^{-t\delta(j,Q)}\, dv(Q),$$

since δ is a point-pair invariant and v is $\mathbf{PSL}(2,\mathbb{C})$-invariant. Hence we obtain

$$\begin{aligned}
\int_{\mathcal{F}} \Theta(P,Q,t)\, dv(Q) &= \int_0^{\infty} \int_{-\infty}^{+\infty} \int_{-\infty}^{+\infty} \exp\left(-t\,\frac{x^2+y^2+r^2+1}{2r}\right) \frac{dx\, dy\, dr}{r^3} \\
&= 2\pi \int_0^{\infty} \int_0^{\infty} \exp\left(-t\,\frac{\rho^2+r^2+1}{2r}\right) \rho\, d\rho\, \frac{dr}{r^3} \\
&= \frac{2\pi}{t} \int_0^{\infty} \exp\left(\frac{-t}{2}\left(r+\frac{1}{r}\right)\right) \frac{dr}{r^2} \\
&= \frac{4\pi}{t} K_1(t).
\end{aligned}$$

For the integral representation of K_1 see Magnus, Oberhettinger, Soni (1966), page 85. □

Recall that $K_1(t) \sim t^{-1}$ as $t \to 0$, so that the Proposition specifies our previous remark.

The function $\Theta(P,Q,t)$ satisfies a noteworthy differential equation of hyperbolic type.

Proposition 1.17. *For $P,\ Q \in \mathbb{H}$, $t > 0$, $\Theta(P,Q,t)$ is a solution of the differential equation*

$$\Delta\, \Theta(P,Q,t) = \left(t^2 \frac{\partial^2}{\partial t^2} + 3t\frac{\partial}{\partial t} - t^2\right) \Theta(P,Q,t) \tag{1.42}$$

where Δ operates on P. The same equation holds when Δ operates on Q.

Proof. Since Θ is symmetric with respect to P and Q, we only need to show (1.42) with Δ operating on P. Now it is easy to see, using our coordinate expression for δ from Chapter 1, Section 1, that differentiation of $\exp(-t\delta(P,MQ))$ with respect to any of the variables x, y, r, t where $P = x + yi + rj$ yields a term which may be estimated by some constant times some power of $\delta(P,MQ)$ times $\exp(-t\delta(P,MQ))$. Hence the differential operators on both sides of (1.42) may be applied to Θ termwise, and this implies that we must only show that

$$\Delta \exp(-t\delta(P,Q)) = \left(t^2 \frac{\partial^2}{\partial t^2} + 3t\frac{\partial}{\partial t} - t^2\right) \exp(-t\delta(P,Q)).$$

This can be done by an elementary computation. □

In our later discussion of the resolvent we shall be concerned with solutions of the differential equation

$$-\Delta\, \varphi(\delta(P,Q)) = \lambda\, \varphi(\delta(P,Q)) \tag{1.43}$$

where Δ operates on P. Using the definitions of Δ and δ, the equation (1.43) reduces to a special case of Riemann's differential equation, namely

$$(\delta^2 - 1)\, \varphi''(\delta) + 3\, \delta\, \varphi'(\delta) + \lambda\, \varphi(\delta) = 0. \tag{1.44}$$

To solve this equation, we introduce a complex parameter s such that

$$\lambda = 1 - s^2. \tag{1.45}$$

For $s \neq 0$, the equation (1.44) has the fundamental system of solutions

$$\varphi_s(\delta) := \frac{(\delta + \sqrt{\delta^2 - 1})^{-s}}{\sqrt{\delta^2 - 1}} \qquad \text{and} \qquad \varphi_{-s}(\delta) \tag{1.46}$$

which are defined for $\delta > 1$. The corresponding solutions of (1.43) are defined for $P \neq Q$ only.

For $\operatorname{Re} s > 1$ the point-pair invariant (1.46) gives rise to a Poincaré series that will be recognized later on as the kernel of the resolvent operator, see Section 2 of Chapter 4. This method for the construction of the resolvent kernel is due to A. Selberg whose investigations were inspired by a pioneering paper of Maaß (1953), see also Hejhal (1976a), page 35. Moreover, Maaß had already constructed the Green function considered in Roelcke (1956a), which served in that work as a substitute for the resolvent kernel. Consequently, we call our Poincaré series a Maaß–Selberg series.

Definition 1.18. For $P, Q \in \mathbb{H}$, $P \not\equiv Q \mod \Gamma$ the Maaß–Selberg series is defined by

$$F(P,Q,s) := \frac{1}{4\pi} \sum_{M \in \Gamma} \varphi_s(\delta(P, MQ)) \tag{1.47}$$

provided that this series converges absolutely.

Proposition 1.19. *The Maaß–Selberg series (1.47) converges uniformly on compact subsets of* $(\mathbb{H} \times \mathbb{H} \setminus \{\, (P, MP) \mid P \in \mathbb{H},\ M \in \Gamma \,\}) \times \{\, s \mid \operatorname{Re} s > 1 \,\}$.

Proof. For $s \in \mathbb{C}$, $\delta > 1$ we have

$$\varphi_s(\delta) = 2^{-s}\, \delta^{-1-s}\, (1 + O(\delta^{-2})) \qquad (\delta \to +\infty) \tag{1.48}$$

uniformly on compact sets with respect to $s \in \mathbb{C}$. Hence the uniform convergence follows from Proposition 1.3. □

It is also clear from (1.48) that the Maaß–Selberg series has the same abscissa of convergence as our function $H(P,Q,s)$. For all cofinite groups Γ this abscissa of convergence is equal to 1, see Corollary 1.6.

Proposition 1.20. *For fixed* $P,\ Q \in \mathbb{H},\ P \not\equiv Q \bmod \Gamma$, *the function*

$$4\pi F(P,Q,s) - 2^{-s} H(P,Q,s) \tag{1.49}$$

has a holomorphic continuation into the half-plane $\{\ s \mid \operatorname{Re} s > -1\ \}$.

Proof. The assertion follows from (1.48) and Proposition 1.3. □

Proposition 1.21. *Let* σ_0 *be the abscissa of convergence of* Γ. *For* $\operatorname{Re} s > \sigma_0,\ P,\ Q \in \mathbb{H},\ P \not\equiv Q \bmod \Gamma$, *the Maaß–Selberg series satisfies the differential equation*

$$-\Delta\ F(P,Q,s) = (1-s^2)\ F(P,Q,s). \tag{1.50}$$

Proof. The individual terms in the series defining F by construction satisfy the differential equation in question. The interchange of summation and differentiation requires some elementary estimates. □

As it will turn out later, the functions H and Θ do not satisfy a functional equation. However, under appropriate assumptions on Γ we shall be able to prove a functional equation for the Maaß–Selberg series F. Hence the inverse Mellin transform of F also satisfies a functional equation. This inverse Mellin transform is defined as follows. For $\operatorname{Re} s > 1$ we have

$$4\pi F(P,Q,s) = \frac{1}{\Gamma(s)} \int_0^\infty \sum_{M\in\Gamma} \frac{1}{\sqrt{\delta_M^2 - 1}}\ e^{-t(\delta_M + \sqrt{\delta_M^2-1})}\ t^{s-1}\ dt, \tag{1.51}$$

where we put

$$\delta_M := \delta(P, MQ). \tag{1.52}$$

Formula (1.51) is justified by a computation as in the proof of Proposition 1.13. Hence we introduce

Definition 1.22. For $P,\ Q \in \mathbb{H},\ P \not\equiv Q \bmod \Gamma,\ t > 0$ define

$$\Lambda(P,Q,t) := \sum_{M\in\Gamma} \frac{1}{\sqrt{\delta_M^2 - 1}} e^{-t(\delta_M + \sqrt{\delta_M^2-1})}, \tag{1.53}$$

where δ_M is given by (1.52).

Our previous considerations on Θ carry over to Λ and yield the following result.

Proposition 1.23. *The following properties hold uniformly on compact subsets of the corresponding ranges of definition.*
The series for Λ converges on

$$(\mathbb{H} \times \mathbb{H} \setminus \{ (P, MP) \mid P \in \mathbb{H},\ M \in \Gamma \}) \times]0, \infty[.$$

There exist constants $c_1, c_2, \epsilon_1, \epsilon_2 > 0$ such that for all $t \geq 1$

$$c_1 e^{-\epsilon_1 t} \leq \Lambda(P, Q, t) \leq c_2 e^{-\epsilon_2 t}. \tag{1.54}$$

There exists a constant $C_1 > 0$ such that for all $t \in]0, 1[$

$$\Lambda(P, Q, t) \leq \frac{C_1}{t}. \tag{1.55}$$

If Γ has finite covolume, there exists a constant $C_2 > 0$ such that for all $t \in]0, 1[$

$$\Lambda(P, Q, t) \geq \frac{C_2}{t}. \tag{1.56}$$

3.2 Definition and Elementary Properties of Eisenstein Series

This section contains a first discussion of automorphic functions which are constructed by averaging the functions on $\mathbb{H}$ defined by

$$z + rj \mapsto r^s.$$

They are called Eisenstein series. They play an important role in the spectral theory of the Laplace operator and in many other parts of what is to come. Our Eisenstein series are sometimes called non-holomorphic, they were first studied by Maaß (1953). Selberg extended their theory shortly later, see Selberg (1989b) for some history.

Let $\Gamma < \mathbf{PSL}(2, \mathbb{C})$ be a discrete group and let $\zeta \in \mathbb{P}^1\mathbb{C} = \mathbb{C} \cup \{\infty\}$. We have defined the stabilizer-group of ζ

$$\Gamma_\zeta := \{ M \in \Gamma \mid M\zeta = \zeta \} \tag{2.1}$$

and its maximal unipotent subgroup

$$\Gamma'_\zeta := \{ M \in \Gamma \mid M\zeta = \zeta, \quad M = I \text{ or } M \text{ is parabolic} \}. \tag{2.2}$$

Choose $A \in \mathbf{PSL}(2, \mathbb{C})$ such that $A\zeta = \infty$, then $(A\Gamma A^{-1})_\infty = A\Gamma_\zeta A^{-1}$, and also $(A\Gamma A^{-1})'_\infty = A\Gamma'_\zeta A^{-1}$. Remember that we call ζ a cusp if there is a full lattice $\Lambda \subset \mathbb{C}$ such that

$$A\Gamma'_\zeta A^{-1} = \left\{ \begin{pmatrix} 1 & \lambda \\ 0 & 1 \end{pmatrix} \;\middle|\; \lambda \in \Lambda \right\}.$$

For all $M \in \mathbf{PSL}(2, \mathbb{C})$, $P \in \mathbb{H}$ let us write

$$MP = z(MP) + r(MP)j. \tag{2.3}$$

There are simple expressions for the $z(MP) \in \mathbb{C}$, $r(MP) > 0$, given in Chapter 1, Section 1.

Let now $\zeta \in \mathbb{P}^1\mathbb{C}$ be a cusp of Γ. For $P \in \mathbb{H}$ consider the formal series

$$E_A^*(P, s) := \sum_{M \in (A\Gamma A^{-1})'_\infty \backslash A\Gamma A^{-1}} r(MP)^{1+s}, \tag{2.4}$$

where the summation extends over a system of representatives of the right cosets of $(A\Gamma A^{-1})'_\infty$ in $A\Gamma A^{-1}$. Then $E_A^*(P, s)$ is termwise independent of the choice of the system of representatives. For every fixed $S \in A\Gamma A^{-1}$, the element MS runs through a system of representatives of the right cosets of $(A\Gamma A^{-1})'_\infty$ in $A\Gamma A^{-1}$ whenever M does. Hence $E_A^*(P, s)$ is $A\Gamma A^{-1}$-invariant, whenever it converges absolutely. Note that E_A^* is a generalized Dirichlet series and hence has an abscissa of (absolute) convergence. We shall show in a moment that $E_A^*(P, s)$ converges absolutely iff the series $H(P, Q, s)$, studied in Section 1, converges absolutely provided $\Gamma \neq \Gamma_\zeta$. In particular, the series (2.4) converges for $\operatorname{Re} s > 1$.

First we discuss some formal properties of $E_A^*(P, s)$ and handle the question of convergence in Propositions 2.2, 2.3. We transform back in order to construct a Γ-invariant function. The $A\Gamma A^{-1}$-invariance of E_A^* implies that $E_A^*(AP, s)$ is Γ-invariant.

Note that $M = ALA^{-1}$ runs through a system of representatives of the right cosets of $(A\Gamma A^{-1})'_\infty$ in $A\Gamma A^{-1}$ if L runs through a corresponding system for Γ'_ζ in Γ. Note further that $r(MAP) = r(ALP)$. Writing again M instead of L in the summation condition we see that the series

$$E_A(P, s) := \sum_{M \in \Gamma'_\zeta \backslash \Gamma} r(AMP)^{1+s} \tag{2.5}$$

agrees with $E_A^*(AP, s)$ termwise and is Γ-invariant whenever it converges absolutely. We call E_A an Eisenstein series for Γ at the cusp ζ.

Observe that E_A depends on the choice of A. If $B \in \mathbf{PSL}(2, \mathbb{C})$ is another element such that $B\zeta = \infty$, we have

$$B = \begin{pmatrix} \alpha & \beta \\ 0 & \alpha^{-1} \end{pmatrix} A \tag{2.6}$$

for some α, $\beta \in \mathbb{C}$, $\alpha \neq 0$. This implies that

$$E_B(P, s) = |\alpha|^{2+2s} E_A(P, s). \tag{2.7}$$

Thus E_B agrees with E_A up to a trivial factor. Hence we feel justified to call E_A the Eisenstein series for Γ, ζ. Whenever appropriate we shall make a suitable choice of A. If we restrict A such that $(A\Gamma A^{-1})'_\infty$ has a fundamental parallelogram of Euclidean area 1 and if we impose the same restriction on B, we necessarily have $|\alpha| = 1$ in (2.6) and hence $E_A(P,s) = E_B(P,s)$. This normalization is suggested by Kubota (1968a). Whenever a normalization of this kind is applied to Eisenstein series this will be stated explicitly.

Some authors prefer to define the Eisenstein series with the summation condition $M \in \Gamma'_\zeta \backslash \Gamma$ replaced by the condition $M \in \Gamma_\zeta \backslash \Gamma$. Obviously this results in a multiplication of E_A by $[\Gamma_\zeta : \Gamma'_\zeta]^{-1}$. Note that the index $[\Gamma_\zeta : \Gamma'_\zeta]$ is restricted to the values 1,2,3,4,6.

There is a simple behaviour of the Eisenstein series if Γ is conjugated within $\mathbf{PSL}(2,\mathbb{C})$. Let $S \in \mathbf{PSL}(2,\mathbb{C})$ and let $G = S^{-1}\Gamma S$. Then $\eta = S^{-1}\zeta$ is a cusp of G and $AS\eta = \infty$. Since L runs through a system of representatives of the right cosets of G'_η in G iff $M = SLS^{-1}$ runs through a corresponding system for Γ'_ζ and Γ, we see that

$$E_{AS}(P,s) = E_A(SP,s), \tag{2.8}$$

where $E_{AS}(\cdot,s)$ is the Eisenstein series for G, η. In the special case $S \in \Gamma$, equation (2.8) amounts to

$$E_{AS}(P,s) = E_A(P,s) \qquad (S \in \Gamma,\ P \in \mathbb{H}). \tag{2.9}$$

This means that the Eisenstein series of Γ-equivalent cusps agree if we choose the corresponding matrices as above. Hence there exist at most as many linearly independent Eisenstein series $E_A(\cdot,s)$ as there are Γ-equivalence classes of cusps of Γ. Our later study will even show that the number of linearly independent Eisenstein series and the number of Γ-equivalence classes of cusps of Γ coincide.

Proposition 2.1. *Let $\Gamma < \mathbf{PSL}(2,\mathbb{C})$ be a discrete group and let $\zeta = A^{-1}\infty$ with $A \in \mathbf{PSL}(2,\mathbb{C})$ be a cusp of Γ.*

(1) If $\Gamma = \Gamma_\zeta$, the Eisenstein series is a finite sum and equal to a constant multiple of $r(AP)^{1+s}$. The abscissa of convergence of the Eisenstein series equals $-\infty$ in this case.

(2) For $\Gamma \neq \Gamma_\zeta$ the Eisenstein series $E_A(P,s)$ converges iff the series $H(P,Q,s)$ (see (1.1)) converges for some $Q \in \mathbb{H}$. A necessary condition for convergence is that $\operatorname{Re} s > 0$.

Proof. Part (1) is obvious. To prove (2) we may restrict ourselves to the case $\zeta = \infty$, $A = I$. Let $\Lambda \subset \mathbb{C}$ be the lattice corresponding to the maximal unipotent subgroup $\Gamma'_\infty \subset \Gamma_\infty$. Using the notation introduced in (2.3), we have for $s \in \mathbb{R}$

$$
\begin{aligned}
H(P,j,s) &= \sum_{M\in\Gamma}\left(\frac{2\,r(MP)}{|z(MP)|^2+r(MP)^2+1}\right)^{1+s} \\
&= 2^{1+s}\sum_{M\in\Gamma'_\infty\backslash\Gamma} r(MP)^{1+s}\sum_{\omega\in\Lambda}(|z(MP)+\omega|^2+r(MP)^2+1)^{-1-s}.
\end{aligned}
\tag{2.10}
$$

Hence convergence of $H(P,j,s)$ (s real) implies that $s>0$. In the inner sum over $\omega\in\Lambda$, the number $z(MP)$ can be replaced by a representative $z^*(MP)$ of $z(MP)$ in a fundamental parallelogram $\mathcal{P}$ of Λ. Choose

$$
\mathcal{P}=\left\{\ \alpha_1\omega_1+\alpha_2\omega_2\ \middle|\ \alpha_1,\ \alpha_2\in\mathbb{R},\ |\alpha_1|\le\frac{1}{2},\ |\alpha_2|\le\frac{1}{2}\ \right\}
$$

where $0\neq\omega_1\in\Lambda$ is such that $|\omega_1|$ is minimal and where $\omega_2\in\Lambda\backslash\mathbb{Z}\omega_1$ is such that $|\omega_2|$ is minimal. Then we have

$$
\frac{|\omega|}{4}\le|z^*(MP)+\omega|\qquad\text{for all}\quad 0\neq\omega\in\Lambda.
\tag{2.11}
$$

There exists a $C>0$ such that for fixed P and all $M\in\Gamma$ we have $r(MP)\le C$ and $|z^*(MP)|\le C$. Hence we obtain for $s>0$

$$
\begin{aligned}
(2C^2+1)^{-1-s} &\le \sum_{\omega\in\Lambda}(|z^*(MP)+\omega|^2+r(MP)^2+1)^{-1-s} \\
&\le 1+\sum_{0\neq\omega\in\Lambda}\left(\frac{|\omega|^2}{16}\right)^{-1-s}.
\end{aligned}
\tag{2.12}
$$

Substituting this estimate into (2.10), we see first: If $H(P,j,s)$ converges for $s\in\mathbb{R}$, we have $s>0$, and the Eisenstein series

$$
E(P,s):=\sum_{M\in\Gamma'_\infty\backslash\Gamma} r(MP)^{1+s}
\tag{2.13}
$$

converges. Second, assume that (2.13) converges. Note that for all $M=\begin{pmatrix}\cdot&\cdot\\c&d\end{pmatrix}\in\Gamma$, $\omega\in\Lambda$

$$
\begin{pmatrix}\cdot&\cdot\\c&d\end{pmatrix}\begin{pmatrix}1&\omega\\0&1\end{pmatrix}=\begin{pmatrix}\cdot&\cdot\\c&c\omega+d\end{pmatrix}\in\Gamma.
$$

Since we assume that $\Gamma\neq\Gamma_\infty$, there exists an $M=\begin{pmatrix}\cdot&\cdot\\c&d\end{pmatrix}\in\Gamma$ such that $c\neq 0$. Writing $P=z+rj$, the series $E(P,s)$ is a convergent majorant of

$$
\sum_{\omega\in\Lambda} r(M(P+\omega))^{1+s}=\sum_{\omega\in\Lambda}\left(\frac{r}{|cz+c\omega+d|^2+|c|^2\,r^2}\right)^{1+s}
$$

which majorizes $\sum_{0\neq\omega\in\Lambda}|\omega|^{-2-2s}$. Hence the convergence of the Eisenstein series (2.13) yields that $s>0$ and substituting (2.12) into (2.10) we see that $H(P,j,s)$ converges. Since $H(P,j,s)$ converges iff $H(P,Q,s)$ converges for all $Q\in\mathbb{H}$ we obtain our assertion. □

Recall that a discrete subgroup of $\mathbf{PSL}(2,\mathbb{C})$ is called elementary if it has at most two limit points in $\mathbb{P}^1\mathbb{C}$. If Γ is non-elementary, the possibility (1) of Proposition 2.1 is excluded and we have the following corollary.

Corollary 2.2. *If $\Gamma<\mathbf{PSL}(2,\mathbb{C})$ is a non-elementary discrete group with cusps, all the Eisenstein series for Γ have abscissa of convergence equal to σ_0, where σ_0 denotes the abscissa of convergence of Γ. Moreover $\sigma_0\geq 0$.*

In the next proposition and its corollary we describe the asymptotic behaviour of the Eisenstein series as cusps are approached.

Proposition 2.3. *Let $\Gamma<\mathbf{PSL}(2,\mathbb{C})$ be a discrete group and for $A,\ B\in\mathbf{PSL}(2,\mathbb{C})$ let $\zeta=A^{-1}\infty,\ \eta=B^{-1}\infty$ be cusps of Γ. Let furthermore $\alpha>0,\ \beta>\sigma_0$ be given. Then, writing $r=r(P)$, the series*

$$r^{-1-s}\,E_A(P,s)=\sum_{M\in\Gamma'_\zeta\backslash\Gamma}\left(\frac{r(AMP)}{r}\right)^{1+s} \tag{2.14}$$

converges uniformly for $(P,s)\in B^{-1}\{\ z+r'j\in\mathbb{H}\ \mid\ r'\geq\alpha\ \}\times\{\ s\ \mid\ \operatorname{Re}s\geq\beta\ \}$. Let

$$\delta_{\eta,\zeta}:=\begin{cases}1 & \text{if}\quad \eta\equiv\zeta \bmod \Gamma,\\ 0 & \text{if}\quad \eta\not\equiv\zeta \bmod \Gamma.\end{cases} \tag{2.15}$$

If $\eta,\ \zeta$ are Γ-equivalent choose $L_0\in\Gamma$ with $\zeta=L_0\eta$ and let

$$AL_0B^{-1}=\begin{pmatrix}\cdot & \cdot\\ 0 & d_0\end{pmatrix}. \tag{2.16}$$

Then

$$E_A(B^{-1}(z+rj),s)=(\delta_{\eta,\zeta}\ [\Gamma_\zeta:\Gamma'_\zeta]\ |d_0|^{-2s-2}+o(1))\ r^{1+s} \tag{2.17}$$

as $r\to\infty$, uniformly in $\{\ z+rj\in\mathbb{H}\ \mid\ r\geq\alpha\ \},\ s\in\mathbb{C},\ \operatorname{Re}s\geq\beta$.

Remember that a function $f:\mathbb{R}\to\mathbb{C}$ is called $o(1)$ as $r\to\infty$ iff $f(r)\to 0$ as r goes to infinity.

Proof. Note that we have

$$r^{-1-s}\, E_A(P,s) = \sum_{\substack{M \in \Gamma'_\zeta \backslash \Gamma \\ AM = \left(\begin{smallmatrix} * & * \\ c & d \end{smallmatrix}\right)}} \|cP+d\|^{-2-2s}$$

where $\|cP+d\|^2 = |cz+d|^2 + |c|^2 r^2$, $(P = z + rj)$. Let $P \in B^{-1}\{\, z + r'j \in \mathbb{H} \mid r' \geq \alpha \,\}$ and $Q \in \{\, z + r'j \in \mathbb{H} \mid r' \geq \alpha \,\}$ so that $P = B^{-1}Q$. We then have

$$E_A(P,s) = E_A(B^{-1}Q, s) = E_{AB^{-1}}(Q,s) \tag{2.18}$$

(see (2.8)), where $E_{AB^{-1}}$ is the Eisenstein series for $B\Gamma B^{-1}$ at its cusp $B\zeta$. This reduces the proof of the first part of our proposition to the special case $\eta = \infty$, $B = I$ which we are going to treat now.

Let $K \subset \mathbb{C}$ be a compact set, put $P = z + rj$, $z \in K$, $r \geq \alpha$ and let $P_0 = z_0 + r_0 j$ be fixed. For $(c,d) \in \mathbb{C}^2$ with $(c,d) \neq (0,0)$ put

$$(\gamma, \delta) := (|c|^2 + |d|^2)^{-\frac{1}{2}}\, (c,d).$$

Then a compactness argument shows that

$$\frac{\|cP_0+d\|^2}{\|cP+d\|^2} = \frac{\|\gamma P_0+\delta\|^2}{\|\gamma P+\delta\|^2} \leq \frac{|\gamma z_0+\delta|^2 + |\gamma|^2\, r_0^2}{|\gamma z+\delta|^2 + |\gamma|^2\, \alpha^2} \tag{2.19}$$

is bounded by a bound independent of (c,d), $z \in K$, $r \geq \alpha$. Hence the series (2.14) converges uniformly for $(P,s) \in \{\, z + rj \in \mathbb{H} \mid z \in K,\ r \geq \alpha \,\} \times \{\, s \mid \operatorname{Re} s \geq \beta \,\}$. But since $\eta = \infty$ is a cusp, the invariance of (2.14) with respect to Γ'_∞ readily implies the uniformity of convergence with K replaced by $\mathbb{C}$.

For the proof of the second part of the proposition we consider the case of arbitrary B, η. The first part implies that

$$\sum_{\substack{M \in \Gamma'_\zeta \backslash \Gamma \\ AMB^{-1} = \left(\begin{smallmatrix} * & * \\ c & d \end{smallmatrix}\right), c \neq 0}} \|c(z+rj)+d\|^{-2-2s} \to 0 \tag{2.20}$$

as $r \to \infty$, uniformly with respect to $z \in \mathbb{C}$, $s \in \mathbb{C}$, $\operatorname{Re} s \geq \beta$. So we have to determine all elements $M \in \Gamma'_\zeta \backslash \Gamma$ such that

$$AMB^{-1} = \begin{pmatrix} \cdot & \cdot \\ 0 & d \end{pmatrix}, \qquad M \in \Gamma'_\zeta \backslash \Gamma. \tag{2.21}$$

Note that AMB^{-1} is of the form (2.21) iff AMB^{-1} fixes ∞, that happens iff $B^{-1}\infty = \eta$ and $A^{-1}\infty = \zeta$ are Γ-equivalent. This proves (2.17) for Γ-inequivalent cusps η, ζ. Assume now that $\eta \equiv \zeta \bmod \Gamma$, choose $L_0 \in \Gamma$ such that $L_0\eta = \zeta$, and put

$$AL_0B^{-1} = \begin{pmatrix} \cdot & \cdot \\ 0 & d_0 \end{pmatrix}.$$

Then for all other $M \in \Gamma$ such that AMB^{-1} has the form (2.21), we have $M \in \Gamma_\zeta L_0$. Hence there are exactly $[\Gamma_\zeta : \Gamma'_\zeta]$ different choices for M in $\Gamma'_\zeta \backslash \Gamma$. For all these M there exists an element

$$S = A^{-1} \begin{pmatrix} \epsilon^{-1} & \gamma \\ 0 & \epsilon \end{pmatrix} A \in \Gamma_\zeta$$

such that $M = SL_0$, where ϵ is a root of unity. This implies

$$\begin{pmatrix} \cdot & \cdot \\ 0 & d \end{pmatrix} = AMB^{-1} = \begin{pmatrix} \epsilon^{-1} & \gamma \\ 0 & \epsilon \end{pmatrix} AL_0B^{-1} = \begin{pmatrix} \cdot & \cdot \\ 0 & \epsilon d_0 \end{pmatrix},$$

and hence $|d| = |d_0|$. Hence summands corresponding to elements of the form (2.21) yield the same contribution, and we obtain (2.17) in full generality. □

Corollary 2.4. *Let Γ be a discrete subgroup of* $\mathbf{PSL}(2, \mathbb{C})$ *with cusps. Then all the Eisenstein series for Γ are of polynomial growth at all cusps of Γ. More precisely, if $\zeta = A^{-1}\infty$, $\eta = B^{-1}\infty$ are cusps of Γ then there is a constant $K \geq 0$ with*

$$E_A(B^{-1}(z + rj), s) = O(r^K) \qquad \text{as } r \to \infty$$

uniformly with respect to $z \in \mathbb{C}$. The number of linearly independent Eisenstein series for Γ and the number of Γ-inequivalent cusps of Γ coincide.

The proof is obvious from Proposition 2.3.

Proposition 2.5. *Let Γ be a discrete subgroup of* $\mathbf{PSL}(2, \mathbb{C})$ *and let $\zeta = A^{-1}\infty$ be a cusp of Γ. Then the Eisenstein series $E_A(P, s)$ is a real analytic function of P and a holomorphic function of s in* $\operatorname{Re} s > \sigma_0$. *In addition it satisfies the differential equation*

$$(-\Delta - (1 - s^2))\, E_A(P, s) = 0. \tag{2.22}$$

Proof. Differentiating

$$(|cz + d|^2 + |c|^2 r^2)^{-1-s} \quad (z \in \mathbb{C}, r > 0) \tag{2.23}$$

any number of times with respect to x, y, r, where $z = x + yi$, yields an expression which is locally majorized by the modulus of the term (2.23). This proves the locally uniform convergence of the differentiated Eisenstein series.

Now r^{1+s} satisfies the differential equation $-\Delta r^{1+s} = (1 - s^2)\, r^{1+s}$ and since Δ commutes with the action of $\mathbf{PSL}(2, \mathbb{C})$ on functions, we obtain (2.22). So the Eisenstein series satisfies an elliptic partial differential equation with real analytic coefficients and hence is real analytic. □

3.3 Fourier Expansion in Cusps and the Maaß–Selberg Relations

In this section we introduce two important tools for the study of automorphic functions corresponding to discrete groups which are cofinite but not cocompact. We start off by discussing the Fourier expansion of an automorphic function at a cusp. The main theorem reads as follows.

Theorem 3.1. *Suppose Λ is a lattice in $\mathbb{C}$. Let $f : \mathbb{H} \to \mathbb{C}$ be a Λ-invariant C^2-function, that is $f(P+\gamma) = f(P)$ for all $\gamma \in \Lambda$, satisfying the differential equation $-\Delta f = \lambda f$. Choose $s \in \mathbb{C}$ with $\lambda = 1 - s^2$. Assume further that $f(z + rj)$ is of polynomial growth as $r \to \infty$, that is*

$$f(z + rj) = O(r^K) \qquad \text{as} \quad r \to \infty \tag{3.1}$$

for some constant K uniformly with respect to $z \in \mathbb{C}$. Then, in case $s \neq 0$, f possesses a Fourier expansion

$$f(z + rj) = a_0\ r^{1+s} + b_0\ r^{1-s} + \sum_{0 \neq \mu \in \Lambda^\circ} a_\mu\ r\ K_s(2\pi|\mu|r)\ e^{2\pi i <\mu, z>} \tag{3.2}$$

whereas in case $s = 0$

$$f(z + rj) = a_0\ r + b_0\ r \log r + \sum_{0 \neq \mu \in \Lambda^\circ} a_\mu\ r\ K_0(2\pi|\mu|r)\ e^{2\pi i <\mu, z>}. \tag{3.3}$$

Here $< \cdot, \cdot >$ is the usual scalar product on $\mathbb{R}^2 = \mathbb{C}$ and Λ° denotes the dual lattice of Λ :

$$\Lambda^\circ = \{\ \mu \ \mid\ <\mu, \gamma> \in \mathbb{Z} \quad \text{for all} \quad \gamma \in \Lambda\ \}.$$

K_s is the usual Bessel function defined in Magnus, Oberhettinger, Soni (1966), page 66.

Proof. Since the function $z \mapsto f(z + rj)$ is real analytic and Λ-invariant, it has a Fourier expansion

$$f(z + rj) = \sum_{\mu \in \Lambda^\circ} g_\mu(r)\ e^{2\pi i <\mu, z>} \tag{3.4}$$

which may be differentiated termwise any number of times. Hence our formula for Δ in the coordinates z, r yields the ordinary differential equation

$$\left(r^2 \frac{d^2}{dr^2} - r \frac{d}{dr} + \lambda - 4\ \pi^2\ |\mu|^2\ r^2 \right) \cdot g_\mu(r) = 0. \tag{3.5}$$

For $\mu = 0$, $s \neq 0$ the functions r^{1+s}, r^{1-s} form a fundamental system of solutions of (3.5). For $\mu = 0$, $s = 0$, we take the functions r, $r \log r$ as fundamental system. This gives the first term on the right-hand side of (3.2), (3.3).

We now discuss (3.5) for $\mu \neq 0$. If $Z_\nu(u)$ is an arbitrary solution of Bessel's differential equation of order ν, the function

$$w = u^\alpha \; Z_\nu(\beta u) \tag{3.6}$$

satisfies the differential equation

$$u^2 \, \frac{d^2 w}{du^2} + (1 - 2\alpha) \; u \, \frac{dw}{du} + ((\beta u)^2 + \alpha^2 - \nu^2) \; w = 0. \tag{3.7}$$

This fact can be found in Magnus, Oberhettinger, Soni (1966), page 77. Comparing (3.5), (3.7) we see that we may choose $\alpha = 1$, $\beta = 2\pi i|\mu|$, $\nu = s$. Hence we arrive at the solution $g_\mu(r) = r \; Z_s(2\pi i|\mu| r)$ of (3.5). The function $Z_s(2\pi i|\mu| r)$ can be written as a linear combination of the fundamental system $K_s(2\pi|\mu|r)$, $I_s(2\pi|\mu|r)$. So the general solution of (3.5) is

$$g_\mu(r) = a_\mu \; r \; K_s(2\pi|\mu|r) + b_\mu \; r \; I_s(2\pi|\mu|r), \tag{3.8}$$

where a_μ, b_μ are constants.

We conclude from (3.4) that

$$g_\mu(r) = \frac{1}{|\mathcal{P}|} \int_{\mathcal{P}} f(z + rj) \; e^{-2\pi i <\mu, z>} \; dx \; dy, \tag{3.9}$$

where $|\mathcal{P}|$ denotes the Euclidean area of a fundamental parallelogram $\mathcal{P}$ of Λ. By assumption (3.1), f is of polynomial growth. Hence $g_\mu(r)$ is of polynomial growth. The function $K_s(x)$ decreases exponentially and $I_s(x)$ increases exponentially as $x \to +\infty$. Hence $b_\mu = 0$ in (3.8), which finishes the proof. $\square$

Theorem 3.2. *Let Λ, f be as in Theorem 3.1, except that polynomial growth (3.1) is no longer assumed. Instead, assume that f is square integrable over a cusp sector at infinity, that is*

$$\int_{\mathcal{P} \times [1, \infty[} |f(Q)|^2 \; dv(Q) < \infty, \tag{3.10}$$

where $\mathcal{P}$ is a fundamental parallelogram for Λ. Then $f(z+rj)$ is of polynomial growth as $r \to \infty$. Hence f has a Fourier expansion of the form (3.2), (3.3). Moreover

(1) $a_0 = b_0 = 0$ if $\lambda \geq 1$, i.e. if $s = it$ for some $t \in \mathbb{R}$,
(2) $b_0 = 0$ if $\operatorname{Re} s < 0$,
(3) $a_0 = 0$ if $\operatorname{Re} s > 0$.

Proof. We use the same notation as in the preceeding proof and obtain from Parseval's theorem

$$\int_{\mathcal{P}\times[1,\infty[} |f(Q)|^2 \, dv(Q) = |\mathcal{P}| \sum_{\mu\in\Lambda^\circ} \int_1^\infty |g_\mu(r)|^2 \, \frac{dr}{r^3}.$$

Since the left-hand side is finite, we conclude that

$$\int_1^\infty |g_\mu(r)|^2 \, \frac{dr}{r^3} < \infty \qquad \text{for all} \;\; \mu \in \Lambda^\circ. \tag{3.11}$$

This implies that for $\mu \neq 0$ the coefficient b_μ in (3.8) must be equal to zero. Hence f has a Fourier expansion of the form (3.2), (3.3). Moreover, (3.11) with $\mu = 0$ readily implies that (1)–(3) are true. Since, given $\mu \neq 0$, the Bessel function $K_s(2\pi|\mu|r)$ decreases exponentially as $r \to \infty$, the series

$$\sum_{0\neq\mu\in\Lambda^\circ} a_\mu \; r \; K_s(2\pi|\mu|r) \; e^{2\pi i<\mu,z>}$$

decreases exponentially as $r \to \infty$ and uniformly with respect to $z \in \mathbb{C}$. This shows that $f(z + rj)$ is of polynomial growth as $r \to \infty$. So our theorem is proved. □

The proof of the above theorem also gives the following useful corollary.

Corollary 3.3. *Let Λ be a lattice in $\mathbb{C}$ and let $f : \mathbb{H} \to \mathbb{C}$ be a Λ-invariant solution of the differential equation $-\Delta f = \lambda f$ with $\lambda \geq 1$. Assume that f is square integrable over a cusp sector at infinity. Then $f(z + rj)$ is of exponential decrease as $r \to \infty$, that is*

$$f(z + rj) = O(e^{-\epsilon r}) \quad as \qquad r \to \infty$$

for some $\epsilon > 0$ uniformly with respect to $z \in \mathbb{C}$.

Corollary 3.4. *Let $\Gamma <$ **PSL**$(2, \mathbb{C})$ be a discrete group having a fundamental domain $\mathcal{F}$ of finite hyperbolic volume. Assume further that Γ has exactly one class of cusps. Then any Γ-invariant solution f of the differential equation $-\Delta f = 0$ satisfying $\int_{\mathcal{F}} |f|^2 \, dv < \infty$ is constant.*

Proof. We may assume that Γ possesses exactly one cusp which is at $\infty \in \mathbb{P}^1\mathbb{C}$. By Theorem 2.3.9 we choose a fundamental domain $\mathcal{F}$ for Γ so that $\mathcal{F}$ consists of a compact set together with a cusp sector at ∞. The stabilizer of ∞ in Γ contains a maximal lattice and f satisfies (3.10) with the fundamental parallelogram of this lattice. We may choose $s = -1$, and then we derive from Theorem 3.2 and its proof that f has an expansion

$$f(z + rj) = a_0 + \sum_{0\neq\mu\in\Lambda^\circ} a_\mu \; r \; K_1(2\pi|\mu|r) \; e^{2\pi i<\mu,z>},$$

hence

$$g(z+rj) := \sum_{0 \neq \mu \in \Lambda^\circ} a_\mu \; r \; K_1(2\pi|\mu|r) \; e^{2\pi i <\mu, z>}$$

is of exponential decrease as $r \to \infty$.

The function $|g|$ restricted to $\mathcal{F}$ must attain its maximum at some point $P_0 \in \mathcal{F}$. Moreover, g is Γ-invariant. Therefore, $|g|$ attains its maximum on $\mathbb{H}$ in P_0. But g is a solution of $-\Delta g = 0$. Hence by the mean-value formula of Theorem 5.4 $|g|$ reduces to a constant, and, in fact, is equal to zero, since it is of exponential decrease, as $r \to \infty$. Summing up, we obtain $f = a_0$ which yields our assertion. □

Let us point out that Corollary 3.4 is actually true for all groups Γ of finite covolume. This will be proved in Theorem 4.1.8. For cocompact groups the result is included in Theorem 5.4.

Definition 3.5. Let $\Gamma < \mathbf{PSL}(2, \mathbb{C})$ be an arbitrary discrete group. An automorphic function for the group Γ with parameter $\lambda \in \mathbb{C}$ is a function $f : \mathbb{H} \to \mathbb{C}$ with the following properties:

(1) f is Γ-invariant,
(2) $f \in C^2(\mathbb{H})$ and f satisfies the differential equation $-\Delta f = \lambda f$,
(3) f is of polynomial growth at all cusps of Γ, that is, if $\zeta = A^{-1}\infty$ ($A \in \mathbf{PSL}(2, \mathbb{C})$) is a cusp of Γ, then there exists a constant $K > 0$ such that

$$f(A^{-1}(z+rj)) = O(r^K) \qquad \text{as } r \to \infty$$

uniformly with respect to $z \in \mathbb{C}$.

The vector space of all automorphic functions for the group Γ with parameter $\lambda \in \mathbb{C}$ is denoted by $\mathcal{A}(\Gamma, \lambda)$.

Obviously it suffices in (3) to require that f is of polynomial growth at all the cusps of a maximal system of Γ-inequivalent cusps of Γ. Suppose that $f \in \mathcal{A}(\Gamma, \lambda)$ and let $A^{-1}\infty$ be a cusp of Γ. Then $f \circ A^{-1}$ has a Fourier expansion of the form (3.2), (3.3). This Fourier expansion is called the Fourier expansion of f at $A^{-1}\infty$. It follows from Theorem 3.2 that every Γ-invariant function $f \in C^2(\mathbb{H})$ satisfying (2) such that $|f|^2$ is v-integrable over a fundamental domain of Γ belongs to $\mathcal{A}(\Gamma, \lambda)$. By way of example we mention that the Eisenstein series $E_A(P, s)$ are automorphic functions for the group Γ with parameter $\lambda = 1 - s^2$ provided that $\operatorname{Re} s > 1$. These are however not square-integrable.

We shall prove in the next chapter that $\mathcal{A}(\Gamma, \lambda)$ is a finite-dimensional vector space if Γ is a discrete group of finite covolume.

We are now going to discuss the second tool, that is the Maaß–Selberg relations. Let Γ be a discrete subgroup of $\mathbf{PSL}(2, \mathbb{C})$ of finite covolume and suppose that $B_1, \cdots, B_h \in \mathbf{PSL}(2, \mathbb{C})$ are chosen so that

$$\eta_1 = B_1^{-1}\infty, \cdots, \eta_h = B_h^{-1}\infty \in \mathbb{P}^1\mathbb{C}$$

are representatives for the Γ-classes of the cusps of Γ. Let Λ_i be the lattice in $\mathbb{C}$ corresponding to η_i. Let $\mathcal{P}_i$ be a closed fundamental domain for the action of $B_i \Gamma_{\eta_i} B_i^{-1}$ on $\mathbb{P}^1\mathbb{C} \setminus \{\infty\} = \mathbb{C}$. For $Y > 0$ put

$$\mathcal{F}_i(Y) := B_i^{-1}\{ \, z + rj \;\;|\;\; z \in \mathcal{P}_i, \; r \geq Y \, \}.$$

Choose the $Y = (Y_1, \cdots, Y_h)$ large enough so that the requirements of Theorem 2.3.9 are satisfied. Then there is a compact set $\mathcal{F}_0 \subset \mathbb{H}$ so that

$$\mathcal{F} := \mathcal{F}_0 \cup \mathcal{F}_1(Y_1) \cup \cdots \cup \mathcal{F}_h(Y_h)$$

is a fundamental domain for Γ, see Theorem 2.3.9.

Suppose now that $f : \mathbb{H} \to \mathbb{C}$ is a Γ-invariant solution of the differential equation $-\Delta f = \lambda f$ with $\lambda \in \mathbb{C}$, $\lambda \neq 1$. Suppose further that f is of polynomial growth at all cusps of Γ. That is to say

$$f(B_i^{-1}(z + rj)) = O(r^K)$$

for $i = 1, \cdots h$ and some constant $K \geq 0$. Then by Theorem 3.1 f has expansions

$$f(B_i^{-1}P) = a_i \; r^{1+s} + b_i \; r^{1-s} + \ldots .$$

where $P = z + rj$ and $s \in \mathbb{C}$ satisfies $\lambda = 1 - s^2$. Define now functions $u_i : \mathbb{H} \to \mathbb{C}$ by

$$u_i(B_i^{-1}P) = a_i \; r^{1+s} + b_i \; r^{1-s}.$$

We now modify f in the cusp sectors $\mathcal{F}_i(Y_i)$ and define

$$f^Y(P) := \begin{cases} f(P) - u_i(P) & \text{if } P \in \mathcal{F}_i(Y_i) \text{ and } P \notin \mathcal{F}_0, \\ f(P) & \text{if } P \in \mathcal{F}_0. \end{cases} \tag{3.12}$$

By our choice of Y the cusp sectors $\mathcal{F}_i(Y_i)$ are disjoint. Since the intersections $\mathcal{F}_0 \cap \mathcal{F}_i(Y_i)$ have Lebesgue measure 0, our definition gives f^Y up to a set of measure 0. This is enough for the following theorem. Note that f^Y decreases exponentially when we approach the cusps in $\mathcal{F}$.

Theorem 3.6. *Let Γ be a cofinite subgroup of* $\mathbf{PSL}(2, \mathbb{C})$ *and suppose that notation has been chosen as just explained. Let $f, g : \mathbb{H} \to \mathbb{C}$ be Γ-invariant solutions of the differential equations $-\Delta f = \lambda f$, $-\Delta g = \mu g$ with $\lambda, \mu \in \mathbb{C}$, $\lambda \neq 1 \neq \mu$. Choose $s, t \in \mathbb{C}$ with $\lambda = 1 - s^2$, $\mu = 1 - t^2$. Assume further that f, g are of polynomial growth at all cusps of Γ. Let*

$$f(B_i^{-1}P) = a_i \; r^{1+s} + b_i \; r^{1-s} + \ldots, \qquad g(B_i^{-1}P) = c_i \; r^{1+t} + d_i \; r^{1-t} + \ldots$$

be the zeroth terms of the Fourier expansions of f and g at the cusps $\eta_1, \cdots, \eta_h$. Define f^Y, g^Y as above. Then

$$(3.13)\qquad \begin{aligned}(s^2-\bar{t}^2)\int_{\mathcal{F}} f^Y \overline{g^Y}\, dv = \sum_{i=1}^{h} \frac{|\Lambda_i|}{[\Gamma_{\eta_i} : \Gamma'_{\eta_i}]} &\big((s-\bar{t})\, a_i\bar{c}_i\, Y_i^{s+\bar{t}} \\ &+ (s+\bar{t})\, a_i\bar{d}_i\, Y_i^{s-\bar{t}} - (s+\bar{t})\, b_i\bar{c}_i\, Y_i^{-s+\bar{t}} \\ &- (s-\bar{t})\, b_i\bar{d}_i\, Y_i^{-s-\bar{t}}\big).\end{aligned}$$

Theorem 3.6 is proved by a simple application of Stokes' Theorem. We shall not give the details. In Roelcke (1966), (1967), Chapter 9 the 2-dimensional version is worked out. The equations of type (3.13) are called the Maaß–Selberg relations. Zagier (1981b) proves the Maaß–Selberg relations for Eisenstein series by means of the Rankin–Selberg method. His proof immediately extends to the 3-dimensional case. They will be used in Chapter 6 and also in certain covolume computations in Chapter 8. They can also be used as it is done in the appendix of Langlands (1976) to get the analytic continuation of the Eisenstein series. The Maaß–Selberg relations are also a useful tool for the computation of spectral asymptotics.

3.4 Fourier Expansion of Eisenstein Series

Combining now our results on Eisenstein series with our previous study of Fourier expansions in cusps we see that all Eisenstein series possess Fourier expansions of the form (3.2). We proceed to determine these Fourier expansions explicitly. We use notations from the two preceding sections.

If $\Gamma < \mathbf{PSL}(2,\mathbb{C})$ is a discrete group and $\zeta = A^{-1}\infty \in \mathbb{P}^1\mathbb{C}$ is one of its cusps, the Eisenstein series of Γ at ζ was, in Section 2, defined as

$$E_A(P,s) := \sum_{M\in\Gamma'_\zeta\backslash\Gamma} r(AMP)^{1+s}.$$

This series converges for $\operatorname{Re} s > \sigma_0$, σ_0 being the abscissa of convergence of Γ, defined in Section 1. If $\eta = B^{-1}\infty \in \mathbb{P}^1\mathbb{C}$ is another cusp of Γ, the function $P \mapsto E_A(B^{-1}P,s)$ is invariant under the action of the lattice Λ corresponding to $(B\Gamma B^{-1})'_\infty = B\Gamma'_\eta B^{-1}$, that is

$$(4.1)\qquad B\Gamma'_\eta B^{-1} = \left\{ \begin{pmatrix} 1 & \omega \\ 0 & 1 \end{pmatrix} \;\middle|\; \omega \in \Lambda \right\}.$$

As before, we write $< \cdot,\cdot >$ for the Euclidean inner product on $\mathbb{R}^2 = \mathbb{C}$ and Λ° for the lattice dual to Λ with respect to this inner product. Writing $P = z + rj \in \mathbb{H}$, the growth conditions and the differential equation established for $E_A(B^{-1}P,s)$ in Section 2 will ensure the existence of an expansion

$$E_A(B^{-1}P,s) = a_0\, r^{1+s} + b_0\, r^{1-s} + \sum_{0\neq\mu\in\Lambda^\circ} a_\mu\, r\, K_s(2\pi|\mu|r)\, e^{2\pi i<\mu,z>}$$

of type (3.2). We shall give here an explicit formula for the coefficients a_μ. In the formulation we shall use the Kronecker symbol

$$\delta_{\eta,\zeta} := \begin{cases} 1 & \text{if} \quad \eta \equiv \zeta \bmod \Gamma, \\ 0 & \text{if} \quad \eta \not\equiv \zeta \bmod \Gamma, \end{cases}$$

for the cusps η, ζ of Γ.

Theorem 4.1. *Let $\Gamma < \mathbf{PSL}(2,\mathbb{C})$ be a non-elementary discrete group and for $A,\ B \in \mathbf{PSL}(2,\mathbb{C})$ let $\zeta = A^{-1}\infty$, $\eta = B^{-1}\infty$ be two cusps of Γ. Let $\Lambda \subset \mathbb{C}$ be the lattice corresponding to $(B\Gamma B^{-1})'_\infty = B\Gamma'_\eta B^{-1}$. Then for $\operatorname{Re} s > \sigma_0$ the Λ-invariant function $E_A(B^{-1}P,s)$ has the Fourier expansion*

$$\begin{aligned} E_A(B^{-1}P,s) &= \delta_{\eta,\zeta}[\Gamma_\zeta : \Gamma'_\zeta]|d_0|^{-2-2s}\, r^{1+s} \\ &\quad + \frac{\pi}{|\mathcal{P}|s}\left(\sum_{\left(\begin{smallmatrix} * & * \\ c & d \end{smallmatrix}\right)\in\mathcal{R}} |c|^{-2-2s}\right) r^{1-s} \\ &\quad + \frac{2\pi^{1+s}}{|\mathcal{P}|\Gamma(1+s)} \sum_{0\neq\mu\in\Lambda^\circ} |\mu|^s \\ &\quad \cdot \left(\sum_{\left(\begin{smallmatrix} * & * \\ c & d \end{smallmatrix}\right)\in\mathcal{R}} \frac{e^{2\pi i<\mu,\frac{d}{c}>}}{|c|^{2+2s}}\right) r\, K_s(2\pi|\mu|r)\, e^{2\pi i<\mu,z>}, \end{aligned} \tag{4.2}$$

where the following notation is used. $\mathcal{R}$ denotes a system of representatives $\left(\begin{smallmatrix} * & * \\ c & d \end{smallmatrix}\right)$ of the double cosets in

$$A\Gamma'_\zeta A^{-1}\backslash A\Gamma B^{-1}/B\Gamma'_\eta B^{-1}, \tag{4.3}$$

such that $c \neq 0$. $\mathcal{P}$ is a period parallelogram for Λ with Euclidean area $|\mathcal{P}|$. If η and ζ are Γ-equivalent, let $L_0 \in \Gamma$ be such that $L_0\eta = \zeta$ and let d_0 be defined by

$$AL_0B^{-1} = \begin{pmatrix} \cdot & \cdot \\ 0 & d_0 \end{pmatrix}. \tag{4.4}$$

Proof. The function $P \mapsto E_A(B^{-1}P,s)$ is a Λ-invariant solution of (2.22) and of polynomial growth and hence has a Fourier expansion of the form (3.2). We need not even know the precise form of this Fourier expansion a priori, since we shall obtain it anew by direct computation. We only use that the function $P \mapsto E_A(B^{-1}P,s)$ $(\operatorname{Re} s > \sigma_0)$ has a Fourier expansion of the form

$$E_A(B^{-1}P,s) = \sum_{\mu\in\Lambda^\circ} a_\mu(r,s)\, e^{2\pi i<\mu,z>} \qquad (P = z + rj). \tag{4.5}$$

Let $\mathcal{P}$ be a period parallelogram for Λ with the Euclidean area $|\mathcal{P}|$. Then we have

$$(4.6) \qquad a_\mu(r,s) = \frac{1}{|\mathcal{P}|}\int_{\mathcal{P}} E_A(B^{-1}(z+rj),s)\ e^{-2\pi i<\mu,z>}\ dx\ dy$$

with $z = x+iy$. It is our task to compute these integrals for all $\mu \in \Lambda^\circ$. Since the Eisenstein series converges uniformly on compact sets, we have

$$(4.7) \quad a_\mu(r,s) = \frac{1}{|\mathcal{P}|}\sum_{M\in\Gamma'_\zeta\backslash\Gamma}\int_{\mathcal{P}} r(AMB^{-1}(z+rj))^{1+s}\ e^{-2\pi i<\mu,z>}\ dx\ dy.$$

First we want to reduce our system of elements AMB^{-1} modulo $B\Gamma'_\eta B^{-1}$ from the right. Consider

$$(4.8) \qquad AMB^{-1} = \begin{pmatrix} \cdot & \cdot \\ c & d \end{pmatrix}.$$

Then there exists an $M \in \Gamma$ in the same class such that $c = 0$ iff AMB^{-1} fixes ∞, that is iff $\eta = B^{-1}\infty$ and $\zeta = A^{-1}\infty$ are equivalent modulo Γ. If ζ and η are Γ-equivalent, let $L_0 \in \Gamma$ be so chosen that $L_0\eta = \zeta$ and put

$$AL_0B^{-1} = \begin{pmatrix} \cdot & \cdot \\ 0 & d_0 \end{pmatrix}.$$

Then for all $M \in \Gamma$ such that AMB^{-1} has the form (4.8) with $c = 0$, we have $|d| = |d_0|$, and there are exactly $[\Gamma_\zeta : \Gamma'_\zeta]$ different elements in $\Gamma'_\zeta\backslash\Gamma$ with this property. Now, if $\mu \neq 0$, the contribution of these terms to the right-hand side of (4.7) equals zero, whereas for $\mu = 0$ the contribution is equal to

$$(4.9) \qquad \delta_{\eta,\zeta}\ [\Gamma_\zeta : \Gamma'_\zeta]|d_0|^{-2-2s}\ r^{1+s}.$$

Hence we are left with the computation of the sum

$$(4.10) \qquad \frac{1}{|\mathcal{P}|}\sum_{\substack{AMB^{-1}=\left(\begin{smallmatrix} * & * \\ c & d\end{smallmatrix}\right) \\ c\neq 0}}\int_{\mathcal{P}}\left(\frac{r}{\|c(z+rj)+d\|^2}\right)^{1+s} e^{-2\pi i<\mu,z>}\ dx\ dy,$$

where the summation extends over $M \in \Gamma'_\zeta\backslash\Gamma$ such that $c \neq 0$, that is over all cosets in $A\Gamma'_\zeta A^{-1}\backslash A\Gamma B^{-1}$ with $c \neq 0$. For $\omega \in \Lambda$ we have

$$\begin{pmatrix} \cdot & \cdot \\ c & d \end{pmatrix}\begin{pmatrix} 1 & \omega \\ 0 & 1 \end{pmatrix} = \begin{pmatrix} \cdot & \cdot \\ c & c\omega + d \end{pmatrix},$$

and for $c \neq 0$ different elements $\omega \in \Lambda$ represent different cosets from $A\Gamma'_\zeta A^{-1}\backslash A\Gamma B^{-1}$. Hence if we let $\mathcal{R}$ denote a system of representatives $\begin{pmatrix} \cdot & \cdot \\ c & d \end{pmatrix}$ of the double cosets (4.3) such that $c \neq 0$, we obtain the following expression for the sum (4.10):

$$\frac{1}{|\mathcal{P}|} \sum_{\left(\begin{smallmatrix} * & * \\ c & d \end{smallmatrix}\right) \in \mathcal{R}} \sum_{\omega \in \Lambda} \int_{\mathcal{P}} \left(\frac{r}{\|c(z+\omega)+d\|^2 + |c|^2 r^2} \right)^{1+s} e^{-2\pi i <\mu, z>} dx\, dy$$

$$(4.11) \quad = \frac{1}{|\mathcal{P}|} \sum_{\left(\begin{smallmatrix} * & * \\ c & d \end{smallmatrix}\right) \in \mathcal{R}} \int_{\mathbb{C}} \left(\frac{r}{|cz+d|^2 + |c|^2\, r^2} \right)^{1+s} e^{-2\pi i <\mu, z>}\, dx\, dy$$

$$= \frac{1}{|\mathcal{P}|} \sum_{\left(\begin{smallmatrix} * & * \\ c & d \end{smallmatrix}\right) \in \mathcal{R}} \frac{e^{2\pi i <\mu, \frac{d}{c}>}}{|c|^{2+2s}} \int_{\mathbb{C}} \left(\frac{r}{|z|^2 + r^2} \right)^{1+s} e^{-2\pi i |\mu| x}\, dx\, dy,$$

where we have applied an orthogonal linear transformation of $\mathbb{R}^2$ sending μ to $(|\mu|, 0)$.

We distinguish between the cases $\mu = 0, \mu \neq 0$. In the case $\mu = 0$ we obtain

$$(4.12) \quad \begin{aligned} \int_{\mathbb{C}} \left(\frac{r}{|z|^2 + r^2} \right)^{1+s} dx\, dy &= r^{1-s} \int_{\mathbb{C}} (|z|^2 + 1)^{-1-s}\, dx\, dy \\ &= 2\pi r^{1-s} \int_0^\infty \frac{\rho\, d\rho}{(\rho^2+1)^{1+s}} = \frac{\pi}{s}\, r^{1-s}. \end{aligned}$$

Collecting terms from (4.6)–(4.12), we finally obtain for $\mu = 0$ the zeroth term in the Fourier expansion (4.2).

In the case $\mu \neq 0$, we perform the following computation:

$$(4.13) \quad \begin{aligned} &\int_{\mathbb{C}} \left(\frac{r}{|z|^2 + r^2} \right)^{1+s} e^{-2\pi i |\mu| x}\, dx\, dy \\ &= r^{1-s} \int_{-\infty}^{+\infty} \int_{-\infty}^{+\infty} \frac{dy}{(y^2 + (1+x^2))^{1+s}}\, e^{-2\pi i r |\mu| x}\, dx \\ &= r^{1-s} \int_{-\infty}^{+\infty} \frac{dt}{(1+t^2)^{1+s}} \int_{-\infty}^{+\infty} \frac{e^{-2\pi i r |\mu| x}}{(1+x^2)^{s+\frac{1}{2}}}\, dx \\ &= \frac{2\pi^{1+s} |\mu|^s}{\Gamma(1+s)}\, r\, K_s(2\pi |\mu| r), \end{aligned}$$

where K_s denotes the modified Bessel function defined in Magnus, Oberhettinger, Soni (1966), page 66. The definite integrals can be found there on pages 6, 85. Gathering up terms we arrive at (4.2). □

3.5 Expansion of Eigenfunctions and the Selberg Transform

The most important result of the present section is an explicit formula for the so-called Selberg transform. For motivation we give a brief explanation of this notion.

Selberg (1956) considers invariant differential operators and invariant integral operators. As mentioned in section 1, the ring of invariant differential operators equals $\mathbb{C}[\Delta]$ in our case. So it is sufficient to consider the operator Δ only. The invariant integral operators are of the form

$$f \mapsto \int_{\mathbb{H}} K(\cdot, Q)\, f(Q)\, dv(Q),$$

where K is a point-pair invariant. Now the fundamental relation between the invariant differential operators and the invariant integral operators is this: Consider a solution f of the equation

$$-\Delta f = \lambda f. \tag{5.1}$$

Then the function f is also an eigenfunction of the invariant integral operators, that is

$$\int_{\mathbb{H}} K(P, Q)\, f(Q)\, dv(Q) = h(\lambda)\, f(P), \tag{5.2}$$

where the eigenvalue $h(\lambda)$ depends only on K and on the eigenvalue λ in (5.1) and not on the particular function f in the eigenspace corresponding to the eigenvalue λ. In fact, $h(\lambda)$ may be computed from K by means of a simple integral transform: Write

$$K = k \circ \delta$$

with an appropriate $k : [1, \infty[\to \mathbb{C}$, see Proposition 1.1.9, and put $\lambda = 1 - s^2$. Then we shall prove

$$h(\lambda) = \frac{\pi}{s} \int_1^\infty k\left(\frac{1}{2}\left(t + \frac{1}{t}\right)\right) (t^s - t^{-s}) \left(t - \frac{1}{t}\right) \frac{dt}{t}. \tag{5.3}$$

Observe that the right-hand side of (5.3) has the same value for either sign of s. For $s = 0$ the limit $s \to 0$ has to be taken. All this holds under suitable conditions which guarantee the existence of the integrals in question. These conditions are given in Theorem 5.3. We do not assume the above facts as known but prefer to give precise proofs valid in our special space $\mathbb{H}$.

Our method of proof proceeds as follows. First, we shall derive an expansion of the solutions of (5.1) in elementary solutions which corresponds to the Fourier expansion of the solutions in the case of the hyperbolic plane, see

Elstrodt (1973a), Section 2. The formulas become especially simple if we use the unit ball model $\mathbb{B}$ of three-dimensional hyperbolic space.

In the present case, a separation of variables by means of spherical harmonics is appropriate. We remind the reader of the following facts concerning spherical harmonics on the unit sphere $S^2 \subset \mathbb{R}^3$, which may be found in the literature, see for example Rellich (1952), Teil I, Kapitel II, or Erdélyi, Magnus, Oberhettinger, Tricomi (1953), Chapter 11.

Choosing coordinates as in Chapter 1, Section 2, the Euclidean Laplace operator Δ_e on $\mathbb{B}$ may be written in the form

$$(5.4) \qquad \Delta_e = \frac{\partial^2}{\partial \xi^2} + \frac{\partial^2}{\partial \eta^2} + \frac{\partial^2}{\partial \zeta^2} = \frac{1}{\rho^2}\frac{\partial}{\partial \rho}\rho^2\frac{\partial}{\partial \rho} + \frac{1}{\rho^2}\Theta,$$

where Θ is defined by

$$(5.5) \qquad \Theta := \frac{1}{2}\sum_{j,k=1}^{3}\left(\xi_j \frac{\partial}{\partial \xi_k} - \xi_k \frac{\partial}{\partial \xi_j}\right)^2 .$$

The coordinates are denoted by

$$\begin{pmatrix} \xi_1 \\ \xi_2 \\ \xi_3 \end{pmatrix} := \begin{pmatrix} \xi \\ \eta \\ \zeta \end{pmatrix} .$$

The operator Θ is the Laplace-Beltrami operator on S^2. It is known that Θ has a complete orthonormal system of eigenfunctions. The eigenvalues of $-\Theta$ are

$$(5.6) \qquad \mu_\ell := \ell(\ell+1) \qquad (\ell = 0, 1, \dots)$$

and the multiplicity of μ_ℓ is equal to $2\ell + 1$. So there exists a complete orthonormal system

$$(5.7) \qquad Y_{\ell j} \qquad (j = 1, 2, \dots, 2\ell + 1)$$

of eigenfunctions of $-\Theta$ corresponding to the eigenvalue μ_ℓ. These functions $Y_{\ell j}$ may be constructed in the following manner. A harmonic form of degree ℓ is a homogeneous polynomial u of degree ℓ in three variables ξ, η, ζ, such that $\Delta_e \, u = 0$. For any harmonic form u of degree ℓ the function

$$Y := \left(\xi^2 + \eta^2 + \zeta^2\right)^{-\frac{\ell}{2}} \; u$$

is a spherical harmonic of degree ℓ, that is an eigenfunction of $-\Theta$ with eigenvalue μ_ℓ. All eigenfunctions of $-\Theta$ with eigenvalue μ_ℓ are obtained in this way, and there are exactly $2\ell + 1$ linearly independent eigenfunctions which we denote by $Y_{\ell j}$.

Theorem 5.1. *Let* $g : \mathbb{B} \to \mathbb{C}$ *be a solution of the differential equation* $-\Delta\, g = \lambda\, g$ *and put* $\lambda = 1 - s^2$. *Then* g *has an expansion of the form*

$$(5.8)\qquad \begin{aligned} &g(x) \\ &= \sum_{\ell=0}^{\infty}\sum_{j=1}^{2\ell+1} \alpha_{\ell j}\, \rho^{\ell}\, (1-\rho^2)^{1+s}\, F\left(\ell+s+1, s+\frac{1}{2}; \ell+\frac{3}{2}; \rho^2\right) Y_{\ell j}\left(\frac{x}{\rho}\right) \end{aligned}$$

where $x \in \mathbb{B}$, $\rho = \|x\|$. *The* $\alpha_{\ell j}$ *are constants and* $F(a,b;c;z)$ *is the hypergeometric function.*

For the definition and properties of the hypergeometric function see Magnus, Oberhettinger, Soni (1966).

Proof. We start with some function $g : \mathbb{B}\backslash\{0\} \to \mathbb{C}$ which satisfies $-\Delta g = \lambda g$ in $\mathbb{B}\backslash\{0\}$. We may expand g in a series with respect to the spherical harmonics $Y_{\ell j}$

$$(5.9)\qquad g(x) = \sum_{\ell=0}^{\infty}\sum_{j=1}^{2\ell+1} g_{\ell j}(\rho)\, Y_{\ell j}\left(\frac{x}{\rho}\right).$$

Observe that g is a real analytic function since g is a solution of an elliptic differential equation with real analytic coefficients. Hence it is easy to show by partial integration, using the symmetry of Θ on S^2, that the expansion (5.9) also converges pointwise absolutely and may be differentiated any number of times under the summation sign.

The differential equation $-\Delta g = \lambda g$ implies an ordinary differential equation for the functions $g_{\ell j}$ occurring in (5.9). The representation (5.4) of Δ_e together with the coordinate expression for the hyperbolic Laplace operator, see Chapter 1, Section 2, and

$$-\Theta\, Y_{\ell j} = \ell(\ell+1)\, Y_{\ell j}$$

yields for $\varphi = g_{\ell j}$:

$$(5.10)\qquad -\frac{1}{4}(1-\rho^2)^2\left(\frac{1}{\rho^2}\frac{d}{d\rho}\rho^2\frac{d}{d\rho} + \frac{2\rho}{1-\rho^2}\frac{d}{d\rho} - \frac{\ell(\ell+1)}{\rho^2}\right)\varphi = \lambda\varphi.$$

Here we substitute $u = \rho^2$, $\varphi(\rho) = \psi(u)$ and obtain after some computations

$$(5.11)\qquad \begin{aligned} &\frac{d^2\psi}{du^2} + \left[\frac{3}{2\, u} - \frac{1}{u-1}\right]\frac{d\psi}{du} \\ &+ \left[\frac{\ell(\ell+1)}{4\, u} + \frac{\lambda}{u-1} - \frac{\ell(\ell+1)}{4}\right]\frac{\psi}{u(u-1)} = 0. \end{aligned}$$

This is a special case of the so-called generalized hypergeometric differential equation, see Magnus, Oberhettinger, Soni (1966), page 57

$$\begin{aligned}(5.12)\qquad &\frac{d^2w}{dz^2} + \left[\frac{1-\alpha-\alpha'}{z} + \frac{1-\gamma-\gamma'}{z-1}\right]\frac{dw}{dz} \\ &+ \left[\frac{-\alpha\alpha'}{z} + \frac{\gamma\gamma'}{z-1} + \beta\beta'\right]\frac{w}{z(z-1)} = 0,\end{aligned}$$

where the parameters $\alpha, \alpha', \ldots, \gamma, \gamma'$ are subject to the condition

$$\alpha + \alpha' + \beta + \beta' + \gamma + \gamma' = 1.$$

In (5.11) the parameters have the special values

$$(5.13)\qquad \alpha = \beta = \frac{\ell}{2}, \qquad \alpha' = \beta' = -\frac{\ell+1}{2}, \qquad \gamma = 1+s, \qquad \gamma' = 1-s.$$

The solution of the generalized hypergeometric differential equation is represented by Riemann's P-function

$$(5.14)\qquad w = P\left\{\begin{matrix} 0 & \infty & 1 & \\ \alpha & \beta & \gamma & z \\ \alpha' & \beta' & \gamma' & \end{matrix}\right\}$$

see Magnus, Oberhettinger, Soni (1966), page 57. Taking the solution W_1 of equation (5.12) from Magnus, Oberhettinger, Soni (1966), page 60 and observing (5.13), we obtain the solution

$$\begin{aligned}(5.15)\qquad \varphi(\rho) &= \rho^{2\alpha}\,(1-\rho^2)^\gamma\, F(\alpha+\beta+\gamma, \alpha+\beta'+\gamma; 1+\alpha-\alpha'; \rho^2) \\ &= \rho^\ell\,(1-\rho^2)^{1+s}\, F\left(\ell+s+1, s+\frac{1}{2}; \ell+\frac{3}{2}; \rho^2\right)\end{aligned}$$

of equation (5.10). Some explanation is in order in connection with the formulas in Magnus, Oberhettinger, Soni (1966), page 60. To obtain the solutions of (5.12) one has to replace all factors of the hypergeometric function which are of the form of a power of $z-b \quad (b=\infty)$ by 1. Furthermore in the argument of F the limit as $b \to \infty$ has to be taken.

Now we are able to finish the proof of our theorem. The equation (5.12) has exponents α, α' corresponding to the singularity at $a = 0$. In our case, α' is negative. This means that there exists a solution of (5.11), hence of (5.10), which is unbounded as $u \to 0$. But, if $g : \mathbb{B} \to \mathbb{C}$ is a solution of $-\Delta g = \lambda g$ in all of $\mathbb{B}$, the functions $g_{\ell j}(\rho)$ occurring in (5.9) are necessarily bounded as $\rho \to 0$. Hence $g_{\ell j}(\rho)$ is a multiple of the solution (5.15) since the solution (5.15) is bounded as $\rho \to 0$. This proves our assertion. Note that the solution (5.15) is invariant with respect to $s \mapsto -s$ since

$$(5.16)\qquad F(a,b;c;z) = (1-z)^{c-a-b}\, F(c-a, c-b; c; z)$$

for $|z| < 1$. □

Lemma 5.2. *Let $f : \,]1,\infty[\to \mathbb{C}$ be a C^2-function and let $f(\delta(P,Q))$ be a solution of the differential equation*

$$-\Delta\, f(\delta(P,Q)) = \lambda\, f(\delta(P,Q)), \tag{5.17}$$

where Δ *acts on* P *and* $P \in \mathbb{H} \setminus \{Q\}$. *Suppose that* $\lambda = 1 - s^2$ *with* $s \neq 0$. *Then* $f(\delta)$ *is a linear combination of* $\varphi_s(\delta)$ *and* $\varphi_{-s}(\delta)$ *(see (1.46)).*

Proof. Since δ is a point-pair invariant, it is sufficient to prove that $f(\delta(P,j))$ is a linear combination of $\varphi_s(\delta(P,j))$ and $\varphi_s(\delta(P,j))$. To apply the first part of the preceding proof, we transform $\mathbb{H}$ to $\mathbb{B}$ such that $j \mapsto 0$, $P \mapsto x$ and put $\rho = \|x\|$. Then $f(\delta(P,j))$ becomes some function $\varphi(\rho)$ for $\rho \in]0,1[$, which satisfies (5.10) with $\ell = 0$. The corresponding function $\psi(u)$, is represented by

$$P\begin{Bmatrix} 0 & \infty & 1 & \\ 0 & 0 & 1+s & z \\ -\frac{1}{2} & -\frac{1}{2} & 1-s & \end{Bmatrix}. \tag{5.18}$$

From Magnus, Oberhettinger, Soni (1966), page 61 (function W_{17}), we obtain the solution

$$\hat{\psi}_s(u) = (1-u)^{1+s}\ F(s+1, s+\frac{1}{2}; 2s+1; 1-u) \tag{5.19}$$

of (5.11). But by Magnus, Oberhettinger, Soni (1966), page 39 this is an elementary function and gives us the solution

$$\hat{\varphi}_s(\rho) := \frac{1}{\rho}\ (1-\rho^2)^{1+s}\ (1+\rho)^{-2s} = \frac{1-\rho^2}{\rho}\ \left(\frac{1-\rho}{1+\rho}\right)^s \tag{5.20}$$

of (5.10). Obviously $\hat{\varphi}_{-s}(\rho)$ also is a solution, and these solutions form a fundamental system of solutions for $s \neq 0$. Hence $\varphi(\rho)$ is a linear combination of $\hat{\varphi}_s(\rho)$ and $\hat{\varphi}_{-s}(\rho)$. We transform back to $\mathbb{H}$. We have

$$\delta(P,j) = \cosh d(P,j) = \cosh d'(x,0) = \cosh\left(\log\frac{1+\rho}{1-\rho}\right) = \frac{1+\rho^2}{1-\rho^2}. \tag{5.21}$$

Putting $\delta = \delta(P,j)$ we also have

$$\rho^2 = \frac{\delta-1}{\delta+1}, \qquad 1-\rho^2 = \frac{2}{\delta+1}, \qquad \frac{1+\rho}{1-\rho} = \delta + \sqrt{\delta^2-1}. \tag{5.22}$$

Substituting (5.22) into (5.20), we arrive at

$$\varphi_s(\delta) = \frac{1-\rho^2}{2\rho}\ \left(\frac{1-\rho}{1+\rho}\right)^s. \tag{5.23}$$

This proves the Lemma. □

The following remark will be used in our proof of Theorem 5.3. We consider the contribution of the summand for $\ell = 0$ in (5.8). Observe that there is only one spherical function for $\ell = 0$ which may be chosen as $Y_{01} = \frac{1}{\sqrt{4\pi}}$.

Moreover, all summands with $\ell > 0$ vanish at $x = 0$. Hence the zeroth summand in (5.8) is equal to

$$g_0(\rho) := g(0)\,(1-\rho^2)^{1+s}\,F\left(s+1, s+\frac{1}{2}; \frac{3}{2}; \rho^2\right). \tag{5.24}$$

The hypergeometric function in (5.24) is an elementary function, see Magnus, Oberhettinger, Soni (1966), page 39. Hence

$$g_0(\rho) = \frac{1}{4}\,g(0)\,\frac{1-\rho^2}{\rho}\,\frac{1}{s}\left(\left(\frac{1+\rho}{1-\rho}\right)^s - \left(\frac{1-\rho}{1+\rho}\right)^s\right), \tag{5.25}$$

where the right-hand side has to be replaced by its $\lim_{s\to 0}$ if s equals zero.

We are now able to prove the formula (5.3) for the Selberg transform.

Theorem 5.3. *Let $k :]1,\infty[\to \mathbb{C}$ be a measurable function and let $f : \mathbb{H} \to \mathbb{C}$ be a solution of the differential equation $-\Delta f = \lambda f$, where $\lambda = 1 - s^2$.*

(1) If k and f are non-negative and if the Lebesgue integral

$$h(\lambda) = \frac{\pi}{s}\int_1^\infty k\left(\frac{1}{2}\left(t+\frac{1}{t}\right)\right)\,(t^s - t^{-s})\,\left(t-\frac{1}{t}\right)\,\frac{dt}{t} \tag{5.26}$$

exists, then the function $k(\delta(P,\cdot))\,f(\cdot)$ is v-integrable over $\mathbb{H}$ for all $P \in \mathbb{H}$, and the equation

$$\int_{\mathbb{H}} k(\delta(P,Q))\,f(Q)\,dv(Q) = h(\lambda)\,f(P) \tag{5.27}$$

holds.

(2) If $k(\delta(P,\cdot))\,f(\cdot)$ is v-integrable over $\mathbb{H}$, then the Lebesgue integral (5.26) exists and (5.27) holds.

For $s = 0$ the factor $\frac{1}{s}(t^s - t^{-s})$ in (5.26) has to be replaced by its limit $2\log t$. The integral transform (5.26) is called the Selberg transform.

Proof. To compute the integral $\int_{\mathbb{H}} k(\delta(P,Q))\,f(Q)\,dv(Q)$, we transform $\mathbb{H}$ to $\mathbb{B}$ so that $P \mapsto 0$, $Q \mapsto x \in \mathbb{B}$. Then f gives rise to a solution $g : \mathbb{B} \to \mathbb{C}$ of the differential equation $-\Delta g = \lambda g$ such that $f(P) = g(0)$, and g is non-negative whenever f is non-negative. A change of variables together with Fubini's theorem and (5.21) yields

$$\begin{aligned} \int_{\mathbb{H}} k(\delta(P,Q))\,f(Q)\,dv(Q) &= \int_{\mathbb{B}} k(\delta'(0,x))\,g(x)\,dv'(x) \\ &= \int_0^1 k\left(\frac{1}{2}\left(\frac{1+\rho}{1-\rho}+\frac{1-\rho}{1+\rho}\right)\right)\cdot\int_{S^2} g(\rho\zeta)\,d\Omega(\zeta)\,\frac{8\rho^2\,d\rho}{(1-\rho^2)^3}, \end{aligned} \tag{5.28}$$

where Ω denotes the Euclidean surface measure on S^2. Because of the uniform convergence of (5.8) on compact subsets of $\mathbb{B}$ the integral over S^2 in (5.28) is readily computed by termwise integration. Observe that only the zeroth summand of (5.8) actually gives a contribution, since all Ω-integrals of the functions $Y_{\ell j}(\zeta)$ with $\ell \geq 1$ vanish. These functions are orthogonal to the constant spherical function $Y_{01} = (4\pi)^{-\frac{1}{2}}$.

Using the formula (5.25) for the zeroth summand in (5.8) and inserting $f(P) = g(0)$, we obtain now

$$
\begin{aligned}
&\int_{\mathbb{H}} k(\delta(P,Q))\, f(Q)\, dv(Q) \\
&= \frac{\pi}{s} \int_0^1 k\left(\frac{1}{2}\left(\frac{1+\rho}{1-\rho} + \frac{1-\rho}{1+\rho}\right)\right) \left(\left(\frac{1+\rho}{1-\rho}\right)^s - \left(\frac{1-\rho}{1+\rho}\right)^s\right) \\
&\qquad \cdot \frac{8\,\rho\, d\rho}{(1-\rho^2)^2}\, f(P) \\
&= \frac{\pi}{s} \int_1^\infty k\left(\frac{1}{2}\left(t + \frac{1}{t}\right)\right) \left(t^s - t^{-s}\right) \left(t - \frac{1}{t}\right) \frac{dt}{t}\, f(P).
\end{aligned}
\tag{5.29}
$$

The following justifications are necessary: If the integrand on the left-hand side is integrable, our computations are justified and imply the existence of (5.26) and the equality (5.27). If k and f are non-negative, our computations are also justified and imply that (5.27) holds, where both sides of (5.27) are allowed to be infinite. But if (5.26) is finite, which is our hypothesis, then the left-hand side of (5.27) is also finite, and (5.27) holds. □

The orthogonality of the spherical harmonics $Y_{\ell j}$ $(\ell \geq 1,\ j = 1, \ldots, 2\ell+1)$ with respect to Y_{01} can also be used to give the following analogue of the Gauß mean-value formula for functions on $\mathbb{H}$, or on $\mathbb{B}$, which are harmonic in the hyperbolic sense.

Theorem 5.4. *Let $f : \mathbb{H} \to \mathbb{C}$ be a C^2-function which is harmonic in the hyperbolic sense and let $B(P,R)$ denote the hyperbolic ball with center P and radius $R > 0$. Then f satisfies the mean-value formula*

$$
f(P) = \frac{1}{v(B(P,R))} \int_{B(P,R)} f(Q)\, dv(Q). \tag{5.30}
$$

Proof. Transform $\mathbb{H}$ to $\mathbb{B}$ such that $P \mapsto 0$. This gives a corresponding function g on $\mathbb{B}$, which is harmonic in the hyperbolic sense such that $g(0) = f(P)$. We apply (5.25) with $s = 1$. This gives $g_0(\rho) = g(0) = f(P)$ and hence by (5.8)

$$
\int_{S^2} g(\rho\zeta)\, d\Omega(\zeta) = 4\pi g(0)
$$

and also

$$\begin{aligned}\int_{B'(0,R)} g(x)\, dv'(x) &= \int_0^{\frac{e^R-1}{e^R+1}} \int_{S^2} g(\rho\zeta)\, \frac{8\,\rho^2\, d\Omega(\zeta)\, d\rho}{(1-\rho^2)^3} \\ &= 4\pi g(0) \int_0^{\frac{e^R-1}{e^R+1}} \frac{8\,\rho^2\, d\rho}{(1-\rho^2)^3} \\ &= v'(B'(0,R))\, g(0),\end{aligned}$$

where $B'(0,R)$ denotes the hyperbolic ball with center 0 and radius R contained in $\mathbb{B}$. This proves our assertion for the unit ball model of hyperbolic 3-space, and a transformation back to $\mathbb{H}$ yields our theorem. □

For later use we report on some formulas and technical points concerning the Selberg transform. We let $\mathcal{S}([1,\infty[)$ be the space of C^∞-functions $f : [1,\infty[\to \mathbb{R}$ which are together with all their derivatives of rapid decrease as $x \to \infty$. Rapid decrease for a function $k : [1,\infty[\to \mathbb{R}$ means that $x^n k(x)$ is bounded for every $n \in \mathbb{N}$ as $x \to \infty$. The Schwartz space $\mathcal{S}(\mathbb{R})$ is defined analoguously.

Lemma 5.5. *Let $k \in \mathcal{S}([1,\infty[)$ then the Selberg transform h of k defined in (5.26) exists for every $s \in \mathbb{C}$. Define for $x \in \mathbb{R}$*

$$\begin{aligned}g(x) &:= 2\int_{\mathbb{R}^2} k\left(u_1^2 + u_2^2 + \frac{e^x + e^{-x}}{2}\right) du_1\, du_2 \\ &= 2\pi \int_0^\infty k\,(u + \cosh(x))\; du.\end{aligned} \tag{5.31}$$

Then g is even and of rapid decrease as $|x| \to \infty$ and

$$h(1+t^2) = \int_{-\infty}^{\infty} g(x)\, e^{itx}\, dx, \quad g(x) = \frac{1}{2\pi}\int_{-\infty}^{\infty} h(1+t^2)\, e^{-ixt}\, dt. \tag{5.32}$$

We also have the formula:

$$\int_{\mathbb{R}^2} k\left(\frac{\|u\|^2 + r_1^2 + r_2^2}{2r_1 r_2}\right) du = r_1\, r_2\, g\left(\log\left(\frac{r_1}{r_2}\right)\right) \qquad (r_1, r_2 > 0). \tag{5.33}$$

Define further for $x \geq 1$

$$Q(x) := 2\pi \int_0^\infty k(x+t)\; dt. \tag{5.34}$$

Then we have

$$g(x) = Q\left(\frac{e^x + e^{-x}}{2}\right), \qquad Q'(x) = -2\pi\, k(x). \tag{5.35}$$

Formula (5.31) follows by introducing polar coordinates. The first of formulas (5.32) follows from

$$\int_{\mathbb{H}} k(\delta(P,Q))\, r_Q^{1+s}\, dv(Q) = h(1-s^2)\, r_P^{1+s} \tag{5.36}$$

the second from Fourier inversion. In (5.36) we use the notation $P = (z_P, r_P)$ and $Q = (z_Q, r_Q)$.

In Chapter 6 we shall make use of the following observation.

Lemma 5.6. *There is a constant* $0 < \epsilon_0 \le 1/16$ *so that for every* $0 < \epsilon \le \epsilon_0$ *and every nonnegative continuous function* $k : [1, \infty[\to \mathbb{R}$ *with support in* $[1, 1+\epsilon]$ *and with* $\int_{\mathbb{H}} k(\delta(j,Q))\, dv = 1$ *we have*

$$|h(1+t^2) - 1| \le \frac{7}{2}(1+|t|)\sqrt{\epsilon} \qquad \textit{for } |t| \le \frac{1}{4\sqrt{\epsilon}}.$$

Proof. Given $\epsilon > 0$ we define $M_\epsilon := \{\, Q \in \mathbb{H} \mid \delta(j,Q) \le 1+\epsilon \,\}$. If $Q \in M_\epsilon$ then

$$1+\epsilon - \sqrt{\epsilon(2+\epsilon)} \le r_Q \le 1+\epsilon + \sqrt{\epsilon(2+\epsilon)}. \tag{5.37}$$

We may choose ϵ_0 so small that (5.37) implies $|\log r_Q| \le 2\sqrt{\epsilon}$ for all $0 < \epsilon \le \epsilon_0$ and $Q \in M_\epsilon$. Note that $|e^z - 1| \le (7/4)|z|$ for all $z \in \mathbb{C}$ with $|z| < 1$, see Abramowitz, Stegun (1972), 4.2.38. Hence

$$|r_Q^{1+it} - 1| \le \frac{7}{2}(1+|t|)\sqrt{\epsilon} \tag{5.38}$$

for $0 < \epsilon \le \epsilon_0$ and $Q \in M_\epsilon$ if $(1+|t|)\sqrt{\epsilon} \le 1/2$. Choosing a priori $\epsilon_0 \le 1/16$, inequality (5.38) holds for $|t| \le 1/(4\sqrt{\epsilon})$ and hence

$$\begin{aligned} |h(1+t^2) - 1| &\le \int_{\mathbb{H}} k(\delta(j,Q))\, |r_Q^{1+it} - 1|\, dv(Q) \\ &= \int_{M_\epsilon} k(\delta(j,Q))\, |r_Q^{1+it} - 1|\, dv(Q) \le \frac{7}{2}(1+|t|)\sqrt{\epsilon}. \end{aligned}$$

□

3.6 Behaviour of the Poincaré Series at the Abscissa of Convergence

Throughout this section we assume that Γ is a discrete subgroup of $\mathbf{PSL}(2, \mathbb{C})$ of finite covolume. We fix a fundamental domain $\mathcal{F}$ such that $v(\mathcal{F}) < \infty$. Sharpening the results of Corollary 1.7, we shall prove that

$$\lim_{s \downarrow 1} (s-1) H(P,Q,s) = \frac{4\pi}{v(\mathcal{F})}. \tag{6.1}$$

The even sharper result

$$\lim_{t \to 0} t^2 \Theta(P, Q, t) = \frac{4\pi}{v(\mathcal{F})}. \tag{6.2}$$

is much more difficult to prove, it is contained in Chapter 6.

Along with our computation of the above limit we shall analyze the positive harmonic Γ-invariant functions. In fact we shall prove that only constants can have this property.

Our proof of (6.1) consists of modification of a method originated by Patterson (1976a). We first outline a general construction, and afterwards we state the results in Lemma 6.1.

Choose arbitrary points $A, B \in \mathbb{H}$ and put

$$H(s) := H(A, B, s) = \sum_{M \in \Gamma} \delta(A, MB)^{-s-1}, \tag{6.3}$$

assume also $s > 1$. We know from Corollary 1.7 that $H(s)$ tends to infinity as $s \downarrow 1$. Let ϵ_P denote the Dirac measure in $P \in \mathbb{H}$. Introduce now the probability measure μ_s by

$$\mu_s := \frac{1}{H(s)} \sum_{M \in \Gamma} \delta(A, MB)^{-s-1} \; \epsilon_{MB}. \tag{6.4}$$

Notice that μ_s depends on the choice of A, B. We put

$$\hat{\mathbb{H}} := \mathbb{H} \cup \mathbb{P}^1\mathbb{C} = \mathbb{H} \cup (\mathbb{C} \times \{0\}) \cup \{\infty\}. \tag{6.5}$$

We topologize $\mathbb{H} \cup (\mathbb{C} \times \{0\})$ by its embedding into $\mathbb{R}^3$. We choose the Alexandroff compactification topology on $\hat{\mathbb{H}}$. Notice that the isometry from $\mathbb{H}$ to $\mathbb{B}$ described in Chapter 1 extends to a homeomorphism of $\hat{\mathbb{H}}$ onto the closed unit ball $\bar{\mathbb{B}} \subset \mathbb{R}^3$. We write $C(\hat{\mathbb{H}})$ for the space of continuous complex valued functions on $\hat{\mathbb{H}}$.

Now we may consider the measures μ_s as probability measures on the compact space $\hat{\mathbb{H}}$. A known compactness theorem from measure theory states: *If X is a compact metric space, the space $\mathcal{M}(X)$ of probability measures on X equipped with its vague topology (weak topology defined by the continuous functions on X) is also compact.* (See Edwards (1965), page 205 or Parthasarathy (1967), page 45). This means the following in the present situation: *For any sequence $(s_\nu)_{\nu \geq 1}$ in $]1, \infty[$ such that $s_\nu \to 1$ as $\nu \to \infty$ there exists a subsequence $(t_\nu)_{\nu \geq 1}$ of $(s_\nu)_{\nu \geq 1}$ and a probability measure μ on $\hat{\mathbb{H}}$ such that*

$$\int_{\hat{\mathbb{H}}} f \, d\mu_{t_\nu} \quad \longrightarrow \quad \int_{\hat{\mathbb{H}}} f \, d\mu \tag{6.6}$$

for all $f \in C(\hat{\mathbb{H}})$ as $\nu \to \infty$.

For all $f \in C(\hat{\mathbb{H}})$ having a compact support contained in $\mathbb{H}$, the limit (6.6) is equal to zero, since $H(t_\nu) \to \infty$. Hence the support of μ is contained in $\mathbb{P}^1\mathbb{C} = \mathbb{C} \cup \{\infty\}$, see Edwards (1965), page 202.

Starting with the initially chosen point $A = a + tj \in \mathbb{H}$ we form the kernel

$$(6.7) \qquad K_s(P,P') := \left(\frac{r}{|z-z'|^2 + r^2 + r'^2} \cdot \frac{|z'-a|^2 + r'^2 + t^2}{t} \right)^{s+1},$$

where we have chosen the coordinates $P = z + rj \in \mathbb{H}$ with $z \in \mathbb{C}$, $r > 0$ and $P' = z' + r'j \in \mathbb{H} \cup \mathbb{C}$ with $z' \in \mathbb{C}$, $r' \geq 0$. We extend this definition by

$$(6.8) \qquad K_s(P,\infty) := \left(\frac{r}{t}\right)^{s+1}.$$

Then $K_s : \mathbb{H} \times \hat{\mathbb{H}} \to \mathbb{C}$ is a continuous function. For $P, Q \in \mathbb{H}$ we simply have

$$(6.9) \qquad K_s(P,Q) = \left(\frac{\delta(A,Q)}{\delta(P,Q)}\right)^{s+1}.$$

This yields for $s > 1, P \in \mathbb{H}$

$$(6.10) \qquad \begin{aligned} G_s(P) &:= \int_{\mathbb{H}} K_s(P,Q)\, d\mu_s(Q) \\ &= \frac{1}{H(s)} \sum_{M \in \Gamma} K_s(P, MB)\, \delta(A, MB)^{-s-1} \\ &= \frac{H(P,B,s)}{H(A,B,s)}. \end{aligned}$$

G_s is a Γ-invariant function on $\mathbb{H}$.

We want to investigate the behaviour of (6.10) as s tends to one. Since we do not know a priori anything about the possible existence of the limit, we start with an arbitrary sequence $(s_\nu)_{\nu \geq 1}$ in $]1, \infty[$ such that $s_\nu \to 1$. Then we conclude from Corollary 1.7 that $(s_\nu - 1)H(A,B,s_\nu)$ is bounded from above and from below by positive constants. Hence there exists a subsequence $(s'_\nu)_{\nu \geq 1}$ of $(s_\nu)_{\nu \geq 1}$ such that the limit

$$(6.11) \qquad \lim_{\nu \to \infty} (s'_\nu - 1) H(A,B,s'_\nu) =: \alpha \in]0, \infty[$$

exists. Now there exist a subsequence $(t_\nu)_{\nu \geq 1}$ of $(s'_\nu)_{\nu \geq 1}$ and a probability measure μ on $\hat{\mathbb{H}}$ such that (6.6) holds.

The function

$$(6.12) \qquad G(P) := \int_{\mathbb{P}^1\mathbb{C}} K_1(P,Z)\, d\mu(Z) \qquad (P \in \mathbb{H})$$

is well defined and satisfies

$$G(P) = \lim_{\nu\to\infty} \int_{\hat{\mathbb{H}}} K_1(P,Z)\, d\mu_{t_\nu}(Z) \tag{6.13}$$

because μ is supported by $\mathbb{P}^1\mathbb{C}$. To relate G with G_s, we want to replace K_1 in (6.13) by K_{t_ν}. An easy estimate shows that for any $P \in \mathbb{H}$ there exists a constant $C(P) > 0$ such that

$$|K_s(P,Z) - K_1(P,Z)| \leq C(P)\,(s-1) \tag{6.14}$$

for all $Z \in \hat{\mathbb{H}}$. Since all measures involved are probability measures, we may replace K_1 in (6.13) by K_{t_ν}, and this yields in view of (6.10), (6.11).

$$\begin{aligned} G(P) &= \lim_{\nu\to\infty} \int_{\hat{\mathbb{H}}} K_{t_\nu}(P,Z) d\mu_{t_\nu}(Z) = \lim_{\nu\to\infty} \frac{H(P,B,t_\nu)}{H(A,B,t_\nu)} \\ &= \frac{1}{\alpha} \lim_{\nu\to\infty} H(P,B,t_\nu)(t_\nu - 1). \end{aligned} \tag{6.15}$$

The remaining part of the proof makes essential use of the fact that G is harmonic in the hyperbolic sense. Observe that $-\Delta\, r^{s+1} = (1-s^2)\, r^{s+1}$. Since Δ is $\mathbf{PSL}(2,\mathbb{C})$-invariant, we also have

$$-\Delta \left(\frac{r}{|z-z'|^2 + r^2} \right)^{s+1} = (1-s^2) \left(\frac{r}{|z-z'|^2 + r^2} \right)^{s+1}. \tag{6.16}$$

Hence

$$-\Delta\, K_s(P,Z) = (1-s^2)\, K_s(P,Z) \tag{6.17}$$

where Δ acts on P and Z is in $\mathbb{P}^1\mathbb{C}$. In particular we have $\Delta K_1(P,Z) = 0$ for all $Z \in \mathbb{P}^1\mathbb{C}$. The application of Δ under the integral sign in (6.12) is justified and yields $\Delta G = 0$, so G is harmonic in the hyperbolic sense. Summing up, we have proved the following lemma.

Lemma 6.1. *Let Γ be a cofinite subgroup of $\mathbf{PSL}(2,\mathbb{C})$. Then for all $A, B \in \mathbb{H}$ and for any sequence $(s_\nu)_{\nu\geq 1}$ in $]1,\infty[$ such that $\lim_{\nu\to\infty} s_\nu = 1$, there exist a subsequence $(t_\nu)_{\nu\geq 1}$ of $(s_\nu)_{\nu\geq 1}$ and a probability measure μ on $\hat{\mathbb{H}}$ such that the following hold:*

(1) The limit $\lim_{\nu\to\infty} (t_\nu - 1)H(A,B,t_\nu) =: \alpha$ exists and $0 < \alpha < \infty$.
(2) The function

$$G(P) = \int_{\mathbb{P}^1\mathbb{C}} K_1(P,Z)\, d\mu(Z)$$

of $P \in \mathbb{H}$ is harmonic on $\mathbb{H}$ in the hyperbolic sense, that is $\Delta G = 0$, and

$$\lim_{\nu\to\infty} (t_\nu - 1)H(P,B,t_\nu) = \alpha \cdot G(P)$$

for all $P \in \mathbb{H}$.

We want to show that G is square-integrable over $\mathcal{F}$. The proof of this fact relies upon the Selberg transform.

Lemma 6.2. *Let g be a Γ-invariant C^2-function on $\mathbb{H}$ such that $g \geq 0$, $\Delta g = 0$. Then for $s \in \mathbb{C}$, $\operatorname{Re} s > 1$, $P \in \mathbb{H}$ the following integral exists and has the given value:*

$$\int_{\mathcal{F}} H(P,Q,s)\; g(Q)\; dv(Q) = 2^s\; \pi\; \frac{\Gamma(\frac{s+1}{2})\Gamma(\frac{s-1}{2})}{\Gamma(s+1)}\; g(P). \tag{6.18}$$

Proof. It suffices to prove the lemma for real $s > 1$, so let us assume that $s > 1$. Then

$$\int_{\mathcal{F}} H(P,Q,s)\; g(Q)\; dv(Q) = \int_{\mathbb{H}} \delta(P,Q)^{-s-1}\; g(Q)\; dv(Q),$$

where a priori the integrals may be infinite. To prove the finiteness of the integral in question and the equation (6.18) it suffices to compute the Selberg transform from Section 5 for our data at hand. In the application of Theorem 5.3 one has to take $s = 1$ and not confuse it with the s used here. The Selberg transform obviously exists and is equal to

$$\begin{aligned}
&\pi \int_1^\infty \left(\frac{1}{2}\left(t+\frac{1}{t}\right)\right)^{-s-1} (t-t^{-1})\left(t-\frac{1}{t}\right)\frac{dt}{t} \\
&= \pi \int_0^\infty \left(\frac{1}{2}\left(t+\frac{1}{t}\right)\right)^{-s-1} \left(t-\frac{1}{t}\right) dt \\
&= 2^{s+1}\pi \int_0^\infty (t^2+1)^{-s-1}\; t^s\; (t^2-1)\; dt \\
&= 2^s\pi \int_0^\infty (u+1)^{-s-1}\; (u-1)\; u^{\frac{s-1}{2}}\; du \qquad (t^2 = u) \\
&= 2^s\pi \left(\mathrm{B}\left(\frac{s+1}{2}+1, \frac{s-1}{2}\right) - \mathrm{B}\left(\frac{s+1}{2}, \frac{s+1}{2}\right)\right) \\
&= 2^s\pi \frac{\Gamma(\frac{s+1}{2})\Gamma(\frac{s-1}{2})}{\Gamma(s+1)}.
\end{aligned}$$

See Magnus, Oberhettinger, Soni (1966), page 7 for help. Since the Selberg transform is finite ($s > 1$), Theorem 5.3 implies the Lemma. Observe that Lemma 6.2 holds for arbitrary discrete subgroups of $\mathbf{PSL}(2,\mathbb{C})$. □

We proceed to prove (6.1).

Theorem 6.3. *If Γ is a discrete subgroup of $\mathbf{PSL}(2,\mathbb{C})$ having a fundamental domain $\mathcal{F}$ of finite hyperbolic volume, then*

$$\lim_{s \downarrow 1} (s-1)H(P,Q,s) = \frac{4\pi}{v(\mathcal{F})}. \tag{6.19}$$

Proof. We continue our considerations which gave Lemma 6.1. Let G be as in Lemma 6.1. Then we obtain from (6.15) and from Fatou's lemma

$$\begin{aligned}\int_{\mathcal{F}} G^2\, dv &= \frac{1}{\alpha} \int_{\mathcal{F}} (\liminf_{\nu \to \infty} (t_\nu - 1)H(P,B,t_\nu))\; G(P)\; dv(P) \\ &\leq \frac{1}{\alpha} \liminf_{\nu \to \infty} (t_\nu - 1) \int_{\mathcal{F}} H(P,B,t_\nu)\; G(P)\; dv(P).\end{aligned}$$

Here the right-hand side is known from (6.18), and evaluating the lim inf, we arrive at

$$\int_{\mathcal{F}} G^2\, dv \leq \frac{4\pi}{\alpha}\; G(B).$$

So G is square-integrable. Since we have $v(\mathcal{F}) < \infty$ by hypothesis, this implies that G is constant. This fact is proved in full generality in Section 1 of the next Chapter. Observe that the constancy of G is clear from Theorem 5.4 if Γ is cocompact. If Γ has exactly one class of cusps we have proved it in Corollary 3.4.

Now we know that G is constant, put $G(P) = \gamma$. Observe that we have

$$\lim_{\nu \to \infty} (t_\nu - 1)H(P,B,t_\nu) = \alpha\gamma$$

for all $P \in \mathbb{H}$ and that for all ν

$$(t_\nu - 1)H(P,B,t_\nu) \leq C_1(B)$$

for all $P \in \mathbb{H}$ with a constant $C_1(B)$ which does not depend on P. Hence an application of Lebesgue's theorem on bounded convergence yields

$$\begin{aligned}\gamma^2 v(\mathcal{F}) = \int_{\mathcal{F}} G^2\, dv &= \frac{1}{\alpha} \int_{\mathcal{F}} (\lim_{\nu \to \infty} (t_\nu - 1)H(P,B,t_\nu))\; G(P)\; dv(P) \\ &= \frac{1}{\alpha} \lim_{\nu \to \infty} (t_\nu - 1) \int_{\mathcal{F}} H(P,B,t_\nu)\; G(P)\; dv(P) = \frac{4\pi}{\alpha}\; G(B) = \frac{4\pi}{\alpha}\gamma.\end{aligned}$$

Hence we obtain $\alpha\gamma = \frac{4\pi}{v(\mathcal{F})}$, that is

$$\lim_{\nu \to \infty} (t_\nu - 1)H(P,B,t_\nu) = \frac{4\pi}{v(\mathcal{F})} \tag{6.20}$$

for all $P \in \mathbb{H}$.

We have now proved: If $B \in \mathbb{H}$ is an arbitrary point and if $(s_\nu)_{\nu \geq 1}$ is an arbitrary sequence in $]1, \infty[$ such that $\lim_{\nu \to \infty} s_\nu = 1$, then there exists a subsequence (t_ν) of (s_ν) such that (6.20) holds for all $P \in \mathbb{H}$. But this immediately implies that $\frac{4\pi}{v(\mathcal{F})}$ is the only accumulation point of $(s-1)H(P,B,s)$

as s tends to one. Hence (6.19) holds with Q replaced by B, and since B was arbitrary the proof is complete. □

A slight modification of the argument of the preceding proof implies that any positive Γ-invariant harmonic function on $\mathbb{H}$ is a constant. This we shall prove in the next theorem.

Theorem 6.4. *Let Γ be a discrete subgroup of* $\mathbf{PSL}(2,\mathbb{C})$ *of finite covolume. Assume that $f : \mathbb{H} \to \mathbb{R}$ is a nonnegative, Γ-invariant function which is harmonic in the hyperbolic sense. Then f is constant.*

Proof. First we shall show that f is v-integrable over a fundamental domain $\mathcal{F}$ of Γ. This is obtained from (6.19) and from Fatou's lemma:

$$\begin{aligned}\frac{4\pi}{v(\mathcal{F})}\int_{\mathcal{F}} f\,dv &= \int_{\mathcal{F}} (\liminf_{s\downarrow 1}\,(s-1)H(P,Q,s))\;f(Q)\;dv(Q)\\ &\le \liminf_{s\downarrow 1}(s-1)\int_{\mathcal{F}} H(P,Q,s)\;f(Q)\;dv(Q)\\ &= 4\pi f(P)\end{aligned}\tag{6.21}$$

by Lemma 6.2. Hence f is v-integrable over $\mathcal{F}$. Therefore, replacing Fatou's lemma by the Lebesgue bounded convergence theorem, we have equality in (6.21), that is

$$\frac{4\pi}{v(\mathcal{F})}\int_{\mathcal{F}} f\,dv = 4\pi f(P)$$

for all $P \in \mathbb{H}$. Hence f is constant. □

Theorem 6.3 has the following interesting consequence.

Corollary 6.5. *If Γ is a discrete subgroup of* $\mathbf{PSL}(2,\mathbb{C})$ *having a fundamental domain $\mathcal{F}$ of finite hyperbolic volume, then for all $P, Q \in \mathbb{H}$*

$$\sum_{\substack{M\in\Gamma\\ \delta(P,MQ)\le T}} \delta^{-2}(P,MQ) \sim \frac{4\pi}{v(\mathcal{F})}\log T.\tag{6.22}$$

Proof. Let us write (6.19) in the form

$$\sum_{M\in\Gamma} \delta^{-2}(P,MQ)\;e^{-t\log\delta(P,MQ)} \sim \frac{4\pi}{v(\mathcal{F})}\cdot\frac{1}{t}\tag{6.23}$$

as $t \downarrow 0$. Since the generalized Dirichlet series appearing on the left-hand side of (6.23) has positive coefficients, an application of Karamata's Tauberian Theorem, see Karamata (1931a), (1931b) yields

$$\sum_{\substack{M\in\Gamma \\ \log\delta(P,MQ)\le x}} \delta^{-2}(P,MQ) \sim \frac{4\pi}{v(\mathcal{F})}x$$

as $x \to \infty$. Substituting $x = \log T$ we obtain (6.22). □

Observe that (6.22) is a precise asymptotic version of the estimates from Lemmas 2.6.1, 2.6.2 and of (1.6), (1.15).

3.7 Notes and Remarks

It seems difficult if not impossible to trace back the origin of the various series encountered in the present chapter, since similar functions have been discussed intensively ever since Poincaré (1916), Tome II introduced his now famous series. The function $H(P,Q,s)$ was discussed in the corresponding two-dimensional situation by Huber (1959), (1961). The case of arbitrary dimension was considered by Bérard-Bergery (1971), (1973). The universal property of discontinuous groups stated in Theorem 1.8 seems to be new. An analogous theorem holds in the case of the hyperbolic plane, and it is unclear, to what extent this universal property is related to other known universal properties of Fuchsian groups or Kleinian groups, for this see Marden (1974), Sturm, Shinnar (1974), Apanasov (1975a), Beardon (1983).

The analogue of Theorem 1.10 for the hyperbolic plane is known, see Tsuji (1959), page 517, Rao (1969), page 636, Elstrodt (1973c). Theorem 1.10 yields another simple proof of Corollary 1.6: Suppose that $\Gamma < \mathbf{PSL}(2,\mathbb{C})$ is a discrete group with a fundamental domain $\mathcal{F}$ of finite hyperbolic volume. We want to show that $H(P,Q,1)$ diverges. Assume to the contrary that $H(P,Q,1)$ converges for some $P,Q \in \mathbb{H}$. Then $H(\cdot,j,1)$ is bounded on $\mathbb{H}$ and hence

$$\int_{\mathcal{F}} H(P,j,1)\,dv(P) < \infty.$$

But the computation (1.3) implies that the integral is infinite, and we arrive at a contradiction. Hence $H(P,Q,1)$ diverges for all $P,Q \in \mathbb{H}$.

Patterson (1976a) invented his method in connection with his work on the limit set of finitely generated Fuchsian groups of the second kind. This work was extended to hyperbolic spaces of arbitrary dimension and discussed in a much broader framework by Sullivan (1978), (1979), (1986); see Ahlfors (1980) for a report on Sullivan's work on ergodic properties of Möbius transformations.

Up to normalization, the kernel (6.7) is a power of the Poisson kernel for the upper half-space $\mathbb{H}$. For an introductory account of the properties of the Poisson kernel in the case of the hyperbolic plane see Eymard (1977).

Chapter 4. Spectral Theory of the Laplace Operator

In the present chapter we analyse the operation of the Laplace-Beltrami operator Δ on its natural domain in the Hilbert space $L^2(\Gamma\backslash\mathbb{H})$ of square-integrable functions on $\Gamma\backslash\mathbb{H}$ for discrete subgroups $\Gamma < \mathbf{PSL}(2,\mathbb{C})$ in the frame-work of Hilbert space theory. Our discussion uses certain well-known facts about self-adjoint operators on Hilbert spaces, they can be found in Dunford, Schwartz (1958), Reed, Simon (1972) or Kato (1976).

The starting point of our discussion is the fact that $-\Delta$ is essentially self-adjoint and positive on the subspace $\mathcal{D} \subset L^2(\Gamma\backslash\mathbb{H})$ consisting of all C^2-functions $f \in L^2(\Gamma\backslash\mathbb{H})$ such that $\Delta f \in L^2(\Gamma\backslash\mathbb{H})$. We prove this result in Section 1 for all discrete groups Γ of finite covolume. Our proof is self-contained and simple. The only tools which we do not develop here are some standard Hilbert space theory and Weyl's lemma. The operator $-\Delta$ has then a unique self-adjoint extension denoted by $-\tilde{\Delta}$. We write $\sigma(-\tilde{\Delta})$, $\rho(-\tilde{\Delta})$ for its spectrum and resolvent set. It follows that $\sigma(-\tilde{\Delta}) \subset [0,\infty[$. The resolvent operator

$$R_\lambda = (-\tilde{\Delta} - \lambda)^{-1} : \rho(-\tilde{\Delta}) \to \mathcal{B}(L^2(\Gamma\backslash\mathbb{H}))$$

is a holomorphic map from the resolvent set to the Banach space $\mathcal{B}(L^2(\Gamma\backslash\mathbb{H}))$ of bounded operators on $L^2(\Gamma\backslash\mathbb{H})$.

A second major step in the discussion of Δ is the explicit construction of the resolvent of $-\tilde{\Delta}$ in Section 2. We prove that the resolvent operator $R_\lambda = (-\tilde{\Delta} - \lambda)^{-1}$ is described by an integral operator whose kernel is the Maaß–Selberg series $F(P,Q,s)$ introduced in Section 1 of Chapter 3. In Section 2 we shall also show that $F(P,Q,s)$ defines an integral operator of Carleman type for any $s \in \mathbb{C}$ with $\mathrm{Re}\, s > 1$. Suppose $\Phi : \mathbb{H} \times \mathbb{H} \to \mathbb{C}$ is a measurable kernel which is Γ-invariant in both variables. Following the usual terminology we call Φ a kernel of *Carleman type* if

$$\int_{\mathcal{F}} |\Phi(P,Q)|^2 \, dv(Q) < \infty$$

for all $P \in \mathbb{H}$. In this case, the corresponding integral operator

$$L^2(\Gamma \setminus \mathbb{H}) \ni f \mapsto \int_{\mathcal{F}} \Phi(\cdot, Q)\, f(Q)\, dv(Q)$$

is called an integral operator of Carleman type. Here $\mathcal{F}$ denotes a fundamental domain of Γ. The Maaß–Selberg series does not necessarily describe the resolvent for all $\lambda = 1 - s^2$ contained in the resolvent set of $-\tilde{\Delta}$ since this series converges only when $\operatorname{Re} s$ is sufficiently large. We will show that $\operatorname{Re} s > 1$ suffices for all groups Γ. This suggests an analytical continuation of the resolvent kernel into the s-plane such that its basic properties are preserved. Starting from the resolvent equation, we carry out this continuation into the set $\{ s \mid \operatorname{Re} s > 0,\ s \notin [0,1] \}$ in Section 4.

The resolvent kernel will be proved to be of Hilbert-Schmidt type if and only if $\Gamma\backslash\mathbb{H}$ is compact in Section 3. From the theory of Hilbert-Schmidt operators we infer that if Γ is a discrete cocompact subgroup of $\mathbf{PSL}(2,\mathbb{C})$ then $-\Delta : \mathcal{D} \to L^2(\Gamma\backslash\mathbb{H})$ possesses a complete orthonormal system $(e_n)_{n\geq 0}$ of eigenfunctions with a discrete set of associated eigenvalues $0 = \lambda_0 < \lambda_1 \leq \lambda_2 \leq \ldots$, such that $\lambda_n \to \infty$ and

$$\sum_{n=1}^{\infty} \lambda_n^{-2} < \infty.$$

The case of cocompact groups Γ is studied in more detail in Chapter 5.

We resume the study of the resolvent kernel for discrete non-cocompact groups of finite covolume in Section 5. In this case the resolvent kernel is no longer of Hilbert-Schmidt type. It is extremely difficult to prove general results on the eigenfunctions of $-\Delta : \mathcal{D} \to L^2(\Gamma\backslash\mathbb{H})$. We show that if $(f_n)_{n\geq 0}$ is the orthonormalized system of eigenfunctions and $(\lambda_n)_{n\geq 0}$ are the corresponding eigenvalues then every λ_n occurs with finite multiplicity. The eigenvalue 0 occurs with multiplicity 1, assuming $\lambda_0 = 0$ we also prove $\sum_{n=1}^{\infty} \lambda_n^{-2} < \infty$. In addition we show that the series $\sum_{n=1}^{\infty} < g, f_n > f_n$ converges also pointwise absolutely and uniformly on compact sets in $\mathbb{H}$. Here $< \cdot, \cdot >$ stands for the scalar product in our Hilbert space. The key to these results is an approximation theorem formulated in Theorem 5.1. This theorem says that there exists a real symmetric kernel $F^*(P,Q,s)$ for $P, Q \in \mathbb{H}$, $P \not\equiv Q \bmod \Gamma$ and $s > 1$ which is of Hilbert-Schmidt type such that the integral operator associated with F^* has the same action as the resolvent operator on a certain subspace which contains all but finitely many of the eigenfunctions of Δ. The kernel F^* is constructed from F by a suitable modification of F near the cusps of Γ.

4.1 Essential Self-Adjointness of the Laplace-Beltrami Operator

Let $\Gamma < \mathbf{PSL}(2,\mathbb{C})$ be a discrete group. We denote by $L^2(\Gamma\backslash\mathbb{H})$ the set of all Γ-invariant Borel-measurable functions $f : \mathbb{H} \to \mathbb{C}$ which satisfy

$$\int_{\mathcal{F}} |f|^2 \, dv < \infty, \tag{1.1}$$

where $\mathcal{F}$ denotes a (measurable) fundamental domain of Γ. Although the elements of $L^2(\Gamma\backslash\mathbb{H})$ are equivalence classes of functions we talk about them as functions and do not distinguish between functions which coincide almost everywhere. For $f, g \in L^2(\Gamma\backslash\mathbb{H})$ the function $f\bar{g}$ is Γ-invariant. Hence the definition

$$< f, g >:= \int_{\mathcal{F}} f\bar{g} \, dv \tag{1.2}$$

makes sense, and $< \cdot, \cdot >$ is an inner product on $L^2(\Gamma\backslash\mathbb{H})$. As is well-known, the space $L^2(\Gamma\backslash\mathbb{H}, dv)$ is a Hilbert space. The norm corresponding to the inner product (1.2) is

$$\|f\| = (< f, f >)^{1/2} \qquad (f \in L^2(\Gamma\backslash\mathbb{H})). \tag{1.3}$$

We want to define Δ on an appropriate domain $\mathcal{D} \subset L^2(\Gamma\backslash\mathbb{H})$ such that $\Delta : \mathcal{D} \to L^2(\Gamma\backslash\mathbb{H})$ is an essentially self-adjoint operator. We infer from Chapter 1 that for any C^2-function $f \in L^2(\Gamma\backslash\mathbb{H})$ we have $(\Delta f) \circ M = \Delta(f \circ M) = \Delta f$ for all $M \in \Gamma$. This means that Δf automatically is Γ-invariant, but possibly does not satisfy (1.1). Hence we are led to introduce the following domains for Δ.

Definition 1.1. Let $\Gamma < \mathbf{PSL}(2, \mathbb{C})$ be a discrete group. The natural domains of definition of Δ are

$$\mathcal{D} := \{ f \in L^2(\Gamma\backslash\mathbb{H}) \cap C^2(\mathbb{H}) \mid \Delta f \in L^2(\Gamma\backslash\mathbb{H}) \}, \tag{1.4}$$

$$\mathcal{D}^\infty := \{ f \in L^2(\Gamma\backslash\mathbb{H}) \cap C^\infty(\mathbb{H}) \mid \pi_\Gamma(\mathrm{supp}(f)) \text{ is compact in } \Gamma\backslash\mathbb{H} \}. \tag{1.5}$$

Here $\pi_\Gamma : \mathbb{H} \to \Gamma\backslash\mathbb{H}$ is the natural projection, and $\mathrm{supp}(f)$ is the support of f.

Notice that $\mathcal{D}^\infty \subset \mathcal{D}$. Moreover, $\mathcal{D}^\infty$ and $\mathcal{D}$ are dense subspaces of $L^2(\Gamma\backslash\mathbb{H})$. We need the following version of the partition of unity lemma.

Lemma 1.2. *Let $\Gamma < \mathbf{PSL}(2, \mathbb{C})$ be a discrete group. Then there exists a Γ-invariant C^∞-partition of unity on $\mathbb{H}$, i.e. there exist C^∞-functions of compact support $h_\nu : \mathbb{H} \to [0, 1]$ for $\nu \in \mathbb{N}$ such that $0 \le h_\nu \le 1$. Furthermore there are relatively compact open neighbourhoods U_ν with $\mathrm{supp}(h_\nu) \subset U_\nu$ such that the sets MU_ν for $M \in \Gamma$, $\nu \in \mathbb{N}$ form a locally finite covering of $\mathbb{H}$, and*

$$1 = \sum_{\substack{M \in \Gamma \\ \nu \in \mathbb{N}}} h_\nu \circ M.$$

Proof. Recall that a family $(A_\iota)_{\iota \in I}$ of subsets of a topological space X is called locally finite if any point $x \in X$ has a neighbourhood U such that $U \cap A_\iota$

is non-empty only for finitely many $\iota \in I$. Assume that we have a family $(U_\nu)_{\nu\in\mathbb{N}}$ of relatively compact open subsets of $\mathbb{H}$, such that $(MU_\nu)_{(M,\nu)\in\Gamma\times\mathbb{N}}$ is a locally finite (open) covering of $\mathbb{H}$. Assume further that each U_ν has an open subset V_ν with $\bar{V}_\nu \subset U_\nu$ such that $(MV_\nu)_{(M,\nu)\in\Gamma\times\mathbb{N}}$ is a covering of $\mathbb{H}$. Then the proof is completed as follows. For every $\nu \in \mathbb{N}$ choose a C^∞-function $g_\nu \geq 0$ such that $g_\nu(x) > 0$ for all $x \in \bar{V}_\nu$ and $\mathrm{supp}(g_\nu) \subset U_\nu$. The existence of such a function g_ν is well-known, see Helgason (1962), pages 2–3. Then

$$g := \sum_{M\in\Gamma,\nu\in\mathbb{N}} g_\nu \circ M$$

is a strictly positive, Γ-invariant C^∞-function on $\mathbb{H}$. Hence $h_\nu := \frac{g_\nu}{g}$ has the required properties.

To complete the proof, we construct families $(U_\nu)_{\nu\in\mathbb{N}}$, $(V_\nu)_{\nu\in\mathbb{N}}$ as required above. Consider a fixed Poincaré normal polyhedron $\mathcal{F}$ of Γ, and let

$$\mathcal{F}_n := \mathcal{F} \cap \overline{B(j,n)} \quad (n \geq 1),$$

where $B(j,n)$ is the hyperbolic ball with centre j and radius n. Cover $\mathcal{F}_1$ by finitely many open hyperbolic balls of radius 1 called $V_1, \ldots, V_{k_1-1}$, and proceed inductively as follows. Suppose that for $n \in \mathbb{N}$ with $n > 1$ hyperbolic balls of radius 1: $V_1, \ldots, V_{k_n-1}$, are chosen such that $\mathcal{F}_n$ is covered by these sets. If $\mathcal{F}_{n+1}\backslash\mathcal{F}_n$ is empty, put

$$V_{k_n} := \emptyset, \qquad k_{n+1} := k_n + 1.$$

If $\mathcal{F}_{n+1}\backslash\mathcal{F}_n$ is non-empty, cover $\mathcal{F}_{n+1}\backslash\mathcal{F}_n$ by finitely many balls of radius 1 denoted by $V_{k_n}, \ldots, V_{k_{n+1}-1}$ such that the centres of these balls belong to $\mathcal{F}_{n+1}\backslash\mathcal{F}_n$. Let U_ν be the hyperbolic ball of radius 2 having the same centre as V_ν and put $U_\nu := \emptyset$ if $V_\nu = \emptyset$. Then the sets U_ν, V_ν are as required. □

The elements of $\mathcal{D}^\infty$ can be represented as follows.

Lemma 1.3. *$\mathcal{D}^\infty$ coincides with the set of functions $g : \mathbb{H} \to \mathbb{C}$ having a representation of the form*

$$g = \sum_{M\in\Gamma} h \circ M, \tag{1.6}$$

where $h \in C_c^\infty(\mathbb{H})$, that is h is a C^∞-function with compact support.

Proof. Clearly any g of the form (1.6) belongs to $\mathcal{D}^\infty$. To prove the converse, let $g \in \mathcal{D}^\infty$ be given. Let h_ν $(\nu \in \mathbb{N})$ be as in Proposition 1.2. Since $\pi_\Gamma(\mathrm{supp}(g))$ is a compact subset of $\Gamma\backslash\mathbb{H}$, there exists a finite subset $F \subset \mathbb{N}$ such that

$$\mathrm{supp}(g) \cap M^{-1}\mathrm{supp}(h_\nu) = \emptyset$$

for all $\nu \in \mathbb{N}\backslash F,\ M \in \Gamma$. Then we have $g\cdot(h_\nu \circ M) = 0$ for all $\nu \in \mathbb{N}\backslash F,\ M \in \Gamma$. This implies

$$g = \sum_{M\in\Gamma,\ \nu\in F} g\cdot(h_\nu \circ M) = \sum_{M\in\Gamma} h\circ M,$$

where

$$h := \sum_{\nu\in F} g\cdot h_\nu \in C_c^\infty(\mathbb{H}).$$

This completes the proof. □

We recall some notions of Hilbert space theory. Let $\mathbf{H}$ be a separable complex Hilbert space, and let $A : \mathcal{D}_A \to \mathbf{H}$ be a linear operator which is defined on a dense linear subspace $\mathcal{D}_A$ of $\mathbf{H}$. Then A is called symmetric if $< Af, g >=< f, Ag >$ for all $f, g \in \mathcal{D}_A$. The adjoint $A^* : \mathcal{D}_{A^*} \to \mathbf{H}$ of the densely defined operator A is defined as follows: $\mathcal{D}_{A^*}$ is the set of all $g \in \mathbf{H}$, such that there exists an $h \in \mathbf{H}$ satisfying $< Af, g >=< f, h >$ for all $f \in \mathcal{D}_A$. In this case, h is uniquely determined, and $A^*g := h$ defines a linear operator $A^* : \mathcal{D}_{A^*} \to \mathbf{H}$. A is symmetric if and only if A^* extends A, that is $\mathcal{D}_A \subseteq \mathcal{D}_{A^*}$ and $A^*|_{\mathcal{D}_A} = A$. A is called self-adjoint if $A = A^*$.

A is called essentially self-adjoint if A^* is self-adjoint. We recall some equivalent characterizations of this notion in Proposition 1.4. An operator $A : \mathcal{D}_A \to \mathbf{H}$ is called closed if its graph $\{\ (f, Af)\ \mid\ f \in \mathcal{D}_A\ \} \subset \mathbf{H}\times\mathbf{H}$ is a closed subset. If A has a closed extension, A is called closable. In this case, there exists a closed extension $\tilde{A} : \mathcal{D}_{\tilde{A}} \to \mathbf{H}$ of A, called the closure of A, with $\mathcal{D}_{\tilde{A}}$ minimal among all closed extensions. An element $g \in \mathbf{H}$ belongs to $\mathcal{D}_{\tilde{A}}$ if and only if there exists a sequence $(f_n)_{n\geq 1}$ in $\mathcal{D}_A$ converging to g such that $(Af_n)_{n\geq 1}$ converges to $h \in \mathbf{H}$, say, and then $\tilde{A}g := h$. A is closable if and only if $\mathcal{D}_{A^*}$ is dense in $\mathbf{H}$, and in this case $\tilde{A} = A^{**} := (A^*)^*$.

Proposition 1.4. *Let $A : \mathcal{D}_A \to \mathbf{H}$ be an operator defined on a dense linear subspace $\mathcal{D}_A$ of the complex Hilbert space $\mathbf{H}$. Then the following statements are equivalent.*

(1) A is essentially self-adjoint.
(2) A is closable, and $\tilde{A}$ is self-adjoint.
(3) A is closable, and $\tilde{A} = A^*$.
(4) A is symmetric, and $(A+i)\mathcal{D}_A$ and $(A-i)\mathcal{D}_A$ are dense in $\mathbf{H}$.

For the proof, we refer to any standard text on Hilbert space theory, see for example Reed, Simon (1972), pages 256–257 or Weidmann (1979), Section 5.3.

The main result of this section will be that the operators

$$-\Delta : \mathcal{D} \to L^2(\Gamma\backslash\mathbb{H}), \qquad -\Delta : \mathcal{D}^\infty \to L^2(\Gamma\backslash\mathbb{H})$$

are essentially self-adjoint. The crucial step is to show that $\Delta : \mathcal{D} \to L^2(\Gamma\backslash\mathbb{H})$ is symmetric. This requires an appropriate version of Green's theorem. We use our usual coordinates $\mathbb{H} = \{ x + yi + rj \mid x, y, r \in \mathbb{R},\ r > 0 \}$. The partial derivatives of a differentiable function $f : \mathbb{H} \to \mathbb{C}$ are denoted by f_x, f_y, f_r.

Lemma 1.5. *Let $f, g \in C^1(\mathbb{H})$ and put*

$$\mathbf{Gr}(f,g) := r^2(f_x\bar{g}_x + f_y\bar{g}_y + f_r\bar{g}_r). \tag{1.7}$$

Then the function $\mathbf{Gr}(f,g)$ has the following transformation property:

$$\mathbf{Gr}(f,g) \circ T = \mathbf{Gr}(f \circ T, g \circ T) \tag{1.8}$$

for all isometries T of $\mathbb{H}$.

The proof requires an elementary matrix calculation based on the explicit formulas for the action of $\mathbf{Iso}(\mathbb{H})$ on $\mathbb{H}$. We skip the details.

A function $f \in C^1(\mathbb{H})$ gives rise to a cotangent vector field on $\mathbb{H}$. Using the hyperbolic metrics on the tangent spaces this can by duality be turned into a tangent vector field, called the hyperbolic gradient field $\mathbf{grad}(f)$. In our coodinate system we find for $P = (x, y, r) \in \mathbb{H}$

$$\mathbf{grad}(f)|_P = r^2 \left(f_x(P) \left.\frac{\partial}{\partial x}\right|_P + f_y(P) \left.\frac{\partial}{\partial y}\right|_P + f_r(P) \left.\frac{\partial}{\partial r}\right|_P \right). \tag{1.9}$$

Hence we find

$$\mathbf{Gr}(f,f)(P) = \|\mathbf{grad}(f)|_P\|^2$$

where the length on the tangent space is computed using the hyperbolic metric.

Lemma 1.6. *Let $T \in \mathbf{PSL}(2,\mathbb{C})$ and let $h \in C^1(\mathbb{H})$ be a T-invariant function, then the differential form*

$$\omega := \frac{1}{r}\frac{\partial h}{\partial x}\, dy \wedge dr + \frac{1}{r}\frac{\partial h}{\partial y}\, dr \wedge dx + \frac{1}{r}\frac{\partial h}{\partial r}\, dx \wedge dy. \tag{1.10}$$

is T-invariant.

This is again proved by a straightforward computation.

Theorem 1.7. *Let $\Gamma < \mathbf{PSL}(2,\mathbb{C})$ be a discrete group and let $\mathcal{F}$ be a fundamental domain of Γ. Then for all $f, g \in \mathcal{D}$ the function*

$$\mathbf{Gr}(f,g) = r^2(f_x\bar{g}_x + f_y\bar{g}_y + f_r\bar{g}_r) \tag{1.11}$$

is Γ-invariant and v-integrable over $\mathcal{F}$ and satisfies

$$< -\Delta f, g >= \int_{\mathcal{F}} r^2 (f_x \bar{g}_x + f_y \bar{g}_y + f_r \bar{g}_r) \, dv. \tag{1.12}$$

In particular, $-\Delta : \mathcal{D} \to L^2(\Gamma \backslash \mathbb{H})$ is symmetric and positive, that is we have $< -\Delta f, g > = < f, -\Delta g >$, $< -\Delta f, f > \geq 0$ for all $f, g \in \mathcal{D}$.

This Theorem actually holds in much greater generality and was proved for arbitrary complete Riemannian manifolds with discontinuous groups of isometries by Roelcke (1960), page 145, Satz 5. Roelcke's proof relies upon a variant of Stokes' theorem which may not be generally known. Since we are mainly interested in the case of cofinite groups, we give a self-contained proof of Theorem 1.7 for this case. The authors are indebted to W. Roelcke for a very helpful correspondence which corrected an error of the corresponding proof in Roelcke (1956a), page 9 and which led to the following version.

Proof of Theorem 1.7 for groups of finite covolume. Since the statement of our theorem does not depend on the fundamental domain we choose a Poincaré normal polyhedron in case Γ is cocompact and in case Γ is of finite covolume but not cocompact we choose $\mathcal{F}$ to be of the kind described in Theorem 2.3.9. We let $B_1, \ldots, B_h \in \mathbf{PSL}(2, \mathbb{C})$ be so that

$$\zeta_1 = B_1^{-1}\infty, \ldots, \zeta_h = B_h^{-1}\infty \in \mathbb{P}^1\mathbb{C}$$

is a complete system of representatives for the Γ-classes of cusps of Γ. We furthermore choose a fundamental domain $\mathcal{P}_\nu \subset \mathbb{C}$ for the action of $B_\nu \Gamma_{\zeta_\nu} B_\nu^{-1}$ on $\mathbb{P}^1\mathbb{C} \setminus \{\infty\} = \mathbb{C}$ and $Y > 0$ all satisfying the requirements of Theorem 2.3.9. We write

$$\mathcal{F}_\nu(Y) := B_\nu^{-1}\{\, z + rj \;\mid\; z \in \mathcal{P}_\nu,\ r \geq Y \,\} \tag{1.13}$$

for the corresponding cusp sector. Then there is a compact polyhedron $\mathcal{F}_0 \subset \mathbb{H}$ so that

$$\mathcal{F} = \mathcal{F}_0 \cup \mathcal{F}_1(Y) \cup \ldots \cup \mathcal{F}_h(Y) \tag{1.14}$$

is a fundamental domain for Γ. For cocompact groups we interpret this equation as saying $\mathcal{F} = \mathcal{F}_0$. For $R > Y$ we let

$$\mathcal{F}_R := \mathcal{F} \cap \bigcup_{\nu=1}^{h} B_\nu^{-1}\{\, z + rj \mid z \in \mathbb{C},\ r \leq R \,\} \tag{1.15}$$

be the fundamental domain $\mathcal{F}$ cut off at height R. Observe that $\mathcal{F}_R$ depends on the choice of $B_1, \ldots, B_h$.

We now turn to the analytic part of the proof. Let $f, g \in \mathcal{D}$. It follows from Lemma 1.5 that $\mathbf{Gr}(f, g)$ is Γ-invariant. We still have to show that (1.11) is v-integrable over $\mathcal{F}$ and that (1.12) holds. It is sufficient to prove this in the special case $f = g$ since the general assertion follows by the Cauchy-Schwarz inequality and by polarization.

Now let $f \in \mathcal{D}$ and consider the Γ-invariant (see Lemma 1.6) differential form

$$\omega := f_x \bar{f}\, \frac{dy \wedge dr}{r} + f_y \bar{f}\, \frac{dr \wedge dx}{r} + f_r \bar{f}\, \frac{dx \wedge dy}{r} \tag{1.16}$$

and integrate $d\omega$ over $\mathcal{F}_R$ where R is sufficiently large. Since

$$d\omega = (\Delta f)\bar{f}\; dv + \mathbf{Gr}(f, f)\; dv, \tag{1.17}$$

we obtain by Stokes' theorem

$$J(R) := \int_{\mathcal{F}_R} ((\Delta f)\bar{f} + \mathbf{Gr}(f, f))\; dv = \int_{\mathcal{F}_R} d\omega = \int_{\partial \mathcal{F}_R} \omega. \tag{1.18}$$

Now the boundary of $\mathcal{F}_R$ consists of parts of hyperbolic planes which are Γ-equivalent in pairs, and of cross-sections at the cusps. Because Γ-equivalent pairs of bounding hyperbolic planes inherit opposite orientations, the corresponding contributions to the above integral over $\partial\mathcal{F}_R$ cancel.

Abbreviating $\mathcal{Q}_\nu(R) := \{\, z + Rj \;\mid\; z \in \mathcal{P}_\nu \,\}$ we thus obtain

$$J(R) = \sum_{\nu=1}^{h} \int_{B_\nu^{-1}\mathcal{Q}_\nu(R)} \omega$$

for all sufficiently large R. We consider a typical term of the sum on the right-hand side and omit the index ν for convenience. Letting $(x', y', r') = P' = BP = B(x, y, r)$ and ω' be ω in the coordinates (x', y', r') we obtain from Lemma 1.6

$$\begin{aligned}
\left| \int_Y^R \left(\int_{B^{-1}\mathcal{Q}(t)} \omega \right) \frac{dt}{t} \right| &= \left| \int_Y^R \left(\int_{\mathcal{Q}(t)} \omega' \right) \frac{dt}{t} \right| \\
&= \left| \int_Y^R \left(\int_{\mathcal{Q}(t)} \left(\frac{\partial}{\partial r'} f(x', y', r') \right) \overline{f(x', y', r')}\; \frac{dx'\; dy'}{r'} \right) \Bigg|_{r'=t} \frac{dt}{t} \right| \\
&\leq \left(\int_{\mathcal{P}\times[Y,R]} |r' f_{r'}(x', y', r')|^2\; dv \right)^{1/2} \cdot \left(\int_{\mathcal{P}\times[Y,R]} |f(x', y', r')|^2\; dv \right)^{1/2} \\
&\leq \left(\int_{\mathcal{P}\times[Y,R]} \mathbf{Gr}(f, f) \circ B\; dv \right)^{1/2} \cdot \left(\int_{\mathcal{P}\times[Y,R]} |f \circ B|^2\; dv \right)^{1/2} \\
&= \left(\int_{\mathcal{F}(Y,R)} \mathbf{Gr}(f, f)\; dv \right)^{1/2} \cdot \left(\int_{\mathcal{F}(Y,R)} |f|^2\; dv \right)^{1/2},
\end{aligned}$$

where we have omitted the index ν from $\mathcal{F}_\nu(Y, R) := B_\nu^{-1}(\mathcal{P}_\nu \times [Y, R])$. We infer from Theorem 2.3.9 that the sets $\mathcal{F}_\nu(Y, R)$, $\nu = 1, \dots, h$ are disjoint subsets of $\mathcal{F}_R$ and obtain from the Cauchy-Schwarz inequality

$$\left|\int_Y^R J(r)\,\frac{dr}{r}\right|$$

$$(1.19)\qquad \leq \sum_{\nu=1}^{h}\left(\int_{\mathcal{F}_\nu(Y,R)} \mathbf{Gr}(f,f)\,dv\right)^{1/2}\cdot\left(\int_{\mathcal{F}_\nu(Y,R)} |f|^2\,dv\right)^{1/2}$$

$$\leq h\left(\int_{\mathcal{F}_R} \mathbf{Gr}(f,f)\,dv\right)^{1/2}\cdot\|f\|.$$

In view of (1.18), all we have to show is that

$$(1.20)\qquad \lim_{R\to\infty} J(R) = 0.$$

If we can show that

$$(1.21)\qquad \int_{\mathcal{F}} \mathbf{Gr}(f,f)\,dv < \infty,$$

equation (1.20) can easily be established as follows. By (1.21) and (1.18), $\lim_{R\to\infty} J(R)$ exists since $f\in\mathcal{D}$. But this limit cannot be different from zero, since otherwise the left-hand side of (1.19) tends to infinity as $R\to\infty$ whereas the right-hand side remains bounded. Hence we only have to show that (1.21) holds.

Assume to the contrary that

$$(1.22)\qquad \int_{\mathcal{F}_R} \mathbf{Gr}(f,f)\,dv \to +\infty$$

as $R\to\infty$. Then

$$\varphi(R) := \operatorname{Re}\int_{R_0}^R J(r)\,\frac{dr}{r}$$

also tends to $+\infty$ as $R\to\infty$ as is obvious from (1.18). We have $R\,\varphi'(R) = \operatorname{Re} J(R)$. In (1.18), the first term on the right-hand side is bounded as $R\to\infty$. Hence there exist constants $c>0,\ R_1\geq Y$ such that

$$R\,\varphi'(R) = \operatorname{Re}\ J(R) \geq c\int_{\mathcal{F}_R} \mathbf{Gr}((f,f)\,dv > 0$$

for all $R\geq R_1$. In particular, we have $\varphi'(R)>0$ for all $R\geq R_1$. From (1.19) we infer that there exists a constant $C>0$ such that $0<\varphi(R)\leq C\sqrt{R\,\varphi'(R)}$ for all $R\geq R_1$, and hence

$$\frac{1}{R} \leq C^2\frac{\varphi'(R)}{(\varphi(R))^2}$$

for all $R\geq R_1$. This implies

$$\log R - \log R_1 \leq C^2\left(\frac{1}{\varphi(R_1)} - \frac{1}{\varphi(R)}\right)$$

for all $R \geq R_1$, contradicting the fact that $\varphi(R)$ tends to infinity as $R \to \infty$. Hence (1.22) leads to a contradiction and we are done. □

Theorem 1.8. *Let $\Gamma < \mathbf{PSL}(2, \mathbb{C})$ be a cofinite subgroup. Then the eigenvalues of the operator $-\Delta : \mathcal{D} \to L^2(\Gamma\backslash\mathbb{H})$ are non-negative real numbers. Zero is an eigenvalue of multiplicity one and the corresponding eigenspace is generated by the constant eigenfunction.*

The proof is obvious from Theorem 1.7. Note that we now have also filled the gap in Theorem 3.6.3.

Theorem 1.9. *Let $\Gamma \in \mathbf{PSL}(2, \mathbb{C})$ be a discrete group. Then the operators*

$$\Delta : \mathcal{D}^\infty \to L^2(\Gamma\backslash\mathbb{H}), \qquad \Delta : \mathcal{D} \to L^2(\Gamma\backslash\mathbb{H}) \tag{1.23}$$

are essentially self-adjoint and have the same self-adjoint extension.

Proof. Since an essentially self-adjoint operator in a Hilbert space has a unique self-adjoint extension, it is clear that Δ with domain $\mathcal{D}^\infty$ and Δ with domain $\mathcal{D}$ have the same self-adjoint extension if both operators are essentially self-adjoint. Since Δ is symmetric on both domains we only have to show that $(\Delta + i)\ \mathcal{D}^\infty$, $(\Delta - i)\ \mathcal{D}^\infty$ are dense subspaces of $L^2(\Gamma\backslash\mathbb{H})$. Now Δ has real coefficients and $\mathcal{D}^\infty$ is invariant under conjugation. Hence we only have to show that $(\Delta + i)\ \mathcal{D}^\infty$ is a dense subspace of $L^2(\Gamma\backslash\mathbb{H})$. Let $u \in L^2(\Gamma\backslash\mathbb{H})$ and assume $< u, \Delta f + if >= 0$ for all $f \in \mathcal{D}^\infty$. We shall show that $u = 0$. This will imply that the closure of $(\Delta + i)\ \mathcal{D}^\infty$ is the whole of $L^2(\Gamma\backslash\mathbb{H})$.

Using Lemma 1.3 we can write every $f \in \mathcal{D}^\infty$ in the form

$$f = \sum_{M \in \Gamma} h \circ M \tag{1.24}$$

for some function $h \in C_c^\infty(\mathbb{H})$. Conversely, for any $h \in C_c^\infty(\mathbb{H})$, (1.24) defines a function $f \in \mathcal{D}^\infty$. Notice that the sum (1.24) is a finite sum if the variable ranges over a compact set. Hence we have

$$\Delta f = \sum_{M \in \Gamma} (\Delta h) \circ M.$$

Using the dominated convergence theorem and the Γ-invariance of u, we may write

$$0 = \sum_{M \in \Gamma} \int_{\mathcal{F}} u(P)\ \overline{(\Delta h + ih) \circ M(P)}\ dv(P) = \int_{\mathbb{H}} u\ \overline{(\Delta + i)h}\ dv. \tag{1.25}$$

By assumption, this holds for all $h \in C_c^\infty(\mathbb{H})$. Hence by Weyl's lemma (cf. Hellwig (1960), page 172) u is almost everywhere equal to a C^2-function. There is another regularity theorem in Agmon (1965), page 66 which even implies that u is almost everywhere equal to a C^∞-function. So we may assume without loss of generality that u is twice continuously differentiable. Now apply Green's formula and Stokes' theorem to a ball B such that $\bar{B} \subset \mathbb{H}$:

$$\begin{aligned}\int_B (-\Delta u\, \bar{h} - u\, \overline{\Delta h})\, dv = &\int_{\partial B} (u\bar{h}_x - u_x\bar{h})\, \frac{dy \wedge dr}{r} \\ &+ (u\bar{h}_y - u_y\bar{h})\, \frac{dr \wedge dx}{r} + (u\bar{h}_r - u_r\bar{h})\, \frac{dx \wedge dy}{r}.\end{aligned} \tag{1.26}$$

Choosing the ball such that it contains the support of h, the right-hand side of (1.26) vanishes and we obtain

$$\int_{\mathbb{H}} u\, \overline{\Delta h}\, dv = \int_{\mathbb{H}} (\Delta u)\, \bar{h}\, dv.$$

We conclude from (1.25)

$$\int_{\mathbb{H}} ((\Delta - i)u)\, \bar{h}\, dv = 0$$

for all $h \in C_c^\infty(\mathbb{H})$. Hence $\Delta u = iu$. But by Theorem 1.7, the operator $\Delta : \mathcal{D} \to L^2(\Gamma\backslash\mathbb{H})$ has only real eigenvalues. This implies that $u = 0$ and we are done. □

Definition 1.10. We define

$$\tilde{\Delta} : \tilde{\mathcal{D}} \to L^2(\Gamma\backslash\mathbb{H}) \tag{1.27}$$

to be the unique self-adjoint extension of $\Delta : \mathcal{D}^\infty \to L^2(\Gamma\backslash\mathbb{H})$ and of $\Delta : \mathcal{D} \to L^2(\Gamma\backslash\mathbb{H})$.

An element $g \in L^2(\Gamma\backslash\mathbb{H})$ belongs to the domain $\tilde{\mathcal{D}}$ of $\tilde{\Delta}$ if and only if there exists a sequence $(f_n)_{n\geq 1}$ in $\mathcal{D}^\infty$ (resp. in $\mathcal{D}$) converging to g, such that $(\Delta f_n)_{n\geq 1}$ converges in $L^2(\Gamma\backslash\mathbb{H})$. In this case

$$\tilde{\Delta} g := \lim_{n\to\infty} \Delta f_n.$$

The following lemma will be used later.

Lemma 1.11. *If $f \in \tilde{\mathcal{D}}$ is twice continuously differentiable, then f belongs to $\mathcal{D}$.*

Proof. Take an arbitrary $g \in \mathcal{D}^\infty$ and write g in the form (1.6) with $h \in C_c^\infty(\mathbb{H})$ and let $\mathcal{F}$ be a fundamental domain of Γ. Then we have

$$\begin{aligned}\int_{\mathcal{F}} (\tilde{\Delta} f)\, \bar{g}\, dv &= < \tilde{\Delta} f, g >= < f, \tilde{\Delta} g >= < f, \Delta g >= \\ &= \int_{\mathbb{H}} f\, (\overline{\Delta h})\, dv = \int_{\mathbb{H}} (\Delta f)\, \bar{h}\, dv = \int_{\mathcal{F}} (\Delta f)\, \bar{g}\, dv.\end{aligned}$$

In this computation we have made use of arguments of the type (1.25), (1.26). Since $g \in \mathcal{D}^\infty$ is arbitrary, this implies $\tilde{\Delta} f = \Delta f \in L^2(\Gamma \backslash \mathbb{H})$. □

The following remark seems useful with respect to Theorem 1.9. Let $\Gamma <$ $\mathbf{PSL}(2,\mathbb{C})$ be a discrete group and consider the linear operators $S := \Delta|_{\mathcal{D}^\infty}$, and $T := \Delta|_{\mathcal{D}}$. Note that the proof of Theorem 1.9 makes essential use of the fact that T is symmetric. Hence in some sense we may say that the symmetry of T is sufficient for the essential self-adjointness of S. We shall show now that the essential self-adjointness of S implies the symmetry of T almost trivially. An elementary application of Stokes' theorem, compare (1.25), (1.26), yields $< Sf, g >=< f, Tg >$ for all $f \in \mathcal{D}^\infty$, $g \in \mathcal{D}$ and hence $S \subset T \subset S^*$, where $S \subset T$ means that T extends S. Assume now that S is essentially self-adjoint. Then we have $S^* = S^{**}$, and we conclude $T \subset S^* = S^{**} \subset T^*$. Hence T is symmetric, which we wanted to show.

Quite general results on the self-adjointness of formally symmetric partial differential operators have been obtained by Chernoff (1973) and Unell (1980).

There is also another approach to define the self-adjoint extension of Δ. Here one uses the Friedrichs extension of a densely defined, positive, symmetric operator on a Hilbert space, see Kato (1976), Chapter 6. This approach is taken by Lax, Phillips (1976). We start off with our operator $\Delta : \mathcal{D}^\infty \to L^2(\Gamma \setminus \mathbb{H})$ with associated closable sesquilinear form

$$Q(f,g) = \int_{\mathcal{F}} \mathbf{Gr}(f,g)\, dv =< -\Delta f, g > \qquad (f,g \in \mathcal{D}^\infty).$$

The closed sesquilinear form associated to Q has the domain

$$H^1(\Gamma \setminus \mathbb{H}) := \{\, f \in L^2(\Gamma \setminus \mathbb{H}) \mid \mathbf{grad}(f) \in L^2(\Gamma \setminus \mathbb{H}) \,\}.$$

The hyperbolic gradient occurring in the definition of $H^1(\Gamma \backslash \mathbb{H})$ is understood in the distributional sense. What this means is explained in Section 1 of Chapter 6, in particular in (6.1.21). $H^1(\Gamma \setminus \mathbb{H})$ is usually called the first Sobolev space of $\Gamma \setminus \mathbb{H}$. Associated to the densely defined, closed sesquilinear form Q with domain $H^1(\Gamma \setminus \mathbb{H})$ we have a self-adjoint operator $\mathfrak{Q}$, with possibly smaller domain so that

$$Q(f,g) =< \mathfrak{Q} f, g >$$

for all f in the domain of $\mathfrak{Q}$ and $g \in H^1(\Gamma \setminus \mathbb{H})$. $\mathfrak{Q}$ is called the Friedrichs extension of $\Delta : \mathcal{D}^\infty \to L^2(\Gamma \setminus \mathbb{H})$.

Theorem 1.12. *Let $\Gamma < \mathbf{PSL}(2,\mathbb{C})$ be a discrete group. The Friedrichs extension of $\Delta : \mathcal{D}^\infty \to L^2(\Gamma \setminus \mathbb{H})$ is $\tilde{\Delta}$. Its domain can also be described as the second Sobolev space*

$$\tilde{\mathcal{D}} = H^2(\Gamma \setminus \mathbb{H}) := \{\, f \in L^2(\Gamma \setminus \mathbb{H}) \mid \Delta f \in L^2(\Gamma \setminus \mathbb{H}) \,\}. \tag{1.28}$$

The differential operator Δ is applied to $f \in L^2(\Gamma \setminus \mathbb{H})$ in the distributional sense, see Chapter 6, Section 1 for explanation. The theorem follows easily from our formula (1.12) and the uniqueness statements in the first representation theorem from Kato (1976), Chapter 6.

4.2 The Resolvent Kernel

We shall prove in this section that the Maaß–Selberg series (3.1.47) is the kernel of the resolvent operator $(-\tilde{\Delta} - \lambda)^{-1}$ for suitable values of λ. We start with a useful technical lemma which will imply later that the resolvent kernel is of Carleman type.

Lemma 2.1. *Let $t > 1$ and $Q \in \mathbb{H}$. Then there exists a constant $C > 0$ independent of Q such that for all $M \in \mathbf{PSL}(2, \mathbb{C})$*

$$\int_{\mathbb{H}} \delta(MP, Q)^{-t} \delta(P, Q)^{-t} \, dv(P) \le C \, \frac{1 + \log \delta(MQ, Q)}{\delta(MQ, Q)^t}. \tag{2.1}$$

C can be chosen as a continuous function of t.

Proof. Transform $\mathbb{H}$ isometrically onto the unit ball $\mathbb{B}$ such that $P \mapsto x$, $Q \mapsto 0$ (cf. Chapter 1). Let $M \in \mathbf{PSL}(2, \mathbb{C})$ correspond to the isometry S of $\mathbb{B}$. Then we have by Chapter 1

$$\delta(P, Q) = \cosh d(P, Q) = \cosh d'(x, 0) = \frac{1 + \|x\|^2}{1 - \|x\|^2} \tag{2.2}$$

and hence

$$\begin{aligned} &\int_{\mathbb{H}} \delta(MP, Q)^{-t} \delta(P, Q)^{-t} \, dv(P) \\ &= \int_{\mathbb{B}} \left(\frac{1 - \|Sx\|^2}{1 + \|Sx\|^2} \right)^t \left(\frac{1 - \|x\|^2}{1 + \|x\|^2} \right)^t dv'(x) \\ &\le \int_{\mathbb{B}} (1 - \|Sx\|^2)^t (1 - \|x\|^2)^t \, dv'(x), \end{aligned} \tag{2.3}$$

where v' is the hyperbolic volume measure on $\mathbb{B}$.

Here we need an expression for $1 - \|Sx\|^2$. For $0 \ne y \in \mathbb{B}$ let $y^* = \frac{y}{\|y\|^2}$ be the reflection of y with respect to the unit sphere S^2 and let $A_{y^*}(x)$ be the reflection of x with respect to the sphere with centre y^* orthogonal to S^2. Then A_{y^*} is given by

$$A_{y^*}(x) = y^* + \frac{\|y^*\|^2 - 1}{\|x - y^*\|^2}(x - y^*). \tag{2.4}$$

Consider also the reflection B_y in the plane containing 0 with normal vector $\frac{y}{\|y\|}$. It is given formally by $B_y(x) = x - 2 < x, y > y^*$. Then $T_y = B_y \circ A_{y^*}$ is an isometry of $\mathbb{B}$. Moreover, we obtain, after some computation

$$\|T_y(x)\| = \|A_{y^*}(x)\| = \frac{\|x-y\|}{\|y\| \, \|x-y^*\|}, \tag{2.5}$$

$$1 - \|T_y(x)\|^2 = \frac{(1-\|x\|^2)(1-\|y\|^2)}{\|y\|^2 \|x-y^*\|^2}. \tag{2.6}$$

Observe that $A_{y^*}(y) = 0$ and hence $T_y(y) = 0$. So T_y is an orientation preserving isometry of $\mathbb{B}$ transforming y to the origin such that the distance of $T_y(x)$ from the unit sphere is given by (2.6). In the case $y = 0$ we define T_y to be the identity and observe that (2.6) remains true if we replace the denominator on the right-hand side by its limit as $y \to 0$, which is equal to one.

Now let S be as in (2.3), put $y := S^{-1}(0)$ and consider $U := T_y S^{-1}$. Then U is an orientation preserving isometry of $\mathbb{B}$, and U leaves 0 fixed. This implies that $U \in \mathbf{SO}(3)$, and we obtain $S = U^{-1} T_y$ and hence by (2.6)

$$1 - \|Sx\|^2 = 1 - \|T_y(x)\|^2 = \frac{(1-\|y\|^2)\,(1-\|x\|^2)}{\|y\|^2 \, \|x-y^*\|^2}. \tag{2.7}$$

Now observe that v' is given by

$$dv' = \frac{8\rho^2 \, d\rho \, d\Omega}{(1-\rho^2)^3} \tag{2.8}$$

where $\rho = \|x\|$ and where Ω denotes the Euclidean surface measure on the unit sphere $S^2 \subset \mathbb{R}^3$. Inserting the expression (2.7) into the right-hand side of (2.3) we find

$$\int_{\mathbb{B}} (1-\|Sx\|^2)^t \, (1-\|x\|^2)^t \, dv'(x) = 8\,(1-\|S^{-1}(0)\|^2)^t$$
$$\cdot \int_0^1 (1-\rho^2)^{2t-3} \left(\int_{S^2} \left\| \|S^{-1}(0)\|x - \frac{S^{-1}(0)}{\|S^{-1}(0)\|} \right\|^{-2t} d\Omega \right) \rho^2 \, d\rho.$$

Here we assume that $S^{-1}(0) \neq 0$. Applying a suitable orthogonal linear mapping V we may replace the vector $\|S^{-1}(0)\|^{-1} S^{-1}(0)$ by $(0,0,-1)$. Since Ω is $\mathbf{O}(3)$-invariant, we may write x instead of Vx. Introducing polar coordinates we have

$$d\Omega = \sin\vartheta \, d\vartheta \, d\varphi, \qquad 0 \le \varphi \le 2\pi, \ 0 \le \vartheta \le \pi,$$

and performing the integration with respect to φ we obtain

$$\begin{aligned}
&\int_{\mathbb{B}} (1-\|Sx\|^2)^t\,(1-\|x\|^2)^t\,dv'(x) \\
&= 16\pi\,(1-\|S^{-1}(0)\|^2)^t \int_0^1 (1-\rho^2)^{2t-3}\,\rho^2 \\
&\cdot \int_0^\pi \left(\|S^{-1}(0)\|^2\rho^2 + 2\rho\|S^{-1}(0)\|\cos\vartheta + 1\right)^{-t} \sin\vartheta\,d\vartheta\,d\rho \\
&= 32\pi\,(1-\|S^{-1}(0)\|^2)^t \\
&\cdot \int_0^1 (1-\rho^2)^{2t-3}\,F\left(t,t-\frac{1}{2};\frac{3}{2};\|S^{-1}(0)\|^2\rho^2\right)\rho^2\,d\rho \qquad (2.9)\\
&= 16\pi\,(1-\|S^{-1}(0)\|^2)^t \\
&\cdot \int_0^1 u^{\frac{1}{2}}\,(1-u)^{2t-3}\,F\left(t,t-\frac{1}{2};\frac{3}{2};\|S^{-1}(0)\|^2u\right)du \\
&= 16\pi\,\frac{\Gamma(\frac{3}{2})\,\Gamma(2t-2)}{\Gamma(2t-\frac{1}{2})} \\
&\cdot F\left(t,t-\frac{1}{2};2t-\frac{1}{2};\|S^{-1}(0)\|^2\right)(1-\|S^{-1}(0)\|^2)^t.
\end{aligned}$$

In the above computation some formulas involving the hypergeometric function are used, see Magnus, Oberhettinger, Soni (1966), page 55. The above result is also true if $S^{-1}(0)=0$. Now observe that there exists a constant $c>0$ such that for all $\xi\in[0,1[$

$$|F(a,b;a+b;\xi)| \le c(1+|\log(1-\xi)|), \tag{2.10}$$

see Magnus, Oberhettinger, Soni (1966), page 49. In the case at hand we have $a=t$, $b=t-\frac{1}{2}$, and c may be chosen as a continuous function of t. Combining (2.3), (2.9), (2.10) we obtain finally

$$\begin{aligned}
&\int_{\mathbb{H}} \delta(MP,Q)^{-t}\delta(P,Q)^{-t}\,dv(P) \\
&\le C_1\,(1-\|S^{-1}(0)\|^2)^t\,(1+|\log(1-\|S^{-1}(0)\|^2)|) \\
&\le C\left(\frac{1-\|S^{-1}(0)\|^2}{1+\|S^{-1}(0)\|^2}\right)^t\left(1+\left|\log\left(\frac{1-\|S^{-1}(0)\|^2}{1+\|S^{-1}(0)\|^2}\right)\right|\right) \\
&= C\,\delta(M^{-1}Q,Q)^{-t}\,(1+\log\delta(M^{-1}Q,Q)) \\
&= C\,\delta(MQ,Q)^{-t}\,(1+\log\delta(MQ,Q))
\end{aligned}$$

for some constant C having the required properties. □

Lemma 2.2. *Let* $s\in\mathbb{C}$,

$$\varphi_s(\delta) := \frac{(\delta+\sqrt{\delta^2-1})^{-s}}{\sqrt{\delta^2-1}} \qquad (\delta>1), \tag{2.11}$$

and suppose that $\lambda = 1 - s^2$ *and* $u \in C_c^2(\mathbb{H})$. *Then*

$$4\pi\, u(Q) = \int_{\mathbb{H}} \varphi_s(\delta(P,Q))\, (-\Delta - \lambda)u(P)\, dv(P). \tag{2.12}$$

Proof. We transform $\mathbb{H}$ isometrically onto $\mathbb{B}$ such that $Q \mapsto 0$, $P \mapsto x = (\xi, \eta, \zeta)$. Put $\rho = \|x\|$. On $\mathbb{B}$, we consider the differential form

$$\begin{aligned} \omega = 2\,(1-\rho^2)^{-1}\,&((f_\xi g - f g_\xi)\, d\eta \wedge d\zeta \\ &+ (f_\eta g - f g_\eta)\, d\zeta \wedge d\xi + (f_\zeta g - f g_\zeta)\, d\xi \wedge d\eta). \end{aligned} \tag{2.13}$$

Then we have $d\omega = ((\Delta f)g - f(\Delta g))\, dv'$. Transform the integral in (2.12) to an integral over $\mathbb{B}$ as indicated above, exclude a small Euclidean ball of radius ϵ and centre 0 and denote the inverse image of $\mathbb{B}_\epsilon := \{\, x \in \mathbb{B} \mid \|x\| \geq \epsilon \,\}$ in $\mathbb{H}$ by $\mathbb{H}_\epsilon$. Then we have

$$\begin{aligned} &\int_{\mathbb{H}} \varphi_s(\delta(P,Q))\, (-\Delta - \lambda)u(P)\, dv(P) \\ &\qquad = \lim_{\epsilon \to 0} \int_{\mathbb{H}_\epsilon} \varphi_s(\delta(P,Q))\, (-\Delta - \lambda)u(P)\, dv(P), \end{aligned} \tag{2.14}$$

and

$$\begin{aligned} &\int_{\mathbb{H}_\epsilon} \varphi_s(\delta(P,Q))\, (-\Delta - \lambda)u(P)\, dv(P) \\ &= \int_{\mathbb{H}_\epsilon} ((\Delta \varphi_s(\delta(P,Q)))\, u(P) - \varphi_s(\delta(P,Q))\, \Delta u(P))\, dv(P) \\ &= \int_{\mathbb{B}_\epsilon} d\omega, \end{aligned} \tag{2.15}$$

where we have to insert

$$f(x) = \frac{1-\rho^2}{2\rho} \left(\frac{1-\rho}{1+\rho} \right)^s = \varphi_s(\delta(P,Q)), \tag{2.16}$$

and $g(x) = u(P)$ in (2.13).

We compute the right-hand side of (2.15) by means of Stokes' theorem. Let S_ϵ be the boundary of $\mathbb{B}_\epsilon$ with orientation such that the normal is directed outside. Introducing spherical coordinates, we obtain the restriction of our differential form to the sphere S_ϵ from

$$f_\xi\, d\eta \wedge d\zeta + f_\eta\, d\zeta \wedge d\xi + f_\zeta\, d\xi \wedge d\eta = \frac{1}{\rho} < (\xi, \eta, \zeta), \mathrm{grad}(f) > \rho^2\, d\Omega, \tag{2.17}$$

with $\rho = \epsilon$. Here $< \cdot\, , \,\cdot >$ denotes the scalar product in $\mathbb{R}^3$, $\mathrm{grad}(f)$ is the Euclidean gradient of f, and Ω is the Euclidean surface measure on S^2 (cf. Spivak (1965), page 128, Theorem 5.6). Hence we obtain

$$\int_{\mathbb{B}_\epsilon} d\omega = -\int_{S_\epsilon} \omega = -\,2(1-\epsilon^2)^{-1}\left(\int_{S^2} \frac{1}{\rho} < (\xi,\eta,\zeta), \mathrm{grad}(f) > g\ \rho^2\ d\Omega \right.$$
$$\left. - \int_{S^2} \frac{1}{\rho} < (\xi,\eta,\zeta), \mathrm{grad}(g) > f\ \rho^2\ d\Omega \right)\Big|_{\rho=\epsilon}.$$

As x tends to zero, we have $f(x) = O(\rho^{-1})$. Hence the second integral converges to zero as $\epsilon \to 0$. However, in the first integral we have

$$\frac{1}{\rho} < (\xi,\eta,\zeta), \mathrm{grad}(f) >= \frac{\partial}{\partial\rho}\frac{1-\rho^2}{2\rho}\left(\frac{1-\rho}{1+\rho}\right)^s = -\frac{1}{2\rho^2} + O\left(\frac{1}{\rho}\right).$$

Hence we finally obtain $\lim_{\epsilon\to 0}\int_{\mathbb{B}_\epsilon} d\omega = 4\pi\ g(0) = 4\pi\ u(Q)$. This completes the proof of the Lemma. □

We go on to prove that the Maaß–Selberg series defines a kernel of Carleman type. For this and later purposes it is convenient to decompose the kernel into a continuous part and a part with singularities, such that the contribution of the singularities has small support.

Choose once and for all a function $\psi \in C_c^\infty([0,\infty[)$, such that ψ is decreasing and

$$\psi(t) = \begin{cases} 1 & \text{for} \quad 0 \le t \le 2, \\ 0 & \text{for} \quad t > 3. \end{cases} \tag{2.18}$$

Define

$$g(t) := \frac{1}{4\pi\sqrt{2}}\ \frac{1}{\sqrt{t-1}}\ \psi(t) \qquad (t > 1) \tag{2.19}$$

Then the function

$$k_s(t) := \frac{1}{4\pi}\ \varphi_s(t) - g(t) \tag{2.20}$$

is continuous for $t \ge 1$, and satisfies an estimate of the form $|k_s(t)| \le C_1\, t^{-1-s}$ for all $t \ge 1$, where C_1 may be chosen as a continuous function of s.

Now let $\Gamma < \mathbf{PSL}(2,\mathbb{C})$ be a discrete group with fundamental domain $\mathcal{F}$ and let σ_0 be the abscissa of convergence of Γ. Define

$$G(P,Q) := \sum_{M\in\Gamma} g(\delta(P,MQ)) \qquad (P,\ Q \in \mathbb{H},\ P \not\equiv Q \bmod \Gamma), \tag{2.21}$$

$$K(P,Q,s) := \sum_{M\in\Gamma} k_s(\delta(P,MQ)) \qquad (P,\ Q \in \mathbb{H},\ \mathrm{Re}\, s > \sigma_0). \tag{2.22}$$

Observe that G has a weak singularity for $Q \to P$. If P and Q belong to compact subsets of $\mathbb{H}$, only finitely many terms on the right-hand side of (2.21) can be different from zero. The same holds if P belongs to a compact subset of $\mathbb{H}$ and Q to a Poincaré normal polyhedron for Γ. The series (2.22) defines a continuous function of $(P, Q, s) \in \mathbb{H} \times \mathbb{H} \times]\sigma_0, \infty[$. Moreover,

$$F(P, Q, s) = G(P, Q) + K(P, Q, s) \tag{2.23}$$

for $P,\ Q \in \mathbb{H},\ P \not\equiv Q \bmod \Gamma,\ \operatorname{Re} s > \sigma_0$. The series of the absolute values for F and K will be denoted by

$$|F|(P, Q, s) := \frac{1}{4\pi} \sum_{M \in \Gamma} |\varphi_s(\delta(P, MQ))| = F(P, Q, \operatorname{Re} s), \tag{2.24}$$

$$|K|(P, Q, s) := \sum_{M \in \Gamma} |k_s(\delta(P, MQ))|. \tag{2.25}$$

Lemma 2.3. *For* $\operatorname{Re} s > \max(0, \sigma_o)$, *the integral*

$$\int_{\mathcal{F}} (|F|(P, Q, s))^2 \, dv(Q) \tag{2.26}$$

converges uniformly on compact sets with respect to $P,\ s$ in the following sense. For any $\epsilon > 0$ and all compact subsets $\mathcal{K} \subset \mathbb{H},\ \mathcal{S} \subset \{ s \mid \operatorname{Re} s > \max(0, \sigma_0) \}$ there exists a compact subset $\mathcal{L} \subset \mathcal{F}$ such that

$$\int_{\mathcal{F} \setminus \mathcal{L}} (|F|(P, Q, s))^2 \, dv(Q) < \epsilon \tag{2.27}$$

for all $P \in \mathcal{K},\ s \in \mathcal{S}$. An analogous statement holds for $G(P, Q)$ and $|K|(P, Q, s)$.

Proof. Suppose $\operatorname{Re} s = \sigma > \max(0, \sigma_0)$, and let $L \subset \mathcal{F}$ be a measurable subset. Then we have

$$\begin{aligned} &\int_{\mathcal{F} \setminus L} (|K|(P, Q, s))^2 \, dv(Q) \\ &\quad = \sum_{M \in \Gamma} \int_{\mathbb{H} \setminus (\Gamma L)} |k_s(\delta(P, Q))| \, |k_s(\delta(P, MQ))| \, dv(Q). \end{aligned} \tag{2.28}$$

Now let $\mathcal{K} \subset \mathbb{H},\ \mathcal{S} \subset \{ s \mid \operatorname{Re} s > \max(0, \sigma_0) \}$ be compact sets, choose $P_0 \in \mathcal{K}$ and put $t := 1 + \min \{ \operatorname{Re} s \mid s \in \mathcal{S} \}$. Let $C > 0$ be such that $|k_s(\delta(P, Q))| \le C \, \delta(P_0, Q)^{-t}$ for all $P \in \mathcal{K},\ Q \in \mathbb{H},\ s \in \mathcal{S}$. This is possible by the obvious estimate for k_s. First choose $L = \emptyset$ in (2.28) to obtain for all $s \in \mathcal{S},\ P \in \mathcal{K}$

$$\begin{aligned} &\int_{\mathcal{F}} (|K|(P, Q, s))^2 \, dv(Q) \\ &\le C^2 \sum_{M \in \Gamma} \int_{\mathbb{H}} \delta(P_0, Q)^{-t} \delta(P_0, MQ)^{-t} \, dv(Q) < \infty \end{aligned} \tag{2.29}$$

by Lemma 2.1. Hence the integral (2.26) converges.

Secondly we prove the uniformity statement. For this we observe that for any finite subset $E \subset \Gamma$ and $s \in \mathcal{S},\ P \in \mathcal{K}$

$$\begin{aligned}
&\int_{\mathcal{F}\setminus L} (|K|(P,Q,s))^2\, dv(Q) \\
&\le C^2 \sum_{M \in E} \int_{\mathbb{H}\setminus(\Gamma L)} \delta(P_0,Q)^{-t}\delta(P_0,MQ)^{-t}\, dv(Q) \qquad (2.30)\\
&+ C^2 \sum_{M \in \Gamma\setminus E} \int_{\mathbb{H}} \delta(P_0,Q)^{-t}\delta(P_0,MQ)^{-t}\, dv(Q).
\end{aligned}$$

If $\epsilon > 0$ is given, choose E such that the second term on the right-hand side of (2.30) is less than $\epsilon/2$. This is possible by Lemma 2.1. Then choose a compact subset $L \subset \mathcal{F}$ such that the first term also is less than $\epsilon/2$. This proves (2.27) for $|K|$.

The corresponding statements for G are easily seen, since G is a finite sum if P ranges over a compact set and Q ranges over a Poincaré normal polyhedron for Γ. The statements for $|K|$ and G imply the truth of the lemma for $|F|$. □

Lemma 2.4. *Let* $P_0 \in \mathbb{H},\ s_0 \in \mathbb{C},\ \operatorname{Re} s_0 > \max(0,\sigma_0)$. *Then*

$$\lim_{\substack{P \to P_0 \\ s \to s_0}} \int_{\mathcal{F}} |F(P,Q,s) - F(P_0,Q,s_0)|^2\, dv(Q) = 0$$

uniformly on every compact subset of $\mathbb{H} \times \{\ s \mid \operatorname{Re} s > \max(0,\sigma_0)\ \}$. *A corresponding statement holds for* $G(P,Q)$ *and* $K(P,Q,s)$.

Proof. Let Φ be any one of the kernels $F(P,Q,s),\ G(P,Q),\ K(P,Q,s)$. Choose $\epsilon > 0$ and let $\mathcal{K} \subset \mathbb{H},\ \mathcal{S} \subset \{\ s \mid \operatorname{Re} s > \max(0,\sigma_0)\ \}$ be compact sets. Use Lemma 2.3 and choose a compact subset $\mathcal{L} \subset \mathcal{F}$ such that

$$\int_{\mathcal{F}\setminus\mathcal{L}} |\Phi(P,Q,s) - \Phi(P',Q,s')|^2\, dv(Q) < \frac{\epsilon}{2}$$

for all $P,\ P' \in \mathcal{K},\ s,\ s' \in \mathcal{S}$. The corresponding integral over $\mathcal{L}$ will also be less than $\frac{\epsilon}{2}$, provided that $P,\ P' \in \mathcal{K},\ s,\ s' \in \mathcal{S},\ d(P,P') < \eta,\ |s-s'| < \eta$, where $\eta > 0$ is sufficiently small. □

Summing up the essential content of Lemmas 2.3, 2.4, we have the following corollary.

Corollary 2.5. *For* $\operatorname{Re} s > \max(0,\sigma_0)$, *the Maaß–Selberg series* $F(P,Q,s)$ *is a kernel of Carleman type, and it is continuous in the mean square.*

Theorem 2.6. *Let $\Gamma < \mathbf{PSL}(2,\mathbb{C})$ be a discrete group with fundamental domain $\mathcal{F}$ and abscissa of convergence σ_0. Suppose $\lambda = 1 - s^2$, $\operatorname{Re} s > \max(0, \sigma_0)$. Then any element $u \in \tilde{\mathcal{D}}$ can be represented by a continuous function which satisfies*

$$u(P) = \int_{\mathcal{F}} F(P,Q,s)\,(-\tilde{\Delta} - \lambda)u(Q)\,dv(Q) \qquad (P \in \mathbb{H}). \tag{2.31}$$

Proof. For arbitrary $u \in \tilde{\mathcal{D}}$ the integral

$$f(P) = \int_{\mathcal{F}} F(P,Q,s)\,(-\tilde{\Delta} - \lambda)u(Q)\,dv(Q) \qquad (P \in \mathbb{H}). \tag{2.32}$$

exists and the right-hand side remains finite if $F(P,Q,s)$ and $(-\tilde{\Delta} - \lambda)u$ are replaced by $|F|(P,Q,s)$ and $|(-\tilde{\Delta} - \lambda)u|$, respectively (cf. Lemma 2.3). Hence the dominated convergence theorem implies that

$$f(P) = \frac{1}{4\pi}\int_{\mathbb{H}} \varphi_s(\delta(P,Q))\,(-\tilde{\Delta} - \lambda)u(Q)\,dv(Q). \tag{2.33}$$

Suppose first that we have $u \in \mathcal{D}^\infty$. Then u has the form $u = \sum_{M\in\Gamma} g\circ M$ with some $g \in C_c^\infty(\mathbb{H})$, see Lemma 1.3. Letting $h := (-\Delta - \lambda)g \in C_c^\infty(\mathbb{H})$ we have $(-\Delta - \lambda)u = \sum_{M\in\Gamma} h\circ M$ and hence by the monotone convergence theorem

$$\begin{aligned}
&\frac{1}{4\pi}\int_{\mathbb{H}} |\varphi_s(\delta(P,Q))|\,|(-\Delta - \lambda)u(Q)|\,dv(Q)\\
&\leq \frac{1}{4\pi}\int_{\mathbb{H}} |\varphi_s(\delta(P,Q))| \sum_{M\in\Gamma} |h(MQ)|\,dv(Q)\\
&= \int_{\mathbb{H}} |F|(P,Q,s)\,|h(Q)|\,dv(Q).
\end{aligned} \tag{2.34}$$

Since the compact support of h is covered by finitely many Γ-images of a Poincaré normal polyhedron for Γ, the right-hand side of (2.34) is finite. This justifies an application of the dominated convergence theorem in (2.33) for our u and we obtain

$$\begin{aligned}
&\frac{1}{4\pi}\sum_{M\in\Gamma}\int_{\mathbb{H}} \varphi_s(\delta(P,Q))\,(-\Delta - \lambda)(g\circ M)(Q)\,dv(Q)\\
&= \sum_{M\in\Gamma} (g\circ M)(P) = u(P).
\end{aligned}$$

In this computation we have used Lemma 2.2. Hence the theorem is true for all $u \in \mathcal{D}^\infty$.

Now assume that $u \in \tilde{\mathcal{D}}$. By Theorem 1.9 and also Definition 1.10 the operator $\tilde{\Delta} : \tilde{\mathcal{D}} \to L^2(\Gamma\setminus\mathbb{H})$ is the closure of $\Delta : \mathcal{D}^\infty \to L^2(\Gamma\setminus\mathbb{H})$. Hence there exists a sequence $(u_n)_{n\geq 1}$ in $\mathcal{D}^\infty$ such that $u = \lim_{n\to\infty} u_n$, $\tilde{\Delta} u = \lim_{n\to\infty} \Delta u_n$, the limits to be understood in the norm sense. The continuity of the inner product implies

$$\begin{aligned} f(P) &= < (-\tilde{\Delta} - \lambda)u, \overline{F(P, \cdot, s)} > = \lim_{n\to\infty} < (-\Delta - \lambda)u_n, \overline{F(P, \cdot, s)} > \\ &= \lim_{n\to\infty} \int_{\mathcal{F}} F(P, Q, s)\ (-\Delta - \lambda)u_n(Q)\ dv(Q) = \lim_{n\to\infty} u_n(P), \end{aligned}$$

since (2.31) holds with u being replaced by $u_n \in \mathcal{D}^\infty$. This means that the sequence (u_n) converges to f pointwise and to u in the mean square. Hence $f = u$ almost everywhere. But f is continuous by (2.32) and Lemma 2.4. So u can be represented by a continuous function on $\mathbb{H}$ which satisfies (2.31). This completes our proof. □

For any $\lambda \in \rho(-\tilde{\Delta})$ the resolvent operator $R_\lambda := (-\tilde{\Delta} - \lambda)^{-1} : L^2(\Gamma \setminus \mathbb{H}) \to \tilde{\mathcal{D}}$ is defined on all of $L^2(\Gamma \setminus \mathbb{H})$ and has its values in $\tilde{\mathcal{D}}$. If $\lambda \in \rho(-\tilde{\Delta})$ has the form $\lambda = 1 - s^2$ for some $s \in \mathbb{C}$ such that $\operatorname{Re} s > \max(0, \sigma_0)$, we may insert $u = (-\tilde{\Delta} - \lambda)^{-1} f$ with $f \in L^2(\Gamma \setminus \mathbb{H})$ in (2.31) to obtain

$$R_\lambda f = \int_{\mathcal{F}} F(\,\cdot\,, Q, s)\ f(Q)\ dv(Q). \tag{2.35}$$

In particular, this is true for all $\lambda \in \mathbb{C} \setminus [0, \infty[$ which satisfy the above conditions. This means that the Maaß–Selberg series is the kernel of the resolvent R_λ. The following considerations show that $\lambda = 1 - s^2$, $\operatorname{Re} s > \max(0, \sigma_0)$ imply that $\lambda \in \rho(-\tilde{\Delta})$, so that (2.35) holds true for all such λ and $f \in L^2(\Gamma \backslash \mathbb{H})$.

Theorem 2.7. *Let $\Gamma < \mathbf{PSL}(2, \mathbb{C})$ be a discrete group with abscissa of convergence σ_0. Then $\sigma(-\tilde{\Delta}) \subset [1 - (\max(0, \sigma_0))^2, \infty[$.*

Some explanations are in order before we prove this theorem. If Γ has finite covolume, we have $\sigma_0 = 1$, and the theorem is trivially true since $-\tilde{\Delta}$ is a positive self-adjoint operator. However, if $\sigma_0 < 1$, it gives us new information on the spectrum of $-\tilde{\Delta}$. In particular, if $\sigma_0 \leq 0$, we have $\sigma(-\tilde{\Delta}) \subset [1, \infty[$.

Proof of Theorem 2.7. Let $(E_\lambda)_{\lambda \in \mathbb{R}}$ be the right continuous spectral family of $-\tilde{\Delta}$ and assume that $\alpha < \beta < 1 - (\max(0, \sigma_0))^2$. We shall show that $E_\alpha = E_\beta$ and this will yield our result because for all $\lambda \in \sigma(-\tilde{\Delta})$ and for all $\epsilon > 0$ we have $E_{\lambda - \epsilon} \neq E_{\lambda + \epsilon}$.

For $\alpha \leq t \leq \beta$, $\epsilon > 0$, $\lambda = t \pm i\epsilon$ we apply formula (2.35). The well-known relation between the resolvent and the spectral family of a self-adjoint operator, see e.g. Dunford, Schwartz (1958), page 1202, Theorem 10, now yields for an arbitrary $f \in L^2(\Gamma \setminus \mathbb{H})$:

$$
\begin{aligned}
\frac{1}{2}\left(E_\beta + E_{\beta-0} - E_\alpha - E_{\alpha-0}\right) f &= \lim_{\epsilon\to 0} \frac{1}{2\pi i}\int_\alpha^\beta (R_{t+i\epsilon} - R_{t-i\epsilon})\, f\, dt \\
&= \lim_{\epsilon\to 0} \frac{1}{2\pi i}\int_\alpha^\beta \int_{\mathcal{F}} (F(\cdot, Q, \sqrt{1-t-i\epsilon}) \\
&\qquad - F(\cdot, Q, \sqrt{1-t+i\epsilon}))f(Q)\, dv(Q)\, dt.
\end{aligned} \tag{2.36}
$$

Here the limits are taken in the norm sense in $L^2(\Gamma \setminus \mathbb{H})$, and the integrals with respect to t are $L^2(\Gamma\setminus\mathbb{H})$-limits of Riemann sums. The square roots are taken with positive real parts. First we consider the pointwise limit on the right-hand side of (2.36). An application of the Cauchy-Schwarz inequality and of Lemma 2.4 yields

$$
\begin{aligned}
\lim_{\epsilon\to 0}\int_\alpha^\beta \int_{\mathcal{F}} &(F(P, Q, \sqrt{1-t-i\epsilon}) \\
&- F(P, Q, \sqrt{1-t+i\epsilon}))f(Q)\, dv(Q)\, dt = 0
\end{aligned} \tag{2.37}
$$

for all $P \in \mathbb{H}$. Here the integral with respect to t simply is a Riemann integral of the corresponding continuous integrand. We show that this Riemann integral is equal to the $L^2(\Gamma \setminus \mathbb{H})$-limit of the corresponding Riemann sums occurring in (2.36). Let $w_\epsilon(\cdot, t) := (R_{t+i\epsilon} - R_{t-i\epsilon})\, f$ for $\epsilon > 0,\ \alpha \le t \le \beta$. Then $w_\epsilon : \mathbb{H}\times[\alpha,\beta] \to \mathbb{C}$ is a continuous function, and $[\alpha,\beta] \ni t \mapsto w_\epsilon(\cdot,t) \in L^2(\Gamma\setminus\mathbb{H})$ is a continuous family of elements from $\tilde{\mathcal{D}}$. Consider now a sequence $T_n(P)$ of Riemann sums for the Riemann integral $g_\epsilon(P) := \int_\alpha^\beta w_\epsilon(P,t)\, dt$. This sequence may also be thought of as a sequence $T_n \in L^2(\Gamma \setminus \mathbb{H})$ of Riemann sums for the $L^2(\Gamma \setminus \mathbb{H})$-Riemann integral $h_\epsilon := \int_\alpha^\beta w_\epsilon(\cdot,t)\, dt$. The sequence $(T_n)_{n\ge 1}$ converges to g_ϵ pointwise and to h_ϵ in the mean square. Hence $h_\epsilon = g_\epsilon$ almost everywhere. Summing up, we obtain from (2.37)

$$
\lim_{\epsilon\to 0}\int_\alpha^\beta (R_{t+i\epsilon} - R_{t-i\epsilon})\, f\, dt = 0
$$

and hence (2.36) yields $(E_\beta + E_{\beta-0} - E_\alpha - E_{\alpha-0})\, f = 0$ for all $f \in L^2(\Gamma\setminus\mathbb{H})$. This implies that $E_\beta - E_\alpha = 0$ which completes the proof. □

Corollary 2.8. *Let $\Gamma < \mathbf{PSL}(2, \mathbb{C})$ be a discrete group with abscissa of convergence σ_0. Suppose that $\lambda \in \mathbb{C} \setminus [1, \infty[$, and assume that $\lambda = 1 - s^2$, $\operatorname{Re} s > \max(0, \sigma_0)$. Then λ belongs to the resolvent set $\rho(-\tilde{\Delta})$, and for all $f \in L^2(\Gamma \setminus \mathbb{H})$ the resolvent operator is given by*

$$
R_\lambda f = \int_{\mathcal{F}} F(\cdot, Q, s)\, f(Q)\, dv(Q). \tag{2.38}
$$

Proof. Theorem 2.7 yields that any λ satisfying the hypotheses of the corollary belongs to $\rho(-\tilde{\Delta})$. Equation (2.38) is clear from the discussion preceding Theorem 2.7. □

Let Γ, σ_0 be as in Corollary 2.8. Then the set of $\lambda \in \mathbb{C}$ satisfying the hypotheses of the Corollary has the following simple description. For $\sigma_0 \leq 0$ the only restriction on λ is that $\lambda \notin [1, \infty[$. In this case the resolvent operator is represented by the formula (2.38) for all $\lambda \in \mathbb{C} \setminus [1, \infty[$. Hence it is possible to compute the spectral family by (2.36). In the case of the hyperbolic plane the corresponding calculations are done in Elstrodt (1973b), Section 6. Since in our present work we are mainly interested in the case of cofinite groups which have $\sigma_0 = 1$, we do not include a discussion of this point here. Now assume that $0 < \sigma_0 \leq 1$. Then the points $\lambda \in \mathbb{C} \setminus [1, \infty[$ such that $\operatorname{Re} \sqrt{1-\lambda} > \sigma_0$ are precisely the complex numbers $\lambda = \zeta + i\eta$ lying to the left of the parabola described by the equation $\eta^2 = 4\sigma_0^2(\xi - (1 - \sigma_0^2))$. In the case of cofinite groups we have $\sigma_0 = 1$, and the parabola is described by the equation $\eta^2 = 4\xi$. With this figure in mind it is natural to ask whether it is possible to represent the resolvent operator in the form (2.35) with a suitable kernel for all $\lambda \in \mathbb{C} \setminus [1 - \sigma_0^2, \infty[$. We shall prove in Section 4 that this is true. The proof of this theorem on analytic continuation of the resolvent kernel relies upon an explicit formula for the norm $\|F(P, \cdot, s)\|$ which we shall develop now.

Theorem 2.9. *Let $\Gamma < \mathbf{PSL}(2, \mathbb{C})$ be a discrete group with fundamental domain $\mathcal{F}$ and abscissa of convergence σ_0, and suppose that $\lambda, \mu \in \mathbb{C} \setminus [1, \infty[$, $\lambda = 1 - s^2$, $\mu = 1 - t^2$ and assume* $\operatorname{Re} s > \max(0, \sigma_0)$, $\operatorname{Re} t > \max(0, \sigma_0)$. *Then*

$$(2.39) \quad (\lambda - \mu) \int_{\mathcal{F}} F(P, Z, s)\, F(Z, Q, t)\, dv(Z) = \lim_{Z \to Q} (F(P, Z, s) - F(P, Z, t))$$

for all P, $Q \in \mathbb{H}$. In particular, for non-real λ the equation

$$(2.40) \qquad \int_{\mathcal{F}} |F(P, Z, s)|^2\, dv(Z) = \frac{1}{\lambda - \bar{\lambda}} \lim_{Z \to P} (F(P, Z, s) - F(P, Z, \bar{s}))$$

holds, whereas for λ real we have

$$(2.41) \quad \begin{aligned} \int_{\mathcal{F}} |F(P, Z, s)|^2\, dv(Z) &= \lim_{Z \to P} \frac{\partial}{\partial \lambda} F(P, Z, s) \\ &= \lim_{\mu \to \lambda} \frac{1}{\lambda - \mu} \left(\lim_{Z \to P} (F(P, Z, s) - F(P, Z, t)) \right). \end{aligned}$$

Proof. Since $\lambda, \mu \in \rho(-\tilde{\Delta})$ we have Hilbert's resolvent equation

$$(\lambda - \mu) R_\lambda R_\mu = R_\lambda - R_\mu$$

which we apply to an arbitrary $f \in \mathcal{D}^\infty$ using (2.38) and obtain after a formal interchange of the succession of integrations for all $P \in \mathbb{H}$

$$
\begin{aligned}
&(\lambda-\mu)\int_{\mathcal{F}} F(P,Z,s)\int_{\mathcal{F}} F(Z,Q,t)\, f(Q)\, dv(Q)\, dv(Z) \\
&= (\lambda-\mu)\int_{\mathcal{F}}\int_{\mathcal{F}} F(P,Z,s)\, F(Z,Q,t)\, dv(Z)\, f(Q)\, dv(Q) \qquad (2.42)\\
&= \int_{\mathcal{F}} (F(P,Q,s)-F(P,Q,t))\, f(Q)\, dv(Q).
\end{aligned}
$$

Observe that both sides are continuous functions of P by Theorem 2.6. To justify the interchange of the order of integrations notice that the function

$$
\mathbb{H} \ni Q \mapsto \int_{\mathcal{F}} |F(P,Z,s)|\, |F(Z,Q,t)|\, dv(Z)
$$

is continuous, see Corollary 2.5. This implies that the integral

$$
\int_{\mathcal{F}}\int_{\mathcal{F}} |F(P,Z,s)|\, |F(Z,Q,t)|\, dv(Z)\, |f(Q)|\, dv(Q)
$$

is finite for all $f \in \mathcal{D}^\infty$. Hence equation (2.42) holds for all $f \in \mathcal{D}^\infty,\ P \in \mathbb{H}$ by Fubini's theorem. This implies that for every $P \in \mathbb{H}$ the equation

$$
\begin{aligned}
(\lambda-\mu)\int_{\mathcal{F}} F(P,Z,s)\, F(Z,Q,t)\, dv(Z) &= F(P,Q,s)-F(P,Q,t) \\
&= \lim_{Z\to Q} (F(P,Z,s)-F(P,Z,t))
\end{aligned}
$$

holds for almost all $Q \in \mathbb{H}$. But both sides of this equation are continuous functions of $Q \in \mathbb{H}$, see Corollary 2.5 and (2.23). Note that $K(P,Q,s)$ is continuous on $\mathbb{H} \times \mathbb{H} \times \{\, s \mid \operatorname{Re} s > \sigma_0 \,\}$. Hence (2.39) holds for all $P,\ Q \in \mathbb{H}$, and the remaining assertions are now obvious since $\overline{F(P,Q,s)} = F(Q,P,\bar{s})$. □

Corollary 2.10. *Let Γ, $\mathcal{F}$, σ_0, λ, s be as in Theorem 2.9. Then for λ non-real and $P \in \mathbb{H}$*

$$
\begin{aligned}
\int_{\mathcal{F}} |F(P,Q,s)|^2\, dv(Q) &= \frac{|\Gamma_P|}{4\pi}\frac{1}{s+\bar{s}} \\
&- \frac{1}{4\pi(s^2-\bar{s}^2)} \sum_{M\in\Gamma\backslash\Gamma_P} (\varphi_s(\delta(P,MP)) - \varphi_{\bar{s}}(\delta(P,MP)))\,,
\end{aligned} \tag{2.43}
$$

and for λ real and $P \in \mathbb{H}$

$$
\begin{aligned}
\int_{\mathcal{F}} |F(P,Q,s)|^2\, dv(Q) &= \frac{|\Gamma_P|}{8\pi s} \\
&+ \frac{1}{8\pi s} \sum_{M\in\Gamma\backslash\Gamma_P} \varphi_s(\delta(P,MP)) \log\left(\delta(P,MP) + \sqrt{\delta(P,MP)^2-1}\right).
\end{aligned} \tag{2.44}
$$

Proof. If λ is non-real, we have for any $P \in \mathbb{H}$ by (2.40)

$$(\lambda - \bar{\lambda}) \int_{\mathcal{F}} |F(P,Q,s)|^2 \, dv(Q) = \lim_{Z \to P} \frac{|\Gamma_P|}{4\pi} \left(\varphi_s(\delta(P,Z)) - \varphi_{\bar{s}}(\delta(P,Z))\right)$$
$$+ \frac{1}{4\pi} \sum_{M \in \Gamma \backslash \Gamma_P} \left(\varphi_s(\delta(P,MP)) - \varphi_{\bar{s}}(\delta(P,MP))\right).$$

Since $\varphi_s(\delta) - \varphi_{\bar{s}}(\delta) = -(s - \bar{s}) + O(\sqrt{\delta - 1})$ as $\delta \to 1$, equation (2.43) is proved, and (2.44) is derived from (2.43) by an obvious limit process. □

Theorem 2.11. *Let $\Gamma < \mathbf{PSL}(2, \mathbb{C})$ be a discrete group and assume that the series*

$$\sum_{M \in \Gamma} \delta(P, MQ)^{-1} \tag{2.45}$$

converges for all $P, Q \in \mathbb{H}$. Then $\tilde{\Delta} : \tilde{\mathcal{D}} \to L^2(\Gamma \setminus \mathbb{H})$ has no eigenvalues.

Of course the assumption of Theorem 2.11 does not hold for cofinite groups. However the trivial group is an example for Theorem 2.11.

Proof. By Theorem 2.7 we have only to show that there are no eigenvalues of $-\tilde{\Delta}$ in $[1, \infty[$. Suppose that $\lambda \geq 0$ and that $f \in \tilde{\mathcal{D}}$ satisfies $-\tilde{\Delta} f = \lambda f$. Then we obtain for all $\zeta \in \mathbb{C}$ the equation $(-\tilde{\Delta} - (\lambda + \zeta)) f = -\zeta f$. We now choose ζ any complex number such that $\lambda + \zeta$ belongs to the region where the resolvent is presented by the Maaß–Selberg series. This implies that

$$f(P) = -\zeta \int_{\mathcal{F}} F(P, Q, s(\zeta)) \, f(Q) \, dv(Q),$$

where $\lambda + \zeta = 1 - s(\zeta)^2$, $\operatorname{Re} s(\zeta) > 0$. Hence we obtain from the Cauchy-Schwarz inequality

$$|f(P)| \leq |\zeta| \left(\int_{\mathcal{F}} |F(P, Q, s(\zeta))|^2 \, dv(Q) \right)^{1/2} \|f\|. \tag{2.46}$$

We shall show that the right-hand side of this inequality tends to zero if ζ tends to zero along a certain curve. This will imply that $f(P) = 0$ for all $P \in \mathbb{H}$, i.e. $f = 0$, so that λ cannot be an eigenvalue for $-\tilde{\Delta}$.

Let $\lambda = 1 + t^2$ for $t \geq 0$ and put $s(\epsilon) := it + \epsilon(1 + i)$ with $\epsilon > 0$. Then

$$1 - s(\epsilon)^2 = 1 + t^2 + 2t\epsilon - 2i\epsilon(t + \epsilon) = \lambda + \zeta(\epsilon),$$

where $\zeta(\epsilon) = 2t\epsilon - 2i\epsilon(t + \epsilon)$ is a non-real complex number. Hence $\lambda + \zeta(\epsilon)$ belongs to the region where the resolvent is represented by the Maaß–Selberg series, and we obtain from (2.43)

$$|\zeta(\epsilon)|^2 \int_{\mathcal{F}} |F(P,Q,s(\epsilon))|^2 \, dv(Q) \leq \frac{t^2\epsilon^2 + \epsilon^2(t+\epsilon)^2}{2\pi\epsilon}$$

$$\cdot \left(|\Gamma_P| + \frac{1}{2(t+\epsilon)} \sum_{M \in \Gamma \setminus \Gamma_P} 2 \frac{\left(\delta(P,MP) + \sqrt{\delta(P,MP)^2 - 1}\right)^{-\epsilon}}{\sqrt{\delta(P,MP)^2 - 1}} \right).$$

Here the right-hand side tends to zero as $\epsilon \to 0$ since the series (2.45) converges by hypothesis. (Consider the cases $t = 0$, $t > 0$ separately.) This proves the theorem. □

Theorem 2.12. *Let $\Gamma < \mathbf{PSL}(2,\mathbb{C})$ be a discrete group and assume that the series $\sum_{M \in \Gamma} \delta(P,MQ)^{-1-t}$ converges for some $t \in [0,1]$. Then $\lambda = 1 - t^2$ is not an eigenvalue for $-\tilde{\Delta} : \tilde{D} \to L^2(\Gamma \setminus \mathbb{H})$.*

This theorem is proved using the same idea as in Theorem 2.11.

4.3 Hilbert-Schmidt Type Resolvents

Let $\Gamma < \mathbf{PSL}(2,\mathbb{C})$ be a discrete group. In this section we shall use the results proved so far to investigate the case when the resolvent of the self-adjoint extension $\tilde{\Delta}$ of our Laplace operator Δ on $L^2(\Gamma \setminus \mathbb{H})$ is of a particularly simple kind. The main result of this section is:

Theorem 3.1. *Let $\Gamma < \mathbf{PSL}(2,\mathbb{C})$ be a discrete group and $\lambda \in \rho(-\tilde{\Delta})$. Then the operator $R_\lambda = (-\tilde{\Delta} - \lambda)^{-1}$ is of Hilbert-Schmidt type if and only if the group Γ is cocompact.*

Proof. Since R_λ is the resolvent of a self-adjoint operator it is clear that R_λ is of Hilbert-Schmidt type for all $\lambda \in \rho(-\tilde{\Delta})$ whenever it is of Hilbert-Schmidt type for some $\lambda_0 \in \rho(-\tilde{\Delta})$. Hence it suffices to prove the Theorem for $\lambda = 1 - s^2$, $\operatorname{Re} s > 1$.

Assume first that Γ is cocompact, let $\mathcal{F}$ be a compact fundamental domain for Γ and let $F(P,Q,s)$ be the corresponding Maaß–Selberg series. Then the function

$$P \mapsto \int_{\mathcal{F}} |F(P,Q,s)|^2 \, dv(Q)$$

is continuous, see Corollary 2.5, and hence

$$\int_{\mathcal{F}} \int_{\mathcal{F}} |F(P,Q,s)|^2 \, dv(Q) \, dv(P) < \infty.$$

This proves that R_λ is of Hilbert-Schmidt type if Γ is cocompact.

Now assume that Γ is a group such that R_λ is of Hilbert-Schmidt type, and let $\lambda = 1 - s^2$ for $s \in \mathbb{R}$, $s > 1$. Then it follows from (2.44) that

$$\int_{\mathcal{F}} |F(P,Q,s)|^2 \, dv(Q) \geq \frac{1}{8\pi s}.$$

Since the left-hand side of this inequality is v-integrable over $\mathcal{F}$, we see that $v(\mathcal{F}) < \infty$. It remains to show that Γ has no cusps. We conclude this from the next Theorem 3.2, which is of interest in itself. Note that there are other possible ways of arriving at this result. One way is to show that $[1, \infty[$ is contained in the essential spectrum of $-\tilde{\Delta}$ whenever Γ has a cusp. This can be proved by the same methods as in Elstrodt and Roelcke (1974), p. 397, Satz 2.1. Another way is to use the spectral decomposition for $-\tilde{\Delta}$ given in Section 5 of this chapter. These results show that $-\tilde{\Delta}$ cannot have a pure point spectrum unless Γ is cocompact. The reader unwilling to investigate the rather tough estimates in the proof of Theorem 3.2 may skip this proof and draw the conclusion from Section 5 of this chapter which is more important than Theorem 3.2. $\square$

Theorem 3.2. *Let $\Gamma < \mathbf{PSL}(2, \mathbb{C})$ be a discrete group with abscissa of convergence σ_0, so that ∞ is a cusp of Γ. Let $\Gamma'_\infty = \left\{ \begin{pmatrix} 1 & \omega \\ 0 & 1 \end{pmatrix} \;\middle|\; \omega \in \Lambda \right\}$ be the maximal lattice contained in the stabilizer Γ_∞ of ∞ in Γ, and let $\mathcal{P}$ be a fundamental parallelogram for Λ with Euclidean area $|\mathcal{P}|$. Suppose that $s \in \mathbb{C} \backslash \mathbb{R}$, $\operatorname{Re} s > \max(0, \sigma_0)$. Then*

$$\|F(z + rj, \,\cdot\,, s)\|^2 = \frac{[\Gamma_\infty : \Gamma'_\infty]}{2(s + \bar{s}) \, |s|^2 \, |\mathcal{P}|} \, r^2 + O(r^{2(1 - \operatorname{Re} s)}) + O(r \log r) \tag{3.1}$$

as $r \to \infty$ uniformly with respect to $z \in \mathbb{C}$.

Proof. The proof rests upon (2.43). By Theorem 2.3.2 there exists $r_0 \geq 1$ such that $\delta(P, MP) \geq 2$ for all $M \in \Gamma \backslash \Gamma_\infty$, $P = z + rj \in \mathbb{H}$, $r \geq r_0$. We first discuss the contribution of the elements $M \in \Gamma \backslash \Gamma_\infty$ to the right-hand side of (2.43). Clearly there exists a constant $C_1 > 0$ such that

$$h(P) := \sum_{M \in \Gamma \backslash \Gamma_\infty} |\varphi_s(\delta(P, MP))| \leq C_1 \sum_{M \in \Gamma \backslash \Gamma_\infty} \delta(P, MP)^{-1 - \operatorname{Re} s} \tag{3.2}$$

for all $P = z + rj \in \mathbb{H}$, $r \geq r_0$. Denoting by $\mathcal{R}$ a representative system of the right cosets $\Gamma'_\infty M$, $M \in \Gamma \backslash \Gamma_\infty$, we obtain further

$$h(P) \leq C_1 \sum_{M \in \mathcal{R}} \sum_{\omega \in \Lambda} \delta(P, MP + \omega)^{-1 - \operatorname{Re} s}. \tag{3.3}$$

Letting $t := 1 + \operatorname{Re} s$ and $P = z + rj$, $MP = z_M + r_M j$, we have

$$\sum_{\omega \in \Lambda} \delta(P, MP + \omega)^{-t} = \sum_{\omega \in \Lambda} \left(\frac{2 r r_M}{|z - z_M + \omega|^2 + r^2 + r_M^2} \right)^t. \tag{3.4}$$

Estimating the right-hand side of this expression, we may assume that $z - z_M$ belongs to a fundamental parallelogram for Λ which we choose as follows. Let $0 \neq \omega_1 \in \Lambda$ be such that $|\omega_1|$ is minimal, and choose $\omega_2 \in \Lambda \setminus \mathbb{Z}\omega_1$ such that $|\omega_2|$ is also minimal. Then $\Lambda = \mathbb{Z}\omega_1 + \mathbb{Z}\omega_2$, and we have to estimate the right-hand side of (3.4) subject to the condition $z - z_M = \alpha_1\omega_1 + \alpha_2\omega_2,\ \alpha_1, \alpha_2 \in \mathbb{R},\ |\alpha_1| \leq \frac{1}{2},\ |\alpha_2| \leq \frac{1}{2}$. By a straightforward computation we have for $0 \neq \omega \in \Lambda,\ \omega = m_1\omega_1 + m_2\omega_2,\ m_1, m_2 \in \mathbb{Z}$

$$|z - z_M + \omega|^2 \geq \epsilon^2\ |m_1 i + m_2|^2,$$

where $\epsilon > 0$ is defined by $\epsilon^2 := \frac{1}{8}\ |\omega_1|^2$. This implies for our sum (3.4) that

$$\begin{aligned}
\sum_{\omega \in \Lambda} \delta(P, MP + \omega)^{-t} &\leq (2rr_M)^t \left(r^{-2t} + \sum_{0 \neq \gamma \in \mathbb{Z}[i]} (\epsilon^2|\gamma|^2 + r^2)^{-t} \right) \\
&\leq (2rr_M)^t \left(r^{-2t} + \sum_{k=1}^{\infty} 8k\ (\epsilon^2 k^2 + r^2)^{-t} \right) \\
&\leq (2rr_M)^t \left(r^{-2t} + \frac{8}{\epsilon} \sum_{k=1}^{\infty} (\epsilon^2 k^2 + r^2)^{-(t-\frac{1}{2})} \right) \\
&\leq (2rr_M)^t \left(r^{-2t} + \frac{8}{\epsilon} \int_0^{\infty} (\epsilon^2 x^2 + r^2)^{-(t-\frac{1}{2})}\ dx \right) \\
&= (2rr_M)^t\ (r^{-2t} + C_2\ r^{2-2t}).
\end{aligned} \tag{3.5}$$

The computation uses that there exist $8k$ lattice points on the boundary of the square with vertices $\pm k \pm ki,\ k \geq 1$. The constant C_2 is defined by

$$C_2 = \frac{8}{\epsilon} \int_0^{\infty} (\epsilon^2 u^2 + 1)^{-(t-\frac{1}{2})}\ du \tag{3.6}$$

and is independent of $P \in \mathbb{H}$ and $M \in \mathcal{R}$. For $r \geq r_0 \geq 1$, the right-hand side of (3.5) has the upper bound $2^t\ (1 + C_2)\ r^{2-t}\ r_M^t$ which implies for our function (3.2) the inequality

$$h(z + rj) \leq C_3\ r^{1-\operatorname{Re} s} \sum_{M \in \mathcal{R}} r_M^{1+\operatorname{Re} s} \tag{3.7}$$

for $r > r_0$, where $C_3 > 0$ is a constant independent of $P \in \mathbb{H}$. Here the sum on the right-hand side is an Eisenstein series in which the sumands belonging to Γ_∞ are missing. From our estimates for Eisenstein series in Chapter 3 we deduce

$$\sum_{M \in \mathcal{R}} r_M^{1+\operatorname{Re} s} = O(r^{1-\operatorname{Re} s}) \tag{3.8}$$

as $r \to \infty$ uniformly with respect to $z \in \mathbb{C}$. Hence we finally see that the contribution of the elements $M \in \Gamma \setminus \Gamma_\infty$ to the right-hand side of (2.43) is absorbed by the term $O(r^{2(1-\operatorname{Re} s)})$ in the statement of the theorem.

Second, we have to analyze the contribution of Γ_∞ to the right-hand side of (2.43). Since Γ'_∞ has finite index in Γ_∞, it is more convenient to start with Γ'_∞. Let s be fixed, $\operatorname{Re} s > \max(0, \sigma_0)$, and observe that $\delta = \cosh d$, and $\varphi_s(\delta) = \frac{e^{-sd}}{\sinh d}$, where d denotes the hyperbolic distance. Hence the function

$$g(\delta) = -\frac{1}{s-\bar{s}} \left(\varphi_s(\delta) - \varphi_{\bar{s}}(\delta)\right) \tag{3.9}$$

is continuous for $\delta \geq 1$ and satisfies

$$g(\delta) = e^{-\sigma d} \frac{\sin td}{t \sinh d} = \frac{1}{t} \sin\left(t \log(\delta + \sqrt{\delta^2 - 1})\right) \varphi_\sigma(\delta), \tag{3.10}$$

where $s = \sigma + it$, $\sigma = \operatorname{Re} s$. The contribution of the elements $\begin{pmatrix} 1 & \omega \\ 0 & 1 \end{pmatrix} \in \Gamma'_\infty$ to the right-hand side of (2.43) is equal to $(4\pi(s+\bar{s}))^{-1}$ times

$$f(P) := \sum_{\omega \in \Lambda} g(\delta(P, P+\omega)). \tag{3.11}$$

We have to estimate this contribution for $P = z + rj$, $z \in \mathbb{C}$, $r \to \infty$.

Let $\mathcal{P}$ be a fundamental parallelogram for Λ, and denote the Euclidean area of $\mathcal{P}$ by $|\mathcal{P}|$. Then we have

$$\begin{aligned} &f(P) \\ &= \frac{1}{|\mathcal{P}|} \sum_{\omega \in \Lambda} \int_{\mathcal{P}} \left(g(\delta(P, P+\omega)) - g(\delta(P, P+\omega+\xi+i\eta))\right) d\xi\, d\eta \\ &+ \frac{1}{|\mathcal{P}|} \int_{\mathbb{R}^2} g(\delta(P, P+\xi+i\eta))\, d\xi\, d\eta. \end{aligned} \tag{3.12}$$

The second integral on the right-hand side of (3.12) is equal to

$$\begin{aligned} \frac{1}{|\mathcal{P}|} \int_{\mathbb{R}^2} g\left(\frac{\xi^2+\eta^2+2r^2}{2r^2}\right) d\xi\, d\eta &= \frac{2r^2}{|\mathcal{P}|} \int_{\mathbb{R}^2} g(\xi^2+\eta^2+1)\, d\xi\, d\eta \\ &= \frac{4\pi r^2}{|\mathcal{P}|} \int_0^\infty g(\rho^2+1)\, \rho\, d\rho = \frac{2\pi r^2}{|\mathcal{P}|} \int_1^\infty g(t)\, dt \\ &= -\frac{2\pi r^2}{(s-\bar{s})\, |\mathcal{P}|} \int_1^\infty (\varphi_s(t) - \varphi_{\bar{s}}(t))\, dt \\ &= -\frac{2\pi r^2}{(s-\bar{s})\, |\mathcal{P}|} \int_0^\infty (\varphi_s(\cosh x) - \varphi_{\bar{s}}(\cosh x)) \sinh x\, dx \\ &= -\frac{2\pi r^2}{(s-\bar{s})\, |\mathcal{P}|} \int_0^\infty (e^{-sx} - e^{-\bar{s}x})\, dx = \frac{2\pi r^2}{|s|^2\, |\mathcal{P}|}. \end{aligned} \tag{3.13}$$

So it remains to estimate the first term on the right-hand side of (3.12). This term is bounded above by

$$(3.14)\quad f_1(P) := \frac{1}{|\mathcal{P}|} \sum_{\omega\in\Lambda} \int_{\mathcal{P}} |\, g(\delta(P,P+\omega)) - g(\delta(P,P+\omega+\xi+i\eta))\,|\; d\xi\, d\eta.$$

Note that $|g(\delta)-g(\delta')| = |\delta-\delta'|\,|g'(\delta'')|$ for some δ'' between δ and δ'. From (3.10) we have

$$\begin{aligned}
|g'(\delta)| &\le \frac{1}{|t|}\,|\varphi_\sigma'(\delta)| + \frac{\varphi_\sigma(\delta)}{\sqrt{\delta^2-1}} \\
&= \varphi_\sigma(\delta)\left(\left(\frac{\sigma}{|t|}+1\right)\frac{1}{\sqrt{\delta^2-1}} + \frac{1}{|t|}\frac{\delta}{\delta^2-1}\right) \qquad (3.15)\\
&< \left(\frac{\sigma+1}{|t|}+1\right)\frac{\delta}{\delta^2-1}\,\varphi_\sigma(\delta) < \left(\frac{\sigma+1}{|t|}+1\right)\frac{1}{\delta-1}\,\varphi_\sigma(\delta).
\end{aligned}$$

Note that the right-hand side of (3.15) is a monotonically decreasing function of δ. We shall substitute a lower bound for δ'' in the right-hand side of (3.15) and produce an upper bound for $|g'(\delta'')|$.

We choose $\delta = \delta(P,P+\omega)$, $\delta' = \delta(P,P+\omega+\xi+i\eta)$ where $P = z+rj$, $\omega\in\Lambda$, $\xi+i\eta\in\mathcal{P}$. Here $\mathcal{P}$ is the fundamental parallelogram described in the lines following equation (3.4). Then we have

$$(3.16)\qquad |\delta-\delta'| \le \frac{\big||\omega|^2 - |\omega+\xi+i\eta|^2\big|}{2r^2} \le C_4\frac{|\omega|}{r^2}$$

for all $0\ne\omega\in\Lambda$, $P\in\mathbb{H}$, $\xi+i\eta\in\mathcal{P}$, where C_4 denotes a positive constant depending only on Λ. Writing $\omega = m_1\omega_1 + m_2\omega_2$, with $m_1,\ m_2\in\mathbb{Z}$ and $\xi+i\eta = \alpha_1\omega_1+\alpha_2\omega_2$ with $|\alpha_1|\le\frac12$, $|\alpha_2|\le\frac12$, we find for $\omega\ne0$ by straightforward computation $|\omega+\xi+i\eta|^2 \ge \frac{1}{16}|\omega|^2$, that is $|\omega|\le 4|\omega+\xi+i\eta|$, and hence from (3.16)

$$(3.17)\qquad |\delta-\delta'| \le 4\,C_4\,\frac{|\omega+\xi+i\eta|}{r^2}$$

for all $0\ne\omega\in\Lambda$, $P\in\mathbb{H}$, $\xi+i\eta\in\mathcal{P}$. We can also estimate $|\omega+\xi+i\eta|^2\le C_5|\omega|^2$ for $\omega\ne0$, where C_5 is a positive constant depending only on Λ. Hence the value of δ'' is no less than

$$\min(\delta,\delta') = \min\left(\frac{|\omega|^2+2r^2}{2r^2},\ \frac{|\omega+\xi+i\eta|^2+2r^2}{2r^2}\right) \ge 1 + C_6\frac{|\omega+\xi+i\eta|^2}{2r^2}$$

for $0\ne\omega\in\Lambda$ and where $C_6>0$ depends only on Λ.

Collecting terms from (3.15), (3.17), and observing that $\omega=0$ contributes a bounded function of $P\in\mathbb{H}$ to the right-hand side of (3.14), we arrive at

$$\begin{aligned}
f_1(P) \le C_7 + C_8 \sum_{0\ne\omega\in\Lambda} \int_{\mathcal{P}} & \frac{|\omega+\xi+i\eta|}{r^2}\left(\frac{|\omega+\xi+i\eta|^2}{r^2}\right)^{-1} \qquad (3.18)\\
&\cdot\varphi_\sigma\left(1+C_6\frac{|\omega+\xi+i\eta|^2}{2r^2}\right)\, d\xi\, d\eta,
\end{aligned}$$

where C_7, C_8 do not depend on $P \in \mathbb{H}$. Note that the translates of $\mathcal{P}$ under $\Lambda \backslash \{0\}$ fill $\mathbb{C} \backslash \mathcal{P}$ exactly once. Choose $R > 0$ so small that the Euclidean disc of radius R with centre 0 is contained in $\mathcal{P}$. Then we have

$$
\begin{aligned}
f_1(P) &\leq C_7 + C_8 \, \frac{1}{r} \int_{|\xi + i\eta| > R} \left(\frac{|\xi + i\eta|}{r} \right)^{-1} \varphi_\sigma \left(1 + C_6 \frac{|\xi + i\eta|^2}{2r^2} \right) \, d\xi \, d\eta \\
&= C_7 + 2\pi \, C_8 \, r \int_{\frac{R}{r}}^{\infty} \rho^{-1} \, \varphi_\sigma \left(1 + \frac{1}{2} \, C_6 \, \rho^2 \right) \, \rho \, d\rho \\
&\leq C_7 + 2\pi \, C_8 \, r \int_{\frac{R}{r}}^{\infty} \left(1 + \frac{1}{2} \, C_6 \, \rho^2 \right)^{-\sigma} (C_6 \, \rho^2)^{-\frac{1}{2}} \, d\rho \\
&= O(r \log r) \qquad \text{as} \quad r \to \infty
\end{aligned}
$$

uniformly with respect to $z \in \mathbb{C}$.

Summing up, we see from (3.9), (3.12), (3.13), (3.14) and the last estimate that the contribution of Γ_∞ to the right-hand side of (2.59) is equal to

$$
[\Gamma_\infty : \Gamma'_\infty] \cdot \frac{1}{4\pi(s + \bar{s})} \frac{2\pi}{|s|^2 |\mathcal{P}|} r^2 + O(r \log r).
$$

This completes the proof of our theorem. □

It would also be possible to prove an analogous estimate for real s using a similar reasoning. Since we have no application for such an estimate, we skip the tedious details.

Corollary 3.3. *Suppose that $\Gamma < \mathbf{PSL}(2, \mathbb{C})$ is a discrete cocompact group. Then the operator $-\Delta : \mathcal{D} \to L^2(\Gamma \backslash \mathbb{H})$ has a complete orthonormal system $(e_n)_{n \geq 0}$ of eigenfunctions with corresponding eigenvalues $0 = \lambda_0 < \lambda_1 \leq \lambda_2 \leq \ldots$ counted according to their multiplicities and $\sum_{n=1}^{\infty} \lambda_n^{-2} < \infty$.*

Proof. The proof follows immediately from the observation that $(-\tilde{\Delta} - \lambda)^{-1}$ is for $(\lambda < 0)$ a self-adjoint linear operator of Hilbert-Schmidt type (cf. Theorem 3.1) and hence has a complete orthonormal system of eigenfunctions $(e_n)_{n \geq 0}$ with associated eigenvalues μ_n such that $\sum_{n=0}^{\infty} \mu_n^2 < \infty$. □

The eigenfunction e_0 is constant and we may choose $e_0 = (v(\mathcal{F}))^{-\frac{1}{2}}$, where $\mathcal{F}$ denotes a fundamental domain for Γ. This follows immediately from Theorem 1.8. A direct proof using the maximum principle for harmonic functions yields the same conclusion in a simpler way.

4.4 Analytic Continuation of the Resolvent Kernel

This section contains results concerning the continuation with respect to the variable s of the Maaß–Selberg series $F(P,Q,s)$. Throughout the section we keep the following notation fixed. $\Gamma < \mathbf{PSL}(2,\mathbb{C})$ is a discrete group with fundamental domain $\mathcal{F}$ and with abscissa of convergence σ_0. Put

$$\sigma_0^+ := \max(0,\sigma_0), \tag{4.1}$$

denote the resolvent set of $-\tilde{\Delta}$ by $\rho(-\tilde{\Delta})$ and let $R_\lambda := (-\tilde{\Delta}-\lambda)^{-1}$ be the resolvent of $-\tilde{\Delta}$.

We know from Corollary 2.8 that every $\lambda \in \mathbb{C} \setminus [1,\infty[$ such that $\operatorname{Re}\sqrt{1-\lambda} > \sigma_0^+$ belongs to $\rho(-\tilde{\Delta})$ and that the resolvent operator is represented by the kernel

$$G_\lambda(P,Q) := F(P,Q,\sqrt{1-\lambda}) \tag{4.2}$$

for these λ. It is our aim to continue $G_\lambda(P,Q)$ as a function of λ into the domain $\mathbb{C} \setminus [1-(\sigma_0^+)^2,\infty[$ in such a way that the analytical properties of G_λ are preserved. This aim will be achieved after some lengthy, though rather elementary discussion. The final results are given in Theorems 4.1, 4.4. We start with a theorem which already contains the crucial representation of the resolvent but which does not yet contain the differentiability properties of the continuation of the resolvent kernel.

Theorem 4.1. *For every $\mu \in \mathbb{C}\setminus[1-(\sigma_0^+)^2,\infty[$, $P \in \mathbb{H}$ there exists a unique element*

$$G_\mu(P,\cdot) \in L^2(\Gamma \setminus \mathbb{H}) \tag{4.3}$$

depending continuously on (P,μ) such that

$$R_\mu f(P) = \int_{\mathcal{F}} G_\mu(P,Z)\, f(Z)\, dv(Z) \tag{4.4}$$

for all $f \in L^2(\Gamma \setminus \mathbb{H})$, $P \in \mathbb{H}$. Equation (4.4) holds both as an equality between two elements of $L^2(\Gamma \setminus \mathbb{H})$ and as a pointwise equality between two continuous functions on $\mathbb{H}$. The function G_μ satisfies

$$G_\mu(P,Q) = G_\mu(Q,P), \qquad \overline{G_\mu(P,Q)} = G_{\bar{\mu}}(P,Q). \tag{4.5}$$

Let $G(P,Q)$ be the function defined in (2.21), then $G_\mu(P,Q) - G(P,Q)$ depends continuously on $(P,Q,\mu) \in \mathbb{H} \times \mathbb{H} \times (\mathbb{C} \setminus [1-(\sigma_0^+)^2,\infty[)$. G_μ solves the integral equation

$$(\lambda-\mu)\int_{\mathcal{F}} G_\lambda(P,Z)\, G_\mu(Z,P)\, dv(Z) = \lim_{Z\to Q}(G_\lambda(P,Z) - G_\mu(P,Z)) \tag{4.6}$$

for all $\lambda, \mu \in \mathbb{C} \setminus [1-(\sigma_0^+)^2, \infty[,\ P,\ Q \in \mathbb{H}$. *In addition,*

$$G_\lambda(P,Q) = F(P,Q,\sqrt{1-\lambda}) \tag{4.7}$$

if $\operatorname{Re}\sqrt{1-\lambda} > \sigma_0^+$.

Proof. The idea of proof is to start with a fixed $\lambda \in \rho(-\tilde{\Delta})$ such that $\operatorname{Re}\sqrt{1-\lambda} > \sigma_0^+$ and to consider (4.6) as an equation for G_μ.

So let us start with a fixed non-real λ such that $\operatorname{Re}\sqrt{1-\lambda} > \sigma_0^+$, and define G_λ by (4.7). Let $\mu \in \mathbb{C} \setminus [1-(\sigma_0^+)^2, \infty[$. We want to define G_μ. If μ also satisfies $\operatorname{Re}\sqrt{1-\lambda} > \sigma_0^+$, we know from Corollary 2.9 that equation (4.6) holds. Write (4.6) in the form $(\lambda-\mu)R_\lambda G_\mu(\cdot,Q) = G_\lambda(\cdot,Q) - G_\mu(\cdot,Q)$, that is

$$(E + (\lambda-\mu)R_\lambda)G_\mu(\cdot,Q) = G_\lambda(\cdot,Q), \tag{4.8}$$

where E denotes the identity map on $L^2(\Gamma \setminus \mathbb{H})$. We now drop our above hypothesis on μ and consider (4.8) as an equation for $G_\mu(\cdot,Q)$. Observe that $\|R_\lambda\| \le |\operatorname{Im}\lambda|^{-1}$. Hence the operator on the left of (4.8) has an inverse whenever $|\mu-\lambda| < |\operatorname{Im}\lambda|$ and there exists a unique solution $G_\mu(\cdot,Q) \in L^2(\Gamma \setminus \mathbb{H})$ of (4.8) whenever μ satisfies $|\mu-\lambda| < |\operatorname{Im}\lambda|$. The uniqueness of the solution immediately implies that (4.7) holds.

By the remarks following Corollary 2.8 λ is allowed to vary unrestrictedly to the left of a certain parabola, this shows that $G_\mu(\cdot,Q)$ is now defined for all $\mu \in \mathbb{C} \setminus [1-(\sigma_0^+)^2, \infty[$. We shall see later that $G_\mu(\cdot,Q)$ does not depend on λ so that our notation is justified.

Under the condition $|\mu-\lambda| < |\operatorname{Im}\lambda|$ the unique solution of (4.8) is given by the Neumann series

$$\begin{aligned} G_\mu(\cdot,Q) &= \sum_{n=0}^{\infty} (\mu-\lambda)^n\ R_\lambda^n G_\lambda(\cdot,Q) \\ &= G_\lambda(\cdot,Q) + \sum_{n=1}^{\infty} (\mu-\lambda)^n\ R_\lambda^n G_\lambda(\cdot,Q). \end{aligned} \tag{4.9}$$

This is an element of $L^2(\Gamma \setminus \mathbb{H})$ which obviously depends continuously on (Q,μ) in the norm sense. In addition, we know from Corollary 2.5 and Theorem 2.6 that for all $n \ge 1$ the function $R_\lambda^n G_\lambda(\cdot,Q)(P)$ depends continuously on $(P,Q) \in \mathbb{H} \times \mathbb{H}$. Since we have

$$\begin{aligned} &|R_\lambda^n G_\lambda(\cdot,Q)(P)| = \left| \int_{\mathcal{F}} G_\lambda(P,Z)\ (R_\lambda^{n-1} G_\lambda(\cdot,Q))(Z)\ dv(Z) \right| \\ &\le \|R_\lambda\|^{n-1} \|G_\lambda(P,\cdot)\|\ \|G_\lambda(\cdot,Q)\| \le \left(\frac{1}{|\operatorname{Im}\lambda|} \right)^{n-1} \|G_\lambda(P,\cdot)\|\ \|G_\lambda(\cdot,Q)\|, \end{aligned}$$

we conclude that the series on the right-hand side of (4.9) converges uniformly on compact sets with respect to (P,Q,μ), $P,\ Q \in \mathbb{H}$, μ subject to the condition $|\mu - \lambda| < |\operatorname{Im} \lambda|$. This shows that $G_\mu(P,Q)$ has the same type of singularity for $P \to Q$ as $G_\lambda(P,Q)$, and we see that $G_\mu(P,Q) - G(P,Q)$ is a continuous function of $(P,Q,\mu) \in \mathbb{H} \times \mathbb{H} \times (\mathbb{C} \setminus [1-(\sigma_0^+)^2, \infty[)$. Obviously, $G_\mu(P,Q)$ is a holomorphic function of $\mu \in \mathbb{C} \setminus [1-(\sigma_0^+)^2, \infty[$ whenever $P, Q \in \mathbb{H}$, $P \not\equiv Q \bmod \Gamma$ are kept fixed.

Equation (4.7) is now obvious. To prove (4.5) we proceed to show that all the terms on the right-hand side of (4.9) are symmetric with respect to P and Q. This is obvious for the first term $G_\lambda(P,Q)$.

In the infinite series $(R_\lambda G_\lambda(\cdot,Q))(P) = \int_{\mathcal{F}} G_\lambda(P,Z)\, G_\lambda(Z,Q)\, dv(Z)$, which is symmetric with respect to P, Q. The term

$$\begin{aligned}(R_\lambda^2 G_\lambda(\cdot Q))(P) \\ = \int_{\mathcal{F}} \left(\int_{\mathcal{F}} G_\lambda(P,Z_1)\, G_\lambda(Z_1,Z_2)\, G_\lambda(Z_2,Q)\, dv(Z_2) \right) dv(Z_1)\end{aligned} \tag{4.10}$$

obviously has the required symmetry property if the order of integration can be interchanged. Now observe that $F(P,Q,\operatorname{Re}\sqrt{1-\lambda})$ is defined by a series with positive terms for which the right-hand side of (4.10) converges. This yields a suitable integrable majorant for the application of Fubini's theorem on the right-hand side of (4.10), hence (4.10) is symmetric with respect to P and Q. The remaining terms in (4.9) are handled in the same way. This proves (4.5).

We proceed to prove (4.4). Once we have proved this, it is clear that $G_\mu(\cdot,Q)$ does not depend on λ. To prove (4.4), let $\mu \in \mathbb{C} \setminus [1-(\sigma_0^+)^2, \infty[$ be given and choose $\lambda = 1 - s^2, \operatorname{Re} s > \sigma_0^+$ such that $|\mu - \lambda| < |\operatorname{Im} \lambda|$. Then G_μ satisfies (4.6) by definition. Hence we obtain from (4.5) for all $f \in \mathcal{D}^\infty, P \in \mathbb{H}$

$$\begin{aligned}& f(P) - \int_{\mathcal{F}} G_\mu(P,Q)\,(-\Delta - \lambda) f(Q)\, dv(Q) \\ &= \int_{\mathcal{F}} (G_\lambda(Q,P) - G_\mu(Q,P))\,(-\Delta - \lambda) f(Q)\, dv(Q) \\ &= (\lambda - \mu) \int_{\mathcal{F}} \left(\int_{\mathcal{F}} G_\lambda(Q,Z) G_\mu(Z,P) dv(Z) \right) (-\Delta - \lambda) f(Q)\, dv(Q).\end{aligned} \tag{4.11}$$

We want to interchange the order of integration here, and we first justify this procedure. We know that $G_\lambda(\cdot,Q)$ depends continuously on Q in the mean-square sense. Hence $Q \mapsto \int_{\mathcal{F}} |G_\lambda(Q,Z)\, G_\mu(Z,P)|\, dv(Z)$ is a continuous function, and this yields that

$$\int_{\mathcal{F}} \int_{\mathcal{F}} |G_\lambda(Q,Z)\, G_\mu(Z,P)|\, dv(Z)\, |(-\Delta - \lambda) f(Q)|\, dv(Q) < \infty$$

since $f \in \mathcal{D}^\infty$. Interchanging the order of integration in (4.11) we arrive at

$$\begin{aligned} & f(P) - \int_{\mathcal{F}} G_\mu(P,Q)\,(-\Delta-\lambda)f(Q)\,dv(Q) \\ (4.12) \quad & = (\lambda-\mu)\int_{\mathcal{F}} G_\mu(Z,P)\left(\int_{\mathcal{F}} G_\lambda(Q,Z)\,(-\Delta-\lambda)f(Q)\,dv(Q)\right)dv(Z) \\ & = (\lambda-\mu)\int_{\mathcal{F}} G_\mu(P,Z)\,f(Z)\,dv(Z). \end{aligned}$$

This yields

$$\int_{\mathcal{F}} G_\lambda(P,Z)\;(-\Delta-\mu)f(Z)\;dv(Z) = f(P) \tag{4.13}$$

for all $f \in \mathcal{D}^\infty,\ P \in \mathbb{H}$. Hence the equation

$$(R_\mu g)(P) = \int_{\mathcal{F}} G_\mu(P,Z)\;g(Z)\;dv(Z) \tag{4.14}$$

is true for all $g \in U$, where $U := (-\Delta-\lambda)\mathcal{D}^\infty$. Note that U is a dense linear subspace of $L^2(\Gamma\setminus\mathbb{H})$ since $\lambda \in \rho(-\tilde{\Delta})$. This allows us to extend (4.14) to all $g \in L^2(\Gamma\setminus\mathbb{H})$: Let $g \in L^2(\Gamma\setminus\mathbb{H})$, then there exists a sequence $(g_n)_{n\geq 1}$ in U such that $g_n \to g$ in $L^2(\Gamma\setminus\mathbb{H})$. Hence

$$\int_{\mathcal{F}} G_\mu(P,Z)\;g_n(Z)\;dv(Z) \to \int_{\mathcal{F}} G_\mu(P,Z)\;g(Z)\;dv(Z), \tag{4.15}$$

where the right-hand side is a continuous function of $P \in \mathbb{H}$. The sequence $(R_\mu g_n)_{n\geq 1}$ converges pointwise to the function on the right-hand side of (4.15) and it converges in the norm to $R_\mu g$. This yields

$$R_\mu g(P) = \int_{\mathcal{F}} G_\mu(P,Z)\;g(Z)\;dv(Z)$$

both as an equality between two elements of $L^2(\Gamma\setminus\mathbb{H})$ and as an equality between two continuous functions on $\mathbb{H}$. This proves (4.4) and this shows that G_μ does not depend on λ.

We still have to show that (4.6) is true for arbitrary $\lambda,\ \mu \in \mathbb{C}\setminus[1-(\sigma_0^+)^2,\infty[$. But since we now have the resolvent equation (4.4) at our disposal in greater generality, we may simply repeat the proof of Theorem 2.9 to obtain (4.6). This proves our theorem. □

Definition 4.2. For $\operatorname{Re} s > 0,\ s \notin [0,\sigma_0^+]$ define the extended Maaß–Selberg series by

$$F(P,Q,s) := G_{1-s^2}(P,Q). \tag{4.16}$$

Here we must require $P, Q \in \mathbb{H},\ P \not\equiv Q \bmod \Gamma$.

We also want to show that the extended Maaß–Selberg series is a real analytic function of $P,\ Q \in \mathbb{H}$ depending holomorphically on s satisfying the differential equation

$$(4.17) \qquad (-\Delta - (1-s^2))\ F(\cdot, Q, s) = 0$$

on $\mathbb{H} \setminus \Gamma Q$. It is convenient here to discuss first the function $H(P,Q,s)$ from Chapter 3, Section 1. Then it will be easy to carry over the result from $H(P,Q,s)$ to $F(P,Q,s)$ by means of Proposition 3.1.20.

Theorem 4.3. *The function $H(P,Q,s)$ has an analytic continuation to the slit half-plane $\operatorname{Re} s > 0, s \notin [0, \sigma_0^+]$. As a function of $P, Q \in \mathbb{H}$, $H(P,Q,s)$ is real analytic and contained in $\mathcal{D}$. It is furthermore symmetric, that is $H(P,Q,s) = H(Q,P,s)$ and satisfies the differential equation*

$$(4.18) \qquad (-\Delta - (1-s^2))\ H(\cdot, Q, s) = (s+1)\ (s+2)\ H(\cdot, Q, s+2).$$

Proof. Since the Theorem is trivially true for $\sigma_0 \le 0$, we may assume that $\sigma_0 > 0$. We know from (3.1.29) that (4.18) is satisfied for $\operatorname{Re} s > \sigma_0$. Note that the right-hand side of (4.18) makes sense for $\operatorname{Re} s > \sigma_0 - 2$. We further know from Lemma 2.1 that $H(\cdot, Q, s+2) \in L^2(\Gamma \setminus \mathbb{H})$ if $\operatorname{Re} s > \sigma_0 - 2$. Note that we assume that $\sigma_0 > 0$. Since we have just constructed the inverse of the operator on the left-hand side of (4.18), we can define

$$(4.19) \qquad H(P,Q,s) := (s+1)\ (s+2) \int_{\mathcal{F}} F(P,Z,s)\ H(Z,Q,s+2)\ dv(Z)$$

for $P, Q \in \mathbb{H}$, $\operatorname{Re} s > 0$, $s \notin [0, \sigma_0^+]$. This is a continuous function of (P,Q,s) on its domain, and this obviously defines an extension of our function $H(P,Q,s)$ given by (3.1.1) for $\operatorname{Re} s > \sigma_0$. As a function of s with $\operatorname{Re} s > 0$, $s \notin [0, \sigma_0^+]$ the extended function (4.19) is holomorphic. Since our extended Maaß–Selberg series (4.16) is symmetric with respect to P and Q, we conclude from (3.1.48), (3.1.49) that the extended function $H(P,Q,s)$ is also symmetric with respect to P and Q.

Now observe that the resolvent operator maps $L^2(\Gamma \setminus \mathbb{H})$ onto $\tilde{\mathcal{D}}$. Hence $H(\cdot, Q, s) \in \tilde{\mathcal{D}}$ for $\operatorname{Re} s > 0$, $s \notin [0, \sigma_0^+]$ and

$$(-\tilde{\Delta} - (1-s^2))\ H(\cdot, Q, s) = (s+1)\ (s+2)\ H(\cdot, Q, s+2).$$

This yields that for all $h \in C_c^\infty(\mathbb{H})$

$$\int_{\mathbb{H}} (-\tilde{\Delta} - (1-s^2)) H(\cdot, Q, s)\ h\ dv = \int_{\mathbb{H}} (s+1)\ (s+2)\ H(\cdot, Q, s+2)\ h\ dv,$$

i.e. $H(\cdot, Q, s)$ is a weak solution of (4.18). But now a known regularity theorem applies (cf. Agmon (1965), page 66, Theorem 6.6) and yields $H(\cdot, Q, s) \in C^\infty(\mathbb{H})$. In particular, we conclude from Lemma 1.11 that $H(\cdot, Q, s)$ belongs to $\mathcal{D}$. Hence (4.18) holds pointwise. But now we have that $H(\cdot, Q, s)$ is a twice continuously differentiable solution of an elliptic partial differential equation with real analytic coefficients. Thus $H(\cdot, Q, s)$ is real analytic in the given domain. □

Theorem 4.4. *For fixed $Q \in \mathbb{H}$, the continuation $G_\lambda(P,Q) = F(P,Q,s)$ where $P \in \mathbb{H} \setminus \Gamma Q$, $\mathrm{Re}\, s > 0$, $s \notin [0, \sigma_0^+]$ of the resolvent kernel given by Theorem 4.1 is a real analytic function of P satisfying the differential equation*

$$(4.20) \qquad (-\Delta - \lambda)\, G_\lambda(P,Q) = (-\Delta - (1-s^2))\, F(P,Q,s) = 0.$$

Proof. We deduce from Chapter 3, equation (3.1.48) that

$$(4.21) \qquad F(P,Q,s) = (2^{s+2}\pi)^{-1} H(P,Q,s) + L(P,Q,s),$$

where $L(P,Q,s)$ is some kernel which is defined for $P \not\equiv Q \bmod \Gamma$ and for $\mathrm{Re}\, s > \sigma_0 - 2$. Equation (4.21) first holds for $\mathrm{Re}\, s > \sigma_0$ and by analytic continuation, (4.21) remains true for $\mathrm{Re}\, s > 0$, $s \notin [0, \sigma_0^+]$. Since the right-hand side of (4.21) is real analytic with respect to $P \in \mathbb{H} \setminus \Gamma Q$, the left-hand side is also a real analytic function of $P \in \mathbb{H} \setminus \Gamma Q$. From the series defining L we have the differential equation

$$(4.22) \qquad (-\Delta - (1-s^2))\, L(P,Q,s) + \frac{(s+1)\,(s+2)}{2^{s+2}\pi}\, H(P,Q,s+2) = 0$$

for all s with $\mathrm{Re}\, s > \sigma_0 - 2$. Hence we obtain (4.20) from (4.18), (4.22). □

The existence of a representation of the form (4.4) for the resolvent operator by means of a kernel may also be proved by a simple application of the Riesz representation theorem:

Theorem 4.5. *Let $\Gamma < \mathbf{PSL}(2,\mathbb{C})$ be a discrete group. Then there is a kernel $G_\mu(P,Q)$ of Carleman type so that $R_\mu f(P) = \int_{\mathcal{F}} G_\mu(P,Z)\, f(Z)\, dv(Z)$ for all $\mu \in \rho(-\tilde{\Delta})$, $f \in L^2(\Gamma \setminus \mathbb{H})$ and $P \in \mathbb{H}$.*

Proof. Choose $\lambda \in \mathbb{C}$, $\lambda = 1 - s^2$, $\mathrm{Re}\, s > \sigma_0^+$. Then $R_\mu = R_\lambda - (\lambda - \mu) R_\lambda R_\mu$. Since R_λ is represented by the Maaß–Selberg series, which we already now to be of Carleman type, we see that $R_\mu f$ is for $f \in L^2(\Gamma \setminus \mathbb{H})$ a continuous function on $\mathbb{H}$ satisfying the estimate $|R_\mu f(P)| \le \|F(P,\cdot,s)\|\ (1 + |\lambda - \mu|\ \|R_\mu\|)\ \|f\|$. Hence $L^2(\Gamma \setminus \mathbb{H}) \ni f \mapsto R_\mu f(P) \in \mathbb{C}$ is a bounded linear form on $L^2(\Gamma \setminus \mathbb{H})$. We deduce our theorem now from the Riesz representation theorem. □

It is known that the resolvent of a self-adjoint elliptic differential operator is an integral operator with a kernel of Carleman type under very general hypotheses, see F. Browder (1959), (1961), (1962).

The technique used to prove Theorem 4.5 offers another approach to Theorem 4.1. Working out the details on the lines of this approach seems to be more difficult than the method given above. Nevertheless, the approach via the Riesz representation theorem seems to be quite useful since it easily generalizes to related questions. As another example consider the map

$$L^2(\Gamma \setminus \mathbb{H}) \ni f \mapsto (E_\beta - E_\alpha)\, f(P), \tag{4.23}$$

where $\alpha,\ \beta \in \mathbb{R},\ \alpha < \beta,\ P \in \mathbb{H}$, and where $(E_\lambda)_{\lambda \in \mathbb{R}}$ is the spectral family of $-\tilde{\Delta}$. Since $(E_\beta - E_\alpha) f \in \tilde{\mathcal{D}}$, we conclude that

$$\begin{aligned} |(E_\beta - E_\alpha) f(P)| &= \left| \int_{\mathcal{F}} F(P,Q,s)\, (-\tilde{\Delta} - \lambda)(E_\beta - E_\alpha) f(Q)\, dv(Q) \right| \\ &\leq \|F(P,\cdot,s)\|\, (|\lambda| + |\beta| + |\alpha|)\, \|f\|. \end{aligned}$$

This shows that the linear form (4.23) is continuous. Hence we may apply once more the Riesz representation theorem to obtain the following: There exists an element $K_{\alpha,\beta}(P,\cdot) \in L^2(\Gamma \setminus \mathbb{H})$ such that

$$(E_\beta - E_\alpha) f(P) = \int_{\mathcal{F}} K_{\alpha,\beta}(P,Q)\, f(Q)\, dv(Q)$$

for all $f \in L^2(\Gamma \setminus \mathbb{H})$.

It is a remarkable fact that Theorems 4.1, 4.3, 4.4 hold for arbitrary discrete groups Γ. For the 2-dimensional analogue of the function $H(P,Q,s)$ this was proved by Patterson (1976b), page 66, using a different method. Patterson applies the methods of spectral theory to solve a problem of W.K. Hayman. It would be interesting to know whether the more elementary methods of the present section suffice to solve the analogue of Hayman's problem in the case of dimension three.

Let now Γ be a cofinite group which is not cocompact. It is a remarkable fact that the resolvent kernel can be be continued meromorphically to all of $\mathbb{C}$. We shall see later that $[1, \infty[\subset \sigma(\tilde{\Delta})$ and hence the resolvent R_{1-s^2} cannot be meromorphically continued beyond the line $\operatorname{Re} s = 0$.

Theorem 4.6. *Let $\Gamma < \mathbf{PSL}(2,\mathbb{C})$ be a discrete group of finite covolume. Then its Maaß–Selberg series can be continued meromorphically to all of $\mathbb{C}$.*

For cocompact groups this is a consequence of our result in Section 3 which shows that the resolvent itself continues meromorphically to $\mathbb{C}$. For non-cocompact groups this result is contained in Proposition 6.4.6. Results like Theorem 4.6 were first proved by L. D. Faddeev (1967) in the 2-dimensional situation. His method can be generalized to dimension 3. The paper (1980a) of W. Müller contains a far reaching generalization of Faddeev's results.

4.5 Approximation by Kernels of Hilbert-Schmidt Type

Suppose that Γ is a discrete subgroup of $\mathbf{PSL}(2,\mathbb{C})$ with fundamental domain $\mathcal{F}$ such that $\Gamma \setminus \mathbb{H}$ is non-compact and of finite volume. Then by Theorem 3.1 the resolvent kernel is no longer of Hilbert-Schmidt type. The spectrum of the Laplacian $-\Delta : \mathcal{D} \to L^2(\Gamma \setminus \mathbb{H})$ turns out to be no longer discrete. In fact, using the arguments of Elstrodt and Roelcke (1974) it is rather easy to show that $[1,\infty[$ belongs to the essential spectrum of $-\Delta$. We shall even be able to describe the non-discrete part of the spectrum rather explicitly in terms of the analytically continued Eisenstein series. This means that the non-discrete part of the spectrum is quite well understood.

Much less is known about the discrete spectrum. The aim of the present section is to show that the situation parallels the results in the cocompact case in so far as the eigenvalues $(\lambda_n)_{n\geq 1}$ belonging to the cusp eigenfunctions of $-\Delta : \mathcal{D} \to L^2(\Gamma \setminus \mathbb{H})$ still satisfy

$$\sum_{n\geq 1} \lambda_n^{-2} < \infty. \tag{5.1}$$

Cusp functions are defined in Definition 5.1. It will turn out in Chapter 6 that all but finitely many eigenvalues belong to cusp eigenfunctions. The proof of (5.1) uses a method of Selberg which is described in Roelcke (1967), paragraph 8 for the case of the hyperbolic plane. The three-dimensional case is slightly more complicated since there arise unbounded functions as upper bounds which are square integrable over $\mathcal{F} \times \mathcal{F}$ whereas in the two-dimensional case the upper bounds are constant.

We introduce some notation. We choose a fundamental domain $\mathcal{F}$ for Γ of the kind described in Proposition 2.3.9. We let $A_1, \dots, A_h \in \mathbf{PSL}(2,\mathbb{C})$ be so that

$$\zeta_1 = A_1^{-1}\infty, \dots, \zeta_h = A_h^{-1}\infty \in \mathbb{P}^1\mathbb{C}$$

is a complete system of representatives for the Γ-classes of cusps of Γ. The stabilizer of the cusp ζ_ν in Γ is denoted by $\Gamma_\nu = \Gamma_{\zeta_\nu}$. We put

$$\Theta_\nu = A_\nu \Gamma A_\nu^{-1}, \qquad \Lambda_\nu := A_\nu \Gamma_{\zeta_\nu} A_\nu^{-1}, \qquad \Lambda'_\nu := A_\nu \Gamma'_{\zeta_\nu} A_\nu^{-1}.$$

Γ'_{ζ_ν} is the unipotent subgroup of the stabilizer Γ_ν of ζ_ν in Γ. Then Λ'_ν corresponds to a lattice in $\mathbb{C}$, and we identify Λ'_ν with that lattice. We furthermore choose fundamental domains $\mathcal{R}_\nu \subset \mathbb{C}$ for the action of $A_\nu \Gamma_{\zeta_\nu} A_\nu^{-1}$ on $\mathbb{P}^1\mathbb{C} \setminus \{\infty\} = \mathbb{C}$ and $Y > 0$ all satisfying the requirements of Proposition 2.3.9. We write

$$\mathcal{F}_\nu(Y) := A_\nu^{-1}\{\, z + rj \;\mid\; z \in \mathcal{R}_\nu,\ r \geq Y \,\}.$$

for the corresponding cusp sector. Then there is a compact polyhedron $\mathcal{F}_0 \subset \mathbb{H}$ so that

$$\mathcal{F} = \mathcal{F}_0 \cup \mathcal{F}_1(Y) \cup \ldots \cup \mathcal{F}_h(Y)$$

is a fundamental domain for Γ. Two points $P = z + rj$, $P' = z' + r'j \in \mathbb{H}$ with r, $r' \geq Y$ are Θ_ν-equivalent if and only if they are equivalent under Λ_ν.

Definition 5.1. Suppose that $f \in L^2(\Gamma \setminus \mathbb{H})$. Then we call the function

$$a_\nu(r) := \frac{1}{|\mathcal{R}_\nu|} \int_{\mathcal{R}_\nu} f(A_\nu^{-1}(x + iy + jr))\, dx\, dy$$

the zeroth Fourier coefficient of f at ζ_ν. Here, $|\mathcal{R}_\nu|$ is the Euclidean area of $\mathcal{R}_\nu$. If $a_\nu(r) = 0$ for almost all $r \geq R$, then we say that the zeroth Fourier coefficient of f vanishes a.e. near ζ_ν, and if this holds for all $\nu = 1, \ldots, h$, then we say that the zeroth Fourier coefficients of f vanish a.e. near all the cusps of Γ and call f a cusp function.

The space of cusp functions will be further analyzed in Section 1 of Chapter 6. The central result of the present section is:

Theorem 5.2. *Let Γ be a discrete subgroup of $\mathbf{PSL}(2, \mathbb{C})$ with fundamental domain $\mathcal{F}$ such that $\Gamma \setminus \mathbb{H}$ is non-compact and of finite volume. Then there exists a function $F^*(P, Q, s)$ for P, $Q \in \mathbb{H}$, $P \not\equiv Q \bmod \Gamma$, $s > 1$ with the following properties.*

(1) $F^*(\cdot, \cdot, s)$ is real-valued, Γ-invariant in both variables, and symmetric in P, Q, that is $F^*(P, Q, s) = F^*(Q, P, s)$.
(2) $F^*(\cdot, \cdot, s)$ is measurable and of Hilbert-Schmidt type, that is

$$\int_{\mathcal{F}} \int_{\mathcal{F}} |F^*(P, Q, s)|^2\, dv(P)\, dv(Q) < \infty. \tag{5.2}$$

(3) The function

$$P \mapsto \int_{\mathcal{F}} |F^*(P, Q, s)|^2\, dv(Q)$$

is bounded on compact subsets of $\mathbb{H}$.
(4) If $f \in L^2(\Gamma \setminus \mathbb{H})$ is a cusp function, then

$$\int_{\mathcal{F}} F^*(P, Q, s)\, f(Q)\, dv(Q) = \int_{\mathcal{F}} F(P, Q, s)\, f(Q)\, dv(Q). \tag{5.3}$$

Proof. We shall construct F^* from F by a certain modification of F near the cusps of Γ. We use again without further explanation the notation of (2.18)–(2.24). The modification of F will be carried out separately in the components G and K of the decomposition (2.23), and we shall show that Theorem 5.2 holds true with the modified functions G^* and K^* and with G and K instead of F^* and F.

For $R > Y$ we put

$$\mathcal{F}_\nu = \mathcal{F}_\nu(R) \tag{5.4}$$

and we choose Y so large that the following condition holds:

(5.5) If $P \in \mathcal{F}_\nu$ for some ν and $Q \in \mathcal{F}$, $M \in \Gamma$ and $\delta(P, MQ) \leq 3$, then $M \in \Gamma_{\zeta_\nu}$.

We have to determine upper bounds for various functions of $(P, Q) \in \mathcal{F} \times \mathcal{F}$ subject to the condition

$$P \in \mathcal{F}_\nu \qquad \text{for some } \nu = 1, \dots, h, \qquad Q \in \mathcal{F}. \tag{5.6}$$

We keep the following notations fixed:

$$\begin{aligned} P &= z + rj, & Q &= w + tj, \\ P_\nu &:= A_\nu P = z_\nu + r_\nu j, & Q_\nu &:= A_\nu Q = w_\nu + t_\nu j, \\ MQ_\nu &= w_{\nu M} + t_{\nu M} j. \end{aligned} \tag{5.7}$$

In the following, $C_1, \dots, C_{18}$ denote suitable positive constants possibly dependent on s but independent of P and Q.

In the first half of the proof we analyze the contribution of $K(P, Q, s)$, see (2.22) to the resolvent kernel; the second half is similarly concerned with $G(P, Q)$. We investigate the growth behaviour of the two functions on the right-hand side of the decomposition

$$K(P, Q, s) = \sum_{M \in \Gamma_\nu} k_s(\delta(P, MQ)) + \sum_{M \in \Gamma \setminus \Gamma_\nu} k_s(\delta(P, MQ)). \tag{5.8}$$

Step 1. The second term on the right-hand side of (5.8) is bounded under the condition (5.6). To prove this, we apply the obvious estimate for the function k_s, see (2.20), and find

$$\begin{aligned} f_\nu(P, Q) &:= \sum_{M \in \Gamma \setminus \Gamma_\nu} |k_s(\delta(P, MQ))| \leq C_1 \sum_{M \in \Gamma \setminus \Gamma_\nu} \delta(P, MQ))^{-1-s} \\ &= C_1 \sum_{M \in \Gamma \setminus \Gamma_\nu} \delta(A_\nu P, (A_\nu M A_\nu^{-1}) A_\nu Q)^{-1-s} \\ &= C_1 \sum_{M \in \Theta_\nu \setminus \Lambda_\nu} \delta(P_\nu, MQ_\nu)^{-1-s}. \end{aligned} \tag{5.9}$$

This reduces the problem to the case $\zeta_\nu = \infty$, $A_\nu = I$. Let V_ν be a representative system of the right cosets $\Lambda'_\nu M$, $M \in \Theta_\nu \setminus \Lambda_\nu$. Then we have

$$\begin{aligned} &\sum_{M \in \Theta_\nu \setminus \Lambda_\nu} \delta(P_\nu, MQ_\nu)^{-1-s} \\ &= \sum_{M \in V_\nu} \sum_{\omega \in \Lambda'_\nu} \left(\frac{2 r_\nu t_{\nu M}}{|z_\nu - w_{\nu M} + \omega|^2 + r_\nu^2 + t_{\nu M}^2} \right)^{-1-s}. \end{aligned}$$

As a function of $z_\nu - w_{\nu M}$, the inner sum is periodic mod Λ'_ν. Hence we may assume without loss of generality that $z_\nu - w_{\nu M}$ belongs to a suitably chosen fundamental parallelogram $\mathcal{P}_\nu$ for Λ'_ν. Let $0 \neq \omega_1 \in \Lambda'_\nu$ be such that $|\omega_1|$ is minimal and choose $\omega_2 \in \Lambda'_\nu \setminus \mathbb{Z}\omega_1$ such that $|\omega_2|$ is minimal. Then $\{\omega_1, \omega_2\}$ is a $\mathbb{Z}$-basis for Λ'_ν and

$$\mathcal{P}_\nu = \left\{ \zeta_1\omega_1 + \zeta_2\omega_2 \;\middle|\; \zeta_1, \zeta_2 \in \left[-\frac{1}{2}, \frac{1}{2}\right] \right\}$$

is a fundamental parallelogram for Λ'_ν. We draw from the inequality following (3.4) that there exists an $\epsilon > 0$ such that $|z_\nu - w_{\nu M} + \omega|^2 \geq \epsilon^2 |m_1 i + m_2|^2$ for all $0 \neq \omega = m_1\omega_1 + m_2\omega_2 \in \Lambda'_\nu$ and for all $z_\nu - w_{\nu M} \in \mathcal{P}_\nu$. Proceeding as in (3.5) we then find

$$\begin{aligned} &\sum_{\omega \in \Lambda'_\nu} \delta(P_\nu, MQ_\nu + \omega)^{-1-s} \\ &\leq (2r_\nu\, t_{\nu M})^{1+s} \left(r_\nu^{-2s-2} + \sum_{0 \neq \gamma \in \mathbb{Z}[i]} (\epsilon^2|\gamma|^2 + r_\nu^2)^{-1-s} \right) \\ &\leq (2r_\nu\, t_{\nu M})^{1+s} (r_\nu^{-2s-2} + C_2\, r_\nu^{-2s}) \leq C_3\, r_\nu^{1-s}\, (t_{\nu M})^{1+s} \end{aligned}$$

since $r_\nu \geq R$ (see (5.6)). We have estimated the sum over $0 \neq \gamma \in \mathbb{Z}[i]$ by the corresponding double integral to obtain the upper bound $C_2 r_\nu^{-2s}$. The above yields

$$f_\nu(P, Q) \leq C_4\, r_\nu^{1-s} \sum_{M \in V_\nu} (t_{\nu M})^{1+s}. \tag{5.10}$$

Here the sum on the right-hand side is an Eisenstein series for the cusp ∞ of the group Θ_ν in which the terms belonging to the stabilizer Λ_ν of ∞ in Θ_ν are missing. Using the notation of Sections 2, 4 of Chapter 3, we have

$$\sum_{M \in V_\nu} (t_{\nu M})^{1+s} = E_I(Q_\nu, s) - [\Lambda_\nu : \Lambda'_\nu]\, t_\nu^{1+s}. \tag{5.11}$$

Using the result of Theorem 3.4.1, we see that this expression is a bounded function of $Q_\nu \in \mathcal{F}_\nu$. If $Q \in \mathcal{F}_\mu$ approaches ζ_μ $(\mu \neq \nu)$, then $Q_\nu \in A_\nu \mathcal{F}_\nu$ approaches the cusp $A_\nu \zeta_\mu \neq \infty$ of Θ_ν, and Theorem 3.4.1 again yields that (5.11) is bounded, and (5.11) is trivially bounded on compact subsets of $\mathbb{H}$. Hence $f_\nu(P, Q)$ is bounded under the condition (5.6), and step 1 is finished.

Step 2. The first term on the right-hand side of (5.8) satisfies

$$\sum_{M \in \Gamma_\nu} k_s(\delta(P, MQ)) = \frac{2\pi\, r_\nu\, t_\nu}{|\mathcal{R}_\nu|}\, \tilde{k}_s\left(\frac{r_\nu}{t_\nu}\right) + O(1 + \sqrt{r_\nu t_\nu}) \tag{5.12}$$

uniformly in P, Q under the condition (5.6). Here,

$$\tilde{k}_s(u) := \int_0^\infty k_s\left(\rho + \frac{1}{2}\left(u + \frac{1}{u}\right)\right)\,d\rho \qquad \text{for} \quad u > 0. \tag{5.13}$$

To show this, we first look at the contribution of Γ_ν'. Let $\mathcal{P}_\nu$ be the fundamental parallelogram for Λ_ν' introduced above. Then we have

$$\begin{aligned} |\mathcal{P}_\nu| \sum_{M \in \Gamma_\nu'} k_s(\delta(P, MQ)) &= |\mathcal{P}_\nu| \sum_{\omega \in \Lambda_\nu'} k_s(\delta(P_\nu, Q_\nu + \omega)) \\ &= \sum_{\omega \in \Lambda_\nu'} \int_{\mathcal{P}_\nu} k_s(\delta(P_\nu, Q_\nu + \omega)) - k_s(\delta(P_\nu, Q_\nu + \omega + \xi + i\eta))\,d\xi\,d\eta \qquad (5.14) \\ &\quad + \int_{\mathbb{R}^2} k_s(\delta(P_\nu, Q_\nu + \xi + i\eta))\,d\xi\,d\eta. \end{aligned}$$

Here, all sums and integrals converge absolutely by virtue of the line following (2.20). The behaviour of the second term on the right-hand side of (5.14) is rather obvious:

$$\begin{aligned} &\int_{\mathbb{R}^2} k_s(\delta(P_\nu, Q_\nu + \xi + i\eta))\,d\xi\,d\eta \\ &= \int_{\mathbb{R}^2} k_s\left(\frac{|z_\nu - w_\nu - \xi - i\eta|^2 + r_\nu^2 + t_\nu^2}{2r_\nu t_\nu}\right)\,d\xi\,d\eta \\ &= 2r_\nu t_\nu \int_{\mathbb{R}^2} k_s\left(\xi^2 + \eta^2 + \frac{r_\nu^2 + t_\nu^2}{2r_\nu t_\nu}\right)\,d\xi\,d\eta \qquad (5.15) \\ &= 4\pi r_\nu t_\nu \int_0^\infty k_s\left(\rho^2 + \frac{1}{2}\left(\frac{r_\nu}{t_\nu} + \frac{t_\nu}{r_\nu}\right)\right)\,\rho\,d\rho \\ &= 2\pi r_\nu t_\nu\,\tilde{k}_s\left(\frac{r_\nu}{t_\nu}\right). \end{aligned}$$

Taking into account that Γ_ν' is of finite index in Γ_ν, we see that the contribution of these terms equals the first term on the right-hand side of (5.12). We go ahead to estimate the error term

$$\begin{aligned} g_\nu(P, Q) := \sum_{\omega \in \Lambda_\nu'} \int_{\mathcal{P}_\nu} &|k_s(\delta(P_\nu, Q_\nu + \omega)) \\ &- k_s(\delta(P_\nu, Q_\nu + \omega + \xi + i\eta))|\,d\xi\,d\eta. \end{aligned} \tag{5.16}$$

For $\delta \geq 3$ we have $k_s(\delta) = \frac{1}{4\pi}\varphi_s(\delta)$ and hence

$$k_s'(\delta) = -k_s(\delta)\left(\frac{s}{\sqrt{\delta^2 - 1}} + \frac{\delta}{\delta^2 - 1}\right),$$

$$k_s''(\delta) = k_s(\delta)\left(\left(\frac{s}{\sqrt{\delta^2 - 1}} + \frac{\delta}{\delta^2 - 1}\right)^2 + \frac{s\,\delta}{(\delta^2 - 1)^{3/2}} + \frac{1 + \delta^2}{(\delta^2 - 1)^2}\right).$$

This implies for $\delta \geq 3$ that $|k_s'(\delta)|$ is monotonically decreasing to zero. We now write

$$k_s(\delta) - k_s(\delta') = (\delta - \delta')\, k_s'(\delta'') \tag{5.17}$$

for some δ'' between δ and δ' and put

$$\delta = \delta(P_\nu, Q_\nu + \omega), \quad \delta' = \delta(P_\nu, Q_\nu + \omega + \zeta), \qquad \zeta = \xi + i\eta. \tag{5.18}$$

Then

$$\delta - \delta' = \frac{2\,\mathrm{Re}((z_\nu - w_\nu - \omega)\bar{\zeta}) - |\zeta|^2}{2r_\nu t_\nu}$$

and hence

$$|\delta - \delta'| \le C_5 \frac{|z_\nu - w_\nu - \omega - \zeta| + 1}{8r_\nu t_\nu} \tag{5.19}$$

for all $P \in \mathcal{F}_\nu,\ Q \in \mathcal{F},\ \zeta \in \mathcal{P}_\nu,\ \omega \in \Lambda_\nu'$.

We want to substitute (5.17) into (5.16) and we want to replace δ'' by δ' in the course of our estimate. This will be possible due to the monotonicity of $|k_s'|$ in $[3, \infty[$ provided δ, δ' are sufficiently large. We first consider those terms, for which δ, δ' is not large enough. The magnitude of δ, δ' depends of course on P and Q. We have the following cases:

(i): $P, Q \in \mathcal{F}_\nu$. Then we have $r_\nu \ge R,\ t_\nu \ge R$, and hence $\delta(P_\nu, Q_\nu + \omega + \zeta) \le 12$ for some $\zeta \in \mathcal{P}_\nu$ and for some P, Q only if $|z_\nu - w_\nu - \omega - \zeta|^2 \le 24R^2$. The number of all $\omega \in \Lambda_\nu'$ satisfying this inequality is bounded by some absolute constant (independent of P, Q, ζ). These finitely many ω contribute only an $O(1)$ to the right-hand side of (5.16). Note that k_s is uniformly bounded on $[1, \infty[$.

(ii): $P \in \mathcal{F}_\nu,\ Q \in \mathcal{F},\ t_\nu \le \frac{R}{24}$. Then we have

$$\frac{r_\nu^2 + t_\nu^2}{2r_\nu t_\nu} > \frac{1}{2}\frac{r_\nu}{t_\nu} \ge 12$$

and hence $\delta \ge 12,\ \delta' \ge 12$.

(iii): $P \in \mathcal{F}_\nu,\ Q \in \mathcal{F},\ \frac{R}{24} \le t_\nu \le R$. Then Q is restricted to a compact subset of $\mathcal{F}$. Again the number of $\omega \in \Lambda_\nu'$ such that $\delta(P_\nu, Q_\nu + \omega + \zeta) \le 12$ for some P, Q, ζ is bounded by some absolute constant independent of P, Q, ζ and the corresponding terms in (5.16) only contribute an $O(1)$ to the right-hand side.

Now we apply (1.1.19) to obtain

$$\delta(P_\nu, Q_\nu + \omega) \ge \frac{1}{4} \frac{\delta(P_\nu, Q_\nu + \omega + \zeta)}{\delta(Q_\nu + \omega, Q_\nu + \omega + \zeta)} \ge \frac{1}{4}\delta(P_\nu, Q_\nu + \omega + \zeta).$$

Hence we have $\delta'' \ge \frac{1}{4}\delta(P_\nu, Q_\nu + \omega + \zeta)$. Exclude the finitely many $\omega \in \Lambda_\nu'$ for which the right-hand side in this inequality is less than 3 and put $k_s^*(u) := |k_s'(\max(3, u))|$. Then our considerations above yield

$$
\begin{aligned}
g_\nu(P,Q) &\le C_6 + C_5 \sum_{\omega\in\Lambda'_\nu} \int_{\mathcal{P}_\nu} k_s^*\left(\frac{\delta(P_\nu, Q_\nu+\omega+\zeta)}{4}\right) \\
&\quad\cdot \frac{|z_\nu - w_\nu - \omega - \zeta| + 1}{8r_\nu t_\nu}\, d\xi\, d\eta \\
&= C_6 + C_5 \int_{\mathbb{R}^2} k_s^*\left(\frac{\xi^2+\eta^2+r_\nu^2+t_\nu^2}{8r_\nu t_\nu}\right) \frac{\sqrt{\xi^2+\eta^2}+1}{8r_\nu t_\nu}\, d\xi\, d\eta \\
&\le C_7 + C_8 \sqrt{r_\nu t_\nu} \int_{\mathbb{R}^2} k_s^*(\xi^2+\eta^2)\, \sqrt{\xi^2+\eta^2}\, d\xi\, d\eta \\
&= C_7 + 2\pi\, C_8 \sqrt{r_\nu t_\nu} \int_0^\infty k_s^*(\rho^2)\, \rho^2\, d\rho \\
&\le C_9\, (1+\sqrt{r_\nu t_\nu}).
\end{aligned}
$$

Since Γ'_ν is of finite index in Γ_ν, our proof of (5.12) is complete.

Step 3. We now define for $P,\ Q \in \mathcal{F}$

$$
(5.20)\quad K^*(P,Q,s) := \begin{cases} K(P,Q,s) - \dfrac{2\pi r_\nu t_\nu}{|\mathcal{R}_\nu|}\, \tilde{k}_s\left(\dfrac{r_\nu}{t_\nu}\right) & \text{if } P,Q \in \mathcal{F}_\nu, \\ K(P,Q,s) & \text{if } (P,Q) \in \mathcal{F}\times\mathcal{F} \setminus \cup_{\nu=1}^h \mathcal{F}_\nu \times \mathcal{F}_\nu. \end{cases}
$$

We define $K^*(\cdot,\cdot,s)$ on $\mathbb{H}\times\mathbb{H}$ by the requirement of Γ-invariance in both variables. We claim that Theorem 5.2 holds true with (K^*, K) instead of (F^*, F).

Statement (1) is trivially true since $\tilde{k}_s(\frac{1}{u}) = \tilde{k}_s(u)$. In order to prove that $K^*(\cdot,\cdot,s)$ is of Hilbert-Schmidt type, we decompose

$$
\begin{aligned}
&\int_{\mathcal{F}}\int_{\mathcal{F}} |K^*(P,Q,s)|^2\, dv(P)\, dv(Q) \\
(5.21)\qquad &= \sum_{\nu=1}^h \int_{\mathcal{F}_\nu}\int_{\mathcal{F}} |K^*(P,Q,s)|^2\, dv(P)\, dv(Q) \\
&\quad + \int_{\mathcal{K}}\int_{\mathcal{F}} |K^*(P,Q,s)|^2\, dv(P)\, dv(Q),
\end{aligned}
$$

where $\mathcal{K}$ is a suitable compact subset of $\mathcal{F}$. Clearly, the second term on the right-hand side is finite by (5.20) and Corollary 2.5. Further, we know from step 1 and step 2 that $K^*(P,Q,s) = O(1+\sqrt{r_\nu t_\nu})$ for $(P,Q) \in \mathcal{F}_\nu \times \mathcal{F}$. Since

$$
\begin{aligned}
\int_{\mathcal{F}_\nu}\int_{\mathcal{F}} (\sqrt{r_\nu t_\nu})^2\, dv(P)\, dv(Q) &= \int_{\mathcal{R}_\nu\times[R,\infty[} r\, \frac{dx\, dy\, dr}{r^3} \int_{\mathcal{F}} t_\nu\, dv(Q) \\
&= \frac{|\mathcal{R}_\nu|}{R} \int_{A_\nu \mathcal{F}} t\, \frac{dx\, dy\, dt}{t^3} < \infty,
\end{aligned}
$$

we see that (5.21) is indeed finite.

In order to prove (3) we have only to show that

$$P \mapsto \int_{\mathcal{F}} (K(P,Q,s) - K^*(P,Q,s))\, dv(Q) =: \Phi(P)$$

is bounded on $\mathbb{H}$. Obviously $\Phi(P) = 0$ for all $P \in \Gamma(\mathcal{F} \setminus \cup_{\nu=1}^{h} \mathcal{F}_\nu)$. Suppose that $P' \in \mathbb{H}$ is Γ-equivalent to $P \in \mathcal{F}_\nu$. Then

$$\begin{aligned}
|\Phi(P')| = |\Phi(P)| &\leq \frac{2\pi\, r_\nu}{|\mathcal{R}_\nu|} \int_{\mathcal{F}_\nu} t_\nu \left| \tilde{k}_s\left(\frac{t_\nu}{r_\nu}\right)\right| \, dv(Q) \\
&= \frac{2\pi\, r_\nu}{|\mathcal{R}_\nu|} \int_{\mathcal{R}_\nu \times [R,\infty[} t \left| \tilde{k}_s\left(\frac{t}{r_\nu}\right)\right| \frac{du\, dv\, dt}{t^3} = 2\pi \int_{\frac{R}{r_\nu}} |\tilde{k}_s(t)| \, \frac{dt}{t^2} \\
&\leq 2\pi \int_1^\infty |\tilde{k}_s(t)| \, \frac{dt}{t^2} < \infty,
\end{aligned}$$

since

$$|\tilde{k}_s(u)| \leq \int_0^\infty |k_s\left(\rho + \frac{1}{2}\left(u + \frac{1}{u}\right)\right)| \, d\rho \leq \int_1^\infty |k_s(\rho)| \, d\rho < \infty.$$

Hence Φ is bounded on all of $\mathbb{H}$.

To prove (4), assume that $f \in L^2(\Gamma \setminus \mathbb{H})$ and

$$\int_{\mathcal{R}_\nu} f(A_\nu^{-1}(\xi + i\eta + jt))\, d\xi\, d\eta = 0 \tag{5.22}$$

for $\nu = 1, ..., h$ and almost all $t > R$. We need only to show that (5.3) holds with (K, K^*) instead of (F, F^*) and with $\mathcal{F}_\nu$ $(\nu = 1, \ldots, h)$ instead of $\mathcal{F}$. In addition, we have to prove our assertion only for $P \in \mathcal{F}_\nu$. Now we have under these conditions by (5.20)

$$\begin{aligned}
&\int_{\mathcal{F}_\nu} (K(P,Q,s) - K^*(P,Q,s))\, f(Q)\, dv(Q) \\
&= \frac{2\pi}{|\mathcal{R}_\nu|} \int_{\mathcal{F}_\nu} r_\nu t_\nu\, \tilde{k}_s\left(\frac{r_\nu}{t_\nu}\right) f(Q)\, dv(Q) \\
&= \frac{2\pi\, r_\nu}{|\mathcal{R}_\nu|} \int_{\mathcal{R}_\nu \times [R,\infty[} t\, \tilde{k}_s\left(\frac{r_\nu}{t}\right) f(A_\nu^{-1}(\xi + i\eta + jt))\, \frac{d\xi\, d\eta\, dt}{t^3} = 0
\end{aligned} \tag{5.23}$$

by (5.22). This finishes the first half of the proof, dealing with the contribution of $K(P,Q,s)$.

In the second half of our proof we shall be similarly concerned with $G(P,Q)$ (cf. (2.19), (2.21)). We start from the decomposition

$$G(P,Q) = \sum_{M \in \Gamma_\nu} g(\delta(P, MQ)) + \sum_{M \in \Gamma \setminus \Gamma_\nu} g(\delta(P, MQ)). \tag{5.24}$$

Step 4. We have

$$(5.25)\qquad \begin{aligned} G(P,Q) = {} & \frac{2\pi r_\nu t_\nu}{|\mathcal{R}_\nu|}\,\tilde{g}\left(\frac{r_\nu}{t_\nu}\right) + O(g(\delta(P,Q))) \\ & + O\left(\sqrt{r_\nu t_\nu}\,(1+\log(1+r_\nu t_\nu))\right) \end{aligned}$$

uniformly in $(P,Q) \in \mathcal{F}_\nu \times \mathcal{F}$, where

$$(5.26)\qquad \tilde{g}(u) := \int_0^\infty g\left(\rho + \frac{1}{2}\left(u + \frac{1}{u}\right)\right)\,d\rho \qquad \text{for} \quad u > 0.$$

In order to show this, we first note that the second sum on the right-hand side of (5.24) is zero on grounds of our choice (5.5) of R and of (5.6). The first sum in (5.24) is again compared with the corresponding integral. We first deal with the contribution of Γ'_{ζ_ν}. Repeating the computations (5.14), (5.15) we find

$$(5.27)\qquad \begin{aligned} & |\mathcal{P}_\nu| \sum_{M\in\Gamma'_\nu} g(\delta(P,MQ)) \\ & = \sum_{\omega\in\Lambda'_\nu} \int_{\mathcal{P}_\nu} (g(\delta(P_\nu, Q_\nu+\omega)) - g(\delta(P_\nu,Q_\nu+\omega+\xi+i\eta)))\,d\xi\,d\eta \\ & + \int_{\mathbb{R}^2}\!\!\int g(\delta(P_\nu,Q_\nu+\xi+i\eta))\,d\xi\,d\eta, \end{aligned}$$

$$(5.28)\qquad \int_{\mathbb{R}^2} g(\delta(P_\nu,Q_\nu+\xi+i\eta))\,d\xi\,d\eta = 2\pi\, r_\nu t_\nu\, \tilde{g}\left(\frac{r_\nu}{t_\nu}\right).$$

Observing that Γ'_ν is of finite index in Γ_ν we see that the second term on the right-hand side of (5.27) gives rise to the first term on the right-hand side of (5.25).

We go ahead to estimate the sum

$$(5.29)\qquad \begin{aligned} h_\nu(P,Q) := \frac{1}{|\mathcal{P}_\nu|} \sum_{\omega\in\Lambda'_\nu} \int_{\mathcal{P}_\nu} & |g(\delta(P_\nu,Q_\nu+\omega)) \\ & - g(\delta(P_\nu,Q_\nu+\omega+\xi+i\eta))|\,d\xi\,d\eta. \end{aligned}$$

The main idea is to replace $g(\delta(P_\nu,Q_\nu+\omega+\xi+i\eta))$ by $g(\delta(P_\nu,Q_\nu+\omega'))$ where $\omega' \in \Lambda'_\nu$ is a lattice point near ω such that $\delta(P_\nu,Q_\nu+\omega') \geq \delta(P_\nu,Q_\nu+\omega+\zeta)$ for all $\zeta \in \mathcal{P}_\nu$. Then the monotonicity of g will imply that many terms in the majorant for (5.29) cancel and this will enable us to estimate (5.29).

We choose a $\mathbb{Z}$-basis $\{\omega_1,\ \omega_2\}$ of Λ'_ν as above and put

$$(5.30)\qquad \mathcal{P}_\nu = \left\{ \xi_1\omega_1 + \xi_2\omega_2 \;\middle|\; \xi_1,\ \xi_2 \in \left[-\frac{1}{2},\frac{1}{2}\right] \right\}.$$

Estimating h_ν we can assume without loss of generality that $z_\nu - w_\nu = \alpha_1\omega_1 + \alpha_2\omega_2$ with $\alpha_1,\alpha_2 \in \left[-\frac{1}{2},\frac{1}{2}\right]$. We put further $\zeta = \xi + i\eta = \xi_1\omega_1 + \xi_2\omega_2$, with

$\xi_1,\ \xi_2 \in [-1/2, 1/2]$. Suppose that $\omega = m_1\omega_1 + m_2\omega_2$ with $m_1,\ m_2 \in \mathbb{Z}\setminus\{0\}$, and let $\epsilon_1 := 2m_1/|m_1|,\ \epsilon_2 := 2m_2/|m_2|$. Then we have

$$\begin{aligned}
&|z_\nu - w_\nu - \omega - (\epsilon_1\omega_1 + \epsilon_2\omega_2)|^2 - |z_\nu - w_\nu - \omega - \zeta|^2 \\
&\quad = 2\,\mathrm{Re}((\omega - (z_\nu - w_\nu))(\epsilon_1\bar{\omega}_1 + \epsilon_2\bar{\omega}_2)) + |\epsilon_1\omega_1 + \epsilon_2\omega_2|^2 \\
&\qquad - 2\,\mathrm{Re}(\omega - (z_\nu - w_\nu))\bar{\zeta} - |\zeta|^2.
\end{aligned} \tag{5.31}$$

Note that by our choice of ω_1, ω_2 we have $|\omega_1 \pm \omega_2|^2 \geq |\omega_2|^2$ and hence $|2\,\mathrm{Re}\,\omega_1\bar{\omega}_2| \leq |\omega_1|^2$. This yields by a simple computation

$$|\epsilon_1\omega_1 + \epsilon_2\omega_2|^2 - |\zeta|^2 \geq \frac{13}{4}|\omega_1|^2 \tag{5.32}$$

since $\xi_1, \xi_2 \in [-\frac{1}{2}, \frac{1}{2}]$. Further we have the estimate

$$\begin{aligned}
&2\,\mathrm{Re}(((m_1 - \alpha_1)\omega_1 + (m_2 - \alpha_2)\omega_2)((\epsilon_1 - \xi_1)\bar{\omega}_1 + (\epsilon_2 - \xi_2)\bar{\omega}_2)) \\
&\geq \frac{1}{2}(|m_1 - \alpha_1| + |m_2 - \alpha_2|)|\omega_1|^2
\end{aligned} \tag{5.33}$$

since by the above

$$2|\epsilon_1 - \xi_1| - |\epsilon_2 - \xi_2| \geq 2(2 - |\xi_1|) - (2 + |\xi_2|) = 2 - 2|\xi_1| - |\xi_2| \geq \frac{1}{2}$$

and similarly $2|\epsilon_2 - \xi_2| - |\epsilon_1 - \xi_1| \geq \frac{1}{2}$. This yields under the above restrictions

$$\delta(P_\nu, Q_\nu + \omega + \epsilon_1\omega_1 + \epsilon_2\omega_2) \geq \delta(P_\nu, Q_\nu + \omega + \zeta)$$

and hence, due to the monotonicity of g

$$\begin{aligned}
&\sum_{\substack{|m_1|\geq 1 \\ |m_2|\geq 1}} \int_{\mathcal{P}_\nu} |g(\delta(P_\nu, Q_\nu + \omega)) - g(\delta(P_\nu, Q_\nu + \omega + \xi + i\eta))|\, d\xi\, d\eta \\
&\leq |\mathcal{P}_\nu| \sum_{(m_1,m_2)\in J} g(\delta(P_\nu, Q_\nu + m_1\omega_1 + m_2\omega_2)),
\end{aligned} \tag{5.34}$$

where J denotes the index set

$$J = \{\ (m_1, m_2)\ \ |\ \ m_1, m_2 \in \mathbb{Z},\ m_1 \in \{\pm 1, \pm 2\} \quad \text{or} \quad m_2 \in \{\pm 1, \pm 2\}\ \}.$$

To estimate (5.29), we still have to take the terms with $m_1 = 0$ or $m_2 = 0$ into consideration. The contribution of the terms with $m_1 = 0$ is

$$\begin{aligned}
&\sum_{m_2\in\mathbb{Z}} \int_{\mathcal{P}_\nu} |g(\delta(P_\nu, Q_\nu + m_2\omega_2)) \\
&\qquad\qquad - g(\delta(P_\nu, Q_\nu + m_2\omega_2 + \xi + i\eta))|\, d\xi\, d\eta \\
&\leq |\mathcal{P}_\nu| \sum_{m_2\in\mathbb{Z}} g(\delta(P_\nu, Q_\nu + m_2\omega_2)) \\
&\quad + C_{10}|\mathcal{P}_\nu| \int_{-\frac{1}{2}}^{\frac{1}{2}} \int_{-\infty}^{+\infty} g(\delta(P_\nu, Q_\nu + \xi_1\omega_1 + \xi_2\omega_2))\, d\xi_2\, d\xi_1.
\end{aligned} \tag{5.35}$$

We have the estimate

$$\begin{aligned}
A &:= \int_{-\frac{1}{2}}^{\frac{1}{2}} \int_{-\infty}^{+\infty} g(\delta(P_\nu, Q_\nu + \xi_1\omega_1 + \xi_2\omega_2))\, d\xi_2\, d\xi_1 \\
&\le \int_{-1}^{1} \int_{-\infty}^{+\infty} g\left(\frac{|\xi_1\omega_1 + \xi_2\omega_2|^2 + r_\nu^2 + t_\nu^2}{2r_\nu t_\nu}\right) d\xi_2\, d\xi_1 \\
&\le \int_{-1}^{1} \int_{-\infty}^{+\infty} g\left(\frac{\frac{1}{2}(\xi_1^2 + \xi_2^2)\, |\omega_1|^2 + r_\nu^2 + t_\nu^2}{2r_\nu t_\nu}\right) d\xi_2\, d\xi_1
\end{aligned}$$

because g is monotonically decreasing and since $|\xi_1\omega_1 + \xi_2\omega_2|^2 \ge \frac{1}{2}(\xi_1^2 + \xi_2^2)|\omega_1|^2$ follows from a simple computation. We put $\gamma := \frac{|\omega_1|}{\sqrt{2}}$ and proceed:

$$\begin{aligned}
A &\le \frac{2}{|\omega_1|^2} \int_{-\gamma}^{\gamma} \int_{-\infty}^{+\infty} g\left(\frac{u^2 + v^2 + r_\nu^2 + t_\nu^2}{2r_\nu t_\nu}\right) du\, dv \\
&= \frac{8}{|\omega_1|^2} \int_{0}^{\gamma} \int_{0}^{\sqrt{6r_\nu t_\nu}} g\left(\frac{u^2 + v^2 + r_\nu^2 + t_\nu^2}{2r_\nu t_\nu}\right) du\, dv \\
&\le \frac{\sqrt{2}}{\pi\, |\omega_1|^2} \int_{0}^{\gamma} \int_{0}^{\sqrt{6r_\nu t_\nu}} \left(\frac{u^2 + v^2 + r_\nu^2 + t_\nu^2}{2r_\nu t_\nu} - 1\right)^{-\frac{1}{2}} du\, dv.
\end{aligned}$$

By some further obvious estimates we find

$$\begin{aligned}
A &\le \frac{2\,\sqrt{r_\nu t_\nu}}{\pi\, |\omega_1|^2} \int_{0}^{\sqrt{6r_\nu t_\nu}} \int_{0}^{\gamma} \frac{du\, dv}{\sqrt{u^2 + v^2}} \\
&\le \frac{2\,\sqrt{r_\nu t_\nu}}{\pi\, |\omega_1|^2} \int_{0}^{\sqrt{6r_\nu t_\nu}} \log\left(1 + \frac{2\gamma}{v}\right) dv \\
&\le C_{11}\, \sqrt{r_\nu t_\nu}\, \left(C_{12} + \int_{1}^{1+6r_\nu t_\nu} \log\left(1 + \frac{2\gamma}{v}\right) dv\right) \\
&\le C_{13}\, \sqrt{r_\nu t_\nu}\, (1 + \log(1 + r_\nu t_\nu)).
\end{aligned} \tag{5.36}$$

The contribution of the terms with $m_2 = 0$ is handled in the same way. Inserting (5.36) and the contribution of the terms with $m_2 = 0$ into (5.29), we find

$$\begin{aligned}
h_\nu(P, Q) \le &\sum_{\substack{|m_1| \le 2 \text{ or} \\ |m_2| \le 2}} g(\delta(P_\nu, Q_\nu + m_1\omega_1 + m_2\omega_2)) \\
&+ C_{14}\sqrt{r_\nu t_\nu}(1 + \log(1 + r_\nu t_\nu)).
\end{aligned} \tag{5.37}$$

The sum in (5.37) is estimated as follows: We single out the term with $m_1 = m_2 = 0$. Then there exists an $\epsilon > 0$ such that for all $z_\nu - w_\nu \in \mathcal{P}_\nu$

$$|z_\nu - w_\nu - m_1\omega_1 - m_2\omega_2|^2 \ge \epsilon^2 m_2^2 \text{ for all } (m_1, m_2) \text{ with } |m_1| \le 2,\ m_2 \ne 0,$$
$$|z_\nu - w_\nu - m_1\omega_1 - m_2\omega_2|^2 \ge \epsilon^2 m_1^2 \text{ for all } (m_1, m_2) \text{ with } m_1 \ne 0,\ |m_2| \le 2.$$

This yields

$$\begin{aligned}
&\sum_{\substack{|m_1|\le 2 \text{ or} \\ |m_2|\le 2}} g(\delta(P_\nu, Q_\nu + m_1\omega_1 + m_2\omega_2)) \\
&\le g(\delta(P,Q)) + 20 \sum_{m=1}^{\infty} g\left(\frac{\epsilon^2 m^2 + r_\nu^2 + t_\nu^2}{2r_\nu t_\nu}\right) \\
&\le g(\delta(P,Q)) + C_{15} \sum_{1\le m\le \epsilon^{-1}\sqrt{6r_\nu t_\nu}} \left(\frac{\epsilon^2 m^2 + r_\nu^2 + t_\nu^2}{2r_\nu t_\nu} - 1\right)^{-\frac{1}{2}} \\
&\le g(\delta(P,Q)) + C_{16}\sqrt{r_\nu t_\nu} \sum_{1\le m\le \epsilon^{-1}\sqrt{6r_\nu t_\nu}} \frac{1}{m} \\
&\le g(\delta(P,Q)) + C_{17}\sqrt{r_\nu t_\nu}\,(1+\log(1+r_\nu t_\nu)).
\end{aligned}$$

This yields our final estimate

$$h_\nu(P,Q) \le g(\delta(P,Q)) + C_{18}\sqrt{r_\nu\, t_\nu}\,(1+\log(1+r_\nu t_\nu)). \tag{5.38}$$

The proof of (5.25) is now complete.

Step 5. We define for $P, Q \in \mathcal{F}$

$$G^*(P,Q) := \begin{cases} G(P,Q) - \dfrac{2\pi r_\nu t_\nu}{|\mathcal{R}_\nu|}\, \tilde{g}\left(\dfrac{r_\nu}{t_\nu}\right) & \text{if } P,Q \in \mathcal{F}_\nu, \\ G(P,Q) & \text{if } (P,Q) \in \mathcal{F}\times\mathcal{F} \setminus \cup_{\nu=1}^{h} \mathcal{F}_\nu \times \mathcal{F}_\nu. \end{cases} \tag{5.39}$$

We further define G^* on all of $\mathbb{H}\times\mathbb{H}$ by Γ-invariance in both variables. Then we claim that Theorem 5.2 holds true with (G^*, G) instead of (F^*, F). A look back at step 3 and at (5.25) yields that all we have to show is statement (3) and that

$$\int_{\mathcal{F}_\nu}\int_{\mathcal{F}} g(\delta(P,Q))^2\, dv(P)\, dv(Q) < \infty. \tag{5.40}$$

Obviously (5.40) is true if

$$\int_{\mathcal{F}_\nu}\int_{\mathbb{H}} g(\delta(P,Q))^2\, dv(P)\, dv(Q) < \infty. \tag{5.41}$$

Now choose $T \in \mathbf{PSL}(2,\mathbb{C})$ such that $TQ = j$. Then we have

$$\int_{\mathbb{H}} g(\delta(P,Q))^2\, dv(P) = \int_{\mathbb{H}} g(\delta(TP,j))^2\, dv(P) = \int_{\mathbb{H}} g(\delta(P,j))^2\, dv(P).$$

The right-hand side is finite and independent of Q. This implies statement (3) and (5.41), and step 5 is complete.

We finally define

$$F^*(P,Q,s) := G^*(P,Q) + K^*(P,Q,s).$$

Then it is immediate that F^* has the required properties since G^* and K^* have. $\square$

Note that the kernel F^* in the preceding construction can be written as

$$F^*(P,Q,s) := \begin{cases} F(P,Q,s) - \dfrac{r_\nu t_\nu}{2|\mathcal{R}_\nu|}\, \tilde{\varphi}_s\left(\dfrac{r_\nu}{t_\nu}\right) & \text{if } P,Q \in \mathcal{F}_\nu, \\ F(P,Q,s) \qquad \text{if } (P,Q) \in \mathcal{F} \times \mathcal{F} \setminus \cup_{\nu=1}^{h} \mathcal{F}_\nu \times \mathcal{F}_\nu, \end{cases} \tag{5.42}$$

where for $u > 0$

$$\tilde{\varphi}_s(u) := \int_0^\infty \varphi_s\left(\rho + \frac{1}{2}\left(u + \frac{1}{u}\right)\right)\, d\rho. \tag{5.43}$$

For $u \geq 1,\ u = e^\xi$ with $\xi \geq 0$, we have

$$\begin{aligned} \tilde{\varphi}_s(u) &= \int_0^\infty \varphi_s(\rho + \cosh\xi) d\rho \\ &= \int_\xi^\infty \varphi_s(\cosh t) \sinh t dt = \int_\xi^\infty e^{-ts} dt = \frac{1}{s} u^{-s}. \end{aligned}$$

We thus obtain

$$\tilde{\varphi}_s(u) = \begin{cases} \dfrac{u^{-s}}{s} & \text{for} \quad u \geq 1, \\ \dfrac{u^s}{s} & \text{for} \quad 0 < u \leq 1, \end{cases} \tag{5.44}$$

and we can write down F^* explicitly by (5.42) and by

$$F^*(P,Q,s) := \begin{cases} F(P,Q,s) - \dfrac{r_\nu^{1-s} t_\nu^{1+s}}{2s|\mathcal{R}_\nu|} & \text{if } P,Q \in \mathcal{F}_\nu,\ r_\nu \geq t_\nu, \\ F(P,Q,s) - \dfrac{r_\nu^{1+s} t_\nu^{1-s}}{2s|\mathcal{R}_\nu|} & \text{if } P,Q \in \mathcal{F}_\nu,\ r_\nu \leq t_\nu. \end{cases} \tag{5.45}$$

These formulae clearly display the terms which destroy the square integrability of $F(\cdot,\cdot,s)$ over $\mathcal{F} \times \mathcal{F}$.

The proof of Theorem 5.2 is admittedly rather tedious but in view of Corollaries 5.3, 5.4, 5.5, 5.6 to come the results are well worth the effort. Comparing our proof with the two-dimensional case as described by Roelcke (1967), paragraph 8 we point out that Roelcke in his case is even able to show that the sum $\sum_{m\geq 1} < g, e_m > e_m$ from Corollary 5.4 converges uniformly on the entire hyperbolic plane, since his analogue of the function

$$P \mapsto \int_{\mathcal{F}} |F^*(P,Q,s)|^2\, dv(Q)$$

is even bounded on the entire hyperbolic plane. This does not result from our estimates, and we do not know if it is possible to define F^* in such a way that this comes out.

Corollary 5.3. *Let $\Gamma < \mathbf{PSL}(2,\mathbb{C})$ be a cofinite but non-cocompact subgroup. Suppose that $-\Delta : \mathcal{D} \to L^2(\Gamma \setminus \mathbb{H})$ has the orthonormalized system $(e_m)_{m\geq 1}$ of cusp eigenfunctions and $-\Delta e_m = \lambda_m e_m$. Then every λ_n occurs with finite multiplicity, and $\sum_{m\geq 1} \lambda_m^{-2} < \infty$.*

We stress that we have no information yet how many cusp eigenfunctions a specific cofinite group might have. See Section 6.6 of for further discussion.

Proof. Choose $s > 1$ and put $\lambda = 1 - s^2$. Then Theorem 5.2 yields

$$\int_{\mathcal{F}} F^*(P,Q,s)\, e_m(Q)\, dv(Q) = \frac{1}{\lambda_m - \lambda} e_m(P).$$

Hence $(e_m)_{n\geq 1}$ is an orthonormalized system of eigenfunctions of a self-adjoint integral operator of Hilbert-Schmidt type whence $\sum_{m\geq 1}^{\infty} (\lambda_m - \lambda)^{-2} < \infty$. This yields the result. □

Corollary 5.4. *Let Γ and $(e_m)_{m\geq 1}$ be as in Corollary 5.3 and suppose that $g \in \tilde{\mathcal{D}}$. Then the contribution $\sum_{m\geq 1} < g, e_m > e_m$ of the system $(e_m)_{m\geq 1}$ to the eigenfunction expansion of g converges also pointwise absolutely and uniformly on compact subsets of $\mathbb{H}$.*

Proof. Choose $s > 1$ and put $\lambda = 1 - s^2$, $h := (-\Delta - \lambda)g$. Let $p, q \in \mathbb{N}, p < q$. Then we have for all $P \in \mathbb{H}$

$$\begin{aligned}
&\sum_{n=p}^{q} | < g, e_n > e_n(P)| \\
&\quad = \sum_{n=p}^{q} | < g, (\lambda_n - \lambda) e_n > | \left| \int_{\mathcal{F}} F^*(P,Q,s)\, e_n(Q)\, dv(Q) \right| \\
&= \sum_{n=p}^{q} | < h, e_n > | \left| \int_{\mathcal{F}} F^*(P,Q,s)\, e_n(Q)\, dv(Q) \right| \\
&\leq \left(\sum_{n=p}^{q} | < h, e_n > |^2 \right)^{\frac{1}{2}} \left(\int_{\mathcal{F}} |F^*(P,Q,s)|^2\, dv(Q) \right)^{\frac{1}{2}}.
\end{aligned}$$

Theorem 5.2, (3) now yields the assertion. □

Corollary 5.5. *Let Γ and $(e_m)_{m\geq 1}$ be as in Corollary 5.3. Then the series*

$$L(P,Q) := \sum_{m\geq 1} \lambda_m^{-2}\, e_m(Q)\, \overline{e_m(P)} \qquad (P, Q \in \mathbb{H}) \tag{5.46}$$

converges absolutely and uniformly on compact subsets of $\mathbb{H} \times \mathbb{H}$. *The continuous function* $L(P, Q)$ *represents an element of* $L^2(\mathcal{F} \times \mathcal{F}, dv \otimes dv)$. *It is furthermore integrable over* $\mathcal{F}$.

Proof. We choose a real number $s > 1$ and put $\lambda = 1 - s^2$. We obtain from Bessel's inequality

$$\sum_{m \geq 1} (\lambda_m - \lambda)^{-2} \, |e_m(P)|^2 = \sum_{m \geq 1} | < F^*(P, \cdot, s) e_m, e_m > |^2 \leq \|F^*(P, \cdot, s)\|^2.$$

From Theorem 5.2 we infer that the left hand side is bounded if P ranges over a compact subset of $\mathbb{H}$. The proof of the corollary is finished by some elementary considerations using the Schwarz inequality and by application of Corollary 5.3. □

Corollary 5.6. *Suppose that* Γ *is a discrete group of finite covolume and let* $\lambda \in \mathbb{C}$. *Then the vector space* $\mathcal{A}(\Gamma, \lambda)$ *of all automorphic functions for the group* Γ *with parameter* $\lambda \in \mathbb{C}$ *is finite-dimensional.*

Proof. If Γ is cocompact, then every automorphic function is an eigenfunction of $-\Delta : \mathcal{D} \to L^2(\Gamma \setminus \mathbb{H})$ and the assertion follows from Corollary 3.3. Suppose now that Γ has cusps and let $\zeta_1, \ldots, \zeta_h$ be a maximal system of Γ-inequivalent cusps of Γ, $\zeta_\nu = A_\nu^{-1}\infty$ with $A_\nu \in \mathbf{PSL}(2, \mathbb{C})$. Every $f \in \mathcal{A}(\Gamma, \lambda)$ has a Fourier expansion of the type (3.3.2) at ζ_ν and hence

$$f \circ A_\nu^{-1}(z + rj) = a_\nu(f) \, r^{1+s} + b_\nu(f) \, r^{1-s} + O(e^{-\epsilon r})$$

as $r \to \infty$ uniformly in $z \in \mathbb{C}$ for some $\epsilon > 0$. For $s = 0$ the term r^{1-s} has to be replaced by r log$r$. With every $f \in \mathcal{A}(\Gamma, \lambda)$ we associate the vector $a(f) := (a_1(f), \ldots, a_h(f), b_1(f), \ldots, b_h(f)) \in \mathbb{C}^{2h}$. Choose $f_1, \ldots, f_k \in \mathcal{A}(\Gamma, \lambda)$ with k maximal such that $a(f_1), \ldots, a(f_k)$ are linearly independent, and suppose that $f \in \mathcal{A}(\Gamma, \lambda)$. Then $a(f)$ is a linear combination of $a(f_1), \ldots, a(f_k)$:

$$a(f) = \alpha_1 a(f_1) + \ldots + \alpha_k a(f_k)$$

with $\alpha_1, \ldots, \alpha_k \in \mathbb{C}$. Hence $f - (\alpha_1 f_1 + \ldots + \alpha_k f_k)$ is a cusp function with eigenvalue λ or vanishes identically. Since the space of cusp functions with fixed eigenvalue is finite-dimensional the assertion follows. □

Chapter 5. Spectral Theory of the Laplace Operator for Cocompact Groups

In the present Chapter we give applications and finer points of the spectral theory of the Laplace-Beltrami operator Δ on $L^2(\Gamma\backslash\mathbb{H})$ in case $\Gamma < \mathbf{PSL}(2,\mathbb{C})$ is a discrete cocompact group. We already know from the preceding Chapter that $-\Delta$ is essentially self-adjoint and positive on the subspace $\mathcal{D} \subset L^2(\Gamma\backslash\mathbb{H})$ consisting of all C^2-functions $f \in L^2(\Gamma\backslash\mathbb{H})$ such that $\Delta f \in L^2(\Gamma\backslash\mathbb{H})$. This means that the closure of the graph of Δ in $L^2(\Gamma\backslash\mathbb{H}) \times L^2(\Gamma\backslash\mathbb{H})$ is the graph of a self-adjoint linear operator $\tilde{\Delta}: \tilde{\mathcal{D}} \to L^2(\Gamma\backslash\mathbb{H})$.

The complex number λ is contained in the resolvent set $\rho(-\tilde{\Delta})$ of the operator $-\tilde{\Delta}$ if and only if the resolvent operator $R_\lambda = (-\tilde{\Delta} - \lambda)^{-1}$ is a bounded linear operator mapping $L^2(\Gamma\backslash\mathbb{H})$ bijectively onto $\tilde{\mathcal{D}}$. For suitable values of λ we have shown in Section 2 of Chapter 4 that the resolvent operator $(-\tilde{\Delta} - \lambda)^{-1}$ is described by an integral operator of Carleman type the kernel of which is the Maaß–Selberg series

$$F(P,Q,s) = \sum_{M\in\Gamma} \varphi_s(\delta(P,MQ)),$$

where $s \in \mathbb{C}$, $\operatorname{Re} s > 1$ and

$$\varphi_s(\delta) = \frac{1}{4\pi} \frac{(\delta + \sqrt{\delta^2 - 1})^{-s}}{\sqrt{\delta^2 - 1}}$$

with the distance-function $\delta(\cdot,\cdot)$ defined in Chapter 1. More specifically we have proved that

$$\int_{\mathcal{F}} |F(P,Q,s)|^2 \, dv(Q) < \infty$$

for $s \in \mathbb{C}$, $\operatorname{Re} s > 1$, $P \in \mathbb{H}$ arbitrary, and $\mathcal{F}$ a fundamental domain for Γ. Furthermore we have

$$R_\lambda f(P) = \int_{\mathcal{F}} F(P,Q,s) f(Q) \, dv(Q)$$

if $\lambda = 1 - s^2$, $\operatorname{Re} s > 1$ and $f \in L^2(\Gamma\backslash\mathbb{H})$. Henceforth we also call the series $F(P,Q,s)$ the resolvent kernel. The resolvent kernel does not necessarily describe the resolvent for all $\lambda = 1 - s^2$ contained in the resolvent set of $-\tilde{\Delta}$

since the Maaß–Selberg series converges only when $\operatorname{Re} s$ is sufficiently large, for example $\operatorname{Re} s > 1$ suffices for all groups Γ.

The resolvent kernel is of Hilbert-Schmidt type if as in the case considered here $\Gamma \backslash \mathbb{H}$ is compact. It follows that the operator $-\Delta : \mathcal{D} \to L^2(\Gamma \backslash \mathbb{H})$ possesses a complete orthonormal system $(e_n)_{n \geq 0}$ of eigenfunctions with associated eigenvalues $0 = \lambda_0 < \lambda_1 \leq \lambda_2 \leq \dots$ such that

$$\sum_{n=1}^{\infty} \lambda_n^{-2} < \infty.$$

Apart from the constant eigenfunction e_0 and its associated eigenvalue $\lambda_0 = 0$ none of the eigenfunctions and the eigenvalues λ_n can be computed explicitly. In order to get some information on the eigenvalues one takes the trace in the eigenfunction expansion of appropriate kernels so that the unknown eigenfunctions disappear. This method cannot be applied to the resolvent kernel itself because this kernel is not of trace class. However, if the resolvent kernel is applied twice, one gets a kernel of trace class, and the trace of this kernel can be evaluated explicitly be means of Selberg's methods. The result is Theorem 2.2 which has several interesting consequences. First we obtain a variant of Huber's theorem on the connection between the eigenvalue and length spectra.

The iterated resolvent kernel seems to be the most interesting kernel for the computation of the trace since the corresponding trace formula leads immediately to the Selberg zeta function and its fascinating properties. We define and discuss the Selberg zeta function $Z_\Gamma(s)$ in Section 4. This function is defined by some kind of Euler product. The trace formula enables us to continue the zeta function as an entire function of s into the whole s-plane and to prove that it satisfies a simple functional equation relating $Z_\Gamma(-s)$ to $Z_\Gamma(s)$. The zeros of Z_Γ are precisely the numbers $\pm s_n$ such that $\lambda_n = 1 - s_n^2$ is an eigenvalue of $-\Delta : \mathcal{D} \to L^2(\Gamma \backslash \mathbb{H})$. This means that the analogue of the Riemann hypothesis is true for the Selberg zeta function save for the finitely many zeros of Z_Γ in $[-1, 1]$. The analogue of the Lindelöf hypothesis is also true for Z_Γ. The Selberg zeta function is an entire function of order precisely 3, and we can prove quite satisfactory results on the canonical factorization of Z_Γ in the sense of Hadamard's theorem on entire functions of finite order.

As a first application of the Selberg zeta function we prove Weyl's asymptotic law on the distribution of the eigenvalues of $-\Delta$. We also produce the standard bound for the error term in this asymptotic law, see Theorem 5.6. A second application is given in Section 7 where we derive the so-called prime geodesic theorem for the asymptotic distribution of the norms of the primitive hyperbolic or loxodromic elements of Γ.

5.1 The Hyperbolic Lattice-Point Problem

We give here an immediate application of the spectral theory of the Laplace operator developed in Chapter 3. Let $\Gamma < \mathbf{PSL}(2, \mathbb{C})$ be a discrete group. The hyperbolic lattice point problem asks for the asymptotics of the number of lattice-points MQ, with $M \in \Gamma$ in a hyperbolic ball with center P

$$|\{ \ MQ \ \mid \ M \in \Gamma, \ d(P, MQ) \leq R \ \}|$$

where the radius R of the ball tends to infinity. We shall give here the principal term for discrete cocompact groups $\Gamma < \mathbf{PSL}(2, \mathbb{C})$.

We start off by considering the eigenfunction expansion of the Maaß–Selberg series $F(P, \cdot, s)$ for $P \in \mathbb{H}$, $\operatorname{Re} s > 1$ with respect to the complete orthonormal system $(\bar{e}_n)_{n \geq 0}$. Since

$$< F(P, \cdot, s), \bar{e}_n >= \int_{\mathcal{F}} F(P, Q, s) \ e_n(Q) \ dv(Q) = \frac{1}{s^2 - s_n^2} \ e_n(P),$$

where we put $\lambda_n = 1 - s_n^2$, we obtain from the completeness relation

$$\int_{\mathcal{F}} |F(P, Q, s)|^2 \ dv(Q) = \sum_{n=0}^{\infty} \frac{|e_n(P)|^2}{|s^2 - s_n^2|^2}.$$

The left-hand side of this equation is a continuous function of (P, s), $(P \in \mathbb{H}, \ \operatorname{Re} s > 1)$. We conclude from Dini's theorem that the series on the right-hand side converges uniformly on $\mathbb{H} \times K$, where $K \subset \{ \ s \ \mid \ \operatorname{Re} s > 1 \ \}$ is an arbitrary compact set. In view of Corollary 4.3.3 the convergence is even uniform on compact subsets of

$$\mathbb{H} \times \{ \ s \in \mathbb{C} \ \mid \ s \neq \pm s_n, \ n \geq 0 \ \}.$$

We apply this to the Parseval relation

$$\int_{\mathcal{F}} F(P, Z, s) \ F(Z, Q, t) \ dv(Z) = \sum_{n=0}^{\infty} \frac{1}{s^2 - s_n^2} \ \frac{1}{t^2 - s_n^2} \ e_n(P) \ \overline{e_n(Q)}$$

with $\operatorname{Re} s > 1$, $\operatorname{Re} t > 1$. Here, the series on the right-hand side converges uniformly on compact sets with respect to (P, Q, s, t) with side conditions $P, \ Q \in \mathbb{H}$, $s, \ t \in \mathbb{C}$, $s \neq \pm s_n$, $t \neq \pm s_n$ for all $n \geq 0$. Hence the right-hand side is a holomorphic function of $s \in \mathbb{C} \setminus \{ \ \pm s_n \ \mid \ n \geq 0 \ \}$. In particular, noting that $\lambda_0 = 0$ is a simple eigenvalue of $-\Delta$, we see that our function is holomorphic in some half-plane $\operatorname{Re} s > 1 - \epsilon$, except for a simple pole at $s = 1$, where $\epsilon > 0$ is sufficiently small.

Using Theorem 4.2.9, we rewrite the last equation in the form

$$\lim_{Z \to Q} (F(P, Z, s) - F(P, Z, t)) = \sum_{n=0}^{\infty} \left(\frac{1}{s^2 - s_n^2} - \frac{1}{t^2 - s_n^2} \right) \ e_n(P) \ \overline{e_n(Q)}.$$

Writing $\delta = \cosh d$, where d denotes the hyperbolic distance, we have

$$F(P,Q,s) = \frac{1}{4\pi} \sum_{M\in\Gamma} \frac{e^{-d(P,MQ)s}}{\sinh d(P,MQ)}.$$

An application of Ikehara's theorem, see Wiener (1933), page 127, theorem 16, Lang (1968), Chapter 15 immediately yields

$$\frac{1}{4\pi} \sum_{\substack{M\in\Gamma \\ 0<d(P,MQ)\le \log T}} \frac{1}{\sinh d(P,MQ)} \sim \frac{1}{2\,v(\mathcal{F})}\,T \qquad \text{for } T\to\infty.$$

This is equivalent to

$$\sum_{\substack{M\in\Gamma \\ \delta(P,MQ)\le T}} \frac{1}{\delta(P,MQ)} \sim \frac{4\pi}{v(\mathcal{F})}\,T \qquad \text{for } T\to\infty.$$

Choosing

$$\alpha(T) := \sum_{\substack{M\in\Gamma \\ \delta(P,MQ)\le T}} \frac{1}{\delta(P,MQ)},$$

we see by partial summation

$$\sum_{\substack{M\in\Gamma \\ \delta(P,MQ)\le T}} \frac{1}{\delta(P,MQ)^2} = \int_{\frac{1}{2}}^{T} \frac{d\alpha(x)}{x} = \left[\frac{\alpha(x)}{x}\right]_{\frac{1}{2}}^{T} + \int_{\frac{1}{2}}^{T} \frac{\alpha(x)}{x^2}\,dx$$
$$\sim \frac{4\pi}{v(\mathcal{F})}\log T.$$

This gives a new proof of (3.6.22) for cocompact groups.

We can even do better and prove the asymptotics for the hyperbolic lattice-point problem for cocompact groups. Let $\alpha(t)$ be as above and put

$$\beta(x) := \int_0^x t\,d\alpha(t).$$

Then we have

$$\beta(x) \sim \frac{1}{2}\frac{4\pi}{v(\mathcal{F})}\,x^2 \qquad \text{for } x\to\infty.$$

Now, using the function (3.1.31) we have for all $t>0$

$$\Theta(P,Q,t) = \sum_{M\in\Gamma} e^{-t\delta(P,MQ)} = \int_0^\infty x\,e^{-tx}\,d\alpha(x) = \int_0^\infty e^{-tx}\,d\beta(x)$$
$$= t\int_0^\infty \beta(x)\,e^{-tx}\,dx.$$

Our knowledge of the asymptotic behaviour of β now implies that

$$\Theta(P, Q, t) \sim \frac{4\pi}{v(\mathcal{F})} t^{-2} \qquad \text{for } t \downarrow 0.$$

This is (3.6.2) in the cocompact case. An application of Karamata's theorem, see Karamata (1931a) or Lang (1968) Chapter 15, to the latter relation immediately yields

$$|\{ M \in \Gamma \mid \delta(P, MQ) \leq T \}| \sim \frac{2\pi}{v(\mathcal{F})} T^2 \qquad \text{for } T \to \infty.$$

This relation is equivalent to the asymptotic result for the number of lattice-points MQ ($M \in \Gamma$) in a hyperbolic ball with center P as the radius of the ball tends to infinity, namely

$$|\{ MQ \mid M \in \Gamma,\ d(P, MQ) \leq R \}| \sim \frac{\pi}{2\, v(\mathcal{F})\, |\Gamma_Q|} e^{2R} \qquad \text{for } R \to \infty.$$

Summing up, we have now proved the following result, which we call the hyperbolic lattice-point theorem.

Theorem 1.1. *Suppose that $\Gamma < \mathbf{PSL}(2, \mathbb{C})$ is a discrete cocompact group with fundamental domain $\mathcal{F}$. Then the following asymptotic relations hold:*

$$(1.1) \qquad \sum_{\delta(P,MQ) \leq T} \frac{1}{\delta(P, MQ)} \sim \frac{4\pi}{v(\mathcal{F})} T \qquad \textit{for } T \to \infty,$$

$$(1.2) \qquad \Theta(P, Q, t) \sim \frac{4\pi}{v(\mathcal{F})} t^{-2} \qquad \textit{for } t \downarrow 0,$$

$$(1.3) \qquad |\{ M \in \Gamma \mid \delta(P, MQ) \leq T \}| \sim \frac{2\pi}{v(\mathcal{F})} T^2 \qquad \textit{for } T \to \infty.$$

The relations (1.1)–(1.3) are all equivalent in the sense that each one of them implies the other two by means of Abelian or Tauberian theorems. For estimates of the error terms and related problems see Fricker (1968), Günther (1979), (1980), Thurnheer (1979), (1980), Lax and Phillips (1981), Levitan (1987). In fact Lax and Phillips prove the following asymptotic expansion.

Theorem 1.2. *Suppose that $\Gamma < \mathbf{PSL}(2, \mathbb{C})$ is a discrete group of finite covolume with fundamental domain $\mathcal{F}$. Let*

$$(1.4) \qquad 0 = \lambda_0 < \lambda_1 \leq \lambda_2 \leq \ldots \leq \lambda_N < 1$$

be the eigenvalues of the Laplace operator $-\Delta$ on $L^2(\Gamma \backslash \mathbb{H})$ in the interval $[0, 1[$ counted with their multiplicities. Put $s_\nu := \sqrt{1 - \lambda_\nu}$ for $\nu = 1, \ldots, N$. Let $\varphi_1, \ldots, \varphi_N$ be the corresponding orthonormal eigenfunctions. Then

$$
\begin{aligned}
|\{ M \in \Gamma \mid d(P, MQ) \le T \}| &= \frac{\pi}{2\, v(\mathcal{F})}\, e^{2T} \\
&+ \pi \sum_{\nu=1}^{N} \frac{\Gamma(s_\nu)}{\Gamma(s_\nu + 2)}\, \varphi_\nu(P)\, \bar{\varphi}_\nu(Q)\, e^{(s_\nu+1)T} + O\left(e^{\frac{3T}{2}}\right).
\end{aligned} \tag{1.5}
$$

The proof of Lax and Phillips makes heavy use of the wave equation for the Laplace operator. We will not include it here. Of course the error term is conjectured to be $O(\exp(T))$. The eigenvalues addressed to in (1.4) are exactly those which are not embedded in the continuous spectrum if Γ is cofinite but not cocompact. They are called the exceptional eigenvalues.

5.2 Computation of the Trace

As always in this Chapter we assume that Γ is a discrete subgroup of $\mathbf{PSL}(2, \mathbb{C})$ such that $\Gamma \backslash \mathbb{H}$ is compact. Let $(e_n)_{n \ge 0}$ be a complete orthonormal system of eigenfunctions of $-\Delta : \mathcal{D} \to L^2(\Gamma \setminus \mathbb{H})$. So we have $-\Delta e_n = \lambda_n e_n$ for $n \ge 0$. Supposing that $\lambda \in \rho(-\tilde{\Delta})$, the resolvent operator R_λ satisfies

$$R_\lambda e_n = \frac{1}{\lambda_n - \lambda}\, e_n.$$

We shall see in Corollary 5.7 that the series $\sum_{n=1}^{\infty} \lambda_n^{-1}$ diverges. Hence R_λ is not a trace-class operator. But if we take another $\mu \in \rho(-\tilde{\Delta})$, the operator $R_\lambda R_\mu$, as a product of two Hilbert-Schmidt operators, is of trace-class, see Weidmann (1979), Section 7.1, and

$$R_\lambda R_\mu e_n = (\lambda_n - \lambda)^{-1} (\lambda_n - \mu)^{-1} e_n.$$

Put $\lambda = 1 - s^2$, $\mu = 1 - t^2$ with $s,\ t \in \mathbb{C}$ and also $\lambda_n = 1 - s_n^2$. Then the trace of $R_\lambda R_\mu$ is given by

$$
\begin{aligned}
(\lambda - \mu) \operatorname{tr}(R_\lambda R_\mu) &= \sum_{n=0}^{\infty} \left(\frac{1}{\lambda_n - \lambda} - \frac{1}{\lambda_n - \mu} \right) \\
&= \sum_{n=0}^{\infty} \left(\frac{1}{s^2 - s_n^2} - \frac{1}{t^2 - s_n^2} \right).
\end{aligned} \tag{2.1}
$$

Computing this trace in another way we shall arrive at the logarithmic derivative of the Selberg zeta function which will be studied in Section 4. Suppose that we have in addition $\operatorname{Re} s > 1$, $\operatorname{Re} t > 1$. Then the operator $(\lambda - \mu)\, R_\lambda R_\mu$ is represented by the Maaß–Selberg series. This yields

$$(\lambda - \mu) \operatorname{tr}(R_\lambda R_\mu) = \int_{\mathcal{F}} \lim_{Z \to P} (F(P, Z, s) - F(P, Z, t))\, dv(P). \tag{2.2}$$

Observe that

$$\varphi_s(\delta) - \varphi_t(\delta) =: h(\delta) \tag{2.3}$$

is continuous for $\delta \geq 1$, and

$$h(1) = -(s-t), \qquad h(\delta) = O(\delta^{-2-\epsilon}) \quad \text{as } \delta \to \infty \text{ for some } \epsilon > 0. \tag{2.4}$$

Since $\mathcal{F}$ is compact, we deduce from (2.4), (3.1.2) and Proposition 1.1.8 that we may interchange the summation with the integration in (2.2) to obtain

$$(\lambda - \mu)\operatorname{tr}(R_\lambda R_\mu) = -\frac{v(\mathcal{F})}{4\pi}(s-t) + \frac{1}{4\pi}\sum_{\substack{M\in\Gamma\\ M\neq I}} \int_{\mathcal{F}} h(\delta(P, MP))\, dv(P). \tag{2.5}$$

Here the expression on the right-hand side remains finite if we replace h by its bound (2.4).

We compute the series (2.5) following a clever method of Selberg (1956), page 63 et seq. Let $\{T\} = \{T\}_\Gamma$ run through the set of Γ-conjugacy classes of the elements $M \in \Gamma$, $M \neq 1$. Observe that $S^{-1}TS = M$ runs through $\{T\}$ exactly once whenever S runs through a system $\mathcal{C}(T) \setminus \Gamma$ of representatives of the right cosets of the centralizer $\mathcal{C}(T)$ of T in Γ. This implies that

$$\begin{aligned} \sum_{\substack{M\in\Gamma\\ M\neq I}} \int_{\mathcal{F}} h(\delta(P, MP))\, dv(P) &= \sum_{\{T\}} \sum_{M\in\{T\}} \int_{\mathcal{F}} h(\delta(P, MP))\, dv(P) \\ &= \sum_{\{T\}} \sum_{S\in\mathcal{C}(T)\setminus\Gamma} \int_{\mathcal{F}} h(\delta(P, S^{-1}TSP))\, dv(P) \\ &= \sum_{\{T\}} \sum_{S\in\mathcal{C}(T)\setminus\Gamma} \int_{S\mathcal{F}} h(\delta(Q, TQ))\, dv(Q), \\ &= \sum_{\{T\}} \int_{\mathcal{F}(\mathcal{C}(T))} h(\delta(Q, TQ))\, dv(Q), \end{aligned} \tag{2.6}$$

where $\mathcal{F}(\mathcal{C}(T))$ denotes a fundamental domain of $\mathcal{C}(T)$.

It is our task to compute the various contributions appearing on the right-hand side of (2.6). First we consider the case of a hyperbolic or loxodromic element. Let $T \in \Gamma$ be hyperbolic or loxodromic. Then T is conjugate in $\mathbf{PSL}(2, \mathbb{C})$ to a unique element

$$D(T) = \begin{pmatrix} a(T) & 0 \\ 0 & a(T)^{-1} \end{pmatrix} \tag{2.7}$$

such that $|a(T)| > 1$. Then $D(T)(z + rj) = K(T)z + N(T)rj$, where $K(T) = a(T)^2$ is the *multiplier* of T and where $N(T) = |a(T)|^2 = |K(T)| > 1$ is the *norm* of T.

Since T is hyperbolic or loxodromic, an easy check using the normal form (2.7) yields that an element of Γ commutes with T if and only if it has the same fixed points in $\mathbb{C} \cup \{\infty\}$ as T. This implies that $\mathcal{C}(T)$ has the following structure: Let $\mathcal{E}(T)$ be the set of elements of finite order contained in $\mathcal{C}(T)$. Then $\mathcal{C}(T)$ is the direct product of $\mathcal{E}(T)$ with an infinite cyclic group $< T_0 >$ generated by some hyperbolic or loxodromic element $T_0 \in \mathcal{C}(T)$. Note that this direct product decomposition is preserved if T_0 is replaced by $T_0 E$ or $T_0^{-1} E$ for some $E \in \mathcal{E}(T)$. All the elements $T_0 E, T_0^{-1} E$ with $E \in \mathcal{E}(T)$ have the same norm $N(T_0)$. Hence although T_0 itself is not uniquely determined by T, the norm $N(T_0)$ is. $N(T_0)$ is the minimal norm of a hyperbolic or loxodromic element from $\mathcal{C}(T)$. Following Selberg (1956), we call T_0 a *primitive hyperbolic or loxodromic element for T in Γ*, respectively. One can also impose the normalization condition $T = T_0^k E$ for some $k \geq 1$ and $E \in \mathcal{E}(T)$ on T_0. This is advantageous if $\mathcal{E}(T) = \{I\}$, since then T_0 itself is uniquely determined by this condition. We shall not require such a normalization in the sequel.

Note that the elements of $\mathcal{E}(T)$ are hyperbolic rotations around the axis of T. $\mathcal{E}(T)$ is a cyclic group generated by a rotation with rotation angle $\frac{2\pi}{|\mathcal{E}(T)|}$ where $|\mathcal{E}(T)|$ denotes the order of $\mathcal{E}(T)$. Choose $V \in \mathbf{PSL}(2, \mathbb{C})$ such that

$$T = V^{-1} D(T) V. \tag{2.8}$$

Then $V\mathcal{C}(T)V^{-1}$ has the fundamental domain

$$\mathcal{F}_0 = \left\{ \rho e^{i\varphi} + rj \;\middle|\; \rho \geq 0,\ 0 \leq \varphi \leq \frac{2\pi}{|\mathcal{E}(T)|},\ 1 \leq r \leq N(T_0) \right\}. \tag{2.9}$$

Hence we obtain

$$\begin{aligned} &\int_{\mathcal{F}(\mathcal{C}(T))} \varphi_s(\delta(Q, TQ))\, dv(Q) \\ &= \int_{\mathcal{F}_0} \varphi_s(\delta(Q, D(T)Q))\, dv(Q) \\ &= \int_{\mathcal{F}_0} \varphi_s\left(\frac{|K(T)-1|^2\ |z|^2 + (N(T)^2+1)\ r^2}{2\ N(T)\ r^2}\right) \frac{dx\ dy\ dr}{r^3} \\ &= \frac{2\pi \log N(T_0)}{|\mathcal{E}(T)|\ |a(T)-a(T)^{-1}|^2} \int_{\alpha(T)}^{\infty} \varphi_s(u)\, du, \end{aligned} \tag{2.10}$$

where

$$\alpha(T) := \frac{1}{2}(N(T) + N(T)^{-1}).$$

But substituting $u = \cosh x$, we find for $\alpha > 1$

$$\begin{aligned} \int_{\alpha}^{\infty} \varphi_s(u)\, du &= \int_{\operatorname{arcosh} \alpha}^{\infty} e^{-sx}\ dx = \frac{1}{s} e^{-s \operatorname{arcosh} \alpha} \\ &= \frac{1}{s}\left(\alpha + \sqrt{\alpha^2 - 1}\right)^{-s}. \end{aligned} \tag{2.11}$$

Since $\alpha(T) + \sqrt{\alpha(T)^2 - 1} = N(T)$, we finally obtain for all hyperbolic or loxodromic elements $T \in \Gamma$

$$(2.12) \quad \int_{\mathcal{F}(\mathcal{C}(T))} \varphi_s(\delta(Q, TQ))\, dv(Q) = \frac{2\pi \log N(T_0)}{s\, |\mathcal{E}(T)|\, |a(T) - a(T)^{-1}|^2} N(T)^{-s}.$$

Here, $\mathcal{E}(T)$ is the subgroup of $\mathcal{C}(T)$ of elements of finite order, and T_0 is a primitive hyperbolic or loxodromic element for T in Γ.

The contribution of the conjugacy classes of the elliptic elements in Γ to the right-hand side of (2.6) is determined in essentially the same way, but the details are slightly more delicate. The final result will be equal to (2.12) verbatim if the corresponding concepts are introduced properly. Assume that $R \in \Gamma$ is elliptic. Then R is a rotation around a hyperbolic line which remains pointwise fixed under R. This hyperbolic line meets $\mathbb{C} \cup \{\infty\}$ in the fixed points of R in $\mathbb{C} \cup \{\infty\}$. The subgroup of Γ containing all the elements of Γ with the same fixed points in $\mathbb{C} \cup \{\infty\}$ as R contains a rotation R_0 with minimal rotation angle, and we have

$$(2.13) \qquad R = R_0^k \qquad \text{for some integer } k \text{ with } 1 \leq k \leq |< R_0 >| - 1.$$

Obviously $< R_0 >$, the cyclic group generated by R_0, is a subgroup of $\mathcal{C}(R)$. We call R_0 a *primitive elliptic element of Γ associated with R*. R_0 is uniquely determined up to inversion. This means that k is uniquely determined up to the change $k \mapsto \operatorname{ord}(R_0) - k$, where $\operatorname{ord}(R_0)$ stands for the order of R_0. It will be apparent from our computations that this change leaves our results unaffected.

We proceed to show that there must exist elements in $\mathcal{C}(R)$ not contained in $< R_0 >$. We prove this in a rather indirect way: Suppose that we have $\mathcal{C}(R) =< R_0 >$, choose $V \in \mathbf{PSL}(2, \mathbb{C})$ such that

$$(2.14) \qquad R_0 = V^{-1} R\left(\frac{2\pi}{m}\right) V,$$

where

$$(2.15) \qquad R(\varphi) := \begin{pmatrix} e^{\frac{i\varphi}{2}} & 0 \\ 0 & e^{-\frac{i\varphi}{2}} \end{pmatrix} \qquad (\varphi \in \mathbb{R}),$$

and $m := \operatorname{ord}(R_0)$. Note that $R(\varphi)$ acts on $\mathbb{H}$ as follows:

$$(2.16) \qquad R(\varphi)(z + rj) = e^{i\varphi} z + rj \qquad (z \in \mathbb{C},\ r > 0).$$

Hence by our assumption the group $V\mathcal{C}(R)V^{-1}$ has the fundamental domain

$$(2.17) \qquad \mathcal{F}_1 = \left\{ \rho e^{i\varphi} + rj \;\middle|\; r > 0,\ \rho \geq 0,\ 0 \leq \varphi \leq \frac{2\pi}{m} \right\}.$$

We know from our deduction of (2.5) that the integral

$$\int_{\mathcal{F}(\mathcal{C}(R))} h(\delta(Q, RQ))\, dv(Q) \tag{2.18}$$

converges (absolutely). But under our assumption we simply have

$$\begin{aligned}
&\int_{\mathcal{F}(\mathcal{C}(R))} h(\delta(Q, RQ))\, dv(Q) \\
&= \int_{V\mathcal{F}(\mathcal{C}(R))} h(\delta(V^{-1}Q, RV^{-1}Q))\, dv(Q) \\
&= \int_{\mathcal{F}_1} h(Q, VRV^{-1}Q)\, dv(Q) = \int_{\mathcal{F}_1} h\left(Q, R\left(\frac{2\pi k}{m}\right) Q\right) dv(Q) \qquad (2.19)\\
&= \int_{\mathcal{F}_1} h\left(\frac{|\exp(\frac{2\pi i k}{m}) - 1|^2 |z|^2 + 2r^2}{2r^2}\right) \frac{dx\, dy\, dr}{r^3} \\
&= \int_{\mathcal{F}_1} h\left(1 + 2\sin^2 \frac{\pi k}{m} |z|^2\right) \frac{dx\, dy\, dr}{r},
\end{aligned}$$

which obviously diverges for the domain (2.17) (look at the dependence on r). Hence in the case of dimension 3, the centralizer of an elliptic element of a group with compact quotient never is a finite cyclic group. This phenomenon contrasts the familiar situation in the case of Fuchsian groups acting on the hyperbolic plane. We can even say more: The integral (2.18) remains divergent whenever $\mathcal{C}(R)$ is assumed to be a finite group since the integrand is invariant under $\mathcal{C}(R)$ and since $< R_0 >$ is of finite index in $\mathcal{C}(R)$ in this case. We conclude that $\mathcal{C}(R)$ necessarily is an infinite group. Hence $\mathcal{C}(R)$ contains a hyperbolic or loxodromic element. We proceed to determine the structure of $\mathcal{C}(R)$ in detail and to compute the integral (2.18). There are two cases to consider.

First case: Assume that all the elements in $\mathcal{C}(R) \setminus < R_0 >$ are hyperbolic or loxodromic. Then $< R_0 >$ coincides with $\mathcal{E}(R)$, the subgroup of $\mathcal{C}(R)$ of all elements of finite order. Every hyperbolic or loxodromic element of $\mathcal{C}(R)$ commutes with R and hence has the same fixed points in $\mathbb{C} \cup \{\infty\}$ as R. Thus the fixed line of R is the axis of every hyperbolic or loxodromic element of $\mathcal{C}(R)$. Choose $T_0 \in \mathcal{C}(R) \setminus \mathcal{E}(R)$ such that $N(T_0)$ is minimal. Then we have the direct product decomposition

$$\mathcal{C}(R) = < T_0 > \times \mathcal{E}(R), \tag{2.20}$$

and we recognize the same type of group theoretic structure of the centralizer as in the case of a hyperbolic or loxodromic element. In particular, $\mathcal{C}(R)$ is an abelian group. The set

$$\mathcal{F}_1 := \left\{ \rho e^{i\varphi} + rj \;\middle|\; \rho \geq 0,\ 0 \leq \varphi \leq \frac{2\pi}{m},\ 1 \leq r < N(T_0) \right\} \tag{2.21}$$

is a fundamental domain of $V\mathcal{C}(R)V^{-1}$ in our first case. Substituting (2.21) into (2.19) we find that

$$
(2.22) \qquad
\begin{aligned}
&\int_{\mathcal{F}(\mathcal{C}(R))} h(\delta(Q, RQ))\, dv(Q) \\
&= \frac{2\pi}{m} \log N(T_0) \int_0^\infty h\left(1 + 2\sin^2\left(\frac{\pi k}{m}\right)\rho^2\right) \rho d\rho \\
&= \frac{\pi \ \log N(T_0)}{2m \ \sin^2(\frac{\pi k}{m})} \int_1^\infty h(u)\ du \\
&= \frac{\pi \ \log N(T_0)}{2m \ \sin^2(\frac{\pi k}{m})} \left(\frac{1}{s} - \frac{1}{t}\right)
\end{aligned}
$$

(see (2.3), (2.11)). For the sake of unification with the results in the second case it is advantageous to write (2.22) in a slightly different form. Note that in the present case $m = |\mathcal{E}(R)|$, where $\mathcal{E}(R)$ is the maximal finite subgroup of $\mathcal{C}(R)$. On the other hand, we also have

$$
(2.23) \qquad m = m(R),
$$

where $m(R)$ denotes the order of a primitive elliptic element associated with R. The numbers $|\mathcal{E}(R)|$ and $m(R)$ coincide in the present case but will be different in the second case. In view of (2.23), we now write (2.22) in the form

$$
(2.24) \qquad \int_{\mathcal{F}(\mathcal{C}(R))} h(\delta(Q, RQ))\ dv(Q) = \frac{\pi \log N(T_0)}{2\ |\mathcal{E}(R)|\ \sin^2(\frac{\pi k}{m(R)})} \left(\frac{1}{s} - \frac{1}{t}\right).
$$

This finishes our computation in the first case.

Second case: There exists another elliptic element $S \in \mathcal{C}(R) \setminus < R_0 >$. Transform R to normal form (2.15). Then an elementary computation shows that $R(\varphi)$ commutes with $g = \begin{pmatrix} a & b \\ c & d \end{pmatrix} \in \mathbf{PSL}(2, \mathbb{C})$ if and only if either $b = c = 0$ (i.e. if $R(\varphi)$ and g both have the fixed points 0 and ∞ in $\mathbb{C} \cup \{\infty\}$) or $\exp \frac{i\varphi}{2} = \pm i$ and $a = d = 0$. Since we assume that R and S do not have the same fixed points in $\mathbb{C} \cup \{\infty\}$, the second case occurs here, i.e., R is a hyperbolic rotation with rotation angle π, and $S \neq R$ is an elliptic element of order 2 commuting with R, hence we find a $V \in \mathbf{PSL}(2, \mathbb{C})$ with

$$
(2.25) \qquad VRV^{-1} = \begin{pmatrix} i & 0 \\ 0 & -i \end{pmatrix}, \qquad VSV^{-1} = \begin{pmatrix} 0 & b \\ -b^{-1} & 0 \end{pmatrix}.
$$

Let $R_0 \in \Gamma$ be a primitive elliptic element associated with R. Then $S^{-1}R_0S = R_0^{-1}$. Hence

$$
(2.26) \qquad \mathcal{E}(R) := < R_0 > \cup < R_0 > S
$$

is a finite subgroup of $\mathcal{C}(R)$ of dihedral type, and $|\mathcal{E}(R)| = 2 \operatorname{ord} R_0$. The group $\{I, R, S, RS\}$ is a subgroup of $\mathcal{E}(R)$ isomorphic to the Klein four group.

We proceed to show that $\mathcal{E}(R)$ is a maximal subgroup of $\mathcal{C}(R)$ containing only elements of finite order. To prove this, suppose that $\mathcal{G} \subset \mathcal{C}(R)$ is a subgroup containing only elements of finite order such that $\mathcal{E}(R) \subset \mathcal{G}$. Let $A \in \mathcal{G}, A \notin < R_0 >$. Then it follows from our proof of (2.25) that

$$VAV^{-1} = \begin{pmatrix} 0 & \beta \\ -\beta^{-1} & 0 \end{pmatrix} \in \mathbf{PSL}(2, \mathbb{C})$$

for some $\beta \neq 0$. Hence $VASV^{-1}$ is described by a diagonal matrix, and since $AS \in \mathcal{G}$ is an element of finite order having the same fixed points in $\mathbb{C} \cup \{\infty\}$ as R, we conclude that $AS \in < R_0 >$, i.e., $\mathcal{G} \subset \mathcal{E}(R)$. This proves the maximality of $\mathcal{E}(R)$. It is well-known that a discrete subgroup of $\mathbf{PSL}(2, \mathbb{C})$ contains only elements of finite order if and only if it is finite (cf. Beardon (1977), page 62, Theorem 5.1). Hence we may equivalently say that $\mathcal{E}(R)$ is a maximal finite subgroup of $\mathcal{C}(R)$.

We know from our previous discussion of (2.19) that $\mathcal{C}(R)$ contains a hyperbolic or loxodromic element. The fixed line of R is the axis of every hyperbolic or loxodromic element of $\mathcal{C}(R)$. Note that the elements of the coset $< R_0 > S$ map the fixed line of R onto itself interchanging the endpoints. If $X \in \mathcal{C}(R)$ is hyperbolic or loxodromic, we have

$$\begin{aligned} &AXA^{-1} = X \quad \text{for all } A \in < R_0 >, \\ &BXB^{-1} = X^{-1} \quad \text{for all } B \in < R_0 > S. \end{aligned} \tag{2.27}$$

Choose a hyperbolic or loxodromic element $T_0 \in \mathcal{C}(R)$ such that $N(T_0)$ is minimal. We show that

$$\mathcal{C}(R) = \{ \, T_0^n E \ \mid \ E \in \mathcal{E}(R), \ n \in \mathbb{Z} \, \}. \tag{2.28}$$

To see this, assume first that $T \in \mathcal{C}(R)$ is elliptic and that $T \notin < R_0 >$. Then

$$VTV^{-1} = \begin{pmatrix} 0 & \alpha \\ -\alpha^{-1} & 0 \end{pmatrix} \in \mathbf{PSL}(2, \mathbb{C})$$

has the property that $VTSV^{-1}$ is described by a diagonal matrix. Hence $TS \in \mathcal{C}(R)$ has the same fixed points in $\mathbb{C} \cup \{\infty\}$ as R. This implies that $TS = T_0^n R_0^\nu$ for some integers, ν, n with $0 \leq \nu < \operatorname{ord} R_0$. Hence $T = T_0^n R_0^\nu S$ belongs to the set on the right-hand side of (2.28).

Second, consider the case that $T \in \mathcal{C}(R)$ is hyperbolic or loxodromic. Then T has the same fixed points as T_0. Hence there exists an integer n such that $T_0^{-n} T$ is of finite order. Since $T_0^{-n} T$ has the same fixed points in $\mathbb{C} \cup \{\infty\}$ as R, we conclude that $T_0^{-n} T = R_0^\nu$ for some integers n, ν. This proves (2.28). Note that $< T_0 > \times < R_0 >$ is an Abelian subgroup of index 2 in $\mathcal{C}(R)$ in the present case. The set $\{I, S\}$ is a representative system of the cosets of $< T_0 > \times < R_0 >$ in $\mathcal{C}(R)$.

We prove that all the maximal finite subgroups of $\mathcal{C}(R)$ are conjugate in $\mathbf{PSL}(2, \mathbb{C})$. Let $\mathcal{G}$ be a maximal finite subgroup of $\mathcal{C}(R)$. We know from our previous discussion that there exists an elliptic element $A \in \mathcal{C}(R)$ of order two such that R and A have different pairs of fixed points in $\mathbb{C} \cup \{\infty\}$ and such that $\mathcal{G} =< R_0 > \cup < R_0 > A$. It follows from (2.28) that $A = T_0^n R_0^\nu S$ for some integers n, ν. Let $T_1 \in \mathbf{PSL}(2, \mathbb{C})$ be a hyperbolic or loxodromic element such that $T_0 = T_1^2$. Then T_1 commutes with R_0, and $T_1^{-n} R_0^\ell A T_1^n = R_0^{\ell+\nu} S$ for all $\ell \in \mathbb{Z}$. This yields $T_1^{-n} \mathcal{G} T_1^n = \mathcal{E}(R)$, i.e., $\mathcal{G}$ is conjugate to $\mathcal{E}(R)$ in $\mathbf{PSL}(2, \mathbb{C})$.

Combining the results on $\mathcal{C}(R)$ in both cases, we have now proved the following theorem.

Theorem 2.1. *Let $\Gamma < \mathbf{PSL}(2, \mathbb{C})$ be a discrete cocompact group. Suppose that $R \in \Gamma$ is elliptic, and let $R_0 \in \Gamma$ be a primitive elliptic transformation associated with R. Then the centralizer $\mathcal{C}(R)$ of R in Γ contains hyperbolic or loxodromic elements. There are two possibilities.*

(1) Either all the elliptic elements of $\mathcal{C}(R)$ are contained in $\mathcal{E}(R) :=< R_0 >$, the cyclic group generated by R_0. Then $\mathcal{C}(R)$ is abelian, and

$$\mathcal{C}(R) =< T_0 > \times < R_0 >, \tag{2.29}$$

where $T_0 \in \mathcal{C}(R)$ is a hyperbolic or loxodromic element such that the norm $N(T_0)$ is minimal. In this case, all the elements in $\mathcal{C}(R) \setminus \{I\}$ have the same pair of fixed points in $\mathbb{C} \cup \{\infty\}$.

(2) Or R is elliptic of order 2, and there exists an elliptic element $S \in \mathcal{C}(R)$ also of order 2 whose fixed line meets the fixed line of R orthogonally in a common point. Then for every such S,

$$S^{-1} R_0 S = R_0^{-1}, \tag{2.30}$$

and

$$\mathcal{E}(R) :=< R_0 > \cup < R_0 > S \tag{2.31}$$

is a maximal finite subgroup of $\mathcal{C}(R)$. $\mathcal{E}(R)$ is of dihedral type. All the maximal finite subgroups of $\mathcal{C}(R)$ are conjugate in $\mathbf{PSL}(2, \mathbb{C})$. The centralizer $\mathcal{C}(R)$ has the presentation

$$\begin{aligned} \mathcal{C}(R) =< R_0,\ T_0,\ S \ \mid \ & R_0 T_0 = T_0 R_0,\ S R_0 S^{-1} = R_0^{-1}, \\ & S T_0 S^{-1} = T_0^{-1},\ R_0^m = S^2 = 1 >, \end{aligned}$$

where $T_0 \in \mathcal{C}(R)$ is a hyperbolic or loxodromic element such that $N(T_0)$ is minimal. The group $< T_0 > \times < R_0 >$ is an abelian subgroup of index 2 in $\mathcal{C}(R)$, and $\{I, S\}$ constitutes a representative system of the cosets of $< T_0 > \times < R_0 >$ in $\mathcal{C}(R)$.

We proceed with our discussion of the second case and compute the contribution of the Γ-conjugacy class $\{R\}$ to the right-hand side of (2.6). The group $< T_0 > \times < R_0 >$ is a subgroup of index 2 in $\mathcal{C}(R)$. Putting $m := \operatorname{ord} R_0$, we see that $\mathcal{F}_1$ (cf. (2.21)) is a fundamental domain for $V(< T_0 > \times < R_0 >)V^{-1}$. Note that the function

$$\mathbb{H} \ni Q \mapsto h(\delta(Q, RQ)) \tag{2.32}$$

is invariant under $\mathcal{C}(R)$. Hence we can use our previous computations (2.19), (2.22) to obtain

$$\begin{aligned} \int_{\mathcal{F}(\mathcal{C}(R))} h(\delta(Q, RQ))\, dv(Q) &= \frac{1}{2} \int_{\mathcal{F}_1} h(\delta(Q, VRV^{-1}Q))\, dv(Q) \\ &= \frac{\pi \log N(T_0)}{4m \sin^2(\frac{\pi k}{m})} \left(\frac{1}{s} - \frac{1}{t} \right). \end{aligned} \tag{2.33}$$

Here, $m = m(R) = \operatorname{ord} R_0$ is an even integer, and we may choose $k = \frac{m}{2}$, since $R = R_0^k$ is of order 2. Since $|\mathcal{E}(R)| = 2m$ we finally arrive at

$$\int_{\mathcal{F}(\mathcal{C}(R))} h(\delta(Q, RQ))\, dv(Q) = \frac{\pi \log N(T_0)}{2\, |\mathcal{E}(R)|\, \sin^2\left(\frac{\pi k}{m}\right)} \left(\frac{1}{s} - \frac{1}{t} \right), \tag{2.34}$$

which is in perfect agreement with (2.24).

Collecting terms from (2.5), (2.6), (2.12), (2.24), (2.34), we arrive at the following result.

Theorem 2.2. *Suppose that $\Gamma < \mathbf{PSL}(2, \mathbb{C})$ is a discrete group with compact fundamental domain $\mathcal{F}$, and let $0 = \lambda_0 < \lambda_1 \leq \lambda_2 \leq \ldots$ be the eigenvalues of $-\Delta : \mathcal{D} \to L^2(\Gamma \setminus \mathbb{H})$. Put $\lambda_n = 1 - s_n^2$, with $s_n \in \mathbb{C}$. Assume that $\lambda = 1 - s^2$, $\mu = 1 - t^2$, with* $\operatorname{Re} s$, $\operatorname{Re} t > 1$. *Then*

$$\begin{aligned} (\lambda - \mu)\, \operatorname{tr}(R_\lambda R_\mu) &= \sum_{n=0}^{\infty} \left(\frac{1}{s^2 - s_n^2} - \frac{1}{t^2 - s_n^2} \right) \\ &= -\frac{v(\mathcal{F})}{4\pi}(s - t) + \left(\frac{1}{2s} - \frac{1}{2t} \right) \sum_{\{R\} \text{ ellipt.}} \frac{\log N(T_0)}{|\mathcal{E}(R)|\, 4 \sin^2(\frac{\pi k}{m(R)})} \\ &\quad + \frac{1}{2s} \sum_{\{T\} \text{ lox.}} \frac{\log N(T_0)}{|\mathcal{E}(T)|\, |a(T) - a(T)^{-1}|^2} N(T)^{-s} \\ &\quad - \frac{1}{2t} \sum_{\{T\} \text{ lox.}} \frac{\log N(T_0)}{|\mathcal{E}(T)|\, |a(T) - a(T)^{-1}|^2} N(T)^{-t}, \end{aligned} \tag{2.35}$$

where the following notation is used. The summation with respect to $\{R\}$ extends over the finitely many Γ-conjugacy classes of the elliptic elements $R \in \Gamma$, and for such a class $N(T_0)$ is the minimal norm of a hyperbolic or loxodromic element of the centralizer $\mathcal{C}(R)$. It is further understood that R is the k-th power of a primitive elliptic element $R_0 \in \mathcal{C}(R)$ describing a hyperbolic rotation around the fixed axis of R with minimal rotation angle $\frac{2\pi}{m(R)}$. Further, $\mathcal{E}(R)$, $\mathcal{E}(T)$ are maximal finite subgroups contained in $\mathcal{C}(R)$, $\mathcal{C}(T)$, respectively. The summation with respect to $\{T\}$ lox. *extends over the Γ-conjugacy classes of hyperbolic or loxodromic elements of Γ, and T_0 denotes* *a primitive hyperbolic or loxodromic element associated with T. Finally, T is conjugate in* $\mathbf{PSL}(2,\mathbb{C})$ *to the transformation described by the diagonal matrix with diagonal entries $a(T), a(T)^{-1}$, where $|a(T)| > 1$.*

We can transform the right-hand side of (2.35) into an even more concise form such that the distinction between the elliptic and the hyperbolic or loxodromic elements vanishes. Note that the square of the trace makes sense for every element $M \in \mathbf{PSL}(2,\mathbb{C})$. Now, if R is an elliptic element of the form (2.15), we have

$$4\sin^2\left(\frac{\pi k}{m(R)}\right) = |(\operatorname{tr}(R))^2 - 4|. \tag{2.36}$$

Since the elliptic elements have eigenvalues of absolute value one, it is natural to define

$$N(R) := 1, \quad \text{if } R \text{ is elliptic.} \tag{2.37}$$

If T is hyperbolic or loxodromic, we draw from (2.7)

$$|a(T) - a(T)^{-1}|^2 = |(\operatorname{tr}(T))^2 - 4|. \tag{2.38}$$

Inserting (2.36) - (2.38) into (2.35), we obtain under the hypotheses of Theorem 2.2

$$\begin{aligned}(\lambda-\mu)\operatorname{tr}(R_\lambda R_\mu) &= \sum_{n=0}^{\infty}\left(\frac{1}{s^2-s_n^2} - \frac{1}{t^2-s_n^2}\right)\\ &= -\frac{v(\mathcal{F})}{4\pi}(s-t) + \frac{1}{2s}\sum_{\{T\}}\frac{\log N(T_0)}{|\mathcal{E}(T)|\,|\operatorname{tr}(T)^2-4|}N(T)^{-s}\\ &\quad -\frac{1}{2t}\sum_{\{T\}}\frac{\log N(T_0)}{|\mathcal{E}(T)|\,|\operatorname{tr}(T)^2-4|}N(T)^{-t}\end{aligned} \tag{2.39}$$

Here, the summation extends over all Γ-conjugacy classes of elements $T \in \Gamma, T \neq I$, and $N(T_0)$ is the minimal norm of a hyperbolic or loxodromic element contained in $\mathcal{C}(T)$. Equation (2.39) looks more satisfying from the aesthetic point of view but actually is less advantageous than (2.35) if one wants to introduce the Selberg zeta function.

It is also possible to write the contribution of the elliptic elements to the right-hand side of (2.35) in a different way. Suppose that R_0 is a primitive elliptic element of Γ of order $m = m(R_0)$, and assume first that $\{R_0\} \neq \{R_0^{-1}\}$. Then $\{R_0^k\} \neq \{R_0^\ell\}$ for $k \neq \ell \bmod m$, and we can sum with respect to k in (2.35) using the partial fraction expansion of $\pi^2/\sin^2 \pi z$. This yields

$$\begin{aligned}\sum_{k=1}^{m-1} \sin^{-2}\left(\frac{\pi k}{m}\right) &= \frac{1}{\pi^2} \sum_{n=-\infty}^{+\infty} \sum_{k=1}^{m-1} \left(\frac{k}{m}+n\right)^{-2} \\ &= \frac{m^2}{\pi^2}\left(\sum_{\ell\neq 0} \ell^{-2} - \sum_{\ell \neq 0} (\ell m)^{-2}\right) \\ &= \frac{m^2-1}{3}.\end{aligned}$$

Similarly, if $\{R_0\} = \{R_0^{-1}\}$, the conjugacy classes $\{R_0^k\}$ for $(1 \leq k \leq [\frac{m}{2}])$ are precisely the distinct conjugacy classes of the powers of R_0, and we draw from the above result

$$\begin{aligned}\sum_{1\leq k \leq [\frac{m}{2}]} \sin^{-2}\left(\frac{\pi k}{m}\right) &= \frac{1}{2}\sum_{k=1}^{m-1} \sin^{-2}\left(\frac{\pi k}{m}\right) + \frac{m+1}{2} - \left[\frac{m+1}{2}\right] \\ &= \frac{m^2-1}{6} + \frac{m+1}{2} - \left[\frac{m+1}{2}\right].\end{aligned}$$

Since

$$|\mathcal{E}(R_0)| = \begin{cases} m(R_0) & \text{for} \quad \{R_0\} \neq \{R_0^{-1}\}, \\ 2m(R_0) & \text{for} \quad \{R_0\} = \{R_0^{-1}\}, \end{cases}$$

we obtain in (2.35)

$$\begin{aligned}\sum_{\{R\}\ \text{ellipt.}} \frac{\log N(T_0)}{|\mathcal{E}(R)|\, 4\sin^2(\frac{\pi k}{m(R)})} &= \sum_{\substack{\{R_0\}\ \text{ellipt.} \\ \{R_0\}\neq\{R_0^{-1}\}}} \frac{m(R_0) - m(R_0)^{-1}}{12} \log N(T_0) \\ &+ \sum_{\substack{\{R_0\}\ \text{ellipt.} \\ \{R_0\}=\{R_0^{-1}\}}} \frac{m(R_0) - m(R_0)^{-1}}{48} \log N(T_0) \\ &+ \sum_{\substack{\{R_0\}\ \text{ellipt.} \\ \{R_0\}=\{R_0^{-1}\}}} \left(\frac{m(R_0)+1}{2} - \left[\frac{m(R_0)+1}{2}\right]\right) \frac{\log N(T_0)}{8\, m(R_0)}.\end{aligned}$$

We found in (2.1) that the equation

$$(2.40) \qquad (\lambda - \mu)\operatorname{tr}(R_\lambda R_\mu) = \sum_{n=0}^{\infty}\left(\frac{1}{s^2 - s_n^2} - \frac{1}{t^2 - s_n^2}\right)$$

holds for all $s, t \in \mathbb{C}$ such that $s \neq \pm s_n, t \neq \pm s_n$ for all n. Now keep t fixed, $\operatorname{Re} t > 1$. Then (2.40) is the meromorphic continuation into the entire s-plane of the holomorphic function of $s, \operatorname{Re} s > 1$, appearing on the right-hand side of (2.39). This is the key to the analytic continuation and to the marvellous properties of the Selberg zeta function which we shall investigate in more detail in Section 4.

5.3 Huber's Theorem

We digress into uniqueness questions concerning the eigenvalue and length spectra of the manifold associated with a discrete cocompact group $\Gamma < \mathbf{PSL}(2, \mathbb{C})$. These uniqueness questions were discussed first by H. Huber (1959), (1961) in the case of the hyperbolic plane. Bérard-Bergery (1971), (1973) and Riggenbach (1975) elaborated on Huber's proof for the case of hyperbolic spaces of arbitrary dimension and for cocompact groups without fixed points. The case of groups with fixed points is treated in Parnovski (1992). We choose a slightly different approach based on Theorem 2.2.

Let the hypothesis and notations be as in Theorem 2.2, and assume for the moment that Γ contains no elliptic elements. Then (2.35) simply reads

$$\begin{aligned} \sum_{n=0}^{\infty} \left(\frac{1}{s^2 - s_n^2} - \frac{1}{t^2 - s_n^2} \right) &= -\frac{v(\mathcal{F})}{4\pi}(s-t) \\ &+ \frac{1}{2s} \sum_{\{T\} \text{ lox.}} \frac{\log N(T_0)}{|\mathcal{E}(T)|\, |a(T) - a(T)^{-1}|^2} N(T)^{-s} \\ &- \frac{1}{2t} \sum_{\{T\} \text{ lox.}} \frac{\log N(T_0)}{|\mathcal{E}(T)|\, |a(T) - a(T)^{-1}|^2} N(T)^{-t}. \end{aligned} \tag{3.1}$$

This equation holds for $\operatorname{Re} s > 1, \operatorname{Re} t > 1$.

We interpret the various summands in (3.1) in geometric terms associated with the compact manifold $M = \Gamma \backslash \mathbb{H}$ equipped with its Riemannian metric inherited from $\mathbb{H}$. The left-hand side of (3.1) is determined by the sequence $(s_n)_{n \geq 0}$, i.e., by the eigenvalue spectrum $(\lambda_n)_{n \geq 0}$ of the Laplacian on M. The first term on the right-hand side has an obvious geometric meaning. It is simply given by the volume $v(\mathcal{F})$ of M. The norms $N(T)$ appearing on the right-hand side of (3.1) can also be explained in geometric terms. Recall that $\mathbb{H}$ is the universal covering of M and that Γ is isomorphic to the fundamental group of M. The conjugacy classes of Γ are in a natural bijective correspondence with the free homotopy classes of closed continuous paths on M. This correspondence is established as follows. Consider a free homotopy class W of M. It can be shown that this class contains a curve which is locally of minimal length, a closed geodesic γ of M. This geodesic γ is uniquely determined up to the choice of its initial point. Consider a lift of γ to $\mathbb{H}$. This lift is a hyperbolic line segment L in $\mathbb{H}$. The projections of the endpoints A, B of L to M coincide. Hence there exists an element $T \in \Gamma$ such that $TA = B$, and T is uniquely determined by this equation because Γ has no fixed points. On the other hand, L itself is not uniquely determined since all the hyperbolic line segments $SL \quad (S \in \Gamma)$ are also lifts of γ. The element of Γ matching the endpoints of SL is equal to STS^{-1}. We associate the Γ-conjugacy class $\{T\}$ with γ, i.e., with the free homotopy class W. This correspondence is bijective.

We show that the length of γ can be recovered from $\{T\}$. To avoid trivialities, assume that γ is not a point. Then $T \neq I$, and the hyperbolic line segment mentioned above is part of the axis of T. Take any point A on this axis and consider the segment L on the axis between A and TA. Then L projects to γ on M, and L has length $\log N(T)$. Hence γ also has length $\log N(T)$. This implies that the family $(\log N(T))_{\{T\}}(T \neq I)$ is the set of lengths of non-trivial closed geodesics on M counted with proper multiplicities.

We now abandon our hypothesis that Γ acts fixed-point freely on H and admit that Γ may contain elliptic elements. Bearing in mind the above geometrical description, we define the length spectrum of Γ as follows:

Definition 3.1. Let $\Gamma < \mathbf{PSL}(2,\mathbb{C})$ be a discrete cocompact group (possibly containing elliptic elements). Suppose that $\mu_j = \log N(T_j)$ $(j \geq 1)$ are the logarithms of the norms of the hyperbolic or loxodromic elements of Γ, arranged in strictly increasing order. Then the family of ordered pairs

$$\left(\mu_j, \sum_{\substack{\{T\}\text{lox.} \\ \log N(T)=\mu_j}} \frac{\log N(T_0)}{|\mathcal{E}(T)|\; |a(T)-a(T)^{-1}|^2} \right)_{(j\geq 1)} \tag{3.2}$$

is called the length spectrum of Γ.

This notion of length spectrum imitates the corresponding definition in the fixed-point free case as employed by Huber (1959), (1961) Bérard-Bergery (1971), (1973), and Riggenbach (1975). Observe that our notion of length spectrum really is a group theoretic concept although we maintain the geometric language from the fixed-point free case.

The contribution of the elliptic elements of Γ to the right-hand side of (2.35) leads us to introduce the following quantity.

Definition 3.2. If $\Gamma < \mathbf{PSL}(2,\mathbb{C})$ is a discrete cocompact group, the real number

$$E := \sum_{\{R\}\text{ ellipt.}} \frac{\log N(T_0)}{|\mathcal{E}(R)|\; 4\sin^2\left(\frac{\pi k}{m(R)}\right)} = \sum_{\{R\}\text{ ellipt.}} \frac{\log N(T_0)}{|\mathcal{E}(R)|\; |(\mathrm{tr} R)^2-4|} \tag{3.3}$$

is called the elliptic number of Γ.

We do not know if the elliptic number of Γ has an interpretation in terms of geometric data of the quotient $\Gamma \setminus \mathbb{H}$. The quotient $\Gamma \setminus \mathbb{H}$ inherits the structure of a topological manifold but not necessarily the structure of a Riemannian manifold of constant sectional curvature -1 (cf. Grunewald, Mennicke (1980)). The following result is our analogue of Huber's theorem.

Following a suggestion of M.-F. Vignéras, we merely impose the hypothesis that the eigenvalue or length spectra coincide up to finitely many terms.

Theorem 3.3. *Let Γ_1, Γ_2 be discrete cocompact groups. Then the following hold:*

(1) Suppose that the eigenvalue spectra for Γ_1 and Γ_2 agree up to at most finitely many terms. Then the eigenvalue spectra, the length spectra, the volumes and the elliptic numbers for Γ_1 and Γ_2 coincide.
(2) Suppose that the length spectra for Γ_1 and Γ_2 agree up to at most finitely many terms. Then the length spectra, the eigenvalue spectra, the volumes and the elliptic numbers for Γ_1 and Γ_2 coincide.

Proof. (1) By assumption, the left-hand sides of (2.35) for Γ_1 and Γ_2 agree up to finitely many terms. Letting s tend to infinity in the corresponding equation, we see first that the volumes for Γ_1 and Γ_2 coincide. Omit the contribution of the volumes and let $t \to \infty$ in the new equation. This yields an equation of the form

$$\begin{aligned}\sum_{\text{fin.}} (\pm)\frac{2s}{s^2 - s_n^2} = E_1 - E_2 + \sum_{\{T\}_1 \text{ lox.}} c(T)\, N(T)^{-s} \\ - \sum_{\{T\}_2 \text{ lox.}} c(T)\, N(T)^{-s},\end{aligned} \tag{3.4}$$

where fin. indicates a certain finite sum involving the numbers s_n for Γ_1, Γ_2, where E_1, E_2 are the elliptic numbers for Γ_1, Γ_2, and where $\{T\}_1$, $\{T\}_2$ means summation over the conjugacy classes $\{T\}$ of hyperbolic or loxodromic elements of Γ_1 and Γ_2, respectively. Letting s tend to infinity, we have $E_1 = E_2$. Hence (3.4) can be written in the form

$$\sum_{\text{fin.}} (\pm)\frac{2s}{s^2 - s_n^2} = \sum_{N(T)} \gamma(T)\, N(T)^{-s}, \tag{3.5}$$

where the summation extends over all norms of hyperbolic or loxodromic elements from $\Gamma_1 \cup \Gamma_2$, and where $\gamma(T)$ indicates the difference of the corresponding weights for Γ_1 and Γ_2 occurring in (3.1). We have to show that $\gamma(T) = 0$ for all T. Assume that this is false and let $T^* \in \Gamma_1 \cup \Gamma_2$ be such that

$$N(T^*) := \min \{\ N(T) \ \mid\ \gamma(T) \neq 0\ \}.$$

Then we have

$$N(T^*)^s \sum_{\text{fin.}} (\pm)\frac{2s}{s^2 - s_n^2} = \gamma(T^*) + \sum_{N(T) > N(T^*)} \gamma(T) \left(\frac{N(T)}{N(T^*)}\right)^{-s}. \tag{3.6}$$

Fix $\sigma > 0$ so large that

$$\sum_{N(T)>N(T^*)} |\gamma(T)| \left(\frac{N(T)}{N(T^*)}\right)^{-\sigma} < \frac{|\gamma(T^*)|}{2}$$

and let $s = \sigma + i\tau$, $\tau \to +\infty$. Then the left-hand side of (3.6) tends to zero whereas the right-hand side does not. This contradiction yields $\gamma(T) = 0$ for all T, i.e., the length spectra for Γ_1 and Γ_2 coincide. Hence the right-hand side of (3.4) vanishes, and the terms on the left of (3.4) must cancel as well. This means that eigenvalue spectra for Γ_1 and Γ_2 coincide completely.

b) Assume that the length spectra for Γ_1 and Γ_2 are the same up to at most finitely many terms. Then (2.35) yields that the volumes for Γ_1 and Γ_2 are the same. Omit the contribution of the volumes, multiply the corresponding equation by $2s$ and compare the poles in the s-plane. This readily yields that the eigenvalue spectra for Γ_1 and Γ_2 coincide (including the multiplicities). Omitting the term from the eigenvalues, we see that the elliptic numbers and the length spectra for Γ_1 and Γ_2 are the same. □

Two discrete cocompact groups are said to be *isospectral* if their eigenvalue spectra (or length spectra) coincide. Theorem 3.3 has the following remarkable consequence.

Corollary 3.4. *Suppose that Γ_1, Γ_2 are discrete cocompact isospectral groups. Then either both Γ_1 and Γ_2 contain elliptic elements or neither of them contains elliptic elements.*

The following problem seems to be unsolved. Suppose that Γ_1, Γ_2 are discrete cocompact groups such that the eigenvalue spectra (or length spectra, respectively) for Γ_1 and Γ_2 agree up to a sequence which is of lower density in some sense. Are Γ_1 and Γ_2 isospectral? For groups Γ_1 and Γ_2 which act discontinuously and cocompactly on the hyperbolic plane there are bounds $c(\Gamma_1, \Gamma_2)$ so that the agreement of the first $c(\Gamma_1, \Gamma_2)$ eigenvalues of the Laplacian implies isospectrality. These bounds depend on the deformation spaces of Γ_1 and Γ_2, see Buser (1992), Chapter 14. In dimension 3 it seems hard to even formulate a similar conjecture.

It was a long standing problem first formulated by Gelfand to find cocompact subgroups Γ_1, $\Gamma_2 < \mathbf{PSL}(2, \mathbb{C})$ which are isospectral but non-conjugate. The first examples were given by Vignéras (1978), (1980b). We shall describe some arithmetic examples in Corollary 10.1.8. Somewhat simpler examples can be given by the method of Sunada, see Reid (1992). In this paper it is also proved that isospectral arithmetic groups are commensurable. This is unknown in general.

5.4 The Selberg Zeta Function

The right-hand side of our trace formula (2.35) is for $\operatorname{Re} s > 1$ the logarithmic derivative of an infinite product, the so-called Selberg ζ-function. The trace formula is then the key to the investigation of the amazing properties of this function. After having given the precise definition we shall obtain in this section its analytical continuation, the location of its zeros and various other analytical facts.

Throughout the present section we keep the following notation and hypotheses: Let $\Gamma < \mathbf{PSL}(2, \mathbb{C})$ be a discrete cocompact group with fundamental domain $\mathcal{F}$. Suppose that

$$0 = \lambda_0 < \lambda_1 \leq \lambda_2 \leq \dots \tag{4.1}$$

are the eigenvalues of the operator $-\Delta : \mathcal{D} \to L^2(\Gamma \setminus \mathbb{H})$, and put

$$N := \max \{ n \geq 0 \mid \lambda_n < 1 \}, \tag{4.2}$$

$$s_n := \sqrt{1 - \lambda_n} \qquad \text{for } n = 0, \dots, N, \tag{4.3}$$

$$s_n := i\sqrt{\lambda_n - 1} \qquad \text{for } n \geq N + 1, \tag{4.4}$$

where the square roots are non-negative real numbers. Then

$$\lambda_n = 1 - s_n^2 \qquad (n \geq 0), \tag{4.5}$$

$$s_n = it_n, \ t_n \geq 0 \qquad \text{for } n \geq N + 1. \tag{4.6}$$

We use the same notation as in Theorem 2.2. If $R \in \Gamma$ is elliptic, we let T_0 be a hyperbolic or loxodromic element from the centralizer $\mathcal{C}(R)$ of R in Γ such that $N(T_0)$ is minimal. Let $\mathcal{E}(R)$ be a maximal subgroup of elements of finite order contained in $\mathcal{C}(R)$. If $T \in \Gamma$ is hyperbolic or loxodromic, we let $\mathcal{E}(T)$ be the (cyclic) subgroup of elements of finite order of $\mathcal{C}(T)$, and we denote a primitive hyperbolic or loxodromic element of Γ associated with T by T_0. Sums or products with respect to $\{T_0\}$ extend over all Γ-conjugacy classes of primitive hyperbolic or loxodromic elements of Γ. The summation condition "$\{T\}$ lox." means that the summation extends over all Γ-conjugacy classes of hyperbolic or loxodromic elements of Γ, and similarly "$\{R\}$ ellipt." means summation over all Γ-conjugacy classes of elliptic elements of Γ.

Every hyperbolic or loxodromic element $T \in \Gamma$ is conjugate in $\mathbf{PSL}(2, \mathbb{C})$ to a unique element

$$D(T) = \begin{pmatrix} a(T) & 0 \\ 0 & a(T)^{-1} \end{pmatrix}, \qquad |a(T)| > 1. \tag{4.7}$$

For brevity, we put

$$E := \sum_{\{R\}\,\text{ellipt.}} \frac{\log N(T_0)}{|\mathcal{E}(R)|\,|(\text{tr}R)^2 - 4|}, \tag{4.8}$$

$$m(T) := |\mathcal{E}(T)|, \qquad \text{if } T \text{ is hyperbolic or loxodromic.} \tag{4.9}$$

Then we can write the trace formula from Theorem 2.2 in the form

$$\begin{aligned} \sum_{n=0}^{\infty}\left(\frac{1}{s^2 - s_n^2} - \frac{1}{t^2 - s_n^2}\right) &= -\frac{v(\mathcal{F})}{4\pi}\,(s-t) + \left(\frac{1}{2s} - \frac{1}{2t}\right)E \\ &+ \frac{1}{2s}\sum_{\{T\}\,\text{lox.}} \frac{\log N(T_0)}{m(T)\,|a(T) - a(T)^{-1}|^2}\,N(T)^{-s} \\ &- \frac{1}{2t}\sum_{\{T\}\,\text{lox.}} \frac{\log N(T_0)}{m(T)\,|a(T) - a(T)^{-1}|^2}\,N(T)^{-t}. \end{aligned} \tag{4.10}$$

This equation holds under the condition

$$\text{Re}\, s > 1, \qquad \text{Re}\, t > 1, \tag{4.11}$$

and the left-hand side of (4.10) converges uniformly on compact sets with respect to (s,t) provided $s \neq \pm s_n$, $t \neq \pm s_n$ for all $n \geq 0$.

Definition 4.1. For $\text{Re}\, s > 1$, the Selberg zeta function for Γ is defined by

$$Z(s) := \prod_{\substack{\{T_0\} \in \mathcal{R} \\ k,\ell \geq 0 \\ k \equiv \ell \bmod m(T_0)}} \left(1 - a(T_0)^{-2k}\,\overline{a(T_0)}^{-2\ell}\,N(T_0)^{-s-1}\right) \tag{4.12}$$

where the product with respect to $\{T_0\}$ extends over a maximal reduced system $\mathcal{R}$ of Γ-conjugacy classes of the primitive hyperbolic or loxodromic elements of Γ. $\mathcal{R}$ is called reduced if no two of its elements have representatives with the same centralizer. The corresponding Selberg xi-function is

$$\Xi(s) := \exp\left(-\frac{v(\mathcal{F})}{6\pi}s^3 + Es\right) Z(s), \tag{4.13}$$

where E is given by (4.8).

Note that for $\sigma = \text{Re}\, s > 1$

$$\begin{aligned} &\sum_{\substack{\{T_0\} \\ k,\ell \geq 0}} |a(T_0)^{-2k}\,\overline{a(T_0)}^{-2\ell}\,N(T_0)^{-s-1}| \\ &= \sum_{\{T_0\}} \left(1 - |a(T_0)|^{-2}\right)^{-1} N(T_0)^{-\sigma-1}. \end{aligned} \tag{4.14}$$

Since we know from Lemma 2.6.3 that the counting function

$$\text{(4.15)} \qquad \pi_0(x) := \left| \left\{ \{T\} \;\middle|\; \begin{array}{l} T \in \Gamma,\ T \text{ hyperbolic or loxodromic,} \\ N(T) \leq x \end{array} \right\} \right|$$

satisfies

$$\text{(4.16)} \qquad \pi_0(x) = O(x^2) \qquad \text{for } x \to \infty,$$

we conclude by partial summation that

$$\text{(4.17)} \qquad \sum_{\{T\}\,\text{lox.}} N(T)^{-(s+1)}$$

converges absolutely for $\sigma = \operatorname{Re} s > 1$. Hence the series on the right-hand side of (4.14) converges for $\sigma > 1$, and we conclude that Z and Ξ are nowhere vanishing, holomorphic functions defined in the half-plane $\operatorname{Re} s > 1$.

We deduce from (4.10) that the abscissa of convergence of (4.17) is precisely one. To see this, note that (4.10) implies

$$\sum_{\{T\}\,\text{lox.}} \frac{\log N(T_0)}{m(T)\,|a(T) - a(T)^{-1}|^2}\, N(T)^{-s} \sim \frac{1}{s-1} \qquad \text{for } s \downarrow 1.$$

Hence we deduce from Karamata's theorem, see Karamata (1931a) or Lang (1968), Chapter 15, that

$$\sum_{\substack{\{T\}\,\text{lox.}\\ N(T)\leq x}} \frac{\log N(T_0)}{m(T)\,|1 - a(T)^{-2}|^2}\, N(T)^{-2} \sim \log x \qquad \text{for } x \to \infty.$$

Now $m(T)$ is bounded because there exist only finitely many Γ- conjugacy classes of elliptic elements contained in Γ. Since $|a(T)| \to \infty$ as $\{T\}$ runs through the conjugacy classes of the hyperbolic or loxodromic elements of Γ, we see that (4.17) has abscissa of convergence one. We can even conclude that

$$\text{(4.18)} \qquad \sum_{\substack{\{T\}\,\text{lox.}\\ N(T)\leq x}} \frac{\log N(T_0)}{m(T)\, N(T)^2} \sim \log x \qquad \text{for } x \to \infty.$$

If one uses Ikehara's theorem instead of Karamata's theorem, one obtains an even sharper result, see Theorem 7.3.

It is obvious from the definition that

$$\text{(4.19)} \qquad Z(\bar{s}) = \overline{Z(s)}, \qquad \Xi(\bar{s}) = \overline{\Xi(s)}.$$

The logarithmic derivative of the Selberg ζ-function is computed in the following Lemma.

Lemma 4.2. *For* $\operatorname{Re} s > 1$, *we have*

$$\frac{Z'}{Z}(s) = \sum_{\{T\}\text{ lox.}} \frac{\log N(T_0)}{m(T)\ |a(T) - a(T)^{-1}|^2}\ N(T)^{-s}. \tag{4.20}$$

Proof. We consider the sum on the right-hand side of (4.20) and put

$$T = T_0^{n+1} E_0^{\nu}, \qquad n \geq 0,\ \ 1 \leq \nu \leq m(T_0),$$

where T_0 is a primitive element associated with T and where E_0 generates the cyclic group $\mathcal{E}(T) = \mathcal{E}(T_0)$. Note that all the elements $T_0 E_0^{\nu}$, $1 \leq \nu \leq m(T_0)$ are precisely the primitive elements associated with T. Hence, if $\{T_0\}$ runs through a maximal reduced system $\mathcal{R}$ of Γ-conjugacy classes of primitive hyperbolic or loxodromic elements of Γ, then $T = T_0^{n+1} E_0^{\nu}$, $n \geq 0$, $1 \leq \nu \leq m(T_0)$ runs through a representative system of all Γ-conjugacy classes $\{T\}$ of all hyperbolic or loxodromic elements of Γ. Transforming the fixed points of T_0 to 0 and ∞, we conjugate E_0 in $\mathbf{PSL}(2, \mathbb{C})$ to $\pm \begin{pmatrix} \zeta(T_0) & 0 \\ 0 & \zeta(T_0)^{-1} \end{pmatrix}$, where $\zeta(T_0)$ is a primitive $2m(T_0)$-th root of unity.

Hence we have

$$\begin{aligned}
&\sum_{\{T\}\text{ lox.}} \frac{\log N(T_0)}{m(T)\ |a(T) - a(T)^{-1}|^2} N(T)^{-s} \\
&= \sum_{\substack{\{T_0\}\in\mathcal{R} \\ n\geq 0 \\ 1\leq\nu\leq m(T_0)}} \frac{\log N(T_0)}{m(T_0)\ |\zeta(T_0)^{\nu}\ a(T_0)^{n+1} - \zeta(T_0)^{-\nu}\ a(T_0)^{-n-1}|^2}\ N(T_0)^{-s(n+1)} \\
&= \sum_{\substack{\{T_0\}\in\mathcal{R} \\ n\geq 0 \\ 1\leq\nu\leq m(T_0)}} \frac{N(T_0)^{-(s+1)(n+1)}\ \log N(T_0)}{m(T_0)(1 - \zeta(T_0)^{-2\nu}\ a(T_0)^{-2(n+1)})(1 - \overline{\zeta(T_0)}^{-2\nu}\ \overline{a(T_0)}^{-2(n+1)})} \\
&= \sum_{\substack{\{T_0\}\in\mathcal{R} \\ k,\ell,n\geq 0 \\ k\equiv\ell \bmod m(T_0)}} a(T_0)^{-2k(n+1)}\ \overline{a(T_0)}^{-2\ell(n+1)}\ N(T_0)^{-(s+1)(n+1)}\ \log N(T_0) \\
&= \sum_{\substack{\{T_0\}\in\mathcal{R} \\ k,\ell\geq 0 \\ k\equiv\ell \bmod m(T_0)}} \frac{a(T_0)^{-2k}\ \overline{a(T_0)}^{-2\ell}\ N(T_0)^{-(s+1)}\ \log N(T_0)}{1 - a(T_0)^{-2k}\ \overline{a(T_0)}^{-2\ell}\ N(T_0)^{-(s+1)}} = \frac{Z'}{Z}(s).
\end{aligned}$$

This proves the Lemma. □

Theorem 4.3. *Let the hypotheses and notations be as in Definition 4.1. Then the Selberg zeta function and the xi-function, defined for* $\operatorname{Re} s > 1$ *by (4.12), (4.13), are entire functions of* s*, and*

$$\frac{1}{2s}\frac{\Xi'}{\Xi}(s) - \frac{1}{2t}\frac{\Xi'}{\Xi}(t) = \sum_{n=0}^{\infty} \left(\frac{1}{s^2 - s_n^2} - \frac{1}{t^2 - s_n^2} \right) \tag{4.21}$$

for all $s,\ t \in \mathbb{C} \setminus \{\ \pm s_n \ \mid\ n \geq 0\ \}$. *The zeros of* Z *and* Ξ *are the numbers* $\pm s_n$, $n \geq 0$. *If* $\lambda_n \neq 1$, *both the numbers* s_n, $-s_n$ *are zeros whose multiplicities are equal to the multiplicity of the eigenvalue* λ_n. *If* $\lambda_k = 1$ *is an eigenvalue of* $-\Delta : \mathcal{D} \to L^2(\Gamma \setminus \mathbb{H})$, *then* $s_k = 0$ *is a zero whose multiplicity equals twice the multiplicity of the eigenvalue* $\lambda_k = 1$.

Proof. It is obvious from (4.10), (4.12), (4.13), (4.20) that (4.21) holds for $\operatorname{Re} s > 1$, $\operatorname{Re} t > 1$. Now let t be fixed, $\operatorname{Re} t > 1$. Then the right-hand side of (4.21) is a meromorphic function of $s \in \mathbb{C}$ with simple poles only at the points $\pm s_n$. Hence $\frac{\Xi'}{\Xi}$ is a meromorphic function with simple poles only at the points $\pm s_n$. Since

$$\frac{2s}{s^2 - s_n^2} = \frac{1}{s + s_n} + \frac{1}{s - s_n},$$

we conclude: For $\lambda_n \neq 1$, the residue of $\frac{\Xi'}{\Xi}$ at $\pm s_n$ is equal to the multiplicity of λ_n, and if $\lambda_k = 1$ is an eigenvalue, the residue of the pole $s_k = 0$ equals twice the multiplicity of the eigenvalue $\lambda_k = 1$. Since all the residues are non-negative integers, we see that Ξ is an entire function whose zeros are as described in the Theorem. This implies our assertion. □

Corollary 4.4. *The Selberg zeta function and the xi-function satisfy the functional equations*

$$Z(-s) = \exp\left(-\frac{v(\mathcal{F})}{3\pi} s^3 + 2Es \right) Z(s), \tag{4.22}$$

$$\Xi(-s) = \Xi(s). \tag{4.23}$$

Proof. In view of (4.13), we only have to prove (4.23). We see from (4.21) that

$$\frac{1}{2s} \frac{\Xi'}{\Xi}(s) \tag{4.24}$$

is invariant with respect to $s \to -s$.

Since $\frac{\Xi'}{\Xi}$ is an odd function, it follows that the logarithmic derivative of $\frac{\Xi(s)}{\Xi(-s)}$ equals zero. Hence $\frac{\Xi(s)}{\Xi(-s)}$ is constant. Now we know from Theorem 4.3 that the order of Ξ at zero is even. Hence

$$\lim_{s \to 0} \frac{\Xi(s)}{\Xi(-s)} = 1.$$

This proves the Corollary. □

The Selberg zeta function closely resembles in its properties the usual zeta- or L-functions of number theory. Since the primitive elements of Γ

can be thought of as substitutes of prime numbers, (4.12) is an analogue of the Euler product expansion. Theorem 4.3 and Corollary 4.4 say that $Z(s)$ has an analytic continuation to the whole s-plane and satisfies a simple functional equation which takes its simplest form in terms of the function Ξ. The notations Z, Ξ are analogous to the usual notations ζ, ξ for the Riemann zeta function ζ and its associated function ξ,

$$\xi(s) = \frac{s(s-1)}{2} \pi^{-\frac{s}{2}} \Gamma\left(\frac{s}{2}\right) \zeta(s).$$

The line $\operatorname{Re} s = 0$ is the "critical line" for Z, and with the only exception of $s_0 = 1, \ldots, s_N$, $-s_0 = -1, \ldots, -s_N \in [-1, 1]$ all the zeros of Z are on the critical line. Thus the analogue of the Riemann hypothesis is true for the Selberg zeta function save for the zeros in $[-1, 1]$ just mentioned. There exist no series of trivial zeros of the Selberg zeta function in our case in contrast to the zeta or L-functions of analytic number theory and in contrast to the properties of the Selberg zeta function for a Fuchsian group with compact fundamental domain acting on the hyperbolic plane, see Selberg (1956), Hejhal (1976a), (1976b), Venkov (1979c), Elstrodt (1981).

The reason for the absence of trivial zeros is explained in Gangolli (1977b). It is also known that the function ζ is an entire function of order one and that the Selberg zeta function for a Fuchsian group with compact quotient is an entire function of order 2. We shall show later that in the case of dimension 3, Z is an entire function of order 3, see Theorem 5.8.

5.5 Weyl's Asymptotic Law and the Hadamard Factorisation of the Zeta Function

Suppose that $s > 1$ is real. Then the factors in (4.12) corresponding to the triplets of indices

$$(\{T_0\}, k, \ell), \quad (\{T_0\}, \ell, k)$$

combined yield a positive real factor. Hence

$$Z(s) > 0 \qquad \text{for } s > 1. \tag{5.1}$$

Remember that we know all the zeros of Z from Theorem 4.3. In particular, there are no zeros in the region

$$\{\, s \in \mathbb{C} \mid \operatorname{Re} s > 0, s \notin]0, 1] \,\}. \tag{5.2}$$

Hence there exists a holomorphic logarithm $\log Z$ of Z in this region which is uniquely determined by the requirement that

$$\log Z(s) \in \mathbb{R} \qquad \text{for real} \quad s > 1. \tag{5.3}$$

We follow the awkward classical notion denoting our holomorphic logarithm by $\log Z$. Suppose now that $0 \neq t \in \mathbb{R}$ is such that ti is not a zero of Z, say

$$t_n < |t| < t_{n+1}, \quad n \geq N+1. \tag{5.4}$$

Then $\log Z$ has a unique continuous extension to $\{ \, it \mid t_n < |t| < t_{n+1} \, \}$ which we also denote by $\log Z$. Imitating the well-established notation for the Riemann zeta function we put for $s \in \mathbb{C}$ with $\operatorname{Re} s \geq 0$, $s \notin [0,1]$ and $s \neq \pm i t_n$ for all $n \geq N+1$

$$\arg Z(s) := \operatorname{Im}\,(\log Z(s)), \tag{5.5}$$

see Titchmarsh (1951). The aim of the present subsection is to deduce the asymptotic behaviour of the number of eigenvalues less than T

$$|\{ \, n \geq 0 \mid \lambda_n \leq T \, \}| \qquad \text{for} \quad T \to \infty. \tag{5.6}$$

Since we are concerned with the asymptotic behaviour only, we can likewise consider

$$A(T) := |\{ \, n \mid n \geq N+1, \; t_n < T \, \}|. \tag{5.7}$$

The argument principle gives us a relation of $A(T)$ with the values of $\log Z$ on the critical line.

Theorem 5.1. *Suppose that $T > 0$, $T \neq t_n$ for all $n \geq N+1$. Then*

$$A(T) = \frac{v(\mathcal{F})}{6\pi^2} T^3 + \frac{E}{\pi} T + \frac{1}{\pi} \arg Z(iT) - N. \tag{5.8}$$

Proof. We infer from Theorem 4.3 and from the argument principle that

$$2(A(T) + N) = \frac{1}{2\pi i} \int_{\partial R(T)} \frac{Z'}{Z}(s)\, ds, \tag{5.9}$$

where $R(T)$ is the rectangle with vertices $2+iT$, $-2+iT$, $-2-iT$, $2-iT$. The boundary of $R(T)$ splits into two parts, $R^+(T)$, $R^-(T)$, situated in the half-planes $\operatorname{Re} s \geq 0$ and $\operatorname{Re} s \leq 0$, respectively. Note that the transformation $s \to -s$ maps $R^-(T)$ onto $R^+(T)$ such that the orientation is preserved. Hence we deduce from the functional equation (4.22)

$$\begin{aligned}
&\frac{1}{2\pi i} \int_{\partial R(T)} \frac{Z'}{Z}(s)\, ds \\
&= \frac{1}{2\pi i} \int_{R^+(T)} \frac{Z'}{Z}(s)\, ds - \frac{1}{2\pi i} \int_{R^+(T)} \frac{Z'}{Z}(-s)\, ds \\
&= \frac{1}{2\pi i} \int_{R^+(T)} \left(-\frac{v(\mathcal{F})}{\pi} s^2 + 2E \right) ds + \frac{1}{\pi i} \int_{R^+(T)} \frac{Z'}{Z}(s)\, ds \\
&= \frac{v(\mathcal{F})}{3\pi^2} T^3 + \frac{2E}{\pi} T + \frac{2}{\pi} \arg Z(iT).
\end{aligned} \tag{5.10}$$

In view of (5.9), this proves our assertion. □

We proceed to prove that the term $\arg Z(iT)$ in (5.8) grows less rapidly as $T \to \infty$ than the leading term $\frac{v(\mathcal{F})}{6\pi^2}T^3$.

Lemma 5.2. *Suppose that $\sigma < 0$. Then there exists a bounded function $f_\sigma : \mathbb{R} \to \mathbb{C}$ such that*

$$Z(\sigma + it) = f_\sigma(t) \exp\left(\frac{v(\mathcal{F})}{\pi}|\sigma|t^2\right) Z(-\sigma - it) \qquad \textit{for all } t \in \mathbb{R}. \tag{5.11}$$

Proof. The assertion is obvious from the functional equation (4.22). □

Lemma 5.3. *Z is an entire function of order at most 4.*

We hasten to state that this is a preliminary result only since we show in Theorem 5.8 that Z is an entire function of order 3.

Proof. We know from Corollary 4.3.3 that

$$\sum_{n=0}^{\infty}{}' \; |s_n|^{-4} < \infty,$$

the prime indicating that terms with $s_n = 0$ (if any) must be omitted. Let $p \in \{0, 1, 2, 3\}$ be the minimal integer such that

$$\sum_{n=0}^{\infty}{}' \; |s_n|^{-p-1} < \infty, \tag{5.12}$$

and let $k \geq 0$ denote the multiplicity of the eigenvalue 1 of $-\Delta : \mathcal{D} \to L^2(\Gamma \setminus \mathbb{H})$. Then the canonical product, see Titchmarsh (1939), section 8.23

$$\begin{aligned} \Phi(s) := \; s^{2k} & \prod_{n=0}^{\infty}{}' \left(1 - \frac{s}{s_n}\right) \exp\left(\frac{s}{s_n} + \ldots + \frac{1}{p}\left(\frac{s}{s_n}\right)^p\right) \\ & \cdot \prod_{n=0}^{\infty}{}' \left(1 + \frac{s}{s_n}\right) \exp\left(-\frac{s}{s_n} + \ldots + \frac{1}{p}\left(-\frac{s}{s_n}\right)^p\right) \end{aligned} \tag{5.13}$$

is an entire function of order equal to the exponent of convergence of the series

$$\sum_{n=0}^{\infty}{}' \; |s_n|^{-\alpha}$$

which is at most equal to 4 (cf. Titchmarsh (1939), section 8.25). Φ has the same zeros at Z, and an elementary computation based on (4.21) yields that

$$\frac{Z'}{Z} - \frac{\Phi'}{\Phi} \tag{5.14}$$

is a polynomial of degree at most 2. This implies that

$$Z(s) = \Phi(s)e^{q(s)}, \tag{5.15}$$

where q is a polynomial of degree at most 3. Hence Z is an entire function of order at most of 4. □

Corollary 5.4. *Suppose that $\alpha, \beta \in \mathbb{R}, \alpha < \beta$. Then there exists a constant $C > 0$ such that*

$$Z(\sigma + it) = O(e^{Ct^2}) \quad \text{as} \quad |t| \to \infty \tag{5.16}$$

uniformly with respect to $\sigma \in [\alpha, \beta]$.

Proof. Without loss of generality we may assume that $\alpha < -1 < 1 < \beta$. Then $Z(\beta + it)$ is a bounded function of $t \in \mathbb{R}$. Together with Lemma 5.2 this implies that (5.16) is true for $\sigma = \alpha$ and for $\sigma = \beta$. Since Z is of finite order by Lemma 5.3, the Phragmén-Lindelöf theorem (cf. e.g. Rudin (1974), section 12.7) yields the assertion. □

Theorem 5.5. *Suppose that $T > 0$, $T \neq t_n$ for all $n \geq N + 1$. Then*

$$\arg\ Z(iT) = O(T^2) \qquad \text{as } T \to \infty. \tag{5.17}$$

Proof. Let $P(T)$ be the polygonal curve consisting of the line segments $Q(T)$ from 2 to $2 + iT$ and $R(T)$ from $2 + iT$ to iT. Then

$$\arg\ Z(iT) = \mathrm{Im} \int_{P(T)} \frac{Z'}{Z}(s)\, ds = \arg\ Z(2+iT) + \mathrm{Im} \int_{R(T)} \frac{Z'}{Z}(s)\, ds. \tag{5.18}$$

Recall that for $\mathrm{Re}\, s > 1$ the function $Z(s)$ is represented by the product (4.12). Since the series on the right-hand side of the following equation converges absolutely for $\mathrm{Re}\, s > 1$ and attains real values for real $s > 1$, we see that for $\mathrm{Re}\, s > 1$

$$\begin{aligned} &- \log Z(s) \\ &= \sum_{\substack{\{T_0\} \in \mathcal{R} \\ k,\ell \geq 0,\ n \geq 1 \\ k \equiv \ell \bmod m(T_0)}} \frac{1}{n}\, a(T_0)^{-2nk}\, \overline{a(T_0)}^{-2n\ell}\, N(T_0)^{-n(s+1)} \\ &= \sum_{\substack{\{T_0\} \in \mathcal{R} \\ k,\ell \geq 0,\ n \geq 1 \\ 1 \leq \nu \leq m(T_0)}} \frac{\zeta(T_0)^{-2\nu(k-\ell)}\, a(T_0)^{-2nk} \overline{a(T_0)}^{-2n\ell}\, N(T_0)^{-n(s+1)}}{n\, m(T_0)} \\ &= \sum_{\substack{\{T_0\} \in \mathcal{R} \\ n \geq 1 \\ 1 \leq \nu \leq m(T_0)}} \frac{1}{n} \frac{1}{m(T_0)} \frac{1}{|1 - a(T_0)^{-2n}\, \zeta(T_0)^{-2\nu}|^2}\, N(T_0)^{-n(s+1)} \end{aligned} \tag{5.19}$$

where $\zeta(T_0)$ is a primitive $(2m(T_0))$-th root of unity. Hence if we choose $0 < \alpha < 1$ such that $|a(T_0)|^{-2} \leq \alpha$ for all primitive hyperbolic or loxodromic elements of Γ, we see that for $\sigma = \operatorname{Re} s > 1$

$$
\begin{aligned}
|\log Z(s)| &\leq (1-\alpha)^{-2} \sum_{\substack{\{T_0\}\in\mathcal{R} \\ n\geq 1}} \frac{1}{n} N(T_0)^{-n(\sigma+1)} \\
&= (1-\alpha)^{-2} \sum_{\{T_0\}\in\mathcal{R}} \log(1-N(T_0)^{-\sigma-1})^{-1}.
\end{aligned}
\tag{5.20}
$$

The right-hand side of (5.20) is finite (cf. (4.17)). Hence

$$
\arg\ Z(2+iT) = O(1) \qquad \text{as } T \to \infty. \tag{5.21}
$$

It remains to prove that the second term on the right-hand side of (5.18) is $O(T^2)$ as $T \to \infty$. The proof of this estimate employs the same method as the corresponding proof for the Riemann zeta function (cf. Titchmarsh (1951), pages 180-181). Note that

$$
\operatorname{Im} \int_{R(T)} \frac{Z'}{Z}(s)ds
$$

equals the increment of the argument of $Z(s)$ as s runs through the line segment from $2+iT$ to iT. Now imagine s runnung this way. Each time the argument of $Z(s)$ changes by a quantity of absolute value at least π, the real part of $Z(s)$ undergoes a change of sign. Let $c(T)$ be the number of times $\operatorname{Re} Z(\sigma + iT)$ changes sign as σ decreases from 2 to zero. Then

$$
\left| \operatorname{Im} \int_{R(T)} \frac{Z'}{Z}(s)\ ds \right| \leq \pi(c(T)+2). \tag{5.22}
$$

Now (4.19) implies that $c(T)$ equals the number $\nu(T)$ of zeroes of the entire function

$$
\varphi_T(w) = Z(w+iT) + Z(w-iT) \tag{5.23}
$$

in the interval $[0,2]$ up to an error term not exceeding 2 due to the possible zeros at 0 and 2. The number $\nu(T)$ of zeroes of φ_T in $[0,2]$ is estimated by means of Jensen's formula (cf. Rudin (1974), Theorem 15.18) applied to the disc of center 0 and radius 3. If 0 is not a zero of $\varphi(T)$, we obtain

$$
\nu(T) \log \frac{3}{2} \leq \frac{1}{2\pi} \int_0^{2\pi} \log |\varphi_T(3e^{it})|\ dt - \log |\varphi_T(0)|.
$$

Inserting (5.16), we have $\nu(T) = O(T^2)$ as $T \to \infty$. If 0 is a zero of φ_T, a slight move of the center of our disc yields the same conclusion, and we are finished. □

Theorem 5.6. *Suppose that $T > 0, T \neq t_n$ for all $n \geq N+1$. Then the counting function $A(T) := |\{\, n \mid n \geq N+1,\ t_n < T \,\}$ satisfies*

$$A(T) = \frac{v(\mathcal{F})}{6\pi^2}\, T^3 + O(T^2) \qquad \textit{as } T \to \infty. \tag{5.24}$$

Proof. The proof is now obvious from (5.8) and (5.17). □

Corollary 5.7. *The series*

$$\sum_{n=0}^{\infty}{}' |s_n|^{-\alpha}, \qquad (\alpha \in \mathbb{R})$$

converges if and only if $\alpha > 3$. We further have

$$\sum_{0<t_n\leq T} |s_n|^{-3} = \frac{v(\mathcal{F})}{2\pi^2} \log T + O(1) \qquad \textit{as } T \to \infty. \tag{5.25}$$

Proof. We apply partial summation: For $n > m \geq N+1, t_m \neq 0, t_n < T < t_{n+1}$, we have

$$\sum_{\nu=m}^{n} |s_\nu|^{-\alpha} = [x^{-\alpha}A(x)]_{t_m}^{T} + \alpha \int_{t_m}^{T} x^{-\alpha-1}A(x)dx.$$

The asymptotic law (5.24) now yields both assertions. It is even possible to sharpen the $O(1)$ in (5.25) to $C + O(T^{-1})$, where C is some constant. □

Theorem 5.8. *The functions Z and Ξ are entire functions of order 3.*

Proof. We repeat the argument of the proof of Lemma 5.3 using the result on the exponent of convergence of the zeros of Z and Ξ given in Corollary 5.6. Then we see that we must choose $p = 3$ in (5.12), (5.13) and that the canonical product (5.13) is an entire function of order precisely 3. Now we have the representation (5.15) where q is a polynomial of degree at most 3. Hence Z and Ξ are entire functions or order ρ at most 3. But since the exponent γ of convergence of the sequence of zeros for Z and Ξ is precisely 3, it follows from the inequality $3 = \gamma \leq \rho \leq 3$ (cf. Titchmarsh (1939), section 8.22) that the order of Z and Ξ is precisely equal to 3. □

We mention in passing that (5.25) implies that Z and Ξ are both not of finite type. This follows from a theorem of Lindelöf (cf. Boas (1954), page 27).

We can even be more explicit on the Hadamard factorisation of Z and Ξ. Since we now know that $p = 3$ in (5.13), we can write down the canonical product for the sequence of zeros of Z and Ξ in the form

$$\Phi(s) = s^{2k} \prod_{n=0}^{\infty}{}' \left(1 - \left(\frac{s}{s_n}\right)^2\right) \exp\left(\left(\frac{s}{s_n}\right)^2\right). \tag{5.26}$$

Here $k \geq 0$ is the multiplicity of the eigenvalue 1 of $-\Delta : \mathcal{D} \to L^2(\Gamma \setminus \mathbb{H})$, and the prime indicates that factors with $s_n = 0$ (if any) must be omitted. We now deduce from (4.21), (5.26) that for an arbitrary $t \neq \pm s_n (n \geq 0)$

$$\begin{aligned}
\frac{1}{2s}\frac{\Xi'}{\Xi}(s) - \frac{1}{2s}\frac{\Phi'}{\Phi}(s) &= \sum_{n=0}^{\infty}\left(\frac{1}{s^2 - s_n^2} - \frac{1}{t^2 - s_n^2}\right) + \frac{1}{2t}\frac{\Xi'}{\Xi}(t) \\
&\quad - \frac{1}{2s}\left(\frac{2k}{s} + \sum_{n=0}^{\infty}{}'\left(\frac{2s}{s^2 - s_n^2} + \frac{2s}{s_n^2}\right)\right) \\
&= \frac{1}{2t}\frac{\Xi'}{\Xi}(t) - \frac{k}{t^2} - \sum_{n=0}^{\infty}{}'\left(\frac{1}{t^2 - s_n^2} + \frac{1}{s_n^2}\right) \\
&= \frac{1}{2t}\frac{\Xi'}{\Xi}(t) - \frac{1}{2t}\frac{\Phi'}{\Phi}(t).
\end{aligned} \tag{5.27}$$

Hence there exist constants $\alpha, \beta \in \mathbb{C}, \alpha \neq 0$, such that

$$\Xi(s) = \alpha e^{\beta s^2} \Phi(s). \tag{5.28}$$

Since $s_n^2 \in \mathbb{R}$ for all n, we see also that α and β are real numbers. The corresponding factorisation of Z is now clear from (4.13), and we have proved:

Corollary 5.9. *There exist real constants α, β such that Ξ and Z have canonical factorisations of the form*

$$\Xi(s) = \alpha\, e^{\beta s^2}\, \Phi(s), \tag{5.29}$$

$$Z(s) = \alpha \exp\left(\frac{v(\mathcal{F})}{6\pi} s^3 + \beta s^2 - Es\right) \Phi(s) \tag{5.30}$$

where $\Phi(s)$ denotes the canonical product (5.26) formed with the zeros of Ξ and Z.

Note that the coefficients of the odd powers of s in the polynomial in (5.30) are known explicitly, whereas α and β are not known explicitly. This agrees with the functional equations (4.22), (4.23). Of course, we can express α, β in terms of the Taylor coefficients of Ξ. Obviously

$$\alpha = \lim_{s \to 0} s^{-2k}\, \Xi(s) = \lim_{s \to 0} s^{-2k}\, Z(s) \tag{5.31}$$

is the first nonvanishing coefficient in the Taylor series for Ξ or Z at 0. Similarly, we can extract β from the Taylor series of Ξ at 0. Note that Φ has a Taylor expansion at 0 of the form

$$\Phi(s) = s^{2k}\ (1 + a_4 s^4 + \ldots) \tag{5.32}$$

and hence

$$\begin{aligned} \Xi(s) &= \alpha s^{2k}\ (1 + \beta s^2 + \frac{1}{2}\beta^2 s^4 + \ldots)\ (1 + a_4 s^4 + \ldots) \\ &= \alpha s^{2k}\ (1 + \beta s^2 + (a_4 + \frac{1}{2}\beta^2) s^4 + \ldots). \end{aligned} \tag{5.33}$$

This yields

$$\beta = \lim_{s\to 0} \frac{\alpha^{-1} s^{-2k}\ \Xi(s) - 1}{s^2}. \tag{5.34}$$

In view of

$$\frac{\Xi'}{\Xi}(s) = \frac{2k}{s} + \frac{2\beta s + \ldots}{1 + \beta s^2 + \ldots}, \tag{5.35}$$

formula (5.34) can also be written in the form

$$\beta = \lim_{s\to 0} \frac{1}{2s}\left(\frac{\Xi'}{\Xi}(s) - \frac{2k}{s}\right). \tag{5.36}$$

The latter formula can also be deduced from (5.27). Formula (5.26) has some noteworthy consequences. Obviously,

$$\frac{1}{2s}\frac{\Phi'}{\Phi}(s) = \frac{k}{s^2} + \sum_{n=0}^{\infty}{}' \left(\frac{1}{s^2 - s_n^2} + \frac{1}{s_n^2}\right) \tag{5.37}$$

and hence putting $s = \sigma + it$ we have

$$\operatorname{Im}\left(\frac{1}{2s}\frac{\Phi'}{\Phi}(s)\right) = -2\sigma t \sum_{n=0}^{\infty} |s^2 - s_n^2|^{-2}. \tag{5.38}$$

This implies that the zeros of Φ' are all either real or purely imaginary. Differentiating (5.37) once more, we obtain

$$\frac{1}{2s}\left(\frac{1}{2s}\frac{\Phi'}{\Phi}\right)'(s) = -\sum_{n=0}^{\infty} (s^2 - s_n^2)^{-2}. \tag{5.39}$$

Here the right-hand side is negative whenever s is real or purely imaginary, and $s \neq \pm s_n$ for all n. Suppose $s > 0$ is real and between two positive zeros $0 < s_n < s_{n-1} (1 \leq n \leq N)$. Then it follows from (5.37), (5.39), that $\frac{1}{2s}\frac{\Phi'}{\Phi}(s)$ decreases strictly from $+\infty$ to $-\infty$ as s increases from s_n to s_{n-1}. Hence Φ' vanishes precisely once in $]s_n, s_{n-1}[$, and since Φ' is odd, Φ' also vanishes precisely once in $]-s_{n-1}, -s_n[$. The same sort of behaviour is shown by Φ' between any two consecutive zeros of Φ on the positive or negative imaginary axis. Of course, $\Phi'(0) = 0$, because Φ' is odd, and there exist no zeros of Φ' in $]-s_N, s_N[$ which are different from 0. There are no zeros of Φ' in

$]-\infty,-1]\cup[1,\infty[$ as is obvious from (5.37). Needless to say, there are very difficult open problems connected with the explicit determination of the λ_n or s_n for a given group. Not a single example of a discrete cocompact group Γ is known where the sequence $(\lambda_n)_{n\geq 0}$ is known explicitly, not even a single eigenvalue different from 0 is known explicitly for any particular group Γ.

5.6 Analogue of the Lindelöf Hypothesis for the Selberg Zeta Function

The following subsection is motivated by the far-reaching formal analogies between the Selberg zeta function and the Riemann zeta function. We briefly review the relevant facts about $\zeta(s)$ from Titchmarsh (1951), pages 81–82 and page 276 et seq. It is known that $\zeta(s)$ is of polynomial growth in vertical strips. Hence it follows from the general theory of Dirichlet series, see Titchmarsh (1939), section 9.41, that

$$\mu(\sigma) := \inf \{ \gamma \in \mathbb{R} \mid \zeta(\sigma+it) = O(|t|^{\gamma}) \quad \text{for } |t| \to \infty \} \tag{6.1}$$

is a well-defined non-negative monotonically decreasing function of σ. Moreover, $\mu(\cdot)$ is convex downwards. Hence $\mu(\cdot)$ is continuous. For $\sigma > 1$, the function $t \mapsto \zeta(\sigma+it)$ is bounded. Hence

$$\mu(\sigma) = 0 \qquad \text{for } \sigma > 1. \tag{6.2}$$

The functional equation for ζ then yields

$$\mu(\sigma) = \frac{1}{2} - \sigma \qquad \text{for } \sigma < 0. \tag{6.3}$$

These equations also hold by continuity for $\sigma = 1$ and for $\sigma = 0$ respectively.

The exact value of $\mu(\sigma)$ is unknown for any value of $\sigma \in]0,1[$. Since $\mu(\cdot)$ is convex, the simplest possible hypothesis is that

$$\mu(\sigma) = \begin{cases} 0 & \text{for} \quad \sigma \geq 1/2, \\ \frac{1}{2} - \sigma & \text{for} \quad \sigma \leq 1/2, \end{cases}. \tag{6.4}$$

This is the Lindelöf hypothesis for ζ. It is equivalent to the statement that

$$\zeta\left(\frac{1}{2} + it\right) = O(|t|^{\epsilon}) \qquad \text{for } |t| \to \infty \tag{6.5}$$

for every $\epsilon > 0$. It is still unknown if the Lindelöf hypothesis is true, but it can be shown that the truth of the Lindelöf hypothesis follows from that of the Riemann hypothesis, see Titchmarsh (1951), page 282–283. But the Lindelöf hypothesis is not equivalent to the Riemann hypothesis. In fact, the Lindelöf hypothesis is equivalent to a condition on the distribution of the zeros of the Riemann zeta function which is much less restrictive than the Riemann hypothesis, see Titchmarsh (1951), page 279.

We turn to the corresponding problem for the Selberg zeta function. Neglecting the finitely many zeros of Z in $[-1,1]$, we may say that the analogue of the Riemann hypothesis is true for Z. Hence we may expect the analogue of the Lindelöf hypothesis to be true as well. To find out what the analogue of the Lindelöf hypothesis is, recall the formal correspondence between Riemann's and Selberg's zeta functions. Write down the O-condition in (6.1) in the form

$$\zeta(\sigma+it) = O(\exp(\gamma \log t)) \qquad \text{for } t \to \infty.$$

Now it is known that many estimates for the Selberg zeta function in the case of the hyperbolic plane are formally equal to the corresponding estimates for ζ, if we replace $\log t$ by t in the estimates for ζ, see Hejhal (1976b). In the case of three-dimensional hyperbolic space, we have to replace $\log t$ by t^2. Hence, in view of (5.11) and Corollary 5.4 it is natural to define

$$M(\sigma) := \inf \left\{ \gamma \in \mathbb{R} \;\middle|\; Z(\sigma+it) = O\left(\exp(\gamma \frac{v(\mathcal{F})}{\pi} t^2)\right) \quad \text{for } t \to \infty \right\}. \tag{6.6}$$

Lemma 6.1. *M is a non-negative, downwards convex and monotonically decreasing continuous function satisfying*

$$M(\sigma) = 0 \qquad \textit{for } \sigma \geq 1, \tag{6.7}$$

$$M(\sigma) = |\sigma| \qquad \textit{for } \sigma \leq -1. \tag{6.8}$$

Proof. We know from Corollary 5.4 that $-\infty \leq M(\sigma) < \infty$, and it will turn out in a moment that M is actually finite. First we will show that M is convex downwards. Let $\sigma_1 < \sigma_2$ and assume that γ_1, γ_2 are real and

$$M(\sigma_1) < \gamma_1, \qquad M(\sigma_2) < \gamma_2.$$

Put

$$f(s) = \exp\left(\frac{v(\mathcal{F})}{\pi}(\alpha s^3 + \beta s^2)\right) Z(s),$$

where α, β are real numbers still to be chosen suitably. Note that

$$\mathrm{Re}(\alpha s^3 + \beta s^2) = \alpha\sigma^3 + \beta\sigma^2 - (3\alpha\sigma + \beta)t^2,$$

where $s = \sigma + it$. Choosing

$$3\alpha = \frac{\gamma_2 - \gamma_1}{\sigma_2 - \sigma_1}, \qquad \beta = \gamma_1 - 3\alpha\sigma_1,$$

we see that

$$3\alpha\sigma + \beta = \frac{\sigma - \sigma_1}{\sigma_2 - \sigma_1}(\gamma_2 - \gamma_1) + \gamma_1$$

and is equal to γ_j for $\sigma = \sigma_j, \ j = 1, 2$. Hence

$$f(\sigma_j + it) = O(1) \qquad \text{for } |t| \to \infty \qquad (j = 1, 2).$$

Note that Z is an entire function of finite order. Hence the Phragmén–Lindelöf theorem can be applied to f. It follows that f is bounded in the strip $\sigma_1 \leq \sigma \leq \sigma_2$, and hence we have in this range

$$M(\sigma) \leq \frac{\sigma - \sigma_1}{\sigma_2 - \sigma_1}(\gamma_2 - \gamma_1) + \gamma_1 = \frac{\sigma_2 - \sigma}{\sigma_2 - \sigma_1}\gamma_1 + \frac{\sigma - \sigma_1}{\sigma_2 - \sigma_1}\gamma_2.$$

Taking the inf with respect to $\gamma_1, \ \gamma_2$, we obtain in the same range for σ

$$M(\sigma) \leq \frac{\sigma_2 - \sigma}{\sigma_2 - \sigma_1} M(\sigma_1) + \frac{\sigma - \sigma_1}{\sigma_2 - \sigma_1} M(\sigma_2).$$

Hence M is convex.

Obviously $Z(s)$ is bounded in every half-plane $\operatorname{Re} s \geq 1 + \delta$ $(\delta > 0)$. Therefore, $M(\sigma) \leq 0$ for $\sigma > 1$. It follows from the product expansion of Z that $|Z(\sigma + it)|$ $(t \in \mathbb{R})$ is bounded from below by some positive constant if $\sigma > 1$ is sufficiently large. Hence $M(\sigma) = 0$ for σ sufficiently large. Applying the functional equation of Z, we see that $M(\sigma) = |\sigma|$ for $-\sigma$ sufficiently large. Now the convexity of M yields $M(\sigma) \geq 0$ for all $\sigma \in \mathbb{R}$. Hence M is finite everywhere, and M is a non-negative downwards convex function. In particular, M is continuous. Summing up, we have proved (6.7). Another application of the functional equation yields (6.8). Now take $\sigma_1 < \sigma < \sigma_2$, and let $\sigma_2 > 1$. Then we have

$$M(\sigma) \leq \frac{\sigma_2 - \sigma}{\sigma_2 - \sigma_1} M(\sigma_1) \leq M(\sigma_1).$$

Hence M is monotonically decreasing as well. □

The obvious analogue of the Lindelöf hypothesis for the Selberg zeta function is that

$$M(\sigma) = \begin{cases} 0 & \text{for } \sigma \geq 0, \\ |\sigma| & \text{for } \sigma \leq 0. \end{cases} \tag{6.9}$$

This is actually true and will be proved in Corollary 6.3. Adapting the corresponding proof for the Riemann zeta function, which is based on the assumption that the Riemann hypothesis is true, to the present situation, we even can prove a slightly stronger result given in Theorem 6.2. The proof is an elaboration of ideas of Littlewood, see Titchmarsh (1951), pages 282–283.

Theorem 6.2. *For $|t| \to \infty$, the asymptotic relation*

$$\log Z(\sigma + it) = O(|t|^{2(1-\sigma)} \log |t|) \tag{6.10}$$

holds uniformly with respect to $\frac{1}{\log|t|} \le \sigma \le 1$.

Proof. It follows from Corollary 5.4 and from the boundedness of $Z(s)$ in the half-plane $\operatorname{Re} s \ge 2$, that there exists a constant $C > 0$ such that

$$\operatorname{Re}\log Z(u+iv) \le Cv^2 \qquad \text{for all} \quad u > 0,\ |v| \ge 1. \tag{6.11}$$

We want to estimate $\log Z(w)$ for $\operatorname{Re} w \ge \delta > 0$. Apply the Borel-Carathéodory theorem, see Titchmarsh (1939), section 5.5, to the circles with centers $2+it$ and with radii $R = 2 - \frac{\delta}{2}$, $r = 2 - \delta$. This gives in view of (6.11)

$$|\log Z(w)| \le \frac{4(2-\delta)}{\delta}\ C\ (|t|+2)^2 + \frac{8-3\delta}{\delta}\ \log|Z(2+it)|,$$

where $|t| \ge 3$, $|w-(2+it)| \le r$. Hence there exists some constant A, independent of δ, such that

$$|\log Z(u+iv)| \le \frac{A}{\delta} v^2 \qquad \text{for all} \quad u \ge \delta,\ |v| \ge 3. \tag{6.12}$$

Now let $0 < \delta < \sigma < 1$, $a := \delta^{-1}$, $s = \sigma + it$, $|t| \ge 3$. We shall estimate $\log Z(\sigma+it)$ by means of Hadamard's three-circles theorem, see Titchmarsh (1939), section 5.3. Let

$$r_1 = a - 1 - \delta, \qquad r_2 = a - \sigma, \qquad r_3 = a - \delta,$$

and let C_j denote the circle of radius r_j and center $a + it$,

$$M_j := \max_{w \in C_j} |\log Z(w)| \qquad (j = 1,2,3).$$

Then Hadamard's theorem says

$$M_2 \le M_1^{1-\alpha}\ M_3^{\alpha}, \tag{6.13}$$

where

$$\alpha = \frac{\log\frac{r_2}{r_1}}{\log\frac{r_3}{r_1}} = \frac{\log(1+\frac{1+\delta-\sigma}{a-1-\delta})}{\log(1+\frac{1}{a-1-\delta})} = 1 - \sigma + O(\delta) \qquad \text{for } \delta \to +0 \tag{6.14}$$

uniformly with respect to σ. We see from (6.12) that

$$M_3 \le \frac{A}{\delta}\ (|t|+a)^2 \qquad \text{for } |t| - a \ge 3. \tag{6.15}$$

Choosing

$$\beta = \inf_{\{T_0\}} \log N(T_0),$$

we see from (5.19) that for $u > 1$

$$|\log Z(u+iv)| \le \frac{1}{\beta} \sum_{\{T\}\,\text{lox.}} \frac{\log N(T_0)}{m(T_0)\ |a(T) - a(T)^{-1}|^2}\ N(T)^{-u} \le \frac{B}{u-1}$$

for some constant $B > 0$. The last step uses the trace formula (4.10). Choosing the absolute constant $A > 0$ sufficiently large, we have

$$|\log Z(w)| \le \frac{A}{\delta} \qquad \text{for } \operatorname{Re} w \ge 1+\delta \tag{6.16}$$

and hence

$$M_1 \le \frac{A}{\delta}. \tag{6.17}$$

Summing up, we see from (6.13)–(6.17) that

$$|\log Z(\sigma+it)| \;\le\; M_2 \;\le\; M_1^{1-\alpha} M_3^{\alpha} \;\le\; \frac{A}{\delta}\,(|t|+a)^{2(1-\sigma+O(\delta))} \tag{6.18}$$

where A is some absolute constant and where the O-constant does not depend on σ, t. Hence we are free to choose $a = \delta^{-1} = \log|t|$, where $|t|$ is sufficiently large. Then

$$(|t|+a)^{O(\delta)} = O(1) \qquad \text{for } |t| \to \infty$$

and our Theorem follows from (6.18). □

Corollary 6.3. *The analogue of the Lindelöf hypothesis holds for the Selberg zeta function, that is*

$$M(\sigma) = \begin{cases} 0 & \text{for } \sigma \ge 0, \\ |\sigma| & \text{for } \sigma \le 0. \end{cases} \tag{6.19}$$

Proof. Formula (6.19) holds for $|\sigma| \ge 1$ by Lemma 6.1, and we deduce from (6.10) and the continuity of M that $M(\sigma) = 0$ for $0 \le \sigma \le 1$ as well. Using the convexity of M, we conclude from these equalities that (6.19) holds unrestrictedly. □

5.7 The Prime Geodesic Theorem

The primitive elements of Γ play a rôle comparable to that of the prime numbers. Hence we call the theorem describing the asymptotic behaviour of

$$\pi_{00}(x) := \left| \left\{ \{T_0\}_\Gamma \;\middle|\; \begin{array}{l} T_0 \text{ primitive hyperbolic or loxodromic,} \\ N(T_0) \le x \end{array} \right\} \right| \tag{7.1}$$

the prime geodesic theorem.

The problems met in the present situation closely parallel those in classical analytic number theory, in so far that we have quite natural analogues of the number theoretic data. Analytic number theory takes a special interest in the function

(7.2) $$\pi(x) := |\{\, p \mid p \le x,\ p \text{ a prime } \}|$$

counting the number of primes less than or equal to x. The asymptotic behaviour of the Chebyshev functions

(7.3) $$\vartheta(x) := \sum_{p \le x} \log p,$$

and

(7.4) $$\psi(x) := \sum_{n \le x} \Lambda(n),$$

where

(7.5) $$\Lambda(n) = \begin{cases} \log p & \text{for } n = p^k,\ p \text{ a prime}, \\ 0 & \text{otherwise}, \end{cases}$$

is closely tied up with the behaviour of $\pi(x)$ as $x \to \infty$. So, for example it is well known that the prime number theorem

(7.6) $$\pi(x) \sim \frac{x}{\log x} \qquad \text{for } x \to \infty$$

and both the relations

(7.7) $$\psi(x) \sim x \qquad \text{for } x \to \infty$$

and

(7.8) $$\vartheta(x) \sim x \qquad \text{for } x \to \infty$$

are equivalent. This means that each one of the relations (7.6)–(7.8) yields the other two in an elementary way. We shall meet an analogous situation for our function π_{00}.

We look for natural analogues of ψ and ϑ. Recall that the numbers $\Lambda(n)$ appear in the logarithmic derivative of the Riemann ζ-function

(7.9) $$-\frac{\zeta'}{\zeta}(s) = \sum_{n=1}^{\infty} \Lambda(n)\, n^{-s}.$$

Comparing (7.9) and (4.20) it is natural to define

(7.10) $$\Lambda(T) := \frac{\log N(T_0)}{m(T)\, |a(T) - a(T)^{-1}|^2}$$

for T a hyperbolic or loxodromic element of Γ. Then we have

(7.11) $$\frac{Z'}{Z}(s) = \sum_{\{T\} \text{ lox.}} \Lambda(T)\, N(T)^{-s}.$$

We take

$$\Psi(x) := \sum_{\substack{\{T\}\text{ lox.}\\ N(T)\le x}} \Lambda(T) \tag{7.12}$$

as a substitute for the Chebyshev function ψ.

We proceed to define a substitute Θ for ϑ. Note that ϑ and ψ are only slightly different. We want this to be true for Θ and Ψ as well. The denominator in (7.10) is asymptotically equal to $m(T)N(T)$ if $\{T\}$ runs through the Γ-conjugacy classes of the hyperbolic or loxodromic elements of Γ. There exist only finitely many primitive classes $\{T_0\}$ such that $m(T_0) \neq 1$. Hence we define

$$\Theta(x) := \sum_{\substack{\{T_0\}\text{ lox.}\\ N(T_0)\le x}} \frac{\log N(T_0)}{N(T_0)}, \tag{7.13}$$

where the summation extends over all Γ-conjugacy classes $\{T_0\}$ of primitive hyperbolic or loxodromic elements of Γ with norm $N(T_0) \le x$.

Lemma 7.1. $\Theta(x) - \Psi(x) = O(\log x)$ *for* $x \to \infty$.

Proof. Splitting off the contribution of the primitive elements to the right-hand side of (7.12), we infer that there exists a constant $C > 0$ such that

$$\begin{aligned} &\Big|\Psi(x) - \sum_{\substack{\{T_0\}\text{ lox.}\\ N(T_0)\le x}} \Lambda(T_0)\Big| \\ &\le C\left(\sum_{\substack{\{T_0\}\text{ lox.}\\ N(T_0)^2\le x}} \frac{\log N(T_0)}{N(T_0)^2} + \sum_{\substack{\{T_0\}\text{ lox.}\\ N(T_0)^3\le x}} \frac{\log N(T_0)}{N(T_0)^3} + \dots \right). \end{aligned} \tag{7.14}$$

In fact, we may take

$$C := \max_{\{T\}\text{ lox.}} \{\, |1 - a(T)^{-2}|^{-1} \,\}.$$

The finite sum of sums on the right-hand side of (7.14) has at most $O(\log x)$ terms. We know from (4.18) that the biggest term in this sum satisfies

$$\sum_{\substack{\{T_0\}\text{ lox.}\\ N(T_0)\le x}} \frac{\log N(T_0)}{N(T_0)^2} = O(\log x) \qquad \text{for } x \to \infty. \tag{7.15}$$

Moreover, we have for all $k \ge 3$

$$\sum_{\substack{\{T_0\}\text{ lox.}\\ N(T_0)^k\le x}} \frac{\log N(T_0)}{N(T_0)^k} \le \sum_{\{T_0\}\text{ lox.}} \frac{\log N(T_0)}{N(T_0)^3} < \infty.$$

Hence

$$\Psi(x) - \sum_{\substack{\{T_0\} \text{ lox.} \\ N(T_0) \leq x}} \Lambda(T_0) = O(\log x) \qquad \text{for } x \to \infty. \tag{7.16}$$

Recall that there exist only finitely many primitive Γ-conjugacy classes $\{T_0\}$ such that $m(T_0) \neq 1$. Thus it suffices to prove that

$$\sum_{\substack{\{T_0\} \text{ lox.} \\ N(T_0) \leq x}} \frac{\log N(T_0)}{|a(T_0) - a(T_0)^{-1}|^2} - \Theta(x) = O(\log x) \qquad \text{for } x \to \infty. \tag{7.17}$$

This is done as follows. Note that there exists a constant $C_1 > 0$ such that

$$||a(T_0) - a(T_0)^{-1}|^{-2} - N(T_0)^{-1}| = N(T_0)^{-1} ||1 - a(T_0)^{-2}|^{-2} - 1| \leq C_1 \, N(T_0)^{-2}$$

for all primitive hyperbolic or loxodromic elements of Γ. This yields by (7.15)

$$\left| \sum_{\substack{\{T_0\} \text{ lox.} \\ N(T_0) \leq x}} \frac{\log N(T_0)}{|a(T_0) - a(T_0)^{-1}|^2} - \Theta(x) \right| \leq C_1 \sum_{\substack{\{T_0\} \text{ lox.} \\ N(T_0) \leq x}} \frac{\log N(T_0)}{N(T_0)^2} = O(\log x) \tag{7.18}$$

for $x \to +\infty$. This proves our assertion. □

Our proof of Lemma 7.1 is not completely elementary since it rests upon (7.15) which was drawn from the formula for the trace of the iterated resolvent kernel. However, the following completely elementary estimate yields a result which is only slightly worse than (7.15). Apply partial summation

$$\sum_{\substack{\{T_0\} \text{ lox.} \\ N(T_0) \leq x}} \frac{\log N(T_0)}{N(T_0)^2} = \left[\frac{\log t}{t^2} \pi_{00}(t) \right]_{1+\delta}^{x} + \int_{1+\delta}^{x} \pi_{00}(t) \, \frac{2 \log t - 1}{t^3} \, dt,$$

where $\delta > 0$ is sufficiently small. Since we know from Lemma 2.6.3 that

$$\pi_{00}(x) = O(x^2) \qquad \text{for } x \to \infty,$$

we infer that

$$\sum_{\substack{\{T_0\} \text{ lox.} \\ N(T_0) \leq x}} \frac{\log N(T_0)}{N(T_0)^2} = O(\log^2 x) \qquad \text{for } x \to \infty. \tag{7.19}$$

Using (7.19) in place of (7.15), we deduce that

$$\Theta(x) - \Psi(x) = O(\log^2 x) \qquad \text{for } x \to \infty. \tag{7.20}$$

This is now proved by elementary means only.

Theorem 7.2. *The following asymptotic relations hold:*

$$\Psi(x) \sim x \qquad \text{for } x \to \infty, \tag{7.21}$$

$$\Theta(x) \sim x \qquad \text{for } x \to \infty. \tag{7.22}$$

Proof. It is clear from (4.10) and (4.20) that

$$\frac{Z'}{Z}(s) - \frac{1}{s-1}$$

converges to a finite limit for $\operatorname{Re} s \to 1$. This convergence is uniform in every interval $|\operatorname{Im} s| \leq \alpha$. Hence the hypotheses of Ikehara's theorem are satisfied, see Lang (1968), Chapter 15. This fact immediately implies that

$$\Psi(x) \sim x \qquad \text{for } x \to \infty.$$

The asymptotic relation (7.22) is now obvious from (7.21) and from Lemma 7.1. □

Theorem 7.3 (Prime Geodesic Theorem).

$$\pi_{00}(x) \sim \frac{x^2}{2\log x} \qquad \text{for } x \to +\infty. \tag{7.23}$$

Proof. Choose $\alpha > 1$ so small that

$$\Theta(t) = 0 \qquad \text{for } 1 \leq t \leq \alpha,$$

and apply partial summation

$$\pi_{00}(x) = \int_\alpha^x \frac{t}{\log t}\, d\Theta(t) = \left[\frac{t\,\Theta(t)}{\log t}\right]_\alpha^x - \int_\alpha^x \frac{\Theta(t)}{\log t}\, dt + \int_\alpha^x \frac{\Theta(t)}{\log^2 t}\, dt. \tag{7.24}$$

The asymptotic law (7.22) yields

$$\begin{aligned} \int_\alpha^x \frac{\Theta(t)}{\log t}\, dt &\sim \int_\alpha^x \frac{t}{\log t}\, dt = \int_{\alpha^2}^{x^2} \frac{du}{\log u} \sim \ell i(x^2) \\ &\sim \frac{x^2}{\log x^2} = \frac{x^2}{2\log x} \qquad \text{for } x \to \infty. \end{aligned} \tag{7.25}$$

Here the integral logarithm is defined for $x > 1$ by

$$\ell i(x) := \lim_{\delta \to 0} \left(\int_0^{1-\delta} \frac{dt}{\log t} + \int_{1+\delta}^x \frac{dt}{\log t} \right).$$

We use the well-known fact that

$$\ell i x \sim \frac{x}{\log x} \qquad \text{for } x \to \infty.$$

On the other hand, we find for the second integral on the right-hand side of (7.24)

$$\begin{aligned}
\int_\alpha^x \frac{\Theta(t)}{\log^2 t}\,dt &\sim \int_\alpha^x \frac{t}{\log^2 t} dt = 2\int_{\alpha^2}^{x^2} \frac{du}{\log^2 u} \\
&= 2\int_{\alpha^2}^{x} \frac{du}{\log^2 u} + 2\int_{x}^{x^2} \frac{du}{\log^2 u} \\
&= O(\ell i(x)) + O\left(\frac{1}{\log x}\int_x^{x^2} \frac{du}{\log u}\right) \\
&= O(\ell i(x)) + O\left(\frac{\ell i(x^2)}{\log x}\right) = O\left(\frac{x^2}{\log^2 x}\right).
\end{aligned} \tag{7.26}$$

Inserting (7.22), (7.25), (7.26) in (7.24), we obtain (7.23). This proves our assertion. □

Lemma 7.4. *The counting function π_0 of Lemma 2.6.3 satisfies*

$$\pi_0(x) = \pi_{00}(x) + O\left(\frac{x}{\log x}\right) \qquad \textit{for } x \to \infty. \tag{7.27}$$

Proof. We have

$$\pi_0(x) = \pi_{00}(x) + \pi_{00}\left(x^{\frac{1}{2}}\right) + \ldots + \pi_{00}\left(x^{\frac{1}{n(x)}}\right)$$

where $n(x)+1$ is the smallest integer such that $\pi_{00}(x^{\frac{1}{n(x)+1}}) = 0$. Then

$$n(x) = O(\log x), \qquad \pi_{00}\left(x^{\frac{1}{2}}\right) = O\left(\frac{x}{\log x}\right),$$

and for $3 \le k \le n(x)$

$$\pi_{00}\left(x^{\frac{1}{k}}\right) \le \pi_{00}\left(x^{\frac{1}{3}}\right) = O\left(\frac{x^{\frac{2}{3}}}{\log x}\right).$$

Hence

$$\pi_0(x) = \pi_{00}(x) + O\left(\frac{x}{\log x}\right) \qquad \text{for } x \to \infty$$

as was to be shown. □

Corollary 7.5. *The counting function π_0 satisfies*

(7.28)
$$\pi_0(x) \sim \frac{x^2}{2\,\log x} \qquad \textit{for } x \to \infty.$$

Proof. The result is obvious from (7.27) and (7.23). □

The proof of the prime geodesic theorem (7.23) imitates Wiener's proof of the prime number theorem for $\pi(x)$. But in the present situation we are more fortunate than in classical analytic number theory because we have rather good qualitative knowledge of the zeros of the corresponding zeta function. Hence one may sharpen (7.23) considerably and give a good error term; compare Brüdern (1995). We leave the details to the reader.

5.8 Notes and Remarks

The Selberg zeta function was introduced in Selberg's pioneering work Selberg (1956). Selberg communicates the properties of his zeta function in the case of the hyperbolic plane, and there are brief indications concerning the 3-dimensional case, see page 79. The two-dimensional case is now easily accessible in the literature, see Hejhal (1976a), Venkov (1979c), Elstrodt (1981). These papers quote extensive lists of references. The Selberg zeta function for discrete subgroups $\Gamma < \mathbf{PSL}(2,\mathbb{C})$ without elliptic elements such that $\Gamma\backslash\mathbb{H}$ is compact was also introduced by Vishik (1976) in different notations. Vishik's paper contains no proofs, and there seems to be an error in his version of the functional equation, see Vishik (1976), page 256, Theorem 1. For a discussion of the Selberg zeta function for compact space forms of symmetric spaces of rank one see Gangolli (1977b), for the non-compact case see Gangolli and Warner (1980).

Our introduction of the Selberg zeta function follows an approach suggested by Elstrodt (1981), section 10. This approach was developed in detail by Fischer (1987). A noteworthy feature of our discussion is that we allow Γ to contain elliptic elements. It turns out that the results are essentially the same as for groups without fixed points in $\mathbb{H}$.

The version (5.24) of Weyl's asymptotic law obviously agrees with the known asymptotic result

$$|\{\, n \mid \lambda_n \le T \,\}| = \frac{\mathrm{vol}(M)}{(4\pi)^{\frac{m}{2}}\Gamma(\frac{m}{2}+1)}\, T^{\frac{m}{2}} + O(T^{\frac{m-1}{2}}) \qquad \text{for } T \to \infty$$

which holds for the eigenvalues of $-\Delta$ on an arbitrary compact Riemannian manifold of dimension m, see Avakumović (1956), Hörmander (1968). It is also known that this estimate of the error term is best possible in general, see Avakumović (1956), Hörmander (1968). But for the special types of eigenvalue problems on compact quotients considered here, it is possible to improve on the above asymptotics slightly. This was done by Hejhal (1976a), page 119 et seq. and Randol (1978) in the case of the corresponding eigenvalue problem for fixed point free cocompact subgroups of $\mathbf{PSL}(2, \mathbb{R})$ acting on the hyperbolic plane. They proved that a term $\log T$ can be introduced in the denominator of the error term. The same improvement actually is possible in much more generality as shown in Bérard (1976), (1977), Kolk (1977a), (1977b) and Duistermaat, Kolk and Varadarajan (1979), page 89, Theorem 9.1.

The constants α and β in Corollary 5.7 may possibly be expressible in terms of the eigenvalue spectrum or in terms of geometric data in a more direct way than we have done. We do not know whether this is in fact the case. For the case of the hyperbolic plane see Fischer (1987).

Chapter 6. Spectral Theory of the Laplace Operator for Cofinite Groups

This chapter is a continuation of Chapter 4. Having established there some fundamental facts about the spectral theory of the Laplace operator on the Hilbert spaces $L^2(\Gamma\backslash\mathbb{H})$ for discrete subgroups $\Gamma < \mathbf{PSL}(2,\mathbb{C})$ and having treated the case of cocompact groups in Chapter 5, we turn here to the finer properties of the Laplace operator for groups Γ which are of finite covolume but not cocompact. We already know that Δ defined on an appropriate domain in $L^2(\Gamma\backslash\mathbb{H})$ has a unique self-adjoint extension $\tilde{\Delta}$. If Γ is not cocompact the spectrum of $\tilde{\Delta}$ is no longer purely discrete. We aim here for an explicit description of the non-discrete part of the spectrum of $\tilde{\Delta}$ and also for an explicit decomposition formula for L^2-functions. We shall give two versions of this theory in Section 3. The first, more in the spirit of Selberg (1989b), gives $\tilde{\Delta}$ as a certain multiplication operator. The second describes the eigenpackets for $\tilde{\Delta}$ following Roelcke (1956a), (1966), (1967). Section 2 contains a quick treatment of the theory of eigenpackets. It is included here since nowadays textbooks treat the spectral theory of unbounded operators usually via spectral families.

An important ingredient to this chapter is the meromorphic continuation of the Eisenstein series defined in Chapter 3, Section 2. We establish it in Section 1, using an adaptation of a method of Colin de Verdière (1981), (1982), (1983). In Section 4 we study the spectral expansions of certain integral kernels. Finally in Section 5 we derive the Selberg trace formula. We use in principle the approach of Selberg (1989b). Certain ideas in the proofs of Sections 4,5 are taken from Cohen, Sarnak (1979).

6.1 Meromorphic Continuation of the Eisenstein Series

Let $\Gamma < \mathbf{PSL}(2,\mathbb{C})$ be a non-cocompact discrete subgroup which has finite covolume. The Eisenstein series were introduced in Section 2 of Chapter 3 as certain Poincaré series depending on the parameter $s \in \mathbb{C}$. In the range of their absolute convergence ($\operatorname{Re} s > 1$) they define Γ-invariant functions depending holomorphically on the parameter s. We shall prove in this section that the Eisenstein series can be continued meromorphically to all of $\mathbb{C}$. We shall also give some information on the poles and on the growth behaviour of the continued functions. We thank Leonid Parnovski for advice on the subject.

To formulate our results precisely, we choose once and for all in this chapter $B_1, \ldots, B_h \in \mathbf{PSL}(2, \mathbb{C})$ so that

$$\eta_1 = B_1^{-1}\infty, \ldots, \eta_h = B_h^{-1}\infty \in \mathbb{P}^1\mathbb{C} \tag{1.1}$$

are representatives for the Γ-classes of cusps of Γ. Recalling our notation from Chapter 3, Section 2 we put for $\nu = 1, \ldots, h$ and $P \in \mathbb{H}$

$$E_{B_\nu}(P, s) = \sum_{M \in \Gamma'_{\eta_\nu} \backslash \Gamma} r(B_\nu M P)^{1+s}. \tag{1.2}$$

This series was proved to converge absolutely and locally uniformly for $\operatorname{Re} s > 1$. We furthermore put

$$E_\nu(P, s) := \frac{1}{[\Gamma_{\eta_\nu} : \Gamma'_{\eta_\nu}]} E_{B_\nu}(P, s). \tag{1.3}$$

As proved in Chapter 3, Section 4 the $E_\nu(P, s)$ have Fourier expansions in the cusps of the form

$$E_\nu(B_\nu^{-1}P, s) = r^{1+s} + \phi_{\nu\nu}(s)\, r^{1-s} + \ldots \tag{1.4}$$

for $\nu = 1, \ldots, h$ and in case $\nu \neq \mu$

$$E_\nu(B_\mu^{-1}P, s) = \phi_{\nu\mu}(s)\, r^{1-s} + \ldots \tag{1.5}$$

The functions $\phi_{\nu\mu}(s)$ are certain Dirichlet series described in Theorem 3.4.1.

Definition 1.1. Let $\Gamma < \mathbf{PSL}(2, \mathbb{C}$ be a cofinite subgroup. Using the notation (1.3), (1.4), (1.5) we define

$$\mathcal{E}(P, s) := \begin{pmatrix} E_1(P, s) \\ \vdots \\ E_h(P, s) \end{pmatrix}, \tag{1.6}$$

$$\Phi(s) := (\phi_{\nu\mu}(s)) \tag{1.7}$$

where ν is the row index and μ the column index. The matrix (1.7) is called the scattering matrix for Γ.

We shall prove the following results on the meromorphic continuation of the matrix valued functions $\Phi(s)$, $\mathcal{E}(P, s)$.

Theorem 1.2. *Let $\Gamma < \mathbf{PSL}(2, \mathbb{C})$ be a cofinite non-cocompact subgroup. Then both $\Phi(s)$ and $\mathcal{E}(P, s)$ have meromorphic continuations to all of $\mathbb{C}$ in the following sense: There is a holomorphic function $g : \mathbb{C} \to \mathbb{C}$ with $g \neq 0$*

such that for every $\nu = 1, ..., h$ *the product* $g(s)E_\nu(P,s)$ *can be continued to a function on* $\mathbb{H} \times \mathbb{C}$ *which is real analytic in* P *and holomorphic in* s. *The continued functions satisfy*

$$(1.8) \qquad \mathcal{E}(P,-s) = \Phi(-s)\ \mathcal{E}(P,s), \qquad \Phi(s)\ \Phi(-s) = I.$$

The function $\det \Phi(s)$ *is a meromorphic function of order* ≤ 4. *The continued components* $E_\nu(P,s)$ *of* $\mathcal{E}(P,s)$ *satisfy*

$$(1.9) \qquad (-\Delta - (1-s^2))\ E_\nu(P,s) = 0$$

if s *is not a pole of* $E_\nu(P,s)$.

A meromorphic function f on $\mathbb{C}$ is defined to be of order $\leq n$ if it can be represented as a quotient $f = g/h$ of two entire functions g, h which satisfy

$$\max_{|z|=r} \{\, |g(z)|\, \},\ \max_{|z|=r} \{\, |h(z)|\, \} \leq \kappa\, e^{\kappa_1 r^n}$$

for all $r \geq 0$ and some constants κ, κ_1. See Levin (1996) for explanations.

The existence of a meromorphic continuation of $\mathcal{E}(P,s)$ of the type indicated above implies that for fixed $P \in \mathbb{H}$ the holomorphic function $s \to \mathcal{E}(P,s)$ $(\operatorname{Re} s > 1)$ can be meromorphically continued and that the poles of all of these continuations do not depend on P. To formulate our result precisely we would have to define a Banach space of functions on $\mathbb{H}$ which contains all the $E(\cdot, s)$ for $\operatorname{Re} s > 1$ and perform the continuation in this space. It will be clear from the proof of Theorem 1.2 that this is possible, but we leave this last step in precision to the reader. Additionally we remark that weak holomorphicity is often equivalent to the strong version, see Reed, Simon (1972), Chapter 6 for comments.

We shall give here an adapted version of a proof due to Colin de Verdière (1981), (1982), (1983). Results like Theorem 1.2, but for discrete subgroups of $\mathbf{PSL}(2, \mathbb{R})$ were first proved by Selberg, see Selberg (1989b) for comments. The problem is also treated in Roelcke (1956b), (1966), (1967) and Faddeev (1967). Detailed expositions are given in the books of Neunhöffer (1973), Kubota (1973), and Hejhal (1976a). The ultimate result on the meromorphic continuation of much more general Eisenstein series is due to Langlands (1976).

Our proof starts off by constructing a self-adjoint extension of the Laplace operator Δ restricted to a certain subspace of $L^2(\Gamma \backslash \mathbb{H})$. This extension will have the advantage of having a Hilbert–Schmidt resolvent. The construction is due to Lax and Phillips (1976), who consider a similar operator in the 2-dimensional situation.

To avoid notational difficulties we shall give the proof in case Γ has only one class of cusps. We assume that this class is represented by $\infty \in \mathbb{P}^1\mathbb{C}$. We then choose a fundamental domain $\mathcal{F}$ of the kind described in Proposition 2.3.9 for Γ. Let $\mathcal{P} \subset \mathbb{C}$ be a fundamental polygon for the action of Γ_∞ on

$\mathbb{C} = \mathbb{P}^1\mathbb{C}\setminus\{\infty\}$ and $Y > 0$ from Proposition 2.3.9. The fundamental domain $\mathcal{F}$ is then the union $\mathcal{F} = \mathcal{F}_0 \cup \mathcal{F}_\infty(Y)$ of a compact polyhedron $\mathcal{F}_0$ and the cusp sector $\mathcal{F}_\infty(Y) = \{\ z + rj \in \mathbb{H} \ \mid\ z \in \mathcal{P},\ r > Y\ \}$. Following (1.3) we write

$$E(P, s) := \frac{1}{[\Gamma_\infty : \Gamma'_\infty]}\ E_I(P, s). \tag{1.10}$$

We also leave off the index in the function ϕ arising in (1.4). We furthermore put

$$\Omega := \{\ s \in \mathbb{C}\ \mid\ \operatorname{Re} s > 0,\ 1 - s^2 \notin \sigma(-\tilde{\Delta})\ \}. \tag{1.11}$$

By Section 1 of Chapter 4 the operator $-\tilde{\Delta}$ is self-adjoint and positive, hence

$$\{s \in \mathbb{C}\ \mid\ \operatorname{Re} s > 0,\ s \notin [0, 1]\ \} \subset \Omega.$$

The particular structure of $\mathcal{F}$ may be used to define the following Γ-invariant functions on $\mathbb{H}$.

Definition 1.3. Let $Y < Y' < Y''$ and $h : [Y, \infty[\to \mathbb{R}$ be a C^∞-function which satisfies $h(y) = 1$ for $Y'' \le y$ and $h(y) = 0$ for $y \le Y'$. For $s \in \mathbb{C},\ P \in \mathcal{F}$ we define

$$\tilde{h}_s(P) := \begin{cases} h(r)\ r^{1+s} & \textit{if } P = z + rj \in \mathcal{F}_\infty(Y), \\ 0 & \textit{if } P \in \mathcal{F}_0. \end{cases} \tag{1.12}$$

By Theorems 2.3.2–2.3.4 the function $\tilde{h}_s$ extends to a Γ-invariant C^∞-function

$$\tilde{h}_s : \mathbb{H} \to \mathbb{C}. \tag{1.13}$$

For $s \in \mathbb{C},\ P \in \mathbb{H}$ we define

$$H(P, s) := (-\Delta - (1 - s^2))\ \tilde{h}_s(P), \tag{1.14}$$

and for $s \in \Omega$ we put

$$\tilde{E}(P, s) := \tilde{h}_s(P) - (-\tilde{\Delta} - (1 - s^2))^{-1} H(P, s). \tag{1.15}$$

Note that $H(P, s)$ is a Γ-invariant C^∞-function in the variable P. Being equal to 0 for all $P = (z, r)$ with $r > Y''$ it is contained in $L^2(\Gamma \setminus \mathbb{H})$.

Lemma 1.4. *The function $H(P, s)$ depends holomorphically on $s \in \Omega$ and as a function of $P \in \mathbb{H}$ it is of compact support when restricted to $\mathcal{F}$. The function $\tilde{E}(P, s)$ also depends holomorphically on $s \in \Omega$. It is the unique Γ-invariant function which satisfies the conditions*

$$(-\Delta - (1 - s^2))\ \tilde{E}(P, s) = 0, \qquad \tilde{E}(P, s) - \tilde{h}_s(P) \in L^2(\Gamma \setminus \mathbb{H}). \tag{1.16}$$

For $\operatorname{Re} s > 1$ *we have* $\tilde{E}(P,s) = E(P,s)$ *for all* $P \in \mathbb{H}$.

Lemma 1.4 is evident from the construction of $\tilde{E}$ and the definition of Ω. From what we have proved about the Eisenstein series in Chapter 3 we infer that $E(P,s)$ satisfies (1.16) for $\operatorname{Re} s > 1$.

We add some remarks on distributions which will be of importance in our arguments. We define as before $\pi_\Gamma : \mathbb{H} \to \Gamma \setminus \mathbb{H}$ to be the projection map. For our purposes here it is convenient to slightly vary Definition 4.1.1. We put:

$$\mathcal{D}^\infty := \left\{ \varphi : \mathcal{F} \to \mathbb{R} \;\middle|\; \begin{array}{l} \varphi \text{ is the restriction to } \mathcal{F} \text{ of a} \\ \Gamma\text{-invariant } C^\infty\text{-function on } \mathbb{H}, \\ \pi_\Gamma(\operatorname{supp}(\varphi)) \text{ is compact in } \Gamma \setminus \mathbb{H} \end{array} \right\}. \tag{1.17}$$

On $\mathcal{D}^\infty$ we consider the usual topology of uniform convergence of all derivatives of φ on compact subsets of $\mathcal{F}$, see Rudin (1973). The corresponding space of distributions, that is continuous, complex valued linear functionals on $\mathcal{D}^\infty$ is denoted by $\mathcal{D}'$. If Γ happens to be torsion-free then $\Gamma \setminus \mathbb{H}$ is a Riemannian manifold and $\mathcal{D}^\infty$, $\mathcal{D}'$ can be identified with the spaces of C_c^∞-functions and distributions on $\Gamma \setminus \mathbb{H}$.

We write $L^1_{\text{loc}}(\Gamma \setminus \mathbb{H})$ for the space of measurable, Γ-invariant functions which are locally integrable modulo functions which are 0 almost everywhere. Note that every $f \in L^2(\Gamma \setminus \mathbb{H})$ is also contained in $L^1_{\text{loc}}(\Gamma \setminus \mathbb{H})$. An element $f \in L^1_{\text{loc}}(\Gamma \setminus \mathbb{H})$ gives rise to an element $f \in \mathcal{D}'$ by the formula

$$f(\varphi) := \int_{\mathcal{F}} f(P)\,\varphi(P)\,dv(P) \qquad (\varphi \in \mathcal{D}^\infty). \tag{1.18}$$

A distribution is said to be *contained in* $L^2(\Gamma \setminus \mathbb{H})$ if it arises through formula (1.18) for some $f \in L^2(\Gamma \setminus \mathbb{H})$. The distribution U is said to be *represented by* $f \in L^2(\Gamma \setminus \mathbb{H})$ if $U(\varphi) = f(\varphi)$ for all $\varphi \in \mathcal{D}^\infty$. When we apply differential operators to distributions some care has to be taken since we work with the hyperbolic volume measure. We define for $U \in \mathcal{D}'$, $i \in \mathbb{N} \cup \{0\}$ and $\varphi \in \mathcal{D}^\infty$

$$\begin{gathered} \ell_x^{[i]} U(\varphi) := (-1)^i\, U(r^i \frac{\partial^i}{\partial x^i} \varphi), \qquad \ell_y^{[i]} U(\varphi) := (-1)^i\; U(r^i \frac{\partial^i}{\partial y^i} \varphi), \\ \ell_r^{[i]} U(\varphi) := (-1)^i\; U(r^3 \frac{\partial^i}{\partial r^i} (r^{i-3} \varphi)). \end{gathered} \tag{1.19}$$

Here x, y, r are our usual coordinates on $\mathbb{H}$. The formulas (1.19) give new distributions $\ell_x^{[i]} U$, $\ell_y^{[i]} U$, $\ell_r^{[i]} U$. Let $f \in L^2(\Gamma \setminus \mathbb{H})$ be a C^∞-function and t be one of x, y, r then integration by parts using Stokes' theorem shows that the distribution $\ell_t^{[i]} f$ is represented by $r^i \frac{\partial^i}{\partial t^i} f$. If we apply one of our operators $\ell_t^{[i]}$ to a function $f \in L^2(\Gamma \setminus \mathbb{H})$ we understand applying it to the corresponding distribution. We define

$$\Delta U := (\ell_x^{[2]} + \ell_y^{[2]} + \ell_r^{[2]} - \ell_r^{[1]})\, U \qquad (U \in \mathcal{D}'). \tag{1.20}$$

A small computation shows that $(\Delta U)(\varphi) = U(\Delta\varphi)$ for all $\varphi \in \mathcal{D}^\infty$. We furthermore define for $f \in L^2(\Gamma \setminus \mathbb{H})$

$$\mathbf{grad}(f) \in L^2(\Gamma \setminus \mathbb{H}) \qquad \text{iff} \qquad \ell_x^{[1]} f,\ \ell_y^{[1]} f,\ \ell_r^{[1]} f \in L^2(\Gamma \setminus \mathbb{H}). \tag{1.21}$$

Here $\mathbf{grad}(f)$ is the hyperbolic gradient of f mentioned in Chapter 4, Section 1. If f is also a C^2-function then (1.21) amounts to our condition $\|\mathbf{grad}(f)\|^2 \in L^2(\Gamma \setminus \mathbb{H})$ from Chapter 4, Section 1. If $f, g \in L^2(\Gamma \setminus \mathbb{H})$ are elements with $\mathbf{grad}(f)$, $\mathbf{grad}(g) \in L^2(\Gamma \setminus \mathbb{H})$ we may define

$$Q(f,g) := \int_{\mathcal{F}} \left(\ell_x^{[1]} f \cdot \overline{\ell_x^{[1]} g} + \ell_y^{[1]} f \cdot \overline{\ell_y^{[1]} g} + \ell_r^{[1]} f \cdot \overline{\ell_r^{[1]} g} \right) dv. \tag{1.22}$$

The product of the distributions occuring under the integral stands for the product of the L^2-functions representing them. If $g = \varphi$ happens to be in $\mathcal{D}^\infty$ then again a computation transforms (1.22) into

$$Q(f,\varphi) = \int_{\mathcal{F}} f\,(-\Delta\varphi)\, dv = (-\Delta f)(\varphi). \tag{1.23}$$

For $f \in L^2(\Gamma \setminus \mathbb{H})$ the zeroth Fourier coefficient is defined by

$$f_0(r) := \frac{1}{|\mathcal{P}|} \int_{\mathcal{P}} f(x,y,r)\, dx\, dy \qquad (r \in]Y,\infty[). \tag{1.24}$$

This definition is justified: since f is contained in $L^1_{\text{loc}}(\Gamma \setminus \mathbb{H})$ formula (1.24) defines f_0 almost everywhere and Fubini's theorem tells us that $f_0 \in L^1_{\text{loc}}(]Y,\infty[, r^{-3}dr)$. For any function $\psi :]Y,\infty[\to \mathbb{C}$ we define $\tilde{\psi} : \mathcal{F} \to \mathbb{C}$ by

$$\tilde{\psi}(P) := \begin{cases} \psi(r) & \text{if } P = z + rj \in \mathcal{F}_\infty(Y), \\ 0 & \text{if } P \in \mathcal{F}_0 \end{cases} \tag{1.25}$$

and extend $\tilde{\psi}$ to all of $\mathbb{H}$ by Γ-automorphy. Given $f \in L^2(\Gamma \setminus \mathbb{H})$ it is easy to see that $\tilde{f}_0 \in L^2(\Gamma \setminus \mathbb{H})$ and

$$\int_{\mathcal{F}} \tilde{\psi}\, \bar{f}\, dv = \int_{\mathcal{F}} \tilde{\psi}\, \overline{\tilde{f}_0}\, dv = \int_Y^\infty \psi\, \bar{f}_0\, \frac{dr}{r^3}. \tag{1.26}$$

Corresponding to the projection map $\pi : \mathcal{F}_\infty(Y) \to]Y,\infty[,\ \pi(x,y,r) = r$ there are two maps between spaces of distributions

$$\pi_* : \mathcal{D}' \to \mathcal{D}'(]Y,\infty[), \qquad \pi^* : \mathcal{D}'(]Y,\infty[) \to \mathcal{D}' \tag{1.27}$$

defined by

$$\pi_*(U)(\psi) := U(\tilde{\psi}) \qquad \text{for } U \in \mathcal{D}',\ \psi \in \mathcal{D}(]Y,\infty[),$$

$$\pi^*(u)(\varphi) := u(\varphi_0) \qquad \text{for } u \in \mathcal{D}'(]Y,\infty[),\ \varphi \in \mathcal{D}^\infty.$$

Here $\mathcal{D}(]Y,\infty[)$ is the space of real valued C^∞-functions on $]Y,\infty[$ of compact support whereas $\mathcal{D}'(]Y,\infty[)$ consists of the complex valued distributions on $\mathcal{D}(]Y,\infty[)$. We have $\pi_* \circ \pi^* = \mathrm{id}_{\mathcal{D}'(]Y,\infty[)}$, $\pi^*(\Delta_0 u) = \Delta\pi^*(u)$ and $\pi_*(\Delta U) = \Delta_0\pi_*(U)$, where Δ_0 is defined by

$$(1.28) \qquad (\Delta_0 u)(\psi) := u\left(\left(r^2\frac{d^2}{dr^2} - r\frac{d}{dr}\right)\psi\right).$$

We associate to $h \in L^1_{\text{loc}}(]Y,\infty[, r^{-3}dr)$ the distribution

$$h(\psi) := \int h\,\psi\,\frac{dr}{r^3} \qquad (\psi \in \mathcal{D}(]Y,\infty[)).$$

Formula (1.26) reads now as $\pi_*(f) = f_0$. We may now define the distribution $T_a \in \mathcal{D}'$ for $a > Y$ by

$$(1.29) \qquad T_a(\varphi) := \varphi_0(a) \qquad (\varphi \in \mathcal{D}^\infty).$$

We have $T_a = \pi^*(\delta_a)$ where δ_a is the Dirac distribution at a, that is $\delta_a(\psi) = \psi(a)$. It will be important to relate the distribution T_a to the operator Δ. To do this we introduce for $s \in \mathbb{C}$ and $a > Y$ the continuous function $\eta_{a,s} :]Y,\infty[\to \mathbb{C}$ by

$$(1.30) \qquad \eta_{a,s}(r) := \begin{cases} r^{1+s} - a^{2s}\,r^{1-s} & \text{for } r \geq a, \\ 0 & \text{for } r < a. \end{cases}$$

Note that $\tilde{\eta}_{a,s}$ is a continuous function on $\mathbb{H}$. A simple computation of integrals shows

$$(1.31) \qquad (-\Delta_0 - (1-s^2))\,\eta_{a,s} = -2sa^{s-1}\,\delta_a.$$

By application of π^* we find

$$(1.32) \qquad (-\Delta - (1-s^2))\,\tilde{\eta}_{a,s} = -2sa^{s-1}\,T_a.$$

Lemma 1.5. *Let $a \in \mathbb{R}$, $a > Y''$ and define*

$$(1.33) \qquad \mathfrak{H}_a := \{\, f \in L^2(\Gamma\setminus\mathbb{H}) \ \mid\ f_0|_{[a,\infty[} = 0 \,\}.$$

Then $\mathfrak{H}_a$ is a closed subspace of $L^2(\Gamma\setminus\mathbb{H})$. We further put

$$(1.34) \qquad \mathfrak{D}_a := \{\, f \in \mathfrak{H}_a \ \mid\ \mathbf{grad}(f) \in L^2(\Gamma\setminus\mathbb{H}) \,\}.$$

For $f, g \in \mathfrak{D}_a$ we define

$$(1.35) \qquad Q_a(f,g) := Q(f,g)$$

using our construction (1.22). Then Q_a is a closed, symmetric, non-negative sesquilinear form on $\mathfrak{H}_a$ with dense domain $\mathfrak{D}_a$. We further define

(1.36) $\mathfrak{C}_a := \{ \varphi - \tilde{\psi} \mid \varphi \in \mathcal{D}^\infty,\ \psi \in \mathcal{D}(]Y, \infty[),\ \varphi_0(r) = \psi(r) \text{ for } r \geq a \}.$

Then the linear subspace $\mathfrak{C}_a \subset \mathfrak{D}_a$ *is a core of* Q_a.

Proof. By formula (1.26) $\mathfrak{H}_a$ is the orthogonal complement of the subspace $\{ \tilde{\psi} \mid \psi \in \mathcal{D}(]a, \infty[) \}$ of $L^2(\Gamma \setminus \mathbb{H})$ and hence is closed. For the further arguments we use the terminology of Kato (1976), Chapter 6. We remind the reader of the concept of a core. Since Q_a is closed on $\mathfrak{D}_a$ the sesqilinear form

$$< f, g >_a := Q_a(f, g) + < f, g > \qquad (f, g \in \mathfrak{D}_a)$$

turns $\mathfrak{D}_a$ into a Hilbert space. A linear subspace of $\mathfrak{D}_a$ is a core if it is dense in this new Hilbert space. The claims in the above are proved by simple approximation arguments. The following facts have to be used. We put

$$H^1(\Gamma \setminus \mathbb{H}) := \{ f \in L^2(\Gamma \setminus \mathbb{H}) \mid \mathbf{grad}(f) \in L^2(\Gamma \setminus \mathbb{H}) \}.$$

If Γ is torsion-free, then $H^1(\Gamma \setminus \mathbb{H})$ is the first Sobolev space of $\Gamma \setminus \mathbb{H}$, see Hörmander (1963). We already know from Chapter 4 that $H^1(\Gamma \setminus \mathbb{H})$ is dense in $L^2(\Gamma \setminus \mathbb{H})$. The subspace $\mathcal{D}^\infty \subset H^1(\Gamma \setminus \mathbb{H})$ is dense with respect to the scalar product

$$< f, g >_1 := Q(f, g) + < f, g > \qquad (f, g \in H^1(\Gamma \setminus \mathbb{H})).$$

This fact can be inferred from Hörmander (1963) in case Γ is torsionfree. A slight change of the argument given there establishes the general case. Note also that for a sequence (f_n) in $L^2(\Gamma \setminus \mathbb{H})$ which converges to 0 the sequence $((f_n)_0)\tilde{}$ of zeroth Fourier coefficents also converges to 0. □

To any closed, symmetric, densely defined sesquilinear form B bounded from below on a Hilbert space $\mathbf{H}$ with domain $\mathbf{D}_B$ there is associated a self-adjoint operator T (the Friedrichs extension) with domain $\mathbf{D}_T \subset \mathbf{D}_B$ so that $B(v, w) = < Tv, w >$ for all $v \in \mathbf{D}_T,\ w \in \mathbf{D}_B$, see Kato (1976), Chapter 6, Paragraph 2 for the construction. We study this process for our Hilbert space $\mathfrak{H}_a$ and the sesquilinear form Q_a with domain $\mathfrak{D}_a$. The operator Δ_a was first considered in the analogous 2-dimensional situation by Lax and Phillips (1976), pages 206–207.

Proposition 1.6. *Let* Δ_a *be the self-adjoint operator on* $\mathfrak{H}_a$ *associated to the sesquilinear form* Q_a. *Then the domain* $\mathfrak{D}_{\Delta_a}$ *of* Δ_a *consists of those* $f \in \mathfrak{D}_a$ *which satisfy*

$$-\Delta f = g + cT_a \tag{1.37}$$

for some $g \in \mathfrak{H}_a$ *and some* $c \in \mathbb{C}$. *If (1.37) is satisfied we have* $-\Delta_a f = g$.

Proof. Suppose $f \in \mathfrak{D}_a$ satisfies the requirement (1.37) that is

$$\int_{\mathcal{F}} f\,(-\Delta\varphi)\;dv = < g, \varphi > + c\varphi_0(a) \qquad \text{for all } \varphi \in \mathcal{D}^\infty. \tag{1.38}$$

To prove that $f \in \mathfrak{D}_{\Delta_a}$ and $\Delta_a f = g$ we need, by the first representation theorem of Kato (1976), Chapter 6, Paragraph 2, only prove the formula $Q_a(f, \Phi) = < g, \Phi >$ for all elements Φ in our core $\mathfrak{C}_a$ and this is obviuos.

Suppose conversely that $f \in \mathfrak{D}_{\Delta_a}$ and $g \in \mathfrak{H}_a$ satisfy $-\Delta_a f = g$. We need to check (1.38). For $\varphi \in \mathcal{D}^\infty$ we find $\Phi := \varphi - \tilde{\varphi}_0 \in \mathfrak{C}_a$ and hence $Q_a(f, \Phi) = < g, \Phi >$. This results in

$$-\Delta f(\varphi) - g(\varphi) = -\Delta f(\tilde{\varphi}_0) - g(\tilde{\varphi}_0) = \pi^*(\pi_*(-\Delta f - g))(\varphi). \tag{1.39}$$

We determine now the distribution $u := \pi_*(-\Delta f - g) = -\Delta_0 f_0 - g_0$. Since f_0 and g_0 are 0 from a onwards, u vanishes on functions $\psi \in \mathcal{D}(]Y, \infty[)$ with $\mathrm{supp}(\psi) \subset [a, \infty[$. On the other hand u vanishes also on all $\psi \in \mathcal{D}(]Y, \infty[)$ with $\mathrm{supp}(\psi) \subset]Y, a]$ because for them $\Phi_1 := \Phi + \tilde{\psi}$ is in $\mathfrak{C}_a$ and computation (1.39) may be applied to Φ_1. Hence u has support in $\{a\}$. It follows that there is an expansion of u in terms of δ_a and its derivatives

$$u = c\delta_a + \sum_{k=1}^{n} c_k \delta_a^{(k)}.$$

From the fact that $\mathbf{grad}(f) \in L^2(\Gamma \setminus \mathbb{H})$ we infer that the distribution v defined by $v(\psi) = f_0(r^3(r^{-2}\psi)')$ is represented by an element of the space $L^1_{\mathrm{loc}}(]Y, \infty[, r^{-3}dr)$. It is an elementary exercise to deduce that $c_1 = \ldots = c_n = 0$. □

Proposition 1.6 shows that Δ_a is a self-adjoint extension of the restriction $\Delta|_{\mathfrak{H}_a}$ of Δ to $\mathfrak{H}_a$. It is not difficult to see that $\Delta|_{\mathfrak{H}_a}$ has many, in fact a family parametrized by $\mathbf{U}(1)$, of self-adjoint extensions. We will now relate the resolvents of Δ and Δ_a and use the cut-off kernel $F^*(P, Q, s)$ from Chapter 4, Section 5 to prove that Δ_a has a compact resolvent. To do this we turn our Eisenstein series into an L^2-function by cut-off. The following is obvious.

Lemma 1.7. *For $a > Y,\ s \in \mathbb{C},\ \mathrm{Re}\, s > 1$ we introduce*

$$E_a(\cdot, s) := E(\cdot, s) - \tilde{\eta}_{a,s}, \tag{1.40}$$

and call it the cut-off Eisenstein series. We have $E_a(\cdot, s) \in L^2(\Gamma \setminus \mathbb{H})$ and

$$(-\Delta - (1 - s^2))\; E_a(\cdot, s) = 2sa^{s-1}\, T_a.$$

$E_a(\cdot, s)$ is contained in $\mathfrak{H}_a$ iff $\phi(s) = -a^{2s}$. If this is the case then $E_a(\cdot, s)$ is an eigenfunction for $-\Delta_a$ with eigenvalue $1 - s^2$.

Note that $E_a(\cdot, s) \neq 0$ since it isn't smooth when $r = r(P) = a$.

Lemma 1.7 has the following interesting consequence. If $\phi(s) = -a^{2s}$ has a solution with $\mathrm{Re}\, s > 1$ then $1 - s^2$ is a positive real number. This follows since $-\Delta_a$ is self-adjoint and positive. Once we have the meromorphic continuation of the Eisenstein series we get Lemma 1.7 for all s which are not poles of the Eisenstein series and our remark similarly applies. Worked out for the Eisenstein series for arithmetic groups treated in Chapter 8 our remark amounts to a deep fact related to the Riemann hypothesis.

Proposition 1.8. *For $s \in \mathbb{C}$ with $\mathrm{Re}\, s > 1$ we consider the operators*

$$\mathcal{R}_s := (-\tilde{\Delta} - (1-s^2))^{-1}, \quad \mathcal{R}_{a,s} := (-\Delta_a - (1-s^2))^{-1} \tag{1.41}$$

from $\mathfrak{H}_a$ to $L^2(\Gamma \setminus \mathbb{H})$. If s has the property $\phi(s) \neq -a^{2s}$ then the range of $\mathcal{R}_s - \mathcal{R}_{a,s}$ is generated by $E_a(\cdot, s)$ and hence is 1-dimensional. The self-adjoint operator Δ_a has a Hilbert–Schmidt resolvent and hence a purely discrete spectrum.

Proof. We define for $g \in \mathfrak{H}_a$: $f := \mathcal{R}_s\, g$, $\hat{f} := \mathcal{R}_{a,s}\, g$. The zeroth Fourier coefficient of f satisfies $(-\Delta_0 - (1-s^2))\, f_0 = 0$ in $]Y, \infty[$, hence by elliptic regularity, see Rudin (1973), 8.12, it is C^∞ on this interval and we find $f_0(r) = c_1\, r^{1-s}$ because f was in $L^2(\Gamma \setminus \mathbb{H})$. By our assumption $\phi(s) \neq -a^{2s}$ we find a $c_2 \in \mathbb{C}$ so that $(f - c_2\, E_a(\cdot, s))_0 = 0$ in $]a, \infty[$. Hence $f_0|_{[a,\infty[} = 0$ as a distribution and by Proposition 1.6, Lemma 1.7 $f - c_2\, E_a(\cdot, s)$ is in the domain of Δ_a. By our choice of s we find $\hat{f} = f - c_2\, E_a(\cdot, s)$ and we are done with the first part of the proposition.

Since ϕ is an infinite Dirichlet series for $\mathrm{Re}\, s > 1$ we can find an $s \in \mathbb{R}$ with $s > 1$, so that $\phi(s) \neq -a^{2s}$. By the above $\mathcal{R}_s - \mathcal{R}_{a,s}$ is Hilbert–Schmidt and by Theorem 4.5.2 $\mathcal{R}_s$ is also a Hilbert–Schmidt operator. Hilbert's resolvent equation then shows that $(-\Delta_a - (1-s^2))^{-1}$ is Hilbert–Schmidt for all s so that $1 - s^2$ is not contained in the spectrum of Δ_a. □

There is the following direct argument which shows that Δ_a has a compact resolvent. We use a decomposition $\mathcal{F} = \mathcal{F}_1 \cup \mathcal{F}_\infty(b)$ where $b > a$ and $\mathcal{F}_1$ is compact. We then restrict our form Q_a to both $\mathcal{F}_1$ and $\mathcal{F}_\infty(b)$. The corresponding self-adjoint operators have compact resolvents. On $\mathcal{F}_1$ this is standard theory and on $\mathcal{F}_\infty(b)$ it follows by a computation using separation of variables. Finally we use the Neumann comparison theorem, see Chavel (1984), to obtain the result on $\mathcal{F}$.

Before proving Theorem 1.2 we need the following result from operator theory. We thank R. Meise for helping with the proof.

Lemma 1.9. *Let $\mathbf{H}$ be a separable Hilbert space and Θ a densily defined self-adjoint, positive operator on $\mathbf{H}$. Assume that the spectrum $\sigma(\Theta)$ of Θ is a discrete subset of $[0, \infty[$. Let $\sigma_1, \sigma_2, \ldots$ be the nonzero eigenvalues of Θ counted without multiplicities. Assume further that*

$$\sum_{n=1}^{\infty} \frac{1}{\sigma_n^2} < \infty. \tag{1.42}$$

Then there are holomorphic functions $D : \mathbb{C} \to \mathcal{B}(\mathbf{H})$ *and* $d : \mathbb{C} \to \mathbb{C}$ *which satisfy*

$$\max\{\, |d(z)|,\ \|D(z)\|\,\} \le \kappa\, e^{\kappa_1 |z|^2} \qquad \textit{for all } z \in \mathbb{C} \tag{1.43}$$

with some constants κ, κ_1 *and*

$$(\Theta - z)^{-1} = \frac{D(z)}{d(z)} \qquad \textit{for all } z \in \rho(\Theta) = \mathbb{C} \setminus \sigma(\Theta). \tag{1.44}$$

Note that assumption (1.42) is satisfied once the resolvent $(\Theta - z)^{-1}$ is a Hilbert–Schmidt operator for some $z \in \rho(\Theta)$.

Proof. Replacing Θ by a suitable shift $\Theta - \mu$ we may without loss of generality assume that $\sigma(\Theta) \subset]0, \infty[$. By a well known formula (Kato (1976), paragraph 5) we have

$$\|(\Theta - z)^{-1}\| = \frac{1}{\operatorname{dist}(z, \sigma(\Theta))} \qquad \text{for } z \in \rho(\Theta). \tag{1.45}$$

Here the distance of two nonempty subsets $S, T \subset \mathbb{C}$ is defined as $\operatorname{dist}(S,T) := \inf\{\, |s - t| \mid s \in S, t \in T\,\}$. Formula (1.45) shows that the resolvent $(\Theta - z)^{-1}$ has poles of order 1 in the points of $\sigma(\Theta)$. It also shows that $\|(\Theta - z)^{-1}\|$ is bounded in every subset of $\rho(\Theta)$ which lies in a distance bounded from below away from $\sigma(\Theta)$.

Choose now a holomorphic function $d : \mathbb{C} \to \mathbb{C}$ which has zeroes of order 1 in the points $\sigma_i \in \sigma(\Theta)$ and which satisfies

$$|d(z)| \le \kappa_2\, e^{\kappa_3 |z|^2} \qquad \text{for all } z \in \mathbb{C}.$$

It follows from (1.42) and a theorem of Lindelöf (see Boas (1954), 2.10.1 and Rubel, Taylor (1968)) that this is possible. Note that (1.42) implies that the counting function $n(r, \sigma(\Theta)) := |\{\, \sigma \in \sigma(\Theta) \mid |\sigma| \le r\,\}$ satisfies $n(r, \sigma(\Theta)) \le \kappa_4\, r^2$ for some constant κ_4.

Put $D(z) := d(z)\,(\Theta - z)^{-1}$. This defines a holomorphic map $D : \mathbb{C} \to \mathcal{B}(\mathbf{H})$. Consider for $\beta > 2$ the cone

$$\Sigma := \left\{ r\, e^{it} \;\middle|\; r, t \in \mathbb{R},\ r > 0,\ |t| < \frac{\pi}{2\beta} \right\}. \tag{1.46}$$

For $z \in \mathbb{C} \setminus \Sigma$ an estimate of the form (1.43) holds. For $z \in \partial\Sigma$ we have $z = r\, e^{\pm i \frac{\pi}{2\beta}}$ and hence

$$\|D(z)\| \le \frac{1}{\operatorname{dist}(\sigma(\Theta), \partial\Sigma)}\, \kappa_2\, e^{\kappa_3 r^2}$$

Choose now κ_5 so large that $\kappa_3 < \kappa_5 \cos(\pi/\beta)$. This is possible because $\beta > 2$. Define $g(z) := e^{-\kappa_5 z^2}$. For $z = r\, e^{\pm i \frac{\pi}{2\beta}} \in \partial\Sigma$ we have

$$\|gD(z)\| \leq \kappa_2\, e^{\kappa_3 r^2 - \kappa_5 r^2 \cos(\frac{\pi}{\beta})} \leq \kappa_6.$$

Hence the function $gD : \mathbb{C} \to \mathcal{B}(\mathbf{H})$ is bounded on $\partial\Sigma$ and satisfies $\log\|(gD)(z)\| = O(|z|^\beta)$ on Σ. Since β is the angle at the vertex of Σ we may apply the Phragmén–Lindelöf principle for cones (see Koosis (1992)). The applicability of this principle to our (Banach space valued) function $gD : \mathbb{C} \to \mathcal{B}(\mathbf{H})$ is justified by the theorem of Hahn–Banach. □

There are various versions of the above lemma where the assumption (1.42) can be replaced by assumptions on the counting function $n(r, \sigma(\Theta))$ and on the partial sums of powers of the elements of $\sigma(\Theta)$. To prove these the results of Rubel, Taylor (1986) can be used.

Proof of Theorem 1.2. Going back to Definition 1.3 we choose Y', Y'' so that $Y < Y' < Y'' < a$ and make a choice of the function h. Since $H(\cdot, s) \in \mathfrak{H}_a$ we may define for all $s \in \mathbb{C}$ with $1 - s^2 \notin \sigma(-\Delta_a)$

$$\Lambda(P, s) := (-\Delta_a - (1 - s^2))^{-1} H(P, s). \tag{1.47}$$

Since the resolvent of Δ_a is a meromorphic function from $\mathbb{C}$ to the space of bounded operators on $\mathfrak{H}_a$, the function $\Lambda(P, s)$ varies meromorphically in the parameter s. Evidently we have $(-\Delta_a - (1 - s^2))\, \Lambda(P, s) = H(P, s)$. By Proposition 1.6 we find

$$(-\Delta - (1 - s^2))\, \Lambda(P, s) = H(P, s) + c(s)\, T_a \tag{1.48}$$

with some meromorphic function c. We choose the meromorphic function $A(s)$ so that $(-\Delta - (1 - s^2))\, A(s)\, \tilde{\eta}_{a,s} = -c(s) T_a$. In fact we have to take $A(s) = \frac{1}{2} s^{-1} a^{1-s} c(s)$ and we put

$$\Lambda_1(\cdot, s) := \Lambda(\cdot, s) + A(s)\, \tilde{\eta}_{a,s}(\cdot) + \tilde{h}_s(\cdot).$$

From (1.48) and the definition of $H(\cdot, s)$ we get

$$(-\Delta - (1 - s^2))\, \Lambda_1(\cdot, s) = 0. \tag{1.49}$$

Now we deduce from the elliptic regularity theorem, Rudin (1973), 8.12, that $\Lambda_1(P, s)$ is C^∞ in $P \in \mathbb{H}$. We have now a family of functions $\Lambda_1(P, s)$ which is meromorphic in s and C^∞ in the variable P. For large r the zeroth Fourier coefficient of Λ_1 is

$$(\Lambda_1)_0(r, s) = (A(s) + 1)\, r^{1+s} - A(s)\, a^{2s}\, r^{1-s}. \tag{1.50}$$

Suppose that $\Lambda_1(P, s)$ is identically 0 then its zeroth Fourier coefficient is also 0, which is clearly impossible. If we choose $s \in \Omega$ or $-s \in \Omega$, see (1.11), we get from the uniqueness part of Lemma 1.4 the two following equations

$$\Lambda_1(P,s) = (A(s)+1)\,\tilde{E}(P,s), \qquad \Lambda_1(P,-s) = A(-s)\,a^{-2s}\,\tilde{E}(P,s).$$

From them everything is clear except for the fact that the function ϕ arising in the Fourier expansion

$$E(P,s) := \frac{1}{[\Gamma_\infty : \Gamma'_\infty]}\,E_I(P,s) = r^{1+s} + \phi(s)\,r^{1-s} + \dots$$

is a meromorphic function of order ≤ 4. To prove this note that by the above reasoning ϕ can be rationally expressed in the function A from (1.50). We have $A(s) = \frac{1}{2}s^{-1}a^{1-s}c(s)$, hence it is enough to show that $c(s)$ is of order ≤ 4. To do this note that (1.48) means that

$$c(s)\varphi_0(a) = \int_{\mathcal{F}} \Lambda(P,s)\,((-\Delta-(1-s^2))\varphi)\,dv(P) - \int_{\mathcal{F}} H(P,s)\varphi\,dv(P)$$

for every $\varphi \in \mathcal{D}^\infty$. Put $\varphi_1 := (-\Delta-(1-s^2))\varphi$. Using (1.44) for $\Theta = \Delta_a$ we find

$$c(s)\varphi_0(a) = \left\langle \frac{D(1-s^2)}{d(1-s^2)} H(P,s), \varphi_1 \right\rangle - < H(P,s), \varphi > .$$

Choose now $\varphi \in \mathcal{D}^\infty$ with $\varphi_0(a) = 1$, estimate $H(P,s)$ elementarily from its definition (1.14), use (1.43) and the result follows. □

We add some information on the eigenvalues of $-\Delta_a$. Before doing this we consider the definition of E_a in (1.40) to be extended to all of $\mathbb{C}$.

Proposition 1.10. *The eigenfunctions and eigenvalues of $-\Delta_a$ are of the following three types:*

(1) The eigenvalues of Δ which belong to eigenfunctions in $\mathfrak{H}_a$. (All but finitely many of the eigenvalues of Δ have this property.)
(2) The numbers $\lambda = 1-s^2$ with $s \neq 0$ so that the continued Eisenstein series has no pole in s which have the properties $\phi(s) = -a^{2s}$. The eigenfunction is $E_a(\cdot, s) \neq 0$.
(3) Suppose that $\phi(0) = -1$ and $\phi'(0) = -2\log a$. Then $\frac{d}{ds}E_a(\cdot,s)|_0$ is an eigenfunction with eigenvalue 1.

From the above it is easy to see that the given eigenfunctions and eigenvalues really occur. We leave as an exercise using the spectral decomposition developed in Section 3 to prove that Proposition 1.10 is a complete description of the spectrum of $-\Delta_a$.

We need some further properties of our continued Eisenstein series and also of the scattering matrix. We return to our notations (1.3)–(1.7) and abandon our previous assumption that Γ has only one cusp.

Theorem 1.11. *Let $\Gamma < \mathbf{PSL}(2,\mathbb{C})$ be a cofinite subgroup. Then the following hold.*

(1) $\Phi(s)$ *is symmetric where defined.*
(2) *If the functions in the* ν*-th row of* $\Phi(s)$ *have the maximal pole order* n *at a point* $s_0 \in \mathbb{C}$ *then* $E_\nu(\cdot, s)$ *has a pole of order* n *at* s_0.
(3) *The Eisenstein series* $E_\nu(P,s)$ *and the entries of* $\Phi(s)$ *have no poles in* $\{ s \in \mathbb{C} \mid \operatorname{Re} s > 0 \}$, *except possibly finitely many in the segment* $]0,1]$ *on the real line. These poles are simple.*
(4) *Define for* $\sigma \in]0,1]$ *the residue function*

$$RE_\nu(P,\sigma) := \operatorname*{Res}_{s=\sigma} E_\nu(P,s), \tag{1.51}$$

then $RE_\nu(\cdot,\sigma) \in L^2(\Gamma\backslash\mathbb{H})$ *and*

$$< RE_\nu(\cdot,\sigma), RE_\nu(\cdot,\sigma) >= \frac{|\Lambda_\nu|}{[\Gamma_\nu : \Gamma'_\nu]} \operatorname*{Res}_{s=\sigma} \phi_{\nu\nu}(s). \tag{1.52}$$

(5) $E_\nu(P,s)$ *has a simple pole at* $s=1$ *and its residue there is the constant function* $|\mathcal{P}_\nu|\operatorname{vol}(\Gamma)^{-1}$.
(6) *Both the entries of* $\mathcal{E}(\cdot,s)$ *and of* $\Phi(s)$ *have no poles for* $s \in \mathbb{C}$ *with* $\operatorname{Re} s = 0$. $\Phi(s)$ *is a unitary matrix on this line.*

We leave the simple proof as an exercise. See Selberg (1989b), page 653, Theorem 7.3 or Roelcke (1966), (1967) for hints. The use of the Maaß–Selberg relations from Theorem 3.3.6 (particularly for part (4)) is essential.

6.2 Generalities on Eigenfunctions and Eigenpackets

Throughout this section, $\Gamma < \mathbf{PSL}(2,\mathbb{C})$ will denote a discontinuous group on $\mathbb{H}$ and $\mathcal{F}$ a fundamental domain of Γ. We write $\Delta : \mathcal{D} \to L^2(\Gamma \setminus \mathbb{H})$ for the Laplace operator and $\tilde{\Delta} : \tilde{\mathcal{D}} \to L^2(\Gamma \setminus \mathbb{H})$ for its self-adjoint extension, see Chapter 4. Since $\tilde{\Delta}$ is self-adjoint, the Hilbert space $L^2(\Gamma \setminus \mathbb{H})$ splits into the direct orthogonal sum of the subspaces generated by the eigenfunctions and the eigenpackets of $\tilde{\Delta}$. The main subject of the present section is a study of these eigenfunctions and eigenpackets in general. First we shall deal with the eigenfunctions.

Let $f \in \mathcal{D}$ be an eigenfunction of $-\Delta : \mathcal{D} \to L^2(\Gamma \setminus \mathbb{H})$:

$$-\Delta f = \lambda f, \qquad \lambda = 1 - s^2, \qquad s \in \mathbb{C}. \tag{2.1}$$

Assume that Γ has the cusp $\zeta = A\infty$ ($A \in \mathbf{PSL}(2,\mathbb{C})$) and that

$$\left\{ \begin{pmatrix} 1 & \omega \\ 0 & 1 \end{pmatrix} \;\middle|\; \omega \in \Lambda \right\}$$

is a maximal lattice in $(A^{-1}\Gamma A)_\infty$. Then $g = f \circ A$ is a Λ-periodic solution of the differential equation $-\Delta g = \lambda g$. Hence we conclude from Chapter 3 that $f \circ A$ has a Fourier expansion of the form

$$(2.2)\quad f \circ A(z+rj) = a_0 r^{1+s} + b_0 r^{1-s} + \sum_{0 \neq \mu \in \Lambda^\circ} a_\mu \, r \, K_s(2\pi|\mu| r) \, e^{2\pi i <\mu, z>},$$

where Λ° denotes the dual lattice of Λ and where $< \cdot, \cdot >$ stands for the inner product on $\mathbb{R}^2 = \mathbb{C}$. For $s = 0$ the term $b_0 r^{1-s}$ has to be replaced by $b_0 r \log r$. Since $f \circ A$ is square integrable over a cusp sector at infinity, we also know from Chapter 3 that

$$(2.3)\qquad a_0 = b_0 = 0 \qquad \text{if } \lambda \geq 1, \text{ i.e. if } s \text{ is purely imaginary},$$

$$(2.4)\qquad b_0 = 0 \qquad \text{if } \operatorname{Re} s < 0, \qquad a_0 = 0 \qquad \text{if } \operatorname{Re} s > 0.$$

Definition 2.1. Assume $v(\mathcal{F}) < \infty$ and let $f \in \mathcal{D}$ be an eigenfunction of $-\Delta : \mathcal{D} \to L^2(\Gamma \setminus \mathbb{H})$. Then f is called a cusp eigenfunction if for every cusp of Γ the coefficients a_0, b_0 in the corresponding Fourier expansion (2.2) both vanish.

Notice that the vanishing of a_0, b_0 does not depend on the choice of A. Also, if a_0, b_0 vanish in a cusp ζ, then they vanish in all cusps Γ-equivalent to ζ. Hence it suffices in the definition of a cusp eigenfunction, to require the vanishing of the zeroth Fourier coefficients for a maximal set of Γ-inequivalent cusps. Note that the concept of cusp eigenfunction is in various respects stronger than the notion of a cusp function defined in Definition 4.5.1.

The cusp functions for the eigenvalue theory of the hyperbolic Laplacian acting on the upper half-plane are studied in Kubota (1973), Maaß (1949a), (1953), (1964), Roelcke (1956a), (1966), (1967), Selberg (1954), (1965), (1989b), Venkov (1979c).

Proposition 2.2. *Let $\Gamma < \mathbf{PSL}(2, \mathbb{C})$ be a cofinite group. Then the following hold.*

(1) Every eigenfunction f of $-\Delta : \mathcal{D} \to L^2(\Gamma \setminus \mathbb{H})$ corresponding to an eigenvalue $\lambda \geq 1$ is a cusp eigenfunction.
(2) The eigenspaces of $-\Delta$ for eigenvalues $0 \leq \lambda < 1$ are spanned by cusp eigenfunctions and the possible residues of the Eisenstein series at $s = \sqrt{1-\lambda}$.

Proof. Item (1) is obvious from (2.3). For the proof of (2) we assume that Γ has only one cusp as the general case creates only notational difficulties. We keep the notations used in the definition of the Eisenstein series (1.10). Assume that $g \in \mathcal{D} \subset L^2(\Gamma \setminus \mathbb{H})$ is an eigenfuction of $-\Delta$ with an expansion (2.2) of the form $g = r^{1-s} + \dots$. With an appropriate choice of the cut-off height $Y \in \mathbb{R}$ the Maaß–Selberg relation from Theorem 3.3.6 reads for $t \neq s$ as

$$\int_{\mathcal{F}} E(\cdot,t)^Y \, \overline{g^Y} \, dv = C \left(\frac{Y^{t-s}}{t-s} - \frac{\overline{\phi(t)} \, Y^{-t-s}}{t+s} \right) \tag{2.5}$$

with some non zero constant C. If ϕ was holomorphic at s then by Theorem 1.11 (2) the Eisenstein series E would also have this property and the left-hand side of (2.5) would stay bounded as $t \to s$, which is impossible. Hence by Theorem 1.11 (2), (3) both $E(\cdot,t)$ and $\phi(t)$ have poles of order 1 at s. We may then subtract a suitable multiple of the residue at s of $E(\cdot,t)$ from g to obtain a cusp function. □

We now study the orthogonal complement of the subspace of $L^2(\Gamma \setminus \mathbb{H})$ spanned by the eigenfunction of $\tilde{\Delta}$, called the continuous part of $L^2(\Gamma \setminus \mathbb{H})$ with respect to $\tilde{\Delta}$. This space is generated by certain elements called eigenpackets. The notion of an eigenpacket was used by Hellinger (1907), (1909) and Hahn (1912) in the early days of spectral theory and undeservedly received less attention in the following decades, notable exceptions being the work of Rellich (1951), (1952) and Roelcke (1956a), (1966), (1967). In the case of cofinite groups Γ eigenpackets are particularly useful since in this case the continuous part of $L^2(\Gamma \setminus \mathbb{H})$ can be described rather explicitly in terms of eigenpackets defined by Eisenstein series. We shall discuss this in detail in Section 3. For the reader's convenience we offer a brief general discussion of eigenpackets following essentially Roelcke (1956a), paragraph 9, and Jörgens–Rellich (1976), Kapitel II, and Hellwig (1964).

Throughout the rest of this section up to Corollary 2.13,

$$A : \mathcal{D}_A \to \mathbf{H}, \tag{2.6}$$

will denote a symmetric linear operator in the Hilbert space $\mathbf{H}$, and $< \cdot, \cdot >$ will denote the inner product in $\mathbf{H}$.

Definition 2.3. An eigenpacket of A is a map $v : \mathbb{R} \to \mathbf{H}$, $\lambda \mapsto v_\lambda$ having the following properties:

(1) $v_0 = 0$ and $v_\lambda \in \mathcal{D}_A$ for all $\lambda \in \mathbb{R}$.
(2) v is continuous in the norm sense.
(3) $Av_\lambda = \int_0^\lambda \mu \, dv_\mu$ for all $\lambda \in \mathbb{R}$, where the integral is the limit in the norm sense of the corresponding Stieltjes sums.

Applying partial summation to the Stieltjes sums for the integral in Definition 2.3, (3) we have

$$Av_\lambda = \lambda v_\lambda - \int_0^\lambda v_\mu \, d\mu, \tag{2.7}$$

where the integral is the limit of Riemann sums in the norm sense.

Theorem 2.4. *(1) If f is an eigenvector and v an eigenpacket of A, then*

(2.8) $$< f, v_\lambda >= 0 \qquad \textit{for all } \lambda \in \mathbb{R}.$$

(2) If v and w are two eigenpackets of A and $[\alpha, \beta]$, $[\gamma, \delta]$ are two intervals in $\mathbb{R}$ having at most one point in common, the orthogonality relation

(2.9) $$< v_\beta - v_\alpha, w_\delta - w_\gamma >= 0$$

holds.

Proof. (1): Suppose we have $Af = \mu f$. Then the symmetry of A implies that

$$\mu < f, v_\lambda >=< Af, v_\lambda >=< f, Av_\lambda >$$
$$=< f, \int_0^\lambda t\ dv_t >= \lambda < f, v_\lambda > - \int_0^\lambda < f, v_t >\ dt.$$

Hence the continuous function $\varphi(\lambda) :=< f, v_\lambda >$ satisfies the functional equation

(2.10) $$(\lambda - \mu)\varphi(\lambda) = \int_0^\lambda \varphi(t)\ dt.$$

This implies that $\varphi'(t) = 0$ for all $t \neq \mu$, and since φ is continuous and $\varphi(0) = 0$ we conclude that $\varphi(\lambda) = 0$ for all $\lambda \in \mathbb{R}$, which proves (1).

(2): Without loss of generality we may assume that $\alpha < \beta \leq \gamma < \delta$, and since v, w are continuous, it is even sufficient to prove (2.8) in the case $\alpha < \beta < \gamma < \delta$. The symmetry of A implies

$$0 =< v_s - v_\alpha, A(w_t - w_\gamma) > - < A(v_s - v_\alpha), w_t - w_\gamma) >$$
$$=< v_s - v_\alpha, \int_\gamma^t \lambda\ d(w_\lambda - w_\gamma) > - < \int_\alpha^s \lambda\ d(v_\lambda - v_\alpha), w_t - w_\gamma >$$
$$= t < v_s - v_\alpha, w_t - w_\gamma > - < v_s - v_\alpha, \int_\gamma^t (w_\lambda - w_\gamma)\ d\lambda >$$
$$- s < v_s - v_\alpha, w_t - w_\gamma > + < \int_\alpha^s (v_\lambda - v_\alpha)\ d\lambda, w_t - w_\gamma > .$$

Hence the continuous function $\varphi(s,t) :=< v_s - v_\alpha, w_t - w_\gamma >$ satisfies the functional equation

(2.11) $$(t - s)\varphi(s,t) - \int_\gamma^t \varphi(s,\lambda)\ d\lambda + \int_\alpha^s \varphi(\lambda, t)\ d\lambda = 0.$$

We show that every continuous solution of this functional equation vanishes identically in $[\alpha, \beta] \times [\gamma, \delta]$. Let $s \in [\alpha, \beta], t \in [\gamma, \delta]$ and define

$$M := \max\{\ |\varphi(s,t)|\ \mid\ \alpha \leq s \leq \beta,\ \gamma \leq t \leq \delta\ \}.$$

Then we have

$$t - s \geq \gamma - \beta =: \epsilon > 0, \tag{2.12}$$

and hence our functional equation implies

$$|\varphi(s,t)| \leq \frac{M}{\epsilon}(t - \gamma + s - \alpha). \tag{2.13}$$

We show by induction that

$$|\varphi(s,t)| \leq \frac{2^{n-1}M}{\epsilon^n n!}(t - \gamma + s - \alpha)^n \qquad \text{for all } n \geq 1. \tag{2.14}$$

By (2.13), inequality (2.14) is satisfied if $n = 1$. Assume that (2.14) holds for n. Then (2.11) implies

$$\begin{aligned} |\varphi(s,t)| &\leq \frac{1}{\epsilon}\frac{2^{n-1}M}{\epsilon^n n!}\left\{\int_\gamma^t (\lambda - \gamma + s - \alpha)^n \, d\lambda + \int_\alpha^s (t - \gamma + \lambda - \alpha)^n \, d\lambda\right\} \\ &\leq \frac{2^n M}{\epsilon^{n+1}(n+1)!}(t - \gamma + s - \alpha)^{n+1}. \end{aligned}$$

Hence (2.14) holds for all $n \geq 1$. Letting $n \to \infty$ in (2.14), we obtain $\varphi(s,t) = 0$ for all $(s,t) \in [\alpha, \beta] \times [\gamma, \delta]$. In particular, we have

$$0 = \varphi(\beta, \delta) = < v_\beta - v_\alpha, w_\delta - w_\gamma > .$$

This proves the theorem. □

The proof of Theorem 2.4 follows the lines of thought of Rellich (1951), (1952) Teil I. Essentially the same proof is given by Hellwig (1964), Jörgens and Rellich (1976), Roelcke (1956a).

If A is a self-adjoint linear operator in $\mathbf{H}$, there is a simple relation between the eigenpackets of A and the spectral family of A.

Theorem 2.5. *Let $A : \mathcal{D}_A \to \mathbf{H}$ be a self-adjoint linear operator in the Hilbert space $\mathbf{H}$ and let $(E_\lambda)_{\lambda \in \mathbb{R}}$ be the (right continuous) spectral family of A. Denote the closed linear subspace of $\mathbf{H}$ generated by the eigenvectors of A by $\mathbf{H}_{\text{eig}}$, and let $\mathbf{H}_{\text{cont}}$ be the orthogonal complement of $\mathbf{H}_{\text{eig}}$ in $\mathbf{H}$. Then the following holds: A map $v : \mathbb{R} \to \mathbf{H}$ is an eigenpacket of A if and only if for any $\alpha > 0$ there exists an element $f_\alpha \in \mathbf{H}_{\text{cont}}$ such that*

$$v_\lambda = (E_\lambda - E_0) f_\alpha \qquad \text{for all } \lambda \in [-\alpha, \alpha]. \tag{2.15}$$

Proof. If for any $\alpha > 0$, $v|_{[-\alpha,\alpha]}$ is of the form (2.15) for some $f_\alpha \in \mathbf{H}_{\text{cont}}$, then v fulfills the conditions (1)–(3) of Definition 2.3. This is easily seen from the properties of the spectral family associated with a self-adjoint operator.

To prove the converse, let v be an eigenpacket, $\alpha > 0$ and put

$$f_\alpha := v_\alpha - v_{-\alpha}. \tag{2.16}$$

Theorem 2.4, (1) implies that $f_\alpha \in \mathbf{H}_{\text{cont}}$. Hence

$$w_\lambda := (E_\lambda - E_0) f_\alpha \qquad (\lambda \in \mathbb{R}) \tag{2.17}$$

defines an eigenpacket of A as was shown above. We want to show that

$$w_\lambda = v_\lambda \quad \text{if} \quad \lambda \in [-\alpha, \alpha].$$

This will be proved if we show that for $\lambda, \mu \in \mathbb{R}$

$$E_\lambda v_\mu = \begin{cases} 0 & \text{for} \quad \lambda \le \mu \le 0, \\ v_\mu - v_\lambda & \text{for} \quad \mu \le \lambda \le 0, \\ v_\mu & \text{for} \quad \mu \le 0 \le \lambda, \\ 0 & \text{for} \quad \lambda \le 0 \le \mu, \\ v_\lambda & \text{for} \quad 0 \le \lambda \le \mu, \\ v_\mu & \text{for} \quad 0 \le \mu \le \lambda. \end{cases} \tag{2.18}$$

To prove (2.18) we first observe that

$$E_\lambda v_\mu \in \mathbf{H}_{\text{cont}} \tag{2.19}$$

since $v_\mu \in \mathbf{H}_{\text{cont}}$ and $\mathbf{H}_{\text{cont}}$ is E_λ-invariant. Now let $g \in \mathbf{H}_{\text{cont}}$ be arbitrary. Then we have for $\lambda \le \mu \le 0$

$$\begin{aligned} &< E_\lambda v_\mu, g > = < v_\mu, E_\lambda g > \\ &\qquad = \lim_{t \to -\infty} < v_\mu - v_0, (E_\lambda - E_0) g - (E_t - E_0) g > = 0 \end{aligned} \tag{2.20}$$

by Theorem 2.4, (2). Since $g \in \mathbf{H}_{\text{cont}}$ is arbitrary, we conclude from (2.20) that $E_\lambda v_\mu = 0$ for $\lambda \le \mu \le 0$. Similarly, for $\mu \le \lambda \le 0$ we have

$$\begin{aligned} < E_\lambda v_\mu, g > &= < v_\mu, E_\lambda g > \\ &= < (v_\mu - v_\lambda) + (v_\lambda - v_0), (E_\lambda - E_\mu) g > + < v_\mu, E_\mu g > \\ &= < v_\mu - v_\lambda, (E_\lambda - E_\mu) g > + < v_\mu, E_\mu g > \quad \text{(Theorem 2.4, (2))} \\ &= < v_\mu - v_\lambda, (E_\lambda - E_\mu) g > \qquad \text{(by (2.20))} \\ &= \lim_{t \to \infty} < v_\mu - v_\lambda, (E_t - E_{-t}) g > = < v_\mu - v_\lambda, g > . \end{aligned}$$

This implies that $E_\lambda v_\mu = v_\mu - v_\lambda$ for $\mu \le \lambda \le 0$. For $\mu \le 0 \le \lambda$ we have similarly

$$< E_\lambda v_\mu, g > = < v_\mu, E_\lambda g > = \lim_{t \to \infty} < v_\mu - v_0, ((E_t - E_\lambda) + E_\lambda) g > = < v_\mu, g >$$

and hence $E_\lambda v_\mu = v_\mu$ for $\mu \le 0 \le \lambda$. In the case $\lambda \le 0 \le \mu$ we have

$$< E_\lambda v_\mu, g > = < v_\mu, E_\lambda g > = \lim_{t \to -\infty} < v_\mu - v_0, (E_\lambda - E_t) g > = 0$$

and hence $E_\lambda v_\mu = 0$ for $\lambda \le 0 \le \mu$. For $0 \le \lambda \le \mu$ we obtain

$$< E_\lambda v_\mu, g > = < v_\mu, E_\lambda g > = \lim_{t \to -\infty} < v_\mu - v_\lambda, (E_\lambda - E_t) g > + < v_\lambda, E_\lambda g >$$
$$= < v_\lambda, E_\lambda g > = \lim_{t \to \infty} < v_\lambda - v_0, (E_t - (E_t - E_\lambda)) g > = < v_\lambda, g >$$

and hence $E_\lambda v_\mu = v_\lambda$ for $0 \leq \lambda \leq \mu$. In the last case $0 \leq \lambda \leq \mu$ we have

$$< E_\lambda v_\mu, g > = < v_\mu, E_\lambda g >$$
$$= \lim_{t \to \infty} < v_\mu - v_0, (E_\lambda - E_t) g > + < v_\mu, g > = < v_\mu, g >$$

and hence $E_\lambda v_\mu = v_\mu$ for $0 \leq \mu \leq \lambda$. $\square$

Theorem 2.5 is taken from Roelcke (1956a), page 58, Satz 24, see Hellwig (1964) for a slightly different proof. The meaning of Theorem 2.5 can be explained as follows: The use of eigenpackets and eigenvectors transfers the spectral representation of self-adjoint operators in terms of spectral families, as expressed in the spectral theorem into a decomposition of elements of the Hilbert space in terms of elements of the Hilbert space itself. Classically, the relevant formulas are based on the notion of a Hellinger integral, see Hellinger (1907), (1909), Hahn (1912). Roelcke (1956a), (1966) also follows this approach, but he mentions in (1966), page 322, footnote 11 that it is also possible to use Lebesgue–Stieltjes integrals and Radon–Nikodým derivatives. This aspect was worked out by Masani (1968), (1970), (1972) in an abstract setting; the notion of an integral with respect to an orthogonally scattered vector-valued measure already appears in Doob (1953), pages 426–433. We follow the ideas of Doob and Masani restricting our attention to the case of the real line. First, we outline the construction of the (relevant) vector-valued integrals, and we summarize the results of our construction in Theorem 2.6.

If v is an eigenpacket, we intend to define

$$\int_{\mathbb{R}} f(\lambda)\, dv_\lambda \tag{2.21}$$

for a reasonably large class of functions $f : \mathbb{R} \to \mathbb{C}$. It turns out that the integral (2.21) can be defined for vector-valued functions $v : \mathbb{R} \to \mathbf{H}$ which are more general than eigenpackets of symmetric operators. Our definition of the integral (2.21) follows the same pattern as Doob's introduction of stochastic integrals (see Doob (1953), page 426 et seq.). Masani (1968) employs the same method for orthogonally scattered vector valued measures on a pre-ring of subsets of an arbitrary abstract set. As Masani (1968), (1970) points out this theory emerged from Wiener's epoch-making work on the mathematical analysis of Brownian motion. There are many illuminating general remarks on the decomposition theory we need here in Masani's papers (1968), (1970), (1972) to which we refer the less experienced reader.

The crucial property for the definition of (2.21) is the orthogonality property of Theorem 2.4, (2). Hence we start anew with the following assumption:

Let $\mathbf{H}$ be a Hilbert space, and let $v : \mathbb{R} \to \mathbf{H},\ \lambda \mapsto v_\lambda$ be a right continuous function with orthogonal increments, i.e.

$$(2.22)\quad \begin{array}{ll} < v_\beta - v_\alpha, v_\mu - v_\lambda >= 0 & \text{for all} \\ \alpha < \beta,\ \lambda < \mu \quad \text{such that} &]\alpha, \beta] \cap]\lambda, \mu] = \emptyset. \end{array}$$

Then the equation

$$(2.23)\qquad F(\beta) - F(\alpha) = \|v_\beta - v_\alpha\|^2 \qquad (\alpha < \beta)$$

defines a non-decreasing right continuous function $F : \mathbb{R} \to \mathbb{R}$ which is unique up to an additive constant. Hence there exists a unique measure

$$(2.24)\qquad \mu : \mathcal{B} \to [0, \infty]$$

on the σ-algebra $\mathcal{B}$ of all Borel subsets of $\mathbb{R}$ such that

$$(2.25)\qquad \mu(]\alpha, \beta]) = \|v_\beta - v_\alpha\|^2 \qquad (\alpha < \beta).$$

There is a corresponding Hilbert space $L^2(\mathbb{R}, \mathcal{B}, \mu) = L^2(\mu)$ associated with μ,

$$(2.26)\qquad < f, g >= \int_{\mathbb{R}} f\bar{g}\, d\mu \qquad \text{for } f, g \in L^2(\mu)$$

and the space $\mathcal{S}$ of all step functions f of the form

$$(2.27)\qquad f = \sum_{\nu=1}^{n} \xi_\nu \chi_{I_\nu},$$

where $\xi_1, \ldots, \xi_n \in \mathbb{C},\ I_\nu =]\alpha_\nu, \beta_\nu]$ are disjoint intervals in $\mathbb{R}$ half-open to the left, is a dense linear subspace of $L^2(\mu)$. For a subset A of a set M, χ_A denotes the characteristic function of A which is equal to one on A and to zero on the complement of A.

The definition of the integral (2.21) in the case $f \in \mathcal{S}$ is straightforward: For all $f \in \mathcal{S}$ of the form (2.27) define

$$(2.28)\qquad \int_{\mathbb{R}} f(\lambda)\, dv_\lambda := \sum_{\nu=1}^{n} \xi_\nu (v_{\beta_\nu} - v_{\alpha_\nu}).$$

This notion of an integral makes sense, since it does not depend on the particular choice of the representation (2.27) of f, and it has the obvious linearity properties with respect to f and v. Moreover, the following inner-product formula is immediate from (2.22):

$$(2.29)\quad < \int_{\mathbb{R}} f(\lambda)\, dv_\lambda, \int_{\mathbb{R}} g(\lambda)\, dv_\lambda >= \int_{\mathbb{R}} f\bar{g}\, d\mu \qquad \text{for all } f, g \in \mathcal{S},$$

where μ is given by (2.24), (2.25). In particular, we have

$$\left\| \int_{\mathbb{R}} f(\lambda)\, dv_\lambda \right\| = \|f\|_{L^2(\mu)} \qquad \text{for all } f \in \mathcal{S}. \tag{2.30}$$

Hence the map

$$\phi : \mathcal{S} \to \mathbf{H}, \qquad \phi(f) = \int_{\mathbb{R}} f(\lambda)\, dv_\lambda \tag{2.31}$$

is an isometric linear map from $\mathcal{S}$ to a linear subspace of $\mathbf{H}$. Since $\mathbf{H}$ is complete, ϕ can be continued in the obvious way to an isometry from the closure of $\mathcal{S}$ (which is $L^2(\mu)$) to a closed linear subspace of $\mathbf{H}$. Let $f \in L^2(\mu)$. Then there exists a sequence $(f_n)_{n\geq 1}$ in $\mathcal{S}$ such that $f_n \to f$ in $L^2(\mu)$. Formula (2.30) with $f_m - f_n$ instead of f implies that the sequence

$$\left(\int_{\mathbb{R}} f_n(\lambda)\, dv_\lambda \right)_{n\geq 1}$$

is a Cauchy sequence in $\mathbf{H}$ and hence converges. The limit of this Cauchy sequence depends only on f and not on the choice of $(f_n)_{n\geq 1}$; for if $(g_n)_{n\geq 1}$ is a second sequence in $\mathcal{S}$ converging to f, these two sequences can be combined to form a single sequence in $\mathcal{S}$ converging to f in $L^2(\mu)$ whose corresponding sequence of integrals converges in $\mathbf{H}$. Hence

$$\lim_{n\to\infty} \int_{\mathbb{R}} f_n(\lambda)\, dv_\lambda = \lim_{n\to\infty} \int_{\mathbb{R}} g_n(\lambda)\, dv_\lambda,$$

and hence the definition

$$\int_{\mathbb{R}} f(\lambda)\, dv_\lambda := \lim_{n\to\infty} \int_{\mathbb{R}} f_n(\lambda)\, dv_\lambda \tag{2.32}$$

for all $f \in L^2(\mu)$, $f_n \in \mathcal{S}$, $f_n \to f$ in $L^2(\mu)$ makes sense. Now we have defined the integral (2.21) for all $f \in L^2(\mu)$, and equations (2.29), (2.30) hold for all $f, g \in L^2(\mu)$, and the map ϕ extends to an isometry

$$\phi : L^2(\mu) \to \mathbf{H}, \qquad \phi(f) = \int_{\mathbb{R}} f(\lambda)\, dv_\lambda \tag{2.33}$$

from $L^2(\mu)$ to a closed linear subspace of $\mathbf{H}$. Summing up, we obtain the following theorem.

Theorem 2.6. *Let* $\mathbf{H}$ *be a Hilbert space, let* $v : \mathbb{R} \to \mathbf{H}$ *be a right continuous function with orthogonal increments, and let* $\mu : \mathcal{B} \to [0, \infty]$ *be the associated measure on the* σ*-algebra* $\mathcal{B}$ *of Borel subsets of* $\mathbb{R}$ *(see (2.24), (2.25)). Then the definition (2.28) extends by continuity to yield an isometry*

$$\phi : L^2(\mu) \to \mathbf{H}, \qquad \phi(f) = \int_{\mathbb{R}} f(\lambda)\, dv_\lambda \qquad (f \in L^2(\mu)).$$

In particular, $f_n \to f$ *in* $L^2(\mu)$ *if and only if*

$$\int_{\mathbb{R}} f_n(\lambda)\, dv_\lambda \to \int_{\mathbb{R}} f(\lambda)\, dv_\lambda \qquad \textit{in } \mathbf{H}. \tag{2.34}$$

We maintain our assumptions on v and our notations (2.24), (2.25). Let $f \in L^2(\mu)$. Then $f\chi_A \in L^2(\mu)$ for all $A \in \mathcal{B}$, and we define

$$\int_A f(\lambda)\, dv_\lambda := \int_{\mathbb{R}} f(\lambda)\chi_A(\lambda)\, dv_\lambda. \tag{2.35}$$

Since ϕ is an isometry, we have

$$< \int_A f(\lambda)\, dv_\lambda, \int_B g(\lambda)\, dv_\lambda >= \int_{A\cap B} f\bar{g}\, d\mu \tag{2.36}$$

for all $f, g \in L^2(\mu),\ A, B \in \mathcal{B}$.

Lemma 2.7. *Let $\mathbf{H}, v, \mu$ be as in Theorem 2.6 and define*

$$\mathcal{B}_v := \{\ A \in \mathcal{B}\ \mid\ \chi_A \in L^2(\mu)\ \}, \tag{2.37}$$

$$v : \mathcal{B}_v \to \mathbf{H}, \qquad v(A) := \int_{\mathbb{R}} \chi_A(\lambda)\, dv_\lambda \tag{2.38}$$

for $A \in \mathcal{B}_v$. Then the following hold:

(1) $\mathcal{B}_v$ is a δ-ring of subsets of $\mathbb{R}$, and $\mathcal{B}_v$ contains all bounded Borel subsets of $\mathbb{R}$.
(2) v is a countably additive orthogonally scattered measure on $\mathcal{B}_v$, that is

$$v(A) = \sum_{n=1}^{\infty} v(A_n) \tag{2.39}$$

whenever $A \in \mathcal{B}_v$ is the disjoint union of the sets $A_n \in \mathcal{B}_v$ $(n \geq 1)$ and

$$< v(A), v(B) >= 0 \tag{2.40}$$

for all $A, B \in \mathcal{B}_v$ such that $A \cap B = \emptyset$. Moreover, we have

$$\|v(A)\|^2 = \mu(A) \qquad \textit{for all } A \in \mathcal{B}_v. \tag{2.41}$$

Proof. (1): is obvious from the definition.
(2): If $A \in \mathcal{B}_v$ is the disjoint union of the sets $A_n \in \mathcal{B}_v (n \geq 1)$, we have

$$\sum_{\nu=1}^{n} \chi_{A_\nu} \to \chi_A \quad \text{in} \quad L^2(\mu)$$

by the dominated convergence theorem. Hence (2.39) follows from (2.34) and (2.38). Properties (2.40), (2.41) are obvious from (2.36). □

Theorem 2.6 is our version of Masani's isomorphism theorem (cf. Masani (1968), page 79). Lemma 2.7 is a variant of the classical correspondence between measures on the real line and distribution functions. A more general extension theory for countably additive orthogonally scattered measures defined on an abstract pre-ring is in Masani (1968), sect. 2.

The following theorem is of basic importance for our decomposition theory.

Theorem 2.8. *Let* $\mathbf{H}$ *be a Hilbert space, let* $v : \mathbb{R} \to \mathbf{H}$ *be a right continuous function with orthogonal increments and with associated measure* μ *on the* σ*-algebra* $\mathcal{B}$ *of Borel subsets of* $\mathbb{R}$ *(cf. (2.25)), and let*

$$\mathcal{B}_v := \{ \ A \in \mathcal{B} \ \ | \ \ \chi_A \in L^2(\mu) \ \}.$$

Let $\mathbf{H}_v$ *be the image of the isometry* ϕ *of Theorem 2.6 and denote the orthogonal projection from* $\mathbf{H}$ *onto* $\mathbf{H}_v$ *by* P_v*. Then the following hold:*

(1) For any $x \in \mathbf{H}$*, the set function (cf. (2.38))*

$$\nu_x : \mathcal{B}_v \to \mathbb{C}, \qquad \nu_x(A) := < x, v(A) > \qquad for\ A \in \mathcal{B}_v \tag{2.42}$$

is countably additive. In particular, the restriction of ν_x *to the Borel subsets of any compact interval is a complex measure. This measure is absolutely continuous with respect to the restricted measure* μ*, and hence (putting pieces together) there exists a (Borel measurable) Radon–Nikodým derivative* $d\nu_x/d\mu$ *defined on all of* $\mathbb{R}$*.*

(2) $d\nu_x/d\mu \in L^2(\mu)$ *and*

$$P_v x = \int_{\mathbb{R}} \frac{d\nu_x}{d\mu}(\lambda)\ dv_\lambda. \tag{2.43}$$

(3) For all $x, y \in \mathbf{H}$*:*

$$< P_v x, P_v y >= \int_{\mathbb{R}} \frac{d\nu_x}{d\mu} \overline{\frac{d\nu_y}{d\mu}}\ d\mu, \tag{2.44}$$

$$\|P_v x\|^2 = \int_{\mathbb{R}} \left| \frac{d\nu_x}{d\mu} \right|^2 \ d\mu. \tag{2.45}$$

Proof. (1): The countable additivity of ν_x is obvious from Lemma 2.7. We proceed to prove absolute continuity: Let $A \in \mathcal{B}_v$, and let $A_1, \ldots, A_n \in \mathcal{B}_v$ be disjoint sets with union A. For convenience, we agree that in the sequel all summands with vanishing denominators are zero by definition. Then we have

$$
\begin{aligned}
&\sum_{k=1}^{n} |\nu_x(A_k)| = \sum_{k=1}^{n} \left| < x, \frac{1}{\|v(A_k)\|} v(A_k) > \right| \; \|v(A_k)\| \\
&\le \left(\sum_{k=1}^{n} | < x, \frac{1}{\|v(A_k)\|} v(A_k) > |^2 \right)^{1/2} \left(\sum_{k=1}^{n} \|v(A_k)\|^2 \right)^{1/2} \\
&\le \|x\| (\mu(A))^{1/2},
\end{aligned} \tag{2.46}
$$

where we have applied the Cauchy–Schwarz inequality, Bessel's inequality, and Lemma 2.7. If we now restrict ν_x to the Borel sets of a compact interval, we obtain a complex measure, whose total variation is absolutely continuous with respect to the restricted μ, by (2.46). Putting pieces together, we arrive at a Radon–Nikodým derivative $d\nu_x/d\mu$ which is a Borel measurable function defined on all of $\mathbb{R}$ and which is uniquely determined almost everywhere with respect to μ.

(2): Since $\phi : L^2(\mu) \to \mathbf{H}_v$ is bijective, there exists a unique element $f_x \in L^2(\mu)$ such that

$$
P_v x = \int_{\mathbb{R}} f_x(\lambda) \, dv_\lambda. \tag{2.47}
$$

Note that $v(A) \in \mathbf{H}_v$ for all $A \in \mathcal{B}_v$ and hence

$$
\begin{aligned}
\nu_x(A) &= < x, v(A) > = < P_v x, v(A) > \\
&= < \int_{\mathbb{R}} f_x(\lambda) \, dv_\lambda, \int_{\mathbb{R}} \chi_A(\lambda) \, dv_\lambda > = \int_A f_x(\lambda) \, d\mu(\lambda)
\end{aligned} \tag{2.48}
$$

for all $A \in \mathcal{B}_v$, see (2.36). Since (2.48) holds for all bounded Borel subsets of $\mathbb{R}$, we conclude that μ-almost everywhere

$$
f_x = \frac{d\nu_x}{d\mu}. \tag{2.49}
$$

If we consider a representative of the equivalence class $f_x \in L^2(\mu)$, we first observe that equation (2.49) holds almost everywhere with respect to μ. Hence we are justified to consider the Radon–Nikodým derivative as an element of $L^2(\mu)$, and then (2.49) holds as an equality between two elements of $L^2(\mu)$. This proves (2).

(3): is immediate from (2) and from Theorem 2.6. □

Theorem 2.8 is crucial for our decomposition theory. This theorem and its proof correspond to Masani's (1968) projection theorem 5.10, page 80. We want to point out here, that in Theorem 2.8 the Radon–Nikodým derivative $d\nu_x/d\mu$ can also be replaced by a Besicovitch derivative. This is shown in a more general context in Masani (1968), sect. 6.

We explain the relation of our concept of integral with the classical notions of Riemann type integrals and Hellinger type integrals. Let $\mathbf{H}, v, \mu, \mathbf{H}_v, P_v, \nu_x$

be as in Theorem 2.8. Assume that $f \in C_c(\mathbb{R})$ is a continuous function on $\mathbb{R}$ with compact support $\operatorname{supp}(f) \subset]\alpha, \beta[$. Let

$$(2.50) \qquad T: \alpha = \lambda_0 < \lambda_1 < \ldots < \lambda_n = \beta$$

be a subdivision of $[\alpha, \beta]$ and approximate f by the step function

$$(2.51) \qquad f_T = \sum_{\nu=1}^{n} f(\lambda'_n)\, \chi_{]\lambda_{\nu-1},\lambda_\nu]},$$

where the $\lambda'_\nu \in [\lambda_{\nu-1}, \lambda_\nu]$ are intermediate points. If we choose a sequence $(T_m)_{m\geq 1}$ of subdivisions (2.50) such that the maximal length of the intervals of the subdivisions tends to zero as $m \to \infty$, the corresponding sequence f_{T_m} converges to f uniformly on $\mathbb{R}$ and hence also converges to f in $L^2(\mu)$ because all the functions involved have their supports contained in $[\alpha, \beta]$. Hence (2.28) and (2.34) imply that the Riemann sums

$$(2.52) \qquad \sum_{\nu=1}^{n} f(\lambda'_\nu)(v_{\lambda_\nu} - v_{\lambda_{\nu-1}})$$

converge to

$$(2.53) \qquad \int_{\mathbb{R}} f(\lambda)\, dv_\lambda$$

as the width of the subdivision shrinks to zero. Hence (2.53) is equal to the corresponding Riemann type integral if f is a continuous function with compact support. The same is true for all integrals

$$(2.54) \qquad \int_{]\alpha,\beta]} f(\lambda)\, dv_\lambda \qquad (\alpha, \beta \in \mathbb{R},\ \alpha < \beta)$$

if f is a continuous function on $\mathbb{R}$. This implies, in particular, that our previous notation

$$\int_0^\lambda \mu\, dv_\mu$$

(cf. Definition 2.3) with the usual sign convention in the case $\lambda < 0$ agrees with our present notion of integral.

Note that the Riemann sum (2.52) has the norm square

$$\sum_{\nu=1}^{n} |f(\lambda'_\nu)|^2 \|v_{\lambda_\nu} - v_{\lambda_{\nu-1}}\|^2 = \sum_{\nu=1}^{n} |f(\lambda'_\nu)|^2 (F(\lambda_\nu) - F(\lambda_{\nu-1})),$$

where F is given by (2.23). Hence we obtain that

$$(2.55) \qquad \left\| \int_{\mathbb{R}} f(\lambda)\, dv \right\|^2 = \int_{\mathbb{R}} |f(\lambda)|^2\, dF(\lambda) \qquad \text{for all } f \in C_c(\mathbb{R}),$$

where the right-hand side is a Riemann–Stieltjes integral. An analogous equality holds for the integral (2.54).

We proceed to establish the relation to Hellinger-type integrals. Let $x \in \mathbf{H}$ and $\alpha, \beta \in \mathbb{R}$, $\alpha < \beta$, and let $Q_{\alpha,\beta}$ be the orthogonal projection from $\mathbf{H}$ onto the closed linear subspace $\mathbf{H}_{\alpha,\beta}$ of $\mathbf{H}$ generated by all elements

$$v_{\lambda'} - v_{\lambda''}. \qquad (\alpha \leq \lambda' < \lambda'' \leq \beta) \tag{2.56}$$

Recall that $\mathbf{H}_v$ is the closed linear subspace of $\mathbf{H}$ generated by all vectors

$$v_{\lambda'} - v_{\lambda''} \quad \text{with} \quad \lambda',\, \lambda'' \in \mathbb{R}. \tag{2.57}$$

Hence we deduce from the orthogonality properties of v and from (2.43) that

$$Q_{\alpha,\beta}x = \int_{]\alpha,\beta]} \frac{d\nu_x}{d\mu}\, dv_\lambda. \tag{2.58}$$

Going back to the definition of our integral we see that the element (2.58) is a norm limit of a sequence of linear combinations of the elements (2.56). Let T be a subdivision of the form (2.50). Then the orthogonality properties of v imply that the best approximation (in the norm sense) of $Q_{\alpha,\beta}x$ by an element in the span of

$$v_{\lambda_\nu} - v_{\lambda_{\nu-1}}, \qquad \nu = 1, \ldots, n \tag{2.59}$$

is given by

$$x_T = \sum_{\nu=1}^{n} \frac{< x, v_{\lambda_\nu} - v_{\lambda_{\nu-1}} >}{\|v_{\lambda_\nu} - v_{\lambda_{\nu-1}}\|^2} (v_{\lambda_\nu} - v_{\lambda_{\nu-1}}), \tag{2.60}$$

where summands with vanishing denominators are understood to be equal to zero, as above. We introduce the functions

$$\varphi : \mathbb{R} \to \mathbb{C}, \qquad \varphi(\lambda) := < x, v_\lambda > \quad (\lambda \in \mathbb{R}) \tag{2.61}$$

and F (cf. (2.23)) and have

$$x_T = \sum_{\nu=1}^{n} \frac{\varphi(\lambda_\nu) - \varphi(\lambda_{\nu-1})}{F(\lambda_\nu) - F(\lambda_{\nu-1})} (v_{\lambda_\nu} - v_{\lambda_{\nu-1}}). \tag{2.62}$$

The considerations above yield that $Q_{\alpha,\beta}x$ can be approximated in the norm by elements of the form (2.62). Assume now that v is also continuous (which is true if v is an eigenpacket of a symmetric operator). Let $T_m, m \geq 1$ be an arbitrary sequence of subdivisions of the form (2.50) such that the width of T_m tends to zero as $m \to \infty$. Let $\epsilon > 0$ and choose T such that the corresponding x_T (see (2.60)) satisfies $\|Q_{\alpha,\beta}x - x_T\| < \epsilon$. As $m \to \infty$, the division points $\lambda_0, \ldots, \lambda_n$ of T are approximated by suitable division points occurring in T_m. Hence there is an $m_0 \in \mathbb{N}$ such that for all $m \geq m_0$ there is an element y_m in the span of the set of difference vectors belonging to

T_m (cf. (2.59)) such that $\|x_T - y_m\| < \epsilon$ for all $m \geq m_0$. Hence we have $\|Q_{\alpha,\beta}x - y_m\| < 2\epsilon$ for all $m \geq m_0$. But the best approximation to $Q_{\alpha,\beta}x$ is given by elements of the form (2.60), (2.62). Denoting the corresponding element for T_m by x_{T_m}, we conclude that

$$\lim_{m\to\infty} x_{T_m} = Q_{\alpha,\beta}x. \tag{2.63}$$

We call the sum (2.62) a Hellinger sum and denote the limit of Hellinger sums by the Hellinger integral

$$\int_\alpha^\beta \frac{d\varphi(\lambda)\, dv_\lambda}{dF(\lambda)}. \tag{2.64}$$

We have now proved: If v is continuous and if T_m is an arbitrary sequence of subdivisions of $[\alpha, \beta]$ with width converging to zero as $m \to \infty$, the corresponding sequence of Hellinger sums (2.62) converges to the Hellinger integral (2.64). The projection of $x \in \mathbf{H}$ to the closed linear subspace $\mathbf{H}_{\alpha,\beta}$ is given by

$$Q_{\alpha,\beta}x = \int_{]\alpha,\beta]} \frac{d\nu_x}{d\mu}(\lambda)\, dv_\lambda = \int_\alpha^\beta \frac{d\varphi(\lambda)\, dv_\lambda}{dF(\lambda)} \tag{2.65}$$

(see (2.61), (2.23)).

The norm of the element (2.62) is easily computed:

$$\|x_T\|^2 = \sum_{\nu=1}^n \frac{|\varphi(\lambda_\nu) - \varphi(\lambda_{\nu-1})|^2}{F(\lambda_\nu) - F(\lambda_{\nu-1})}. \tag{2.66}$$

If T runs through a sequence T_m as above, the corresponding Hellinger sums (2.66) approach a limit which is written as the Hellinger integral

$$\int_\alpha^\beta \frac{|d\varphi(\lambda)|^2}{dF(\lambda)}. \tag{2.67}$$

Thus we have proved: For any $x \in \mathbf{H}$ the norm of the projection $Q_{\alpha,\beta}x$ is given by

$$\|Q_{\alpha,\beta}x\|^2 = \int_{]\alpha,\beta]} \left|\frac{d\nu_x}{d\mu}\right|^2 d\mu = \int_\alpha^\beta \frac{|d\varphi(\lambda)|^2}{dF(\lambda)}. \tag{2.68}$$

Letting $\alpha \to -\infty, \beta \to +\infty$, we obtain: If v is continuous, the projection of $x \in \mathbf{H}$ onto the closed linear subspace $\mathbf{H}_v$ of $\mathbf{H}$ generated by the elements (2.57) is given by

$$P_v x = \int_{\mathbb{R}} \frac{d\nu_x}{d\mu}(\lambda)\, dv_\lambda = \int_{-\infty}^{+\infty} \frac{d\varphi(\lambda)\, dv_\lambda}{dF(\lambda)}. \tag{2.69}$$

The norm square of this projection is equal to

$$\|P_v x\|^2 = \int_{\mathbb{R}} \left| \frac{d\nu_x}{d\mu} \right|^2 d\mu = \int_{-\infty}^{+\infty} \frac{|d\varphi(\lambda)|^2}{dF(\lambda)}. \tag{2.70}$$

The improper Hellinger integrals in (2.69), (2.70) obviously exist.

We now resume the discussion of spectral theory in terms of eigenpackets and eigenvectors. Before we write down the proper version of the spectral theorem, we make the definition: Two eigenpackets $(v_\lambda)_{\lambda\in\mathbb{R}}, (w_\lambda)_{\lambda\in\mathbb{R}}$ are called orthogonal whenever $< v_\alpha, w_\beta >= 0$ for all $\alpha, \beta \in \mathbb{R}$. In view of Theorem 2.4 this is equivalent to $< v_\lambda, w_\lambda >= 0$ for all $\lambda \in \mathbb{R}$. A system of orthogonal eigenpackets of a symmetric operator A in $\mathbf{H}$ is called a maximal orthogonal system of eigenpackets, whenever every eigenpacket of A which is orthogonal to all eigenpackets in the system vanishes identically.

Theorem 2.9. *Let $A : \mathcal{D}_A \to \mathbf{H}$ be a self-adjoint linear operator in the separable Hilbert space $\mathbf{H}$. Then:*

(1) There exists a countable orthonormal system e_m $(m \geq 1)$ of eigenvectors of A (possibly finite or empty) and a countable orthogonal system $v_n : \mathbb{R} \to \mathbf{H}$ $(n \geq 1)$ of eigenpackets of A (possibly finite or empty) such that the e_m $(m \geq 1)$ and the $v_{n,\lambda}$ $(n \geq 1,\ \lambda \in \mathbb{R})$ are complete in $\mathbf{H}$ (i.e. span a dense linear subspace of $\mathbf{H}$).

(2) Every orthonormal system of eigenvectors of A and every orthogonal system of eigenpackets of A can be enlarged to a complete system of orthonormal eigenvectors and orthogonal eigenpackets.

Proof. It suffices to prove (2). The space $\mathbf{H}$ splits into the direct orthogonal sum of the space $\mathbf{H}_{\text{eig}}$ generated by the eigenvectors of A and its orthogonal complement $\mathbf{H}_{\text{cont}}$. The spaces $\mathbf{H}_{\text{eig}}$ and $\mathbf{H}_{\text{cont}}$ both reduce A (cf. Weidmann (1979), Satz 7.28 and Satz 7.29). Every orthonormal system of eigenvectors of A can be embedded into a complete orthonormal system of eigenvectors of the restriction $A : \mathcal{D}_A \cap \mathbf{H}_{\text{eig}} \to \mathbf{H}_{\text{eig}}$. To see this, consider the direct orthogonal decomposition of $\mathbf{H}_{\text{eig}}$ into the sum of the eigenspaces of A. It remains to be shown that every finite or countably infinite orthogonal system $v_n : \mathbb{R} \to \mathbf{H}$ of eigenpackets can be embedded into an orthogonal system of eigenpackets of A which is complete in $\mathbf{H}_{\text{cont}}$. To do this, let U denote the closed linear subspace of $\mathbf{H}_{\text{cont}}$ which is generated by the given eigenpackets $v_{n,\lambda}$, $n \geq 1$, $\lambda \in \mathbb{R}$. Because $\mathbf{H}_{\text{cont}}$ is invariant with respect to the projections E_λ $(\lambda \in \mathbb{R})$ of the spectral family of A, it is clear from (2.15) that U is E_λ-invariant for all λ. Hence the space $U^\perp$ also is E_λ-invariant for all λ. Choose a complete orthonormal system $(f_n)_{n\geq 1}$ in $U^\perp$ and form the eigenpackets w_n, $w_{n,\lambda} := (E_\lambda - E_0)f_n$ (see Theorem 2.5). Taking the union of the v_n and the w_m, we arrive at an orthogonal system of eigenpackets of A which is complete in $\mathbf{H}_{\text{cont}}$. This proves our theorem. □

Our proof of Theorem 2.9, part (2) also yields the following result.

Theorem 2.10. *Let $A : \mathcal{D}_A \to \mathbf{H}$ be a self-adjoint operator and let e_m $(m \geq 1)$ be a maximal (= non-extendable) orthonormal system of eigenvectors of A. Let v_n $(n \geq 1)$ be a maximal orthogonal system of eigenpackets of A none of which vanishes identically. Then the combined systems of e_m $(m \geq 1)$ and v_n $(n \geq 1)$ are complete in $\mathbf{H}$ (i.e. the vectors e_m $(m \geq 1)$ and $v_{n,\lambda}$ $(n \geq 1,\ \lambda \in \mathbb{R})$ span a dense linear subspace of $\mathbf{H}$).*

Theorem 2.11 (Completeness and Expansion Theorem). *Let $A : \mathcal{D}_A \to \mathbf{H}$ be a self-adjoint operator in the separable Hilbert space $\mathbf{H}$, and let $(e_m)_{m\geq 1}$ be an orthonormal system of eigenvectors and v_n $(n \geq 1)$ be an orthogonal system of eigenpackets of A. Then the system $(e_m)_{m>1}$ together with the system v_n $(n \geq 1)$ is complete if and only if for any $x \in \mathbf{H}$ the following completeness relation is satisfied:*

$$\|x\|^2 = \sum_{m\geq 1} |< x, e_m >|^2 + \sum_{n\geq 1} \int_{\mathbb{R}} \left|\frac{d\nu_{n,x}}{d\mu_n}\right|^2 d\mu_n. \tag{2.71}$$

Here $\nu_{n,x}$ is defined by (2.42) with v replaced by v_n and μ_n is defined by (2.24) and (2.25) with v replaced by v_n. Equivalently, the completeness relation (2.71) can be written in terms of Hellinger integrals in the form

$$\|x\|^2 = \sum_{m\geq 1} |< x, e_m >|^2 + \sum_{n\geq 1} \int_{-\infty}^{\infty} \frac{|d\varphi_n(\lambda)|^2}{dF_n(\lambda)}, \tag{2.72}$$

where φ_n, F_n are defined by (2.61), (2.23) with v replaced by v_n. If completeness holds, every $x \in \mathbf{H}$ has an expansion of the form

$$x = \sum_{m\geq 1} < x, e_m > e_m + \sum_{n\geq 1} \int_{\mathbb{R}} \frac{d\nu_{n,x}}{d\mu_n}(\lambda)\, dv_{n,\lambda}. \tag{2.73}$$

Equivalently, this expansion formula can be written in terms of Hellinger integrals in the form

$$x = \sum_{m\geq 1} < x, e_m > e_m + \sum_{n\geq 1} \int_{-\infty}^{\infty} \frac{d\varphi_n(\lambda)\, dv_{n,\lambda}}{dF_n(\lambda)}. \tag{2.74}$$

Proof. Let $\mathbf{H}_m := \mathbb{C}e_m$ and put $\mathbf{K}_n := \overline{\text{span}\{\ v_{n,\lambda}\ \mid\ \lambda \in \mathbb{R}\ \}}$. Then the completeness property holds if and only if $\mathbf{H}$ splits into the direct sum

$$\mathbf{H} = \bigoplus_{m\geq 1} \mathbf{H}_m \oplus \bigoplus_{n\geq 1} \mathbf{K}_n.$$

This holds true if and only if every $x \in \mathbf{H}$ is the sum of its projections to all these subspaces. Collecting the contributions of the projections on the spaces $\mathbf{K}_n$ from Theorem 2.8, we see that completeness holds iff the completeness relation (2.71) is satisfied for all $x \in \mathbf{H}$, and in the case of completeness, the expansion formula (2.73) is valid. The reformulation of this in terms of Hellinger integrals can be drawn from the discussion above, see formulas (2.69) and (2.70). □

Corollary 2.12. *Let $A : \mathcal{D}_A \to \mathbf{H}$ be a self-adjoint operator in the separable Hilbert space $\mathbf{H}$ with associated spectral family $(E_\lambda)_{\lambda \in \mathbb{R}}$, let $(e_m)_{m\geq 1}$, v_n $(n \geq 1)$ be as in Theorem 2.11 and assume that $(e_m)_{m>1}$ together with v_n $(n \geq 1)$ is complete in $\mathbf{H}$. Then the spectral family of A satisfies*

$$(2.75) \qquad E_\alpha x = \sum_{\lambda_m \leq \alpha} < x, e_m > e_m + \sum_{n\geq 1} \int_{-\infty}^{\alpha} \frac{d\nu_{n,x}}{d\mu_n}(\lambda)\, dv_n(\lambda)$$

where $x \in \mathbf{H}, \alpha \in \mathbb{R}$ Here $Ae_m = \lambda_m e_m$ and $\nu_{n,x}$, μ_n have the same meaning as in Theorem 2.11.

The proof is obvious from Theorem 2.11 and (2.18).

This finishes our general discussion of decomposition theory. We come back to our self-adjoint operators $-\tilde{\Delta} : \tilde{\mathcal{D}} \to L^2(\Gamma \setminus \mathbb{H})$ for groups $\Gamma <$ $\mathbf{PSL}(2, \mathbb{C})$ which are cofinite but not cocompact. A major task for the next sections is to recast this abstract expansion theory into a form as concrete as possible. We collect now what we know so far. First we show that the process of closing Δ does not produce any new eigenfunctions or eigenpackets. More precisely, we have the following theorem.

Theorem 2.13. *Let $\Gamma < \mathbf{PSL}(2, \mathbb{C})$ be a cofinite group and let $(E_\lambda)_{\lambda \in \mathbb{R}}$ be the spectral family of the self-adjoint operator $\tilde{\Delta} : \tilde{\mathcal{D}} \to \mathbb{H}$. Then for all $f \in L^2(\Gamma \setminus \mathbb{H})$, $\lambda, \mu \in \mathbb{R}$ the element $(E_\mu - E_\lambda)f$ is almost everywhere equal to a real analytic function belonging to $\mathcal{D}$. In particular, all eigenfunctions and eigenpackets of $\tilde{\Delta}$ are real analytic and hence are eigenfunctions or eigenpackets of Δ, respectively. Furthermore we have $v_\lambda = 0$ for every eigenpacket and every $\lambda < 0$.*

Proof. First we show that

$$(2.76) \qquad u := (E_\mu - E_\lambda)f \qquad (\lambda \leq \mu)$$

is almost everywhere equal to a C^∞-function on $\mathbb{H}$. Take an arbitrary $g \in \mathcal{D}^\infty$ and write g in the form

$$(2.77) \qquad g = \sum_{M \in \Gamma} h \circ M,$$

where $h \in C_c^\infty(\mathbb{H})$. Then g belongs to the domain of definition of A^n for all $n \geq 1$, where $A := \Delta|_{\mathcal{D}}$, and $\tilde{\Delta}^n g = \Delta^n g$. Moreover, it is known from Hilbert space theory that u belongs to the domain of definition of $\tilde{\Delta}^n$ for all $n \geq 1$.

Hence the self-adjointness of $\tilde{\Delta}^n$ implies $< \tilde{\Delta}^n u, g >=< u, \Delta^n g >$. Inserting the Poincaré series (2.77) for g we obtain by dominated convergence

$$\int_{\mathbb{H}} (\tilde{\Delta}^n u)\, \bar{h}\, dv = \int_{\mathbb{H}} u\, (\Delta^n \bar{h})\, dv.$$

This equation holds for all $h \in C_c^\infty(\mathbb{H})$. Since Δ has real-analytic coefficients, the regularity theorem for the solutions of elliptic partial differential equations implies that u is almost everywhere equal to a function in $C^{2n-2}(\mathbb{H})$, see Dunford, Schwartz (1958), page 1708, corollary 4. Hence we conclude that u coincides almost everywhere with a C^∞-function on $\mathbb{H}$. Thus we may assume that $u \in C^\infty(\mathbb{H})$.

We proceed to prove that u (cf. (2.76)) is real analytic. Lemma 4.1.11 and the spectral representation of $\tilde{\Delta}^n$ yield

$$\|\Delta^n u\| = \|\tilde{\Delta}^n u\| = \left(\int_\lambda^\mu t^{2n}\, d < E_t f, f > \right)^{1/2} \leq (|\lambda| + |\mu|)^n \|f\|.$$

Now a theorem of Kotaké and Narasimhan, see Komatsu (1962), implies that u is real analytic, and $u \in \mathcal{D}$ by Lemma 4.1.11. The analyticity of the eigenpackets of $\tilde{\Delta}$ follows from Theorem 2.5. For the last statement note that $-\tilde{\Delta}$ is positive and hence $E_\lambda = 0$ for all $\lambda < 0$. □

Theorem 2.13 holds analogously for arbitrary discrete subgroups $\Gamma <$ **PSL**$(2, \mathbb{C})$, not only for cofinite subgroups; see Roelcke (1966), p. 323, Satz 5.6. The same remark applies to Corollary 2.14, Propositions 2.15, 2.16 and Theorem 2.17.

Corollary 2.14. *Let $\Gamma <$ **PSL**$(2, \mathbb{C})$ be a cofinite group. There are subsets*

(2.78) $$\mathfrak{D} = \mathfrak{D}(\Gamma) \subset \mathbb{N}, \qquad \mathfrak{C} = \mathfrak{C}(\Gamma) \subset \mathbb{N},$$

an orthonormal system $(e_m)_{m \in \mathfrak{D}}$ of eigenfunctions and an orthogonal system $(v_n)_{n \in \mathfrak{C}}$ of eigenpackets of the essentially self-adjoint linear operator $-\Delta : \mathcal{D} \to L^2(\Gamma \setminus \mathbb{H})$, such that the system $(e_m)_{m \in \mathfrak{D}}$ together with $(v_n)_{n \in \mathfrak{C}}$ is complete in $L^2(\Gamma \setminus \mathbb{H})$. We define

(2.79) $$L^2_{\text{disc}}(\Gamma \setminus \mathbb{H}), \qquad L^2_{\text{cont}}(\Gamma \setminus \mathbb{H})$$

to be the closed subspaces generated by the $(e_m)_{m \in \mathfrak{D}}$ or $(v_n)_{n \in \mathfrak{C}}$ respectively. Then $L^2_{\text{disc}}(\Gamma \setminus \mathbb{H})$ and $L^2_{\text{cont}}(\Gamma \setminus \mathbb{H})$ are orthogonal complements of each other in $L^2(\Gamma \setminus \mathbb{H})$.

This follows immediately from Theorem 2.13. Note that the restriction of the operator $-\Delta : \mathcal{D} \to L^2(\Gamma \setminus \mathbb{H})$ to the set of all real analytic functions belonging to $\mathcal{D}$ also is an essentially self-adjoint linear operator with closure $-\tilde{\Delta} : \tilde{\mathcal{D}} \to L^2(\Gamma \setminus \mathbb{H})$.

Together with Corollary 2.14 come the expansions (2.73), (2.74). We prove now pointwise versions of these. Let v_λ be one of the eigenpackets addressed to in Corollary 2.14. For a function f from the domain of definition $\tilde{\mathcal{D}}$ of $\tilde{\Delta}$ we define as in (2.61) $\varphi_f(\lambda) :=< f, v_\lambda >$. Let F be defined as in (2.23). Given real numbers $\alpha < \beta$ and $P \in \mathbb{H}$ we define similarly to (2.62)

$$f_T(P) = \sum_{\nu=1}^{n} \frac{\varphi_f(\lambda_\nu) - \varphi_f(\lambda_{\nu-1})}{F(\lambda_\nu) - F(\lambda_{\nu-1})}(v_{\lambda_\nu}(P) - v_{\lambda_{\nu-1}}(P)) \tag{2.80}$$

for a sudivision T as in (2.50). Let (T_m) be a sequence of subdivisions as in (2.63). We may now try to define a function on $\mathbb{H}$ by the limit

$$\lim_{m\to\infty} f_{T_m}(P). \tag{2.81}$$

Proposition 2.15. *Let $\Gamma < \mathbf{PSL}(2,\mathbb{C})$ be a cofinite group, v_λ an eigenpacket of $-\Delta$ in $L^2(\Gamma \setminus \mathbb{H})$ and $f \in \tilde{\mathcal{D}}$. Then the limit (2.81) exists, is independent of the sequence (T_m) and depends continuously on $P \in \mathbb{H}$. We define*

$$\int_\alpha^\beta \frac{d\varphi_f(\lambda)\, dv_\lambda(P)}{dF(\lambda)} := \lim_{m\to\infty} f_{T_m}(P). \tag{2.82}$$

The limit

$$\lim_{\substack{\beta\to\infty \\ \alpha\to-\infty}} \int_\alpha^\beta \frac{d\varphi_f(\lambda)\, dv_\lambda(P)}{dF(\lambda)} =: \int_{-\infty}^{\infty} \frac{d\varphi_f(\lambda)\, dv_\lambda(P)}{dF(\lambda)} \tag{2.83}$$

also exist and depends continuously on $P \in \mathbb{H}$. The function (2.83) represents the element (2.69) in $L^2(\Gamma \setminus \mathbb{H})$. Its norm in $L^2(\Gamma \setminus \mathbb{H})$ may be computed as

$$\left\| \int_{-\infty}^{\infty} \frac{d\varphi_f(\lambda)\, dv_\lambda(P)}{dF(\lambda)} \right\|^2 = \int_{-\infty}^{\infty} \frac{|d\varphi_f(\lambda)|^2}{dF(\lambda)}. \tag{2.84}$$

We have

$$\begin{aligned} \int_{-\infty}^{\infty} \frac{d\varphi_f(\lambda)\, dv_\lambda(P)}{dF(\lambda)} &= \int_{-\infty}^{\infty} \frac{d\varphi_f(\lambda)}{dF(\lambda)} \frac{dv_\lambda(P)}{dF(\lambda)}\, dF(\lambda) \\ &= \int_0^{\infty} \frac{d\varphi_f(\lambda)}{dF(\lambda)} \frac{dv_\lambda(P)}{dF(\lambda)}\, dF(\lambda) \end{aligned} \tag{2.85}$$

where the two latter integrals are usual Lebesgue-Stieltjes integrals. The derivatives are Radon–Nikodým derivatives with respect to dF. A similar statement holds for the improper Hellinger integral on the right-hand side of (2.84).

Before commenting on the proof we add some formulas which describe the action of the resolvent kernel $F(P, Q, s)$, see Chapter 4, Sections 2–5, on eigenpackets. We fix $P \in \mathbb{H}$ and a complex number s_0 with $\operatorname{Re} s_0 > 1$ and put $\lambda_0 = 1 - s_0^2$. Since v_λ is an eigenpacket for $-\Delta$ we obtain from Corollary 4.2.8

$$< F(P,\cdot,s_0), \int_0^\mu (\alpha - \lambda_0)\, dv_\alpha >= \overline{v_\mu(P)}. \tag{2.86}$$

Going back to the definitions we easily deduce

$$\begin{aligned} < F(P,\cdot,s_0), v_\lambda > &=< F(P,\cdot,s_0), \int_0^\lambda \frac{1}{\mu-\lambda_0}\, d\int_0^\mu (\alpha - \lambda_0)\, dv_\alpha > \\ &= \int_0^\lambda \frac{1}{\mu-\lambda_0}\, d < F(P,\cdot,s_0), \int_0^\mu (\alpha - \lambda_0)\, dv_\alpha > \\ &= \overline{\int_0^\lambda \frac{1}{\mu-\lambda_0}\, dv_\mu(P)}. \end{aligned} \tag{2.87}$$

The integral in the last line and the outer integral in the line before are limits of Stieltjes sums in $\mathbb{C}$. Note that according to our choice of λ_0 there are no zeroes of $\mu - \lambda_0$ in the relevant ranges. From formula (2.87) we may now more explicitly determine the norm of the square integrable function $F(P,\cdot,s_0)$. We obviously find from Theorem 2.11 and (2.87)

$$\|F(P,\cdot,s_0)\|^2 = \sum_{m\in\mathfrak{D}} \frac{|e_m(P)|^2}{|\lambda_m - \lambda_0|^2} + \sum_{n\in\mathfrak{C}} \int_0^\infty \frac{|dv_{n,\lambda}(P)|^2}{|\lambda - \lambda_0|^2\, dF_n(\lambda)} \tag{2.88}$$

The notation is as in Corollary 2.14. The lower bound 0 for the integrals follows from Theorem 2.13.

From (2.88) and Corollary 4.2.5 we immediately obtain the following consequence which may be seen in analogy with part of the statement in Corollary 4.5.5. For the comparison of Hellinger and Lebesgue–Stieltjes integrals see Theorem 2.8.

Proposition 2.16. *Let* $\Gamma < \mathbf{PSL}(2,\mathbb{C})$ *be a cofinite group,* v_λ *an eigenpacket of* $-\Delta$ *in* $L^2(\Gamma\setminus\mathbb{H})$ *and* $\lambda_0 < 0$. *Then the integral*

$$\int_0^\infty \frac{|dv_\lambda(P)|^2}{(\lambda-\lambda_0)^2\, dF(\lambda)} = \int_0^\infty \frac{1}{(\lambda-\lambda_0)^2} \left|\frac{dv_\lambda(P)}{dF(\lambda)}\right|^2 dF(\lambda) \tag{2.89}$$

exists and defines a function on $\mathbb{H}$ *which is bounded on compact subsets of* $\mathbb{H}$. *The integral on the left is an improper Hellinger integral whereas the integral on the right is a usual Lebesgue–Stieltjes integral. The derivatives are Radon–Nikodým derivatives with respect to* dF.

Proof of Proposition 2.15. The proof consists of a straightforward application of Cauchy–Schwarz. It is word for word the same as the argument on pages 74–77 of Roelcke (1956a). Only Roelcke's Green's function has to be replaced by our resolvent kernel $F(P,Q,s)$. The relevant properties of $F(P,Q,s)$ are (2.87), (2.88). We skip the details. □

We are now able to give the pointwise version of (2.74). The straightforward proof proceeds as in the argument for Satz 34 of Roelcke (1956a); see also Roelcke (1967), p. 271–272, Satz 7.2. Again we replace Roelcke's Green's function by the resolvent kernel $F(P,Q,s)$.

Theorem 2.17. *Let $\Gamma < \mathbf{PSL}(2,\mathbb{C})$ be a cofinite group. Let $(e_m)_{m\in\mathfrak{D}}$ be a complete orthonormal system of eigenfunctions and $(v_n)_{n\in\mathfrak{C}}$ a complete orthogonal system of eigenpackets of the essentially self-adjoint linear operator $-\Delta : \mathcal{D} \to L^2(\Gamma\setminus\mathbb{H})$. Assume that $f \in \tilde{\mathcal{D}}$ and $P \in \mathbb{H}$. Then the expansion*

$$f(P) = \sum_{m\in\mathfrak{D}} < f, e_m > e_m(P) + \sum_{n\in\mathfrak{C}} \int_{-\infty}^{\infty} \frac{d\varphi_{n,f}(\lambda)\, dv_{n,\lambda}(P)}{dF_n(\lambda)} \tag{2.90}$$

converges locally uniformly and absolutely in $\mathbb{H}$. This means that for every compact set $\mathcal{K} \subset \mathbb{H}$ and every $\epsilon > 0$ there are finite subsets $\mathfrak{D}' \subset \mathfrak{D}$ and $\mathfrak{C}' \subset \mathfrak{C}$ and a $\gamma > 0$ such that

$$\sum_{m\in\mathfrak{D}\setminus\mathfrak{D}'} |< f, e_m > e_m(P)| + \sum_{n\in\mathfrak{C}\setminus\mathfrak{C}'} \int_{-\infty}^{\infty} \frac{|d\varphi_{n,f}(\lambda)|\, |dv_{n,\lambda}(P)|}{dF_n(\lambda)} \tag{2.91}$$

and

$$\sum_{m\in\mathfrak{D}\setminus\mathfrak{D}'} |< f, e_m > e_m(P)| + \sum_{n\in\mathfrak{C}} \int_{\gamma}^{\infty} \frac{|d\varphi_{n,f}(\lambda)|\, |dv_{n,\lambda}(P)|}{dF_n(\lambda)} \tag{2.92}$$

are smaller than ϵ for every $P \in \mathcal{K}$.

6.3 Spectral Decomposition Theory

In this section we explicitly describe the spectral decomposition for the self-adjoint extension $\tilde{\Delta}$ of the Laplace operator on $L^2(\Gamma\setminus\mathbb{H})$ in case $\Gamma < \mathbf{PSL}(2,\mathbb{C})$ is a cofinite but not cocompact subgroup. In fact we give two approaches. In the first, which was followed by Roelcke (1966) in the 2-dimensional situation, we describe a maximal system of eigenpackets. In the second we give unitary maps

$$\Theta : L^2(]0,\infty[, dt) \to L^2(\Gamma\setminus\mathbb{H})$$

so that the pullback of $\tilde{\Delta}$ is a multiplication operator. This version is adapted from Selberg (1989b). For both approaches the Eisenstein series, in fact their continuation to the line $\operatorname{Re} s = 0$ play an essential role.

To fix notation we as in Section 1 choose $B_1, \ldots, B_h \in \mathbf{PSL}(2,\mathbb{C})$ so that

$$\eta_1 = B_1^{-1}\infty, \ldots, \eta_h = B_h^{-1}\infty \in \mathbb{P}^1\mathbb{C} \tag{3.1}$$

are representatives for the Γ-classes of cusps of Γ. In our notation from Chapter 3, Section 2 we put for $\nu = 1, \ldots, h$ and $P \in \mathbb{H}$

$$E_\nu(P,s) := \frac{1}{[\Gamma_\nu : \Gamma'_\nu]} E_{B_\nu}(P,s) = \frac{1}{[\Gamma_\nu : \Gamma'_\nu]} \sum_{M \in \Gamma'_\nu \backslash \Gamma} r(B_\nu M P)^{1+s}. \tag{3.2}$$

Here $\Gamma_\nu = \Gamma_{\eta_\nu}$ stands for the stabilizer in Γ of the cusp η_ν whereas Γ'_ν is the unipotent part of Γ_ν. We write $\Lambda_\nu \subset \mathbb{C}$ for the lattice corresponding to Γ'_ν. By the results of Section 1 we can continue these series to meromorphic families of Γ-invariant C^∞-functions on $\mathbb{H}$. We also found that $E_\nu(P,s)$ is actually holomorphic on the imaginary axis $s = it$, $t \in \mathbb{R}$. We also know that if $s \neq 1$ is not a pole of $E_\nu(\cdot, s)$ then it has a Fourier expansion in the cusps of the form

$$E_\nu(B_\nu^{-1}P, s) = r^{1+s} + \phi_{\nu\nu}(s)\, r^{1-s} + \ldots \tag{3.3}$$

for $\nu = 1, \ldots, h$ and in case $\nu \neq \mu$

$$E_\nu(B_\mu^{-1}P, s) = \phi_{\nu\mu}(s)\, r^{1-s} + \ldots \tag{3.4}$$

The meromorphic functions $\phi_{\nu\mu}(s)$ are continuations of certain Dirichlet series. They do not have a pole in s if $E_\nu(\cdot, s)$ doesn't.

Theorem 3.1. *For $P \in \mathbb{H}$ and $\nu = 1, \ldots h$, we define*

$$V_\lambda^{[\nu]}(P) := \begin{cases} 0 & \text{for } \lambda < 1, \\ \displaystyle\int_0^{\sqrt{\lambda - 1}} E_\nu(P, it)\, dt & \text{for } \lambda \geq 1 \end{cases} \tag{3.5}$$

with positive choice of the square root. Then $V_\lambda^{[\nu]}(\cdot) \in L^2(\Gamma \setminus \mathbb{H})$ and the families $(V_\lambda^{[\nu]})_{\lambda \in \mathbb{R}}$ are a maximal system of mutually orthogonal eigenpackets for $-\tilde{\Delta}$. We have

$$\|V_{\lambda_1}^{[\nu]} - V_{\lambda_2}^{[\nu]}\|^2 = 2\pi \frac{|\Lambda_\nu|}{[\Gamma_\nu : \Gamma'_\nu]} (T_1 - T_2) \tag{3.6}$$

for $\lambda_1 \geq \lambda_2 \geq 1$ and $T_1 = \sqrt{\lambda_1 - 1}$, $T_2 = \sqrt{\lambda_2 - 1}$.

Theorem 3.1 shows that the index set $\mathfrak{C}(\Gamma)$ from Corollary 2.14 may be taken as $\mathfrak{C}(\Gamma) = \{1, \ldots, h\}$.

Before proceeding we define

$$\tau : \{\, \varphi \in L^2(]0, \infty[, dt) \mid \alpha\varphi \in L^2(]0, \infty[, dt) \,\} \to L^2(]0, \infty[, dt) \tag{3.7}$$

to be the operator given by multiplication with the function $\alpha :]0,\infty[\to \mathbb{C}$, $\alpha(t) = 1+t^2$. This is an unbounded, self-adjoint operator on $L^2(]0,\infty[, dt)$. Its spectrum is absolutely continuous and consists of the interval $[1,\infty[$.

Theorem 3.2. *For* $\nu = 1, \dots h$, $P \in \mathbb{H}$, $\varphi \in C_c^\infty(]0,\infty[)$ *we define*

$$\Theta^{[\nu]}(\varphi)(P) = \sqrt{\frac{[\Gamma_\nu : \Gamma'_\nu]}{2\pi|\Lambda_\nu|}} \int_0^\infty E_\nu(P, it)\, \varphi(t)\, dt. \tag{3.8}$$

Then $\Theta^{[\nu]}(\varphi) \in L^2(\Gamma \setminus \mathbb{H})$, *and the map* $\Theta^{[\nu]}$ *extends to a unitary map*

$$\Theta^{[\nu]} : L^2(]0,\infty[, dt) \to L^2(\Gamma \setminus \mathbb{H}). \tag{3.9}$$

The maps $\Theta^{[1]}, \dots, \Theta^{[h]}$ *thus obtained can be orthogonally composed to give a unitary map*

$$\Theta := \Theta^{[1]} \perp \dots \perp \Theta^{[h]} :\ L^2(]0,\infty[, dt) \perp \dots \perp L^2(]0,\infty[, dt) \ \to\ L^2(\Gamma \setminus \mathbb{H}).$$

The image of Θ *is* $L^2_{\text{cont}}(\Gamma \setminus \mathbb{H})$. *The pullback of the restriction of* $-\tilde{\Delta}$ *to* $L^2_{\text{cont}}(\Gamma \setminus \mathbb{H})$ *under* Θ *is the operator* $\tau \perp \dots \perp \tau$.

Theorem 3.2 has the following obvious corollary.

Corollary 3.3. *Let* $\Gamma < \mathbf{PSL}(2,\mathbb{C})$ *be a cofinite subgroup. The spectrum of* $-\Delta$ *on* $L^2(\Gamma\setminus\mathbb{H})$ *consists of a discrete part and an absolutely continuous part. The absolutely continuous part consists of the interval* $[1,\infty[$ *with multiplicity equal to the number of classes of cusps of* Γ.

The results of Section 2 imply the following theorem which gives the spectral expansion of elements in $L^2(\Gamma \setminus \mathbb{H})$.

Theorem 3.4. *Let as in Corollary 2.14* $\mathfrak{D} \subset \mathbb{N}$ *be an index set for a complete orthonormal set of eigenfunctions* $(e_n)_{n\in\mathfrak{D}}$ *for* $-\Delta$ *in* $L^2(\Gamma \setminus \mathbb{H})$. *Then the following hold.*

(1) Every $f \in L^2(\Gamma \setminus \mathbb{H})$ *has an* L^2*-convergent expansion*

$$\begin{aligned} f = \sum_{m\in\mathfrak{D}} < f, e_m > e_m + \frac{1}{2\pi} \sum_{\nu=1}^{h} \frac{[\Gamma_\nu : \Gamma'_\nu]}{|\Lambda_\nu|} \\ \cdot \int_0^\infty \frac{d < f, \int_0^T E_\nu(\cdot, it)\, dt > \ d\int_0^T E_\nu(\cdot, it)\, dt}{dT}. \end{aligned} \tag{3.10}$$

The integral is defined as a limit of Hellinger sums in $L^2(\Gamma \setminus \mathbb{H})$.

(2) If f *lies in the domain of definition* $\tilde{\mathcal{D}}$ *of* $-\tilde{\Delta}$ *then*

$$f(P) = \sum_{m \in \mathfrak{D}} <f, e_m> e_m(P) + \frac{1}{2\pi} \sum_{\nu=1}^{h} \frac{[\Gamma_\nu : \Gamma'_\nu]}{|\Lambda_\nu|} \tag{3.11}$$
$$\int_0^\infty E_\nu(P, iT) \frac{d}{dT} < f, \int_0^T E_\nu(\cdot, it)\, dt > dT.$$

This series converges absolutely and locally uniformly in the sense of Theorem 2.17. The function

$$T \mapsto < f, \int_0^T E_\nu(\cdot, it)\, dt >$$

is absolutely continuous, hence differentiable almost everywhere. Its derivative belongs to $L^2([0,\infty[)$. *The integrals from* 0 *to* ∞ *in (3.11) converge locally uniformly in the sense that the integrals* $\lim_{A\to\infty} \int_0^A |\ldots|\, dT$ *converge uniformly on compact subsets of* $\mathbb{H}$.

(3) If $f \in \tilde{\mathcal{D}}$ *and if* $f(\cdot)\overline{E_\nu(\cdot, it)}$ *is absolutely integrable over* $\mathcal{F} \times [0, T]$ *for every* $T \geq 0$ *and every* $\nu = 1, ..., h$ *then (3.11) may be replaced by*

$$f(P) = \sum_{m \in \mathfrak{D}} < f, e_m > e_m(P) \tag{3.12}$$
$$+ \frac{1}{4\pi} \sum_{\nu=1}^{h} \frac{[\Gamma_\nu : \Gamma'_\nu]}{|\Lambda_\nu|} \int_{-\infty}^{\infty} < f, E_\nu(\cdot, it) > E_\nu(P, it)\, dt$$

where $< f\,,\, E_\nu(\cdot, it) >$ *is defined as* $\int_{\mathcal{F}} f(Q)\, \overline{E_\nu(Q, it)}\, dv(Q)$. *The series (3.12) converges absolutely and locally uniformly in the sense of Theorem 2.17. We also obtain the following convergent series for the norm of* f

$$\|f\|^2 = \sum_{m \in \mathfrak{D}} | < f, e_m > |^2 \tag{3.13}$$
$$+ \frac{1}{4\pi} \sum_{\nu=1}^{h} \frac{[\Gamma_\nu : \Gamma'_\nu]}{|\Lambda_\nu|} \int_{-\infty}^{\infty} | < f, E_\nu(\cdot, it) > |^2\, dt.$$

The contribution of the eigenpacket $V_\lambda^{[\nu]}$ to the expansion (2.74) is

$$\int_{-\infty}^{\infty} \frac{d < f, V_\lambda^{[\nu]} > dV_\lambda^{[\nu]}}{dF_\nu(\lambda)}$$
$$= \frac{[\Gamma_\nu : \Gamma'_\nu]}{2\pi |\Lambda_\nu|} \int_0^\infty \frac{d < f, \int_0^T E_\nu(\cdot, it)\, dt > d \int_0^T E_\nu(\cdot, it)\, dt}{dT}.$$

This can be seen by putting in the definition of $V_\lambda^{[\nu]}$, switching variables from λ to $T = \sqrt{\lambda - 1}$ and using formula (3.6).

If $f \in \tilde{\mathcal{D}}$, the contribution of $V_\lambda^{[\nu]}$ to the expansion (2.90) can by Proposition 2.15 be computed as a Hellinger integral and by formulas (2.85) and (3.6) as a Lebesgue integral as follows:

$$\int_1^\infty \frac{d < f, V_\lambda^{[\nu]} > \; dV_\lambda^{[\nu]}(P)}{dF_\nu(\lambda)}$$
$$= \frac{[\Gamma_\nu : \Gamma_\nu']}{2\pi|\Lambda_\nu|} \int_0^\infty \frac{d < f, \int_0^T E_\nu(\cdot, it)\, dt >}{dT} \; \frac{d \int_0^T E_\nu(P, it)\, dt}{dT} \, dT.$$

The Radon–Nikodým derivatives occurring in the above are in fact usual derivatives since $\int_0^T E_\nu(P, it)\, dt$ is differentiable with respect to the variable T. The function $< f, \int_0^T E_\nu(\cdot, it)\, dt >$ is absolutely continuous and hence differentiable almost everywhere with respect to T. By similar reasoning we may compute the contribution of $V_\lambda^{[\nu]}$ to the completeness relation (2.72) as

$$\int_{-\infty}^\infty \frac{|d < f, V_\lambda^{[\nu]} > |^2}{dF_\nu(\lambda)}$$
$$= \frac{1}{2\pi} \frac{[\Gamma_\nu : \Gamma_\nu']}{|\Lambda_\nu|} \int_0^\infty \left| \frac{d}{dT} < f, \int_0^T E_\nu(\cdot, it)\, dt > \right|^2 dT. \tag{3.14}$$

If $f \in \tilde{\mathcal{D}}$ and if $f(\cdot)\overline{E_\nu(\cdot, it)}$ is absolutely integrable over $\mathcal{F} \times [0, T]$ for every $T \geq 0$ then by Fubini and the Fundamental Theorem of Calculus in the Lebesgue version

$$\begin{aligned} \frac{d}{dT} < f, \int_0^T E_\nu(\cdot, it)\, dt > &= \frac{d}{dT} \int_{\mathcal{F}} f(Q) \int_0^T \overline{E_\nu(Q, it)}\, dt\, dv(Q) \\ &= \frac{d}{dT} \int_0^T \int_{\mathcal{F}} f(Q)\, \overline{E_\nu(Q, it)}\, dv(Q)\, dt \\ &= \int_{\mathcal{F}} f(Q)\, \overline{E_\nu(Q, iT)}\, dv(Q). \end{aligned}$$

Of course this equation makes sense only almost everywhere. We then put formally

$$< f, E_\nu(\cdot, iT) > := \int_{\mathcal{F}} f(Q)\, \overline{E_\nu(Q, iT)}\, dv(Q).$$

A simple computation using the functional equation of the $E_\nu(\cdot, iT)$ under $T \to -T$ and that $\Phi(iT)$ is unitary produces the form of the contribution of $V_\lambda^{[\nu]}$ given in (3.12).

These computations show that Theorem 3.4 follows immediately from Theorem 2.11 once the completeness part of Theorem 3.1 is known. We do not enter into the somewhat delicate arguments for the proof of (2) and refer to the careful exposition in Roelcke (1967), paragraph 12, in particular pages 314–319. For most practical purposes statement (3) will suffice anyhow.

As in Section 1 we shall give the proofs in case Γ has only one class of cusps with an occasional comment if in the several cusps case a non-notational difficulty arises. We assume that our class of cusps is represented by $\infty \in \mathbb{P}^1\mathbb{C}$. We then choose a fundamental domain $\mathcal{F}$ of the kind described in Proposition 2.3.9 for Γ. Let $\mathcal{P} \subset \mathbb{C}$ be a fundamental polygon for the action of Γ_∞ on $\mathbb{C} = \mathbb{P}^1\mathbb{C}\backslash\{\infty\}$ and $Y > 0$ from Proposition 2.3.9. The fundamental domain $\mathcal{F}$ is then the union $\mathcal{F} = \mathcal{F}_0 \cup \mathcal{F}_\infty(Y)$ of a compact polyhedron $\mathcal{F}_0$ and the cusp sector $\mathcal{F}_\infty(Y) = \{\, z + rj \in \mathbb{H} \mid z \in \mathcal{P},\ r > Y \,\}$. For an $A > Y$ we also use the notation

$$(3.15)\quad \mathcal{F}^A := \mathcal{F}_\infty(A) = \{\, z + rj \in \mathbb{H} \mid z \in \mathcal{P},\ r > A \,\}, \quad \mathcal{F}_A := \mathcal{F} \backslash \mathcal{F}^A.$$

We keep the notation (1.10) and introduce the cut-off Eisenstein series

$$(3.16)\qquad E^A(\cdot, s) := E(\cdot, s) - \alpha(A, \cdot, s)$$

where $\alpha(A, \cdot, s)$ is defined for $P = (x, y, r) \in \mathbb{H}$ by

$$(3.17)\qquad \alpha(A, P, s) := \begin{cases} r^{1+s} + \phi(s)\, r^{1-s} & \text{for } r \geq A, \\ 0 & \text{for } r < A. \end{cases}$$

The function ϕ is the coefficient of r^{1-s} in (1.4). In the computations which will follow we will make use of the Maaß–Selberg relations from Theorem 3.3.6. Putting

$$(3.18)\qquad C_\infty := \frac{|\Lambda_\infty|}{[\Gamma_\infty : \Gamma'_\infty]}$$

they read, assuming $s \neq \pm\bar{s}$, for cut-off Eisenstein series as

$$(3.19)\qquad \begin{aligned} &\int_{\mathcal{F}} |E^A(P, s)|^2 \, dv \\ &= C_\infty \left(\frac{A^{s+\bar{s}} - \phi(s)\overline{\phi(s)} A^{-s-\bar{s}}}{s + \bar{s}} + \frac{\overline{\phi(s)} A^{s-\bar{s}} - \phi(s) A^{-s+\bar{s}}}{s - \bar{s}} \right). \end{aligned}$$

We also introduce the zeroth Fourier coefficient of the Eisenstein series

$$(3.20)\qquad \alpha(P, s) := r^{1+s} + \phi(s)\, r^{1-s} \qquad (P = (x, y, r) \in \mathbb{H}).$$

Similarly to Corollary 3.3.3 the exponential decrease of the Bessel functions for $r \to \infty$ implies that

$$(3.21)\qquad E(P, s) - \alpha(P, s) = O(e^{-\kappa r}) \qquad \text{as } r \to \infty$$

whenever s is not a pole of $E(\cdot, s)$ and $P = (x, y, r) \in \mathbb{H}$. The estimate (3.21) is uniform in x, y and for $s = it$ also uniform in any finite interval for t. This follows from the estimates for the Bessel functions $K_s(r)$ on page 139 of Magnus, Oberhettinger, Soni (1966).

We draw a first consequence from the properties of the Eisenstein series that we have just listed.

Lemma 3.5. *Let $|c_\Gamma|$ be the smallest of the absolute values of the nonzero left-hand lower entries of the elements of Γ. Then the function $|c_\Gamma|^{2s}\phi(s)$ is for every $\epsilon > 0$ bounded in the set*

$$\{\, s \in \mathbb{C} \;\mid\; 0 \leq \operatorname{Re} s \leq 2,\ |\operatorname{Im} s| \geq \epsilon \,\} \cup \{\, s \in \mathbb{C} \;\mid\; 2 \leq \operatorname{Re} s \,\}.$$

It tends uniformly to 0 as $\operatorname{Re} s \to \infty$.

Proof. From the expression for $\phi(s)$ in Theorem 3.4.1 we see that $|c_\Gamma|^{2s}\phi(s)$ is bounded in the second of the above sets. We take $s = a + ib$ with $a > 0$ and $b \neq 0$ in (3.19). Since the left-hand side is nonnegative we get

$$|\phi(s)|^2 \leq A^{4a} + \frac{2aA^{2a}}{|b|}|\phi(s)|.$$

From here the result follows. □

In the arguments which will come we need the following elementary lemma.

Lemma 3.6. *Let $W : \mathbb{H} \times \mathbb{R} \to \mathbb{C}$ be a continuous function so that $(W(\cdot,\lambda))_{\lambda\in\mathbb{R}}$ is a norm continuous family of elements of $L^2(\Gamma\backslash\mathbb{H})$. Then for all $a, b \in \mathbb{R}$:*

$$(R-)\int_a^b W(\cdot,t)\,dt = (L^2-)\int_a^b W(\cdot,t)\,dt,$$

$$(RS-)\int_a^b t\,dW(\cdot,t) = (L^2-)\int_a^b t\,dW(\cdot,t),$$

where the prefixes $(R-)$, $(RS-)$ indicate that the integrals are ordinary Riemann or Riemann–Stieltjes integrals, whereas the prefix $(L^2$-$)$ stands for the L^2-limit of the associated Riemann or Riemann–Stieltjes sums.

Proof. Consider a sequence of Riemann sums $Z_n(P)$ $(P \in \mathbb{H},\ n \in \mathbb{N})$ for

$$g(P) = (R-)\int_a^b W(P,t)\,dt.$$

This sequence can also be regarded as a sequence of Riemann sums $Z_n \in L^2(\Gamma\backslash\mathbb{H})$ for

$$f := (L^2-)\int_a^b W(\cdot,t)\,dt.$$

Hence we simultanuously have the equations $\lim_{n\to\infty} Z_n(P) = g(P)$ and $\lim_{n\to\infty} \|f - Z_n\| = 0$. Then $(Z_n)_{n\in\mathbb{N}}$ has a subsequence converging to f

almost everywhere, hence $f = g$ almost everywhere. As usual we identify elements of $L^2(\Gamma\backslash\mathbb{H})$ having a continuous representative with their unique continuous representative. This proves the first equation, the second follows by partial integration. □

Proof of Theorem 3.1. We have first of all to show that the functions $V_\lambda : \mathbb{H} \to \mathbb{C}$ defined by

$$V_\lambda(P) := \begin{cases} 0 & \text{for } \lambda < 1, \\ \displaystyle\int_0^{\sqrt{\lambda-1}} E(P, it)\, dt & \text{for } \lambda \geq 1 \end{cases}$$

are in $L^2(\Gamma \setminus \mathbb{H})$. During the following computations we put $T := \sqrt{\lambda - 1}$ whenever $\lambda \geq 1$. By (3.21) we only have to show that the functions $\beta(\cdot, T) : \mathbb{H} \to \mathbb{C}$ defined by

$$\beta(P, T) := \int_0^T \alpha(P, it)\, dt$$

are in $L^2(\mathcal{F})$. An obvious partial integration shows that

$$\beta(P, T) = O\left(\frac{r}{\log r}\right) \qquad \text{as } r \to \infty \tag{3.22}$$

uniformly in any finite interval for T. This shows that $V_\lambda(P)$ is square integrable over $\mathcal{F}$ and does depend continuously on λ (in the norm sense). The estimates (3.21) and (3.22) can be used to prove that differentiation with respect to x, y, r can be interchanged with the integral in the definition of $V_\lambda(P)$. We get for $\lambda \geq 1$

$$-\Delta V_\lambda(P) = \int_0^T -\Delta E(P, it)\, dt = \int_0^T (1 + t^2)\, E(P, it)\, dt. \tag{3.23}$$

An argument similar to the above shows then that $-\Delta V_\lambda(P)$ is square integrable over $\mathcal{F}$. Putting $\mu = 1 + t^2$ in (3.23) and remembering Lemma 3.6 we get

$$-\Delta V_\lambda = \int_0^\lambda \mu\, dV_\mu.$$

This proves that $V_\lambda(P)$ is an eigenpacket.

We turn to the proof of formula (3.6). We have for $1 < \lambda_2 = 1 + T_2^2 \leq \lambda_1 = 1 + T_1^2$ with $0 < T_2 \leq T_1$

$$\|V_{\lambda_1} - V_{\lambda_2}\|^2 = \int_{\mathcal{F}} \int_{T_2}^{T_1} \int_{T_2}^{T_1} E(P, it_1)\, \overline{E(P, it_2)}\, dt_1\, dt_2\, dv. \tag{3.24}$$

Inserting the decomposition (3.16) we find

$$\begin{aligned}\|V_{\lambda_1}-V_{\lambda_2}\|^2 &= \int_{\mathcal{F}}\int_{T_2}^{T_1}\int_{T_2}^{T_1} \alpha(A,P,it_1)\,\overline{\alpha(A,P,it_2)}\,dt_1\,dt_2\,dv \\ &+ \int_{\mathcal{F}}\int_{T_2}^{T_1}\int_{T_2}^{T_1} E^A(P,it_1)\,\overline{E^A(P,it_2)}\,dt_1\,dt_2\,dv =: L_1+L_2.\end{aligned} \tag{3.25}$$

We compute the first integral elementarily and the second by means of (3.19). We put

$$\begin{aligned}\psi(r,t_1,t_2) :&= r^{i(t_1-t_2)} + \phi(it_1)r^{-i(t_1+t_2)} + \phi(-it_2)r^{i(t_1+t_2)} \\ &+ \phi(it_1)\phi(-it_2)r^{-i(t_1-t_2)}\end{aligned}$$

and get

$$\begin{aligned}L_1 &= C_\infty \int_A^\infty \int_{T_2}^{T_1}\int_{T_2}^{T_1} \psi(r,t_1,t_2)\,dt_1\,dt_2\,\frac{dr}{r} \\ &= \lim_{B\to\infty} C_\infty \int_A^B \int_{T_2}^{T_1}\int_{T_2}^{T_1} \psi(r,t_1,t_2)\,dt_1\,dt_2\,\frac{dr}{r} \\ &= \lim_{B\to\infty} C_\infty \int_{T_2}^{T_1}\int_{T_2}^{T_1} (\psi_1(B,t_1,t_2)-\psi_1(A,t_1,t_2))\,dt_1\,dt_2\end{aligned} \tag{3.26}$$

where, by an elementary consideration

$$\begin{aligned}\psi_1(r,t_1,t_2) :&= \frac{r^{i(t_1-t_2)}}{i(t_1-t_2)} - \frac{\phi(it_1)r^{-i(t_1+t_2)}}{i(t_1+t_2)} \\ &+ \frac{\phi(-it_2)r^{i(t_1+t_2)}}{i(t_1+t_2)} - \frac{\phi(it_1)\phi(-it_2)r^{-i(t_1-t_2)}}{i(t_1-t_2)}.\end{aligned}$$

Note that $C_\infty\psi_1(A,t_1,t_2)$ is exactly the term on the right-hand side of (3.19) for $s = it_1$, $t = it_2$. In the range of integration of (3.26) we may write $\psi_1(B,t_1,t_2)$ as the sum

$$\psi_1(B,t_1,t_2) = \rho_1(B,t_1,t_2) + \rho_2(B,t_1,t_2) + \rho_3(B,t_1,t_2) \tag{3.27}$$

of the following three terms

$$\begin{aligned}\rho_1(B,t_1,t_2) &= 2\,\frac{\sin((t_1-t_2)\log B)}{(t_1-t_2)}, \\ \rho_2(B,t_1,t_2) &= -\frac{\phi(it_1)B^{-i(t_1+t_2)}}{i(t_1+t_2)} + \frac{\phi(-it_2)B^{i(t_1+t_2)}}{i(t_1+t_2)}, \\ \rho_3(B,t_1,t_2) &= \frac{(1-\phi(it_1)\phi(-it_2))B^{-i(t_1-t_2)}}{i(t_1-t_2)}.\end{aligned} \tag{3.28}$$

Considering the integrals

$$\tilde{\rho}_j(t_2) := \lim_{B\to\infty} C_\infty \int_{T_2}^{T_1} \rho_j(B,t_1,t_2)\,dt_1$$

for $j = 1,2,3$ we find that $\tilde{\rho}_2(t_2) = \tilde{\rho}_3(t_2) = 0$ for $t_2 \in [T_2, T_1]$. This follows from the Riemann–Lebesgue lemma since both $(t_1+t_2)^{-1}$ and $(1-\phi(it_1)\phi(-it_2))/i(t_1-t_2)$ are bounded in this range. For the latter fact we employ $\phi(it)\phi(-it) = 1$ from Theorem 1.2. Finally we have for $t_2 \in]T_2, T_1[$

$$\tilde{\rho}_1(t_2) = \lim_{B\to\infty} C_\infty \int_{T_2}^{T_1} 2\,\frac{\sin(\log B(t_1-t_2))}{(t_1-t_2)}\,dt_1 = 2C_\infty\pi$$

by a well known formula on Dirichlet integrals, see Bochner (1932), page 24, Satz 5. For $t_2 \notin [T_2, T_1]$ we have $\tilde{\rho}_1(t_2) = 0$ by the Riemann–Lebesgue lemma. Putting everything together we have shown that

$$L_1 = 2C_\infty\pi(T_1 - T_2) - C_\infty \int_{T_2}^{T_1}\int_{T_2}^{T_1} \psi_1(A, t_1, t_2)\,dt_1\,dt_2. \tag{3.29}$$

To finish the proof of formula (3.6) we note that in the definition of L_2 above the integrals may be interchanged and we get from (3.19)

$$L_2 = C_\infty \int_{T_2}^{T_1}\int_{T_2}^{T_1} \psi_1(A, t_1, t_2)\,dt_1\,dt_2.$$

This finishes the proof of formula (3.6). The orthogonality of $V^{[\nu]}$ and $V^{[\mu]}$ for $\nu \neq \mu$ in the several cusps case is proved by essentially the same computation.

It remains to prove the completeness of our systems of eigenfunctions and eigenpackets. To do this we consider for a large $A \in \mathbb{R}$ the function $f_A : \mathcal{F} \to \mathbb{C}$ defined by

$$f_A(P) := \begin{cases} 1 & \text{for } r \geq A, \\ 0 & \text{for } r < A. \end{cases} \tag{3.30}$$

Its prolongation to $\mathbb{H}$ via Γ is in $L^2(\Gamma\backslash\mathbb{H})$ and we may compute its Hilbert space norm $\|f_A\|^2$. By direct computation we find

$$\|f_A\|^2 = \frac{|\Lambda_\infty|}{2[\Gamma_\infty : \Gamma'_\infty]}\,A^{-2} = \frac{C_\infty A^{-2}}{2}. \tag{3.31}$$

We shall now evaluate the contributions of the various eigenfunctions and eigenpackets of $-\Delta$ to the completeness relation (2.72) for f_A. Let $0 < \sigma_1 < \sigma_2 < \ldots < \sigma_N = 1$ be the poles of $E(P,s)$ or $\phi(s)$ in the segment $[0,1]$. Following Theorem 1.11 we put for $j = 1, ..., N$

$$e_j(P) := \left(C_\infty \operatorname*{Res}_{s=\sigma_j} \phi(s)\right)^{-\frac{1}{2}} RE(P, \sigma_j). \tag{3.32}$$

Note that $\mathrm{Res}_{s=\sigma_j}\phi(s)$ is positive by Theorem 1.11. By Theorem 1.11 the $e_1, ..., e_N$ are an orthonormal system of eigenfunctions of $-\Delta$ in $L^2(\Gamma\setminus\mathbb{H})$ and

$$|< f_A, e_j >|^2 = C_\infty \frac{A^{-2-2\sigma_j}}{(1+\sigma_j)^2} \operatorname*{Res}_{s=\sigma_j} \phi(s). \tag{3.33}$$

By Theorem 2.9 the system (3.32) may be complemented by cusp eigenfunctions to a complete system of eigenfunctions of $-\Delta$. Note that all cusp eigenfunctions are orthogonal to f_A.

Since $f_A \in \tilde{\mathcal{D}}$ we may use formula (3.14) to evaluate the contribution of V_λ to the completeness relation (2.72). We may also use the computation just after (3.14) to obtain

$$\begin{aligned} B := \int_{-\infty}^{\infty} \frac{|d < f_A, V_\lambda >|^2}{dF(\lambda)} &= \frac{1}{2\pi C_\infty} \int_0^\infty \left| \frac{d}{dT} < f_A, \int_0^T E(\cdot, it)\, dt > \right|^2 dT \\ &= \frac{1}{2\pi C_\infty} \int_0^\infty \left| \int_{\mathcal{F}} f_A(P)\, \overline{E(P, iT)}\, dv(P) \right|^2 dT. \end{aligned}$$

Since $E(P, iT) - \alpha(P, iT)$ has no zeroth Fourier coefficient we may proceed by

$$\begin{aligned} B &= \frac{C_\infty}{2\pi} \int_0^\infty \left| \int_A^\infty \overline{\alpha(P, iT)}\, \frac{dr}{r^3} \right|^2 dT \\ &= \frac{C_\infty}{2\pi} \int_0^\infty \left| \frac{A^{-1-iT}}{1+iT} + \phi(-iT) \frac{A^{-1+iT}}{1-iT} \right|^2 dT. \end{aligned}$$

Since $\phi(iT)\phi(-iT) = 1$ and by a variable change we get

$$B = \frac{C_\infty A^{-2}}{\pi} \int_0^\infty \frac{1}{1+T^2}\, dT + \frac{C_\infty A^{-2}}{2\,\pi} \int_{-\infty}^{+\infty} \phi(iT)\, \frac{A^{-2iT}}{(1+iT)^2}\, dT.$$

We evaluate the second integral by shifting the line of integration to the right and using the residue theorem. With the usual precautions and the help of Lemma 3.5 we find, if A was large enough

$$B = \frac{C_\infty A^{-2}}{2} - C_\infty A^{-2} \sum_{j=1}^{N} \frac{A^{-2\sigma_j}}{(1+\sigma_j)^2} \operatorname*{Res}_{s=\sigma_j} \phi(s). \tag{3.34}$$

Let v_λ be an eigenpacket orthogonal to V_λ. We conclude from (3.31), (3.33) and (3.34) that

$$\int_{-\infty}^{\infty} \frac{|d < f_A, v_\lambda >|^2}{dF(\lambda)} = 0. \tag{3.35}$$

Let U be the closed subspace generated by the $v_{\lambda_2} - v_{\lambda_1}$. We see from the definition (2.66), (2.67) of the symbol in (3.35) that $< f_A, u >= 0$ for every $u \in U$. Hence the function u is a cusp function in the sense of Definition 4.5.1. From Theorem 4.5.2 we conclude that the integral operator Φ^* corresponding to $F^*(P, Q, \lambda_0)$ (λ_0 a large real number) acts on u with the same image as the integral operator Φ corresponding to $F(P, Q, \lambda_0)$. Hence Φ^* leaves U

invariant. Since Φ^* is Hilbert–Schmidt U is the closure of the span of the eigenfunctions of Φ^* in U. Such an eigenfunction would be also an eigenfunction of Φ and hence of $-\Delta$. This shows that $U = \{0\}$. From this we conclude $v_\lambda = 0$ for all $\lambda \in \mathbb{R}$.

This finishes the proof of Theorem 3.1. In the several cusps case the function f_A has to be replaced by a set of functions which are 0 on $\mathcal{F}$ except for the vicinity of a single cusp where they are 1. □

Proof of Theorem 3.2. We first prove $\Theta^{[\nu]}(\varphi) \in L^2(\Gamma \setminus \mathbb{H})$ for $\varphi \in C_c^\infty(]0,\infty[)$ by the same considerations from the proof of Theorem 3.1 which show that $V_\lambda^{[\nu]} \in L^2(\Gamma \setminus \mathbb{H})$. The various inner product relations needed between the $\Theta^{[\nu]}(\varphi)$ and $\Theta^{[\mu]}(\psi)$ are proved by the considerations leading to formula (3.6). The compatibility of the $\Theta^{[\nu]}$ with $-\Delta$ also follows as in the proof of Theorem 3.1. The completeness follows from the analogous statement in Theorem 3.1. □

Finally for this section we draw a conclusion from Proposition 2.16. The result follows by polarization from formula (2.88) analogously to Corollary 4.5.5.

Proposition 3.7. *Let $\Gamma < \mathbf{PSL}(2,\mathbb{C})$ be a cofinite subgroup and $E_\nu(P,it)$ one of its Eisenstein series. The integral*

$$H(P,Q) = \int_{-\infty}^{\infty} (1+t^2)^{-2}\ E_\nu(Q,it)\ \overline{E_\nu(P,it)}\ dt \qquad (P,Q \in \mathbb{H})$$

exists and defines a continuous function on $\mathbb{H} \times \mathbb{H}$.

6.4 Spectral Expansions of Integral Kernels and Poincaré Series

We study here the integral kernels $K(P,Q)$ $(P,\ Q \in \mathbb{H})$ that arose in Chapter 3, Sections 1, 5. Their spectral expansion theory will lead to further analytical properties of our Eisenstein series. The considerations in this section lead up to the trace formula discussed in the next section.

We fix for this section a cofinite subgroup $\Gamma < \mathbf{PSL}(2,\mathbb{C})$ and keep here the relevant notations concerning Eisenstein series from the beginning of Section 1, in particular (1.3) and in the one cusp case (1.10). The fundamental domain $\mathcal{F}$ is always chosen as in Proposition 2.3.9. We also use the notations introduced in Corollary 2.14.

To explain the logic of our arguments we leave at the beginning all questions of convergence or existence of integrals aside. We start off with a function $k : [1,\infty[\to \mathbb{R}$ satisfying suitable growth-hypotheses and consider as in Section 5 of Chapter 3 the point pair invariant $K := k \circ \delta$ where δ is defined in Proposition 1.1.6. We obtain an integral operator

$$\tilde{K}f(P) = \int_{\mathbb{H}} K(P,Q)\ f(Q)\ dv(Q) \qquad (P,\ Q \in \mathbb{H}). \tag{4.1}$$

If f happens to be an eigenfunction of the Laplace operator $-\Delta$, that is $-\Delta f = \lambda f$ then

$$\tilde{K}f(P) = \int_{\mathbb{H}} K(P,Q)\ f(Q)\ dv(Q) = h(\lambda)\ f(P) \tag{4.2}$$

where h is the Selberg transform of k given in Theorem 3.5.3, that is

$$h(1-s^2) := \frac{\pi}{s} \int_1^{\infty} k\left(\frac{1}{2}\left(u + \frac{1}{u}\right)\right)\ (u^s - u^{-s})\left(u - \frac{1}{u}\right)\frac{du}{u}. \tag{4.3}$$

Let now $\Gamma < \mathbf{PSL}(2, \mathbb{C})$ be our cofinite group. We define

$$K_{\Gamma}(P,Q) = \sum_{\gamma \in \Gamma} K(P, \gamma Q) \qquad (P,\ Q \in \mathbb{H}). \tag{4.4}$$

The function K_{Γ} is Γ-invariant in both variables. For fixed $P \in \mathbb{H}$ we will have $K_{\Gamma}(P, \cdot) \in L^2(\Gamma \setminus \mathbb{H})$. We also get an integral operator on Γ-invariant functions by

$$\check{K}_{\Gamma}f(P) = \int_{\mathcal{F}} K_{\Gamma}(P,Q)\ f(Q)\ dv(Q) \qquad (P,\ Q \in \mathbb{H}). \tag{4.5}$$

The operator $\check{K}_{\Gamma}$ will be bounded and self-adjoint. If f is a Γ-invariant eigenfunction of $-\Delta$ with eigenvalue λ then we obtain by unfolding the integral

$$\begin{aligned} \check{K}_{\Gamma}f(P) &= \int_{\mathcal{F}} K_{\Gamma}(P,Q)\ f(Q)\ dv(Q) \\ &= \int_{\mathbb{H}} K(P,Q)\ f(Q)\ dv(Q) = h(\lambda)\ f(P). \end{aligned} \tag{4.6}$$

If $e_m \in L^2(\Gamma \setminus \mathbb{H})$ is one of the eigenfunctions of $-\Delta$ with eigenvalue λ_m and if $K_{\Gamma}(P, \cdot) \in L^2(\Gamma \setminus \mathbb{H})$ then (4.6) reads as

$$< K_{\Gamma}(P, \cdot), e_m >= h(\lambda_m)\ \overline{e_m(P)}. \tag{4.7}$$

This shows that $L^2_{\text{disc}}(\Gamma \setminus \mathbb{H})$ is invariant under $\check{K}_{\Gamma}$. We may also apply $\check{K}_{\Gamma}$ to the Eisenstein series $E_{\nu}(P, it)$. We find

$$\int_{\mathcal{F}} K_{\Gamma}(P,Q)\ E_{\nu}(Q, it)\ dv(Q) = h(1+t^2)\ E_{\nu}(P, it). \tag{4.8}$$

This equation may then be applied to our eigenpackets from Theorem 3.1 to give for $\lambda \geq 1$

$$\check{K}_{\Gamma}\ V_{\lambda}^{[\nu]}(P) = \int_0^{\sqrt{\lambda - 1}} h(1+t^2)\ E_{\nu}(P, it)\ dt. \tag{4.9}$$

By inserting equations (4.7) and (4.9) into the expansions from Theorem 3.4 we get

$$K_\Gamma(P,Q) = \sum_{m\in\mathfrak{D}} h(\lambda_m)\, e_m(Q)\, \overline{e_m(P)} + \frac{1}{4\pi}\sum_{\nu=1}^{h} \frac{[\Gamma_\nu : \Gamma_\nu']}{|\Lambda_\nu|} \int_{-\infty}^{\infty} h(1+t^2)\, E_\nu(Q,it)\, \overline{E_\nu(P,it)}\, dt \tag{4.10}$$

and correspondingly the L^2-norm equation

$$\|K_\Gamma(P,\cdot)\|^2 = \sum_{m\in\mathfrak{D}} |h(\lambda_m)\, e_m(P)|^2 + \frac{1}{4\pi}\sum_{\nu=1}^{h} \frac{[\Gamma_\nu : \Gamma_\nu']}{|\Lambda_\nu|} \int_{-\infty}^{\infty} |h(1+t^2)|^2\, |E_\nu(P,it)|^2\, dt. \tag{4.11}$$

These last series will give interesting expansions and prolongation results for the specific K_Γ discussed in Section 1 of Chapter 3. It will also imply important growth results for the Eisenstein series. Finally we define

$$H_\Gamma(P,Q) := \frac{1}{4\pi}\sum_{\nu=1}^{h} \frac{[\Gamma_\nu : \Gamma_\nu']}{|\Lambda_\nu|} \int_{-\infty}^{\infty} h(1+t^2)\, E_\nu(Q,it)\, \overline{E_\nu(P,it)}\, dt. \tag{4.12}$$

Defining $L_\Gamma(P,Q) := K_\Gamma(P,Q) - H_\Gamma(P,Q)$ we have

$$L_\Gamma(P,Q) = \sum_{m\in\mathfrak{D}} h(\lambda_m)\, e_m(Q)\, \overline{e_m(P)}. \tag{4.13}$$

The associated operator $\check{L}_\Gamma$ will turn out to be of trace-class with an obvious trace. The computation of the trace of $\check{K}_\Gamma - \check{H}_\Gamma$ in the next section will then result in the trace formula.

To make our results precise we write $\mathcal{S}([1,\infty[)$ for the space of C^∞-functions $k : [1,\infty[\to \mathbb{C}$ which are together with all their derivatives of rapid decay as $x \to \infty$. Rapid decay for the function k means as usually that $x^n k(x)$ tends to 0 for $x \to \infty$ and all $n \in \mathbb{N}$. The most important class of functions k to which all of the above can be applied to is $\mathcal{S}([1,\infty[)$.

Theorem 4.1. *Let $\Gamma < \mathbf{PSL}(2,\mathbb{C})$ be a cofinite subgroup and $k \in \mathcal{S}([1,\infty[)$. Then the following hold.*

(1) The series in (4.4) converges absolutely and uniformly on compact subsets of $\mathbb{H}\times\mathbb{H}$. The resulting function K_Γ is symmetric and Γ-invariant, that is we have $K_\Gamma(P,Q) = K_\Gamma(Q,P) = K_\Gamma(P,\gamma Q)$ for all $P,Q \in \mathbb{H}$ and $\gamma \in \Gamma$.

(2) For fixed $P \in \mathbb{H}$ we have $K_\Gamma(P,\cdot) \in L^2(\Gamma\backslash\mathbb{H})$ and $K_\Gamma(P,\cdot)$ is contained in the domain of definition of $-\Delta$. The function $K_\Gamma(P,\cdot)\overline{E_\nu(\cdot,i\cdot)}$ is absolutely integrable over $\mathcal{F}\times[0,T]$ for every $T \geq 0$ and every $\nu = 1,\dots,h$.

(3) The relations (4.7), (4.8), (4.9) hold.
(4) The relations (4.10), (4.11) hold pointwise. The series in (4.10) converges absolutely and uniformly on compact subsets of $\mathbb{H}\times\mathbb{H}$. *The series in (4.11) converges uniformly on compact subsets of* $\mathbb{H}$.
(5) $\check{K}_\Gamma$ *is a bounded self-adjoint operator on* $L^2(\Gamma\setminus\mathbb{H})$.
(6) The (continuous) function $L_\Gamma(\cdot,\cdot)$ *is in* $L^2(\mathcal{F}\times\mathcal{F}, dv\otimes dv)$ *and the operator* $\check{L}_\Gamma$ *on* $L^2(\Gamma\setminus\mathbb{H})$ *is of trace-class.*

Before we come to the proof of Theorem 4.1 we have to describe the behaviour of $K_\Gamma(P,Q)$ when $P\in\mathbb{H}$ is fixed and Q moves towards a cusp of Γ. Let us mention beforehand that the convergence statement in (1) of Theorem 4.1 follows from the corresponding statement for the function $H(P,Q,s)$ from Proposition 3.1.3.

Lemma 4.2. *Let* Γ *be a cofinite group so that* ∞ *is a cusp of* Γ *and let* $C=\Gamma'_\infty\gamma_0$ *be a right coset of* Γ'_∞ *in* Γ_∞. *Define for* $P=(z_P,r_P)$, $Q=(z_Q,r_Q)\in\mathbb{H}$ *and* $k\in\mathcal{S}([1,\infty[)$

$$K^*_\Gamma(P,Q)=\sum_{\gamma\in\Gamma\setminus\Gamma_\infty}K(P,\gamma Q),\quad K^C_\Gamma(P,Q)=\sum_{\gamma\in C}K(P,\gamma Q). \tag{4.14}$$

Then the following hold.

(1) For P *fixed the function* $K^*_\Gamma(P,Q)$ *of* Q *is bounded on* $\mathcal{F}$ *and*

$$K^*_\Gamma(P,Q)=O(r_Q^{-N})\qquad \text{as } r_Q\to\infty \tag{4.15}$$

for every $N\in\mathbb{N}$.
(2) Let g *be the function from Definition 3.5.5. Then for* P, $Q\in\mathbb{H}$

$$K^C_\Gamma(P,Q)=\frac{2}{|\Lambda_\infty|}\,r_P\,r_Q\,g\left(\log\left(\frac{r_P}{r_Q}\right)\right)+O\left((r_Pr_Q)^{-N}\right) \tag{4.16}$$

for every $N\in\mathbb{N}$.

Proof. (1): We choose a constant κ_1 so that $|k(x)|\le\kappa_1/x^{1+N}$ for all $x\in[1,\infty[$. We further let R be a representative system for the right cosets of Γ'_∞ in $\Gamma\setminus\Gamma_\infty$. We obtain from formula (1.1.15)

$$\begin{aligned}|K^*_\Gamma(P,Q)|&\le\sum_{\gamma\in R}\sum_{\omega\in\Lambda_\infty}\left|k\left(\frac{|z_{\gamma Q}-z_P+\omega|^2+r_P^2+r_{\gamma Q}^2}{2r_Pr_{\gamma Q}}\right)\right|\\&\le\kappa_1\sum_{\gamma\in R}r_{\gamma Q}^{1+N}\sum_{\omega\in\Lambda_\infty}\left(\frac{|z_{\gamma Q}-z_P+\omega|^2+r_P^2}{2r_P}\right)^{-1-N}\le\kappa_2\sum_{\gamma\in R}r_{\gamma Q}^{1+N}\\&=\kappa_2\left(E_\infty(Q,N)-[\Gamma_\infty:\Gamma'_\infty]r_Q^{1+N}\right)\end{aligned}$$

with a suitable constant κ_2. The result follows by the usual properties (1.4), (1.5), (3.21) of the Eisenstein series $E_\infty(Q,s)$.

(2): We treat only $K_{\Gamma_\infty}^{C_1}$ where C_1 is the coset of the identity. The non-trivial cosets of Γ'_∞ in Γ_∞ can be treated similarly. If we define

$$f(u) := k\left(\frac{|z_Q - z_P + u|^2 + r_P^2 + r_Q^2}{2r_P r_Q}\right)$$

then we obtain from our formula (1.1.15) and by application of the Poisson summation formula

$$K_\Gamma^{C_1}(P,Q) = \sum_{\omega\in\Lambda_\infty} f(\omega) = |\Lambda_\infty|^{-1} \sum_{\omega\in\Lambda_\infty^\circ} \hat{f}(\omega).$$

The Fourier transform $\hat{f}$ of f is formed with the standard Lebesgue measure on $\mathbb{C} = \mathbb{R}^2$ with respect to the Euclidean scalar product. It can be computed, using (3.5.33), as

$$\hat{f}(v) = 2\,r_P\,r_Q\;e^{-2\pi i<v,z_Q-z_P>} \int_{\mathbb{C}} k\left(|u|^2 + \frac{r_P^2 + r_Q^2}{2r_P r_Q}\right) e^{-2\pi i\sqrt{2r_P r_Q}<v,u>}\,du$$

Since f is of rapid decay also $\hat{f}$ has this property. In fact the speed of decay of $\hat{f}$ is determined by the smoothness of f. This means that for $N \in \mathbb{N}$

$$\hat{f}(v) = O\left((r_P r_Q)^{-N}|v|^{-2N}\right) \qquad \text{as } |v| \to \infty.$$

Summing this over the nontrivial elements of Λ_∞° we get

$$\begin{aligned} K_\Gamma^{C_1}(P,Q) &= |\Lambda_\infty|^{-1}\hat{f}(0) + O\left((r_P r_Q)^{-N}\right) \\ &= \frac{2\,r_P\,r_Q}{|\Lambda_\infty|}\, g\left(\log\left(\frac{r_P}{r_Q}\right)\right) + O\left((r_P r_Q)^{-N}\right). \end{aligned}$$

This proves the result. □

Relations (4.15), (4.16) determine the growth behaviour of $K_\Gamma(P,Q)$ in the cusps of Γ. They show that $K_\Gamma(P,Q)$ is for fixed P square intgrable as a function of Q, but $K_\Gamma(P,P)$ will in general not be square integrable over $\mathcal{F}$.

Proof of Theorem 4.1. (1): This follows as already mentioned from Proposition 3.1.3.

(2): Note first of all that for $j = (0,0,1)$, $Q \in \mathbb{H}$

$$\Delta_Q\; k(\delta(j,Q)) = 3\;\delta(j,Q)\;k'(\delta(j,Q)) + \delta(j,Q)^2\;k''(\delta(j,Q)) - k''(\delta(j,Q)).$$

Hence Lemma 4.2 implies using the Fourier expansions (1.4), (1.5), (3.21) of the Eisenstein series that both $K_\Gamma(P,\cdot)$ and $\Delta K_\Gamma(P,\cdot)$ are in $L^2(\Gamma\backslash\mathbb{H})$. The rest follows similarly.

(3): Follows from (2) and Theorem 3.5.3.

(4): Follows from Theorem 3.3 in the pointwise version and from Corollary 4.5.5 we infer uniformity. Note that by Lemma 3.5.5 the function h is rapidly decreasing.

(5): Follows from the rapid decrease and continuity of the function h.

(6): Follows from formula (4.13) and the rapid decrease of h. □

Our next Proposition is a strong version of Mercer's theorem, see Dieudonné (1960), (11.6.7) or Petrowskij (1953),p. 81. Its proof is obvious from Theorem 3.4, Corollary 4.5.5, Proposition 3.6.

Proposition 4.3. *Let $\Gamma < \mathbf{PSL}(2,\mathbb{C})$ be a cofinite subgroup and $k : [1,\infty[\to \mathbb{R}$ be a C^∞-function. Assume that (4.4) converges absolutely and uniformly on compact subsets of $\mathbb{H}\times\mathbb{H}$, that for fixed $P \in \mathbb{H}$ we have $K_\Gamma(P,\cdot) \in L^2(\Gamma\setminus\mathbb{H})$. In addition let $\check{K}_\Gamma$ be a positive operator. Assume further that $K_\Gamma(P,\cdot)$ lies in the domain of $-\Delta$ and that $K_\Gamma(P,\cdot)\overline{E_\nu(\cdot,i\cdot)}$ is integrable over $\mathcal{F}\times[0,T]$ for every $T\geq 0$. Let the Selberg transform h of k defined by (4.3) exist and satisfy $h(s) = O(|s|^{-2})$ as $|s|\to\infty$. Then the expansion (4.10) converges absolutely and uniformly on compact subsets of $\mathbb{H}\times\mathbb{H}$.*

We apply now the technique explained in the beginning of this section to specific point-pair invariants and deduce explicit formulas for their expansion (4.10). For the arguments the following asymptotic relation for Bessel functions which we learned from F.W.J. Olver is of importance:

(4.17)
$$K_{i\mu}(t) = \sqrt{\frac{2\pi}{\mu}}e^{-\frac{\pi\mu}{2}}\left(\cos\left(\mu\log\left(\frac{t}{2}\right) - \mu\log(\mu) + \mu + \frac{\pi}{4}\right) + o(1)\right).$$

This holds for $\mu,\ t\in\mathbb{R},\ \mu>0$ as t or μ are fixed and $\mu\to\infty$ respectively $t\to 0$. The relation also holds if simultanuously $\mu\to\infty, t\to 0$. Formula (4.17) can be deduced using (9.6.2), (9.6.7), (9.6.10) from Abramowitz, Stegun (1970).

First of all we consider for a cofinite subgroup $\Gamma < \mathbf{PSL}_2(\mathbb{C})$ and $t > 0$ the Poincaré series (3.1.31), that is

$$\Theta(P,Q,t) = \sum_{\gamma\in\Gamma} e^{-t\delta(P,\gamma Q)} \qquad (P,Q\in\mathbb{H}).$$

The hypotheses of Theorem 4.1 are satisfied.

Proposition 4.4. *Let $\Gamma \in \mathbf{PSL}_2(\mathbb{C})$ be a cofinite subgroup and $t > 0$. We define $\mathfrak{D}_{>0}$ to be the index set for the nonconstant eigenfunctions of $-\Delta$. For $m\in\mathfrak{D}_{>0}$ with corresponding eigenvalue λ_m we define $\mu_m := \sqrt{1-\lambda_m}$. Let $0 < \lambda_1 \leq \ldots \leq \lambda_M < 1$ be all the eigenvalues of $-\Delta$ in $]0,1[$ counted with their multiplicity. For them let μ_m be chosen positively. The function $\Theta(P,Q,t)$ has the following properties.*

(1) We have

$$
(4.18)\quad \begin{aligned}\Theta(P,Q,t) &= \frac{4\pi}{t\ \mathrm{vol}(\Gamma)} K_1(t) + \frac{4\pi}{t} \sum_{m\in\mathfrak{D}_{>0}} K_{\mu_m}(t)\, e_m(Q)\, \overline{e_m(P)} \\ &+ \sum_{\nu=1}^{h} \frac{[\Gamma_\nu : \Gamma'_\nu]}{t|\Lambda_\nu|} \int_{-\infty}^{\infty} K_{i\mu}(t)\, E_\nu(Q,i\mu)\, \overline{E_\nu(P,i\mu)}\, d\mu\end{aligned}
$$

where the series converges absolutely and uniformly on compact subsets of $\mathbb{H}\times\mathbb{H}\times]0,\infty[$.

(2) *The function* $\Theta(P,Q,t)$ *satisfies the asymptotic relations*

$$
(4.19)\quad t\Theta(P,Q,t) = \frac{4\pi}{\mathrm{vol}(\Gamma)}\frac{1}{t} + 2\pi \sum_{m=1}^{M} \frac{2^{\mu_m}\,\Gamma(\mu_m)\, e_m(Q)\, \overline{e_m(P)}}{t^{\mu_m}} + O(1)
$$

for $t\to 0$ *and* $\lim_{t\to 0} t^2\Theta(P,Q,t) = \frac{4\pi}{\mathrm{vol}(\Gamma)}$.

Proof. The first problem is to compute the Selberg transform h of $k(x) = e^{-tx}$. Using the integral representations on page 85 of Magnus, Oberhettinger, Soni (1966) we find $h(1-s^2) = \frac{4\pi}{t} K_s(t)$. Part (1) follows now from Theorem 4.1 and (4.17). Of the asymptotic relations the first is the stronger. We deduce it from (4.18) by the help of (4.17), Proposition 3.7, Corollary 4.5.5 and the asymptotic relation for $K_\mu(t)$ $(t\to 0)$ from Magnus, Oberhettinger, Soni (1966), page 66. □

The next Poincaré series we treat is from Definition 3.1.1

$$
H(P,Q,s) = \sum_{\gamma\in\Gamma} \frac{1}{\delta(P,\gamma Q)^{s+1}} \qquad (P,Q\in\mathbb{H},\ \mathrm{Re}\, s>1).
$$

Proposition 4.5. *Let* $\Gamma \in \mathbf{PSL}_2(\mathbb{C})$ *be a cofinite subgroup and* $s\in\mathbb{C}$ *with* $\mathrm{Re}\, s > 1$. *We keep the notations concerning eigenfunctions introduced in the last Proposition. The function* $H(P,Q,s)$ *has the following properties.*

(1) *For* $P\in\mathbb{H}$ *fixed and* $\mathrm{Re}\, s>1$ *we have* $H(P,\cdot,s)\in L^2(\Gamma\setminus\mathbb{H})$.

(2) *We have*

$$
(4.20)\quad \begin{aligned} H(P,Q,s) &= \frac{\pi}{\mathrm{vol}(\Gamma)} \frac{2^s}{\Gamma(s+1)} \Gamma\left(\frac{s-1}{2}\right)\Gamma\left(\frac{s+1}{2}\right) \\ &+ \pi \frac{2^s}{\Gamma(s+1)} \sum_{m\in\mathfrak{D}_{>0}} \Gamma\left(\frac{s-\mu_m}{2}\right)\Gamma\left(\frac{s+\mu_m}{2}\right) e_m(Q)\,\overline{e_m(P)} \\ &+ \sum_{\nu=1}^{h} \frac{[\Gamma_\nu:\Gamma'_\nu]\, 2^s}{4\,|\Lambda_\nu|\, \Gamma(s+1)} \cdot \\ &\cdot \int_{-\infty}^{+\infty} \Gamma\left(\frac{s-i\mu}{2}\right)\Gamma\left(\frac{s+i\mu}{2}\right) E_\nu(Q,i\mu)\,\overline{E_\nu(P,i\mu)}\, d\mu \end{aligned}
$$

where the series converges absolutely and uniformly on compact subsets of $\mathbb{H} \times \mathbb{H} \times \{ s \in \mathbb{C} \mid \operatorname{Re} s > 1 \}$.

(3) *The function* $H(P,Q,s)$ *has a meromorphic continuation into the whole complex s-plane. The continued function is holomorphic in the half-plane* $\operatorname{Re} s > 0$ *up to a pole of order 1 at* $s = 1$. *There are further poles of order 1 at the odd negative integers. If we have* $e_m(Q)\overline{e_m(P)} \neq 0$ *for* $m \in \mathfrak{D}_{>0}$ *there are also poles of order* 1 *at* $s = \pm\mu_m - 2\ell$ $(\ell \in \mathbb{Z},\ \ell \geq 0)$.

Proof. Let first $s > 1$ be real. Then we have by Lemma 4.2.1

$$\int_{\mathcal{F}} H(P,\cdot,s)^2 \, dv(Q) = \int_{\mathbb{H}} \sum_{\gamma \in \Gamma} \delta(\gamma P, Q)^{-s-1} \delta(P,Q)^{-s-1} \, dv(Q)$$
$$\leq C \sum_{\gamma \in \Gamma} \frac{1 + \log \delta(\gamma P, P)}{\delta(\gamma P, P)^{s+1}}$$

with a suitable constant $C > 0$. The last sum converges absolutely by Proposition 3.1.4. Admitting now also complex values of s we see that $H(P,\cdot,s) \in L^2(\Gamma \setminus \mathbb{H})$ for $s \in \mathbb{C}$, $\operatorname{Re} s > 1$.

In order to compute the Selberg transform h of $k(x) = x^{-1-s}$ we rename the former variable s of h to $\tilde{s}$ and find from well-known integral formulas for the Γ-function

$$h(1 - \tilde{s}^2) = 2^s \, \pi \, \frac{\Gamma\left(\frac{s+\tilde{s}}{2}\right) \Gamma\left(\frac{s-\tilde{s}}{2}\right)}{\Gamma(s+1)}.$$

This yields the expansion (4.20) for $H(P,\cdot,s)$ ($\operatorname{Re} s > 1$) in the L^2-sense and hence

$$\|H(P,\cdot,s)\|^2 = \left| \frac{\pi}{\operatorname{vol}(\Gamma)^{1/2}} \frac{2^s}{\Gamma(s+1)} \Gamma\left(\frac{s-1}{2}\right) \Gamma\left(\frac{s+1}{2}\right) \right|^2$$
$$+ \sum_{m \in \mathfrak{D}_{>0}} \left| \frac{2^s \, \pi}{\Gamma(s+1)} \Gamma\left(\frac{s-\mu_m}{2}\right) \Gamma\left(\frac{s+\mu_m}{2}\right) \, e_m(P) \right|^2$$
$$+ \frac{1}{4\pi} \sum_{\nu=1}^{h} \frac{[\Gamma_\nu : \Gamma'_\nu]}{|\Lambda_\nu|} \int_{-\infty}^{+\infty} \left| \frac{2^s \pi \, \Gamma\left(\frac{s-i\mu}{2}\right) \Gamma\left(\frac{s+i\mu}{2}\right)}{\Gamma(s+1)} E_\nu(P, i\mu) \right|^2 \, d\mu.$$

The argument at the beginning of this proof shows that the left-hand side of this equation is a continuous function of (P,s) for $\operatorname{Re} s > 1$. The same holds for the integrals on the right-hand side by Proposition 3.7. Dini's theorem now implies that the series over the eigenvalues and the integrals on the right-hand side of the last equation converge locally uniformly with respect to $(P,s) \in \mathbb{H} \times \{ s \mid \operatorname{Re} s > 1 \}$. Hence by the Cauchy–Schwarz inequality and the Cauchy criterion for uniform convergence the series and the integrals on the right-hand side of (4.20) converge absolutely and locally uniformly with respect to $(P,Q,s) \in \mathbb{H} \times \mathbb{H} \times \{ s \mid \operatorname{Re} s > 1 \}$. Hence (4.20) also holds pointwise. Now Stirling's formula readily implies that the series in (4.20) even is meromorphic in the entire s-plane. Poles may occur only at the points $\pm\mu_m - 2l$ where $m \geq 0$ and $l \geq 0$, $l \in \mathbb{Z}$. These are the possible poles enumerated under (3).

Now we see that the integrals

$$\int_{-\infty}^{+\infty} \Gamma\left(\frac{s-i\mu}{2}\right)\Gamma\left(\frac{s+i\mu}{2}\right) E_\nu(Q,i\mu)\,\overline{E_\nu(P,i\mu)}\,d\mu$$

are holomorphic for $\operatorname{Re} s > 0$ and we are left to prove that these integrals in fact define meromorphic functions in the entire s-plane. To do that we consider a fixed ν and write the integral in question as a complex line integral:

$$I_0(s) := \frac{1}{i}\int_{\operatorname{Re} z=0} \Gamma\left(\frac{s+z}{2}\right)\Gamma\left(\frac{s-z}{2}\right) E_\nu(Q,z)\,E_\nu(P,-z)\,dz.$$

The integral I_0 is holomorphic for $\operatorname{Re} s > 0$. In order to meromorphically continue I_0 we indent the path of integration to the left and choose a large real number $T > 0$ such that the horizontal lines $\operatorname{Im} z = \pm T$ are free of poles of the integrand. We also choose a sufficiently large odd natural number k such that $E_\nu(Q,z)$ has no poles with $\operatorname{Re} z \leq k$. Then we form the integral

$$I_1(s) := \frac{1}{i}\int_W \Gamma\left(\frac{s+z}{2}\right)\Gamma\left(\frac{s-z}{2}\right) E_\nu(Q,z)\,E_\nu(P,-z)\,dz,$$

where W denotes the following piecewise linear path: The part of the imaginary axis from $-i\infty$ to $-iT$, the horizontal line segment from $-iT$ to $-iT-k$, the vertical line segment from $-iT-k$ to $iT-k$, the horizontal line segment from $iT-k$ to iT and then the part of the imaginary axis from iT to $i\infty$. Since k was chosen to be odd, the function $I_1(s)$ is holomorphic in the box $-1 < \operatorname{Re} s < 1$, $\operatorname{Im} s < T$. In order to continue $I_0(s)$ we assume that s satisfies $0 < \operatorname{Re} s < 1$, $|\operatorname{Im} s| < T$. Then by the residue theorem $I_0(s) - I_1(s)$ is equal to 2π times the sum of the residues of the function

$$z \mapsto \Gamma\left(\frac{s+z}{2}\right)\Gamma\left(\frac{s-z}{2}\right) E_\nu(Q,z)\,E_\nu(P,-z)$$

at its singularities contained in the box $-k < \operatorname{Re} z < 0$, $|\operatorname{Im} z| < T$. Denote the sum of residues by $R_T(s)$. It is clear that $R_T(s)$ is a meromorphic function of s in the entire s-plane. We have now shown that

$$I_0(s) = I_1(s) + 2\pi\, R_T(s)$$

for $0 < \operatorname{Re} s < 1$, $|\operatorname{Im} s| < T$. But $I_1(s)$ is even holomorphic for $-1 < \operatorname{Re} s < 1$, $|\operatorname{Im} s| < T$. Since T can be chosen arbitrarily large, this shows that $I_0(s)$ admits a meromorphic continuation to the half-plane $\operatorname{Re} s > -1$.

The preceding method may be iterated. We let k, T be as before and we denote by W' the path W with k replaced by $k+1$. The function

$$I_2(s) := \frac{1}{i} \int_{W'} \Gamma\left(\frac{s+z}{2}\right) \Gamma\left(\frac{s-z}{2}\right) E_\nu(Q,z)\, E_\nu(P,-z)\, dz$$

is holomorphic in the box $-2 < \operatorname{Re} s < 0$, $|\operatorname{Im} s| < T$ since $k+1$ is even. Assuming now $-1 < \operatorname{Re} s < 0$, $|\operatorname{Im} s| < T$, we se that $I_1(s) - I_2(s)$ is equal to 2π times the sum of the residues of the integrand at its singularities contained in the box $-k-1 < \operatorname{Re} z < -k$, $|\operatorname{Im} z| < T$. Using again the freedom of the choice of T, we see that $I_0(s)$ is meromorphic for $\operatorname{Re} s > -2$. The argument may be repeated and yields (3). □

We now treat the Maaß–Selberg series from Definition 3.1.18, that is

$$F(P,Q,s) = \frac{1}{4\pi} \sum_{\gamma \in \Gamma} \varphi_s(P, \gamma Q) \qquad (P, Q \in \mathbb{H},\ \Gamma P \neq \Gamma Q,\ \operatorname{Re} s > 1).$$

Proposition 4.6. *Let $\Gamma \in \mathbf{PSL}_2(\mathbb{C})$ be a cofinite subgroup. We keep the notation from Proposition 4.4. Then the Maaß–Selberg series has the following properties.*

(1) For $\Gamma P \neq \Gamma Q$, $\operatorname{Re} s > 1$ we have the expansion

$$\begin{aligned} F(P,Q,s) &= \frac{1}{\operatorname{vol}(\Gamma)} \frac{1}{s^2-1} + \sum_{m \in \mathfrak{D}_{>0}} \frac{1}{s^2 - \mu_m^2}\, e_m(Q)\, \overline{e_m(P)} \\ &\quad + \frac{1}{4\pi} \sum_{\nu=1}^{h} \frac{[\Gamma_\nu : \Gamma'_\nu]}{|\Lambda_\nu|} \int_{-\infty}^{+\infty} \frac{1}{s^2+\mu^2}\, E_\nu(Q, i\mu)\, \overline{E_\nu(P, i\mu)}\, d\mu. \end{aligned} \tag{4.21}$$

This series and the integrals converge for $P \in \mathbb{H}$ fixed in $L^2(\Gamma \setminus \mathbb{H})$.

(2) For $\operatorname{Re} s > 1$, $\operatorname{Re} w > 1$ the function $(P,Q) \mapsto F(P,Q,s) - F(P,Q,w)$ is continuous on $\mathbb{H} \times \mathbb{H}$ and has the expansion

$$\begin{aligned} F(P,Q,s) - F(P,Q,w) &= (w^2 - s^2) \left(\frac{1}{\operatorname{vol}(\Gamma)} \frac{1}{(s^2-1)(w^2-1)} \right. \\ &+ \sum_{m \in \mathfrak{D}_{>0}} \frac{1}{(s^2-\mu_m^2)(w^2-\mu_m^2)}\, e_m(Q)\, \overline{e_m(P)} \\ &+ \left. \frac{1}{4\pi} \sum_{\nu=1}^{h} \frac{[\Gamma_\nu : \Gamma'_\nu]}{|\Lambda_\nu|} \int_{-\infty}^{+\infty} \frac{1}{(s^2+\mu^2)(w^2+\mu^2)}\, E_\nu(Q, i\mu)\, \overline{E_\nu(P, i\mu)}\, d\mu \right). \end{aligned}$$

For fixed w with $\operatorname{Re} w > 1$ the expansion on the right-hand side converges locally uniformly on $\mathbb{H} \times \mathbb{H} \times \{ s \mid \operatorname{Re} s > 0,\ s \neq s_0, \dots, s_N \}$ and defines a meromorphic continuation of the function on the left-hand side to this domain. Here, $0 = \lambda_0 < \lambda_1 \leq \lambda_2 \leq \dots \leq \lambda_N < 1$ are the exceptional eigenvalues of the operator $-\tilde{\Delta}$ on $\mathcal{D}$ and $\lambda_j = 1 - s_j^2$ $(j = 0, \dots, N)$, $1 = s_0 > s_1 \geq \dots \geq s_N > 0$. The function $s \mapsto F(P,Q,s)$ has a meromorphic continuation to $\{ s \mid \operatorname{Re} s > 0 \}$ such that

$$F(P,Q,s) - \sum_{k=0}^{N} \frac{1}{s^2 - s_k^2}\, e_k(Q)\, \overline{e_k(P)}$$

is a holomorphic function of s in this half-plane for $P, Q \in \mathbb{H}$ with $\Gamma P \neq \Gamma Q$.

(3) *For $P, Q \in \mathbb{H}$ with $\Gamma P \neq \Gamma Q$ the function $s \mapsto F(P,Q,s)$ admits a meromorphic continuation to the entire complex plane and satisfies the functional equation*

$$(4.22)\quad F(P,Q,s) - F(P,Q,-s) = \frac{1}{2s} \sum_{\nu=1}^{h} \frac{[\Gamma_\nu : \Gamma_\nu']}{|\Lambda_\nu|}\, E_\nu(Q,s)\, E_\nu(P,-s).$$

Proof. (1): The L^2-expansion (4.21) holds because $F(P,Q,s)$ is the resolvent kernel for $\operatorname{Re} s > 1$, see also (2.87), (2.88). The normalization factor $1/4\pi$ arises as in Theorem 3.4.

(2): Let $\operatorname{Re} s > 1$ and consider the expansion (2.88):

$$\|F(P,\cdot,s)\|^2 = \frac{1}{\operatorname{vol}(\Gamma)} \frac{1}{|s^2-1|^2} + \sum_{m \in \mathfrak{D}_{>0}} \left| \frac{1}{s^2 - \mu_m^2} e_m(P) \right|^2$$
$$+ \frac{1}{4\pi} \sum_{\nu=1}^{h} \frac{[\Gamma_\nu : \Gamma_\nu']}{|\Lambda_\nu|} \int_{-\infty}^{+\infty} \left| \frac{1}{s^2+\mu^2} E_\nu(P, i\mu) \right|^2 d\mu.$$

We know from Lemma 4.2.4 that the left-hand side is a continuous function of (P,s) ($\operatorname{Re} s > 1$); the same holds for the integrals on the right-hand side by Proposition 3.7. Dini's theorem now shows that

$$\sum_{m \in \mathfrak{D}_{>0}} \left| \frac{1}{s^2 - \mu_m^2} e_m(P) \right|^2, \qquad \int_{-\infty}^{+\infty} \left| \frac{1}{s^2+\mu^2} E_\nu(P, i\mu) \right|^2 d\mu$$

converge locally uniformly with respect to $P \in \mathbb{H}$, and this implies that the convergence is locally uniform for $(P,s) \in \mathbb{H} \times \{ s \mid \operatorname{Re} s > 1 \}$. Now we can go further and see that the right-hand side converges locally uniformly on $\mathbf{H} \times \{ s \mid \operatorname{Re} s > 0,\ s \neq s_0, \ldots, s_N \}$.

Now let $\operatorname{Re} s > 1$, $\operatorname{Re} w > 1$ and consider the continuous function

$$(P,Q) \mapsto F(P,Q,s) - F(P,Q,w)$$

(compare Theorem 4.2.9). This function has the L^2-expansion shown under item (2). The Cauchy-Schwarz inequality combined with the Cauchy criterion for uniform convergence and the aforementioned facts on locally uniform convergence of the expansion for $\|F(P,\cdot,s)\|^2$ shows that this expansion converges for fixed w ($\operatorname{Re} w > 1$) locally uniformly with respect to $(P,Q,s) \in \mathbb{H} \times \mathbb{H} \times \{ s \mid \operatorname{Re} s > 1 \}$. The expression on the right hand side even converges locally uniformly on $\mathbb{H}\times\mathbb{H}\times\{\, s \mid \operatorname{Re} s > 0\, s \neq s_0, \ldots, s_N \,\}$ and defines a meromorphic continuation of $s \mapsto F(P,Q,s) - F(P,Q,w)$ in this region. Removing the constant contribution $F(P,Q,w)$ we see that $s \mapsto F(P,Q,s)$ has a meromorphic continuation to the region $\{\, s \mid \operatorname{Re} s > 0 \,\}$ such that

$$F(P,Q,s) - \sum_{k=0}^{N} \frac{1}{s^2 - s_k^2}\, e_k(Q)\, \overline{e_k(P)}$$

is holomorphic in this half-plane for $P, Q \in \mathbb{H}$ with $\Gamma P \neq \Gamma Q$. The residues of $s \mapsto F(P,Q,s)$ in the half-plane $\operatorname{Re} s > 0$ can be read off from this.

(3): By what has been said the contribution of the eigenfunctions to the right-hand side of the formula under item (2) is a meromorphic function of $s \in \mathbb{C}$ with poles at the points $\pm\mu_m$ $(m \in \mathfrak{D})$. The associated residues are obvious. There may be cancellation of poles due to higher multiplicities or vanishing of the eigenfunctions in special points. To effect the meromorphic continuation of the integral

$$\begin{aligned} I_\nu(s) :&= \int_{-\infty}^{+\infty} \frac{w^2 - s^2}{(s^2+\mu^2)(w^2+\mu^2)}\, E_\nu(Q, i\mu)\, \overline{E_\nu(P, i\mu)}\, d\mu \\ &= \frac{1}{i} \int_{\operatorname{Re}\zeta = 0} \frac{w^2 - s^2}{(s^2-\zeta^2)(w^2-\zeta^2)}\, E_\nu(Q,\zeta)\, E_\nu(P,-\zeta)\, d\zeta \end{aligned}$$

to the half-plane $\{\, s \mid \operatorname{Re} s \leq 0 \,\}$ we alter the path of integration. Let w be fixed with $\operatorname{Re} w > 1$ and let $A > \operatorname{Re} w$, $B > 0$ be chosen so that no poles of the integrand lie on the rectilinear path C starting at $-i\infty$ and connecting $-i\infty$, $-iB$, $A - iB$, $A + iB$, iB, $i\infty$. Let $\zeta_1, \ldots, \zeta_r$ be the poles of the function $\zeta \mapsto E_\nu(Q,\zeta)\, E_\nu(P,-\zeta)$ in the box $0 < \operatorname{Re}\zeta < A$, $|\operatorname{Im}\zeta| < B$. The integral

$$J_\nu(s) := \frac{1}{i} \int_C \frac{w^2 - s^2}{(s^2-\zeta^2)(w^2-\zeta^2)}\, E_\nu(Q,\zeta)\, E_\nu(P,-\zeta)\, d\zeta$$

defines a holomorphic function of s in the box $|\operatorname{Re} s| < A$, $|\operatorname{Im} s| < B$, and for s in this box with $\operatorname{Re} s > 0$ we have by the residue theorem

$$\begin{aligned} I_\nu(s) = J_\nu(s) - 2\pi \sum_{j=1}^{r} \frac{w^2 - s^2}{(s^2-\zeta_j^2)(w^2-\zeta_j^2)} \operatorname*{Res}_{\zeta=\zeta_j} \left(E_\nu(Q,\zeta)\, E_\nu(P,-\zeta) \right) \\ + \frac{\pi}{s}\, E_\nu(Q,s)\, E_\nu(P,-s) - \frac{\pi}{w} E_\nu(Q,w) E_\nu(P,-w) \end{aligned}$$

provided that $s \neq \zeta_1, \ldots, \zeta_r$. This gives the meromorphic continuation of I_ν to the box $|\operatorname{Re} s| < A$, $|\operatorname{Im} s| < B$, and since A, B can be chosen arbitrarily large, the meromorphic continuation of the left-hand side of the eqation under item (2) to the entire s-plane is proved. Removing the term $F(P,Q,w)$ we see that $s \mapsto F(P,Q,s)$ admits a meromorphic continuation to all of $\mathbb{C}$ provided that $\Gamma P \neq \Gamma Q$. We even have a functional equation: Since $J_\nu(s)$ is invariant under $s \mapsto -s$ for s in the box $|\operatorname{Re} s| < A$, $|\operatorname{Im} s| < B$ we find

$$I_\nu(s) - I_\nu(-s) = \frac{\pi}{s}\,(E_\nu(Q,s)\,E_\nu(P,-s) + E_\nu(Q,-s)\,E_\nu(P,s))\,.$$

Since the contribution of the eigenfunctions is invariant under $s \mapsto -s$, we obtain for $\Gamma P \neq \Gamma Q$:

$$\begin{aligned} &F(P,Q,s) - F(P,Q,-s) \\ &\qquad = \frac{1}{4s}\sum_{\nu=1}^{h} \frac{[\Gamma_\nu : \Gamma'_\nu]}{|\Lambda_\nu|}\,(E_\nu(Q,s)\,E_\nu(P,-s) + E_\nu(Q,-s)\,E_\nu(P,s))\,. \end{aligned}$$

The functional equation of the Eisenstein series and $(\overline{\Phi(\overline{s})})^t = \Phi(s)$ yield the functional equation (4.22). □

For the analogue of Proposition 4.6 in the case of dimension 2 see Faddeev (1967), Fay (1977), Fischer (1987), Hejhal (1983), Neunhöffer (1973). The result is contained as a special case in Müller (1980b). The position of the poles of $F(P,Q,s)$, $H(P,Q,s)$ and the functional equation (4.22) are stated erroneously in Elstrodt, Grunewald, Mennicke (1982a).

We shall now apply our results to obtain growth properties of Eisenstein series. Finally we shall investigate the properties of the kernel $H_\Gamma(P,Q)$ from (4.12). We study here only the case of a cofinite group $\Gamma < \mathbf{PSL}(2,\mathbb{C})$ having one class of cusps represented by ∞. We keep the notation from the beginning of Section 1 of this chapter.

Let $|c_\Gamma|$ be the number introduced in Lemma 3.5. It follows from Theorem 1.8 and Lemma 3.5 that $|c_\Gamma|^{2s}\phi(s)$ is bounded in $\mathrm{Re}(s) \geq 0$ except for possible poles in the interval $]0,1]$. Here ϕ is the function from (1.4) and (3.17), it is of crucial importance in the following. The further properties of ϕ that we use are contained in Theorem 1.11.

Definition 4.7. Let $\Gamma < \mathbf{PSL}(2,\mathbb{C})$ be a cofinite subgroup having one class of cusps. Let $0 < \sigma_1 < \sigma_2 \ldots < \sigma_N = 1$ be the poles of $\phi(s)$ in the segment $]0,1]$. We define

$$\phi^*(s) = |c_\Gamma|^{2s}\,\phi(s)\prod_{i=1}^{N}\frac{s-\sigma_i}{s+\sigma_i}\,, \qquad \omega(t) = 1 - \frac{\phi^{*\,\prime}}{\phi^*}(it) \tag{4.23}$$

for $s \in \mathbb{C},\ t \in \mathbb{R}$.

The function ϕ^* is holomorphic for $\mathrm{Re}(s) \geq 0$ and the identities

$$\phi^*(s)\,\phi^*(-s) = 1, \qquad |\phi^*(it)| = 1$$

hold. It is furthermore bounded for $\mathrm{Re}\, s > 0$ and tends uniformly to 0 as $\mathrm{Re}\, s \to \infty$. Therefore we have by Phragmén–Lindelöf $|\phi^*(s)| \leq 1$ for $\mathrm{Re}(s) \geq$

0. Note that the Phragmén–Lindelöf Theorem may be applied here because ϕ^* is of finite order by Theorem 1.2.

We write $\mathbf{P}(\phi^*)$ for the set of poles (with multiplicities) of ϕ^* and similarly $\mathbf{N}(\phi^*)$ for the zeroes of ϕ^*. Note that because of $\overline{\phi^*(s)} = \phi^*(\bar{s})$ poles and zeroes come symmetrically to the real axis. We also have $\rho \in \mathbf{N}(\phi^*)$ if and only if $-\rho \in \mathbf{P}(\phi^*)$. Moreover there is a $B > 0$ so that all poles of ϕ^* lie in the strip $-B < \operatorname{Re} s < 0$. Changing variables to $z := \varphi(s) := (s-1)/(s+1)$ the function $f := \phi^* \circ \varphi^{-1}$ is holomorphic and bounded on the unit disc. Hence by a well-known theorem (see Rudin (1987), Theorem 15.23) the sequence $(z_n)_{(n\geq 1)}$ of zeroes of f satisfies $\sum_{n\geq 1}(1-|z_n|^2) < \infty$, that is

$$\sum_{\rho\in\mathbf{N}(\phi^*)} \frac{\operatorname{Re}\rho}{|\rho+1|^2} < \infty.$$

Hence the product $\prod_{\rho\in\mathbf{P}(\phi^*)}(s+\overline{\rho})(s-\rho)^{-1}$ converges absolutely if we combine the terms with fixed ρ and $\bar{\rho}$ for the nonreal elements $\rho \in \mathbf{P}(\phi^*)$. Alternatively, we could avooid the combination of terms by introducing suitable exponential factors which enforce the convergence in the familiar way of Weierstraß. This product is a meromorphic function with the same zeroes and poles as ϕ^* and we obtain

$$\phi^*(s) = e^{g(s)} \prod_{\rho\in\mathbf{P}(\phi^*)} \frac{s+\bar{\rho}}{s-\rho} \tag{4.24}$$

where g is a polynomial of degree at most 4. Here we use that ϕ is a meromorphic function of order ≤ 4 which is proved in Theorem 1.2. The growth properties ($|\phi^*|$ is bounded for $\operatorname{Re} s > 0$ and $\phi^*(s) \to 0$ as $\operatorname{Re} s \to \infty$) and symmetry of ϕ^* imply that e^g is constant. For the logarithmic derivative we obtain writing every $\rho \in \mathbf{P}(\phi^*)$ as $\rho = \beta + i\gamma$

$$-\frac{\phi^{*\prime}}{\phi^*}(it) = \sum_{\rho\in\mathbf{P}(\phi^*)} \frac{-2\beta}{\beta^2+(t-\gamma)^2}. \tag{4.25}$$

The series on the right-hand side converges absolutely by a standard theorem of complex analysis. Since all poles of ϕ^* have negative β it follows that $\omega(t) \geq 1$ for all t.

The function $\omega(t)$ will serve us as a reference function in our estimates. It controls the sizes of many other functions of importance. To begin with we have for $s = \sigma + it$ with $\sigma > 0$

$$\begin{aligned}
1-|\phi^*(s)|^2 &= 1 - \prod_{\rho\in\mathbf{P}(\phi^*)} \left|\frac{s+\bar{\rho}}{s-\rho}\right|^2 = 1 - \prod_{\rho\in\mathbf{P}(\phi^*)} \left(1 - \frac{-4\sigma\beta}{(\sigma-\beta)^2+(t-\gamma)^2}\right) \\
&\leq \sum_{\rho\in\mathbf{P}(\phi^*)} \frac{-4\sigma\beta}{(\sigma-\beta)^2+(t-\gamma)^2} \leq \sum_{\rho\in\mathbf{P}(\phi^*)} \frac{-4\sigma\beta}{\beta^2+(t-\gamma)^2} \\
&= -2\sigma\frac{\phi^{*\prime}}{\phi^*}(it) = O(\sigma\omega(t)).
\end{aligned}$$

From this and

$$1-|\phi(s)|^2 = 1-|\phi^*(s)|^2 + |\phi^*(s)|^2\left(1-\frac{|\phi(s)|^2}{|\phi^*(s)|^2}\right)$$

and some elementary estimates we find

$$1-|\phi(s)|^2 = O(\sigma\,\omega(t)) \qquad \text{for } 0\le\sigma<\sigma_0<1,\ |t|\ge 1. \tag{4.26}$$

Here as always in this section $s=\sigma+it$. Directly from (3.19) and (4.26) we obtain

$$\int_{\mathcal{F}} |E^A(P,s)|^2\, dv(P) = O(\omega(t)) \tag{4.27}$$

uniformly for $0<\sigma_1<\sigma<\sigma_0<1$. The Maaß–Selberg relation (3.19) holds for $s=\sigma+it$ with $\sigma\neq 0$. If we let σ go to 0 we get

$$\begin{aligned}&\int_{\mathcal{F}} |E^A(P,it)|^2\, dv(P)\\ &= C_\infty\left(2\log A - \frac{\phi'}{\phi}(it) + \frac{\overline{\phi(it)}A^{2it}-\phi(it)A^{-2it}}{2it}\right).\end{aligned} \tag{4.28}$$

From this we immediately infer for $|t|\to\infty$:

$$\int_{\mathcal{F}} |E^A(P,it)|^2\, dv(P) = C_\infty\,\omega(t)+O(1) = -C_\infty\,\frac{\phi'}{\phi}(it)+O(1). \tag{4.29}$$

The various implied constants depend on A which was considered fixed.

For the further development we need three facts about Bessel functions.

Lemma 4.8. *Let $K_s(r)$ ($s\in\mathbb{C}$, $r\in\mathbb{R}$) be the Bessel function defined in Magnus, Oberhettinger, Soni (1966) Chapter 3. Then the following hold.*

(1) For every s with $0\le \operatorname{Re} s\le 1$ and $r>0$ we have

$$|K_s(r)| \le \frac{3e^{-r}}{\sqrt{r}}\left(1+\frac{1}{\sqrt{r}}\right). \tag{4.30}$$

(2) There is a constant $\kappa>0$ so that for every $t\in\mathbb{R}$ and $A>0$

$$\int_A^\infty |K_{it}(r)|^2\,\frac{dr}{r} \ge \kappa\, e^{-5|t|-5A}. \tag{4.31}$$

(3) Let $\alpha_0>0$ be given. Then there are constants $\kappa_0,\ \kappa>0$ so that

$$|K_{it}(\alpha r)| \le\ |K_{it}(\alpha r_0)|\; e^{-\kappa(r-r_0)} \tag{4.32}$$

for all $T\ge 1/2$, $0\le t\le T$, $r\ge r_0\ge\kappa_0 T$, $\alpha\ge\alpha_0$.

Proof. (1): Relation (4.30) can easily be read off from any of the standard integral representations (Magnus, Oberhettinger, Soni (1966), page 85) for $K_s(r)$, see Hejhal (1983), Lemma 12.1 for hints.

(2): By an obvious argument we see that (4.31) follows once it is proved for all $A \geq A_0$. To prove (4.31) for sufficiently large A we put, as is done in an analogous situation in Hejhal (1983), Lemma 12.2, $z = A$, $t^2 + A^2 = y^2$, $\mu = it$ in the last formula on page 104 in Magnus, Oberhettinger, Soni (1966) and obtain

$$\int_A^\infty K_{it}(y)\, y^{1-it}\, (y^2 - A^2)^{it}\, dy = 2^{it} A\, \Gamma(1+it)\, K_1(A). \tag{4.33}$$

There is a constant $\kappa_1 > 0$ so that $|K_1(r)| \geq \kappa_1 \sqrt{r}^{-1} e^{-r}$ for all sufficiently large r, see Magnus, Oberhettinger, Soni (1966), page 139. There also is a constant $\kappa_2 > 0$ so that $|\Gamma(1+it)| \geq \kappa_2 e^{-\pi\,|t|/2}$ for all $t \in \mathbb{R}$. By (4.33) there is $\kappa_3 > 0$ so that

$$\kappa_3 \sqrt{A}\, e^{-A-\frac{\pi}{2}|t|} \leq \int_A^\infty |K_{it}(y)|\, y\, dy. \tag{4.34}$$

We take now $4 \leq B > A$ and get from (4.30)

$$\kappa_3 \sqrt{A}\, e^{-A-\frac{\pi}{2}|t|} - 6 \int_B^\infty e^{-y}\, \sqrt{y}\, dy \leq \int_A^B |K_{it}(y)|\, y\, dy. \tag{4.35}$$

Since $B \geq 4$, a partial integration gives

$$\int_B^\infty e^{-y} \sqrt{y}\, dy = \sqrt{B} e^{-B} + \int_B^\infty e^{-y} \frac{dy}{2\sqrt{y}} < (\sqrt{B} + \frac{1}{4}) e^{-B} < 2\sqrt{B} e^{-B}.$$

We now put $B = 2(A + |t| + \kappa_5)$ and take κ_5 large enough so that there is a constant $\kappa_6 > 0$ with $\kappa_6 e^{-2(A+|t|+\kappa_5)} \leq \kappa_3 \sqrt{A} e^{-A-\pi\,|t|/2} - 2\sqrt{B} e^{-B}$ for all $A \geq 1$ and all t. The Cauchy–Schwarz inequality implies that

$$\int_A^B |K_{it}(y)|^2\, y^{-1}\, dy \cdot \int_A^B y^3\, dy \geq \left(\int_A^B |K_{it}(y)|\, y\, dy \right)^2 \geq \kappa_6^2 e^{-4(A+|t|+\kappa_5)}.$$

Computing the second integral on the left-hand side we find a κ as required but only for $A \geq 1$. If $A < 1$ we increase the lower parameter in the integral (4.31) to $A + 1$ and the result follows by making κ a little bit smaller.

(3): Choose $0 < \kappa < \min\{\, \alpha_0^2,\ 1/4 \,\}$ and put $\kappa_0 := \sqrt{2}/\sqrt{\alpha_0^2 - \kappa}$. Let $T \geq 1/2$ and $r_0 \geq \kappa_0 T$.

For $t \in [0, T]$ we define $G_{it}(r) := \operatorname{sign}(K_{it}(\alpha_0 r_0)) \sqrt{r} K_{it}(\alpha_0 r)$. From Bessel's differential equation for K_{it} we find that G_{it} satisfies the differential equation

$$F''(r) = \left(\alpha_0^2 - \frac{t^2 + \frac{1}{4}}{r^2} \right) F(r) = g(t, r)\, F(r) \qquad (r > 0). \tag{4.36}$$

For $r \geq r_0$ our choices imply $g(t,r) \geq \kappa$, that is, $G''_{it}(r) \geq \kappa G_{it}(r)$ if $G_{it}(r) \geq 0$ and $G''_{it}(r) \leq \kappa G_{it}(r)$ if $G_{it}(r) \leq 0$.

By looking at the derivative we infer that $G_{it}G'_{it}$ is monotonically increasing from r_0 on. Since both $G_{it}(r)$ and $G'_{it}(r)$ tend to 0 as $r \to \infty$ we find that $G_{it}(r)G'_{it}(r) < 0$ for all $r \geq r_0$ and $t \in [0,T]$. It also follows that for fixed t the $G_{it}(r)$ and also the $G'_{it}(r)$ have the same sign and hence no zeroes, for $r \geq r_0$. That is, we have $G_{it}(r) > 0$, $G'_{it}(r) < 0$ and $G''_{it}(r) \geq \kappa G_{it}(r)$ for all $t \in [0,T]$, and $r \geq r_0$.

Define $h(r) := e^{-\sqrt{\kappa}r}$ and consider the (positive) function $f(r) := h(r)/G_{it}(r)$ in $r \geq r_0$. Its derivative is

$$f'(r) = \frac{h'(r)G_{it}(r) - h(r)G'_{it}(r)}{G^2_{it}(r)} = \frac{h(r)(-\sqrt{\kappa}G_{it}(r) - G'_{it}(r))}{G^2_{it}(r)}.$$

The derivative of the numerator is $h(r)(\kappa G_{it}(r) - G''_{it}(r)) \leq 0$. This shows that if $f'(r_1) \leq 0$ for some $r_1 \geq r_0$ then $f'(r) \leq 0$ for all $r \geq r_1$. This means that f is decreasing from r_1 onward. Since $\sqrt{\kappa} \leq 1/2$ we conclude from Magnus, Oberhettinger, Soni (1966), page 139 that $f(r) \to \infty$ for $r \to \infty$. This implies that $f'(r) \geq 0$ for all $r \geq r_0$. Hence f is increasing from r_0 onward. Going back from the G_{it} to the K_{it}, part (3) of the Lemma is proved for α_0, it follows easily for bigger α.

The argument used here is a variant of what appears in Watson (1966), 15.8 under the name of Sturm's method. □

The following lemma will also be needed for our final estimates. We remark that given $\epsilon > 0$ there is a C^∞-function $k_\epsilon : [1,\infty[\to \mathbb{R}$ so that

$$\text{(4.37)} \quad \operatorname{supp}(k_\epsilon) \subset [1, 1+\epsilon], \ \max(|k_\epsilon|) \leq \kappa\epsilon^{-\frac{3}{2}}, \quad \int_{\mathbb{H}} k_\epsilon(\delta(P,Q))\, dv(Q) = 1$$

with some absolute constant κ. This can be checked for $P = j = (0,1) \in \mathbb{H}$. Take an $\epsilon_0 < \epsilon$ but near to ϵ. Let k be the characteristic function of $[1, 1+\epsilon_0]$ then $k \circ \delta$ is the characteristic function of the set $M_{\epsilon_0} = \{\, Q \in \mathbb{H} \mid \delta(j,Q) \leq 1+\epsilon_0 \,\}$. Smooth off k so that it becomes C^∞ with support in $[1,1+\epsilon]$. Then apply our volume formula (1.2.6) for closed balls in $\mathbb{H}$.

Lemma 4.9. *Let $\Gamma < \mathbf{PSL}(2,\mathbb{C})$ be a cofinite subgroup so that ∞ represents the only class of cusps of Γ. Let $0 < \epsilon < 1/16$ and let the C^∞-function k_ϵ be chosen according to (4.37). Let K_Γ be the corresponding averaged function. There is a constant κ_1 depending only on Γ so that*

$$\text{(4.38)} \qquad \int_{\mathcal{F}} |K_\Gamma(P,Q)|^2 \, dv(Q) \leq \kappa_1 \left(r_P^2 \epsilon^{-\frac{1}{2}} + \epsilon^{-\frac{3}{2}} \right)$$

for all $P = (z_P, r_P) \in \mathcal{F}$.

Proof. Writing $Q = (z_Q, r_Q)$ there is a constant $B > 0$ so that $r_Q \geq B$ for all $Q \in \mathcal{F}$, where $\mathcal{F}$ is as before. We now choose a nonzero element $\omega_0 \in \Lambda_\infty$ and a constant A with

$$A \geq \frac{2(1+\epsilon)|\omega_0|^2}{B}$$

and also large enough so that our decomposition (3.15), that is $\mathcal{F} = \mathcal{F}_A \cup \mathcal{F}^A$, exists. Let $P \in \mathcal{F}_A$ then by Lemma 2.6.1 there are at most $C_1(\mathcal{F}_A)(1+\epsilon)^2$ elements γ in the summation for K_Γ to consider. Hence we find a constant κ_2 so that

$$|K_\Gamma(P,Q)|^2 \leq \kappa_2 \sum_{\gamma \in \Gamma} |k_\epsilon(\delta(P,\gamma Q))|^2 \tag{4.39}$$

for all $P \in \mathcal{F}_A$. Put $M_\epsilon(P) := \{\, Q \in \mathbb{H} \mid \delta(P,Q) \leq 1+\epsilon \,\}$. Using (4.39) we may unfold the integral in (4.38) and get the estimate

$$\int_{\mathcal{F}} |K_\Gamma(P,Q)|^2 \, dv(Q) \leq \kappa_2 \int_{\mathcal{F}} \sum_{\gamma \in \Gamma} |k_\epsilon(\delta(P,\gamma Q))|^2 \, dv(Q)$$
$$= \kappa_2 \int_{\mathbb{H}} |k_\epsilon(\delta(P,Q))|^2 \, dv(Q) \leq \kappa^2 \kappa_2 \epsilon^{-3} \, v(M_\epsilon(P)).$$

Since for $0 < \epsilon < 1/16$

$$\operatorname{arcosh}(1+\epsilon) = \log(1+\epsilon+\sqrt{\epsilon(2+\epsilon)}) \leq 2\sqrt{\epsilon}$$

we infer from our volume formula (1.2.6) the existence of a constant κ_3 so that $v(M_\epsilon(P)) \leq \kappa_3 \epsilon^{3/2}$. This implies the assertion for $P \in \mathcal{F}_A$.

So only the case $P \in \mathcal{F}^A$, that is $r_P \geq A$ remains. Let now $r_P \geq A$. Let $\gamma = \begin{pmatrix} a & b \\ c & d \end{pmatrix} \in \Gamma$ with $c \neq 0$. Then by application of Theorem 2.3.1 (Shimizu's lemma) we find

$$\delta(P,\gamma Q) \geq \frac{1}{2}|c|^2 \, r_P \, r_Q \geq \frac{1}{2}\frac{A\,B}{|\omega_0|^2} > 1+\epsilon$$

for all $Q \in \mathcal{F}$. Hence only the elements in Γ_∞ give a contribution to the summation for K_Γ. We only treat the summation over Γ'_∞, the other cosets of Γ'_∞ in Γ_∞ can be handled similarly. Let now $\gamma \in \Gamma'_\infty$ be an element corresponding to $\omega \in \Lambda_\infty$. Then

$$\delta(P,\gamma Q) = \frac{|z_Q - z_P + \omega|^2 + r_P^2 + r_Q^2}{2 r_P \, r_Q}.$$

If γ is to give a contribution to K_Γ then

$$\frac{|z_Q - z_P + \omega|^2}{2 r_P \, r_Q} < \epsilon, \qquad \frac{(r_P - r_Q)^2}{2 r_P \, r_Q} < \epsilon.$$

This implies $\tau_0\, r_P \leq r_Q \leq \tau_1\, r_P$ with $\tau_0 = 1+\epsilon-\sqrt{\epsilon(\epsilon+2)}$, $\tau_1 = 1+\epsilon+\sqrt{\epsilon(\epsilon+2)}$. From this we infer $|z_Q - z_P + \omega|^2 < 2\,\epsilon\,\tau_1\, r_P^2$. Since z_P, z_Q range over a compact domain of $\mathbb{C}$ we find by using the Euclidean lattice point theorem, see Fricker (1982), a constant κ_4 so that only $\kappa_4\,\epsilon\, r_P^2$ such ω can give a contribution to K_Γ. Each of them will give a contribution of at most $\kappa\,\epsilon^{-3/2}$ to $|K_\Gamma|$. From $\tau_0\, r_P \leq r_Q \leq \tau_1\, r_P$ we obtain the estimate

$$\int_{\mathcal{F}} |K_\Gamma(P,Q)|^2 \, dv(Q) \leq \kappa^2 \epsilon^{-3} \cdot \kappa_4 \epsilon^2 r_P^4 \cdot \kappa_5 \int_{\tau_0 r_P}^{\tau_1 r_P} \frac{dr}{r^3}$$

with some constant κ_5. Trivial estimates finish the proof of the lemma. □

Before proceeding with our estimates let us remark that following from Theorem 3.3.1 we have the expansion

$$(4.40) \qquad E(P,s) = r^{1+s} + \phi(s)\, r^{1-s} + \sum_{\mu\in\Lambda^\circ}{}' \phi_\mu(s)\; r\; K_s(2\pi|\mu| r)\; e^{2\pi i <z,\mu>}$$

for the Eisenstein series. Here $P = (z,r) \in \mathbb{H}$ and $s \in \mathbb{C}$ is neither 0 nor a pole of $E(P,s)$. The dash at the sum indicates summation over the non-zero elements. The ϕ_μ are meromorphic functions with no poles on the imaginary axis. Although it has long been clear that $\phi(it)$ is bounded for $t \in \mathbb{R}$ it is difficult to estimate the $\phi_\mu(it)$. From (4.40) we deduce for A large enough by using Parseval's equality (with C_∞ from (3.18))

$$(4.41) \quad \int_{\mathcal{F}A} |E^A(P,it)|^2 \, dv(P) = C_\infty \int_A^\infty \sum_{\mu\in\Lambda^\circ}{}' |\phi_\mu(it)|^2\; |K_{it}(2\pi|\mu|r)|^2 \,\frac{dr}{r}.$$

We are now ready for our final and most important estimates.

Theorem 4.10. *Let $\Gamma <$ **PSL**$(2,\mathbb{C})$ be a cofinite group with one class of cusps represented by ∞. We use the notation $P = (z_P, r_P) \in \mathbb{H}$. The following estimates hold.*

(1) $\int_{-T}^T |E(P,it)|^2\, dt = O(r_P^2 T + T^3)$ for all $P \in \mathcal{F}$ and $T \to \infty$.
(2) $\int_{-T}^T \int_{\mathcal{F}} |E^A(P,it)|^2\, dv(P)\, dt = O(T^3)$ for $T \to \infty$.
(3) $\int_{-T}^T \omega(t)\, dt = O(T^3)$ for $T \to \infty$.
(4) $|\phi_\mu(it)|^2 = O(\omega(t)\, e^{5|t|+\kappa|\mu|})$ with some constant κ for all $\mu \in \Lambda^\circ$, $\mu \neq 0$, $t \in \mathbb{R}$.

Proof. (1): We assume that T is already so large that the requirements of Lemmas 4.9 and 3.5.6 are satisfied for $\epsilon := (16T^2)^{-1}$. We choose the function k_ϵ from Lemma 4.9 with its Selberg transform h and let K_Γ be the corresponding averaged function. From (4.11) and (4.38) we deduce

$$\kappa_1\,(r_P^2 T + T^3) \geq \|K_\Gamma(P,\cdot)\|^2 \geq \frac{C_\infty}{4\,\pi} \int_{-T}^T |h(1+t^2)|^2\; |E(P,it)|^2 \, dt.$$

From Lemma 3.5.6 we deduce a positive absolute lower bound for $h(1+t^2)$ in the range of the last integral.

(2): Let $0 \neq \mu_0 \in \Lambda_\infty$ be an element of minimal length. Let κ_0, κ be the constants from Lemma 4.8, (3) corresponding to $\alpha_0 = |\mu_0|$. Assume that $\kappa_0 T \geq A$ and let B be one of $\kappa_0 T$, $\kappa_0 T + 1$. We find by using (1)

$$\int_{-T}^{T} \int_{\mathcal{F}_B} |E(P, it)|^2 \, dv \, dt = \int_{\mathcal{F}_B} \int_{-T}^{T} |E(P, it)|^2 \, dt \, dv = O(T^3). \tag{4.42}$$

This shows first of all

$$\int_{-T}^{T} \int_{\mathcal{F}_A} |E^A(P, it)|^2 \, dv \, dt = O(T^3) \tag{4.43}$$

because $E = E^A$ in $\mathcal{F}_A$. Let us introduce the notation $\mathcal{F}_A^B = \mathcal{F}_B \setminus \mathcal{F}_A$. By using Parseval's equality we find

$$\begin{aligned} \int_{-T}^{T} \int_{\mathcal{F}_A^B} |E(P, it)|^2 \, dv \, dt &= \int_{-T}^{T} \int_{\mathcal{F}_A^B} |E^A(P, it)|^2 \, dv \, dt \\ &\quad + C_\infty \int_{-T}^{T} \int_{A}^{B} |r^{1+it} + \phi(it) \, r^{1-it}|^2 \, \frac{dr}{r^3} \, dt. \end{aligned} \tag{4.44}$$

Doing an elementary computation with the integral in the last line, we deduce

$$\int_{-T}^{T} \int_{\mathcal{F}_B} |E^A(P, it)|^2 \, dv \, dt = O(T^3). \tag{4.45}$$

from (4.42), (4.43) and (4.44). We use Parseval's equality and after cutting up the integral from $\kappa_0 T$ to ∞ into parts of length 1 we find by (4.32) a constant κ_2 so that

$$\int_{-T}^{T} \int_{\mathcal{F}^{\kappa_0 T}} |E^A(P, it)|^2 \, dv \, dt \leq \kappa_2 \int_{-T}^{T} \int_{\mathcal{F}_{\kappa_0 T}^{\kappa_0 T + 1}} |E^A(P, it)|^2 \, dv \, dt. \tag{4.46}$$

Putting everything together we have proved (2).

(3): Follows from (2) and (4.29).

(4): Follows from (4.41), (4.29) and (4.31) together with some elementary estimates. □

6.5 The Trace Formula and some Applications

This section discusses Selberg's trace formula for a cofinite subgroup $\Gamma < \mathbf{PSL}(2,\mathbb{C})$. We start off with a C^∞-function $k : [1,\infty[\to \mathbb{R}$ of rapid decay and consider the kernels K_Γ, H_Γ, L_Γ introduced in the beginning of the last section. Since k is of rapid decay we conclude from (3.5.34) for any $n \geq 2$

$$Q(x) = \pi \int_x^\infty k(t)\,dt = \pi \int_x^\infty t^{-n}\left(t^n k(t)\right)dt = O(x^{-n+1}) \quad \text{for } x \to \infty.$$

Hence $g(x) := Q(\cosh(x))$ $(x \in \mathbb{R})$ vanishes exponentially as $|x| \to \infty$, and this implies that

$$h(1+t^2) = \int_{-\infty}^{+\infty} g(x)\,e^{itx}\,dt$$

is of rapid decay. Therefore (by (4.21))

$$L_\Gamma(P,Q) = \sum_{m\in\mathfrak{D}} h(\lambda_m)\,e_m(Q)\,\overline{e_m(P)} \qquad (P,Q \in \mathbb{H}) \tag{5.1}$$

is a product of two Hilbert–Schmidt kernels and hence of trace-class. We may put $P = Q$ and integrate over the fundamental domain $\mathcal{F}$ thus obtaining $\sum_{m\in\mathfrak{D}} h(\lambda_m)$ which is the trace of the corresponding operator $\check{L}_\Gamma$. To do the same to the kernel $K_\Gamma - H_\Gamma$ is not directly possible. We define

$$K_\Gamma^{\text{id}},\ K_\Gamma^{\text{cusp}},\ K_\Gamma^{\text{par}},\ K_\Gamma^{\text{ce}},\ K_\Gamma^{\text{nce}},\ K_\Gamma^{\text{lox}} \tag{5.2}$$

similarly to K_Γ only that the summation is extended over the identity element, the non-identity elements of the Γ_ζ (ζ a cusp of Γ), all parabolic elements, the elements of the $\Gamma_\zeta \setminus \Gamma_\zeta'$, all elliptic elements not stabilizing a cusp, and all hyperbolic or loxodromic elements in Γ, respectively. The upper index in K_Γ^{ce} stands for cuspidal elliptic, it is motivated by the fact that all elements of $\Gamma_\zeta \setminus \Gamma_\zeta'$ are elliptic (and stabilize the cusp ζ). The sums involved in (5.2) converge locally uniformly on $\mathbb{H} \times \mathbb{H}$. We have

$$K_\Gamma = K_\Gamma^{\text{id}} + K_\Gamma^{\text{cusp}} + K_\Gamma^{\text{nce}} + K_\Gamma^{\text{lox}}, \qquad K_\Gamma^{\text{cusp}} = K_\Gamma^{\text{par}} + K_\Gamma^{\text{ce}}. \tag{5.3}$$

We shall show that $K_\Gamma^{\text{cusp}}(P,P) - H_\Gamma(P,P)$ is integrable over $\mathcal{F}$ and also compute the integral. The corresponding integrals for the kernels K_Γ^{id}, K_Γ^{nce}, K_Γ^{lox} from (5.1) will also be computed. Here exactly the same method as in Section 2 of Chapter 5 is applicable. The result will be the trace formula which takes after little more work the following form.

Theorem 5.1. *Let $\Gamma < \mathbf{PSL}(2,\mathbb{C})$ be a cofinite group. Assume that Γ has $\kappa > 0$ classes of cusps represented by $\zeta_1, \ldots, \zeta_\kappa$. Let $\mathfrak{D} = \mathfrak{D}(\Gamma)$ the index set for the eigenfunctions of $-\Delta$ in $L^2(\Gamma\backslash\mathbb{H})$ and $\phi := \det\Phi$ the determinant of the scattering matrix introduced in (1.7).*

Let h be a function holomorphic in a strip of width strictly greater then 2 around the real axis satisfying the growth condition $h(1+z^2) = O((1+|z|^2)^{-3/2-\epsilon})$ for $|z| \to \infty$ uniformly in the strip. Let g be the cosine transform

$$g(x) := \frac{1}{2\pi} \int_{-\infty}^{\infty} h(1+t^2)\, e^{-itx}\, dt = \frac{1}{2\pi} \int_{-\infty}^{\infty} h(1+t^2)\, \cos(xt)\, dt.$$

Then there is for each cusp ζ_i a number $\ell_i \in \mathbb{N}$ $(i = 1, \ldots, \kappa)$ and constants

$$c_\Gamma, \quad \tilde{c}_\Gamma, \quad d_\Gamma, \quad d(i,j),\ \alpha(i,j) > 0 \quad (i = 1, \ldots, \kappa\ j = 1, \ldots, \ell_i)$$

so that the following identity holds with all sums being absolutely convergent:

$$\begin{aligned}
&\sum_{m \in \mathfrak{D}} h(\lambda_m) \\
&= \frac{\operatorname{vol}(\Gamma)}{4\pi^2} \int_{-\infty}^{\infty} h(1+t^2)\, t^2\, dt \\
&+ \sum_{\{R\}\,\mathrm{nce}} \frac{\pi g(0) \log N(T_0)}{|\mathcal{E}(R)|\ \sin^2\left(\frac{\pi k}{m(R)}\right)} + \sum_{\{T\}\,\mathrm{lox}} \frac{4\pi g(\log N(T)) \log N(T_0)}{|\mathcal{E}(T)|\ |a(T) - a(T)^{-1}|^2} \\
&+ c_\Gamma\, g(0) + \tilde{c}_\Gamma\, h(1) - \frac{\operatorname{tr}\Phi(0)\, h(1)}{4} \\
&+ \frac{1}{4\pi} \int_{-\infty}^{\infty} h(1+t^2)\, \frac{\phi'}{\phi}(it)\, dt - d_\Gamma \int_{-\infty}^{\infty} h(1+t^2)\, \frac{\Gamma'}{\Gamma}(1+it)\, dt \\
&+ \sum_{i=1}^{\kappa} \sum_{j=1}^{\ell_i} d(i,j) \int_0^{\infty} g(x)\, \frac{\sinh x}{\cosh x - 1 + \alpha(i,j)}\, dx.
\end{aligned} \tag{5.4}$$

The first sum in the third line extends over all Γ-conjugacy classes of elliptic elements in Γ which do not stabilize a cusp the notation is as in (5.2.22) and (5.2.33). The second sum extends over all hyperbolic or loxodromic conjugacy classes, the notation is as in (5.2.10).

If the stabilzer Γ_{ζ_i} of the cusp ζ_i is torsion free then $d(i,j) = 0$ for $j = 1, \ldots, \ell_i$.

The expression (5.4) is called Selberg's trace formula. From our formulation the function k which gives the point-pair invariant has disappeared. We consider here the function h as the primary function. The formalism in Lemma 3.5.5 gives the possibility to formulate (5.4) so that either k or g take the place of h.

The constants c_Γ, etc. appearing in (5.4) can be evaluated in terms of invariants of Γ which use information about the stabilizers of the cusps of Γ. For groups with one class of cusps their value can easily be read off from the proof of Proposition 5.3.

We treat here only the case of a cofinite group $\Gamma < \mathbf{PSL}(2,\mathbb{C})$ having one class of cusps represented by ∞. We keep the notation from the beginning of Sections 1, 4 of this chapter. For the adjustments needed to prove the general trace formula see the end of this section.

Before beginning with the computations we give the following necessary lemma.

Lemma 5.2. *Let $\Lambda \subset \mathbb{C}$ be a lattice. Then there is a constant κ_Λ so that*

$$(5.5)\quad Z_\Lambda(x) = \sum_{\substack{\mu\in\Lambda \\ |\mu|^2\le x}}{}' |\mu|^{-2} = \frac{\pi}{|\Lambda|}\left(\log x + \kappa_\Lambda\right) + O\left(x^{-\frac{1}{2}}\right) \qquad \textit{as } x \to \infty.$$

The constant κ_Λ may be thought of as an analogue of the Euler constant γ for the lattice $\mathbb{Z}$ in $\mathbb{R}$.

Proof. If (5.5) holds for a lattice Λ then the same statement is true for all lattices $r\Lambda$ where $0 \ne r \in \mathbb{R}$. Hence we may assume that $|\Lambda| = 1$. We now use the usual setup from the proof of the Euclidean lattice point theorem, see Fricker (1982). Let μ_1, μ_2 then be $\mathbb{Z}$-module generators for Λ. Define

$$\mathcal{Q} := \{\ r_1\mu_1 + r_2\mu_2 \ \mid \ -1/2 \le r_1,\, r_2 \le 1/2\ \}$$

and for $\mu \in \Lambda$ put $\mathcal{Q}_\mu := \mu + \mathcal{Q}$. The Euclidean measure of all the $\mathcal{Q}_\mu$ is $|\Lambda| = 1$. Let $D(x) := \{\ u \in \mathbb{C} \ \mid \ |u|^2 \le x\ \}$. Then there is a constant $\kappa_1 > 0$ depending only on Λ so that $D((\sqrt{x}-\kappa_1)^2) \subset \bigcup_{|\mu|^2\le x}\mathcal{Q}_\mu \subset D((\sqrt{x}+\kappa_1)^2)$. Define $F : \mathbb{C}\setminus\{0\} \to \mathbb{R}$ by $F(u) := 1/|u|^2$. Identifying $\mathbb{C}$ with $\mathbb{R}^2$ by $u = u_1 + u_2 i$ we find

$$DF = \left(\frac{-2u_1}{(u_1^2+u_2^2)^2}, \frac{-2u_2}{(u_1^2+u_2^2)^2}\right)$$

for the gradient of F. Let $D_1(x) := \{\ u \in \mathbb{C} \ \mid \ 1 \le |u|^2 \le x\ \}$. Then we see by introducing polar coordinates that $\int_{D_1(x)} F(u)\,du = \pi \log x$. We use now for large x an estimate of the form

$$(5.6)\quad \begin{aligned} \sum_{\substack{\mu\in\Lambda \\ |\mu|^2\le x}}{}' |\mu|^{-2} - \pi \log x &= \sum_{\substack{\mu\in\Lambda \\ |\mu|^2\le x}}{}' |\mu|^{-2} - \int_{D_1(x)} F(u)\,du \\ &= \sum_{\mu\in\Lambda}{}' \int_{\mathcal{Q}_\mu} (F(\mu) - F(u))\,du + \mathrm{E}_0 + \mathrm{E}_{\mathrm{x}} + \mathrm{E}_\infty \end{aligned}$$

with three types of error E_0, E_{x}, E_∞ which we explain in the sequel. Note first of all that by the mean value theorem

$$(5.7)\qquad |F(\mu) - F(u)| \le \sup\{\ \|DF(v)\| \ \mid \ v \in \mathcal{Q}_\mu\ \}\ \|\mu - u\| \le \kappa_2 \frac{1}{|\mu|^3}$$

with a suitable constant κ_2 for all $u \in \mathcal{Q}_\mu$. Hence the sum in the last line of (5.6) converges. The error E_0 occurs around 0 and is easily seen to be independent of x. The error E_x occurs in the ring

$$R(x) := \{ \, u \in \mathbb{C} \ \mid \ (\sqrt{x} - \kappa_1)^2 \leq |u|^2 \leq (\sqrt{x} + \kappa_1)^2 \, \}.$$

It is of size

$$|\mathrm{E}_x| \leq \kappa_3 \left(\sum_{\substack{\mu \in \Lambda \\ \mu \in R(x)}}{}' \, |\mu|^{-2} + \int_{R(x)} F(u) \, du \right) \tag{5.8}$$

with suitable κ_3. The sum in (5.8) is seen to be $O(\sqrt{x}^{-1})$ by an obvious estimate, the integral is also $O(\sqrt{x}^{-1})$ by elementary integration. Finally there is the error E_∞ which comes from summing over all μ in the last line of (5.6). It is of size

$$|\mathrm{E}_\infty| \leq \sum_{\substack{\mu \in \Lambda \\ |\mu|^2 \geq x}}{}' \int_{\mathcal{Q}_\mu} (F(\mu) - F(u)) \, du = O(x^{-1})$$

by (5.7). Putting everything together we have proved the lemma. □

We shall show that $K_\Gamma^{\mathrm{par}}(P,P) - H_\Gamma(P,P)$ is integrable over $\mathcal{F}$ by actually computing $\int_{\mathcal{F}_A} K_\Gamma^{\mathrm{par}}(P,P) \, dv(P)$ and $\int_{\mathcal{F}_A} H_\Gamma(P,P) \, dv(P)$ for larger and larger A. It will be clear that the limit as $A \to \infty$ of the difference of these two integrals exists.

Proposition 5.3. *Let $\Gamma < \mathbf{PSL}(2,\mathbb{C})$ be a cofinite group with one class of cusps represented by ∞. Define $C_\infty = |\Lambda_\infty| / [\Gamma_\infty : \Gamma'_\infty]$. Let k be in $\mathcal{S}([1,\infty[)$ and g be the function introduced in Lemma 3.5.5. We always assume that $A \in \mathbb{R}$ is large enough so that $\mathcal{F}_A$ and $\mathcal{F}^A$ (see (3.15)) are defined. The following hold.*

(1) For $A \to \infty$:

$$\begin{aligned} [\Gamma_\infty : \Gamma'_\infty] & \int_{\mathcal{F}_A} K_\Gamma^{\mathrm{par}}(P,P) \, dv(P) \\ & = g(0) \log A + \frac{h(1)}{4} + g(0) \left(\frac{1}{2} \kappa_{\Lambda_\infty} - \gamma \right) \\ & \quad - \frac{1}{2\pi} \int_{-\infty}^{\infty} h(1+t^2) \, \frac{\Gamma'}{\Gamma}(1+it) \, dt + o(1). \end{aligned} \tag{5.9}$$

Here γ is Euler's constant and κ_{Λ_∞} is from (5.5).

(2) There is an $\ell \in \mathbb{N} \cup \{0\}$ and constants c, $d_1, \ldots, d_\ell$, $\alpha_1 > 0, \ldots, \alpha_\ell > 0$ so that for $A \to \infty$:

$$\int_{\mathcal{F}_A} K_\Gamma^{\mathrm{ce}}(P,P)\,dv(P) \tag{5.10}$$
$$= \left(1 - \frac{1}{[\Gamma_\infty : \Gamma'_\infty]}\right) g(0)\,\log A + c\,g(0)$$
$$+ \sum_{i=1}^{\ell} d_i \int_0^\infty g(x) \frac{\sinh x}{\cosh x - 1 + \alpha_i}\,dx + o(1).$$

(3) Assume further that the Selberg transform h of k satisfies $h(1+t^2) = O(e^{-6|t|})$ as $|t| \to \infty$. Then as $A \to \infty$:

$$\int_{\mathcal{F}_A} H_\Gamma(P,P)\,dv(P) = g(0)\,\log A + \frac{\phi(0)\,h(1)}{4} \tag{5.11}$$
$$- \frac{1}{4\pi} \int_{-\infty}^{\infty} h(1+t^2)\,\frac{\phi'}{\phi}(it)\,dt + o(1).$$

Proof. (1): Since the centralizer of every non-identity element of Γ'_∞ in Γ is Γ'_∞ and since $\Gamma_\infty/\Gamma'_\infty$ acts (by conjugation) without fixed points on $\Gamma'_\infty \setminus \{1\}$ we may write (using as before the abreviation $K := k \circ \delta$):

$$K_\Gamma^{\mathrm{par}}(P,Q) = \sum_{\substack{\gamma \in \Gamma \\ \gamma \text{ parabolic}}} K(P,\gamma Q) = \sum_{\gamma \in \Gamma/\Gamma_\infty} \sum_{\mu \in \Lambda_\infty}{}' K(P, \gamma\hat{\mu}\gamma^{-1}Q) \tag{5.12}$$
$$= \sum_{\gamma \in \Gamma/\Gamma_\infty} \sum_{\mu \in \Lambda_\infty}{}' K(\gamma^{-1}P, \hat{\mu}\gamma^{-1}Q).$$

Here the dash at the sum stands for summation over the non-zero elements. If μ is in Λ_∞ we write $\hat{\mu}$ for the corresponding element of Γ'_∞. Let $\mathcal{P}$ be the fundamental domain of Γ_∞ acting on $\mathbb{C}$ used in the construction of $\mathcal{F}$. The Euclidean measure of $\mathcal{P}$ is C_∞. Define

$$\tilde{\mathcal{P}} := \{\, (z,r) \in \mathbb{H} \mid z \in \mathcal{P}\}, \qquad \mathcal{P}_A := \{\, (z,r) \in \mathbb{H} \mid z \in \mathcal{P},\, r \le A\,\}.$$

Let $\mathbf{R}$ be a system of representatives for Γ/Γ_∞. Then both $\tilde{\mathcal{P}}$ and $\bigcup_{\gamma \in \mathbf{R}} \gamma^{-1}\mathcal{F}$ are fundamental domains for the action of Γ_∞ on $\mathbb{H}$. Put furthermore $\mathcal{M} := \bigcup_{\gamma \in \mathbf{R}} \gamma^{-1}\mathcal{F}_A$. It is an exercise using the rapid decrease of k to prove

$$\int_{\mathcal{F}_A} K_\Gamma^{\mathrm{par}}(P,P)\,dv(P) = \sum_{\mu \in \Lambda_\infty}{}' \int_{\mathcal{F}_A} \sum_{\gamma \in \mathbf{R}} K(\gamma^{-1}P, \hat{\mu}\gamma^{-1}P)\,dv(P) \tag{5.13}$$
$$= \sum_{\mu \in \Lambda_\infty}{}' \int_{\mathcal{M}} K(P,\hat{\mu}P)\,dv = \sum_{\mu \in \Lambda_\infty}{}' \int_{\mathcal{P}_A} K(P,\hat{\mu}P)\,dv + o(1)$$
$$=: I + o(1)$$

as $A \to \infty$. Here the dash at the first two sums denotes leaving off the representative of the identity coset, the dash at the last sum stands for leaving off $\mu = 0$. We proceed with

$$\begin{aligned} I &= C_\infty \sum_{\mu\in\Lambda_\infty}{}' \int_0^A k\left(\frac{|\mu|^2}{2\,r^2}+1\right)\frac{dr}{r^3} \\ &= C_\infty \sum_{\mu\in\Lambda_\infty}{}' |\mu|^{-2} \int_{\frac{|\mu|^2}{2A^2}}^\infty k(u+1)\,du. \end{aligned} \tag{5.14}$$

Since the function Z_{Λ_∞} defined in (5.5) is a step function of the correct calibration we have

$$I = C_\infty \int_0^\infty k(u+1)\; Z_{\Lambda_\infty}(2A^2u)\,du. \tag{5.15}$$

We now put in (5.5). Writing $\log(2A^2u) = \log 2 + 2\log A + \log u$ we get

$$\begin{aligned} &[\Gamma_\infty : \Gamma'_\infty]I \\ &= |\Lambda_\infty| \int_0^\infty k(u+1)\,\frac{\pi}{|\Lambda_\infty|}(\log 2 + 2\log A + \log u + \kappa_{\Lambda_\infty})\,du + o(1) \\ &= g(0)\log A + \frac{\log 2}{2}g(0) + \frac{1}{2}\kappa_{\Lambda_\infty}g(0) + \pi\int_0^\infty k(u+1)\log u\,du + o(1). \end{aligned}$$

For this computation the relations between k and g of Lemma 3.5.5 are crucial. What remains to be done is to determine the last integral, that is

$$I_1 = \int_0^\infty k(u+1)\log u\,du = \frac{\partial}{\partial x}\int_0^\infty k(u+1)\,u^x\,du\bigg|_{x=0}.$$

We have

$$\begin{aligned} \int_0^\infty k(u+1)\,u^x\,du &= \int_1^\infty k(u)\,(u-1)^x\,du = -\frac{1}{2\pi}\int_1^\infty Q'(u)(u-1)^x\,du \\ &= -\frac{1}{2\pi}\int_0^\infty Q'(\cosh t)\sinh t\left(\frac{e^t+e^{-t}}{2}-1\right)^x dt \\ &= -\frac{1}{2\pi}\int_0^\infty g'(t)\left(\frac{\left(e^{\frac{t}{2}}-e^{-\frac{t}{2}}\right)^2}{2}\right)^x dt. \end{aligned}$$

Here we used the following formulas from Lemma 3.5.5:

$$k(x) = -\frac{Q'(x)}{2\pi}, \qquad Q\left(\frac{e^x+e^{-x}}{2}\right) = g(x).$$

Carrying out the differentiation with respect to x we find

$$I_1 = -\frac{\log(2)\, g(0)}{2\pi} - \frac{1}{2\pi}\int_0^\infty g'(x)\,(2\log(e^x - 1) - x)\,dx. \tag{5.16}$$

We have by Lemma 3.5.5

$$\frac{h(1)}{2} = -\int_0^\infty t\, g'(t)\,dt, \qquad g'(x) = \frac{1}{2\pi}\int_{-\infty}^\infty it\, h(1+t^2)\, e^{ixt}\,dt.$$

Inserting this into (5.16) and using

$$-it\int_0^\infty e^{ixt}\log(1-e^{-x})\,dx = -\frac{\Gamma'}{\Gamma}(1-it) - \gamma \tag{5.17}$$

we get

$$I_1 = -\frac{\log(2)\, g(0)}{2\pi} + \frac{h(1)}{4\pi} - \frac{\gamma\, g(0)}{\pi} - \frac{1}{2\pi^2}\int_{-\infty}^\infty h(1+t^2)\,\frac{\Gamma'}{\Gamma}(1-it)\,dt. \tag{5.18}$$

Formula (5.17) can be found in Oberhettinger (1957). It is also just a special case of the formula

$$\frac{\Gamma'}{\Gamma}(z) = -\gamma + \int_0^\infty \frac{e^{-t} - e^{-tz}}{1-e^{-t}}\,dt \qquad (\operatorname{Re} z > 0)$$

which can be found in Magnus, Oberhettinger, Soni (1966), p. 16. Inserting (5.18) into (5.15) and using the obvious computations mentioned above we have proved part (1).

(2): Let $\mathcal{CE}$ be the set of elements of Γ which are Γ-conjugate to an element of $\Gamma_\infty \setminus \Gamma'_\infty$. It is easy to see that $\mathcal{CE}$ falls into finitely many Γ-conjugacy classes. We may take representatives of the form

$$g_1 = \begin{pmatrix} \epsilon_1 & \epsilon_1\omega_1 \\ 0 & \epsilon_1^{-1} \end{pmatrix}, \ldots, g_\ell = \begin{pmatrix} \epsilon_\ell & \epsilon_\ell\omega_\ell \\ 0 & \epsilon_\ell^{-1} \end{pmatrix}.$$

We write $\mathcal{C}(g)$ for the centralizer in Γ of an element $g \in \mathcal{CE}$. It is again easy to see that this centralizer is finite. We have

$$K_\Gamma^{\mathrm{ce}}(P,Q) = \sum_{\gamma\in\mathcal{CE}} K(P,\gamma Q) = \sum_{i=1}^{\ell} \frac{1}{|\mathcal{C}(g_i)|}\sum_{\gamma\in\Gamma} K(P,\gamma g_i \gamma^{-1} Q).$$

Let us define for $i = 1,\ldots,\ell$ and A large

$$I_i(A) := \sum_{\gamma\in\Gamma}\int_{\mathcal{F}A} K(P,\gamma g_i\gamma^{-1}P)\,dv(P).$$

Then we have

$$\int_{\mathcal{F}_A} K_\Gamma^{\mathrm{ce}}(P,P)\,dv(P) = \sum_{i=1}^{\ell}\frac{1}{|\mathcal{C}(g_i)|}\, I_i(A).$$

We are left with the task of computing the integrals $I_i(A)$. Define

$$\mathfrak{M}_A := \bigcup_{\gamma\in\Gamma} \gamma\,\mathcal{F}_A = \mathbb{H} \setminus \left(\bigcup_{\gamma\in\Gamma} \gamma \mathbb{H}_A \right).$$

The second equality follows from the definition $\mathbb{H}_A := \{ (z,r) \mid r \geq A \}$ of the horoballs in Corollary 2.3.3 and the fact that $\mathcal{F}$ is a fundamental domain for Γ. Writing out everything explicitly we have

$$I_i(A) = \int_{\mathfrak{M}_A} k\left(\frac{|(1-\epsilon_i^{-2})z_P + \omega_i|^2}{2r_P^2} + 1 \right) dv(P). \tag{5.19}$$

Let $p_i \in \mathbb{C} \subset \mathbb{P}^1\mathbb{C}$ be the second fixed point of g_i. We have $p_i = \omega_i/(\epsilon_i^{-2} - 1)$. By Corollary 2.3.11 we know that p_i is a cusp of Γ choose a $\gamma_i \in \Gamma$ with $\gamma_i\infty = p_i$. Let c_i lower left-hand entry of γ_i. Notice that the absolute value $|c_i|$ of c_i is independent of the choice of γ_i. We have

$$\gamma_i\mathbb{H}_A = \left\{ (z,r) \;\middle|\; |z-p_i|^2 + \left(r - \frac{1}{2A|c_i|^2} \right)^2 \leq \left(\frac{1}{2A|c_i|^2} \right)^2 \right\}. \tag{5.20}$$

From the explicit version (5.19) and the rapid decrease of k we see that

$$I_i(A) = \int_{\mathbb{H}\setminus(\mathbb{H}_A\cup\gamma_i\mathbb{H}_A)} k\left(\frac{|(1-\epsilon_i^{-2})z_P + \omega_i|^2}{2r_P^2} + 1 \right) dv(P) + o(1). \tag{5.21}$$

If the function k has compact support (5.21) holds for large A without the $o(1)$.

Let $\tilde{I}_i(A)$ be the integral in this equation. We shall evaluate $\tilde{I}_i(A)$ in the following. Define

$$\mathbb{H}_A^B := \{ (z,r) \in \mathbb{H} \mid A \leq r \leq B \}.$$

Since $\gamma_i\mathbb{H}_A$ is contained up to a set of measure zero in $\mathbb{H} \setminus \mathbb{H}_{|c_i|^{-2}A^{-1}}$ the domain of integration for the integral in (5.21) can be decomposed as

$$\mathbb{H} \setminus (\mathbb{H}_A \cup \gamma_i\mathbb{H}_A) = \mathbb{H}^A_{|c_i|^{-2}A^{-1}} \cup ((\mathbb{H} \setminus \mathbb{H}_{|c_i|^{-2}A^{-1}}) \setminus \gamma_i\mathbb{H}_A).$$

This splits the integral in (5.21) into two parts: $\tilde{I}_i(A) = J_{i,1}(A) + J_{i,2}(A)$. From 3.5.33 we obtain

$$\begin{aligned} J_{i,1}(A) &= \int_{\mathbb{H}^A_{|c_i|^{-2}A^{-1}}} k\left(\frac{|(1-\epsilon_i^{-2})z_P + \omega_i|^2}{2r_P^2} + 1 \right) dv(P) \\ &= \frac{2g(0)\,(\log|c_i| + \log A)}{|1-\epsilon_i^2|^2}. \end{aligned} \tag{5.22}$$

To complete the result we finally need to compute the integral

$$J_{i,2}(A) := \int_{\mathfrak{L}_A} k\left(\frac{|(1-\epsilon_i^{-2})z_P + \omega_i|^2}{2r_P^2} + 1\right) dv(P)$$

where

$$\mathfrak{L}_A = \left\{ (z,r) \in \mathbb{H} \;\middle|\; \begin{array}{l} r \leq |c_i|^{-2}A^{-1} \\ |z-p_i|^2 + \left(r - \frac{1}{2A|c_i|^2}\right)^2 \geq \left(\frac{1}{2A|c_i|^2}\right)^2 \end{array} \right\}.$$

Introducing polar coordinates for z_P we obtain

$$J_{i,2}(A) = 2\pi \int_0^{|c_i|^{-2}A^{-1}} \int_{\sqrt{|c_i|^{-2}A^{-1}r - r^2}}^{\infty} k\left(\frac{|1-\epsilon_i^{-2}|^2}{2}\frac{t^2}{r^2} + 1\right) t\,dt\,\frac{dr}{r^3}.$$

Put now $T := 1 + 2^{-1}|1-\epsilon_i^{-2}|^2 r^{-2} t^2$ and use (3.5.34) to find

$$J_{i,2}(A) = \frac{1}{|1-\epsilon_i^{-2}|^2} \int_0^{|c_i|^{-2}A^{-1}} Q\left(1 - \frac{|1-\epsilon_i^{-2}|^2}{2} + \frac{|1-\epsilon_i^{-2}|^2}{2|c_i|^2 A}\frac{1}{r}\right)\frac{dr}{r}.$$

Put

$$w := 1 - \frac{|1-\epsilon_i^{-2}|^2}{2} + \frac{|1-\epsilon_i^{-2}|^2}{2|c_i|^2 A}\frac{1}{r}$$

to get

$$J_{i,2}(A) = \frac{1}{|1-\epsilon_i^{-2}|^2} \int_1^{\infty} Q(w)\,\frac{1}{w - 1 + \frac{|1-\epsilon_i^{-2}|^2}{2}}\,dw.$$

The last equation shows that the function $J_{i,2}(A)$ does (surprisingly) not depend on A. Substituting $w := \cosh x$ we finally obtain

$$J_{i,2}(A) = \frac{1}{|1-\epsilon_i^{-2}|^2} \int_0^{\infty} g(x)\,\frac{\sinh x}{\cosh x - 1 + \frac{|1-\epsilon_i^{-2}|^2}{2}}\,dx.$$

Writing $\epsilon_i = e^{2\pi\varphi_i}$ this takes the form

$$\begin{aligned} J_{i,2}(A) &= \frac{1}{4\sin^2\varphi_i} \int_0^{\infty} g(x)\,\frac{\sinh x}{\cosh x - 1 + 2\sin^2\varphi_i}\,dx \\ &= \frac{1}{8\sin^2\varphi_i} \int_0^{\infty} g(x)\,\frac{\sinh x}{\sinh\frac{x}{2} + \sin^2\varphi_i}\,dx. \end{aligned} \tag{5.23}$$

To get the final we combine (5.22), (5.23) and

$$2\sum_{i=1}^{n} \frac{1}{|1-\epsilon_i|^2 |\mathcal{C}(g_i)|} = 1 - \frac{1}{[\Gamma_\infty : \Gamma_\infty']}. \tag{5.24}$$

This last equation is proved by using Theorem 2.1.8 and a detailed analysis of the conjugacy classes and centralizers of cuspidal elliptic elements. We skip the details.

(3): Consider the following computation, all applications of Fubini's theorem being easily verified in retrospect.

$$
\begin{aligned}
4\pi C_\infty \int_{\mathcal{F}_A} H_\Gamma(P,P)\, dv(P) &= \int_{\mathcal{F}_A} \int_{-\infty}^{\infty} h(1+t^2)\, |E(P,it)|^2\, dt\, dv \\
&= \int_{-\infty}^{\infty} h(1+t^2) \int_{\mathcal{F}_A} |E(P,it)|^2\, dv\, dt \\
&= \int_{-\infty}^{\infty} h(1+t^2) \int_{\mathcal{F}_A} |E^A(P,it)|^2\, dv\, dt \\
&= \int_{-\infty}^{\infty} h(1+t^2) \int_{\mathcal{F}} |E^A(P,it)|^2\, dv\, dt \\
&\qquad - \int_{-\infty}^{\infty} h(1+t^2) \int_{\mathcal{F}^A} |E^A(P,it)|^2\, dv\, dt.
\end{aligned}
$$

The second of the terms in the last line is $o(e^{-A})$ as $A \to \infty$ using Theorem 4.10, part (4) and a straightforward argument with Parseval's identity. For the first of the terms we use (4.28) and find it to be equal to

$$
\begin{aligned}
& C_\infty \int_{-\infty}^{\infty} h(1+t^2) \left(2\log A - \frac{\phi'}{\phi}(it) + \frac{\overline{\phi(it)}\, A^{2it} - \phi(it)\, A^{-2it}}{2it} \right) dt \\
&= 2C_\infty \log A \int_{-\infty}^{\infty} h(1+t^2)\, dt - C_\infty \int_{-\infty}^{\infty} h(1+t^2)\, \frac{\phi'}{\phi}(it)\, dt \\
&+ C_\infty \int_{-\infty}^{\infty} h(1+t^2)\, A^{2it}\, \frac{\overline{\phi(it)} - \phi(it)}{2it}\, dt \\
&+ C_\infty \int_{-\infty}^{\infty} h(1+t^2)\, \phi(it)\, \frac{\sin(2t \log A)}{t}\, dt.
\end{aligned}
$$

As already in Section 3 we see that the first of the last two integrals goes to 0 as $A \to \infty$ by the Riemann Lebesgue lemma. The second integral gives up to an $o(1)$-term a contribution of $C_\infty\, \pi\, h(1)\, \phi(0)$ by a formula on Dirichlet integrals, see Bochner (1932), page 24. This finishes part (3). □

Proof of Theorem 5.1, the one cusp case. Let us first prove (5.4) in case h satisfies $h(1+t^2) = O(e^{-6|t|})$ as $|t| \to \infty$ and $h(1+t^2) \in \mathbb{R}$ for all $t \in \mathbb{R}$. Define for $x \geq 1$:

$$
Q(x) := g(\operatorname{arcosh}(x)) \qquad k(x) := -\frac{Q'(x)}{2\pi}.
$$

Then k is of rapid decrease and its h coming via Lemma 3.5.5 coincides with the h we started with. We may now form the function K_Γ corresponding to k and apply Proposition 5.3. By Proposition 5.3 we have

$$\lim_{A\to\infty}\left(\int_{\mathcal{F}_A} K_\Gamma^{\text{cusp}}(P,P)\,dv(P) - \int_{\mathcal{F}_A} H_\Gamma(P,P)\,dv(P)\right)$$
$$= \frac{1}{[\Gamma_\infty:\Gamma'_\infty]}\frac{h(1)}{4} + \frac{g(0)}{[\Gamma_\infty:\Gamma'_\infty]}\left(\frac{1}{2}\kappa_{\Lambda_\infty} - \gamma\right) - \frac{\phi(0)\,h(1)}{4}$$
$$- \frac{1}{2\pi[\Gamma_\infty:\Gamma'_\infty]}\int_{-\infty}^{\infty} h(1+t^2)\,\frac{\Gamma'}{\Gamma}(1+it)\,dt + \frac{1}{4\pi}\int_{-\infty}^{\infty} h(1+t^2)\,\frac{\phi'}{\phi}(it)\,dt$$
$$+ c\,g(0) + \sum_{i=1}^{\ell} d_i \int_0^\infty g(x)\,\frac{\sinh x}{\cosh x - 1 + \alpha_i}\,dx.$$

The terms in the last line being those from (5.10). What remains to be done to prove (5.4) is to evaluate the integrals of $K_\Gamma^{\text{id}}(P,P)$, $K_\Gamma^{\text{nce}}(P,P)$, $K_\Gamma^{\text{lox}}(P,P)$ over $\mathcal{F}$. For the last 2 we use exactly the same method as in Chapter 5. We do not go into the straightforward details. We have

$$\int_{\mathcal{F}} K_\Gamma^{\text{nce}}(P,P)\,dv(P) = \sum_{\{R\}\,\text{nce}} \frac{\pi g(0)\log N(T_0)}{|\mathcal{E}(R)|\,\sin^2\left(\frac{\pi k}{m(R)}\right)},$$

$$\int_{\mathcal{F}} K_\Gamma^{\text{lox}}(P,P)\,dv(P) = \sum_{\{T\}\,\text{lox}} \frac{4\pi g(\log N(T))\log N(T_0)}{|\mathcal{E}(T)|\,|a(T)-a(T)^{-1}|^2}.$$

To treat the remaining integral note that

$$\int_{\mathcal{F}} K_\Gamma^{\text{id}}(P,P)\,dv(P) = k(1)\,\text{vol}(\Gamma). \tag{5.25}$$

Since we want to express everything in terms of h and g alone we have to find $k(1)$. Since $k(x) = -Q'(x)/2\pi$ and from Lemma 3.5.5

$$g'(x) = Q'\left(\frac{e^x+e^{-x}}{2}\right)\frac{e^x-e^{-x}}{2}$$

we have

$$k(1) = \frac{1}{4\pi^2}\lim_{x\to 0}\left(\frac{e^x-e^{-x}}{2}\right)^{-1}\int_{-\infty}^{\infty} h(1+t^2)\,t\,\sin(xt)\,dt$$
$$= \frac{1}{4\pi^2}\int_{-\infty}^{\infty} h(1+t^2)\lim_{x\to 0}\left(\left(\frac{e^x-e^{-x}}{2}\right)^{-1}\sin(xt)\right)t\,dt$$

and the result, that is (5.4), follows.

Having proved (5.4) under growth hypotheses on h we may use easy aproximation arguments, without mentioning k any more, to obtain the result for general h, see Hejhal (1976a), pages 32–34 for hints. □

Proof of Theorem 5.1, the multi-cusp case. We can only add a few remarks here. Of course our basic results that is Theorem 4.10 and Proposition 5.3 have to be adapted. When using the general Maaß–Selberg relations (they are best understood as a matrix equation) in the proof of Proposition 5.3, part (3) the determinant of the unitary matrix $\Phi(it)$ drops out. □

We now turn to the discussion of some more or less immediate applications of Selberg's trace formula. We start off with the existence problem for eigenvalues of the Laplace operator. Given a cofinite group $\Gamma < \mathbf{PSL}(2,\mathbb{C})$ we put similarly to Theorem 5.5.1

$$(5.26) \qquad A(\Gamma, T) := |\{\, n \in \mathfrak{D}(\Gamma) \;\; | \;\; \lambda_n \leq T \,\}|,$$

the counting being done with multiplicities. The following result is obtained analogously to Theorem 5.5.1 from the trace formula.

Theorem 5.4. *Let $\Gamma < \mathbf{PSL}(2,\mathbb{C})$ be a cofinite group and let the notation be chosen as in Theorem 5.1. Then*

$$(5.27) \qquad A(\Gamma, T) - \frac{1}{4\pi} \int_{-\sqrt{T}}^{\sqrt{T}} \frac{\phi'}{\phi}(it)\, dt \sim \frac{\mathrm{vol}(\Gamma)}{6\,\pi^2} T^{\frac{3}{2}}.$$

as $T \to +\infty$.

Proof. Let $R > 0$ and put $h(1+t^2) := e^{-(1+t^2)R}$. Then h may be used in the trace formula (5.4). We have

$$g(x) = \frac{1}{2\,\pi} \int_{-\infty}^{+\infty} e^{-(1+t^2)R + itx}\, dt = \frac{1}{\sqrt{4\pi R}}\, e^{-R - x^2/4R}$$

and

$$\int_{-\infty}^{+\infty} h(1+t^2)\, t^2\, dt = \frac{\sqrt{\pi}}{2} R^{-\frac{3}{2}} e^{-R}.$$

All the sums and integrals in the trace formula are absolutely convergent. A quick check reveals that the sums over the conjugacy classes in (5.4) are $O(R^{-1/2})$ as $R \to +0$. The same is true for the terms involving $g(0)$ or $h(1)$ and for the terms coming from the last sum on the right-hand side of 5.4. Moreover,

$$\frac{\Gamma'}{\Gamma}(z) = \log z + \int_0^{\infty} e^{-uz} \left(\frac{1}{u} - \frac{1}{1 - e^{-u}} \right) du \qquad (\mathrm{Re}\, z > 0)$$

by Gradshteyn, Ryzhik (1994), p. 943, 8.361, formula 8. The integral in this formula gives a contribution of the size $O(R^{-1/2})$ to the integral

$$\int_{-\infty}^{+\infty} h(1+t^2)\, \frac{\Gamma'}{\Gamma}(1+it)\, dt$$

whereas the term $\log z$ contributes a term $O(R^{-\frac{1}{2}-\epsilon})$ as $R \to 0$ for any $\epsilon > 0$. Collecting terms we get

$$\sum_{m\in\mathfrak{D}} e^{-\lambda_m R} - \frac{1}{4\pi}\int_{-\infty}^{+\infty} e^{-R(1+t^2)}\,\frac{\phi'}{\phi}(it)\,dt = \frac{\mathrm{vol}(\Gamma)}{8\pi^{3/2}}R^{-\frac{3}{2}} + O\left(R^{-\frac{1}{2}-\epsilon}\right)$$

as $R \to +0$ for every $\epsilon > 0$. Here we write the left hand side as a Laplace transform in the form $\int_0^\infty e^{-uR}\,d\alpha(u)$. Looking back at our considerations about $\omega(t)$ following Definition 4.7 and noting that $\omega(t) \geq 0$, we find that the conditions of a Tauberian theorem are satisfied, see Postnikov (1979), p. 51, Theorem 5 or Widder (1941), p. 197. This theorem yields

$$\alpha(u) \sim \frac{\mathrm{vol}(\Gamma)}{8\pi^{3/2}\Gamma(3/2+1)}\,u^{\frac{3}{2}} \qquad \text{as } u \to +\infty.$$

Writing out everything in terms of our data we get

$$\alpha(u) = A(\Gamma, u) - \frac{1}{4\pi}\int_{-\sqrt{u-1}}^{\sqrt{u-1}} \frac{\phi'}{\phi}(it)\,dt + O(1)$$

and this gives the assertion. □

Applying sharper methods it is possible to prove (5.27) with an error term. As it stands (5.27) gives a much weaker result on the $A(\Gamma,T)$ than Theorem 5.5.1 does in the cocompact case. Comparing ϕ'/ϕ with ω and using Theorem 4.10, part (3) we only get $A(\Gamma,T) = O(T^{3/2})$ as $T \to \infty$. This is near to a result of Donnelly (1982) but falls short of proving Weyl's law which would say

$$A(\Gamma,T) \sim \frac{\mathrm{vol}(\Gamma)}{6\,\pi^2}\,T^{\frac{3}{2}} \qquad \text{as } T \to \infty. \tag{5.28}$$

For the cofinite groups constructed in the next chapter or more generally for arithmetic congruence groups we shall prove (5.28) by computing the function ϕ as a zeta function of a number field. The analogue of this result for the hyperbolic plane was proved by Huxley (1984). A well known result of analytic number theory fortunately tells us then that the integral on the left-hand side of (5.27) is of strictly lower order than the right-hand side. For general cofinite groups Γ we can not even conclude that $A(\Gamma,T)$ is unbounded. By an argument similar to Venkov (1978b) it can be proved that $A(\Gamma,T)$ is unbounded if Γ is strictly smaller than its commensurator in $\mathbf{Iso}(\mathbb{H})$.

Let us mention that there are certain computer experiments by Grunewald, Huntebrinker (1996) finding heuristically eigenvalues for the Laplace operator for certain cofinite groups. Take for example the group $\mathbf{CT}(28)$ from Chapter 10. It has one class of cusps, is nonarithmetic and equal to its commensurator. The following table of eigenvalues below 220 was found.

(5.29) 45.1 95 112 118 145 163 173 189 196 199 216.

The group $\mathbf{CT}(28)$ having a fundamental domain of approximate volume 0.1732... the list (5.29) seems even to hint at the validity of Weyl's law. The above mentioned experimental paper also contains tables of eigenvalues for some of the groups constructed in the next chapter, we shall reproduce some.

This being what is known in general for our 3-dimensional hyperbolic space we digress to report on a fascinating story for the case of the 2-dimensional hyperbolic space. Here $\mathbb{H}^2$ is the hyperbolic plane and the groups Γ are the classical cofinite Fuchsian groups. The theory of the Laplace operator as far as the proof of the analogue of Theorem 5.4 was carried through in pioneering work of Selberg (1989b). After an older conjecture of Roelcke and Selberg that any such group should satisfy Weyl's asymptotic law, a better understanding due to Phillips and Sarnak (see Phillips, Sarnak (1985a), (1985b), Deshouillers, Iwaniec, Phillips, Sarnak (1985) and Sarnak (1986)) led to the conjecture that for a generic cofinite but not cocompact group Γ the discrete spectrum of the Laplace operator should only consist of finitely many elements. The new conjecture derives its credibility from Fermi's golden rule, a principle well known in physics (see Phillips, Sarnak (1992), (1991)). A cofinite group Γ with infinite discrete spectrum might be deformable in a suitable Teichmüller space. A small deformation should then lead to a new group for which all eigenvalues embedded in the continuous spectrum have disappeared. The term generic is at the moment still somewhat unclear. It is hoped to mean something like element in general position in a Teichmüller space. The new conjecture has subsequently been heuristically checked. There were also important steps toward a proof, see Hejhal, Rackner (1992), Sarnak (1986), (1990) and also Wolpert (1992), (1994) for the state of the art. For the special case of subgroups of the modular group see Phillips, Sarnak (1994), Venkov (1990a), Balslev, Venkov (1997).

There is the effort of P. Sarnak (1986), (1990) to get a conjectural understanding of the situation for more general symmetric spaces. Sarnak somewhat excludes the hyperbolic spaces of dimensions $n \geq 3$. It is known that there are many nonarithmetic or arithmetic noncongruence cofinite groups on these spaces, but due to Mostow's rigidity theorem there are no deformations of these groups. A heuristic argument using Fermi's golden rule is then of no help.

The next immediate application of the trace formula is to the prime geodesic theorem (analogue of Theorem 5.7.3). Defining $\pi_{00}(x)$ exactly as in (5.7.1) we have

$$\pi_{00}(x) \sim \frac{x^2}{\log x} \qquad \text{as } x \to \infty. \tag{5.30}$$

This is proved by again taking $h(1+t^2) = e^{-(1+t^2)R}$ in the trace formula, but this time analyzing the situation when $R \to \infty$.

The Selberg zeta function $Z(s)$ may be formed out of the noncuspidal elliptic and the loxodromic conjugacy classes just as in the cocompact case. The analytic properties are of $Z(s)$ are largely analogous to the cocompact theory.

6.6 Notes and Remarks

The theory developed in this chapter can be carried through on more general symmetric spaces. This was initiated in Selberg (1956). Venkov (1971) and Cohen, Sarnak (1979) consider the case of a hyperbolic space of arbitrary dimension. Venkov (1973) treats the classical symmetric spaces of rank 1. For a still more general situation see Sarnak (1986), (1990). Explicit versions of the Selberg trace formula for $\Gamma = \mathbf{PSL}(2, \mathcal{O})$ where $\mathcal{O}$ is the ring of integers of an imaginary quadratic number field were developed by Tanigawa (1977), Szmidt (1983), Bauer (1991), (1993). In this case the constants in Theorem 5.1 are explicitly known, see also Elstrodt (1985).

A very detailed exposition of the Selberg trace formula for cofinite subgroups of $\mathbf{SL}(2, \mathbb{R})$ was given by Hejhal (1976a), (1983). An alternative approach to the Selberg trace formula via an explicit computation of the trace of the iterated resolvent kernel and the Selberg zeta function was accomplished by Fischer (1987) in the case of the hyperbolic plane. The same method works in the case of dimension 3. For higher dimensions the resolvent kernel must be iterated suitably many times.

Chapter 7. PSL(2) over Rings of Imaginary Quadratic Integers

Having discussed the general theory of discontinuous groups on 3-dimensional hyperbolic space, we shall start in this chapter with the first of several series of arithmetic examples of such groups. This chapter will also fix the notation concerning imaginary quadratic number fields to be used in the following chapters. For the usual facts about algebraic number theory we refer to Lang (1993), Hecke (1923) or Hasse (1964). A useful little table of class numbers and ideal class groups is contained in Cohn (1980).

7.1 Introduction of the Groups

Let $K = \mathbb{Q}(\sqrt{D}) \subset \mathbb{C}$ be an imaginary quadratic number field. We always suppose that $D < 0$ is a squarefree integer. We write $\mathcal{O} = \mathcal{O}_K$ for the ring of integers in K. The discriminant of K is denoted by d_K. The negative integer d_K always satisfies $d_K \equiv 0$ or $d_K \equiv 1$ modulo 4. In fact we have $d_K = 4D$ or $d_K = D$ if $D \equiv 2, 3$ or $D \equiv 1$ modulo 4 respectively. The ring of integers $\mathcal{O}$ has the $\mathbb{Z}$-basis consisting of 1 and ω where

$$\omega = \frac{d_K + \sqrt{d_K}}{2}. \tag{1.1}$$

The zeta function ζ_K of K is for $\operatorname{Re} s > 1$ defined by

$$\zeta_K(s) = \sum_{\mathfrak{a}} \frac{1}{N(\mathfrak{a})^s}, \tag{1.2}$$

where the sum is extended over the non-zero ideals of $\mathcal{O}$. $N(\mathfrak{a})$ is the norm of $\mathfrak{a}$, it is equal to $|\mathcal{O}/\mathfrak{a}|$. We consider here the group $\mathbf{PSL}(2, \mathcal{O}) < \mathbf{PSL}(2, \mathbb{C})$. We have

Theorem 1.1. *Let K be an imaginary quadratic field of discriminant $d_K < 0$, and let $\mathcal{O}$ be its ring of integers. Then the group $\mathbf{PSL}(2, \mathcal{O})$ has the following properties:*

(1) $\mathbf{PSL}(2, \mathcal{O})$ is a discrete subgroup of $\mathbf{PSL}(2, \mathbb{C})$.
(2) $\mathbf{PSL}(2, \mathcal{O})$ has finite covolume, but is not cocompact.

(3) The covolume of $\mathbf{PSL}(2,\mathcal{O})$ *is*

$$\text{(1.3)} \qquad \mathrm{vol}(\mathbf{PSL}(2,\mathcal{O})) = \frac{|d_K|^{3/2}}{4\pi^2}\,\zeta_K(2).$$

(4) $\mathbf{PSL}(2,\mathcal{O})$ *has a fundamental domain* $\mathcal{F}_K$ *bounded by finitely many geodesic surfaces.*
(5) The covering of $\mathbb{H}$ *by the* $\sigma\mathcal{F}_K$, $\sigma \in \mathbf{PSL}(2,\mathcal{O})$, *is locally finite.*
(6) The set $\{\, \sigma \in \mathbf{PSL}(2,\mathcal{O}) \mid \sigma\mathcal{F}_K \cap \mathcal{F}_K \neq \emptyset \,\}$ *is finite.*
(7) $\mathbf{PSL}(2,\mathcal{O})$ *is a geometrically finite group.*
(8) $\mathbf{PSL}(2,\mathcal{O})$ *is finitely presented.*

Item (1) of this theorem is of course obvious. Apart from (3) everything else depends here on the construction of a specific fundamental domain, denoted by $\mathcal{F}_K$. This construction is due to Poincaré and Bianchi (1893), and will be given in the sequel. Our treatment shall in parts follow closely that of Swan (1971). The proof of Theorem 1.1 except for (1.3) will be given in the end of Section 3.

There are several methods to prove formula (1.3), one using Eisenstein series is contained in Chapter 8. We include another depending on the theory of binary hermitian forms in Chapter 9. Still another method using an adelic volume computation is contained in Borel (1981).

For a specific K the volume formula can be easily numerically evaluated. To do this we introduce the character χ_K of K. It is the completely multiplicative function $\chi_K : \mathbb{N} \to \mathbb{C}$ which is 0 on the primes dividing d_K and which is defined on primes p coprime to d_K by

$$\chi_K(p) = \begin{cases} 1 & \text{for} \quad p = 2 \text{ and } d_K \equiv 1 \bmod 8, \\ -1 & \text{for} \quad p = 2 \text{ and } d_K \equiv 5 \bmod 8, \\ 1 & \text{for} \quad p \neq 2 \text{ and } d_K \equiv \text{ a square mod } p, \\ -1 & \text{for} \quad p \neq 2 \text{ and } d_K \equiv \text{ a nonsquare mod } p. \end{cases}$$

Traditionally the notation $\chi_K(n) = (\frac{d_K}{n})$ is used. It is known that χ_K in fact defines a so-called primitive Dirichlet character $\chi_K : \mathbb{Z}/d_K\mathbb{Z} \to \mathbb{C}$ such that $\chi_K(-1) = -1$, see for example Zagier (1981a). The L-function of χ_K is defined as

$$\text{(1.4)} \qquad L(s,\chi_K) = \sum_{n\in\mathbb{N}} \chi_K(n)\, n^{-s} = \sum_{n\in\mathbb{N}} \left(\frac{d_K}{n}\right) n^{-s}.$$

The zeta function of K can then be expressed as $\zeta_K(s) = \zeta_{\mathbb{Q}}(s)\; L(s,\chi_K)$. Putting this into the above we get the equivalent formula

$$\text{(1.5)} \qquad \mathrm{vol}(\mathbf{PSL}(2,\mathcal{O})) = \frac{|d_K|^{3/2}}{24}\; L(2,\chi_K).$$

A direct summation gives then the following little table.

d_K	$\mathrm{vol}(\mathbf{PSL}(2,\mathcal{O}))$	d_K	$\mathrm{vol}(\mathbf{PSL}(2,\mathcal{O}))$
-3	0,169156 ...	-11	1,165895 ...
-4	0,305321 ...	-15	3,138613 ...
-7	0,888914 ...	-19	2,653148 ...
-8	1,003841 ...	-20	4,203969 ...

If $\mathcal{L}$ is the Lobachevski function defined as

$$\mathcal{L}(\Theta) = \int_0^{\Theta} \log |2 \sin u| \; du, \tag{1.6}$$

then (1.5) can be rewritten as

$$\mathrm{vol}(\mathbf{PSL}(2,\mathcal{O})) = \frac{|d_K|}{12} \sum_{k \bmod d_K} \chi(k) \; \mathcal{L}\left(\frac{\pi k}{|d_K|}\right). \tag{1.7}$$

See Milnor (1982) for a discussion of this formula.

7.2 The Cusps

This paragraph describes the set of classes of cusps for the group $\mathbf{PSL}(2,\mathcal{O})$. We need a certain number of facts from number theory. The following notation will be fixed in the sequel. We write

$$\mathcal{M} = \mathcal{M}_K \tag{2.1}$$

for the group of fractional ideals of $\mathcal{O}$. A fractional ideal is a non-zero finitely generated $\mathcal{O}$-submodule of K. Every fractional ideal is a free $\mathbb{Z}$-module of rank 2. The product on $\mathcal{M}$ comes from the usual complex product of modules. The neutral element of the group $\mathcal{M}$ is $\mathcal{O}$ itself. If $\mathfrak{m} \in \mathcal{M}$, then its inverse is $\mathfrak{m}^{-1} = \{ \lambda \in K \mid \lambda \mathfrak{m} \subset \mathcal{O} \}$. In particular we have $\mathfrak{m}^{-1} \cdot \mathfrak{n} = \{ \lambda \in K \mid \lambda \mathfrak{m} \subset \mathfrak{n} \}$. We remark that for an integral non-zero ideal $\mathfrak{a} \subset \mathcal{O}$ the $\mathcal{O}$ module $\mathfrak{a}^{-1}$ is equal to the unique fractional ideal $\mathfrak{c}$ satisfying $\mathfrak{a} \cdot \mathfrak{c} = \mathcal{O}$. For $\mathfrak{m} \in \mathcal{M}$ there are always two integral ideals $\mathfrak{a}, \mathfrak{b} \subset \mathcal{O}$ such that $\mathfrak{m} = \mathfrak{a} \cdot \mathfrak{b}^{-1}$. If $T \subset K$ is a subset we write $< T >$ for the $\mathcal{O}$-module generated by T. If T is a finite subset of K and contains a non-zero element then $< T > \in \mathcal{M}$. We use the notation $< \lambda > = < \{\lambda\} > = \lambda \cdot \mathcal{O}$ and also $< \lambda, \mu > = < \{\lambda, \mu\} > = \lambda \cdot \mathcal{O} + \mu \cdot \mathcal{O}$ for elements $\lambda, \mu \in K$.

The multiplicative group K^* acts by multiplication on $\mathcal{M}$. The quotient

$$\mathcal{J}_K = \mathcal{M}/K^* \tag{2.2}$$

inherits a group structure from that of $\mathcal{M}$. $\mathcal{J}_K$ is the ideal class group of K or $\mathcal{O}$. $\mathcal{J}_K$ is a finite abelian group. Its order

$$h = h_K = |\mathcal{J}_K| \tag{2.3}$$

is called the class number of K or $\mathcal{O}$. We write $\mathfrak{m}^\#$ for the class of $\mathfrak{m} \in \mathcal{M}$ in $\mathcal{J}_K$. For every $\mathfrak{m}^\# \in \mathcal{J}_K$ there is an integral ideal $\mathfrak{a} \subset \mathcal{O}$ with $\mathfrak{a}^\# = \mathfrak{m}^\#$. We shall need a certain number of facts from arithmetic, the first is:

Lemma 2.1. *Let $(x_1, x_2), (y_1, y_2) \in K \times K$. Then the following statements are equivalent:*

(1) $< x_1, x_2 >=< y_1, y_2 >$.

(2) There exists $\sigma \in \mathbf{SL}(2, \mathcal{O})$ such that $\sigma \begin{pmatrix} x_1 \\ x_2 \end{pmatrix} = \begin{pmatrix} y_1 \\ y_2 \end{pmatrix}$.

Proof. (2) $\Rightarrow$ (1) is clear.
(1) $\Rightarrow$ (2). Put $\mathfrak{a} =< x_1, x_2 >=< y_1, y_2 >$. If $\mathfrak{a} =< 0 >$ then there is nothing to prove. Otherwise choose an $n \in \mathbb{N}$ and $\theta \in K^*$ such that $\mathfrak{a}^n =< \theta >$. This is possible because $\mathcal{J}_K$ is a finite group. The equation $\mathfrak{a} \cdot \mathfrak{a}^{n-1} = \mathfrak{a}^n =< \theta >$ shows that there are $\alpha_1, \alpha_2, \beta_1, \beta_2 \in \mathfrak{a}^{n-1}$ with $\theta = \alpha_1 x_1 + \alpha_2 x_2$ and $\theta = \beta_1 y_1 + \beta_2 y_2$. Put

$$\sigma = \begin{pmatrix} \frac{y_1\alpha_1 + x_2\beta_2}{\theta} & \frac{y_1\alpha_2 - x_1\beta_2}{\theta} \\ \frac{y_2\alpha_1 - x_2\beta_1}{\theta} & \frac{y_2\alpha_2 + x_1\beta_1}{\theta} \end{pmatrix}.$$

From the definition of $\alpha_1, \alpha_2, \beta_1, \beta_2$ it is clear that

$$\sigma \begin{pmatrix} x_1 \\ x_2 \end{pmatrix} = \begin{pmatrix} y_1 \\ y_2 \end{pmatrix}$$

and that $\det \sigma = 1$. Since the numerators of the entries of σ are by choice in $\mathfrak{a}^n$, σ has in fact integral entries. $\square$

By Definition 2.1.10 the set of cusps C_Γ of a discrete subgroup $\Gamma < \mathbf{PSL}(2, \mathbb{C})$ is the set of those $[z_1, z_2] \in \mathbb{P}^1\mathbb{C}$ such that the stabilizer in Γ under the action

$$\begin{pmatrix} a & b \\ c & d \end{pmatrix} [z_1, z_2] = [az_1 + bz_2, cz_1 + dz_2]$$

contains a free abelian group of rank 2.

Proposition 2.2. *Let $\Gamma < \mathbf{PSL}(2, \mathbb{C})$ be a subgroup commensurable with $\mathbf{PSL}(2, \mathcal{O})$, then $C_\Gamma = \mathbb{P}^1 K \subset \mathbb{P}^1\mathbb{C}$.*

Proof. Clearly $\infty \in C_\Gamma$ for every such group Γ. Here $\infty = [1, 0]$. If $\eta \in \mathbb{P}^1 K$ there is a $\sigma \in \mathbf{PSL}(2, K)$ with $\sigma\eta = \infty$. The group $\sigma\Gamma\sigma^{-1}$ is still commensurable with $\mathbf{PSL}(2, \mathcal{O})$ by Proposition 2.8.5. Furthermore $(\sigma\Gamma\sigma^{-1})_\infty = \Gamma_\eta$. Hence $\eta \in C_\Gamma$.

To prove the converse let $[z_1, z_2] \in C_\Gamma$. Then there is a non-trivial parabolic element $\sigma \in \Gamma \cap \mathbf{PSL}(2, \mathcal{O})$ with $\sigma[z_1, z_2] = [tz_1, tz_2]$ where $t \in \mathbb{C}^*$. Taking a $\gamma \in \mathbf{PSL}(2, K)$ with $\gamma\sigma\gamma^{-1} = \begin{pmatrix} 1 & \lambda \\ 0 & 1 \end{pmatrix}$ where $\lambda \in K,\ \lambda \neq 0$ we see that $\gamma[z_1, z_2] = [1, 0] = \infty$. □

If $\Gamma < \mathbf{PSL}(2, \mathbb{C})$ is a discrete subgroup of finite covolume we know now by Theorem 2.5.1 that $\Gamma \backslash C_\Gamma$ is a finite set. For our groups $\Gamma = \mathbf{PSL}(2, \mathcal{O})$ we establish this result beforehand and in fact compute $\Gamma \backslash C_\Gamma$.

Definition 2.3. Define j to be the map

$$j: \mathbb{P}^1 K \to \mathcal{J}_K \qquad j: [z_1, z_2] \mapsto < z_1, z_2 >^\# .$$

Clearly j is well defined. The following is clear from Lemma 2.1.

Theorem 2.4. *The induced map*

$$j: \mathbf{PSL}(2, \mathcal{O}) \backslash \mathbb{P}^1 K = \mathbf{PSL}(2, \mathcal{O}) \backslash C_{\mathrm{PSL}(2,\mathcal{O})} \to \mathcal{J}_K$$

is a bijection.

The last result was first observed by Bianchi (1892). It was proved in greater generality by Hurwitz in a letter to Bianchi, see Bianchi (1892), pages 103–105. We also need the following easy consequence of Lemma 2.1.

Corollary 2.5. *Let $\mathfrak{a} \subset \mathcal{O}$ be an ideal and $\gamma \in \mathfrak{a} \backslash \{0\}$. Then there is a constant $\kappa \in]0, \infty[$ depending only on $\mathfrak{a}$ and γ so that if $\mathfrak{a} = < \alpha, \beta >$ with $\beta \neq 0$, there are $\lambda, \mu \in \mathcal{O}$ with $< \lambda, \mu > = \mathcal{O}$, $\mu\alpha - \lambda\beta = \gamma$ and $|\mu| \leq \kappa|\beta|$.*

Proof. Since $\mathcal{O}$ is a Dedekind ring there is a $\delta \in \mathfrak{a}$ such that $< \gamma, \delta > = \mathfrak{a}$. Making the choices of $\alpha_1, \alpha_2, \beta_1, \beta_2$ and θ as in Lemma 2.1, we see that a possible μ is

$$\mu = \frac{\gamma\alpha_1 + \beta\beta_2}{\theta}.$$

Here α_1 only has to satisfy $\alpha\alpha_1 + \beta\alpha_2 = \theta$. So we may replace α_1 by $\alpha_1 + \beta\lambda$ for any $\lambda \in \mathfrak{a}^{n-1}$, doing this we enforce that $|\alpha_1| \leq \kappa|\beta|$ with a constant κ only depending on $\mathfrak{a}$. The triangle inequality applied to the above μ proves the result. □

Next we need the following result from diophantine approximation.

Proposition 2.6. *There is a constant $\kappa_1 \in]0, \infty[$ only depending on K, so that for any $z \in \mathbb{C} \backslash K$ there are infinitely many $\lambda, \mu \in \mathcal{O}$ with*

$$\left| z - \frac{\lambda}{\mu} \right| \leq \frac{\kappa_1}{|\mu|^2}. \tag{2.4}$$

Proof. Any complex number is congruent mod $\mathcal{O}$, to a number in the parallelogram

$$\mathcal{P} = \{ \ x + y\omega \ | \ 0 \leq x \leq 1, \ 0 \leq y \leq 1 \ \}.$$

Let d be the diameter of $\mathcal{P}$. For any natural number M divide $\mathcal{P}$ into M^2 little parallelograms $\mathcal{P}_k$ by dividing the intervals $0 \leq x \leq 1$ and $0 \leq y \leq 1$ into M equal parts. The parallelograms have diameter dM^{-1}. The set

$$S = \{ \ a + b\omega \ | \ a, b \in \mathbb{Z}, \ 0 \leq a, b \leq M \ \}$$

has cardinality $(M+1)^2$. By the pigeon hole principle there are distinct $\mu_1, \mu_2 \in S$ so that $\mu_1 z$ and $\mu_2 z$ are modulo $\mathcal{O}$ contained in the same $\mathcal{P}_k$. Put $\mu = \mu_1 - \mu_2$. Then there is a $\lambda \in \mathcal{O}$ so that $|\mu z - \lambda| \leq dM^{-1}$. We also have $|\mu| \leq (1 + |\omega|) \cdot M$. Hence

$$\left| z - \frac{\lambda}{\mu} \right| \leq \frac{d}{M|\mu|} \leq \frac{\kappa_1}{|\mu|^2}$$

with $\kappa_1 = (1 + |\omega|) \cdot d$. Since $z \notin K$ and since the number $d/M|\mu|$ can be made arbitrarily small by taking M large, there will be infinitely many μ, λ satisfying the required inequality. □

We shall need the following stronger version of Proposition 2.6.

Proposition 2.7. *There is a constant $\kappa_2 \in]0, \infty[$ only depending on K so that for any $z \in \mathbb{C} \backslash K$ there are infinitely many $\lambda, \mu \in \mathcal{O}$ with*

$$\left| z - \frac{\lambda}{\mu} \right| \leq \frac{\kappa_2}{|\mu|^2} \tag{2.5}$$

and $< \mu, \lambda >= \mathcal{O}$.

Proof. Choose a set of representatives $\mathfrak{a}_1, \ldots, \mathfrak{a}_h \subset \mathcal{O}$ for the ideal classes. Put $N_0 = \max\{ \ N\mathfrak{a}_1, \ldots, N\mathfrak{a}_h \ \}$. By Proposition 2.6, there is a constant $\kappa_1 \in]0, \infty[$ and infinitely many $\alpha, \beta \in \mathcal{O}$ with

$$\left| z - \frac{\alpha}{\beta} \right| \leq \frac{\kappa_1}{|\beta|^2}.$$

For every such pair α, β there is a $\lambda \in K^*$ and an $i \in \{1, \ldots, h\}$ such that $\lambda < \alpha, \beta >= \mathfrak{a}_i$. We have $|\lambda|^2 \leq N(\mathfrak{a}_i) \cdot (N < \alpha, \beta >)^{-1} \leq N_0$. Putting $\alpha_1 = \lambda\alpha$ and $\beta_1 = \lambda\beta$ we obtain

$$\left| z - \frac{\alpha_1}{\beta_1} \right| = \left| z - \frac{\alpha}{\beta} \right| \leq \frac{\kappa_1}{|\beta|^2} = \frac{\kappa_1}{|\beta_1|^2} |\lambda|^2 \leq \frac{\kappa_3}{|\beta_1|^2}$$

where $\kappa_3 = N_0 \cdot \kappa_1$. Since

$$\left|z - \frac{\alpha_1}{\beta_1}\right| = \left|z - \frac{\alpha}{\beta}\right|$$

gets arbitrarily small as the α, β vary, there are infinitely many $\alpha, \beta \in \mathcal{O}$ satisfying

$$\left|z - \frac{\alpha}{\beta}\right| \leq \frac{\kappa_3}{|\beta|^2} \quad \text{and} \quad <\alpha, \beta>= \mathfrak{a}_i \tag{2.6}$$

for a fixed $i \in \{1, \ldots, h\}$.

Choose a $\gamma \in \mathfrak{a}_i \backslash \{0\}$. For every pair α, β satisfying (2.6) we find $\lambda, \mu \in \mathcal{O}$ with $<\lambda, \mu>= \mathcal{O}$, $\mu\alpha - \lambda\beta = \gamma$ and $|\mu| \leq \kappa|\beta|$ with the constant κ from Corollary 2.5. We consider now only pairs α, β satisfying $|\beta| > |\gamma|$. There are still infinitely many pairs (α, β) satisfying (2.6). We have $\mu \neq 0$ and

$$\left|z - \frac{\lambda}{\mu}\right| \leq \left|z - \frac{\alpha}{\beta}\right| + \left|\frac{\alpha}{\beta} - \frac{\lambda}{\mu}\right| \leq \frac{\kappa_3}{|\beta|^2} + \left|\frac{\gamma}{\beta\mu}\right| \leq \frac{\kappa_2}{|\mu|^2},$$

where $\kappa_2 = \kappa_3\kappa^2 + |\gamma|\kappa$. Since $|\mu| \geq 1$ we infer

$$\left|z - \frac{\lambda}{\mu}\right| \leq \frac{\kappa_3}{|\beta|^2} + \frac{\kappa_1}{|\beta|}.$$

By choosing $|\beta|$ large, an infinite number of $\lambda, \mu \in \mathcal{O}$ satisfying

$$\left|z - \frac{\lambda}{\mu}\right| \leq \frac{\kappa_2}{|\mu|^2} \quad \text{and} \quad <\lambda, \mu>= \mathcal{O}$$

has to occur. □

The following points will play a special rôle in the construction of the fundamental domains.

Definition 2.8. A complex number z is called singular for K if $|\mu z + \lambda| \geq 1$ for all $\mu, \lambda \in \mathcal{O}$ with $<\mu, \lambda>= \mathcal{O}$. We define $\mathcal{S} = \mathcal{S}_K = \{\ z \in \mathbb{C} \mid z \text{ is singular for } K\ \}$.

If $z \in \mathcal{S}_K$ and $\lambda \in \mathcal{O}$ then clearly $z + \lambda \in \mathcal{S}_K$. By an application of our number theoretic results we shall show that the set $\mathcal{S}$ is finite up to addition of elements from $\mathcal{O}$.

Proposition 2.9. *The set $\mathcal{S}_K$ of singular points for K satisfies $\mathcal{S}_K \subset K$. There are finitely many $s_1, \ldots, s_r \in \mathcal{S}_K$ such that $\mathcal{S}_K = \cup_{i=1}^r (s_i + \mathcal{O})$.*

Proof. If $z \notin K$ it follows from Proposition 2.7 that z is not singular for K. Choose a system of representatives $\mathfrak{a}_1, \ldots, \mathfrak{a}_h \subset \mathcal{O}$ for the ideal classes of K. For every $i \in \{1, \ldots, h\}$ put

$$M_i = \{ \gamma \in \mathfrak{a}_i \backslash \{0\} \mid |\gamma| \le |\lambda| \quad \text{for all} \quad \lambda \in \mathfrak{a}_i \backslash \{0\} \}.$$

Each M_i is a finite non-empty set. Let $s \in \mathcal{S}_K$ be a singular point for K. We write $s = \alpha/\beta$ such that $< \alpha, \beta >= \mathfrak{a}_i$ for some $i \in \{1, \ldots, h\}$. If $\gamma \in M_i$ there is a $\delta \in \mathfrak{a}_i$ with $< \gamma, \delta >= \mathfrak{a}_i$. By Corollary 2.5 there are $\mu, \lambda \in \mathcal{O}$ with

$$\mu\alpha + \lambda\beta = \gamma \quad \text{and} \quad < \mu, \lambda >= \mathcal{O}.$$

We have $|\mu s + \lambda| = |\gamma|/|\beta|$. Since s is singular we find $|\gamma| \ge |\beta|$, hence $\beta \in M_i$. Because $\mathfrak{a}_i / < \beta >$ is finite, we can restrict α by adding an element from $\mathcal{O}$ to s, to a finite set. □

7.3 Description of a Fundamental Domain

We fix an imaginary quadratic number field K and shall construct now a fundamental domain $\mathcal{F}_K \subset \mathbb{H}$ for our group $\mathbf{PSL}(2, \mathcal{O})$ where $\mathcal{O} = \mathcal{O}_K$ is the ring of integers in K.

Definition 3.1. Let $K = \mathbb{Q}(\sqrt{D})$ with $D < 0$ a squarefree integer and d_K the discriminant of K. We define

$$\mathcal{B}_K = \left\{ z + rj \in \mathbb{H} \;\middle|\; \begin{array}{l} |cz+d|^2 + |d|^2 r^2 \ge 1 \text{ for all } c, d \in \mathcal{O} \\ \text{with } < c, d >= \mathcal{O} \end{array} \right\},$$

$$\mathcal{P}_K = \{ z \in \mathbb{C} \mid 0 \le \operatorname{Re} z \le 1, \quad 0 \le \operatorname{Im} z \le \sqrt{|d_K|}/2 \},$$

$$F_K = \mathcal{P}_K \quad \text{for } D \ne -3, -1 \text{ and}$$

$$F_{\mathbb{Q}(i)} = \{ z \in \mathbb{C} \mid 0 \le |\operatorname{Re} z| \le 1/2, \quad 0 \le \operatorname{Im} z \le 1/2 \},$$

$$F_{\mathbb{Q}(\sqrt{-3})} = \left\{ z \in \mathbb{C} \;\middle|\; 0 \le \operatorname{Re} z,\ \frac{\sqrt{3}}{3}\operatorname{Re} z \le \operatorname{Im} z,\ \operatorname{Im} z \le \frac{\sqrt{3}}{3}(1 - \operatorname{Re} z) \right\}$$

$$\cup \left\{ z \in \mathbb{C} \;\middle|\; 0 \le \operatorname{Re} z \le \frac{1}{2},\ -\frac{\sqrt{3}}{3}\operatorname{Re} z \le \operatorname{Im} z \le \frac{\sqrt{3}}{3}\operatorname{Re} z \right\},$$

$$\mathcal{F}_K = \{ z + rj \in \mathcal{B}_K \mid z \in F_K \}.$$

The picture of $F_{\mathbb{Q}(\sqrt{-3})}$ is a closed quadrangle with corners in the points 0, $1/2 - \sqrt{-3}/6$, $1/2 + \sqrt{-3}/6$, $\sqrt{-3}/3$. It is easy to see that the $\mathcal{P}_K$ are fundamental parallelograms for the lattices $\mathcal{O} \subset \mathbb{C}$ and that the F_K are fundamental domains for the groups

$$\mathbf{PSL}(2, \mathcal{O})_\infty = \left\{ \begin{pmatrix} \alpha & \beta \\ 0 & \lambda \end{pmatrix} \in \mathbf{PSL}(2, \mathcal{O}) \right\}$$

acting by $z \to (\alpha z + \beta)/\lambda$ on $\mathbb{C}$.

Definition 3.2. For $\mu, \lambda \in \mathcal{O}$ with $<\mu, \lambda>= \mathcal{O}$ and $\mu \neq 0$ we define

$$S(\mu, \lambda) = \{ \ z + rj \in \mathbb{H} \ \mid \ |\mu z + \lambda|^2 + |\mu|^2 r^2 = 1 \ \}.$$

The $S(\mu, \lambda)$ are hyperbolic planes, that is Euclidean hemispheres having their center in $-\lambda/\mu \in \mathbb{C}$ and having radius $1/|\mu|$. The following lemma will be used to prove that $\mathcal{F}_K$ is a fundamental domain for $\mathbf{PSL}(2, \mathcal{O})$.

Lemma 3.3. *The set $\mathcal{B}_K$ has the following properties:*

(1) A point $P = z + rj \in \mathbb{H}$ lies in $\mathcal{B}_K$ if and only if for every $\sigma \in \mathbf{PSL}(2, \mathcal{O})$ the point $z' + r'j = \sigma P$ satisfies $r' \leq r$.
(2) For every $P \in \mathbb{H}$ there is a $\sigma \in \mathbf{PSL}(2, \mathcal{O})$ with $\sigma P \in \mathcal{B}_K$.
(3) Let $\sigma = \begin{pmatrix} \alpha & \beta \\ \mu & \lambda \end{pmatrix} \in \mathbf{PSL}(2, \mathcal{O})$ with $\mu \neq 0$, then $\mathcal{B}_K \cap \sigma\mathcal{B}_K = \mathcal{B}_K \cap S(\mu, \lambda)$.
(4) For every $\sigma \in \mathbf{PSL}(2, \mathcal{O})_\infty$ we have $\sigma\mathcal{B}_K = \mathcal{B}_K$.

Proof. (1) is clear from the transformation formula for the j-component.

(2) Given a constant $\kappa \in]0, \infty[$ there are only finitely many $c, d \in \mathcal{O}$ with

$$|cz + d|^2 + r^2|c|^2 \leq \kappa.$$

Hence there is a $\sigma = \begin{pmatrix} \alpha & \beta \\ \mu & \lambda \end{pmatrix} \in \mathbf{PSL}(2, \mathcal{O})$ such that

$$|\mu z - \lambda|^2 + r^2|\mu|^2 \leq |cz - d|^2 + r^2|c|^2$$

for all $c, d \in \mathcal{O}$ with $<c, d>= \mathcal{O}$. The point σP has then a maximal j-component in the $\mathbf{PSL}(2, \mathcal{O})$ orbit of P. By (1) we obtain $\sigma P \in \mathcal{B}_K$.

(3) Let $P = z + rj$ be in $\mathcal{B}_K$. Put $\sigma P = z' + r'j.$, then

$$r' = \frac{r}{|\mu z + \lambda|^2 + |\mu|^2 r^2}.$$

If $\sigma P \in \mathcal{B}_K$ then by (1) we have $r = r'$ and $P \in S(\mu, \lambda)$ and conversely.

(4) It is clear that $\mathbf{PSL}(2, \mathcal{O})_\infty$ only permutes the defining equations for $\mathcal{B}_K$. □

Theorem 3.4. *The set $\mathcal{F}_K$ is a fundamental domain for $\mathbf{PSL}(2, \mathcal{O})$.*

Proof. We have to check the requirements in Definition 2.2.1. $\mathcal{B}_K$ is the intersection of closed halfspaces, hence $\mathcal{B}_K$ and $\mathcal{F}_K$ are closed sets. The set F_K was chosen to be a fundamental domain for the action of $\mathbf{PSL}(2, \mathcal{O})_\infty$ on $\mathbb{C}$, hence Lemma 3.3 implies

$$\mathbb{H} = \bigcup_{\sigma \in \mathrm{PSL}(2, \mathcal{O})} \sigma\mathcal{F}_K.$$

It is clear that $\partial\mathcal{B}_K = \cup_{<\mu,\lambda>=\mathcal{O}} \mathcal{B}_K \cap S(\mu,\lambda)$, since points underneath $S(\mu,\lambda)$ do not lie in $\mathcal{B}_K$. By Lemma 3.3 we have $\sigma\mathcal{B}_K^\circ \cap \mathcal{B}_K^\circ = \emptyset$ if $\sigma = \begin{pmatrix} \alpha & \beta \\ \mu & \lambda \end{pmatrix} \in \mathbf{PSL}(2,\mathcal{O})$ with $\mu \neq 0$. We infer from Lemma 3.3 $\sigma\mathcal{F}_K^\circ \cap \mathcal{F}_K^\circ = \emptyset$ if $\sigma \neq I$. This establishes property (2) of fundamental domains. By our definition

$$\partial\mathcal{F}_K \subset \left(\bigcup_{<\mu,\lambda>=\mathcal{O}} S(\mu,\lambda) \right) \cup M$$

where M is the union of 4 vertical planes in $\mathbb{H}$. Hence $\partial\mathcal{F}_K$ is contained in the union of countably many sets of measure 0. This implies that $\partial\mathcal{F}_K$ has measure 0. □

Next we shall establish some finiteness properties of the set $\mathcal{F}_K$.

Definition 3.5. Let $s \in \mathcal{S}_K$ be a singular point for K. If $s = \frac{\alpha}{\beta}$ with $\alpha, \beta \in \mathcal{O}$, then we have $< \beta^2, -\alpha\beta \pm 1 >= \mathcal{O}$ and $< \omega\beta^2, -\omega\alpha\beta \pm 1 >= \mathcal{O}$. We define

$$S_1^{\pm}(s) = S(\beta^2, -\alpha\beta \pm 1), \qquad S_2^{\pm}(s) = S(\omega\beta^2, -\omega\alpha\beta \pm 1).$$

The two boundary circles of $S_1^{\pm}(s)$, namely $\{ z \in \mathbb{C} \mid |\beta^2 z - \alpha\beta \pm 1| = 1 \}$ pass through $s \in \mathbb{C}$ and are tangent along the line $\mathrm{Re}(\beta^2 z - \alpha\beta) = 0$. The boundary circles of $S_2^{\pm}(s)$ also pass through s and are tangent along the line $\mathrm{Re}(\omega\beta^2 z - \omega\alpha\beta) = 0$. Note that these two lines are distinct.

Lemma 3.6.

(1) If $z_0 \in \mathbb{C}\backslash\mathcal{S}_K$, then there is an $\epsilon \in]0,\infty[$ and a neighbourhood U of z_0 in $\mathbb{C}$ such that every $z + rj \in \mathcal{B}_K$ with $z \in U$ satisfies $r > \epsilon$.
(2) If $s \in \mathcal{S}_K$ then there is an $\epsilon \in]0,\infty[$ and a neighbourhood U of s in $\mathbb{C}$ such that every $z + rj \in \mathcal{B}_K$ with $z \in U$ satisfies $r^2 > \epsilon \cdot |z - s|$.

Proof. (1) There are $\mu, \lambda \in \mathbb{C}$ with $\mu \neq 0, < \mu, \lambda >= \mathcal{O}$ and $|\mu z_0 + \lambda| < 1$. Take a $\theta \in]0,1[$ and a neighbourhood U of z_0 such that $|\mu z + \lambda| < \theta$ for all $z \in U$. If $z + rj \in \mathcal{B}_K$ then $|\mu z + \lambda|^2 + r^2|\mu|^2 \geq 1$, hence $\epsilon = |\mu|^{-1}\sqrt{(1-\theta^2)}$ is suitable.

(2) We choose a pair of lines ℓ_1, ℓ_2 in $\mathbb{C}$ which divide every small enough neighbourhood of s into 4 parts, each of which lies in one of the circles $S_1^{\pm}(s), S_2^{\pm}(s)$. For each of these 4 parts we apply a translation and a rotation so that $s = 0$ and that the equation of the corresponding circle is $(\mathrm{Re}\, z - t_0)^2 + (\mathrm{Im}\, z)^2 = t_0^2$ for some $t_0 \in \mathbb{R}$. The equation for the part above the hemisphere through the circle becomes

$$(\mathrm{Re}\, z - t_0)^2 + (\mathrm{Im}\, z)^2 + r^2 \geq t_0^2. \tag{3.1}$$

In the new coordinates the lines ℓ_1, ℓ_2 have equations $\operatorname{Im} z = k_1 \cdot \operatorname{Re} z$, $\operatorname{Im} z = k_2 \cdot \operatorname{Re} z$ for some $k_1, k_2 \in \mathbb{R}$. In the area between the two new lines we have $|\operatorname{Im} z| \leq k \cdot |\operatorname{Re} z|$ with $k = \max\{|k_1|, |k_2|\}$. From (3.1) we deduce

$$r^2 \geq 2r|\operatorname{Re} z| - (\operatorname{Re} z)^2 - k^2 \cdot (\operatorname{Re} z)^2 > \epsilon^2 |\operatorname{Re} z|$$

if $2r > \epsilon^2$ and $\operatorname{Re} z$ is close enough to 0. □

Definition 3.7. Let $\mu, \lambda \in \mathcal{O}$ and $\Delta \in \mathbb{Z}$ with $\mu \neq 0$ and $\Delta > 0$. We define

$$S(\mu, \lambda; \Delta) = \{\ z + rj \ \mid \ |\mu z + \lambda|^2 + |\mu|^2 r^2 = \Delta\ \}.$$

$S(\mu, \lambda; \Delta)$ is called admissible if $|s + \lambda/\mu| \geq \sqrt{\Delta}/|\mu|$ for every singular point $s \in \mathcal{S}_K$.

The set $S(\mu, \lambda; \Delta)$ is a geodesic hemisphere in $\mathbb{H}$, it has center $-\lambda/\mu$ and radius $\sqrt{\Delta}/|\mu|$. The geodesic hemisphere $S(\mu, \lambda; \Delta)$ is admissible if it covers no singular point of K. For $\mu, \lambda, \in \mathcal{O}$ with $\mu \neq 0$ and $< \mu, \lambda >= \mathcal{O}$ we have $S(\mu, \lambda; 1) = S(\mu, \lambda)$. By the definition of singular points (i.e. Definition 2.8), $S(\mu, \lambda; 1)$ is always admissible.

The central finiteness theorem, which will be used in the proof of Theorem 1.1 is:

Theorem 3.8. *Let $\Delta \in \mathbb{Z}$ with $\Delta > 0$. There are only finitely many $\mu, \lambda \in \mathcal{O}$ with $\mu \neq 0$ such that $S(\mu, \lambda; \Delta)$ is admissible and $S(\mu, \lambda; \Delta) \cap \mathcal{F}_K \neq \emptyset$.*

Proof. We assume that an admissible $S(\mu, \lambda; \Delta)$ is given with $S(\mu, \lambda; \Delta) \cap \mathcal{F}_K \neq \emptyset$. Since F_K is compact, there is a $\delta \in]0, \infty[$ such that $|z| < \delta$ for all $z \in F_K$. If $z + rj \in S(\mu, \lambda; \Delta) \cap \mathcal{F}_K$ then

$$\left|\frac{\lambda}{\mu}\right| \leq \left|z + \frac{\lambda}{\mu}\right| + |z| \leq \frac{\sqrt{\Delta}}{|\mu|} + \delta.$$

This shows that it is enough to find a bound for $|\mu|$ if $S(\mu, \lambda; \Delta) \cap \mathcal{F}_K \neq \emptyset$.

It follows from Proposition 2.9 that the set of singular points $\mathcal{S}_K$ is discrete in $\mathbb{C}$. Hence $\mathcal{S}_K \cap F_K$ is finite, say $\mathcal{S}_K \cap F_K = \{s_1, \ldots, s_n\}$. By Lemma 3.6 there is an $\epsilon \in]0, \infty[$ and neighbourhoods U_i of s_i such that $|z - s_i| \leq \epsilon/2$ for all $z \in U_i$ and $r^2 > \epsilon|z - s_i|$ for every $z + rj \in \mathcal{B}_K$ with $z \in U_i$. Put

$$M = F_K \backslash \left(\bigcup_{i=1}^{n} U_i \right).$$

M is a compact set. By making ϵ possibly a little smaller we may by Lemma 3.6 assume that every $z + rj \in \mathcal{B}_K$ with $z \in M$ has $r > \epsilon$. If $z + rj \in S(\mu, \lambda; \Delta)$ then $r \leq \sqrt{\Delta}/|\mu|$, hence $|\mu| > \sqrt{\Delta}/\epsilon$ implies that every $z + rj \in S(\mu, \lambda; \Delta) \cap \mathcal{F}_K$ has $z \notin M$. Suppose now $z \in U_i$. Then $r^2 > \epsilon|z - s_i|$ and

$$\left|z + \frac{\lambda}{\mu}\right|^2 + r^2 = \frac{\Delta}{|\mu|^2}.$$

We infer that

$$\left|z + \frac{\lambda}{\mu}\right|^2 + \epsilon|z - s_i| < \frac{\Delta}{|\mu|^2}.$$

We put $u = z - s_i$ and $t = s_i + \frac{\lambda}{\mu}$. We have $|u + t|^2 + \epsilon|u| < \Delta/|\mu|^2$ and $|t| \geq \sqrt{\Delta}/|\mu|$, since $S(\mu, \lambda; \Delta)$ is admissible. We get $|u + t| < \sqrt{\Delta}/|\mu|$ and

$$\frac{\sqrt{\Delta}}{|\mu|} \leq |t| \leq |u + t| + |u|.$$

From this we conclude

$$\frac{\Delta}{|\mu|^2} \leq |u + t|^2 + 2|u||u + t| + |u|^2 < \frac{\Delta}{|\mu|^2} - \epsilon|u| + 2\frac{\sqrt{\Delta}}{|\mu|} \cdot |u| + |u|^2.$$

Hence $0 < -\epsilon|u| + 2\sqrt{\Delta}|u|/|\mu| + |u|^2$ and $|u| \neq 0$ follows. We conclude that $0 < -\epsilon + 2\sqrt{\Delta}/|\mu| + |u|$ and from $|u| \leq \epsilon/2$ we find $|\mu| \leq 4\sqrt{\Delta} \cdot \epsilon^{-1}$. This is the final bound for $|\mu|$. □

We are now ready to prove Theorem 1.1 apart from (3).

Proof of Theorem 1.1. (1) is obvious.

(2) From the construction of the fundamental domain $\mathcal{F}_K$, see Definition 3.1, it is clear that $\mathbf{PSL}(2, \mathcal{O})$ is not cocompact. We shall prove that $\mathcal{F}_K$ has finite hyperbolic volume. For a singular point $s \in \mathcal{S}_K \cap F_K$ we consider the 4 hemispheres $S_1^{\pm}(s), S_2^{\pm}(s)$ defined in Definition 3.5, and choose an $\epsilon_s \in]0, \infty[$ which is smaller than half the radius of any of these hemispheres. We define

$$(3.2) \quad \mathcal{F}(s) = \left\{ P = z + rj \in \mathbb{H} \;\middle|\; \begin{array}{l} r \leq \epsilon_s,\ |z - s| < \epsilon_s \text{ and} \\ P \text{ lies above the } S_1^{\pm}(s), S_2^{\pm}(s) \end{array} \right\},$$

$$(3.3) \qquad A(s) = \{\ z \in \mathbb{C} \ \mid \ z + rj \in \mathcal{F}(s) \text{ for some } r \in]0, \infty[\ \}.$$

Note that s is the only singular point of K contained in $A(s)$ and that $A(s)$ contains a neighbourhood of s. $\mathcal{F}(s)$ is called a cusp sector for the cusp s. Put

$$(3.4) \qquad M = F_K \backslash \overline{\bigcup_{s \in \mathcal{S}_K \cap F_K} A(s)}.$$

M is a compact set which contains no singular point of K. Hence by Lemma 3.6 there is an $\epsilon \in]0, \infty[$ such that any $z + rj \in \mathcal{F}_K$ with $z \in M$ satisfies $r \geq \epsilon$. Choose $\epsilon_\infty \in]0, \infty[$ satisfying $\epsilon_\infty \leq \epsilon$ and $\epsilon_\infty \leq \epsilon_s$ for all $s \in \mathcal{S}_K \cap F_K$. This is possible since $\mathcal{S}_K \cap F_K$ is finite. We define

$$\mathcal{F}(\infty) = \{\ z + rj \in \mathbb{H} \ \mid \ z \in F_K,\ r \geq \epsilon_\infty\ \}. \tag{3.5}$$

By our definitions we have

$$\mathcal{F}_K \subseteq \mathcal{F}(\infty) \cup \bigcup_{s \in \mathcal{S}_K \cap F_K} \mathcal{F}(s).$$

Since $\mathcal{F}_K$ is closed it is Lebesgue measurable and it is now enough to show that the $\mathcal{F}(s)$ and $\mathcal{F}(\infty)$ are of finite hyperbolic measure. For $\mathcal{F}(\infty)$ we have

$$v(\mathcal{F}(\infty)) = \mu(F_K) \int_{\epsilon_\infty}^{\infty} \frac{dr}{r^3} < \infty \tag{3.6}$$

where μ is the Lebesgue measure in $\mathbb{C}$. If $s \in \mathcal{S}_K \cap F_K$ we choose a $\gamma_s \in \mathbf{PSL}(2, K)$ with $\gamma_s(s) = \infty$. Clearly there is a compact set $T \subset \mathbb{C}$ and a $\epsilon'_s \in]0, \infty[$ such that

$$\gamma_s(\mathcal{F}(s)) \subset \{\ z + rj \in \mathbb{H} \ \mid \ z \in T,\ r \geq \epsilon'_s\ \}.$$

By computation (3.6) the hyperbolic volume of the right-hand set is finite. Note that γ_s transforms any of the hemispheres $S_1^{\pm}(s), S_2^{\pm}(s)$ into a vertical plane in $\mathbb{H}$.

(4) The boundary of $\mathcal{F}_K$ is contained in the union of four vertical planes together with the union of those of $S(\mu, \lambda)$ with $< \mu, \lambda >= \mathcal{O}$ and $S(\mu, \lambda) \cap \mathcal{F}_K \neq \emptyset$. By Theorem 3.8 there are only finitely many such $S(\mu, \lambda)$.

(5) We have to show that each point $z_0 + r_0 j \in \mathbb{H}$ has a neighbourhood U meeting $\sigma\mathcal{F}_K$ for only finitely many $\sigma \in \mathbf{PSL}(2, \mathcal{O})$. Choose an $\epsilon \in]0, r_0[$ and take a bounded neighbourhood U of $z_0 + r_0 j$ so that any $z + rj \in U$ satisfies $r > \epsilon$. Consider

$$\sigma = \begin{pmatrix} a & b \\ c & d \end{pmatrix} \in \mathbf{PSL}(2, \mathcal{O})$$

with $U \cap \sigma\mathcal{F}_K \neq \emptyset$. Since U is bounded there are only finitely many such σ with $c = 0$. From now on we assume that $c \neq 0$. There is $z + rj \in \mathcal{F}_K$ with $\sigma(z + rj) = z' + r'j \in U$. This implies $r' > \epsilon$. By Lemma 3.3 we have

$$r \geq r' = \frac{r}{|cz + d|^2 + r^2|c|^2} \leq \frac{r}{r^2|c|^2} = \frac{1}{r|c|^2}. \tag{3.7}$$

This implies $|c|^2 \leq \epsilon^{-2}$. Hence there are only finitely many possible $c \in \mathcal{O}$. Next we prove that for a fixed $c \in \mathcal{O}$ there are only finitely many possible d. Choose a $\kappa \in]0, \infty[$ such that $|z| \leq \kappa$, this is possible since F_K is compact. We have, from (3.7)

$$\epsilon < r' \leq r|cz + d|^{-2} \quad \text{and} \quad 1 \geq rr'|c|^2 \geq r\epsilon|c|^2.$$

We infer $r \leq \epsilon^{-1}|c|^{-2}$. This implies

$$|cz + d|^2 \leq \epsilon^{-1} r \leq \epsilon^{-2}|c|^{-2} \quad \text{and} \quad |d| \leq |cz + d| + |cz| \leq \epsilon^{-1}|c|^{-1} + |c| \cdot \kappa.$$

If $\sigma' = \begin{pmatrix} a' & b' \\ c & d \end{pmatrix} \in \mathbf{PSL}(2, \mathcal{O})$ is an element with $U \cap \sigma' \mathcal{F}_K \neq \emptyset$, then $\sigma\sigma'^{-1} = \begin{pmatrix} 1 & e \\ 0 & 1 \end{pmatrix}$ with some $e \in \mathcal{O}$. Given σ, e comes from a finite set since $\sigma\sigma'^{-1} U \cap \sigma\mathcal{F}_K \neq \emptyset$.

(6) follows easily from (4), (5).

(7) follows by definition.

(8) follows from Theorem 2.7.3. □

For a given imaginary quadratic number field K it is always effectively possible to find equations for the finitely many hypersurfaces bounding the fundamental domain $\mathcal{F}_K$. See Swan (1971), Chapter 8 for remarks on how to do this in practice. Many examples are contained in Swan (1971) and Bianchi (1892). The $\mathcal{F}_K$ are hyperbolic polyhedra. The translates $\sigma\mathcal{F}_K$, $\sigma \in \mathbf{PSL}(2, \mathcal{O})$ give a locally finite tesselation of $\mathbb{H}$. For small d_K the $\mathcal{F}_K$ or slight modifications of them are regular polyhedra in $\mathbb{H}$ leading to regular tesselations of $\mathbb{H}$. Some examples are described in Grunewald, Gushoff, Mennicke (1982). It is also possible to find effectively the set

$$\{ \sigma \in \mathbf{PSL}(2, \mathcal{O}) \mid \sigma\mathcal{F}_K \cap \mathcal{F}_K \neq \emptyset \}.$$

This leads to a procedure to find explicit presentations for $\mathbf{PSL}(2, \mathcal{O})$. Many examples are contained in Swan (1971).

It is useful to note that $\mathbb{H}$ contains a 2-dimensional complex which is a $\mathbf{PSL}(2, \mathcal{O})$-equivariant retract of $\mathbb{H}$, see Mendoza (1980), Flöge (1983). Using this, Flöge (1983) determines nice presentations of $\mathbf{PSL}(2, \mathcal{O})$.

We shall now give three examples.

Example 1. $K = \mathbb{Q}(i)$, $d_K = -4$.

Here $\mathcal{O} = \mathbb{Z}[i]$ is the ring of Gaußian integers. We have

Proposition 3.9. *The fundamental domain $\mathcal{F}_{\mathbb{Q}(i)}$ for $\mathbf{PSL}(2, \mathcal{O})$ is described by*

$$\mathcal{F}_{\mathbb{Q}(i)} = \left\{ z + rj \in \mathbb{H} \;\middle|\; 0 \leq |\operatorname{Re} z| \leq \frac{1}{2},\ 0 \leq \operatorname{Im} z \leq \frac{1}{2},\ z\bar{z} + r^2 \geq 1 \right\}.$$

$\mathcal{F}_{\mathbb{Q}(i)}$ is a hyperbolic pyramid with one vertex at ∞ and the other four vertices in the points $P_1 = -\frac{1}{2} + \frac{\sqrt{3}}{2}j$, $P_2 = \frac{1}{2} + \frac{\sqrt{3}}{2}j$, $P_3 = \frac{1}{2} + \frac{1}{2}i + \frac{\sqrt{2}}{2}j$, $P_4 = -\frac{1}{2} + \frac{1}{2}i + \frac{\sqrt{2}}{2}j$. Let

$$A = \begin{pmatrix} 1 & 0 \\ 1 & 1 \end{pmatrix}, \quad B = \begin{pmatrix} 0 & -1 \\ 1 & 0 \end{pmatrix}, \quad U = \begin{pmatrix} 1 & 0 \\ i & 1 \end{pmatrix}.$$

Then the following is a presentation for $\mathbf{PSL}(2, \mathbb{Z}[i])$.

$$\mathbf{PSL}(2,\mathbb{Z}[i]) = \left\langle A, B, U \;\middle|\; \begin{array}{l} (AB)^3 = B^2 = AUA^{-1}U^{-1} \\ = (BUBU^{-1})^3 = (BU^2BU^{-1})^2 \\ = (AUBAU^{-1}B)^2 = 1 \end{array} \right\rangle .$$

Proof. Easy estimates show that $S(\mu,\lambda) \cap \mathcal{F}_{\mathbb{Q}(i)} = \emptyset$ unless $(\mu,\lambda) = (1,0)$, $(1,1)$, $(1,-1)$, $(1,i)$. The description of the fundamental domain then easily follows. The presentation can be derived with the help of Corollary 2.7.5, this involves a lot of work. □

Note that the (hyperbolic) convex hull of the points ∞, j, P_2, P_3 (this is one quarter of the above pyramid) is isometric to the tetrahedron under the heading of $\mathbf{CT}(1)$ in the table of Chapter 10.

Example 2. $K = \mathbb{Q}(\sqrt{-3})$, $d_K = -3$.

Here $\mathcal{O} = \mathbb{Z}[\zeta]$ with $\zeta = -\frac{1}{2} + \frac{\sqrt{-3}}{2}$ is the ring of Eisenstein integers. We have

Proposition 3.10. *The fundamental domain $\mathcal{F}_{\mathbb{Q}(\sqrt{-3})}$ for $\mathbf{PSL}(2,\mathcal{O})$ is described by*

$$\mathcal{F}_{\mathbb{Q}(\sqrt{-3})} = \left\{ z + rj \in \mathbb{H} \;\middle|\; z \in F_{\mathbb{Q}(\sqrt{-3})},\;\; z\bar{z} + r^2 \geq 1 \right\}.$$

$\mathcal{F}_{\mathbb{Q}(\sqrt{-3})}$ is a hyperbolic pyramid with one vertex at ∞ and the other four vertices in the points $Q_1 = j$, $Q_2 = \frac{1}{2} - \frac{\sqrt{3}}{6}i + \sqrt{\frac{2}{3}}j$, $Q_3 = \frac{1}{2} + \frac{\sqrt{3}}{6}i + \sqrt{\frac{2}{3}}j$, $Q_4 = \frac{\sqrt{3}}{3}i + \sqrt{\frac{2}{3}}j$. Let

$$A = \begin{pmatrix} 1 & 0 \\ 1 & 1 \end{pmatrix}, \quad B = \begin{pmatrix} 0 & -1 \\ 1 & 0 \end{pmatrix}, \quad U = \begin{pmatrix} 1 & 0 \\ \zeta & 1 \end{pmatrix}.$$

Then the following is a presentation for $\mathbf{PSL}(2,\mathbb{Z}[\zeta])$.

$$\mathbf{PSL}(2,\mathbb{Z}[\zeta]) = \left\langle A, B, U \;\middle|\; \begin{array}{l} (AB)^3 = B^2 = AUA^{-1}U^{-1} = 1, \\ (AUBU^{-2}B)^2 = (AUBU^{-1}B)^3 = 1, \\ A^2UBU^{-1}BUBUBU^{-1}B = 1 \end{array} \right\rangle .$$

The proof goes along the same lines as in the example before. Note that the (hyperbolic) convex hull of the points ∞, Q_1, Q_2, Q_3 (this is one half of the above pyramid) is isometric to the tetrahedron under the heading of $\mathbf{CT}(7)$ in the table of Chapter 10.

Example 3. $K = \mathbb{Q}(\sqrt{-5})$, $d_K = -20$.

Here the ring of integers $\mathcal{O} = \mathbb{Z}[\sqrt{-5}]$ has class number 2. We put $\tilde{\omega} = \sqrt{-5} = \omega + 10$ and

$$\gamma = \begin{pmatrix} 1 & -\frac{1}{2} - \frac{\sqrt{-5}}{2} \\ 0 & 1 \end{pmatrix} \in \mathbf{PSL}(2, \mathbb{Q}(\sqrt{-5})$$

and $\tilde{\mathcal{F}}_{\mathbb{Q}(\sqrt{-5})} = \gamma \mathcal{F}_{\mathbb{Q}(\sqrt{-5})}$. Then $\tilde{\mathcal{F}}_{\mathbb{Q}(\sqrt{-5})}$ is again a fundamental domain for $\mathbf{PSL}(2, \mathcal{O})$. We have translated $\mathcal{F}_{\mathbb{Q}(\sqrt{-5})}$, since $\tilde{\mathcal{F}}_{\mathbb{Q}(\sqrt{-5})}$ is a highly symmetric polyhedron described in the following.

Proposition 3.11. *The fundamental domain $\tilde{\mathcal{F}}_{\mathbb{Q}(\sqrt{-5})}$ for $\mathbf{PSL}(2, \mathbb{Z}[\sqrt{-5}])$ consists of all points $z + rj \in \mathbb{H}$ satisfying the equations:*

$$0 \le |\operatorname{Re} z| \le \frac{1}{2}, \qquad 0 \le |\operatorname{Im} z| \le \frac{\sqrt{5}}{2}, \qquad |z|^2 + r^2 \ge 1,$$
$$|2z \pm \sqrt{-5}|^2 + 4r^2 \ge 1, \qquad |2\sqrt{-5}z \pm 4 \pm \sqrt{-5}|^2 + 20r^2 \ge 1.$$

Let

$$A = \begin{pmatrix} 1 & 1 \\ 0 & 1 \end{pmatrix}, \; B = \begin{pmatrix} 0 & 1 \\ -1 & 0 \end{pmatrix}, \; U = \begin{pmatrix} 1 & \tilde{\omega} \\ 0 & 1 \end{pmatrix}, \; R = \begin{pmatrix} -4 + \tilde{\omega} & -2\tilde{\omega} \\ 2\tilde{\omega} & -4 - \tilde{\omega} \end{pmatrix},$$
$$S = \begin{pmatrix} 4 - \tilde{\omega} & -2\tilde{\omega} \\ 2\tilde{\omega} & 4 + \tilde{\omega} \end{pmatrix}, \quad T = \begin{pmatrix} \tilde{\omega} & 2 \\ 2 & -\tilde{\omega} \end{pmatrix}, \quad C = \begin{pmatrix} \tilde{\omega} & -2 \\ -2 & -\tilde{\omega} \end{pmatrix}.$$

Then $\mathbf{PSL}(2, \mathbb{Z}[\sqrt{-5}])$ has the presentation

$$\mathbf{PSL}(2, \mathbb{Z}[\sqrt{-5}]) = \left\langle \begin{array}{l} A, B, U, R, \\ S, T, C \end{array} \;\middle|\; \begin{array}{l} (AB)^3 = B^2 = T^2 = C^2 \\ = (TB)^2 = AUA^{-1}U^{-1} = RCST \\ = BSBR = CBU^{-1}TUB \\ = S^{-1}U^{-1}A^{-1}TUSTA = 1 \end{array} \right\rangle.$$

Proof. Here it is necessary to make a detailed examination of the finiteness Theorem 3.8 and determine the bounding hemispheres of $\mathcal{F}_{\mathbb{Q}(\sqrt{-5})}$. The details can be found in Swan (1971) or Bianchi (1892). The polyhedron $\tilde{\mathcal{F}}_{\mathbb{Q}(\sqrt{-5})}$ has 13 vertices:

$$P_1 = -\frac{1}{2} - \frac{\sqrt{-5}}{2}, \quad P_2 = \frac{1}{2} - \frac{\sqrt{-5}}{2}, \quad P_3 = \frac{1}{2} + \frac{\sqrt{-5}}{2}, \quad P_4 = -\frac{1}{2} + \frac{\sqrt{-5}}{2},$$
$$P_5 = -\frac{1}{2} - \frac{3\sqrt{-5}}{8} + \frac{\sqrt{3}}{8}j, \; P_6 = -\frac{1}{2} + \frac{3\sqrt{-5}}{8} + \frac{\sqrt{3}}{8}j, \; P_7 = -\frac{2}{5} + \frac{2\sqrt{-5}}{5} + \frac{1}{5}j,$$
$$P_8 = \frac{2}{5} + \frac{2\sqrt{-5}}{5} + \frac{1}{5}j, \quad P_9 = \frac{1}{2} + \frac{3\sqrt{-5}}{8} + \frac{\sqrt{3}}{8}j, \quad P_{10} = \frac{1}{2} - \frac{3\sqrt{-5}}{8} + \frac{\sqrt{3}}{8}j,$$
$$P_{11} = \frac{2}{5} - \frac{2\sqrt{-5}}{5} + \frac{1}{5}j, \quad P_{12} = -\frac{2}{5} - \frac{2\sqrt{-5}}{5} + \frac{1}{5}j, \quad P_{13} = \infty.$$

The singular points of $\mathbb{Q}(\sqrt{-5})$ contained in the boundary of $\tilde{\mathcal{F}}_{\mathbb{Q}(\sqrt{-5})}$ are P_1, P_2, P_3, P_4. The polyhedron $\tilde{\mathcal{F}}_{\mathbb{Q}(\sqrt{-5})}$ is bounded by 11 faces, they are:

I : $S(1,0) \cap \tilde{\mathcal{F}}_{\mathbb{Q}(\sqrt{-5})}$. This is an octagon with right angles.
II : $\{\ z + rj \ \mid \ \operatorname{Re} z = \frac{1}{2}\ \} \cap \tilde{\mathcal{F}}_{\mathbb{Q}(\sqrt{-5})}$,
III : $\{\ z + rj \ \mid \ \operatorname{Re} z = -\frac{1}{2}\ \} \cap \tilde{\mathcal{F}}_{\mathbb{Q}(\sqrt{-5})}$,
these are two pentagons with two right angles and three angles 0.
IV : $\{z + rj \ \mid \ \operatorname{Im} z = \frac{\sqrt{5}}{2}\} \cap \tilde{\mathcal{F}}_{\mathbb{Q}(\sqrt{-5})}$,
V : $\{z + rj \ \mid \ \operatorname{Im} z = -\frac{\sqrt{5}}{2}\} \cap \tilde{\mathcal{F}}_{\mathbb{Q}(\sqrt{-5})}$,
these are two triangles with angles 0.
VI : $S(2, \sqrt{-5}) \cap \tilde{\mathcal{F}}_{\mathbb{Q}(\sqrt{-5})}$, **VII** : $S(2, -\sqrt{-5}) \cap \tilde{\mathcal{F}}_{\mathbb{Q}(\sqrt{-5})}$,
these are two quadrangles with two angles 0 and two right angles.
VIII : $S(2\sqrt{-5}, 4 + \sqrt{-5}) \cap \tilde{\mathcal{F}}_{\mathbb{Q}(\sqrt{-5})}$,
IX : $S(2\sqrt{-5}, 4 - \sqrt{-5}) \cap \tilde{\mathcal{F}}_{\mathbb{Q}(\sqrt{-5})}$,
X : $S(2\sqrt{-5}, -4 + \sqrt{-5}) \cap \tilde{\mathcal{F}}_{\mathbb{Q}(\sqrt{-5})}$,
XI : $S(2\sqrt{-5}, -4 - \sqrt{-5}) \cap \tilde{\mathcal{F}}_{\mathbb{Q}(\sqrt{-5})}$,
these are four triangles with angles $0, \pi/2, \pi/3$.

The generators are those elements $\sigma \in \mathbf{PSL}(2, \mathcal{O})$ so that $\sigma\tilde{\mathcal{F}}_{\mathbb{Q}(\sqrt{-5})} \cap \tilde{\mathcal{F}}_{\mathbb{Q}(\sqrt{-5})}$ is a two dimensional face of the polyhedron $\tilde{F}_{\mathbb{Q}(\sqrt{-5})}$. The resulting identifications are:

A: identifies **II** with **III**, *B:* identifies **I** with itself,
U: identifies **IV** with **V**, *R:* identifies **IX** with **XI**,
S: identifies **VIII** with **X**, *T:* identifies **VI** with itself,
C: identifies **VII** with itself.

By an application of Corollary 2.7.5 the presentation is found. □

7.4 Groups Commensurable with PSL(2,$\mathcal{O}$)

This chapter contains the study of certain important groups which are commensurable with the groups $\mathbf{PSL}(2, \mathcal{O}) < \mathbf{PSL}(2, \mathbb{C})$. Our notation is as before, $K = \mathbb{Q}(\sqrt{d_K})$ being an imaginary quadratic number field with discriminant $d_K < 0$ and with ring of integers $\mathcal{O} = \mathbb{Z} + \mathbb{Z}\omega$.

As our first examples we consider groups H with $\mathbf{PSL}(2, \mathcal{O}) < H < \mathbf{PSL}(2, \mathbb{C})$ which are still discrete in $PSL(2, \mathbb{C})$. It turns out that there is a unique maximal such group. Our construction is reminiscent of one given by Hurwitz and Maaß (1971) in the Hilbert modular case. We shall use the following general notations:

$$Z(R) = \left\{ \begin{pmatrix} r & 0 \\ 0 & r \end{pmatrix} \ \middle| \ r \in R^* \right\} \tag{4.1}$$

denotes the invertible scalar matrices for any subring in $R \subset \mathbb{C}$,

$$\Theta : \mathbf{GL}(2, \mathbb{C}) \to \mathbf{PSL}(2, \mathbb{C}), \qquad \Theta : g \to \frac{1}{\sqrt{\det g}} \cdot g. \tag{4.2}$$

Note that, although the square root is only defined up to a sign, Θ defines a homomorphism which gives an isomorphism between the groups $\mathbf{PGL}(2,\mathbb{C})$ and $\mathbf{PSL}(2,\mathbb{C})$.

Definition 4.1. For $g = \begin{pmatrix} \alpha & \beta \\ \gamma & \delta \end{pmatrix} \in M(2,\mathcal{O})$ let $I(g) =< \alpha, \beta, \gamma, \delta >$ be the ideal generated by its entries. Put

$$G(\mathcal{O}) = \{ \ g \in M(2,\mathcal{O}) \ \mid \ I(g)^2 =< \det g > \neq 0 \ \}, \tag{4.3}$$

$$PG(\mathcal{O}) = Z(K) \cdot G(\mathcal{O})/Z(K) < \mathbf{PGL}(2,K), \tag{4.4}$$

$$H(\mathcal{O}) = \Theta(PG(\mathcal{O})). \tag{4.5}$$

The ideal $I(g)^2$ is the ideal generated by the products $\alpha^2, \alpha\beta, \ldots, \gamma\delta, \delta^2$. Note that we have $\mathbf{GL}(2,\mathcal{O}) \subset G(\mathcal{O})$ and $\mathbf{PSL}(2,\mathcal{O}) \subset H(\mathcal{O})$. The next proposition will give some properties of the above constructions.

Proposition 4.2. *With Definition 4.1 we have:*

(1) $I(g \cdot h) = I(g) \cdot I(h)$ *for all* $g, h \in G(\mathcal{O})$.
(2) $PG(\mathcal{O})$ *is a subgroup of* $\mathbf{PGL}(2,K)$.
(3) The map

$$\varphi : G(\mathcal{O}) \to \mathcal{J}_K, \qquad \varphi : g \mapsto I(g)^{\#} \tag{4.6}$$

defines a homomorphism $\varphi : PG(\mathcal{O}) \to \mathcal{J}_K$ *which is surjective onto the subgroup* $\mathcal{J}_K^{(2)}$ *of elements of order dividing 2 and has kernel* $\mathbf{PGL}(2,\mathcal{O})$.
(4) $H(\mathcal{O})$ *is a discrete subgroup of* $\mathbf{PSL}(2,\mathbb{C})$*, its subgroup* $\mathbf{PSL}(2,\mathcal{O})$ *is normal and* $[H(\mathcal{O}) : \mathbf{PSL}(2,\mathcal{O})] = 2^t$ *where* t *is the number of prime divisors of* d_K.

Proof. (1): We have the evident inclusions:

$$< \det(g \cdot h) > \subseteq I(g \cdot h)^2 \subseteq I(g)^2 \cdot I(h)^2 = < \det(g \cdot h) > .$$

Hence we may infer $I(g \cdot h)^2 = I(g)^2 \cdot I(h)^2$. Since $\mathcal{M}$ is free abelian the assertion follows.

(2): From (1) we obtain that $G(\mathcal{O})$ is a subsemigroup of $\mathbf{GL}(2,K)$. Inverses are supplied by adjoint matrices.

(3): Everything is easy except the surjectivity of φ onto $\mathcal{J}_K^{(2)}$. For this let p be an odd prime divisor of d_K. Put $\mathfrak{a}_p = p\mathbb{Z} + \omega\mathbb{Z}$. It is easy to see that $\mathfrak{a}_p$ is an ideal in $\mathcal{O}$ with $\mathfrak{a}_p^2 = p \cdot \mathcal{O}$. From Hecke (1923), paragraph 48 we obtain that the $\mathfrak{a}_p^{\#}$ generate $\mathcal{J}_K^{(2)}$. Since p and $\omega\bar{\omega}/p = (d_K^2 - d_K)/(4p)$

are coprime we may choose $r, s \in \mathbb{Z}$ with $rp^2 - s\omega\bar{\omega} = p$. This implies that $g_p := \begin{pmatrix} p & \omega \\ s\bar{\omega} & rp \end{pmatrix} \in G(\mathcal{O})$ and $\varphi(g_p) = \mathfrak{a}_p$.

(4): $H(\mathcal{O})$ is discrete since it is of finite index over $\mathbf{PSL}(2, \mathcal{O})$. In fact the index is

$$[H(\mathcal{O}) : \mathbf{PSL}(2, \mathcal{O})] = 2 \cdot [PG(\mathcal{O}) : \mathbf{PGL}(2, \mathcal{O})] = 2 \cdot |\mathcal{J}_K^{(2)}|.$$

By Hecke (1923) we know that $|\mathcal{J}_K^{(2)}| = 2^{t-1}$ where t is the number of prime divisors of d_K. □

Proposition 4.3. *Let $H < PSL(2, \mathbb{C})$ be a discrete subgroup satisfying $\mathbf{PSL}(2, \mathcal{O}) < H$, then $H < H(\mathcal{O})$.*

To prove Proposition 4.3 we treat first the case that $H < \mathbf{PSL}(2, K)$.

Lemma 4.4. *Let $H < \mathbf{PSL}(2, K)$ be a subgroup which is discrete in $\mathbf{PSL}(2, \mathbb{C})$ with $\mathbf{PSL}(2, \mathcal{O}) < H$, then $H = \mathbf{PSL}(2, \mathcal{O})$.*

Proof. If Lemma 4.4 is false, we find a $g = \begin{pmatrix} a & b \\ c & d \end{pmatrix} \in H$ with at least one entry not in $\mathcal{O}$. If $\wp$ is a prime ideal of $\mathcal{O}$ and $x \in K$ is non-zero we write $e(x, \wp) \in \mathbb{Z}$ for the exact order of occurence of $\wp$ in the prime ideal decomposition of $< x >$. We infer that we may find a prime ideal $\wp$ so that $e(x, \wp) < 0$ for at least one of $x = a, b, c, d$. By multiplying g by matrices of the form

$$\begin{pmatrix} 0 & 1 \\ -1 & 0 \end{pmatrix}, \begin{pmatrix} 1 & 0 \\ x & 1 \end{pmatrix}, \begin{pmatrix} 1 & x \\ 0 & 1 \end{pmatrix}$$

with $x \in \mathcal{O}$ we may enforce that $0 > e(a, \wp) < e(c, \wp), e(b, \wp)$ and $e(a, \wp) \leq e(d, \wp)$. Looking at $g^i \begin{pmatrix} 1 & 1 \\ 0 & 1 \end{pmatrix} g^{-i}$ for $i \in \mathbb{N}$ we then see that the additive subgroup A of K generated by the entries of all unipotent matrices of H is not finitely generated. Note that a finitely generated subgroup of the additive subgroup is contained in $\mathfrak{a}^{-1}$ for some non-zero ideal $\mathfrak{a} \subset \mathcal{O}$.

This is contradictory because we can now argue that A is finitely generated. First of all the subgroup of A generated by the entries of the unipotent elements in the stabilizer of a cusp of H is finitely generated. Furthermore by Proposition 2.2 we have $C_H = C_{\mathrm{PSL}(2,\mathcal{O})} = \mathbb{P}^1 K$ and by Theorem 2.4 we may also infer the finiteness of $\mathbf{PSL}(2, \mathcal{O}) \backslash C_H$. Since $\mathcal{O}$ is additively finitely generated and since any unipotent element stabilizes a cusp the proof is complete. □

Proof of Proposition 4.3. Let $g = \begin{pmatrix} a & b \\ c & d \end{pmatrix} \in H$ be any element. We first prove that there are $r \in \mathbb{C}, a_0, \ldots, d_0 \in K$ so that

$$g = r \cdot \begin{pmatrix} a_0 & b_0 \\ c_0 & d_0 \end{pmatrix}. \tag{4.7}$$

To do this let Λ be the lattice in $\mathbb{C}$ corresponding to the translations in H, that is

$$H'_\infty = \left\{ h = \begin{pmatrix} 1 & x \\ 0 & 1 \end{pmatrix} \;\middle|\; h \in H \right\} = \left\{ \begin{pmatrix} 1 & x \\ 0 & 1 \end{pmatrix} \;\middle|\; x \in \Lambda \right\}.$$

Since $\mathcal{O} \subset \Lambda$ and Λ is of finite index over $\mathcal{O}$ we get $\Lambda \subseteq K$.

If g is diagonal ($b = c = 0$) then $a^2\Lambda = \Lambda$ and we have $a^2 \in K$. From $g = a^{-1}\begin{pmatrix} a^2 & 0 \\ 0 & 1 \end{pmatrix}$ we get (4.7).

For a general g we get from Proposition 2.7 and Shimizu's lemma that there are $r, s, u, v \in \mathbb{C}$ and $a_1, \ldots, d_1, a_2, \ldots, d_2 \in K$ so that

$$g = \begin{pmatrix} a_1 r & b_1 s \\ c_1 r & d_1 s \end{pmatrix} = \begin{pmatrix} a_2 u & b_2 u \\ c_2 v & d_2 v \end{pmatrix}.$$

If the coefficients of one row of g are both non-zero we find that we may take $r = s$.

Since we have already treated diagonal elements and consequently also those of the form $g = \begin{pmatrix} 0 & b \\ c & 0 \end{pmatrix}$ we conclude (4.7).

Clearly we may in the representation (4.7) assume that $a_0, \ldots, d_0 \in \mathcal{O}$. We then define

$$\tilde{g} = \begin{pmatrix} a_0 & b_0 \\ c_0 & d_0 \end{pmatrix} \in M(2, \mathcal{O}).$$

We prove now that $\tilde{g} \in G(\mathcal{O})$. Notice first that $g\mathbf{PSL}(2,\mathcal{O})g^{-1} = \mathbf{PSL}(2,\mathcal{O})$, because the subgroup of H generated by $\mathbf{PSL}(2,\mathcal{O})$ and $g\mathbf{PSL}(2,\mathcal{O})g^{-1}$ satisfies the hypothesis of Lemma 4.4. In fact we have proved now that $\mathbf{PSL}(2,\mathcal{O})$ is normal in H. Putting $t = \det\tilde{g}$ and looking at

$$g \begin{pmatrix} 1 & 1 \\ 0 & 1 \end{pmatrix} g^{-1} = t^{-1} \begin{pmatrix} t - a_0 c_0 & a_0^2 \\ c_0^2 & t + a_0 c_0 \end{pmatrix} \in M(2, \mathcal{O})$$

we find that $a_0 c_0, a_0^2, c_0^2 \in t \cdot \mathcal{O}$. Multiplying g by $\begin{pmatrix} 0 & 1 \\ -1 & 0 \end{pmatrix}$ from both sides we obtain the remaining conditions. □

Sometimes the maximality of the groups $H(\mathcal{O})$ can be proved by volume considerations. Take for example $d_K = -3$ and $\mathcal{O}$ the ring of integers in $\mathbb{Q}(\sqrt{-3})$, then the group $\mathbf{PSL}(2,\mathcal{O})$ has 1 class of cusps. Let $\Gamma < \mathbf{PSL}(2,\mathbb{C})$ be a discrete group which contains $\mathbf{PSL}(2,\mathcal{O})$, then we get from Theorem 5.1 of Chapter 2 that

$$[\Gamma : \mathbf{PSL}(2,\mathcal{O})] \leq \frac{24}{\sqrt{3}} \cdot \mathrm{vol}(\mathbf{PSL}(2,\mathcal{O})).$$

Putting in the value of vol(**PSL**$(2,\mathcal{O})$) from our table at the beginning of this chapter we get $[\Gamma : \mathbf{PSL}(2,\mathcal{O})] \leq 2$. This proves that

$$H(\mathcal{O}) = \left\langle \mathbf{PSL}(2,\mathcal{O}), \begin{pmatrix} i & 0 \\ 0 & -i \end{pmatrix} \right\rangle$$

is maximal discrete in $\mathbf{PSL}(2,\mathbb{C})$.

Our second example for groups of interest which are commensurable with $\mathbf{PSL}(2,\mathcal{O})$ arises from the fact that the 4-dimensional K-algebra $M(2,K)$ sometimes has several $\mathbf{GL}(2,K)$-conjugacy classes of maximal $\mathcal{O}$-orders. An $\mathcal{O}$-order is a subring of $M(2,K)$ which is also an $\mathcal{O}$-submodule. It is further required to contain a K-basis of $M(2,K)$ and be finitely generated as $\mathcal{O}$-module. An example is

$$(4.8) \quad M(\mathcal{O},\mathfrak{a}) := \left\{ \begin{pmatrix} a & b \\ c & d \end{pmatrix} \in M(2,K) \;\middle|\; a, d \in \mathcal{O},\ c \in \mathfrak{a},\ b \in \mathfrak{a}^{-1} \right\}$$

whenever $\mathfrak{a} \subset \mathcal{O}$ is a non-zero ideal. The following is elementary, we skip the proof.

Proposition 4.5. *Let $\mathfrak{a} \subset \mathcal{O}$ be a non-zero ideal then $M(\mathcal{O},\mathfrak{a})$ is a maximal order in $M(2,K)$. Every maximal order in $M(2,K)$ is* $\mathbf{GL}(2,K)$*-conjugate to an $M(\mathcal{O},\mathfrak{a})$. If $\mathfrak{b} \subset \mathcal{O}$ is another non-zero ideal then $M(\mathcal{O},\mathfrak{a})$ and $M(\mathcal{O},\mathfrak{b})$ are* $\mathbf{GL}(2,K)$*-conjugate if and only if $\mathfrak{a}$ and $\mathfrak{b}$ have the same image in $\mathcal{J}_K/\mathcal{J}_K^2$.*

Each of the orders $M(\mathcal{O},\mathfrak{a})$ leads to the cofinite subgroup

$$(4.9) \qquad \mathbf{PSL}(\mathcal{O},\mathfrak{a}) := \{\, g \in M(\mathcal{O},\mathfrak{a}) \mid \det g = 1 \,\}/\{\pm 1\}$$

of $\mathbf{PSL}(2,\mathbb{C})$. It is easy to see that each group $\mathbf{PSL}(\mathcal{O},\mathfrak{a})$ is commensurable with $\mathbf{PSL}(2,\mathcal{O})$. For number theoretic purposes it is usually necessary to consider not only $\mathbf{PSL}(2,\mathcal{O})$ but the $\mathbf{PSL}(\mathcal{O},\mathfrak{a})$ where $\mathfrak{a}$ runs through a system of representatives for $\mathcal{J}_K/\mathcal{J}_K^2$.

It is possible to develop a theory of explicit fundamental domains analoguous to Section 3 also for the $\mathbf{PSL}(\mathcal{O},\mathfrak{a})$, see Schneider (1985). These fundamental domains can then be used to compute generators and relations for the groups $\mathbf{PSL}(\mathcal{O},\mathfrak{a})$. We give one example. Consider the case $K = \mathbb{Q}(\sqrt{-10})$. Its ring of integers has class number 2. The non-trivial ideal class is represented by $\mathfrak{b} =< 2, \sqrt{-10} >$. We put $\tilde{\omega} := \sqrt{-10}$ and

$$B := \begin{pmatrix} 0 & -1 \\ 1 & 0 \end{pmatrix},\ T := \begin{pmatrix} 1 & 1 \\ 0 & 1 \end{pmatrix},\ U := \begin{pmatrix} 1 & \tilde{\omega} \\ 0 & 1 \end{pmatrix},\ C := \begin{pmatrix} -\tilde{\omega} & 3 \\ 3 & \tilde{\omega} \end{pmatrix},$$

$$L := \begin{pmatrix} \tilde{\omega} & 3 \\ 3 & -\tilde{\omega} \end{pmatrix},\ D := \begin{pmatrix} \tilde{\omega} - 1 & -4 \\ 3 & 1 + \tilde{\omega} \end{pmatrix},\ W := \begin{pmatrix} 11 & 5\tilde{\omega} \\ 2\tilde{\omega} & -9 \end{pmatrix}.$$

Then $\mathbf{PSL}(2,\mathcal{O})$ is generated by these elements and we have

$$\mathbf{PSL}(2,\mathcal{O}) = \left\langle \begin{matrix} B,T,U,C, \\ L,D,W \end{matrix} \;\middle|\; \begin{matrix} B^2 = C^2 = L^2 = (BT)^3 = 1, \\ (BC)^2 = (BL)^2 = 1, \\ D^{-1}LTD = T^{-1}C,\, WCW^{-1} = U^{-1}LU, \\ U^{-1}DBCD^{-1}U = D^{-1}BLD, \\ U^{-1}LDT^{-1}BD^{-1}U = CD^{-1}T^{-1}BD, \\ WD^{-1}BTDW^{-1} = U^{-1}DTBD^{-1}U \end{matrix} \right\rangle.$$

This presentation is contained in Flöge (1983). Put further

$$V := \begin{pmatrix} 1 & \frac{\tilde{\omega}}{2} \\ 0 & 1 \end{pmatrix},\ A_1 := \begin{pmatrix} 1 & 0 \\ 2 & 1 \end{pmatrix},\ A_2 := \begin{pmatrix} 1 & 0 \\ -\tilde{\omega} & 1 \end{pmatrix},$$

$$A_3 := \begin{pmatrix} -2 & -\frac{\tilde{\omega}}{2} \\ -\tilde{\omega} & 2 \end{pmatrix},\ A_4 := \begin{pmatrix} -3 & -1-\frac{\tilde{\omega}}{2} \\ 2-\tilde{\omega} & 2 \end{pmatrix}.$$

Then by Schneider (1985) we have the presentation

$$\mathbf{PSL}(\mathcal{O},\mathfrak{b}) = \left\langle \begin{matrix} V,T,A_1, \\ A_2,A_3,A_4 \end{matrix} \;\middle|\; \begin{matrix} A_3^2 = (A_1T^{-1})^2 = A_4^3 = (A_1A_4^{-1}T^{-1})^3 \\ = (A_3A_4)^2 = (A_2A_3U^{-1})^2 = 1, \\ (A_4A_1^{-1}A_3T)^2 = (A_2A_4^{-1}U^{-1})^3 = 1, \\ (A_1A_2A_4^{-1}T^{-1}U^{-1})^3 = 1, \\ TU = UT,\, A_1A_2 = A_2A_1 \end{matrix} \right\rangle.$$

Writing G^{ab} for the commutator quotient group of a group G we obtain from the above two presentations

$$\mathbf{PSL}(2,\mathcal{O})^{\mathrm{ab}} = (\mathbb{Z}/2\mathbb{Z})^2 \times \mathbb{Z}^3, \qquad \mathbf{PSL}(\mathcal{O},\mathfrak{b})^{\mathrm{ab}} = \mathbb{Z}/2\mathbb{Z} \times \mathbb{Z}^2$$

which is somewhat surprising.

Finally for this section we have to discuss the general concept of a congruence subgroup. Given a non-zero ideal $\mathfrak{a} \subset \mathcal{O}$ the principal congruence group of level $\mathfrak{a}$ is defined by

$$\Gamma(\mathfrak{a}) := \left\{ \begin{pmatrix} a & b \\ c & d \end{pmatrix} \in \mathbf{SL}(2,\mathcal{O}) \;\middle|\; \begin{pmatrix} a & b \\ c & d \end{pmatrix} \equiv \begin{pmatrix} 1 & 0 \\ 0 & 1 \end{pmatrix} \mod \mathfrak{a} \right\}.$$

It is easy to see that $\Gamma(\mathfrak{a})$ is a normal subgroup of finite index in $\mathbf{SL}(2,\mathcal{O})$, it is in fact the kernel of the mod $\mathfrak{a}$ reduction homomorphism $\mathbf{SL}(2,\mathcal{O}) \to \mathbf{SL}(2,\mathcal{O}/\mathfrak{a})$. Note that this homomorphism is surjective. This is proved in Bass (1964), Corollary 5.2. We similarly define

$$P\Gamma(\mathfrak{a}) := \left\{ \begin{pmatrix} a & b \\ c & d \end{pmatrix} \in \mathbf{PSL}(2,\mathcal{O}) \;\middle|\; \begin{pmatrix} a & b \\ c & d \end{pmatrix} \equiv \begin{pmatrix} \pm 1 & 0 \\ 0 & \pm 1 \end{pmatrix} \mod \mathfrak{a} \right\}$$

to be the full congruence group of level $\mathfrak{a}$ in $\mathbf{PSL}(2,\mathcal{O})$.

Definition 4.6. A discrete subgroup $\Gamma < \mathbf{SL}(2,\mathbb{C})$ which is $\mathbf{SL}(2,\mathbb{C})$-conjugate to a group containing $\Gamma(\mathfrak{a})$ for some non-zero ideal $\mathfrak{a} \subset \mathcal{O}$ is called a congruence subgroup with respect to $\mathbf{SL}(2,\mathcal{O})$. If Γ is $\mathbf{SL}(2,\mathbb{C})$-conjugate to a group not containing any $\Gamma(\mathfrak{a})$ but commensurable with $\mathbf{SL}(2,\mathcal{O})$ it is called a non-congruence group. Congruence groups of $\mathbf{PSL}(2,\mathbb{C})$ are similarly defined with the $P\Gamma(\mathfrak{a})$ as reference groups.

If $\Gamma < \mathbf{PSL}(2,\mathbb{C})$ is a congruence subgroup then there is a $\gamma \in \mathbf{PSL}(2,\mathbb{C})$ so that the index of $\gamma\Gamma\gamma^{-1} \cap \mathbf{PSL}(2,\mathcal{O})$ in $\mathbf{PSL}(2,\mathcal{O})$ is finite. An example of a congruence subgroup appearing again in the next chapter is given for any non-zero ideal $\mathfrak{a} \subset \mathcal{O}$ by

$$\Gamma_0(\mathfrak{a}) := \left\{ \begin{pmatrix} a & b \\ c & d \end{pmatrix} \in \mathbf{PSL}(2,\mathcal{O}) \;\middle|\; c \in \mathfrak{a} \right\}. \tag{4.10}$$

In the following the subgroup

$$\mathbf{Q}(\mathfrak{a}) :=<< \left\{ \begin{pmatrix} 1 & a \\ 0 & 1 \end{pmatrix} \;\middle|\; a \in \mathfrak{a} \right\} >>_{\mathrm{SL}(2,\mathcal{O})} \tag{4.11}$$

given for any non-zero ideal $\mathfrak{a} \subset \mathcal{O}$ plays an important role. Here $<< M >>_G$ stands for the normal closure of a subset M of a group G. We have $\mathbf{Q}(\mathfrak{a}) < \Gamma(\mathfrak{a})$ but in general $\mathbf{Q}(\mathfrak{a})$ does not have finite index in $\mathbf{SL}(2,\mathcal{O})$.

We shall see in the next section that every $\mathbf{PSL}(2,\mathcal{O})$ has many subgroups of finite index which are not congruence subgroups. It is useful to have the following criterion for Γ to be a congruence subgroup. The result is taken from Grunewald, Schwermer (1996).

Proposition 4.7. *Let $\Gamma < \mathbf{SL}(2,\mathcal{O})$ be a subgroup of finite index, then there is a non-zero ideal $\mathfrak{a} \subset \mathcal{O}$ so that $\mathbf{Q}(\mathfrak{a}) < \Gamma$. The ideal $\mathfrak{a}_\Gamma$ which is maximal with this property is called the level of Γ. The subgroup $\Gamma < \mathbf{SL}(2,\mathcal{O})$ is a congruence subgroup if and only if $\Gamma(\mathfrak{a}_\Gamma) < \Gamma$.*

Proof. Let $\mathfrak{a} \subset \mathfrak{b}$ be two non-zero ideals of $\mathcal{O}$, then

$$< \Gamma(\mathfrak{b}),\ \mathbf{Q}(\mathfrak{a}) >= \Gamma(\mathfrak{a}). \tag{4.12}$$

This is a special case of Proposition 5.1 in Bass (1964). From (4.12) the statement follows using some straightforward arguments. □

7.5 The Group Theoretic Structure of PSL(2,$\mathcal{O}$)

Using an explicit presentation obtained from the knowledge of the fundamental domains $\mathcal{F}_K$ from Section 3 it is possible to describe some of the groups $\mathbf{PSL}(2,\mathcal{O})$ neatly as an amalgamated product or an HNN-extension of simpler groups. One instance is the formula

$$\mathbf{PSL}(2,\mathbb{Z}[i]) \cong (\mathbf{V}_4 *_{\mathbb{Z}/2\mathbb{Z}} \mathbf{S}_3) *_{\mathbb{Z}/2\mathbb{Z}*\mathbb{Z}/3\mathbb{Z}} (\mathbf{A}_4 *_{\mathbb{Z}/3\mathbb{Z}} \mathbf{S}_3) \tag{5.1}$$

given by Flöge (1983). Here $\mathbf{V}_4$ is Klein's group of order 4 and $\mathbf{S}_3$, $\mathbf{A}_4$ are the symmetric group on 3 letters, respectively the alternating group on 4 letters. Formulas like (5.1) can only be found if the absolute value of the discriminant d_K of the imaginary quadratic number field K is small, otherwise the combinatorial structure of the fundamental domains $\mathcal{F}_K$ gets too complicated. The representation (5.1) of $\mathbf{PSL}(2,\mathbb{Z}[i])$ as an iterated amalgamated product sheds some light on the group theoretic structure of this group. It is for example possible to explicitly find a subgroup of finite index with a free non-abelian quotient, see Fine (1989). A general result of this nature is given in the following theorem which is taken from Grunewald, Schwermer (1981a). The arguments use a modification of a topological method invented by Zimmert (1973).

Theorem 5.1. *Let $K = \mathbb{Q}(\sqrt{d_K})$ be an imaginary quadratic number field of discriminant $d_K < 0$, let $\mathcal{R}$ be a $\mathbb{Z}$-order in K. Then there exists a subgroup Γ of finite index of $\mathbf{PSL}(2,\mathcal{R})$ such that Γ has a free non-abelian quotient.*

A $\mathbb{Z}$-order $\mathcal{R}$ in K is a subring of K which is finitely generated as $\mathbb{Z}$-module and contains a $\mathbb{Q}$-basis of K. Such a subring of K is always contained in the ring of integers $\mathcal{O}$ of K. Taking the $\mathbb{Z}$-basis 1, ω of $\mathcal{O}$ with ω as in (1.1) we have for every $m \in \mathbb{N}$ the $\mathbb{Z}$-order $\mathcal{O}(m) := \mathbb{Z} + \mathbb{Z}\, m\omega$. As m runs through all natural numbers we get all $\mathbb{Z}$-orders in this way, see Lang (1973).

Decomposition (5.1) is reminiscent of the formula $\mathbf{PSL}(2,\mathbb{Z}) \cong \mathbb{Z}/2\mathbb{Z} * \mathbb{Z}/3\mathbb{Z}$. The latter immediately implies that $\mathbf{PSL}(2,\mathbb{Z})$ contains a subgroup of finite index which is a free group. This cannot happen for $\mathbf{PSL}(2,\mathcal{O})$ because these groups contain free abelian subgroups of rank 2. The paper of Lubotzky (1996) contains an extension of Theorem 5.1 to a more general class of cofinite groups acting on hyperbolic spaces of arbitrary dimension. Lubotzky uses a more group theoretic approach, in contrast to the explicit description of parts of a fundamental domain that is given here.

We define with respect to the order $\mathcal{O}(m)$ the set

$$\mathfrak{W}(m) = \left\{ n \in \mathbb{N} \;\middle|\; \begin{array}{l} (1)\ 4n^2 \leq m^2|d_K| - 3, \\ (2)\ d_K \text{ is a quadratic non-residue mod} \\ \quad\ \text{all the odd prime divisors of } n \\ \quad\ \text{and if } d_K \not\equiv 5 \bmod 8 \text{ then } n \text{ is odd,} \\ (3)\ n > 0,\ (n,m) = 1 \text{ and } n \neq 2 \end{array} \right\}. \tag{5.2}$$

Note that $\mathfrak{W}(m)$ is a finite set, possibly empty. We shall deduce Theorem 5.1 from the following proposition which gives, under certain circumstances, explicit free quotients of the groups $\mathbf{PSL}(2, \mathcal{O}(m))$.

Proposition 5.2. *Let $w(m)$ be the cardinality of the set $\mathfrak{W}(m)$. Then there is a surjective homomorphism from $\mathbf{PSL}(2, \mathcal{O}(m))$ to the free group $\mathbf{F}_{w(m)}$ on $w(m)$ symbols.*

It is a simple matter to compute $\mathfrak{W}(1)$, and thereby prove the existence of a free quotient of $\mathbf{PSL}(2, \mathcal{O})$ if d_K is given. For small $|d_K|$ it often happens that $\mathfrak{W}(1) = \{1\}$. Proposition 5.2 only gives an infinite cyclic quotient in this case. By application of deep number theoretic facts it has been proved in Grunewald, Schwermer (1981c) that there is a constant $\kappa > 0$ so that $|\mathfrak{W}(1)| > 1$ for all $|d_K| > \kappa$.

Before starting the proof of Proposition 5.2 some definitions are necessary. In analogy with Definition 3.1 we put

$$\begin{aligned}\mathcal{B}_K(m) = \{\ (z, r) \in \mathbb{H} \ \mid\ & |cz + d|^2 + |d|^2 r^2 \geq 1 \text{ for all } c, d \in \mathcal{O}(m) \\ & \text{with } < c, d >= \mathcal{O}(m)\ \}.\end{aligned}$$

Here the brackets $< c, d >$ stand for the $\mathcal{O}(m)$-ideal generated by c, d. It is a simple matter to show by a case distinction that

$$\left(\frac{t(a + m\omega)}{n}, \frac{1}{n}\right) \in \mathcal{B}_K(m)$$

for every $a \in \mathbb{Z}$, $n \in \mathfrak{W}(m)$ and $t \in \mathbb{Z}$ coprime to n, see Zimmert (1973), page 81 for a hint. For each $n \in \mathfrak{W}(m)$ and $t \in \mathbb{Z}$ with $(n, t) = 1$ we define

$$(5.3)\quad F_{n,t} := \mathcal{B}_K(m) \cap \left\{ (z, r) \in \mathbb{H} \ \middle|\ \operatorname{Im}\left(z - \frac{tm\omega}{n}\right) \leq (m^4 |d_K|^2)^{-1} \right\}.$$

The $F_{n,t}$ are solid vertical walls built into hyperbolic space, condition (1) of (5.2) implies that the sets $F_{n,t}$ are disjoint for distinct pairs (n, t). We can now describe the identifications between (boundary) points of the $F_{n,t}$ given by elements of $\mathbf{PSL}(2, \mathcal{O}(m))$.

Lemma 5.3. *Assume that $(z, r) \in \mathcal{B}_K(m)$ with $r \geq 5/(2m|d_K|)$. If $\gamma \in \mathbf{PSL}(2, \mathcal{O}(m))$ is an element with $\gamma(z, r) = (z', r') \in \mathcal{B}_K(m)$ then there is a $t' \in \mathbb{Z}$ with*

$$(5.4)\qquad \operatorname{Im}\left(z - \frac{tm\omega}{n}\right) = \operatorname{Im}\left(z' - \frac{t'm\omega}{n}\right).$$

We skip the straightforward proof, see Zimmert (1973), Hilfssatz 1 for the case $m = 1$.

Proof of Proposition 5.2. Using these walls we define for each $n \in \mathfrak{W}(m)$ a continuous map $e_n : \mathcal{B}_K(m) \to S^1 = \{ z \in \mathbb{C} \mid |z| = 1 \}$ by

$$e_n((z,r)) = \begin{cases} 1 & \text{if } (z,r) \notin \bigcup_{t,\ (n,t)=1} F_{n,t}\,, \\ \exp 2\pi i \left(\dfrac{1}{2} + \dfrac{m^4 |d_K|^2}{2} \operatorname{Im}\left(z - \dfrac{tm\omega}{n} \right) \right) & \text{if } (z,r) \in F_{n,t}\,. \end{cases}$$

Clearly the natural projection $\mathcal{B}_K(m) \to \mathbf{PSL}(2,\mathcal{O}(m)) \setminus \mathbb{H}$ is surjective. By Lemma 5.3 there is a unique factorization of e_n over $\mathbf{PSL}(2,\mathcal{O}(m)) \setminus \mathbb{H}$ by a continuous map

$$f_n : \mathbf{PSL}(2,\mathcal{O}(m)) \setminus \mathbb{H} \to S^1\,. \tag{5.5}$$

Define Y as the one-point union of $w(m)$ copies of the sphere S^1 taking as base-point the point 1. We have an isomorphism $H_1(Y,\mathbb{Z}) \cong \Pi H_1(S^1,\mathbb{Z})$ of the homology groups with integer coefficients. Patching the continuous maps f_n together we get a continuous map

$$f : \mathbf{PSL}(2,\mathcal{O}(m)) \setminus \mathbb{H} \to Y, \tag{5.6}$$

which can be viewed by an appropriate choice of a base point P as a map of pointed spaces. Therefore f induces a homomorphism

$$f_* : \pi_1(\mathbf{PSL}(2,\mathcal{O}(m)) \setminus \mathbb{H}, P) \to \pi_1(Y,1) \tag{5.7}$$

on the level of the fundamental group. We remark that $\mathbf{PSL}(2,\mathcal{O}(m)) \setminus \mathbb{H}$ is pathwise connected.

To each $g \in \mathbf{PSL}(2,\mathcal{O}(m))$ we associate now the class in $\pi_1(\mathbf{PSL}(2,\mathcal{O}(m)) \setminus \mathbb{H}, P)$ which is given by the image in $\mathbf{PSL}(2,\mathcal{O}(m)) \setminus \mathbb{H}$ of a path from P to $g \cdot P$ in $\mathbb{H}$. This defines a homomorphism

$$\phi : \mathbf{PSL}(2,\mathcal{O}(m)) \to \pi_1(\mathbf{PSL}(2,\mathcal{O}(m)) \setminus \mathbb{H}, P).$$

If we combine now the homomorphism ϕ with f_* we get homomorphisms

$$F : \mathbf{PSL}(2,\mathcal{O}(m)) \to \pi_1(Y,1), \quad {}_HF : \mathbf{PSL}(2,\mathcal{O}(m)) \to H_1(Y,\mathbb{Z}) \tag{5.8}$$

defined by $F = f_* \circ \phi$ resp. ${}_HF = H \circ F$ where H denotes the Hurewicz-homomorphism. Now we will show that ${}_HF$ is surjective. Once we have proved this we are through. The group $F(\mathbf{PSL}(2,\mathcal{O}(m)))$ is a free group, being a subgroup of the free group $\pi_1(Y,1)$. The image of $F(\mathbf{PSL}(2,\mathcal{O}(m)))$ in $H_1(Y,\mathbb{Z})$ is a free abelian group of rank $w(m)$. Therefore $F(\mathbf{PSL}(2,\mathcal{O}(m)))$ is a free non-abelian group of at least $w(m)$ generators.

By taking the path $x \mapsto \exp 2\pi i x$ as a generator for $H_1(S^1,\mathbb{Z})$ we identify $H_1(S^1,\mathbb{Z})$ with $\mathbb{Z}$. Let g be an element of $\mathbf{PSL}(2,\mathcal{O}(m))$ such that there exist (z,r), $(z',r') \in \mathcal{B}_K(m)$ with $g \cdot (z,r) = (z',r')$. Then the value $F_n(g)$ of

$F_n : \mathbf{PSL}(2, \mathcal{O}(m)) \to H_1(S^1, \mathbb{Z}) \cong \mathbb{Z}$ which is defined in the same way as F (see (5.8)) is given by

$$(5.9)\quad F_n(g) = \begin{cases} \left| \left\{ k \in \mathbb{Z} \;\middle|\; \begin{array}{l} \operatorname{Im} z < \dfrac{km\omega}{n} \le \operatorname{Im} z' \\ (n,k) = 1 \end{array} \right\} \right| & \text{if } \operatorname{Im} z \le \operatorname{Im} z', \\ - \left| \left\{ k \in \mathbb{Z} \;\middle|\; \begin{array}{l} \operatorname{Im} z' < \dfrac{km\omega}{n} \le \operatorname{Im} z \\ (n,k) = 1 \end{array} \right\} \right| & \text{if } \operatorname{Im} z' < \operatorname{Im} z. \end{cases}$$

This gives us a hint how to construct appropriate elements in $\mathbf{PSL}(2, \mathcal{O}(m))$ whose image under ΠF_n span $\Pi H_1(S^1, \mathbb{Z})$.

We imitate the construction given in Zimmert (1973). The elements $n_1, \ldots, n_{w(m)}$ of $\mathfrak{W}(m)$ are indexed in such a way that

$$\frac{r_1}{n_1} > \frac{r_2}{n_2} > \ldots > \frac{r_{w(m)}}{n_{w(m)}},$$

where r_i denotes the greatest natural number which is smaller than $n_i/2$ and satisfies $(n_i, r_i) = 1$. Then the congruences

$$(5.10)\qquad -|a_i + m\omega|^2 \equiv |b_i + m\omega|^2 \bmod n_i$$

are solvable in a_i and b_i for all $i = 1, \ldots, w(m)$, since in a finite field every element is the sum of two squares. Now condition (2) in the definition of $\mathfrak{W}(m)$ implies $(|a + m\omega|^2, n_i) = 1$ for all $a \in \mathbb{Z}$ and $i = 1, \ldots, w(m)$. Therefore we can find integers $s_i \in \mathbb{Z}$ which satisfy the conditions

$$r_i s_i |a_i + m\omega|^2 \equiv 1 \bmod n_i, \quad -r_i s_i |b_i + m\omega|^2 \equiv 1 \bmod n_i, \quad n_i - r_i < s_i$$

for $i = 1, \ldots w(m)$. Put

$$\sigma_i = \begin{pmatrix} s_i(\overline{a_i + m\omega}) & * \\ n_i & r_i(a_i + m\omega) \end{pmatrix},$$

$$\tau_i = \begin{pmatrix} (n_i - r_i)(\overline{b_i + m\omega}) & * \\ n_i & s_i(n + m\omega) \end{pmatrix}$$

where $^-$ denotes complex conjugation as usual. Define

$$z_i = r_i(a_i + m\omega)/n_i, \quad v_i = s_i(b_i + m\omega)/n_i$$

and $t_i = 1/n_i$, for $i = 1, \ldots, w(m)$. Then one gets

$$\sigma_i(z_i, t_i) = (z_i', t_i) \quad \text{and} \quad \tau_i(v_i, t_i) = (v_i', t_i)$$

where $z_i' = -s_i(\overline{a_i + m\omega})/n_i$, $v_i' = (n_i - r_i)(\overline{b_i + m\omega})/n_i$. We have

$$(z_i, t_i),\ (z_i', t_i),\ (v_i, t_i),\ (v_i', t_i) \in \mathcal{B}_K(m)$$

for all $i = 1, \dots, w(m)$. Define $\gamma_i := \tau_i \sigma_i$ for $i = 1, \dots, w(m)$. Then formula (5.9) implies

$$F_{n_j}(\gamma_i) = 0 \quad \text{for} \quad j < i \quad \text{and} \quad F_{n_j}(\gamma_j) = 1.$$

It follows that $\Pi_{n \in W(m)} F_n$ is surjective, this yields the desired surjectivity of ${}_H F$. □

Proof of Theorem 5.1. Now we consider the group $\mathbf{PSL}(2, \mathcal{O}(m))$ for a given order $\mathcal{O}(m)$ in K. If $w(m) > 1$ we are through by Proposition 5.2. So assume $w(m) \leq 1$. By an old result about primes in quadratic number fields it follows that there exist an infinite number of odd primes p such that d_K is a quadratic non-residue modulo p. Therefore one can find three odd prime numbers $p_1 < p_2 < p_3$ which are prime to m so that d_K is a quadratic non-residue modulo p_1. Put $m' = m \cdot p_2 \cdot p_3$, then it is easy to verify that one has $p_1 \in \mathfrak{W}(m')$. An iteration of this construction implies that one can find an order $\mathcal{O}(q)$ with m dividing q and $w(q) > 1$. Now $\mathbf{PSL}(2, \mathcal{O}(q))$ is of finite index in $\mathbf{PSL}(2, \mathcal{O}(m))$ and has a free non-abelian quotient. □

A group G is called SQ-universal if every countable group is isomorphic to a subgroup of a quotient of G. The following is immediate from Theorem 5.1 and Neumann (1973).

Corollary 5.4. *Let K be an imaginary quadratic number field and $\mathcal{O}$ its ring of integers. Then every subgroup of finite index in $\mathbf{PSL}(2, \mathcal{O})$ is* SQ*-universal.*

By application of a theorem of Margulis on normal subgroups in arithmetic groups it turns out that the only number fields for which the group $\mathbf{PSL}(2)$ over their ring of integers is SQ-universal are $\mathbb{Q}$ and imaginary quadratic number fields, see Grunewald, Schwermer (1981a) for comments.

As an application of Theorem 5.1 we solve the congruence subgroup problem for our groups $\mathbf{PSL}(2, \mathcal{O})$. The question is whether every subgroup Γ of finite index in $\mathbf{PSL}(2, \mathcal{O})$ is a congruence subgroup. To give a precise formulation to the solution we introduce the profinite completion $\mathfrak{C}_f(G)$ of a group G. We furthermore write $\mathfrak{C}_c(\mathbf{PSL}(2, \mathcal{O}))$ for the pro-congruence completion of $\mathbf{PSL}(2, \mathcal{O})$. It is the direct limit of the inverse system obtained from the quotients of $\mathbf{PSL}(2, \mathcal{O})$ by normal congruence subgroups. There obviously is an exact sequence

$$(5.11) \qquad <1> \to \mathfrak{C}\mathfrak{K}(\mathcal{O}) \to \mathfrak{C}_f(\mathbf{PSL}(2, \mathcal{O})) \to \mathfrak{C}_c(\mathbf{PSL}(2, \mathcal{O})) \to <1>$$

the profinite group $\mathfrak{C}\mathfrak{K}(\mathcal{O})$ being called the congruence kernel. Serre (1970) contains more explanations concerning the concepts used here.

Proposition 5.5. *Let K be an imaginary quadratic number field and $\mathcal{O}$ its ring of integers. Then $\mathbf{PSL}(2, \mathcal{O})$ has (many) subgroups of finite index which* *are not congruence subgroups. Let* $\mathbf{F}_\infty$ *be the free group on (countably) infinitely many generators. There is a closed subgroup of* $\mathfrak{CK}(\mathcal{O})$ *topologically isomorphic to* $\mathfrak{C}_f(\mathbf{F}_\infty)$.

Proof. A group H is called a section of a group G if there is a subgroup $G_1 < G$ and a normal subgroup $G_2 < G_1$ so that H is isomorphic to G_1/G_2. In Huppert (1967), Kapitel II we find a table of all subgroups of $\mathbf{PSL}(2,k)$ where k is a finite field. From this table we see that (for example) the finite simple group $\mathbf{A}_6$ is not a subgroup of any of the $\mathbf{PSL}(2,k)$. It is an exercise to infer that the alternating group $\mathbf{A}_6$ is not a section of any of the $\mathbf{PSL}(2,\mathcal{O}/\mathfrak{a})$, $\mathfrak{a}$ a non-zero ideal. Since $\mathbf{A}_6$ is generated by two elements we have a surjective homomorphism $\mathbf{F}_2 \to \mathbf{A}_6$. By Theorem 5.1 there is a subgroup $U < \mathbf{PSL}(2,\mathcal{O})$ and a surjective homomorphism $U \to \mathbf{A}_6$. It follows from the above that the kernel of this homomorphism cannot be a congruence subgroup.

The second statement of the Proposition is proved by similar reasoning, see Lubotzky (1982) for the details. □

It is known from work of Serre (1970) that the correspondingly defined congruence kernel $\mathfrak{CK}(\mathcal{O}_L)$ is finite for every number field L different from $\mathbb{Q}$ and the imaginary quadratic fields. Serre's work is based on prior work of Mennicke (1967) on related problems. By a theorem of Melnikov we have $\mathfrak{CK}(\mathbb{Z}) \cong \mathfrak{C}_f(\mathbf{F}_\infty)$. There is no description of $\mathfrak{CK}(\mathcal{O})$ in the imaginary quadratic case.

Very important as far as the ties to number theory are concerned are the commutator quotient groups Γ^{ab} as Γ is one of the $\mathbf{PSL}(2,\mathcal{O})$ or one of its congruence subgroups. As these groups Γ are all finitely generated the Γ^{ab} are finitely generated abelian groups. From the presentations given in Section 3 the following may easily be read off.

$$\mathbf{PSL}(2,\mathbb{Z}[i])^{\mathrm{ab}} \cong \mathbb{Z}/2\mathbb{Z} \times \mathbb{Z}/2\mathbb{Z}, \quad \mathbf{PSL}\left(2,\mathbb{Z}\left[\frac{1+\sqrt{-3}}{2}\right]\right)^{\mathrm{ab}} \cong \mathbb{Z}/3\mathbb{Z},$$

$$\mathbf{PSL}(2,\mathbb{Z}[\sqrt{-5}])^{\mathrm{ab}} \cong \mathbb{Z}/2\mathbb{Z}\times\mathbb{Z}/6\mathbb{Z}\times\mathbb{Z}^2, \ \mathbf{PSL}(2,\mathbb{Z}[\sqrt{-10}])^{\mathrm{ab}} \cong \mathbb{Z}/2\mathbb{Z}\times\mathbb{Z}^3.$$

To describe some further information on the commutator quotient group let $\Gamma < \mathbf{PSL}(2,\mathbb{C})$ be a cofinite group and $\eta_1,\dots,\eta_h \in C_\Gamma$ be a system of representatives for the Γ-classes of cusps of Γ, see Definition 1.1.10 and Section 1.3 for explanation. Let Γ'_{η_i} be the unipotent part of the stabilizer of η_i. This group is free abelian of rank 2. The inclusions $\Gamma'_{\eta_i} < \Gamma$ define a homomorphism

$$\alpha_\Gamma : \mathbf{U}(\Gamma) = \Gamma'_{\eta_1} \times \ldots \times \Gamma'_{\eta_h} \to \Gamma^{\mathrm{ab}}. \tag{5.12}$$

We now have

Proposition 5.6. *Let $\Gamma < \mathbf{PSL}(2, \mathbb{C})$ be a cofinite group then*

$$\mathrm{rk}_{\mathbb{Z}}(\alpha_\Gamma(\mathbf{U}(\Gamma))) = h,$$

where $\mathrm{rk}_{\mathbb{Z}}$ *denotes the free rank, in each of the following cases:*

(1) Γ *is torsion free.*
(2) $\Gamma < \mathbf{PSL}(2, \mathcal{O})$ *where* $\mathcal{O}$ *is the ring of integers in the imaginary quadratic number field* $K = \mathbb{Q}(\sqrt{d_K})$ *and* $d_K \neq -3, -4$.

Proof. This proposition is taken from Serre (1970). The first case is established by a topological argument. Compactify, as described in the Notes and Remarks of Chapter 2, the manifold $\Gamma\backslash\mathbb{H}$ by the addition of h tori. The compactified manifold has odd dimension and it follows from Poincaré duality that the image of the homology of the boundary has the given rank. Case (2) is implied by case (1) and some simple observations. See Serre (1970), Théorème 9. In the cases $d_K = -3, -4$ we have $\mathrm{rk}_{\mathbb{Z}}(\mathbf{PSL}(2, \mathcal{O})^{\mathrm{ab}}) = 0$ as can be seen from the presentations given in Propositions 3.9, 3.10. □

In case $\Gamma < \mathbf{PSL}(2, \mathcal{O})$ a lot is known on the kernel and the image of α_Γ. This is work of Harder (1975a), (1975b), (1979), (1987). Harder treats these problems for the corresponding homomorphisms between cohomology groups. By Proposition 5.6 we know that

$$\mathrm{rk}_{\mathbb{Z}}(\mathbf{PSL}(2, \mathcal{O})^{\mathrm{ab}}) \geq h \tag{5.13}$$

where h is the class number of $\mathcal{O}$ and $d_K \neq -3, -4$. The question of equality in (5.13) is important in connection with existence problems for cuspidal cohomology classes, see Grunewald, Schwermer (1981c) for some explanation. Equality in (5.13) does not hold in general, this is shown by the example $\Gamma = \mathbf{PSL}(2, \mathbb{Z}[\sqrt{-10}])$, ($\mathbb{Z}[\sqrt{-10}]$ has class number 2 and $\mathrm{rk}_{\mathbb{Z}}(\mathbf{PSL}(2, \mathbb{Z}[\sqrt{-10}])^{\mathrm{ab}}) = 3$). This was first observed by Mennicke around 1970. From Rohlfs (1985) we have

Proposition 5.7. *Let* $\mathcal{O}$ *be the ring of integers in the imaginary quadratic number field* $K = \mathbb{Q}(\sqrt{D})$ *(*$D \in \mathbb{Z}$ *squarefree) and* h *its class number, then*

$$\mathrm{rk}_{\mathbb{Z}}(\mathbf{PSL}(2, \mathcal{O})^{\mathrm{ab}}) \geq \frac{\varphi(D)}{24} - \frac{1}{4} + \frac{h}{2} \tag{5.14}$$

where $\varphi(D)$ *is the number of prime residues modulo* $|D|$.

The proof uses expressions, deduced from the Lefschetz trace formula, for the trace of certain involutions on the cohomology of $\mathbf{PSL}(2, \mathcal{O})\backslash\mathbb{H}$ in terms of values of L-functions. We cannot go into this here. For the method see also Harder (1975b). Comparing the growth of $\varphi(D)$ and of the class number one finds that there are only finitely many D so that equality can hold in (5.13).

A good bound for such D can then be deduced from Grunewald, Schwermer (1981c). Finally Vogtmann (1985) arrives, by explicit considerations of fundamental domains, at the following result.

Proposition 5.8. *Let $\mathcal{O}$ be the ring of integers in the imaginary quadratic number field $K = \mathbb{Q}(\sqrt{D})$ ($D \in \mathbb{Z}$ squarefree) and h its class number. Then* $\mathrm{rk}_{\mathbb{Z}}(\mathbf{PSL}(2,\mathcal{O})^{\mathrm{ab}}) \leq h$ *holds precisely in the following cases*

$$-D \in \{1, 2, 3, 5, 6, 11, 15, 19, 23, 31, 39, 47, 71\}.$$

Although we know a lot on the free rank of $\mathbf{PSL}(2,\mathcal{O})^{\mathrm{ab}}$ not much is known about its torsion subgroup. See Schwermer, Vogtmann (1983) and Vogtmann (1985) for some results.

We shall consider now questions similar to the above but for congruence subgroups of the $\mathbf{PSL}(2,\mathcal{O})$. If A is a finitely generated abelian group we write $\mathbf{T}(A)$ for its torsion subgroup and as before $\mathrm{rk}_{\mathbb{Z}}(A) = \dim_{\mathbb{Q}}(\mathbb{Q} \otimes_{\mathbb{Z}} A)$ for its free rank. We also use the following notation for finite abelian groups

$$[n_1, ..., n_m] = \mathbb{Z}/n_1\mathbb{Z} \times ... \times \mathbb{Z}/n_m\mathbb{Z} \qquad (m, n_1, ..., n_m \in \mathbb{N}). \tag{5.15}$$

Most important for number theoretic reasons amongst the congruence subgroups are the $\Gamma_0(\mathfrak{a})$, where $\mathfrak{a} \subset \mathcal{O}$. See (4.10) for the definition.

Let us discuss some concrete examples. The following table contains the invariants (5.15) in case $\mathcal{O} = \mathbb{Z}[i]$ and $\Gamma = \Gamma_0(\wp)$ where $\wp \subset \mathcal{O}$ runs through the prime ideals of residue degree 1 with $N(\wp) \leq 241$.

Table 5.7.1 Abelian invariants of $\Gamma_0(\wp)$, $\wp \subset \mathbb{Z}[i]$

$N(\wp)$	$\mathrm{rk}_{\mathbb{Z}}(\Gamma_0(\wp)^{\mathrm{ab}})$	$\mathbf{T}(\Gamma_0(\wp)^{\mathrm{ab}})$	$N(\wp)$	$\mathrm{rk}_{\mathbb{Z}}(\Gamma_0(\wp)^{\mathrm{ab}})$	$\mathbf{T}(\Gamma_0(\wp)^{\mathrm{ab}})$
5	0	$[2,2,2]$	109	0	$[2,2,2,27]$
13	0	$[2,2,6]$	113	0	$[2,2,4,8,7]$
17	0	$[2,2,8]$	137	1	$[2,2,2,3,3,17]$
29	0	$[2,2,2,3,7]$	149	0	$[2,2,2,7,37]$
37	0	$[2,2,2,9]$	157	0	$[2,2,2,2,2,3,3,13]$
41	0	$[2,2,2,4,5]$	173	0	$[2,2,2,5,43]$
53	0	$[2,2,2,13]$	181	0	$[2,2,2,3,9,5,31]$
61	0	$[2,2,2,3,5]$	193	0	$[2,2,2,2,4,32,3]$
73	0	$[2,2,4,9]$	197	0	$[2,2,2,3,3,3,49]$
89	0	$[2,2,4,11,11]$	229	0	$[2,2,2,3,27,19,19]$
97	0	$[2,2,16,3,5]$	233	1	$[2,2,2,2,2,4,29]$
101	0	$[2,2,2,25,17]$	241	0	$[2,2,8,3,5,19]$

These data can be computed simply by hand using the presentation given in Section 3 for $\mathbf{PSL}(2, \mathbb{Z}[i])$ and the Reidemeister–Schreier algorithm (see Lyndon, Schupp (1977)) which produces a presentation for $\Gamma_0(\wp)$. Of course nowadays there are effective computer programs at hand by which the above table can considerably be enlarged. The pattern which shows up is that as $N(\wp)$ is increased there are only few, but constantly coming $\wp$ so that $\mathrm{rk}_{\mathbb{Z}}(\Gamma_0(\wp)^{\mathrm{ab}}) \neq 0$. The largest $\mathrm{rk}_{\mathbb{Z}}(\Gamma_0(\wp)^{\mathrm{ab}})$ we found was 3. Nothing is known on the behaviour of $\mathrm{rk}_{\mathbb{Z}}(\Gamma_0(\wp)^{\mathrm{ab}})$ as $N(\wp) \to \infty$.

Let $\Gamma < \mathbf{PSL}(2, \mathcal{O})$ be a congruence subgroup. The groups Γ^{ab} come with an interesting set of endomorphisms, the Hecke operators, which we define now. Given $0 \neq \delta \in \mathcal{O}$ we put $\hat{\delta} := \begin{pmatrix} 1 & 0 \\ 0 & \delta \end{pmatrix}$ and consider the endomorphism T_δ defined by the diagram

$$\begin{array}{ccc} \Gamma^{\mathrm{ab}} & \xrightarrow{T_\delta} & \Gamma^{\mathrm{ab}} \\ \Big\downarrow \mathbf{t} & & \Big\uparrow id \\ (\Gamma \cap \hat{\delta}\,\Gamma\,\hat{\delta}^{-1})^{\mathrm{ab}} & \xrightarrow{\hat{\delta}^{-1}} & (\Gamma \cap \hat{\delta}^{-1}\,\Gamma\,\hat{\delta})^{\mathrm{ab}}. \end{array} \tag{5.16}$$

It is easy to see that $\Gamma \cap \hat{\delta}\,\Gamma\,\hat{\delta}^{-1}$ is of finite index in Γ (use that Γ is a congruence subgroup). The map $\mathbf{t}$ is then the transfer homomorphism, see Huppert (1967). The lower arrow is induced by conjugation with $\hat{\delta}^{-1}$, the right-hand arrow is induced by inclusion. The Hecke operator T_δ is then defined as the composition of the three indicated homomorphisms. This definition is a variation of the construction given in Shimura (1971), Chapter 8 for the Hecke operators for congruence subgroups of $\mathbf{PSL}(2, \mathbb{Z})$.

We now sketch some important constructions which are used in the proof of Proposition 5.9. Let $\Gamma < \mathbf{PSL}(2, \mathcal{O})$ be a congruence subgroup. The first cohomology group (with real coefficients) of Γ is defined by $H^1(\Gamma, \mathbb{R}) = \mathrm{Hom}_{\mathbb{Z}}(\Gamma^{\mathrm{ab}}, \mathbb{R})$. The Hecke operator T_δ constructed above induces a linear map

$$\hat{T}_\delta : H^1(\Gamma, \mathbb{R}) \to H^1(\Gamma, \mathbb{R}).$$

Define

$$H^1_{\mathrm{cusp}}(\Gamma, \mathbb{R}) := \{\ \varphi \in H^1(\Gamma, \mathbb{R}) \ \mid\ \varphi(x) = 0 \text{ for all } x \in \mathbf{U}(\Gamma)\ \}.$$

It is easy to see that $\hat{T}_\delta$ maps this space to itself. Assume now that Γ is torsion free. Then Γ is isomorphic to the fundamental group of the quotient manifold $\Gamma\backslash\mathbb{H}$ and $H^1(\Gamma, \mathbb{R})$ can be identified with the first (simplicial) cohomology group $H^1(\Gamma\backslash\mathbb{H}, \mathbb{R})$. By the theorem of de Rham we have isomorphisms

$$H^1(\Gamma, \mathbb{R}) \cong H^1(\Gamma\backslash\mathbb{H}, \mathbb{R}) \cong H^1_{\mathrm{deRham}}(\Gamma\backslash\mathbb{H}, \mathbb{R}).$$

We write $\hat{T}_\delta$ for the Hecke operator induced on $H^1_{\mathrm{deRham}}(\Gamma\backslash\mathbb{H}, \mathbb{R})$ and $H^1_{\mathrm{cusp}}(\Gamma\backslash\mathbb{H}, \mathbb{R})$ for the subspace corresponding to $H^1_{\mathrm{cusp}}(\Gamma, \mathbb{R})$. By results of Harder (1975a), (1975b), (1979) (see also Grunewald, Schwermer (1981b)) we have

$$H^1_{\text{cusp}}(\Gamma\backslash\mathbb{H},\mathbb{R}) = H^1_!(\Gamma\backslash\mathbb{H},\mathbb{R}) = H^1_{\text{harm}}(\Gamma\backslash\mathbb{H},\mathbb{R}) \tag{5.17}$$

where $H^1_!(\Gamma\backslash\mathbb{H},\mathbb{R})$ is the image of cohomology with compact support in $H^1_{\text{deRham}}(\Gamma\backslash\mathbb{H},\mathbb{R})$ and $H^1_{\text{harm}}(\Gamma\backslash\mathbb{H},\mathbb{R})$ the image of the space of harmonic differential 1-forms. Notice that $\Gamma\backslash\mathbb{H}$ inherits a Riemannian structure from $\mathbb{H}$. It follows from results of Harder that there is a decomposition

$$H^1_{\text{deRham}}(\Gamma\backslash\mathbb{H},\mathbb{R}) = H^1_{\text{cusp}}(\Gamma\backslash\mathbb{H},\mathbb{R}) \oplus H^1_{\text{Eis}}(\Gamma\backslash\mathbb{H},\mathbb{R}) \tag{5.18}$$

so that the Hecke operators $\hat{T}_\delta$ respect $H^1_{\text{Eis}}(\Gamma\backslash\mathbb{H},\mathbb{R})$. There is also a precise description of the $\hat{T}_\delta$ on $H^1_{\text{Eis}}(\Gamma\backslash\mathbb{H},\mathbb{R})$ by certain Hecke characters, for the most general version see Harder (1987).

Proposition 5.9. *Let $\Gamma < \mathbf{PSL}(2,\mathcal{O})$ be a congruence subgroup. There is a non-zero ideal $\mathfrak{a}(\Gamma) < \mathcal{O}$ so that all the Hecke operators T_δ for $\delta \in \mathcal{O}$ prime to $\mathfrak{a}(\Gamma)$ commute. The linear maps induced by the T_δ on $\Gamma^{\text{ab}} \otimes_{\mathbb{Z}} \mathbb{Q}$ are semisimple for δ prime to $\mathfrak{a}(\Gamma)$. We have $T_\delta(\mathbf{U}(\Gamma)) \subset \mathbf{U}(\Gamma)$ for all T_δ. The eigenvalues of T_δ on $\Gamma^{\text{ab}}/\mathbf{U}(\Gamma)$ are totally real algebraic integers.*

Proof. The commutation relations and the stability of $\mathbf{U}(\Gamma)$ follow by a straightforward computation, for the details see Grunewald, Helling, Mennicke (1978). The argument for semisimplicity is more complicated. Notice first of all that Γ may without loss of generality be assumed to be torsion free. It is furthermore sufficient to prove the corresponding results for the Hecke operators $\hat{T}_\delta$. Notice first of all that the $\hat{T}_\delta$ respect a lattice in $H^1(\Gamma,\mathbb{R})$, namely the image of $H^1(\Gamma,\mathbb{Z})$. This implies the integrality statement for the eigenvalues. We now use the decomposition (5.18). Given two differential 1-forms ω_1, ω_2 from $H^1_{\text{cusp}}(\Gamma\backslash\mathbb{H},\mathbb{R})$ we may define the so-called Petersson scalar product given by

$$< \omega_1, \omega_2 >= \int_{\Gamma\backslash\mathbb{H}} \omega_1 \wedge *\omega_2$$

where $*$ is the star-operator derived from the Riemannian structure, see Warner (1970). The existence of the integral is implied by (5.17). It is an exercise to verify that $\hat{T}_\delta$ is symmetric with respect to this scalar product. This together with the explicit description of the Hecke operators on $H^1_{\text{Eis}}(\Gamma\backslash\mathbb{H},\mathbb{R})$ given by Harder (1987) implies the statement. □

The eigenvectors of the Hecke operators are of similar importance for the number theory of the imaginary quadratic number field K as those of the Hecke operators on spaces of holomorphic modular forms. As in the classical situation, see Shimura (1971) we may introduce the L-function of an Hecke eigenvector. In the case that $\mathcal{O}$ has class number 1 this is easily defined as follows. For a congruence group $\Gamma < \mathbf{PSL}(2, \mathcal{O})$ let $\mathfrak{a}(\Gamma)$ be the ideal mentioned in Proposition 5.9. Let $v \in H_1(\Gamma, \mathbb{R})$ be an eigenvector for all T_δ with δ coprime to $\mathfrak{a}(\Gamma)$. Assume further that $T_\epsilon(v) = v$ for all units $\epsilon \in \mathcal{O}^*$. Let $\mathfrak{q}$ be a prime ideal of $\mathcal{O}$ which does not divide $\mathfrak{a}(\Gamma)$. Pick a generator q of $\mathfrak{q}$ then $T_q(v) = a_\mathfrak{q} v$ and the eigenvalue $a_\mathfrak{q}$ does not depend on the choice of q. Define

$$L(v, s) := \prod_{\mathfrak{q}}{}' (1 - a_\mathfrak{q} N(\mathfrak{q})^{-s} + N(\mathfrak{q})^{1-2s})^{-1} \tag{5.19}$$

where the product is extended over all prime ideals $\mathfrak{q}$ not dividing $\mathfrak{a}(\Gamma)$. This is a special case of a Langlands L-function, see Gelbart (1975). It can be shown that (5.19) converges in some half plane and has a meromorphic continuation to $\mathbb{C}$. If $L(v, s)$ is completed by suitable Euler factors at the prime ideals dividing $\mathfrak{a}(\Gamma)$ it satisfies a simple functional equation.

We shall now give a concrete example for $K = \mathbb{Q}(i)$. A short meditation shows that once we have an explicit basis for one of the vector spaces $\Gamma_0(\wp)^{\mathrm{ab}} \otimes_\mathbb{Z} \mathbb{Q}$ or $\Gamma_0(\wp)^{\mathrm{ab}} \otimes_\mathbb{Z} \mathbb{Z}/\ell\mathbb{Z}$ we may compute a matrix representation of the linear map induced by the Hecke operator T_δ if the norm of δ is not too large. We give the result if v is a basis vector of $\Gamma_0(\wp_{233})^{\mathrm{ab}} \otimes_\mathbb{Z} \mathbb{Q}$ where $\wp_{233}$ is the prime ideal generated by $8 + 13i$. It is an exercise to show that T_δ only depends on the ideal generated by δ. We have $T_\delta v = a_{<\delta>} v$ with a rational integer $a_{<\delta>}$. In examples we find

$$a_{<1+i>} = -2,\ a_{<3>} = -4,\ a_{<1+2i>} = -2,\ a_{<1-2i>} = -3,\ a_{<7>} = 1,$$

$$a_{<11>} = -10,\ a_{<-3+2i>} = -3,\ a_{<1+4i>} = -3,\ a_{<1-4i>} = -4,\ a_{<19>} = 35.$$

These numbers seem to have the following connection to arithmetic. Consider the elliptic curve (defined over $\mathbb{Q}(i)$) given by the equation

$$E_{233}:\quad y^2 + iy = x^3 + (1+i)x^2 + ix. \tag{5.20}$$

The conductor ideal of E_{233} is easily seen to be $\wp_{233}$. If $\mathfrak{q}$ is a prime ideal of $\mathcal{O}$ define

$$b_\mathfrak{q} := N(\mathfrak{q}) - |\{\ (x, y) \in (\mathbb{Z}[i]/\mathfrak{q})^2 \ \mid\ (x, y) \text{ lies on } E_{233}\ \}|.$$

For the $a_\mathfrak{q}$ given above and many more it can be simply checked that

$$b_\mathfrak{q} = a_\mathfrak{q}. \tag{5.21}$$

If (5.21) was true in general it would say that the Hasse–Weil L-function of E_{233} would be equal to the corresponding function (5.19) suitably completed by Euler factors at the prime ideals dividing $\mathfrak{a}(\Gamma)$. Many more examples of this nature are contained in Grunewald, Helling, Mennicke (1978), Grunewald, Mennicke (1978), Cremona (1984). There is a similar match-up in case of $N(\wp) = 137$ but here a normalization of the generators for the prime ideals $\mathfrak{q}$ has to be introduced, see Grunewald, Mennicke (1978).

Starting from the automorphic representation corresponding to a Hecke eigenform R. Taylor has proved that (5.21) and many other similar match-ups are true for all primes $\mathfrak{q}$ outside a set of Dirichlet density 0, see Taylor (1994), Harris, Soudry, Taylor (1993). Taylor's argument goes via a lifting construction to $\mathbf{Sp}(4, \mathbb{Z})$ and does not produce the elliptic curve directly (as is done in the classical case of holomorphic modular forms of weight 2).

The Hecke operators also induce linear endomorphisms of the $\mathbb{Z}/\ell\mathbb{Z}$-vector spaces $\mathbf{T}(\Gamma_0(\wp)) \otimes_{\mathbb{Z}} \mathbb{Z}/\ell\mathbb{Z}$. These are in general no longer semisimple, but there are interesting simultaneous eigenspaces for the Hecke operators. Consider the prime ideal $\wp_{157} =< 6 - 11i >$ which has norm 157. From the table above we see that $\mathbf{T}(\Gamma_0(\wp_{157})) \otimes_{\mathbb{Z}} \mathbb{Z}/2\mathbb{Z}$ is a 3-dimensional vector space over $\mathbb{Z}/2\mathbb{Z}$. By a computation we find two linearly independent simultaneous eigenvectors v_0, v_1 for the Hecke operators in these spaces. For the first it can be proved that $T_\delta v_0 = 0$ for all δ. For the second we have $T_\delta v_1 = c_\delta v_1$ with $c_\delta \in \mathbb{Z}/2\mathbb{Z}$. We give some of the easily computed numbers c_δ:

$$c_3 = 1,\ c_{1+2i} = 1,\ c_{1-2i} = 0,\ c_{2+3i} = 1,\ c_{2-3i} = 1,$$

$$c_{11} = 1,\ c_{1+4i} = 1,\ c_{1-4i} = 1,\ a_{2+5i} = 0,\ c_{2-5i} = 0.$$

Consider now the cubic polynomial

$$P(x) = x^3 - ix^2 - (1+i)x - 1 - i. \tag{5.22}$$

It is irreducible over $\mathbb{Q}(i)$ and has discriminant $2(6-11i)$. Let L be the Galois extension over $\mathbb{Q}(i)$ generated by P. Its Galois group is $\mathbf{S}_3 \cong \mathbf{GL}(2, \mathbb{Z}/2\mathbb{Z})$. Let $\mathfrak{q} =< q >$ be a prime ideal of $\mathbb{Z}[i]$ unramified in L, that is $\mathfrak{q}$ has to be prime to $2(6 - 11i)$. Let

$$t_{\mathfrak{q}} := \operatorname{tr}(Fr(\mathfrak{q})) \tag{5.23}$$

be the trace of the Frobenius conjugacy class of $\mathfrak{q}$ taken in $\mathbf{GL}(2, \mathbb{Z}/2\mathbb{Z})$. Using their decomposition behaviour it can be checked that

$$t_{\mathfrak{q}} = c_q \tag{5.24}$$

for all primes $\mathfrak{q}$ of small norm. Examples of this nature are contained in Elstrodt, Grunewald, Mennicke (1982a). These suggest an interesting connection, similar to Serre's conjecture (Serre (1987)), between representations of the absolute Galois group of $\mathbb{Q}(i)$ into $\mathbf{GL}(2, \mathbb{Z}/\ell\mathbb{Z})$ and mod ℓ Hecke eigenvectors.

7.6 Spectral Theory of the Laplace Operator

This section discusses some specific facts known about the spectral theory of the Laplace operator acting on $L^2(\Gamma\backslash\mathbb{H})$, where $\Gamma < \mathbf{PSL}(2,\mathbb{C})$ is a subgroup commensurable with $\mathbf{PSL}(2,\mathcal{O})$. What we report on here all relates to the question how the discrete spectrum of this operator might look as a subset of $[0,\infty[$. First of all we already know that it is a discrete subset (with finite multiplicities) containing $\lambda_0 = 0$. Let us give two numerical examples. Tables 5.6.1 and 5.6.2 contain numerical approximations for the small eigenvalues for two of the groups $\mathbf{PSL}(2,\mathcal{O})$ where $\mathcal{O}$ is the ring of integers in $K = \mathbb{Q}(\sqrt{d_K})$.

Table 5.6.1 Eigenvalues $\neq 0$ and ≤ 525 of $-\Delta$ for $d_K = -4$

44.85247	166.640	236.60	305.57	355.70	401.5	434.6	477.8
74.1927	166.880	253.59	305.7	364.8	413.14	458.2	491.7
104.649	199.25	263.69	317.1	370.9	413.3	458.3	511.2
124.403	201.179	289.84	320.12	378.1	425.67	460.5	511.6
147.781	224.577	289.87	320.12	378.58	429.9	460.9	515.2
147.782	224.58	301.60	333.85	378.6	430.6	477.8	523.5

Table 5.6.2 Eigenvalues $\neq 0$ and ≤ 230 of $-\Delta$ for $d_K = -8$

25.4420	77.40	107.45	132.75	156.0	182.9	195.2	213.5
28.471	77.45	107.71	134.13	160.9	189.2	195.3	213.6
45.097	79.25	111.23	136.18	165.9	189.4	201.0	214.9
55.056	95.78	111.36	143.36	166.5	189.9	201.5	226.7
63.150	104.25	120.01	150.26	168.5	190.3	211.5	228.6
70.205	104.31	132.66	150.33	174.2	192.3	212.5	229.7

These tables can amongst others be found in Grunewald, Huntebrinker (1996). The regular appearence of the eigenvalues seems surprising. In fact we shall establish Weyl's distribution law in Section 8.9 in case Γ is a congruence subgroup. For $\Gamma < \mathbf{SL}(2,\mathbb{C})$ or $\Gamma < \mathbf{PSL}(2,\mathbb{C})$ a cofinite group we define $\lambda_1(\Gamma)$ to be the smallest positive eigenvalue of $-\Delta$ acting on $L^2(\Gamma\backslash\mathbb{H})$. In the tables above λ_1 appears to be quite large. Selberg's conjecture says that

$$\lambda_1(\Gamma) \geq 1 \tag{6.1}$$

whenever Γ is a congruence group with respect to one of the $\mathbf{PSL}(2,\mathcal{O})$. This conjecture is a special case of the so called general Ramanujan–Petersson conjectures, see Satake (1966) and Borel, Casselman (1979) for further developments. It can only be properly understood when automorphic representations are considered, see Gelbart (1975) and Vignéras (1985) for explanation. Later in this section we shall prove the following.

Theorem 6.1. *If $\Gamma < \mathbf{PSL}(2,\mathbb{C})$ is a congruence subgroup with respect to one of the $\mathbf{PSL}(2,\mathcal{O})$ then*

$$\lambda_1(\Gamma) \geq \frac{3}{4}. \tag{6.2}$$

The proof which is reported on later in this section consists of an interesting analysis of certain Poincaré series, see Definition 6.5. These give (roughly speaking) a Hilbert space basis of $L^2(\Gamma\backslash\mathbb{H})$ and also a corresponding simple matrix expression for the Laplace operator. The technique originates from Selberg (1965) where it was used for congruence groups acting on 2-dimensional hyperbolic space. The 3-dimensional case is contained in Sarnak (1983). An n-dimensional version is worked out in Elstrodt, Grunewald, Mennicke (1990) and Cogdell, Li, Piatetski-Shapiro, Sarnak (1991). The bound (6.2) has recently been improved to $\lambda_1(\Gamma) \geq 171/196$ by Luo, Rudnick, Sarnak (1995) by other methods.

For specific congruence groups the bound (6.2) can, by elementary means considerably be improved.

Proposition 6.2. *We have the following lower bounds for the smallest positive eigenvalue of the Laplace operator.*

$$\lambda_1\left(\mathbf{PSL}(2,\mathbb{Z}[i])\right) > \frac{2\pi^2}{3}, \qquad \lambda_1\left(\mathbf{PSL}\left(2,\mathbb{Z}\left[\frac{1+\sqrt{-3}}{2}\right]\right)\right) > \frac{32\pi^2}{27},$$

$$\lambda_1\left(\mathbf{PSL}(2,\mathbb{Z}[\sqrt{-2}])\right) > \frac{\pi^2}{4}, \qquad \lambda_1\left(\mathbf{PSL}\left(2,\mathbb{Z}\left[\frac{1+\sqrt{-7}}{2}\right]\right)\right) > \frac{256\pi^2}{735}.$$

These bounds are proved in Elstrodt, Grunewald, Mennicke (1989) and also Stramm (1994). They use an argument of Roelcke (1956a) and rely on a detailed analysis of a fundamental domain of the group in question.

The hypothesis in Theorem 6.1 that Γ should be a congruence subgroup cannot be omitted, as is shown by the following result.

Proposition 6.3. *Let $\Gamma < \mathbf{PSL}(2,\mathbb{C})$ be a cofinite subgroup and $\varphi : \Gamma \to \mathbb{Z}$ a surjective homomorphism. Define $\Gamma_\ell := \varphi^{-1}(\ell\mathbb{Z})$ for $\ell \in \mathbb{N}$. Then there is a constant $\kappa > 0$ so that*

$$\lambda_1(\Gamma_\ell) \leq \begin{cases} \dfrac{\kappa}{\ell} & \textit{in general,} \\[2ex] \dfrac{\kappa}{\ell^2} & \textit{if } \varphi \textit{ vanishes on parabolics} \end{cases} \tag{6.3}$$

where φ is said to vanish on parabolics if the induced homomorphism $\varphi : \Gamma^{\mathrm{ab}} \to \mathbb{Z}$ satisfies $\varphi(\alpha_\Gamma(\mathbf{U}(\Gamma)) = 0$ (see (5.12)).

This result is contained in Klingholz (1996). The proof uses a simple comparison of Raleigh quotients on $\Gamma\backslash\mathbb{H}$ and $\Gamma_\ell\backslash\mathbb{H}$. A similar but somewhat weaker result can be derived using the Selberg trace formula, see Chavel (1984), Chapter 11. The existence of non-trivial homomorphims $\varphi : \Gamma \to \mathbb{Z}$ for subgroups $\Gamma < \mathbf{PSL}(2, \mathcal{O})$ of finite index is implied by Theorem 5.1. The bound given in the second case of (6.3) is rather sharp since Cheeger's isoperimetric inequality (see Chavel (1984)) implies that there is a $\kappa_1 > 0$ so that $\lambda_1(\Gamma_\ell) \geq \kappa_1/\ell^2$ for all $\ell \in \mathbb{N}$.

The eigenvalue 1 for the Laplace operator is of particular importance when congruence subgroups $\Gamma < \mathbf{SL}(2, \mathbb{C})$ are considered. Eigenfunctions with this eigenvalue for certain congruence groups were described by Maaß (1949b). The construction is similar to that of modular forms and functions for congruence subgroups of $\mathbf{SL}(2, \mathbb{Z})$ using Θ-series, see Hecke (1927) and Vignéras (1985). Another construction of infinitely many subgroups of the group $\mathbf{PSL}(2, \mathbb{Z}[\sqrt{-2}])$ with eigenvalue 1 is contained in Hetrodt (1994), (1996).

We start off with a quadratic extension field $K \subset L \subset \mathbb{C}$ of our imaginary quadratic number field $K = \mathbb{Q}(\sqrt{d_K})$. We write $\mathcal{O}_L$ for its ring of integers and σ for the nontrivial element of the Galois group of L over K. Let further $N_{L/\mathbb{Q}}$, $N_{L/K}$, $N_{K/\mathbb{Q}}$ denote the norms of the field extensions given in the index. We have $N_{L/K}(x) = x\sigma(x)$ for $x \in L$. The notations for the different and the discriminant ideals are $\mathfrak{D}_{L/K} \subset \mathcal{O}_L$, $\mathfrak{d}_{L/K} \subset \mathcal{O}$. By Dirichlet's unit theorem the unit group $\mathcal{O}_L^*$ of $\mathcal{O}_L$ is the direct product of a finite and an infinite cyclic group. If $\mathfrak{C}$ is a non-zero ideal of $\mathcal{O}_L$ we write

$$\mathcal{O}_L^*(\mathfrak{C}) := \{\, x \in \mathcal{O}_L^* \quad | \quad x - 1 \in \mathfrak{C} \,\}$$

for the corresponding congruence subgroup. It is of finite index in $\mathcal{O}_L^*$ and may be cyclic. We further put $\mathcal{O}_L^1(\mathfrak{C}) := \{\, x \in \mathcal{O}_L^*(\mathfrak{C}) \ \mid \ N_{L/K}(x) = 1 \,\}$. For a non-zero $\omega \in \mathcal{O}$, non-zero ideals $\mathfrak{A}$, $\mathfrak{C} \subset \mathfrak{A}\mathfrak{D}_{L/K}$ and $A \in \mathcal{O}_L$ we define the representation number

$$a(A, \mathfrak{A}, \mathfrak{C}, \omega) := \left|\{\, \mu \in \mathcal{O}_L \quad | \quad \mu - A \in \mathfrak{A}\mathfrak{D}_{L/K},\ N_{L/K}(\mu) = \omega \,\}/\mathcal{O}_L^1(\mathfrak{C})\right|.$$

An exercise in algebraic number theory shows that this is finite.

Proposition 6.4. *Choose non-zero ideals $\mathfrak{A}$, $\mathfrak{C} \subset \mathcal{O}_L$, $\mathfrak{b} \subset \mathcal{O}$ with $\mathfrak{C} \subset \mathfrak{A}\mathfrak{b}\mathfrak{D}_{L/K}$. Define for $A \in \mathcal{O}_L$ and $0 \neq \omega \in \mathcal{O}$: $a(\omega) = a(A, \mathfrak{A}\mathfrak{b}, \mathfrak{C}, \omega)$. Suppose that $\mathcal{O}_L^*(\mathfrak{C}) = \mathcal{O}_L^1(\mathfrak{C})$ is a cyclic group generated by $E_\mathfrak{C}$ with $|E_\mathfrak{C}| > 1$. Assume that the following ideal is principal*

$$\mathfrak{A}\,\sigma(\mathfrak{A})\,\mathfrak{b}\,\mathfrak{d}_{L/K} = \left\langle \frac{\alpha}{\sqrt{d_K}} \right\rangle \qquad (\alpha \in \mathcal{O}).$$

Define for $P = z + rj \in \mathbb{H}$

$$f(A, \mathfrak{A}, \mathfrak{b}, \mathfrak{C}, P) := \delta(A, \mathfrak{A}, \mathfrak{b}) \log(|E_{\mathfrak{C}}|)\, r + \sum_{\substack{\omega \in \mathcal{O} \\ \omega \neq 0}} a(\omega)\, r\, K_0\left(\frac{4\pi|\omega|r}{|\alpha|}\right) e^{2\pi i \langle \frac{\omega}{\alpha}, \bar{z} \rangle} \tag{6.4}$$

where

$$\delta(A, \mathfrak{A}, \mathfrak{b}) = \begin{cases} 1 & if\ A \in \mathfrak{A}\,\mathfrak{b}\,\mathfrak{D}_{L/K}, \\ 0 & if\ A \notin \mathfrak{A}\,\mathfrak{b}\,\mathfrak{D}_{L/K}. \end{cases}$$

Then the series (6.4) and all its derivatives converge locally uniformly on $\mathbb{H}$. It defines a C^∞-function which is invariant under elements of the full congruence subgroup $\Gamma(\mathfrak{b}\mathfrak{d}_{L/K}) < \mathbf{SL}(2, \mathcal{O})$ and satisfies $-\Delta f = f$.

Proof. The convergence statement and the differential equation follow from the usual properties of the Bessel function, see Section 3.3. The invariance under the congruence group is implied by a Θ-transformation formula, similarly to Hecke (1927). For the details of the straightforward computation see Maaß (1949b). □

Let $\Gamma < \mathbf{PSL}(2, \mathcal{O})$ be a congruence subgroup and define $\mathrm{Eig}_\Gamma(-\Delta, \lambda)$ ($\lambda \in \mathbb{R}$) to be the λ-eigenspace for the operator $-\Delta$ on $L^2(\Gamma \backslash \mathbb{H})$. By a process similar to (5.16) Hecke operators T_δ ($0 \neq \delta \in \mathcal{O}$) acting on $\mathrm{Eig}_\Gamma(-\Delta, \lambda)$ can be defined. Assume that $\mathcal{O}$ has class number 1. Then the simultaneous eigenvectors for these operators give rise to automorphic representations and Langlands L-functions, see Gelbart (1975). The set of L-functions arising from the spaces $\mathrm{Eig}_\Gamma(-\Delta, 1)$ are conjectured to be precisely the Artin L-functions corresponding to irreducible 2-dimensional complex representations of the absolute Galois group of the imaginary quadratic number field K, see Clozel (1990). Those occurring from the subspaces generated by the functions constructed in Proposition 6.4 correspond to representations with dihedral image.

Finally for this section we report on the proof of Theorem 6.1. Clearly the result needs only to be proved for the full congruence groups

$$\Gamma(\mathfrak{a}) := \left\{ \begin{pmatrix} a & b \\ c & d \end{pmatrix} \in \mathbf{SL}(2, \mathcal{O}) \;\middle|\; \begin{pmatrix} a & b \\ c & d \end{pmatrix} \equiv \begin{pmatrix} 1 & 0 \\ 0 & 1 \end{pmatrix} \bmod \mathfrak{a} \right\}$$

in $\mathbf{SL}(2, \mathcal{O})$ ($\mathfrak{a}$ a non-zero ideal in the ring of integers $\mathcal{O}$) introduced in Section 4. Note that ∞ is a cusp of $\Gamma(\mathfrak{a})$. The lattice of translations corresponding to the unipotent part $\Gamma(\mathfrak{a})'_\infty$ of the stabilizer of ∞ is $\mathfrak{a} \subset \mathbb{C}$. For $u, v \in \mathbb{C}$ we as usually define the scalar product

$$< u, v > = \frac{u\bar{v} + \bar{u}v}{2} = \mathrm{Re}(u\bar{v}).$$

We write $\mathfrak{a}^\circ = \{ z \in \mathbb{C} \mid \ < z, a > \in \mathbb{Z} \text{ for all } a \in \mathfrak{a} \}$ for the dual lattice of $\mathfrak{a}$. In the following we use the abbreviation

$$e(z) := e^{2\pi i z} \qquad (z \in \mathbb{C}). \tag{6.5}$$

Our notation here is similar to that of Section 3.2.

Definition 6.5. For a non-zero ideal $\mathfrak{a} \subset \mathcal{O}$ with $m \in \mathfrak{a}^\circ$, $P = z(P) + r(P)j \in \mathbb{H}$ and $s \in \mathbb{C}$ we define

$$U_m(P,s) = \sum_{A \in \Gamma(\mathfrak{a})'_\infty \backslash \Gamma(\mathfrak{a})} r(AP)^s \; e(i|m|r(AP) + < m, z(AP) >). \tag{6.6}$$

The $U_m(P,s)$ are again Poincaré-type series. For $m = 0$ we see one of the Eisenstein series. Once the series converges it defines a $\Gamma(\mathfrak{a})$-invariant function on $\mathbb{H}$. The following consists of simple exercises.

Lemma 6.6. *Let $\mathfrak{a} \subset \mathcal{O}$ be a non-zero ideal and $m \in \mathfrak{a}^\circ$ and $P \in \mathbb{H}$. Then the following hold.*

(1) The series $U_m(P,s)$ converges absolutely for $s \in \mathbb{C}$, $\operatorname{Re} s > 2$. The convergence is uniform on every compact subset of

$$\mathbb{H} \times \{ s \in \mathbb{C} \mid \operatorname{Re} s > 2 \}.$$

(2) For every $s \in \mathbb{C}$ with $\operatorname{Re} s > 2$ the function $U_m(\cdot, s)$ satisfies the differential equation

$$\Delta U_m(P,s) = 2\pi|m|\,(1-2s)\,U_m(P,s+1) - s(2-s)\,U_m(P,s).$$

(3) If $m \neq 0$ and $s \in \mathbb{C}$ with $\operatorname{Re} s > 2$ then $U_m(\cdot, s) \in L^2(\Gamma(\mathfrak{a})\backslash\mathbb{H})$.

Let now $f \in L^2(\Gamma(\mathfrak{a})\backslash\mathbb{H}) \cap C^2(\Gamma(\mathfrak{a})\backslash\mathbb{H})$ be an eigenfunction of the Laplace operator that is $-\Delta f = \lambda f$ with some $\lambda \in [0, \infty[$. We wish to calculate the scalar product

$$< f, U_m(\cdot, \bar{s}) > = \int_{\Gamma(\mathfrak{a})\backslash\mathbb{H}} f \, \overline{U_m(\cdot, \bar{s})} \, dv. \tag{6.7}$$

To do this we use the Fourier expansion of f from Section 3.3. We put

$$t = 1 + \sqrt{1-\lambda}, \qquad \rho = \sqrt{1-\lambda}.$$

We then find $a(f,\mu) \in \mathbb{C}$ for $\mu \in \mathfrak{a}^\circ$ such that f can be expressed as

$$f(z + rj) = a(f,0)\, r^{2-t} + \sum_{\substack{m \in \mathfrak{a}^\circ \\ m \neq 0}} a(f,m)\; r\; K_\rho(2\pi|m|r)\; e^{2\pi i < m,z >}. \tag{6.8}$$

Proposition 6.7. *Assume that* $f \in L^2(\Gamma(\mathfrak{a})\backslash\mathbb{H}) \cap C^2(\Gamma(\mathfrak{a})\backslash\mathbb{H})$ *satisfies* $-\Delta f = \lambda f$ *and (6.8) is the Fourier expansion of* f*. For* $m \in \mathfrak{a}^\circ \setminus \{0\}$ *and* $s \in \mathbb{C}$ *with* $\mathrm{Re}\, s > 2$ *we have*

$$< f, U_m(\cdot, \bar{s}) >= \epsilon\sqrt{\pi}\ |\mathfrak{a}|\ (4\pi|m|)^{1-s}\ \frac{\Gamma(s-t)\,\Gamma(s+t-2)}{\Gamma(s-\frac{1}{2})}\ a(f,m),$$

where $|\mathfrak{a}|$ *is the covolume of the lattice* $\mathfrak{a} \subset \mathbb{C}$ *and* $\epsilon = |\Gamma(\mathfrak{a}) \cap \{\pm I\}|$.

Proof. We choose a fundamental parallelogram $\mathcal{P}$ for the lattice $\mathfrak{a} \subset \mathbb{C}$. Then $\mathcal{P}\times]0,\infty[$ is a fundamental domain for $\Gamma(\mathfrak{a})'_\infty$ acting on $\mathbb{H}$. Putting $\Gamma = \Gamma(\mathfrak{a})$ we find by unfolding the Poincaré series

$$\begin{aligned}
< f, U_m(\cdot, \bar{s}) > &= \int_{\Gamma\backslash\mathbb{H}} f(P)\ \overline{U_m(P,\bar{s})}\ dv(P) \\
&= \int_{\Gamma\backslash\mathbb{H}} f(P) \sum_{A\in\Gamma'_\infty\backslash\Gamma} r(AP)^s\ e^{-2\pi|m|\,r(AP)-2\pi i<z(AP),m>}\ dv(P) \\
&= \sum_{A\in\Gamma'_\infty\backslash\Gamma} \int_{\Gamma\backslash\mathbb{H}} f(P)\ r(AP)^s\ e^{-2\pi|m|\,r(AP)-2\pi i<z(AP),m>}\ dv(P) \\
&= \epsilon \int_{\mathcal{P}\times]0,\infty[} f(P)\ r^s\ e^{-2\pi|m|\,r(P)-2\pi i<z(P),m>}\ dv(P).
\end{aligned}$$

Putting in the Fourier expansion (6.8) for f and carrying out the integration over $\mathcal{P}$ we find

$$< f, U_m(\cdot, \bar{s}) >= \epsilon\ a(f,m)\ |\mathfrak{a}| \int_0^\infty r^{s-1}\ K_\rho(2\pi|m|r)\ e^{-2\pi|m|r}\ \frac{dr}{r}.$$

Now by Gradshteyn, Ryzhik (1994), 6.621 we have

$$\int_0^\infty r^s\ e^{-r}\ K_\nu(r)\ \frac{dr}{r} = \frac{\sqrt{\pi}}{2^s}\frac{\Gamma(s+\nu)\ \Gamma(s-\nu)}{\Gamma(s+\frac{1}{2})} \tag{6.9}$$

which holds for $\mathrm{Re}\,(s) > |\mathrm{Re}\,(\nu)|$. Using this formula we finish the proof. □

Our aim is now to calculate the scalar product of two Poincaré series

$$< U_m(\cdot, s), U_n(\cdot, \bar{t}) >= \int_{\Gamma(\mathfrak{a})\backslash\mathbb{H}} U_m(P,s)\,\overline{U_n(P,\bar{t})}\ dv(P). \tag{6.10}$$

Before proceeding we have to make the following definition.

Definition 6.8. For a non-zero ideal $\mathfrak{a} \subset \mathcal{O}$ and $m, n \in \mathfrak{a}^\circ\backslash\{0\}$ and $c \in \mathfrak{a}\backslash\{0\}$ we define

$$\mathbf{KS}(m,n,c) = \sum_{\substack{x,y\in\mathcal{O}/c\mathfrak{a} \\ xy\equiv 1 \bmod c\mathfrak{a} \\ x\equiv y\equiv 1 \bmod \mathfrak{a}}} e\left(\left\langle \frac{x}{c}, m\right\rangle + \left\langle \frac{y}{c}, n\right\rangle\right). \tag{6.11}$$

This exponential sum is called a Kloosterman sum. The Dirichlet series

$$\mathbf{Z}(m,n,s) = \sum_{c\in\mathfrak{a}\setminus\{0\}} \frac{\mathbf{KS}(m,n,c)}{|c|^{2s}} \tag{6.12}$$

is called the Linnik–Selberg series.

The Kloosterman sums (6.11) satisfy the obvious bound

$$|\mathbf{KS}(m,n,c)| \leq N(\mathfrak{a})\ |c|^2, \tag{6.13}$$

hence the series $\mathbf{Z}(m,n,s)$ converges absolutely for $\mathrm{Re}\, s > 2$. Using the famous bound for the size of Kloosterman sums over finite fields proved by Weil (1948) we can sharpen this to the following

Theorem 6.9. *Let $\mathfrak{a} \subset \mathcal{O}$ be a non-zero ideal and let $m,\ n \in \mathfrak{a}^\circ$ be non-zero elements. Then the Linnik–Selberg series $\mathbf{Z}(m,n,s)$ converges absolutely for $\mathrm{Re}\, s > \frac{3}{2}$.*

In the proof the following simple observation on Dirichlet series plays a role, we leave its proof as an exercise (or refer to Elstrodt, Grunewald, Mennicke (1990), Lemma 7.15 for a proof).

Lemma 6.10. *Let*

$$D(s) = \sum_{N=1}^{\infty} \frac{a_N}{N^s}$$

be a Dirichlet series so that there are constants $\kappa > 0$ and $r \in \mathbb{R}$ and a finite set of primes B so that

$$|a_N| \leq \prod_{\substack{p\|N\\ p\notin B}} \kappa p^{r-\frac{1}{2}} \cdot \prod_{\substack{p\|N\\ p\in B}} \kappa p^{r} \cdot \prod_{p^2|N} \kappa p^{g_p r}$$

for all natural numbers with the prime number decomposition $N = \prod_p p^{g_p}$. Then $D(s)$ converges for $\mathrm{Re}\, s > r + \frac{1}{2}$.

Proof of Theorem 6.9. We start with the fixed non-zero ideal $\mathfrak{a} \subset \mathcal{O}$ and let $m,\ n \in \mathfrak{a}^\circ$ be fixed non-zero elements. If $\mathfrak{b}$ is a non-zero ideal of $\mathcal{O}$ we define $\Theta(\mathfrak{b})$ to be the kernel of the reduction homomorphism $(\mathcal{O}/\mathfrak{ba})^* \to (\mathcal{O}/\mathfrak{a})^*$. Given two characters $\varphi,\ \psi$ of the additive group $\Theta(\mathfrak{b})$ we generalize the definition of the Kloosterman sums (6.11) to

$$\mathbf{KS}(\varphi,\psi,\mathfrak{b}) := \sum_{\substack{x,\ y\in\Theta(\mathfrak{b})\\ xy\equiv 1 \bmod \mathfrak{ab}}} \varphi(x)\,\psi(y). \tag{6.14}$$

If $\mathfrak{b} = \wp_1^{e_1}...\wp_r^{e_r}$ is the prime ideal decomposition of $\mathfrak{b}$ then the Chinese remainder theorem implies a direct product decomposition

$$\Theta(\mathfrak{b}) \cong \Theta(\wp_1^{e_1}) \times ... \times \Theta(\wp_r^{e_r}) \tag{6.15}$$

induced by reduction. The additive characters φ, ψ decompose accordingly: $\varphi = \varphi_1 \times ... \times \varphi_r$ and $\psi = \psi_1 \times ... \times \psi_r$. It is easy to see that we have the product decomposition

$$\mathbf{KS}(\varphi, \psi, \mathfrak{b}) = \mathbf{KS}(\varphi_1, \psi_1, \wp_1^{e_1}) \cdot \ldots \cdot \mathbf{KS}(\varphi_r, \psi_r, \wp_r^{e_r}). \tag{6.16}$$

If $\wp$ is a prime ideal not dividing $\mathfrak{a}$ we have an isomorphism

$$\Theta(\wp) \cong \mathcal{O}/\wp \tag{6.17}$$

also given by reduction. For $0 \neq c \in \mathcal{O}$ we put $\mathfrak{b} =< c >$ and consider the additive characters of $\Theta(\mathfrak{b})$ given by

$$\varphi_m(x) := \left\langle \frac{x}{c}, m \right\rangle, \qquad \psi_n(y) := \left\langle \frac{y}{c}, n \right\rangle.$$

If $\wp$ is a prime ideal which does not divide $\mathfrak{a}$ and which divides $\mathfrak{b}$ exactly to the first power we get through the isomorphisms (6.15), (6.17) additive characters $\tilde{\varphi}$, $\tilde{\psi}$ of $\mathcal{O}/\wp$. For all but finitely many $\wp$ these will be both non-trivial. The exponential sum

$$\mathbf{KS}(\tilde{\varphi}, \tilde{\psi}, \wp) = \sum_{\substack{x,\, y \in \mathcal{O}/\wp \\ xy = 1 \bmod \mathfrak{a}\wp}} \tilde{\varphi}(x)\, \tilde{\psi}(y)$$

is estimated by Weil (1948) as

$$|\mathbf{KS}(\tilde{\varphi}, \tilde{\psi}, \wp)| \leq 2N(\wp)^{\frac{1}{2}}.$$

This estimate together with a the trivial estimate (6.13) for the remaining factors of (6.16) put into Lemma 6.10 implies the Theorem. □

The Linnik–Selberg series can even be meromorphically continued into all of $\mathbb{C}$, see Cogdell, Li, Piatetski-Shapiro, Sarnak (1992) for this result.

Proposition 6.11. *Suppose that s, t are complex numbers with $\operatorname{Re} t > \operatorname{Re} s > 2$, and $m, n \in \mathfrak{a}^\circ$ are both non-zero. Then we have*

$$\begin{aligned} < U_m(\cdot, s), U_n(\cdot, \bar{t}) > &= (4\pi|m|)^{2-s-t}\ c(m,n)\ |\mathfrak{a}|\ \Gamma(s+t-2) \\ &\quad + 2^{3-2t}\pi^{s-t+\frac{3}{2}}|n|^{s-t}\ \frac{\Gamma(t+s-2)\ \Gamma(t-s)}{\Gamma(s)\ \Gamma(t-\frac{1}{2})}\ \mathbf{Z}(m,n,s) \\ &\quad + \sum_{c \in \mathfrak{a}\setminus\{0\}} \frac{\mathbf{KS}(m,n,c)}{|c|^{2s}}\ R_{m,n}(s,t,c), \end{aligned}$$

where

$$c(m,n) := |\{\ a \in \mathcal{O}^*\ \ |\ \ a - 1 \in \mathfrak{a},\ \bar{a}^2 m = n\ \}|.$$

The function $R_{m,n}(s,t,c)$ satisfies

$$|R_{m,n}(s,t,c)| \leq M|c|^{-2}$$

with some constant M depending only on s, t, m, n provided $\operatorname{Re} t > \operatorname{Re} s + 1$.

Proof. We choose a fundamental parallelogram $\mathcal{P}$ for the lattice $\mathfrak{a} \subset \mathbb{C}$. Putting $\Gamma = \Gamma(\mathfrak{a})$ we find as in the proof of Proposition 6.7 by unfolding the second Poincaré series

$$\begin{aligned} < U_m(\cdot, s), U_n(\cdot, \bar{t}) > &= \int_{\Gamma \backslash \mathbb{H}} U_m(P,s)\, \overline{U_n(P,\bar{t})}\, dv(P) \\ &= \int_0^\infty r^t\, e^{-2\pi|n|r} \int_{\mathcal{P}} U_m(P,s)\, e^{-2\pi i <z,n>}\, dz\, \frac{dr}{r^3}. \end{aligned}$$

The inner integral can be written as

$$\begin{aligned} &\int_{\mathcal{P}} U_m(P,s)\, e^{-2\pi i <z,n>}\, dz \\ &= \sum_{A \in \Gamma(\mathfrak{a})'_\infty \backslash \Gamma(\mathfrak{a})} \int_{\mathcal{P}} r(AP)^s\, e^{-2\pi|m|r(AP) + 2\pi i(<z(AP),m> - <z,n>)}\, dz. \end{aligned}$$

We shall decompose this sum into two parts. To do this we put

$$\Gamma_w := \left\{ \begin{pmatrix} a & b \\ c & d \end{pmatrix} \in \Gamma(\mathfrak{a}) \;\middle|\; c \neq 0 \right\}.$$

We furthermore define

$$H_\infty(r) := \sum_{A \in \Gamma(\mathfrak{a})'_\infty \backslash \Gamma(\mathfrak{a})_\infty} \int_{\mathcal{P}} r(AP)^s\, e^{-2\pi|m|r(AP) + 2\pi i(<z(AP),m> - <z,n>)}\, dz,$$

$$H_w(r) := \sum_{A \in \Gamma(\mathfrak{a})'_\infty \backslash \Gamma_w} \int_{\mathcal{P}} r(AP)^s\, e^{-2\pi|m|r(AP) + 2\pi i(<z(AP),m> - <z,n>)}\, dz.$$

We get

$$< U_m(\cdot, s), U_n(\cdot, \bar{t}) >= \int_0^\infty r^t\, e^{-2\pi|n|r}\, H_\infty(r)\, \frac{dr}{r^3} + \int_0^\infty r^t\, e^{-2\pi|n|r}\, H_w(r)\, \frac{dr}{r^3}.$$

The first integral gives the first summand of the right-hand side of the formula in our proposition by a straightforward computation. The further contributions come from the second integral. We skip the computation here. For details see Sarnak (1983) or Elstrodt, Grunewald, Mennicke (1990). □

Proof of Theorem 6.1. As already remarked it suffices to prove the statement for full congruence groups $\Gamma(\mathfrak{a}) < \mathbf{SL}(2, \mathcal{O})$. Assume that there is an eigenvalue λ_1 of the self-adjoint extension $-\tilde{\Delta}$ of the Laplace operator on $L^2(\Gamma(\mathfrak{a})\backslash\mathbb{H})$ with $0 < \lambda_1 < 3/4$. Note that by Theorem 6.2.13 the operators $-\Delta$ and $-\tilde{\Delta}$ have the same sets of eigenvalues. We write $\operatorname{Eig}_{\Gamma(\mathfrak{a})}(-\tilde{\Delta}, \lambda_1)$ for the corresponding eigenspace and $v_1, \ldots, v_p$ for an orthonormal basis of this space. Then

$$2 > t_1 := 1 + \sqrt{1 - \lambda_1} > \frac{3}{2}.$$

This implies that t_1 belongs to the half-plane of absolute convergence of the Linnik–Selberg series. Let $\rho(-\tilde{\Delta})$ be the resolvent set of $-\tilde{\Delta}$ and

$$R_\lambda := (-\tilde{\Delta} - \lambda)^{-1} \qquad (\lambda \in \rho(-\tilde{\Delta}))$$

the corresponding resolvent operator. We already know that $R_{s(2-s)}$ is holomorphic for $\mathrm{Re}\, s > 2$ and meromorphic for $\mathrm{Re}\, s > 1$ with at most finitely many poles of order 1 in the interval $]1, 2[$. We also have from Proposition 6.6

$$U_m(P, s) = 2\pi|m|(2s - 1)\, R_{s(2-s)}\, U_m(P, s + 1). \tag{6.18}$$

Here $U_m(P, s + 1)$ is holomorphic in s for $\mathrm{Re}\, s > 1$, whereas the resolvent operator is meromorphic in this half-plane. This implies that $U_m(P, s)$ is meromorphic for $\mathrm{Re}\, s > 1$. In particular $U_m(P, s)$ has at most a simple pole at $s = t_1$. Putting $\lambda = s(2 - s)$ we have

$$\lambda - \lambda_1 = (s - t_1)(2 - (s + t_1)).$$

By Kato (1966), chapter 5, paragraph 3, 5 we infer that

$$\operatorname*{Res}_{\lambda=\lambda_1} R_\lambda f = \mathrm{pr}_{\lambda_1}(f)$$

for all $f \in L^2(\Gamma(\mathfrak{a})\backslash\mathbb{H})$, where pr_{λ_1} stands for the orthogonal projection of $L^2(\Gamma(\mathfrak{a})\backslash\mathbb{H})$ onto $\mathrm{Eig}_{\Gamma(\mathfrak{a})}(-\tilde{\Delta}, \lambda_1)$. Computing the residue of $U_m(P, s)$ using (6.18) we find

$$\operatorname*{Res}_{s=t_1} U_m(P, s) = \frac{\pi|m|(2t_1 - 1)}{1 - t_1} \sum_{j=1}^{p} < U_m(\cdot, t_1 + 1), v_j > \ v_j(P).$$

The functions v_i have Fourier expansions as described in (6.8), see Theorem 3.3.2. Using the formula in Proposition 6.7 we find

$$\operatorname*{Res}_{s=t_1} U_m(P, s) = -\epsilon \sqrt{\pi}\, |\mathfrak{a}|\, (4\pi|m|)^{1-t_1} \frac{\Gamma(2t_1 - 2)}{\Gamma(t_1 - \frac{1}{2})} \sum_{j=1}^{p} \overline{a(v_j, m)}\, v_j(P). \tag{6.19}$$

We now fix $t \in \mathbb{C}$ with $\mathrm{Re}\, t > 3$ and fix also $n \in \mathfrak{a}^\circ$ and get from (6.19) and Proposition 6.7

$$< \operatorname*{Res}_{s=t_1} U_m(\cdot, s), U_n(\cdot, \bar{t}) >= -\epsilon^2 \pi\, |\mathfrak{a}|^2\, (4\pi|m|)^{1-t_1} (4\pi|n|)^{1-t}$$
$$\cdot \frac{\Gamma(2t_1 - 2)\Gamma(t - t_1)\Gamma(t + t_1 - 2)}{\Gamma(t_1 - \frac{1}{2})\Gamma(t - \frac{1}{2})} \sum_{j=1}^{p} \overline{a(v_j, m)}\, a(v_j, n).$$

In particular we have for $m = n$

$$
\begin{aligned}
&< \operatorname*{Res}_{s=t_1} U_m(\cdot, s), U_m(\cdot, \bar{t}) >= -\epsilon^2 \pi \, |\mathfrak{a}|^2 \, (4\pi |m|)^{2-t-t_1} \\
&\cdot \frac{\Gamma(2t_1 - 2)\Gamma(t - t_1)\Gamma(t + t_1 - 2)}{\Gamma(t_1 - \frac{1}{2})\Gamma(t - \frac{1}{2})} \sum_{j=1}^{p} |a(v_j, m)|^2 .
\end{aligned} \tag{6.20}
$$

Since λ_1 is an eigenvalue of $-\tilde{\Delta}$ we may choose m so that the right-hand side of the last equation is non-zero. Then $< U_m(\cdot, s), U_m(\cdot, \bar{t}) >$ has a pole at $s = t_1$. On the other hand Theorem 6.9 and Proposition 6.11 imply that the latter scalar product is holomorphic for $\operatorname{Re} s > 3/2$. □

7.7 Notes and Remarks

A subject not touched here is the structure of the manifolds $\Gamma \backslash \mathbb{H}$ where $\Gamma < \mathbf{PSL}(2, \mathbb{C})$ is a torsion free subgroup commensurable with one of the $\mathbf{PSL}(2, \mathcal{O})$. To describe an interesting example we put $\zeta = -\frac{1}{2} + \frac{\sqrt{-3}}{2}$ and write $\mathcal{O}_{-3} = \mathbb{Z}[\zeta]$ for the ring of Eisenstein integers. The subgroup

$$
\Gamma_8 = \left\langle \begin{pmatrix} 1 & 1 \\ 0 & 1 \end{pmatrix}, \begin{pmatrix} 1 & 0 \\ \zeta & 1 \end{pmatrix} \right\rangle < \mathbf{PSL}(2, \mathcal{O}_{-3})
$$

is known to be torsion free and of index 12 (see Riley (1975)). The quotient manifold $\Gamma_8 \backslash \mathbb{H}$ was found out to be homeomorphic to the complement of the figure-8 knot in the 3-sphere S^3 by Riley (1975). See also Thurston (1978) for comments on this phenomenon. The group Γ_8 is known to be a congruence subgroup, in particular $P\Gamma(4) < \Gamma_8$. For general reasons we have $\Gamma_8^{\mathrm{ab}} = \mathbb{Z}$. The eigenvalues produced by the Hecke operators (see Section 6.5) acting on Γ_8^{ab} can be described by a certain Hecke character of $\mathbb{Q}(\sqrt{-3})$.

Many further examples of torsion free subgroups Γ of one of the $\mathbf{PSL}(2, \mathcal{O})$ with $\Gamma \backslash \mathbb{H}$ homeomorphic to a link complement in S^3 are known. See Thurston (1978), Hatcher (1983), Grunewald, Hirsch (1995). It was proved by Reid (1991) that the figure-8 knot is the only knot which arises in this way.

If Γ happens to contain torsion elements then the quotient $\Gamma \backslash \mathbb{H}$ still inherits the structure of a topological 3-manifold from $\mathbb{H}$, see Grunewald, Mennicke (1980). Under suitable circumstances it can happen that $\Gamma \backslash \mathbb{H}$ can be compactified to a closed 3-manifold $M(\Gamma)$ by the addition of finitely many points. This happens if $\Gamma = \mathbf{PSL}(2, \mathbb{Z}[i])$ or $\Gamma = \Gamma_0(\wp)$ where $\wp$ is a degree 1 prime ideal of $\mathbb{Z}[i]$. We get the following homeomorphism types of manifolds

$$
M(\mathbf{PSL}(2, \mathbb{Z}[i])) \cong S^3, \qquad M(\Gamma_0(\wp)) \cong S^3 \text{ if } N(\wp) = 2,\ 5,\ 13.
$$

If we let p be a rational prime congruent to 1 modulo 4 and $\wp_p$ be a prime ideal in $\mathbb{Z}[i]$ with $N(\wp) = p$ then we furthermore have

$$M(\Gamma_0(\wp_{17})) \cong L(4,1), \quad M(\Gamma_0(\wp_{29})) \cong L(21,8), \quad M(\Gamma_0(\wp_{37})) \cong L(3,1)$$

where $L(m,n)$ is the lens-space with invariants (m, n). These and more examples are explained in Grunewald, Mennicke (1980).

The principal results of Chapter 6, that is, the meromorphic continuation of the Eisenstein series and the Selberg trace formula, can in case of one of the $\mathbf{PSL}(2,\mathcal{O})$ be derived by number theoretic techniques. The Eisenstein series are treated in the next chapter. For the trace formula see Tanigawa (1977), Szmidt (1983) and Bauer (1993).

Some of the groups $\mathbf{PSL}(2,\mathcal{O})$ are subgroups of discrete groups which can be generated by hyperbolic reflections, see Chapter 10 for examples. The relationship of the $\mathbf{PSL}(2,\mathcal{O})$ to reflection groups is studied in Shvartsman (1987), Shaikheev (1987), Vinberg (1987) and Ruzmanov (1990).

Chapter 8. Eisenstein Series for PSL(2) over Imaginary Quadratic Integers

Suppose that $K = \mathbb{Q}(\sqrt{D})$ is an imaginary quadratic number field ($D \in \mathbb{Z}$, $D < 0$ and square-free) and let $\mathcal{O}$ be the ring of integers of K. We consider here the group $\mathbf{PSL}(2, \mathcal{O}) \subset \mathbf{PSL}(2, \mathbb{C})$. We already know from Chapter 7 that Γ is a discrete subgroup which is cofinite but not cocompact. We study the Eisenstein series defined in Chapter 3 in detail for the group $\mathbf{PSL}(2, \mathcal{O})$. In fact we shall establish most of the general facts proved in Section 6.1 for the Eisenstein series of general cofinite groups by direct number theoretic methods. We shall for example relate the determinant of the scattering matrix to the zeta function of the Hilbert class field of K. The control we have over the Eisenstein series will also in turn imply many interesting number theoretic results.

Most of the material of this chapter is taken from Elstrodt, Grunewald, Mennicke (1985). We shall use the notation introduced in Chapter 7, particularly in Section 7.2.

8.1 Functions Closely Related to Eisenstein Series

In the present situation, the following direct definition of Eisenstein series is suggestive.

Definition 1.1. For $\mathfrak{m} \in \mathcal{M}$, $P = z + rj \in \mathbb{H}$, $s \in \mathbb{C}$ with $\operatorname{Re} s > 1$ define

$$E_{\mathfrak{m}}(P, s) := N\mathfrak{m}^{1+s} \sum_{\substack{c,d \in K \\ <c,d> = \mathfrak{m}}} \left(\frac{r}{\|cP + d\|^2} \right)^{1+s}, \tag{1.1}$$

where the summation extends over all pairs (c, d) of generators of $\mathfrak{m}$ as an $\mathcal{O}$-module, and

$$\hat{E}_{\mathfrak{m}}(P, s) := N\mathfrak{m}^{1+s} {\sum_{c,d \in \mathfrak{m}}}' \left(\frac{r}{\|cP + d\|^2} \right)^{1+s}, \tag{1.2}$$

where the prime indicates that the summation extends over all c, $d \in \mathfrak{m}$ with $(c, d) \neq (0, 0)$. $E_{\mathfrak{m}}$ and $\hat{E}_{\mathfrak{m}}$ are called Eisenstein series for $\mathbf{PSL}(2, \mathcal{O})$.

The Eisenstein series $E_{\mathfrak{m}}$ and $\hat{E}_{\mathfrak{m}}$ depend only on the class of $\mathfrak{m}$ in $\mathcal{M}/K^*$. They converge uniformly on compact sets for $\operatorname{Re} s > 1$ and diverge for $\operatorname{Re} s \leq 1$. For $\operatorname{Re} s > 1$ the Eisenstein series (1.1), (1.2) are $\mathbf{PSL}(2,\mathcal{O})$-invariant functions that satisfy the differential equation $(-\Delta - (1-s^2))E = 0$. In Theorem 1.7 we shall relate them to the Eisenstein series defined in Chapter 3. First of all we express $\hat{E}_{\mathfrak{m}}$ as a linear combination of the $E_{\mathfrak{n}}$ for $\mathfrak{n} \in \mathcal{M}/K^*$. Here certain zeta functions come up that we define now.

Definition 1.2. For $\mathfrak{m}, \mathfrak{n} \in \mathcal{M}$ let

$$\zeta(\mathfrak{m},\mathfrak{n},s) := (N\mathfrak{m}\mathfrak{n}^{-1})^s \sum_{\lambda\in\mathfrak{m}\mathfrak{n}^{-1}}{}' N\lambda^{-s} \tag{1.3}$$

and for $\mathfrak{m}^\# \in \mathcal{M}/K^*$ define

$$\zeta(\mathfrak{m}^\#, s) := \sum_{\substack{\mathfrak{a}\in\mathfrak{m}^\# \\ \mathfrak{a}\subset\mathcal{O}}} N\mathfrak{a}^{-s}. \tag{1.4}$$

The zeta function (1.4) is the zeta function of the ideal class associated with $\mathfrak{m}^\#$. Hence (1.4) has abscissa of convergence 1, and the same is true for (1.3) by Lemma 1.3 ahead. Note that (1.3) depends only on the classes of $\mathfrak{m}$ and $\mathfrak{n}$ in $\mathcal{M}/K^*$.

Lemma 1.3. *Suppose that* $\mathfrak{m}, \mathfrak{n} \in \mathcal{M}$, $\operatorname{Re} s > 1$. *Then*

$$\zeta(\mathfrak{m},\mathfrak{n},s) = |\mathcal{O}^*|\ \zeta((\mathfrak{m}^{-1}\mathfrak{n})^\#, s) \tag{1.5}$$

where $\mathcal{O}^*$ *stands for the (finite) group of units of* $\mathcal{O}$.

Proof. Since $\zeta(\mathfrak{m},\mathfrak{n},s) = \zeta(\mathfrak{m}\mathfrak{n}^{-1},\mathcal{O},s)$ it suffices to prove the assertion for $\mathfrak{n} = \mathcal{O}$, and since (1.3) depends only on the class of $\mathfrak{m}$ in $\mathcal{M}/K^*$ we may assume that $\mathfrak{m} = \mathfrak{a} \subset \mathcal{O}$ is an integral ideal and $\mathfrak{n} = \mathcal{O}$. Now we are left to prove that $\zeta(\mathfrak{a},\mathcal{O},s) = |\mathcal{O}^*|\zeta((\mathfrak{a}^{-1})^\#, s)$. This result is known, see Borevich, Shafarevich (1966) , p. 310. □

Proposition 1.4. *For* $\mathfrak{m} \in \mathcal{M}$, $P \in \mathbb{H}$, $\operatorname{Re} s > 1$ *we have*

$$|\mathcal{O}^*|\ \hat{E}_{\mathfrak{m}}(P,s) = \sum_{\mathfrak{n}\in\mathcal{M}/K^*} \zeta(\mathfrak{m},\mathfrak{n},s+1)\, E_{\mathfrak{n}}(P,s). \tag{1.6}$$

Proof. Let $\mathfrak{n}$ run through a representative system $\mathcal{V}$ of $\mathcal{M}/K^*$. Consider a pair (γ,δ) of generators of an arbitrary element $\mathfrak{n} \in \mathcal{V}$ and an arbitrary $\lambda \in \mathfrak{m}\mathfrak{n}^{-1}$ and consider the map

$$(\lambda,(\gamma,\delta)) \mapsto (c,d) := (\lambda\gamma,\lambda\delta) \in \mathfrak{m}\oplus\mathfrak{m} \setminus \{0,0\}.$$

This map is surjective, and every (c,d) has precisely $|\mathcal{O}^*|$ different inverse images. This yields the assertion. $\square$

We choose once and for all a representative system $\mathfrak{m}_1, \dots, \mathfrak{m}_h$ of $\mathcal{M}/K^*$ and define the vectors

$$(1.7) \qquad \tilde{\mathcal{E}}(P,s) := \begin{pmatrix} E_{\mathfrak{m}_1}(P,s) \\ \vdots \\ E_{\mathfrak{m}_h}(P,s) \end{pmatrix}, \qquad \hat{\mathcal{E}}(P,s) := \begin{pmatrix} \hat{E}_{\mathfrak{m}_1}(P,s) \\ \vdots \\ \hat{E}_{\mathfrak{m}_h}(P,s) \end{pmatrix},$$

and the matrix

$$(1.8) \qquad \mathcal{A}(s) := \left(\frac{\zeta(\mathfrak{m}_i, \mathfrak{m}_k, s+1)}{|\mathcal{O}^*|} \right)_{i,k=1,\dots,h},$$

where i is the row index and k the column index. Then we arrive at the following result.

Theorem 1.5. *For $P \in \mathbb{H}$, $\operatorname{Re} s > 1$ the equality*

$$(1.9) \qquad \hat{\mathcal{E}}(P,s) = \mathcal{A}(s)\ \tilde{\mathcal{E}}(P,s)$$

holds with the cyclic matrix

$$(1.10) \qquad \mathcal{A}(s) = (\zeta((\mathfrak{m}_i^{-1}\mathfrak{m}_k)^{\#}, s+1))_{i,k=1,\dots,h}.$$

The matrix $\mathcal{A}(s)$ is unitarily equivalent to the diagonal matrix $\mathcal{D}(s)$ with the diagonal entries $L(s+1,\chi_1), \dots, L(s+1,\chi_h)$, where $\chi_1, \dots, \chi_h$ are the characters of the group $\mathcal{M}/K^*$, and where $L(s,\chi)$ is defined by

$$(1.11) \qquad L(s,\chi) = \sum_{\mathfrak{a} \subset \mathcal{O}}{}' \frac{\chi(\mathfrak{a})}{N\mathfrak{a}^s}.$$

More precisely, we have

$$(1.12) \qquad \mathcal{A}(s) = \Omega\ \mathcal{D}(s)\ \Omega^{-1}$$

with the unitary matrix

$$(1.13) \qquad \Omega = \left(\frac{1}{\sqrt{h}}\ \chi_k(\mathfrak{m}_i^{\#}) \right)_{i,k=1,\dots h}.$$

In particular,

$$(1.14) \qquad \det \mathcal{A}(s) = \prod_{k=1}^{h} L(s+1,\chi_k).$$

Proof. (1.9), (1.10) are obvious from (1.5), (1.6). Since

$$\sum_{i=1}^{h} \zeta(\mathfrak{m}_i^{\#}, s+1)\, \chi_k(\mathfrak{m}_i^{\#}) = \sum_{\mathfrak{a} \subset \mathcal{O}} \frac{\chi_k(\mathfrak{a}^{\#})}{N\mathfrak{a}^{s+1}} = L(s+1, \chi_k), \tag{1.15}$$

the remaining assertions are just special cases of the following Lemma 1.6. □

Lemma 1.6. *Let $G = \{\sigma_1, \ldots, \sigma_n\}$ be a finite multiplicative abelian group with character group $\hat{G} = \{\chi_1, \ldots, \chi_n\}$, and let $f : G \to \mathbb{C}$ be a map. Then the cyclic matrix $(f(\sigma_i^{-1}\sigma_k))_{i,k=1,\ldots,n}$ is unitarily equivalent to the diagonal matrix D with diagonal entries $\sum_{i=1}^{n} f(\sigma_i)\, \chi_k(\sigma_i)$ $(k = 1, \ldots, n)$. Explicitly,*

$$(f(\sigma_i^{-1}\sigma_k))_{i,k=1,\ldots,n} = \Omega D \Omega^{-1}$$

with the unitary matrix

$$\Omega = \left(\frac{1}{\sqrt{n}}\, \chi_k(\sigma_i) \right)_{i,k=1,\ldots n}.$$

In particular,

$$\det(f(\sigma_i^{-1}\sigma_k)) = \prod_{k=1}^{n} \sum_{i=1}^{n} f(\sigma_i)\, \chi_k(\sigma_i).$$

This result is essentially due to Dedekind (1931), page 420, who communicated it in a letter to Frobenius (1896). A simple proof is based on the hints in Borevich, Shafarevich (1966), page 421.

It is known from class field theory that the product

$$\prod_{k=1}^{h} L(s, \chi_k) \tag{1.16}$$

in (1.14) is the zeta function of the Hilbert class field of K (see Hasse (1926), Teil I, page 33).

The matrix $\mathcal{A}(s)$ depends meromorphically on $s \in \mathbb{C}$, and since the Euler product

$$L(s, \chi) = \prod_{\mathfrak{p}} (1 - \chi(\mathfrak{p}) N\mathfrak{p}^{-s})^{-1} \tag{1.17}$$

does not vanish for $\operatorname{Re} s > 1$, the matrix $\mathcal{A}(s)$ is invertible for $\operatorname{Re} s > 0$. Hence for $\operatorname{Re} s > 1$ the Eisenstein series $E_{\mathfrak{m}}(\cdot, s)$ $(\mathfrak{m} \in \mathcal{M})$ span the same space of functions on $\mathbb{H}$ as the series $\hat{E}_{\mathfrak{m}}(\cdot, s)$ $(\mathfrak{m} \in \mathcal{M})$.

We show that the functions $E_{\mathfrak{m}}(P, s)$ agree with the Eisenstein series $E_A(P, s)$ of Chapter 3 up to elementary factors. First we introduce the following notation: For $A = \begin{pmatrix} \alpha & \beta \\ \gamma & \delta \end{pmatrix} \in \mathbf{PSL}(2, K)$ let

$$\mathfrak{u}_A := <\gamma, \delta> \in \mathcal{M}, \qquad \mathfrak{v}_A := <\alpha, \beta> \in \mathcal{M}. \tag{1.18}$$

The maps

$$\mathbf{PSL}(2, K) \ni A \mapsto \mathfrak{u}_A \in \mathcal{M}, \qquad \mathbf{PSL}(2, K) \ni A \mapsto \mathfrak{v}_A \in \mathcal{M} \tag{1.19}$$

are surjective.

Theorem 1.7. *If $\zeta \in \mathbb{P}^1(K)$ is a cusp of $\Gamma = \mathbf{PSL}(2, \mathcal{O})$ and*

$$A = \begin{pmatrix} \alpha & \beta \\ \gamma & \delta \end{pmatrix} \in \mathbf{PSL}(2, K)$$

with $A\zeta = \infty$ then

$$E_A(P, s) = \frac{1}{2}(N\mathfrak{u}_A)^{-1-s} E_{\mathfrak{u}_A}(P, s) \tag{1.20}$$

for all $P \in \mathbb{H}$ and $s \in \mathbb{C}$ with $\operatorname{Re} s > 1$.

Proof. We consider the set L of pairs $(c, d) \in K^2$ which generate the $\mathcal{O}$-module $\mathfrak{u}_A$. For every $(c, d) \in L$ there exists an $M \in \mathbf{SL}(2, \mathcal{O})$ such that

$$M \begin{pmatrix} d \\ -c \end{pmatrix} = \begin{pmatrix} \delta \\ -\gamma \end{pmatrix}.$$

We use this fact to construct the map

$$\varphi : L \to \Gamma'_\zeta \backslash \Gamma, \qquad \varphi((c, d)) := \Gamma'_\zeta M.$$

Note that φ is well defined. If $(c, d) \in L$ and $\varphi((c, d)) = \Gamma'_\zeta M$ with $M \in \Gamma$, then $AM = \begin{pmatrix} * & * \\ c & d \end{pmatrix}$. Conversely if $\Gamma'_\zeta M \in \Gamma'_\zeta \backslash \Gamma$ with $AM = \begin{pmatrix} * & * \\ c & d \end{pmatrix}$, then $\varphi^{-1}(\Gamma'_\zeta M) = \{ (c, d), -(c, d) \}$. This implies (1.20). □

8.2 Fourier Expansion of Eisenstein Series for PSL$(2, \mathcal{O})$

The explicit computation of the Fourier expansion of $E_{\mathfrak{m}}(P, s)$ turns out to be rather clumsy, however the Fourier expansion of $\hat{E}_{\mathfrak{m}}(P, s)$ can be determined much more easily. The analogue of Theorem 3.4.1 for $\hat{E}_{\mathfrak{m}}(P, s)$ reads as follows.

Theorem 2.1. *Suppose that $\mathfrak{m} \in \mathcal{M}$ and $\eta = B^{-1}\infty$ ($B \in \mathbf{PSL}(2, K)$) is a cusp of $\mathbf{PSL}(2, \mathcal{O})$. Let Λ be the lattice in $\mathbb{C}$ corresponding to the unipotent stabilizer $\mathbf{PSL}(2, \mathcal{O})'_\eta$ of the cusp η, that is*

$$B\,\mathbf{PSL}(2, \mathcal{O})'_\eta B^{-1} = \left\{ \begin{pmatrix} 1 & \omega \\ 0 & 1 \end{pmatrix} \;\middle|\; \omega \in \Lambda \right\}.$$

Let Λ° be the dual lattice of Λ. Then the Λ-invariant function $\hat{E}_{\mathfrak{m}}(B^{-1}P,s)$ ($P = z + rj \in \mathbb{H}$, $\operatorname{Re} s > 1$) has the Fourier expansion

$$\begin{aligned}\hat{E}_{\mathfrak{m}}(B^{-1}P,s) &= N\mathfrak{u}_B^{1+s}\,\zeta(\mathfrak{m},\mathfrak{u}_B,1+s)\,r^{1+s}\\ &\quad+\frac{\pi N\mathfrak{m}^{1+s}}{|\Lambda|\,s}\left(\sum_{(c,d)\in\mathcal{R}_0}|c|^{-2-2s}\right)r^{1-s}\\ &\quad+\frac{2\pi^{1+s}N\mathfrak{m}^{1+s}}{|\Lambda|\,\Gamma(1+s)}\sum_{0\neq\omega'\in\Lambda^\circ}|\omega'|^s\\ &\quad\cdot\left(\sum_{(c,d)\in\mathcal{R}_0}\frac{e^{2\pi i\langle\omega',\frac{d}{c}\rangle}}{|c|^{2+2s}}\right)r\,K_s(2\pi\,|\omega'|\,r)\,e^{2\pi i<\omega',z>},\end{aligned}\tag{2.1}$$

where the notations are the same as in Theorem 3.4.1 and where $\mathcal{R}_0$ is a maximal system of representatives (c,d) of $(\mathfrak{m}\oplus\mathfrak{m})B^{-1}/B\Gamma'_\eta B^{-1}$ with $c\neq 0$.

Proof. The proof is based on the same method as the proof of Theorem 3.4.1. There is first of all a Fourier expansion of the form

$$\hat{E}_{\mathfrak{m}}(B^{-1}P,s) = \sum_{\omega'\in\Lambda^\circ} a_{\omega'}(r,s)\,e^{2\pi i<\omega',z>}.$$

The Fourier coefficients are computed as

$$\begin{aligned}&a_{\omega'}(r,s)\\ &=\frac{N\mathfrak{m}^{1+s}}{|\Lambda|}\sum_{\substack{(c,d)\in(\mathfrak{m}\oplus\mathfrak{m})B^{-1}\\(c,d)\neq(0,0)}}\int_Q\left(\frac{r}{\|cP+d\|^2}\right)^{1+s}e^{-2\pi i<\omega',z>}\,dx\,dy\\ &=\frac{N\mathfrak{m}^{1+s}}{|\Lambda|}\sum_{\substack{(0,d)\in(\mathfrak{m}\oplus\mathfrak{m})B^{-1}\\d\neq0}}\int_Q\left(\frac{r}{|d|^2}\right)^{1+s}e^{-2\pi i<\omega',x>}\,dx\,dy\\ &\quad+\frac{N\mathfrak{m}^{1+s}}{|\Lambda|}\sum_{\substack{(c,d)\in(\mathfrak{m}\oplus\mathfrak{m})B^{-1}\\c\neq0}}\int_Q\left(\frac{r}{\|cP+d\|^2}\right)^{1+s}e^{-2\pi i<\omega',z>}\,dx\,dy.\end{aligned}\tag{2.2}$$

The first sum on the right-hand side of (2.2) vanishes termwise for $\omega'\neq 0$, and for $\omega' = 0$ it is equal to

$$N\mathfrak{m}^{1+s}\left(\sum_{\substack{(0,d)\in(\mathfrak{m}\oplus\mathfrak{m})B^{-1}\\d\neq0}}|d|^{-2-2s}\right)r^{1+s}.\tag{2.3}$$

If $d\neq 0$, then $(0,d)\in(\mathfrak{m}\oplus\mathfrak{m})B^{-1}$ if and only if $(0,d)B\in(\mathfrak{m}\oplus\mathfrak{m})$, that is, if and only if $d\mathfrak{u}_B\subset\mathfrak{m}$. Hence the sum (2.3) is equal to

$$N\mathfrak{u}_B^{1+s}\,\zeta(\mathfrak{m}, \mathfrak{u}_B, 1+s)\, r^{1+s}.$$

This is the first term on the right-hand side of (2.1). The second sum in (2.2) is treated in the same way as in the proof of Theorem 3.4.1. □

We fix some notations for the rest of this chapter: Suppose that $\mathfrak{m} \in \mathcal{M}$, and let $\eta \in \mathbb{P}^1(K)$ be a cusp of $\Gamma, \eta = B^{-1}\infty$ with

$$B = \begin{pmatrix} \alpha & \beta \\ \gamma & \delta \end{pmatrix} \in \mathbf{PSL}(2, K).$$

Let $\Lambda \subset \mathbb{C}$ be the lattice such that

$$B\,\mathbf{PSL}(2, \mathcal{O})'_\eta\, B^{-1} = \left\{ \begin{pmatrix} 1 & \omega \\ 0 & 1 \end{pmatrix} \;\middle|\; \omega \in \Lambda \right\}. \tag{2.4}$$

We denote the $\mathcal{O}$-modules generated by the row vectors of B by

$$\mathfrak{u} := \mathfrak{u}_B :=< \gamma, \delta >, \qquad \mathfrak{v} := \mathfrak{v}_B :=< \alpha, \beta > . \tag{2.5}$$

The set of row vectors (c, d) in the sum (2.2) is contained in the $\mathcal{O}$-module

$$\mathfrak{L} := (\mathfrak{m} \oplus \mathfrak{m})\, B^{-1}. \tag{2.6}$$

Note that trivially

$$\mathfrak{L} \subset (\mathfrak{m}\mathfrak{u} \oplus \mathfrak{m}\mathfrak{v}). \tag{2.7}$$

For $0 \neq c_0 \in \mathfrak{m}\mathfrak{u}$ let

$$\mathfrak{L}(c_0) := (\{\ c_0\ \} \times K) \cap \mathfrak{L} = \{\ (c, d) \in \mathfrak{L} \ \mid\ c = c_0\ \}. \tag{2.8}$$

Lemma 2.2. *With the preceding notations we have*

(1) $\Lambda = \mathfrak{u}^{-2}$,
(2) if $(c, d) \in \mathfrak{L}$ *and* $\omega \in \Lambda$, *then* $(c, c\omega + d) \in \mathfrak{L}$,
(3) if $0 \neq c_0 \in \mathfrak{m}\mathfrak{u}$, *then* $\mathfrak{L}(c_0) \neq \emptyset$,
(4) $\mathfrak{m}\mathfrak{u}^{-1} \subset \mathfrak{m}\mathfrak{v}$.

Proof. (1): Clearly $\omega \in \Lambda$ if and only if

$$B^{-1} \begin{pmatrix} 1 & \omega \\ 0 & 1 \end{pmatrix} B = \begin{pmatrix} 1 + \gamma\delta\omega & \delta^2\omega \\ -\gamma^2\omega & 1 - \gamma\delta\omega \end{pmatrix} \in \mathbf{PSL}(2, \mathcal{O}). \tag{2.9}$$

This holds if and only if $\gamma^2\omega, \delta^2\omega, \gamma\delta\omega \in \mathcal{O}$. Since $\gamma^2, \delta^2, \gamma\delta$ generate the $\mathcal{O}$-module $\mathfrak{u}^2$, we obtain $\Lambda = \mathfrak{u}^{-2}$.

(2): We have from (2.6) and (2.9)

$$(c, c\omega + d) = (c, d)B \cdot B^{-1} \begin{pmatrix} 1 & \omega \\ 0 & 1 \end{pmatrix} B \cdot B^{-1} \in (\mathfrak{m} \oplus \mathfrak{m})B^{-1} = \mathfrak{L}.$$

(3): We have $c_0 = x\delta - y\gamma$ for some $x, y \in \mathfrak{m}$. Defining $d := -x\beta + y\alpha$ we have $(c_0, d) = (x, y)B^{-1} \in \mathfrak{L}$.

(4): By definition, $\mathcal{O} \subset \mathfrak{u}\mathfrak{v}$. □

The group $\Lambda = \mathfrak{u}^{-2}$ acts on $\mathfrak{L}$ by

$$(c, d) \mapsto (c, c\omega + d) \tag{2.10}$$

where $(c, d) \in \mathfrak{L}$, $\omega \in \Lambda$; see Lemma 2.2, (2). We compute the number of orbits of the restriction of this group action to $\mathfrak{L}(c_0)$.

Lemma 2.3. *If* $0 \neq c_0 \in \mathfrak{m}\mathfrak{u}$, *then*

$$|\mathfrak{L}(c_0)/\Lambda| = \frac{Nc_0}{N\mathfrak{m}\, N\mathfrak{u}}.$$

Proof. We consider the homomorphism of $\mathcal{O}$-modules

$$\begin{aligned} \varphi &: \mathfrak{m}\mathfrak{v} \to (\mathfrak{m}\mathfrak{u}\mathfrak{v} \oplus \mathfrak{m}\mathfrak{u}\mathfrak{v})/(\mathfrak{m} \oplus \mathfrak{m}), \\ \varphi(x) &:= (\gamma x, \delta x) + \mathfrak{m} \oplus \mathfrak{m} \qquad (x \in \mathfrak{m}\mathfrak{v}). \end{aligned}$$

The range of φ is well-defined by Lemma 2.2, (4), and the same result implies

$$\ker \varphi = \mathfrak{m}\mathfrak{v} \cap \mathfrak{m}\mathfrak{u}^{-1} = \mathfrak{m}\mathfrak{u}^{-1}$$

and hence $c_0\mathfrak{u}^{-2} \subset \ker \varphi$. Thus φ induces a homomorphism $\bar{\varphi} : \mathfrak{m}\mathfrak{v}/c_0\mathfrak{u}^{-2} \to (\mathfrak{m}\mathfrak{u}\mathfrak{v} \oplus \mathfrak{m}\mathfrak{u}\mathfrak{v})/\mathfrak{m} \oplus \mathfrak{m}$. The element

$$\lambda_0 := (c_0\alpha, c_0\beta) + \mathfrak{m} \oplus \mathfrak{m} \in (\mathfrak{m}\mathfrak{u}\mathfrak{v} \oplus \mathfrak{m}\mathfrak{u}\mathfrak{v})/\mathfrak{m} \oplus \mathfrak{m}$$

belongs to the image of $\bar{\varphi}$ by Lemma 2.2, (3) and the map

$$\mathfrak{L}(c_0)/\mathfrak{u}^{-2} \to \bar{\varphi}^{-1}(\lambda_0), \qquad \{\, (c_0, c_0\omega + d) \;\mid\; \omega \in \mathfrak{u}^{-2} \,\} \mapsto -d + c_0\mathfrak{u}^{-2}$$

is a bijection. Hence

$$|\mathfrak{L}(c_0)/\mathfrak{u}^{-2}| = |\bar{\varphi}^{-1}(\lambda_0)| = |\ker \bar{\varphi}| = [\mathfrak{m}\mathfrak{u}^{-1} : c_0\mathfrak{u}^{-2}] = \frac{Nc_0}{N\mathfrak{m}\, N\mathfrak{u}}$$

as is required. □

Lemma 2.4. *In Theorem 2.1, we have*

$$\frac{\pi N\mathfrak{m}^{1+s}}{|\Lambda|\, s} \sum_{(c,d)\in\mathcal{R}_0} |c|^{-2-2s} = \frac{2\pi}{\sqrt{|d_K|}\, s} N\mathfrak{u}_B^{1-s}\, \zeta(\mathfrak{m}, \mathfrak{u}_B^{-1}, s).$$

Proof. The set $\mathcal{R}_0$ is a maximal set of representatives $(c, d) \in \mathfrak{L}, c \neq 0$ for the action (2.10), and for a fixed entry c_0 of some element of $\mathcal{R}_0$, the number of different d with $(c_0, d) \in \mathcal{R}_0$ is given by Lemma 2.3. Hence we obtain

$$
\begin{aligned}
(2.11)\qquad N\mathfrak{m}^{1+s} \sum_{(c,d)\in\mathcal{R}_0} |c|^{-2-2s} &= N\mathfrak{m}^{1+s} {\sum_{c\in\mathfrak{m}\mathfrak{u}}}' \frac{|\mathfrak{L}(c)/\Lambda|}{Nc^{1+s}} \\
&= \frac{N\mathfrak{m}^s}{N\mathfrak{u}} {\sum_{c\in\mathfrak{m}\mathfrak{u}}}' Nc^{-s} = N\mathfrak{u}_B^{-1-s}\, \zeta(\mathfrak{m},\mathfrak{u}^{-1},s).
\end{aligned}
$$

Since $\{1, \frac{d_K+\sqrt{d_K}}{2}\}$ is a $\mathbb{Z}$-basis of $\mathcal{O}$, we have $|\mathcal{O}| = \frac{1}{2}\sqrt{|d_k|}$ and hence

$$
(2.12)\qquad |\Lambda| = \frac{1}{2}\sqrt{|d_K|}\, N\mathfrak{u}^{-2}.
$$

Formulas (2.11), (2.12) yield the assertion. □

Our next aim is the explicit computation of the higher Fourier coefficients. This computation is more complicated for the following reason. Note that the choice of B is quite arbitrary; for instance, B may be multiplied from the left by any translation $\begin{pmatrix} 1 & \lambda \\ 0 & 1 \end{pmatrix}$, $\lambda \in K$. Such a change of B leaves Λ unchanged and means that the higher Fourier coefficients are multiplied by $\exp(-2\pi i < \omega', \lambda >)$. We shall circumvent this technical inconvenience later by a suitable choice of B. At the moment a normalization of B is not yet necessary in Lemmas 2.5, 2.6.

For a fixed $0 \neq c_0 \in \mathfrak{m}\mathfrak{u}$ we consider the sum with respect to $(c_0, d) \in \mathcal{R}_0$ in the third term on the right-hand side of (2.1) and define the Kloosterman-like sum

$$
(2.13)\qquad S(\omega', c_0) := \sum_{(c_0,d)\in\mathcal{R}_0} e^{2\pi i \left\langle \omega', \frac{d}{c_0} \right\rangle} \qquad (0 \neq \omega' \in \Lambda^\circ,\ 0 \neq c_0 \in \mathfrak{m}\mathfrak{u}).
$$

Lemma 2.5. *If* $0 \neq \omega' \in \Lambda^\circ$, $0 \neq c_0 \in \mathfrak{m}\mathfrak{u}$, *then*

$$
(2.14)\qquad S(\omega', c_0) = 0 \quad \textit{unless} \quad \frac{\omega'}{\bar{c}_0} \in (\mathfrak{m}\mathfrak{u}^{-1})^\circ.
$$

Proof. Let $(c_0, d) \in \mathfrak{L}$ and $x \in \mathfrak{m}\mathfrak{u}^{-1}$. Then $(c_0, d+x)B = (c_0, d)B + (0, x)B \in \mathfrak{m} \oplus \mathfrak{m}$ because $x\mathfrak{u} \subset \mathfrak{m}$. Hence if (c_0, d) runs through a system of representatives for $\mathfrak{L}(c_0)/\Lambda$, then $(c_0, d+x)$ does the same for every fixed $x \in \mathfrak{m}\mathfrak{u}^{-1}$. This implies

$$
S(\omega', c_0) = e^{2\pi i \left\langle \omega', \frac{x}{c_0} \right\rangle}\, S(\omega', c_0)
$$

for all $x \in \mathfrak{m}\mathfrak{u}^{-1}$. We conclude that $S(\omega', c_0) = 0$ unless the condition $\exp(2\pi i < \omega', x/c_0 >) = 1$ for all $x \in \mathfrak{m}\mathfrak{u}^{-1}$ holds. The latter is equivalent to $\omega'/\bar{c}_0 \in (\mathfrak{m}\mathfrak{u}^{-1})^\circ$. □

Lemma 2.6. *Suppose that* $0 \neq \omega' \in \Lambda^\circ, 0 \neq c_0 \in \mathfrak{m}\mathfrak{u}$, $\frac{\omega'}{\bar{c}_0} \in (\mathfrak{m}\mathfrak{u}^{-1})^\circ$ *and*

$$
(c_0, d_0) \in (\mathfrak{m} \oplus \mathfrak{m})B^{-1}.
$$

Then

$$S(\omega', c_0) = \frac{N c_0}{N\mathfrak{m}\, N\mathfrak{u}}\, e^{2\pi i \left\langle \omega', \frac{d_0}{c_0} \right\rangle},$$

where $\exp(2\pi i < \omega', \frac{d_0}{c_0} >)$ *is a root of unity.*

Proof. If $(c_0, d) \in \mathcal{R}_0$, we have $((c_0, d_0) - (c_0, d))B = (0, d_0 - d)B \in \mathfrak{m} \oplus \mathfrak{m}$, i.e., $d_0 - d \in \mathfrak{m}\mathfrak{u}^{-1}$. Hence all the terms in the sum (2.13) are equal, and the number of terms is given by Lemma 2.3. □

Lemma 2.7. *If* $\mathfrak{n} \in \mathcal{M}$, *then*

$$\mathfrak{n}^\circ = \frac{2}{\sqrt{d_K}}\, \bar{\mathfrak{n}}^{-1}. \tag{2.15}$$

In particular,

$$\Lambda^\circ = \frac{2}{\sqrt{d_K}}\, \bar{\mathfrak{u}}^2. \tag{2.16}$$

Proof. By definition, the dual $\mathbb{Z}$-lattice $\mathfrak{n}^\circ$ is the set of all $\lambda \in K$ such that

$$< \lambda, x >= \frac{\lambda\bar{x} + \bar{\lambda}x}{2} = \frac{\operatorname{tr}(\bar{\lambda}x)}{2}$$

is a rational integer for all $x \in \mathfrak{n}$. Hence $\mathfrak{n}^\circ = 2\bar{\mathfrak{n}}^*$, where $\mathfrak{n}^*$ is the complementary module with respect to the trace form. It is known that $\mathfrak{n}^* = \mathfrak{D}^{-1}\mathfrak{n}^{-1}$, where $\mathfrak{D}$ is the different of K (see Lang (1966), page 57). Here $\mathfrak{D} = \sqrt{d_K}\mathcal{O}$ and the assertion follows. □

We now define a normalization condition on B that will enable us to compute the higher Fourier coefficients explicitly. Remember that $\mathfrak{u}\mathfrak{v} \subset \mathcal{O}$ by Lemma 2.2, (4).

Definition 2.8. A $B \in \mathbf{SL}(2, K)$ or $\mathbf{PSL}(2, K)$ is called quasi-integral if

$$\mathfrak{u}_B \mathfrak{v}_B = \mathcal{O}. \tag{2.17}$$

Maintaining our notation $B = \begin{pmatrix} \alpha & \beta \\ \gamma & \delta \end{pmatrix}$, we see that B is quasi-integral if and only if $\alpha\gamma$, $\alpha\delta$, $\beta\delta \in \mathcal{O}$.

Lemma 2.9.

(1) *For* $\gamma^*, \delta^* \in K$, $(\gamma^*, \delta^*) \neq (0, 0)$, *there exists a quasi-integral matrix* $B^* \in \mathbf{SL}(2, K)$ *such that* $B^* = \begin{pmatrix} \cdot & \cdot \\ \gamma^* & \delta^* \end{pmatrix}$.

(2) *For every $\eta \in \mathbb{P}^1(K)$ there exists a quasi-integral matrix $B \in \mathbf{PSL}(2,K)$ such that $B\eta = \infty$.*
(3) *For every $\mathfrak{n} \in \mathcal{M}$ there exists a quasi-integral matrix $B \in \mathbf{PSL}(2,K)$ such that $\mathfrak{n} = \mathfrak{u}_B$.*

Proof. (1): The $\mathcal{O}$-module $\mathfrak{q} :=< \gamma^*, \delta^* >$ is $\neq \{0\}$ by hypothesis. Hence we have $\mathfrak{q}^h = \lambda\mathcal{O}$ for some $0 \neq \lambda \in K$, where h is the class number of K. We choose a pair α_0, β_0 of generators of $\mathfrak{q}^{h-1}$. Then there exist $\mu_1, \ldots, \mu_4 \in \mathcal{O}$ such that

$$\mu_1\alpha_0\gamma^* + \mu_2\alpha_0\delta^* + \mu_3\beta_0\gamma^* + \mu_4\beta_0\delta^* = \lambda.$$

Hence $B^* = \begin{pmatrix} \alpha^* & \beta^* \\ \gamma^* & \delta^* \end{pmatrix} \in \mathbf{SL}(2,K)$ with

$$\alpha^* := \frac{\mu_2\alpha_0 + \mu_4\beta_0}{\lambda}, \qquad \beta^* := -\frac{\mu_1\alpha_0 + \mu_3\beta_0}{\lambda},$$

satisfies our requirements.

(2): For $\eta = \infty$ we take $B = I$, and for $\eta \in K$ there exists a quasi-integral matrix $B \in \mathbf{PSL}(2,K)$ of the form $B = \begin{pmatrix} \alpha & \beta \\ 1 & -\eta \end{pmatrix}$.

(3): follows immediately from (1). □

The following type of divisor sum will come up in the final formula for the higher Fourier coefficients.

Definition 2.10. For $\mathfrak{a}, \mathfrak{b} \in \mathcal{M}$, $s \in \mathbb{C}$ and $\omega \in K^*$ let

$$\sigma_s(\mathfrak{a}, \mathfrak{b}, \omega) = N\mathfrak{a}^{-s} \sum_{\substack{\lambda \in \mathfrak{a}\mathfrak{b} \\ \omega \in \lambda\mathfrak{a}^{-1}\mathfrak{b}}} N\lambda^s. \tag{2.18}$$

The sum (2.18) is a finite sum. It is empty unless $\omega \in \mathfrak{b}^2$. For $\mathfrak{a} = \mathcal{O}, \mathfrak{b} \subset \mathcal{O}$ an ideal, and $\omega \in \mathfrak{b}^2$, $\omega \neq 0$, the sum (2.18) extends over all divisors λ of ω such that $\lambda \in \mathfrak{b}$ and $\frac{\omega}{\lambda} \in \mathfrak{b}$. If $\mu \in K^*$, then

$$\sigma_s(\mu\mathfrak{a}, \mathfrak{b}, \omega) = \sigma_s(\mathfrak{a}, \mathfrak{b}, \omega). \tag{2.19}$$

The sum (2.18) satisfies the reciprocity formula

$$|\omega|^{-s}\,\sigma_s(\mathfrak{a}, \mathfrak{b}, \omega) = |\omega|^s\,\sigma_{-s}(\mathfrak{a}^{-1}, \mathfrak{b}, \omega). \tag{2.20}$$

Theorem 2.11. *Suppose that $\mathfrak{m} \in \mathcal{M}$ and $\eta \in \mathbb{P}^1(K)$ is a cusp of $\mathbf{PSL}(2,\mathcal{O})$. Choose a quasi-integral matrix $B = \begin{pmatrix} \alpha & \beta \\ \gamma & \delta \end{pmatrix} \in \mathbf{PSL}(2,K)$, such that $\eta = B^{-1}\infty$ and let $\mathfrak{u} :=< \gamma, \delta >$. Then $\hat{E}_{\mathfrak{m}}(B^{-1}P, s)$ $(P = z + rj \in \mathbb{H}$, $\mathrm{Re}\, s > 1)$ has the Fourier expansion*

$$
\begin{aligned}
\hat{E}_{\mathfrak{m}}(B^{-1}P,s) &= N\mathfrak{u}^{1+s}\,\zeta(\mathfrak{m},\mathfrak{u},1+s)\,r^{1+s} \\
&+ \frac{2\pi}{\sqrt{|d_K|}\,s}\,N\mathfrak{u}^{1-s}\,\zeta(\mathfrak{m},\mathfrak{u}^{-1},s)\,r^{1-s} \\
&+ \frac{2^{2+s}\,\pi^{1+s}\,N\mathfrak{u}}{|d_K|^{\frac{s+1}{2}}\,\Gamma(1+s)} \sum_{0\neq\omega\in\mathfrak{u}^2} |\omega|^s\,\sigma_{-s}(\mathfrak{m},\mathfrak{u},\omega) \\
&\cdot r\,K_s\left(\frac{4\pi|\omega|r}{\sqrt{|d_K|}}\right)\, e^{2\pi i\left\langle \frac{2\bar{\omega}}{\sqrt{d_K}},z\right\rangle}.
\end{aligned}
\tag{2.21}
$$

Proof. The coefficients of r^{1+s} and r^{1-s} are given by (2.1) and Lemma 2.4. We compute the higher Fourier coefficients. Let $0\neq\omega'\in\Lambda^\circ$, $0\neq c\in\mathfrak{m}\mathfrak{u}$. If $\frac{\omega'}{\bar{c}}\notin(\mathfrak{m}\mathfrak{u}^{-1})^\circ$, then $S(\omega',c)=0$ by (2.14). Assume now that $\frac{\omega'}{\bar{c}}\in(\mathfrak{m}\mathfrak{u}^{-1})^\circ$, and let $(c,d)\in\mathcal{R}_0$. Then $<\omega',\frac{d}{c}>=<\frac{\omega'}{\bar{c}},d>\in\mathbb{Z}$ because $d\in\mathfrak{m}\mathfrak{v}=\mathfrak{m}\mathfrak{u}^{-1}$ since B is quasi-integral. This means that all terms in the sum (2.13) are equal to one and hence

$$S(\omega',c)=\frac{Nc}{N\mathfrak{m}\,N\mathfrak{u}}$$

(see Lemma 2.6). By (2.16) the map from $\mathfrak{u}^2$ to Λ° defined by

$$\mathfrak{u}\ni\omega\mapsto\omega':=\frac{2}{\sqrt{d_K}}\,\bar{\omega}\in\Lambda^\circ$$

is a bijection. Replacing ω' by $\frac{2}{\sqrt{d_K}}\,\bar{\omega}$, we obtain

$$
\begin{aligned}
N\mathfrak{m}^{1+s}\sum_{(c,d)\in\mathcal{R}_0}\frac{e^{2\pi i<\omega',\frac{d}{c}>}}{|c|^{2+2s}} &= N\mathfrak{m}^{1+s}\sum_{\substack{c\in\mathfrak{m}\mathfrak{u}\\ c\neq 0}} S(\omega',c)\,N(c)^{-1-s} \\
&= \frac{N\mathfrak{m}^s}{N\mathfrak{u}}\sum_{\substack{c\in\mathfrak{m}\mathfrak{u}\\ \frac{\omega'}{\bar{c}}\in(\mathfrak{m}\mathfrak{u}^{-1})^\circ}} Nc^{-s} = \frac{N\mathfrak{m}^s}{N\mathfrak{u}}\sum_{\substack{c\in\mathfrak{m}\mathfrak{u}\\ \omega\in c\mathfrak{m}^{-1}\mathfrak{u}}} Nc^{-s} \\
&= \frac{1}{N\mathfrak{u}}\,\sigma_{-s}(\mathfrak{m},\mathfrak{u},\omega).
\end{aligned}
$$

Inserting this result and (2.12) into (2.1) we obtain (2.21). □

8.3 Meromorphic Continuation by Fourier Expansion and the Kronecker Limit Formula

Formula (2.21) gives the meromorphic continuation of the Eisenstein series $\hat{E}_{\mathfrak{m}}(P,s)$ to the whole s-plane.

Theorem 3.1. *Let the notation be as in Theorem 2.11 with $B = I$. Then the Eisenstein series $\hat{E}_{\mathfrak{m}}(P,s)$ has a meromorphic continuation to the whole s-plane and satisfies the functional equation*

$$\begin{aligned}(3.1)\qquad &\left(\frac{2\pi}{\sqrt{|d_K|}}\right)^{-(1+s)} \Gamma(1+s)\,\hat{E}_{\mathfrak{m}}(P,s)\\ &= \left(\frac{2\pi}{\sqrt{|d_K|}}\right)^{-(1-s)} \Gamma(1-s)\,\hat{E}_{\mathfrak{m}^{-1}}(P,-s)\end{aligned}$$

for every $s \in \mathbb{C}$. The Eisenstein series $\hat{E}_{\mathfrak{m}}(P,s)$ is holomorphic everywhere except for a simple pole at $s = 1$ with residue

$$(3.2)\qquad \operatorname*{Res}_{s=1} \hat{E}_{\mathfrak{m}}(P,s) = \frac{4\pi^2}{|d_K|},$$

and

$$(3.3)\qquad \hat{E}_{\mathfrak{m}}(P,-n) = 0 \qquad \textit{for all } n \in \mathbb{N},\ n \geq 2,$$

whereas

$$(3.4)\qquad \hat{E}_{\mathfrak{m}}(P,-1) = -1.$$

Proof. We rewrite (2.21) in the form

$$\begin{aligned}(3.5)\qquad &\left(\frac{2\pi}{\sqrt{|d_K|}}\right)^{-(1+s)} \Gamma(1+s)\,\hat{E}_{\mathfrak{m}}(B^{-1}P,s)\\ &= Z(\mathfrak{m},\mathfrak{u},1+s)\,N\mathfrak{u}^{1+s}\,r^{1+s} + Z(\mathfrak{m},\mathfrak{u}^{-1},s)\,N\mathfrak{u}^{1-s}\,r^{1-s}\\ &+ 2N\mathfrak{u} \sum_{0\neq\omega\in\mathfrak{u}^2} |\omega|^s\,\sigma_{-s}(\mathfrak{m},\mathfrak{u},\omega)\,r\,K_s\left(\frac{4\pi|\omega|r}{\sqrt{|d_K|}}\right) e^{2\pi i\left\langle \frac{2\bar{\omega}}{\sqrt{d_K}},z\right\rangle}\end{aligned}$$

with

$$(3.6)\qquad Z(\mathfrak{m},\mathfrak{u},s) := \left(\frac{2\pi}{\sqrt{|d_K|}}\right)^{-s} \Gamma(s)\,\zeta(\mathfrak{m},\mathfrak{u},s).$$

The function (3.6) has a meromorphic continuation to the whole s-plane and satisfies the functional equation

$$(3.7)\qquad Z(\mathfrak{m},\mathfrak{u},1-s) = Z(\mathfrak{u},\mathfrak{m},s)$$

(see Lang (1968), page 254). Moreover,

$$(3.8)\qquad Z(\mathfrak{m}^{-1},\mathfrak{u}^{-1},s) = Z(\mathfrak{u},\mathfrak{m},s)$$

by (1.5). Hence the zeroth Fourier coefficient in (3.5) satisfies the functional equation (3.1). Since the Bessel function $K_s(t)$ is an even entire function of $s \in \mathbb{C}$, the reciprocity formula (2.20) implies that the infinite series in (3.5) also satisfies the functional equation (3.1) termwise, and (3.1) follows.

The factor of r^{1-s} in (2.21) has a simple pole at $s = 1$ whereas the remaining terms are holomorphic at $s = 1$. This yields by (1.5)

$$\operatorname*{Res}_{s=1} \hat{E}_{\mathfrak{m}}(P, s) = \frac{2\pi|\mathcal{O}^*|}{\sqrt{|d_K|}} \operatorname*{Res}_{s=1} \zeta((\mathfrak{m}^{-1})^{\#}, s).$$

We now draw from Lang (1968), page 259

$$\operatorname*{Res}_{s=1} \zeta((\mathfrak{m}^{-1})^{\#}, s) = \frac{2\pi}{|\mathcal{O}^*|\sqrt{|d_K|}} \tag{3.9}$$

and (3.2) follows.

The poles of the factors of r^{1+s} and r^{1-s} in (2.21) at $s = 0$ cancel because

$$\zeta((\mathfrak{m}^{-1})^{\#}, 0) = -\frac{1}{|\mathcal{O}^*|} \tag{3.10}$$

as follows from the functional equation of the zeta function. The trivial zeros (3.3) are obvious from (3.1) and (3.4) follows from (3.1), (3.2). $\square$

Theorem 3.1 can also be drawn from Epstein's classical work (Epstein (1903)) on zeta functions. The special value $\hat{E}_{\mathfrak{m}}(B^{-1}P, 0)$ can be computed explicitly by means of the Kronecker limit formula. The result is as follows.

Theorem 3.2. *Let the notation be as in Theorem 2.11, and for* $\mathfrak{a} \in \mathcal{M}$ *let*

$$g(\mathfrak{a}) := (2\pi)^{-12}\, N\mathfrak{a}^6\, |\Delta(\mathfrak{a})| = (2\pi)^{-12}\, N(< 1, \tau >)^6\, |\Delta(\tau)|,$$

where $\mathfrak{a} = \lambda < 1, \tau >$ *with* $\lambda \in K^*$ *and* $\tau \in \mathbb{C}$, $\operatorname{Im}\tau > 0$, *and where* $\Delta = g_2^3 - 27g_3^2$ *is the discriminant from the theory of elliptic functions. Then*

$$\begin{aligned}\hat{E}_{\mathfrak{m}}(B^{-1}P, 0) = \frac{4\pi N\mathfrak{u}}{\sqrt{|d_K|}}\, r\, \{&\log(rN\mathfrak{u}) + \gamma - \log 2\pi \\ &- \log\sqrt{|d_K|} - \frac{1}{12}\log(g(\mathfrak{m}\mathfrak{u}^{-1})\, g(\mathfrak{m}^{-1}\mathfrak{u}^{-1})) \\ &+ \sum_{0 \neq \omega \in \mathfrak{u}^2} \sigma_0(\mathfrak{m}, \mathfrak{u}, \omega)\, K_0\left(\frac{4\pi|\omega|r}{\sqrt{|d_K|}}\right) e^{2\pi i \left\langle \frac{2\bar{\omega}}{\sqrt{d_K}}, z \right\rangle}\},\end{aligned} \tag{3.11}$$

where γ *denotes Euler's constant.*

Proof. We correct a sign error in the Kronecker limit formula in the version of Lang (1973), page 280, (2) and obtain

$$\zeta(\mathfrak{m}, \mathfrak{u}, 1+s) = \frac{2\pi}{\sqrt{|d_K|}} \left(\frac{1}{s} + 2\gamma - \log|d_K| - \frac{1}{6} \log g(\mathfrak{m}\mathfrak{u}^{-1}) \right) + O(s)$$

for $s \longrightarrow 0$. This implies

$$\begin{aligned} & N\mathfrak{u}^{1+s}\, \zeta(\mathfrak{m}, \mathfrak{u}, 1+s)\, r^{1+s} \\ &= \frac{2\pi N\mathfrak{u}}{\sqrt{|d_K|}}\, r \left(\frac{1}{s} + 2\gamma - \log|d_K| + \log(r \cdot N\mathfrak{u}) - \frac{1}{6} \log g(\mathfrak{m}\mathfrak{u}^{-1}) + O(s) \right) \end{aligned}$$

for $s \longrightarrow 0$. In addition, we have from the functional equation of the zeta function (see Lang (1968), page 254)

$$\begin{aligned} & \frac{2\pi}{\sqrt{|d_K|}s} \zeta(\mathfrak{m}, \mathfrak{u}^{-1}, s) = \left(\frac{2\pi}{\sqrt{|d_K|}} \right)^{2s} \frac{\Gamma(1-s)}{\Gamma(1+s)}\, \zeta(\mathfrak{u}^{-1}, \mathfrak{m}, 1-s) \\ &= \left(1 + s \log \frac{(2\pi)^2}{|d_K|} + \dots \right) \cdot \frac{1 + \gamma s + \dots}{1 - \gamma s + \dots} \\ &\cdot \frac{2\pi}{\sqrt{|d_K|}} \left(-\frac{1}{s} + 2\gamma - \log|d_K| - \frac{1}{6} \log g(\mathfrak{m}^{-1}\mathfrak{u}^{-1}) + O(s) \right) \\ &= \frac{2\pi}{\sqrt{|d_K|}} \left(-\frac{1}{s} - 2\log 2\pi - \frac{1}{6} \log g(\mathfrak{m}^{-1}\mathfrak{u}^{-1}) + O(s) \right) \end{aligned}$$

as $s \longrightarrow 0$.This yields

$$\begin{aligned} & \frac{2\pi}{\sqrt{|d_K|}s} N\mathfrak{u}^{1-s}\, \zeta(\mathfrak{m}, \mathfrak{u}^{-1}, s)\, r^{1-s} \\ &= \frac{2\pi N\mathfrak{u}}{\sqrt{|d_K|}}\, r \left(-\frac{1}{s} + \log(rN\mathfrak{u}) - 2\log 2\pi - \frac{1}{6} \log g(\mathfrak{m}^{-1}\mathfrak{u}^{-1}) + O(s) \right) \end{aligned}$$

as $s \longrightarrow 0$, and the assertion follows from (2.21). □

Theorem 3.3. *Let the notation be as in Theorem 3.2. Then*

$$\begin{aligned} & \lim_{s \to 1} \left(\hat{E}_{\mathfrak{m}}(B^{-1}P, s) - \frac{4\pi^2}{|d_K|} \frac{1}{s-1} \right) = |\mathcal{O}^*|\, \zeta((\mathfrak{m}^{-1}\mathfrak{u})^{\#}, 2)\, N\mathfrak{u}^2\, r^2 \\ & + \frac{4\pi^2}{|d_K|} \Bigg(2\gamma - 1 - \log|d_K| - \log(rN\mathfrak{u}) - \frac{1}{6} \log g(\mathfrak{m}\mathfrak{u}) \\ & + 2N\mathfrak{u} \sum_{0 \neq \omega \in \mathfrak{u}^2} |\omega|\, \sigma_{-1}(\mathfrak{m}, \mathfrak{u}, \omega)\, r\, K_1 \left(\frac{4\pi|\omega| r}{\sqrt{|d_K|}} \right) e^{2\pi i \left\langle \frac{2\bar{\omega}}{\sqrt{d_K}}, z \right\rangle} \Bigg). \end{aligned}$$

Proof. We have from the proof of Theorem 3.2

$$\frac{2\pi}{\sqrt{|d_K|}\, s}(rN\mathfrak{u})^{1-s}\,\zeta(\mathfrak{m},\mathfrak{u}^{-1},s) - \frac{4\pi^2}{|d_K|}\frac{1}{s-1}$$
$$= \frac{4\pi^2}{|d_K|}\{(1-(s-1)+\dots)(1-(s-1)\log(rN\mathfrak{u})+\dots)$$
$$\cdot\left(\frac{1}{s-1}+2\gamma-\log|d_K|-\frac{1}{6}\log g(\mathfrak{m}\mathfrak{u})+O(s-1)\right)-\frac{1}{s-1}\}$$
$$= \frac{4\pi^2}{|d_K|}\left(2\gamma-1-\log|d_K|-\log(rN\mathfrak{u})-\frac{1}{6}\log g(\mathfrak{m}\mathfrak{u})+O(s-1)\right)$$

as $s \to 1$. This implies the assertion. □

Theorem 3.3 is an analogue of Kronecker's first limit formula. Asai (1970) studies Kronecker's limit formula for arbitrary number fields of class number one for the cusp ∞.

We briefly discuss the Dirichlet series associated with the Eisenstein series $\hat{E}_{\mathfrak{m}}$. For $\nu \in \mathbb{C}$ and $n \in \mathbb{N}$ let

$$\sigma_\nu(n) := \sum_{\substack{d|n \\ d>0}} d^\nu.$$

Then the Dirichlet series

$$\zeta(s)\zeta(s-\nu) = \sum_{n=1}^{\infty}\frac{\sigma_\nu(n)}{n^s} \qquad (\operatorname{Re} s > \max\{1, \operatorname{Re}\nu+1\}) \tag{3.12}$$

corresponds to the Eisenstein series for the modular group acting on the upper half-plane

$$\sum_{c,d\in\mathbb{Z}}{}' \left(\frac{y}{|cz+d|^2}\right)^s \qquad (y = \operatorname{Im} z > 0) \tag{3.13}$$

by means of the Mellin transform (Maaß (1949a)). An analogous result holds for the Eisenstein series $\hat{E}_{\mathfrak{m}}$.

Theorem 3.4. *Let the notation be as in Theorem 2.11. Then the Mellin transform of the infinite part of the Fourier expansion of $\hat{E}_{\mathfrak{m}}(B^{-1}P,\nu)$ ($\nu \in \mathbb{C}$) is given by*

$$\begin{aligned} &4N\mathfrak{u}^{2s}\int_0^\infty \sum_{0\neq\omega\in\mathfrak{u}^2}|\omega|^\nu\,\sigma_{-\nu}(\mathfrak{m},\mathfrak{u},\omega)\,r^{2s-1}\,K_\nu\left(\frac{4\pi|\omega|r}{\sqrt{|d_K|}}\right)dr \\ &= Z\left(\mathfrak{m},\mathfrak{u}^{-1},s+\frac{\nu}{2}\right)\,Z\left(\mathfrak{m}^{-1},\mathfrak{u}^{-1},s-\frac{\nu}{2}\right), \end{aligned} \tag{3.14}$$

where Z is defined by (3.6).

Proof. We start from the integral

$$\int_0^\infty t^{2s-1} K_\nu(2at)\, dt = \frac{a^{-2s}}{4} \Gamma\left(s + \frac{\nu}{2}\right) \Gamma\left(s - \frac{\nu}{2}\right) \ \left(a > 0, \mathrm{Re}\left(s \pm \frac{\nu}{2}\right) > 0\right)$$

(see Magnus, Oberhettinger, Soni (1966), page 91) and find

$$\begin{aligned} 4 \int_0^\infty & \sum_{0 \neq \omega \in \mathfrak{u}^2} |\omega|^\nu \sigma_{-\nu}(\mathfrak{m}, \mathfrak{u}, \omega)\, r^{2s-1} K_\nu\left(\frac{4\pi|\omega| r}{\sqrt{|d_K|}}\right) dr \\ &= \left(\frac{2\pi}{\sqrt{|d_K|}}\right)^{-2s} \Gamma\left(s + \frac{\nu}{2}\right) \Gamma\left(s - \frac{\nu}{2}\right) \sum_{0 \neq \omega \in \mathfrak{u}^2} \sigma_{-\nu}(\mathfrak{m}, \mathfrak{u}, \omega)\, N\omega^{-s+\frac{\nu}{2}} \end{aligned}$$

whenever $\mathrm{Re}(s \pm \frac{\nu}{2}) > 1$. Now we have from the definitions

$$\begin{aligned} &\zeta\left(\mathfrak{m}^{-1}, \mathfrak{u}^{-1}, s - \frac{\nu}{2}\right) \zeta\left(\mathfrak{m}, \mathfrak{u}^{-1}, s + \frac{\nu}{2}\right) \\ &= N\mathfrak{u}^{2s} \sum_{0 \neq \omega \in \mathfrak{u}^2} \sigma_{-\nu}(\mathfrak{m}, \mathfrak{u}, \omega)\, N\omega^{-s+\frac{\nu}{2}}, \end{aligned}$$

and the assertion follows. □

Corollary 3.5. *The function*

$$\Psi_\nu(\mathfrak{m}, \mathfrak{u}, s) := Z\left(\mathfrak{m}, \mathfrak{u}^{-1}, s + \frac{\nu}{2}\right) Z\left(\mathfrak{m}^{-1}, \mathfrak{u}^{-1}, s - \frac{\nu}{2}\right) \tag{3.15}$$

(see (3.14)) satisfies the functional equation

$$\Psi_\nu(\mathfrak{m}, \mathfrak{u}, 1 - s) = \Psi_\nu(\mathfrak{m}, \mathfrak{u}^{-1}, s). \tag{3.16}$$

Proof. The functional equation (3.16) is a consequence of (3.7), (3.8). □

Note that (3.14)–(3.16) is just a special case of Maaß' (1949b) general theory of Dirichlet series associated with automorphic eigenfunctions of the hyperbolic Laplacian. This remark yields a second proof of (3.16).

Corollary 3.6. *Let the notation be as in Theorem 2.11. Then the function* $g : \mathbb{H} \to \mathbb{C}$ *defined by*

$$\begin{aligned} g(P) &:= \frac{|\mathcal{O}^*||d_K|}{4\pi^2} \zeta((\mathfrak{m}^{-1}\mathfrak{u})^\#, 2)\, N\mathfrak{u}^2\, r^2 \\ &+ 2N\mathfrak{u} \sum_{0 \neq \omega \in \mathfrak{u}^2} |\omega| \sigma_{-1}(\mathfrak{m}, \mathfrak{u}, \omega)\, r\, K_1\left(\frac{4\pi|\omega| r}{\sqrt{|d_K|}}\right) e^{2\pi i \left\langle \frac{2\bar\omega}{\sqrt{d_K}}, z\right\rangle} \end{aligned} \tag{3.17}$$

satisfies the logarithmic transformation formula

$$g(MP) + \log \|cP + d\|^2 = g(P)$$

for all $M = \begin{pmatrix} \cdot & \cdot \\ c & d \end{pmatrix} \in B\,\mathbf{PSL}(2, \mathcal{O})\,B^{-1}$, *and* g *is harmonic for the hyperbolic Laplacian* Δ, *that is* $\Delta g = 0$.

This result follows immediately from Theorem 3.3. The function g of Corollary 3.6 is related with the Dirichlet series $\zeta(\mathfrak{m}^{-1}, \mathfrak{u}^{-1}, s)\,\zeta(\mathfrak{m}, \mathfrak{u}^{-1}, s+1)$ by means of the Mellin transform (see (3.14)). Thus we may consider g as an analogue of the logarithm $\log|\eta|$ of the modulus of the Dedekind η-function since Schoeneberg (1968) and Weil (1968) pointed out that $\zeta(s)\,\zeta(s+1)$ is related with $\log|\eta|$ via the Mellin transform (cf. Asai (1970)).

We also see from Theorem 3.4 that $\hat{E}_{\mathfrak{m}}(B^{-1}P, 0)$ is related with

$$\zeta(\mathfrak{m}^{-1}, \mathfrak{u}^{-1}, s)\;\zeta(\mathfrak{m}, \mathfrak{u}^{-1}, s).$$

The coefficients of this Dirichlet series are given by $\sigma_0(\mathfrak{m}, \mathfrak{u}, \omega)$ $(0 \neq \omega \in \mathfrak{u}^2)$, and this is an analogue of the function $d(n)$ counting the number of divisors of the natural number n. The classical counterpart of our Dirichlet series (for $\nu = 0$) is

$$\zeta^2(s) = \sum_{n=1}^{\infty} \frac{d(n)}{n^s} \qquad (\operatorname{Re} s > 1)$$

which is related with the Fourier expansion of the Eisenstein series (3.13) at $s_0 = 1/2$ via the Mellin transform.

We now develop the results on the meromorphic continuation of the series $E_{\mathfrak{m}}(P, s)$. It is convenient here to use the vector notation (1.7)–(1.10). Let $\sigma \in \mathcal{S}_h$ be the permutation on $\{1, \ldots, h\}$ such that

$$\mathfrak{m}_i^{-1} = \lambda_i\, \mathfrak{m}_{\sigma(i)}, \qquad i = 1, \ldots, h \tag{3.18}$$

for appropriate $\lambda_i \in K^*$, and let $T \in \mathbf{GL}(h, \mathbb{Z})$ be the permutation matrix such that

$$\begin{pmatrix} \alpha_{\sigma(1)} \\ \vdots \\ \alpha_{\sigma(h)} \end{pmatrix} = T \begin{pmatrix} \alpha_1 \\ \vdots \\ \alpha_h \end{pmatrix} \tag{3.19}$$

for all $(\alpha_1, \ldots, \alpha_h)^t \in \mathbb{C}^h$, where the upper index t stands for transposition. Note that T is an orthogonal matrix such that $T^2 = I$. Put

$$R(s) := \left(\frac{2\pi}{\sqrt{|d_K|}}\right)^{-(1+s)} \Gamma(1+s). \tag{3.20}$$

Then (3.1) reads in vector notation

$$R(s)\,\hat{\mathcal{E}}(P, s) = R(-s)\,T\,\hat{\mathcal{E}}(P, -s). \tag{3.21}$$

It follows from Theorem 3.1 and from Theorem 1.5 that $\tilde{\mathcal{E}}(P, s)$ is meromorphic in the whole s-plane and satisfies the functional equation

$$\tilde{\mathcal{E}}(P,s) = \frac{R(-s)}{R(s)}\, \mathcal{A}(s)^{-1}\, T\, \mathcal{A}(-s)\, \tilde{\mathcal{E}}(P,-s). \tag{3.22}$$

The right-hand side of (3.22) can be simplified by means of the following little lemma.

Lemma 3.7. *The matrices* $\mathcal{A}(s)$ *(see (1.10)) and* T *(see (3.18), (3.19)) satisfy*

$$\mathcal{A}(-s) = \frac{R(s-1)}{R(-s)}\, T\, \mathcal{A}(s-1)\, T, \tag{3.23}$$

where $R(s)$ *is defined by (3.20).*

Proof. The zeta function $\zeta(\mathfrak{m}^{\#}, s)$ ($\mathfrak{m} \in \mathcal{M}$) satisfies the functional equation $F(s, \mathfrak{m}^{\#}) = F(1-s, (\mathfrak{m}^{-1})^{\#})$ with $F(s, \mathfrak{m}^{\#}) = R(s-1)\, \zeta(\mathfrak{m}^{\#}, s)$ (see Lang (1968), page 254). Hence we find

$$\begin{aligned}\mathcal{A}(-s) &= (\zeta((\mathfrak{m}_i^{-1}\mathfrak{m}_k)^{\#}, 1-s))_{i,k=1\ldots h} = \frac{R(s-1)}{R(-s)}\, (\zeta((\mathfrak{m}_k^{-1}\mathfrak{m}_i)^{\#}, s)_{i,k=1\ldots h} \\ &= \frac{R(s-1)}{R(-s)}\, T\, (\zeta((\mathfrak{m}_i^{-1}\mathfrak{m}_k)^{\#}, s)_{i,k=1\ldots h}\, T = \frac{R(s-1)}{R(-s)}\, T\, \mathcal{A}(s-1)\, T,\end{aligned}$$

since for every $h \times h$-Matrix (α_{ik}) we have $T\,(\alpha_{ik})\, T = (\alpha_{\sigma(i),\sigma(k)})$. □

Theorem 3.8. *The function* $\tilde{\mathcal{E}}(P,s)$ *has a meromorphic continuation to the whole* s*-plane and satisfies the functional equation*

$$\begin{aligned}\tilde{\mathcal{E}}(P,s) &= \frac{2\pi}{\sqrt{|d_K|}\, s}\, \mathcal{A}(s)^{-1}\, \mathcal{A}(s-1)\, T\, \tilde{\mathcal{E}}(P,-s) \\ &= \frac{2\pi}{\sqrt{|d_K|}\, s}\, \Omega \left(\frac{L(s,\chi_i)}{L(s+1,\chi_i)}\, \delta_{ik} \right) \Omega^t\, \tilde{\mathcal{E}}(P,-s).\end{aligned} \tag{3.24}$$

Every $E_{\mathfrak{m}}(P,s)$ *is holomorphic for* $\operatorname{Re} s > 0$ *except for a simply pole at* $s = 1$ *with residue*

$$\operatorname*{Res}_{s=1} E_{\mathfrak{m}}(P,s) = \frac{4\pi^2}{|d_k|\, \zeta_K(2)}. \tag{3.25}$$

Proof. We deduce from (3.22), (3.23), and (1.10)–(1.12)

$$\begin{aligned}\tilde{\mathcal{E}}(P,s) &= \frac{2\pi}{\sqrt{|d_K|}\, s}\, \mathcal{A}(s)^{-1}\, \mathcal{A}(s-1)\, T\, \tilde{\mathcal{E}}(P,-s) \\ &= \frac{2\pi}{\sqrt{|d_K|}\, s}\, \Omega \left(\frac{L(s,\chi_i)}{L(s+1,\chi_i)}\, \delta_{ik} \right) \Omega^{-1}\, T\, \tilde{\mathcal{E}}(P,-s).\end{aligned}$$

But (1.13) and the definition of T imply $\Omega^{-1}T = \Omega^t$, and (3.24) is proved.

The matrix $\mathcal{A}(s)$ is holomorphic and non-singular for $\operatorname{Re} s > 0$ by Theorem 1.5. Hence $\tilde{\mathcal{E}}(P,s)$ is holomorphic for $\operatorname{Re} s > 0$ except for at most a simple pole at $s = 1$. If $\nu \in \{1, \dots, h\}$ is such that $\chi_\nu = 1$ is the trivial character, we find from (1.12), (1.13) and (3.2):

$$\operatorname*{Res}_{s=1} \tilde{\mathcal{E}}(P,s) = \lim_{s \to 1} \mathcal{A}(s)^{-1}\,(s-1)\,\hat{\mathcal{E}}(P,s) = \frac{4\pi^2}{|d_K|}\,\mathcal{A}(1)^{-1}\begin{pmatrix}1\\ \vdots\\ 1\end{pmatrix}$$

$$= \frac{4\pi^2}{|d_K|}\,\Omega\,\mathcal{D}(1)^{-1}(\sqrt{h}\delta_{\nu k})_{k=1,\dots h} = \frac{4\pi^2}{|d_K|\,\zeta_K(2)}\begin{pmatrix}1\\ \vdots\\ 1\end{pmatrix}. \qquad \square$$

Defining the $h \times h$-matrix

$$\Psi(s) = \frac{2\pi}{\sqrt{|d_K|}\,s}\,\Omega\left(\frac{L(s,\chi_i)}{L(s+1,\chi_i)}\,\delta_{ik}\right)\Omega^t \tag{3.26}$$

we can summarize the following properties of $\Psi(s)$.

Corollary 3.9. *The matrix $\Psi(s)$ satisfies the functional equation*

$$\tilde{\mathcal{E}}(P,s) = \Psi(s)\,\tilde{\mathcal{E}}(P,-s). \tag{3.27}$$

We furthermore have $\Psi(s)\Psi(-s) = I$ where I denotes the $h \times h$ unit matrix and $\overline{\Psi(\bar{s})}^t = \Psi(s)$. In particular, $\Psi(s)$ is Hermitian provided that s is real and Ψ is holomorphic at s. Ψ is holomorphic on the imaginary axis, and $\Psi(it)$ ($t \in \mathbb{R}$) is unitary. The eigenvalues of $\Psi(0)$ are equal to ± 1.

The equation $\overline{\Psi(\bar{s})}^t = \Psi(s)$ follows from the fact that (1.12) remains true with $\bar{\chi}_1, \dots, \bar{\chi}_h$ instead of $\chi_1, \dots \chi_h$. One can deduce the holomorphy of Ψ on the imaginary axis also from the definition (3.26) by means of the non-vanishing of the L-functions on $\operatorname{Re} s = 1$ and the functional equation of the L-functions (see Lang (1968), page 299).

The properties of the matrix $\Psi(s)$ collected in Corollary 3.9 are similar to those satisfied by the scattering matrix Φ of Section 6.1. In fact the two matrices are closely related. To explain the connection we fix a representative system $\eta_1, \dots, \eta_h$ for the $\mathbf{PSL}(2,\mathcal{O})$-classes of cusps. We choose quasi-integral matrices $B_1, \dots, B_h \in \mathbf{PSL}(2,K)$ with $\eta_1 = B_1^{-1}\infty, \dots, \eta_h = B_h^{-1}\infty$. We define the diagonal matrix

$$F(s) := \left(\frac{N\mathfrak{u}_{B_i}^{-1-s}}{|\mathcal{O}^*|}\,\delta_{ik}\right).$$

Note that we have $[\mathbf{PSL}(2,\mathcal{O})_{\eta_i} : \mathbf{PSL}(2,\mathcal{O})'_{\eta_i}] = |\mathcal{O}^*|/2$ for $i = 1,\ldots,h$. Hence we find by (1.20)

$$\mathcal{E}(P,s) = F(s)\,\tilde{\mathcal{E}}(P,s) \tag{3.28}$$

where $\mathcal{E}(P,s)$ is the vector of Eisenstein series from Definition 6.1.1. Let Φ now be the scattering matrix defined with respect to the above data in Definition 6.1.1. From the functional equations (6.1.8) and (3.27) we conclude

$$\Phi(s) = F(s)\,\Psi(s)\,F(-s)^{-1} \tag{3.29}$$

by using (3.28) and the identity theorem for meromorphic functions. In the trace formula of Chapter 6 the determinant of the scattering matrix and also $\operatorname{tr}\Phi(0)$ play an important role. Formula (3.29) gives us the possibility to compute these data.

Theorem 3.10 *Let H be the Hilbert class field of K and t the number of distinct prime divisors of d_K. Then*

$$\det\Phi(s) = (-1)^{\frac{h-2^{t-1}}{2}}\, N(\mathfrak{u}_1 \ldots \mathfrak{u}_h)^{-2s} \left(\frac{2\pi}{\sqrt{|d_K|}\, s}\right)^h \frac{\zeta_H(s)}{\zeta_H(s+1)}, \tag{3.30}$$

$$\operatorname{tr}\Phi(0) = 2^{t-1} - 2. \tag{3.31}$$

Proof. Formula (3.30) comes from (3.26), (3.29). As already mentioned around (1.16) the zeta function of H satisfies

$$\zeta_H(s) = \prod_{k=1}^{h} L(s+1, \chi_k).$$

Since $\Omega^{-1}T = \Omega^t$ it remains to compute the determinant of the matrix T. Obviously it is $\det T = (-1)^r$ where r is the number of elements of order 2 in the ideal class group of K. This number is by genus theory (Hecke (1923)) computed as $r = 2^{t-1}$.

To prove (3.31) we have by (3.29) to determine $\operatorname{tr}\Psi(0)$. Let us first introduce the completed L-function

$$\Lambda(s,\chi) = \left(\frac{\sqrt{|d_K|}}{2\pi}\right)^s \Gamma(s)\, L(s,\chi).$$

By Lang (1967), Chapter 14 the function $\Lambda(s,\chi)$ is meromorphic in $\mathbb{C}$ and satisfies the functional equation $W(\chi)\Lambda(s,\chi) = \Lambda(1-s,\overline{\chi})$ where $W(\chi)$ is the root number and $\overline{\chi}$ is the complex conjugate of χ. In our case χ is totaly unramified in the terminology of Lang (1967). This implies $W(\chi) = 1$. If $\chi = \chi_1$ is the trivial character it has a pole of order 1 at $s = 1$. Otherwise it is holomorphic and non-zero (see Section 6) at $s = 1$. Using this and the functional equation we see that the limit

$$l(\chi) := \lim_{s \to 0} \left(\frac{2\pi}{\sqrt{d_K}\, s} \frac{L(s,\chi)}{L(s+1,\chi)} \right)$$

exists. If χ is a character of order 2, that is $\chi = \overline{\chi}$ we have

$$l(\chi) = \begin{cases} -1 & \chi = \chi_1, \\ 1 & \chi \neq \chi_1,\, \chi^2 = 1. \end{cases}$$

Let $\chi_1, \dots, \chi_h$ be an enumeration of the characters. From (3.26) we infer

$$\operatorname{tr} \Psi(0) = \operatorname{tr}\, \Omega \begin{pmatrix} -1 & 0 & 0 & \dots & 0 \\ 0 & l(\chi_2) & 0 & \dots & 0 \\ \vdots & \vdots & \vdots & \ddots & \vdots \\ 0 & 0 & 0 & \dots & l(\chi_h) \end{pmatrix} \Omega^t$$

From here it is only an exercise in linear algebra to find (3.31). □

Formulas (3.30), (3.31) were proved by Efrat, Sarnak (1985) and by Elstrodt, Grunewald, Mennicke (1985). Formula (3.30) can be simplified if we introduce the completed zeta function of the number field H:

$$\Lambda_H(s) = \left(\frac{\sqrt{|d_K|}}{2\pi} \right)^{hs} \Gamma(s)^h\, \zeta_H(s).$$

This function satisfies $\Lambda_H(1-s) = \Lambda_H(s)$ again by Lang (1968), Chapter 14. We get

$$\det \Phi(s) = (-1)^{\frac{h-2^{t-1}}{2}}\, N(\mathfrak{u}_1 \dots \mathfrak{u}_h)^{-2s} \frac{\Lambda_H(s)}{\Lambda_H(s+1)}. \tag{3.32}$$

8.4 Special Values of Eisenstein Series

In certain cases it is possible to evaluate $\hat{E}_{\mathfrak{m}}(P,s)$ at special points $P \in \mathbb{H}$ in terms of zeta functions. Then the results of Sections 2, 3 yield remarkable relations between various functions and between special values of those functions.

For simplicity we choose $\mathfrak{m} = \mathcal{O}$, and we mostly restrict to the case $P = rj \in \mathbb{H}$ $(r > 0)$. Then we have

$$\|cP + d\|^2 = |d|^2 + r^2 |c|^2 \qquad (c, d \in \mathcal{O}). \tag{4.1}$$

If r^2 is a natural number, then the computation of $\hat{E}_{\mathcal{O}}(rj, s)$ amounts to the summation of the Dirichlet series

$$\hat{E}_{\mathcal{O}}(rj, s) = r^{1+s} \sum_{n=1}^{\infty} \frac{a_n(r^2)}{n^{1+s}}, \tag{4.2}$$

where $a_n(r^2)$ is the number of representations of n by the quaternary quadratic form corresponding to (4.1) when $c, d \in \mathcal{O}$ are expressed in terms of a $\mathbb{Z}$-basis for $\mathcal{O}$.

We define for $n, k \in \mathbb{N}$

$$\sigma_1(n, k) := \sum_{\substack{d|n,\ k \nmid d \\ d>0}} d. \tag{4.3}$$

The associated Dirichlet series is

$$\sum_{n=1}^{\infty} \frac{\sigma_1(n, k)}{n^{s+1}} = \left(1 - \frac{1}{k^s}\right) \zeta(s)\,\zeta(s+1), \tag{4.4}$$

where $\zeta(s)$ denotes the Riemann zeta function.

Theorem 4.1. *For $K = \mathbb{Q}(i)$ we have:*

$$\hat{E}_{\mathbb{Z}[i]}(j, s) = 8(1 - 2^{-2s})\,\zeta(s)\,\zeta(s+1), \tag{4.5}$$

$$\hat{E}_{\mathbb{Z}[i]}(\sqrt{2}j, s) = 2^{\frac{s+3}{2}} \left(2(1 - 2^{-3s}) - (2^{-s} - 2^{-2s})\right) \zeta(s)\,\zeta(s+1), \tag{4.6}$$

$$\begin{aligned} \hat{E}_{\mathbb{Z}[i]}(2j, s) = {} & 2\left(2(2^s - 2^{-3s}) - (1 - 2^{-2s})\right) \zeta(s)\,\zeta(s+1) \\ & + 2^{s+2} L(s)\, L(s+1), \end{aligned} \tag{4.7}$$

where

$$L(s) = \sum_{k=0}^{\infty} \frac{(-1)^k}{(2k+1)^s}, \tag{4.8}$$

$$\hat{E}_{\mathbb{Z}[i]}(\sqrt{3}j, s) = 4 \cdot 3^{\frac{s+1}{2}} (1 - 2^{-s} + 2^{-2s})(1 - 3^{-s})\,\zeta(s)\,\zeta(s+1), \tag{4.9}$$

$$\hat{E}_{\mathbb{Z}[i]}(\sqrt{5}j, s) = 5^{\frac{s+1}{2}} \left(\frac{4(1 - 4^{-s})(1 + 5^{-s})\,\zeta(s)\,\zeta(s+1)}{3} + \frac{8D(s)}{3}\right), \tag{4.10}$$

where

$$D(s) = \sum_{n=1}^{\infty} \frac{\beta(n)}{n^{s+1}} \tag{4.11}$$

is the Dirichlet series associated with the following product of Dedekind η-functions:

$$\eta^2(\tau)\,\eta^2(5\tau) = \sum_{n=1}^{\infty} \beta(n)\, e^{\pi i n \tau} \qquad (\operatorname{Im}\tau > 0). \tag{4.12}$$

The function $D(s)$ is an entire function and satisfies the functional equation $R(s) = R(-s)$ with $R(s) = (\pi/\sqrt{5})^{-s}\,\Gamma(1+s)\,D(s)$. We further have

$$\hat{E}_{\mathbb{Z}[i]}\left(\frac{1}{2} + \frac{1}{2}\sqrt{3}j, s\right) = 6 \cdot 2^{-s}\, 3^{\frac{s+1}{2}}\,(1 - 3^{-s})\,\zeta(s)\,\zeta(s+1). \tag{4.13}$$

Before we enter into the details of the proofs we point out that (4.5), (4.6), (4.7), (4.9), (4.13) are all compatible with the functional equation (3.1). This is due to the fact that $\varphi(s) := (2\pi)^{-s}\,\Gamma(s)\,\zeta(s)\,\zeta(s+1)$ satisfies the functional equation $\varphi(s) = \varphi(-s)$. The remaining elementary factors are even functions of s, and the functional equation of the L-series also concords with (3.1). An analogous remark applies to Corollaries 4.2, 4.3, and to Theorem 4.4.

Proof. We start from (4.2) where $a_n(r^2)$ now is the number of integral representations $n = m_1^2 + m_2^2 + r^2(m_3^2 + m_4^2)$.

(4.5): For $r = 1$ we have Jacobi's formula $a_n(1) = 8\sigma_1(n, 4)$. Hence (4.5) is obvious from (4.4).

(4.6): For $r = \sqrt{2}$ we have

$$a_n(2) = \begin{cases} 4\sigma_1(n) & \text{if} \quad n \text{ is odd}, \\ 8\sigma_1(m) & \text{if} \quad m = n/2 \text{ is odd}, \\ 24\sigma_1(m) & \text{if} \quad n = 2^{\nu} m \text{ with } m \text{ odd and } \nu \geq 2, \end{cases}$$

where $\sigma_1(n)$ equals the sum of the positive divisors of $n \in \mathbb{N}$. This result was stated by Liouville in 1860 (see Dickson (1952), page 227) and proved by Pepin (1890), page 40; for a recent proof see Petersson (1982), page 154, Satz 15.2. An elementary computation yields

$$\hat{E}_{\mathbb{Z}[i]}(\sqrt{2}j, s) = 4 \cdot 2^{\frac{s+1}{2}} \frac{(1 - 2^{-3s}) - 2^{-1}(2^{-s} - 2^{-2s})}{(1 - 2^{-s})(1 - 2^{-1-s})} \sum_{k=0}^{\infty} \frac{\sigma_1(2k+1)}{(2k+1)^{s+1}}.$$

Since

$$\begin{aligned} \zeta(s)\,\zeta(s+1) &= \sum_{n=1}^{\infty} \frac{\sigma_1(n)}{n^{s+1}} = \sum_{k=0}^{\infty}\sum_{\nu=0}^{\infty} \frac{\sigma_1(2^{\nu}(2k+1))}{(2^{\nu}(2k+1))^{s+1}} \\ &= (1 - 2^{-s})^{-1}(1 - 2^{-1-s})^{-1} \sum_{k=0}^{\infty} \frac{\sigma_1(2k+1)}{(2k+1)^{s+1}}, \end{aligned} \tag{4.14}$$

we obtain (4.6).

(4.7): For $r = 2$ it is known that

$$a_n(4) = \begin{cases} 0 & \text{for} \quad n = 4k+3, \\ 4\sigma_1(4k+1) & \text{for} \quad n = 4k+1, \\ 4\sigma_1(m) & \text{for} \quad n/2 = m \text{ odd}, \\ 8\sigma_1(m) & \text{for} \quad n/4 = m \text{ odd}, \\ 24\sigma_1(m) & \text{for} \quad n = 2^\nu m \text{ with } \nu \geq 3 \text{ and } m \text{ odd}. \end{cases}$$

This statement of Liouville was proved by Pepin (1890), pages 41–42, see Dickson (1952), page 227. Looking at (4.6) we find

$$\hat{E}_{\mathbb{Z}[i]}(2j, s) = 2^{s+3} \sum_{k=0}^{\infty} \frac{\sigma_1(4k+1)}{(4k+1)^{s+1}} + 2^{-\frac{s+1}{2}} \hat{E}_{\mathbb{Z}[i]}(\sqrt{2}j, s). \tag{4.15}$$

Hence we have

$$2\sum_{k=0}^{\infty} \frac{\sigma_1(4k+1)}{(4k+1)^{s+1}} = \sum_{k=0}^{\infty} \frac{\sigma_1(2k+1)}{(2k+1)^{s+1}} + \sum_{k=0}^{\infty} \frac{(-1)^k \sigma_1(2k+1)}{(2k+1)^{s+1}}. \tag{4.16}$$

The first sum is known from (4.14), and

$$L(s)\, L(s+1) = \sum_{k=0}^{\infty} \frac{(-1)^k \sigma_1(2k+1)}{(2k+1)^{s+1}}. \tag{4.17}$$

Combining (4.6), (4.14), (4.16) and (4.17) with (4.15) we obtain (4.7).

(4.9): For $r = \sqrt{3}$ we have

$$a_n(3) = \begin{cases} 4\sigma_1(n,3) & \text{if} \quad n \text{ is odd}, \\ 4(2^{\nu+1}-3)\sigma_1(m,3) & \text{if} \quad n = 2^\nu m \text{ with } m \text{ odd and } \nu \geq 1. \end{cases}$$

This result was also communicated by Liouville in 1860 (see Dickson (1952), page 227). It was proved by Dickson (1927), Satz 26, as an application of the theory of quaternion algebras. The formula was recently reproved by Petersson (1982), Satz 11.3, by means of the theory of modular forms. Petersson states

$$a_n(3) = 4 \sum_{\substack{n \equiv 0 \bmod d, \\ d \not\equiv 0 \bmod 3}} (-1)^{n-d}\, d$$

which holds for all $n \in \mathbb{N}$. We find

$$\begin{aligned} \sum_{n=1}^{\infty} \frac{a_n(3)}{n^{s+1}} &= 4 \frac{1 - 2^{-s} + 2^{-2s}}{(1-2^{-s})(1-2^{-1-s})} \sum_{k=0}^{\infty} \frac{\sigma_1(2k+1,3)}{(2k+1)^{s+1}} \\ &= 4(1 - 2^{-s} + 2^{-2s})(1 - 3^{-s})\, \zeta(s)\, \zeta(s+1). \end{aligned}$$

The last identity is obvious on a brief look at (4.4) and (4.14). This yields (4.9).

(4.10): We draw from Petersson (1982), page 107, Satz 11.1 and page 106, (11.9) the formula

(4.18)
$$a_n(5) = \frac{4}{3}\left(\sigma_1(n,4) + 5\sigma_1\left(\frac{n}{5},4\right)\right) + \frac{8}{3}\beta(n)$$

where the integers $\beta(n)$ are defined by (4.12). Note that $\beta(n) = 0$ if n is even. Now (4.10) follows from (4.18) and (4.4). The results on the meromorphic continuation and functional equation of $D(s)$ follow from Theorem 3.1 combined with the remarks after Theorem 4.1.

(4.13): For $P = \frac{1}{2} + \frac{1}{2}\sqrt{3}j$ and $c = m_1 + im_3$, $d = m_2 + im_4 \in \mathbb{Z}[i]$ we have

(4.19)
$$\|cP + d\|^2 = m_1^2 + m_1 m_2 + m_2^2 + m_3^2 + m_3 m_4 + m_4^2.$$

The number a_n of representations of the natural number n by the quaternary quadratic from (4.19) was determined by Dickson (1927), Satz 19. He finds $a_n = 12\sigma_1(n,3)$. This result was confirmed by Petersson (1982), page 80. Hence (4.13) follows from (4.4). □

Corollary 4.2. *For* $K = \mathbb{Q}(\sqrt{-2})$ *we have*

(4.20)
$$\hat{E}_{\mathcal{O}}(j,s) = 2\left(2(1-2^{-3s}) - (2^{-s} - 2^{-2s})\right)\zeta(s)\,\zeta(s+1).$$

Proof. $\{1, \sqrt{-2}\}$ is a $\mathbb{Z}$-basis for $\mathcal{O}$. For $c = m_1 + m_2\sqrt{-2}$, $d = m_3 + m_4\sqrt{-2}$ with $m_1, \ldots, m_4 \in \mathbb{Z}$ we find $\|cj + d\|^2 = m_1^2 + 2m_2^2 + m_3^2 + 2m_4^2$ and hence

$$\hat{E}_{\mathcal{O}}(j,s) = 2^{-\frac{s+1}{2}}\,\hat{E}_{\mathbb{Z}[i]}(\sqrt{2}j, s).$$

The latter function is given by (4.6). □

One can also write down a formula for $\hat{E}_{\mathcal{O}}(2j,s)$ by means of Liouville's result quoted by Dickson (1952), page 227, but here an otherwise unknown Dirichlet series comes up. The same applies to several other cases (see Dickson (1952), page 228).

Corollary 4.3. *For* $K = \mathbb{Q}(\sqrt{-3})$ *we have*

(4.21)
$$\hat{E}_{\mathcal{O}}(j,s) = 12(1-3^{-s})\,\zeta(s)\,\zeta(s+1).$$

Proof. Now $\{1, (1+\sqrt{-3})/2\}$ is a $\mathbb{Z}$-basis for $\mathcal{O}$, and the values of $\|cj+d\|^2$ $(c, d \in \mathcal{O})$ are also given by the form (4.19). Hence we obtain

$$\hat{E}_{\mathcal{O}}(j,s) = \left(\frac{\sqrt{3}}{2}\right)^{-1-s}\hat{E}_{\mathbb{Z}[i]}\left(\frac{1}{2} + \frac{1}{2}\sqrt{3}j, s\right),$$

and (4.13) yields the assertion. □

Similarly, equation (4.10) yields an evaluation of $\hat{E}_{\mathcal{O}}(j,s)$ in the case $K = \mathbb{Q}(\sqrt{-5})$.

Theorem 4.4. *For $K = \mathbb{Q}(\sqrt{-7})$ we have*

$$\hat{E}_{\mathcal{O}}(j,s) = 4(1-7^{-s})\,\zeta(s)\,\zeta(s+1). \tag{4.22}$$

Proof. The ring $\mathcal{O}$ of integers of $\mathbb{Q}(\sqrt{-7})$ has the basis $\{1,(1+\sqrt{-7})/2\}$, and for $c = m_1 + m_2(1+\sqrt{-7})/2$, $d = m_3 + m_4(1+\sqrt{-7})/2$ with $m_1,\ldots,m_4 \in \mathbb{Z}$ we have $|c|^2 + |d|^2 = m_1^2 + m_1m_2 + 2m_2^2 + m_3^2 + m_3m_4 + 2m_4^2$. The number a_n of representations of the natural number n by this form was determined by Dickson (1927), Satz 35, by means of the theory of quaternion algebras. He finds $a_n = 4\sigma_1(n,7)$. This result was recently confirmed by Petersson (1982), Satz 8.3. Now (4.22) is immediate from (4.4). □

The results of Petersson (1982) can also be used to evaluate certain linear combinations of Eisenstein series in terms of $\zeta(s)\,\zeta(s+1)$.

Theorem 4.5. *Let $K = \mathbb{Q}(i)$, and define $L(s)$ by (4.8), $\sigma_{-s}(\mu)$ by (2.18) with $\mathfrak{a} = \mathfrak{b} = \mathcal{O}$. Then the following equations hold:*

$$\begin{aligned} 2(1-2^{-2s})\,\zeta(s)\,\zeta(s+1) &= \zeta(s+1)\,L(s+1) + \frac{\pi}{s}\zeta(s)\,L(s) \\ &\quad + \frac{2\pi^{1+s}}{\Gamma(1+s)} \sum_{0\neq\mu\in\mathbb{Z}[i]} |\mu|^s\,\sigma_{-s}(\mu)\,K_s(2\pi|\mu|), \end{aligned} \tag{4.23}$$

$$\begin{aligned} &\left(1-2^{-3s}-\frac{2^{-s}-2^{-2s}}{2}\right)\zeta(s)\,\zeta(s+1) = \zeta(s+1)\,L(s+1) \\ &+ \frac{\pi}{s}\,2^{-s}\,\zeta(s)\,L(s) + \frac{2^{1-\frac{s}{2}}\pi^{1+s}}{\Gamma(1+s)} \sum_{0\neq\mu\in\mathbb{Z}[i]} |\mu|^s\,\sigma_{-s}(\mu)\,K_s(2\sqrt{2}\pi|\mu|), \end{aligned} \tag{4.24}$$

$$\begin{aligned} &\left(1-2^{-4s}-\frac{2^{-s}-2^{-3s}}{2}\right)\zeta(s)\,\zeta(s+1) + L(s)\,L(s+1) \\ &= 2\zeta(s+1)\,L(s+1) + \frac{\pi}{s}\,2^{1-2s}\,\zeta(s)\,L(s) \\ &+ \frac{2^{2-s}\pi^{1+s}}{\Gamma(1+s)} \sum_{0\neq\mu\in\mathbb{Z}[i]} |\mu|^s\,\sigma_{-s}(\mu)\,K_s(4\pi|\mu|), \end{aligned} \tag{4.25}$$

$$\begin{aligned} &(1-2^{-s}+2^{-2s})\,(1-3^{-s})\,\zeta(s)\,\zeta(s+1) = \zeta(s+1)\,L(s+1) \\ &+ \frac{\pi}{s}\,3^{-s}\,\zeta(s)\,L(s) + \frac{2\cdot 3^{-\frac{s}{2}}\pi^{1+s}}{\Gamma(s+1)} \sum_{0\neq\mu\in\mathbb{Z}[i]} |\mu|^s\,\sigma_{-s}(\mu)\,K_s(2\sqrt{3}\pi|\mu|), \end{aligned} \tag{4.26}$$

$$\begin{aligned}
3(1-3^{-s})\,\zeta(s)\,\zeta(s+1) &= \zeta(s+1)\,L(s+1) \\
&+ \frac{\pi}{s}\,2^{2s}\,3^{-s}\,\zeta(s)\,L(s) \\
&+ \frac{(2\pi)^{1+s}}{\Gamma(s+1)}\,3^{-\frac{s}{2}} \sum_{0\neq\mu\in\mathbb{Z}[i]} |\mu|^s\,\sigma_{-s}(\mu)\,K_s(\sqrt{3}\pi|\mu|)\,e^{-\pi i\,\mathrm{Im}\,\mu}.
\end{aligned} \tag{4.27}$$

Proof. We put the results of Theorem 4.1 into the Fourier expansion (2.21). Then (4.23) is immediate from (4.5), (4.24) follows from (4.6), (4.25) from (4.7), (4.26) from (4.9), and (4.27) from (4.13). □

Other interesting identities come up, if we put (4.20) into (2.21). Remember that the zeta function of $K = \mathbb{Q}(\sqrt{D})$ factors into $\zeta_K(s) = \zeta(s)\,L(s,\chi_K)$ with

$$L(s,\chi_K) = \sum_{n=1}^{\infty} \left(\frac{d_K}{n}\right) n^{-s}.$$

The character χ_K is defined in Section 7.1. Since $\mathbb{Q}(\sqrt{-2})$ has discriminant -8 and class number one, we obtain the following result.

Corollary 4.6. *For $K = \mathbb{Q}(\sqrt{-2})$ we have*

$$\begin{aligned}
&(2(1-2^{-3s}) - 2^{-s} + 2^{-2s})\,\zeta(s)\,\zeta(s+1) \\
&= \zeta(s+1)\,L(s+1,\chi_{\mathbb{Q}(\sqrt{-2})}) + \frac{\pi}{\sqrt{2}s}\,\zeta(s)\,L(s,\chi_{\mathbb{Q}(\sqrt{-2})}) \\
&+ \frac{2^{\frac{1-s}{2}}\pi^{1+s}}{\Gamma(1+s)} \sum_{0\neq\mu\in\mathbb{Z}[\sqrt{-2}]} |\mu|^s\,\sigma_{-s}(\mu)\,K_s(\sqrt{2}\pi|\mu|)
\end{aligned} \tag{4.28}$$

with

$$\sigma_{-s}(\omega) = \sum_{(\mu)|(\omega)} N\mu^{-s}. \tag{4.29}$$

Corollary 4.7. *For $K = \mathbb{Q}(\sqrt{-3})$ we have*

$$\begin{aligned}
&2(1-3^{-s})\,\zeta(s)\,\zeta(s+1) \\
&= \zeta(s+1)\,L(s+1,\chi_{\mathbb{Q}(\sqrt{-3})}) + \frac{2\pi}{\sqrt{3}s}\,\zeta(s)\,L(s,\chi_{\mathbb{Q}(\sqrt{-3})}) \\
&+ \frac{2^{2+s}\pi^{1+s}}{3^{\frac{s+1}{2}}\Gamma(1+s)} \sum_{0\neq\mu\in\mathcal{O}} |\mu|^s\,\sigma_{-s}(\mu)\,K_s\left(\frac{4\pi|\mu|}{\sqrt{3}}\right),
\end{aligned} \tag{4.30}$$

where definition (4.29) applies analogously.

The field $\mathbb{Q}(\sqrt{-7})$ also has class number one and its discriminant is -7. Hence (4.22) and (2.21) yield

Corollary 4.8. *For $K = \mathbb{Q}(\sqrt{-7})$ we have*

$$\begin{aligned} &2(1-7^{-s})\,\zeta(s)\,\zeta(s+1) \\ &= \zeta(s+1)\,L(s+1,\chi_{\mathbb{Q}(\sqrt{-7})}) + \frac{2\pi}{\sqrt{7}s}\,\zeta(s)\,L(s,\chi_{\mathbb{Q}(\sqrt{-7})}) \\ &+ 2\frac{(2\pi)^{1+s}}{7^{\frac{1+s}{2}}\Gamma(1+s)} \sum_{0\neq\mu\in\mathcal{O}} |\mu|^s\,\sigma_{-s}(\mu)\,K_s\left(\frac{4\pi|\mu|}{\sqrt{7}}\right). \end{aligned} \tag{4.31}$$

If we evaluate (4.23)–(4.31) at special points $s \in \mathbb{C}$, interesting identities between special values of zeta- und L-functions arise. We give some simple examples. The evaluation of (4.23) at $s = 2$ is based on the special values $\zeta(2) = \pi^2/6$, $L(3) = \pi^3/32$ which are both due to Euler (Introductio in analysin infinitorum, paragraphs 168 and 175). The number

$$L(2) = \sum_{k=0}^{\infty} \frac{(-1)^k}{(2k+1)^2} = 0.91596559\ldots$$

is known as Catalan's constant. Letting $s = 2$ in (4.23) we obtain

$$\frac{1}{16}\left(\frac{5}{\pi} - \frac{1}{2}\right)\zeta(3) - \frac{L(2)}{12} = \sum_{0\neq\mu\in\mathbb{Z}[i]} |\mu|^2\,\sigma_{-2}(\mu)\,K_2(2\pi|\mu|), \tag{4.32}$$

and the evaluation at $s = 3$ yields

$$\frac{11\,\zeta(3)}{320} - \frac{L(4)}{30} = \sum_{0\neq\mu\in\mathbb{Z}[i]} |\mu|^3\,\sigma_{-3}(\mu)\,K_3(2\pi|\mu|). \tag{4.33}$$

For the evaluation for example of (4.30) at $s = 2$ or $s = 3$ we need the special value

$$L(3,\chi_{\mathbb{Q}(\sqrt{-3})}) = \frac{4\pi^3}{81\sqrt{3}}$$

which is due to Euler, see also Petersson (1982), page 90 or Zagier (1981a), page 53. Now the evaluation of (4.30) at $s = 2$ yields

$$\left(\frac{\sqrt{3}}{\pi} - \frac{1}{6}\right)\frac{\zeta(3)}{9} - \frac{L(2,\chi_{\mathbb{Q}(\sqrt{-3})})}{16} = \sum_{0\neq\mu\in\mathbb{Z}[\rho]} |\mu|^2\,\sigma_{-2}(\mu)\,K_2\left(\frac{4\pi|\mu|)}{\sqrt{3}}\right)$$

with $\rho = \exp(\frac{z\pi i}{3})$, and for $s = 3$ we find

$$\frac{38\,\zeta(3)}{135} - \frac{3\,L(4,\chi_{\mathbb{Q}(\sqrt{-3})})}{10} = 16 \sum_{0\neq\mu\in\mathbb{Z}[\rho]} |\mu|^3\,\sigma_{-3}(\mu)\,K_3\left(\frac{4\pi|\mu|)}{\sqrt{3}}\right).$$

Petersson (1982), page 90 and Zagier (1981a), page 53 also give the value

$$L(3, \chi_{\mathbb{Q}(\sqrt{-7})}) = \frac{32\pi^3}{7^3 \cdot \sqrt{7}},$$

and we find from (4.31) upon evaluation at $s = 2$

$$\begin{aligned} &\frac{2}{7}\left(\frac{\sqrt{7}}{\pi} - \frac{2}{7}\right)\zeta(3) - \frac{7}{48}L(2, \chi_{\mathbb{Q}(\sqrt{-7})}) \\ &= \sum_{0 \neq \mu \in \mathbb{Z}[(1+\sqrt{-7})/2]} |\mu|^2 \, \sigma_{-2}(\mu) \, K_2\left(\frac{4\pi|\mu|)}{\sqrt{7}}\right), \end{aligned} \tag{4.34}$$

and similarly at $s = 3$

$$\begin{aligned} &\frac{239}{2^3 \cdot 5 \cdot 7^2}\zeta(3) - \frac{7^2}{2^5 \cdot 3 \cdot 5}L(4, \chi_{\mathbb{Q}(\sqrt{-7})}) \\ &= \sum_{0 \neq \mu \in \mathbb{Z}[(1+\sqrt{-7})/2]} |\mu|^3 \, \sigma_{-3}(\mu) \, K_3\left(\frac{4\pi|\mu|}{\sqrt{7}}\right). \end{aligned} \tag{4.35}$$

It is also interesting to choose e.g. $s = \frac{1}{2}$ in (4.23)–(4.31). Then the series with the Bessel functions contains elementary functions only. We now evaluate the formulas of Theorem 4.1 at $s = 0$ by means of Theorem 3.2.

Theorem 4.9. *For $0 \neq \mu \in \mathbb{Z}[i]$ let $d(\mu) := \sum_{(\lambda)|(\mu)} 1$. Then the following identities hold:*

$$4 \sum_{0 \neq \mu \in \mathbb{Z}[i]} d(\mu) \, K_0(2\pi|\mu|) = 4 \log \Gamma\left(\frac{1}{4}\right) - \frac{4}{\pi}\log 2 - \gamma - 2\log 2\pi, \tag{4.36}$$

$$\begin{aligned} 4 \sum_{0 \neq \mu \in \mathbb{Z}[i]} d(\mu) \, K_0(\sqrt{8}\pi|\mu|) &= 4 \log \Gamma\left(\frac{1}{4}\right) - \left(\frac{5}{2\pi} + \frac{1}{2}\right)\log 2 \\ &\quad - \gamma - 2\log 2\pi, \end{aligned} \tag{4.37}$$

$$\begin{aligned} 4 \sum_{0 \neq \mu \in \mathbb{Z}[i]} d(\mu) \, K_0(4\pi|\mu|) &= 4 \log \Gamma\left(\frac{1}{4}\right) - \left(\frac{3}{2\pi} + 1\right)\log 2 + \frac{1}{8} \\ &\quad - \gamma - 2\log 2\pi, \end{aligned} \tag{4.38}$$

$$\begin{aligned} 4 \sum_{0 \neq \mu \in \mathbb{Z}[i]} d(\mu) \, K_0(2\sqrt{3}\pi|\mu|) &= 4 \log \Gamma\left(\frac{1}{4}\right) - \left(\frac{1}{\pi} + \frac{1}{2}\right)\log 3 \\ &\quad - \gamma - 2\log 2\pi, \end{aligned} \tag{4.39}$$

$$(4.40)\qquad 4\sum_{0\neq\mu\in\mathbb{Z}[i]} d(\mu)\,K_0(\sqrt{3}\pi|\mu|)\,e^{-\pi i\,\mathrm{Im}\,\mu} = 4\log\Gamma\left(\frac{1}{4}\right) - \left(\frac{3}{\pi}+\frac{1}{2}\right)\log 3 + \log 2 - \gamma - 2\log 2\pi.$$

Proof. We evaluate (3.11) with $B = I$ at the special points occurring in Theorem 4.1. Then the special value

$$g(\mathbb{Z}[i]) = (2\pi)^{-12}|\Delta(i)| = (2\pi)^{-12}\Delta(i)$$

comes up. Since $g_3(i) = 0$, we have

$$\Delta(i) = g_3^2(i) = \left(60\sum_{\substack{m,n\in\mathbb{Z}\\(m,n)\neq(0,0)}}(m+ni)^{-4}\right)^3.$$

Now Hurwitz (1899) proved that

$$(4.41)\qquad \sum_{\substack{m,n\in\mathbb{Z}\\(m,n)\neq(0,0)}}(m+ni)^{-4} = \frac{2^4}{4!}\frac{\omega^4}{10}$$

with the number

$$(4.42)\qquad \omega := 2\int_0^1\frac{dx}{\sqrt{1-x^4}}$$

as an analogue of π. Expressing (4.42) in terms of the gamma-function, we find

$$(4.43)\qquad \omega = \frac{\Gamma(\frac{1}{4})^2}{2\sqrt{2\pi}}$$

and hence

$$(4.44)\qquad \Delta(i) = 2^6\omega^{12} = \frac{\Gamma(\frac{1}{4})^{24}}{2^6(2\pi)^6},$$

$$(4.45)\qquad g\,(\mathbb{Z}[i]) = \frac{\Gamma(\frac{1}{4})^{24}}{2^6(2\pi)^{18}}.$$

Since

$$\zeta(0) = -\frac{1}{2},\qquad \zeta(s) = \frac{1}{s-1}+\gamma+O(s-1)\quad\text{as}\quad s\to 1,$$

equations (4.36), (4.37), (4.39), (4.40) follow from (4.5), (4.6), (4.9), (4.13), respectively. Using the known values $L(1) = \frac{\pi}{4}$, $L(0) = \frac{1}{2}$, (Zagier (1981a)) we obtain (4.38) from (4.7). □

Formulas (4.42)–(4.45) have a long history. The value (4.42) for ω was first obtained by Legendre in 1811, see Whittaker, Watson (1927), page 524. Essentially the same result was almost simultaneously published by Gauß in 1812, see Gauß' Werke, Band III, page 150. The numerical value of ω was already computed by Stirling up to 17 decimal places (the first 15 of which proved to be correct), and Gauß even computed ω up to 26 decimal places by means of the relation of ω with the theta function (see Gauß' Werke, Band III, page 150, page 413 and page 418). Entry [61] in Gauß' diary suggests that (4.41)–(4.43) were already known to Gauß in March 1797.

The result (4.44) was also proved by Ramanujan (1927), page 318, see also page 39 and the remarks on pages 335–336. There is an interesting remark by Ramanujan (1962), page 39 on the special cases in which the complete elliptic integral K can be expressed in terms of gamma-functions. This remark should be looked at in the light of the results of Chowla and Selberg (1967). Formula (4.44) was also communicated by Herglotz (1922). It also follows from the work of Chowla and Selberg (1967). For a disussion of this formula see Weil (1976), page 92.

Theorem 4.10. *For* $0 \neq \mu \in \mathbb{Z}[\sqrt{-2}]$ *let* $d(\mu) := \sum_{(\lambda)|(\mu)} 1$. *Then*

$$
\begin{aligned}
2 \sum_{0 \neq \mu \in \mathbb{Z}[\sqrt{-2}]} d(\mu)\, K_0(\sqrt{2}\pi|\mu|) &= 2 \log \left(\Gamma\left(\frac{1}{8}\right) \Gamma\left(\frac{3}{8}\right) \right) \\
&\quad - \left(1 + \frac{5}{\sqrt{2}\pi}\right) \log 2 - \gamma - 2 \log 2\pi.
\end{aligned}
\tag{4.46}
$$

Proof. For $K = \mathbb{Q}(\sqrt{-2})$ we have from (4.20)

$$
\lim_{s \to 0} \hat{E}_{\mathcal{O}}(j, s) = -5 \log 2. \tag{4.47}
$$

In order to evaluate this limit by means of (3.11) we need the special value

$$
g(\mathcal{O}) = g(\mathbb{Z}[\sqrt{-2}]) = (2\pi)^{-12} \Delta(\sqrt{2}i).
$$

Here we have $\Delta(\sqrt{2}i) = g_2^3(\sqrt{2}i) - 27 g_3^2(\sqrt{2}i)$, and we might use the values of the Eisenstein series as given by Dintzl (1909), (1914):

$$
\sum_{\substack{m,n \in \mathbb{Z} \\ (m,n) \neq (0,0)}} (m + n\sqrt{-2})^{-2k} = \frac{(4\omega)^{2k}}{(2k)!} G_k,
$$

where

$$
\omega := \sqrt{1 - \frac{1}{2}\sqrt{2}} \int_0^1 \frac{dx}{\sqrt{(1-x^2)(1-(\sqrt{2}-1)^2 x^2)}}, \quad G_2 = \frac{1}{3}, \quad G_3 = \frac{2}{3};
$$

see also Herglotz (1922). We are indebted to D. Brümmer (Hagen) for his communication of this value of ω in terms of a Legendre normal integral. E.A. Narishkina (1924) gives ω in terms of the Weierstrass normal form. For the present purpose we prefer the Chowla–Selberg formula (see Chowla, Selberg (1967), page 110, or Weil (1976), page 92) and find

$$\Delta(\sqrt{2}i) = \left(\frac{\pi}{4}\right)^6 \left(\frac{\Gamma(\frac{1}{8})\Gamma(\frac{3}{8})}{\Gamma(\frac{5}{8})\Gamma(\frac{7}{8})}\right)^6 = \frac{\left(\Gamma(\frac{1}{8})\Gamma(\frac{3}{8})\right)^{12}}{2^{15}\,(2\pi)^6}. \tag{4.48}$$

The theorem now follows from (3.11) and (4.47). □

Since

$$\Gamma\left(\frac{3}{8}\right)\Gamma\left(\frac{5}{8}\right) = \frac{\pi}{\cos\frac{\pi}{8}} = 2\pi\sqrt{1-\frac{1}{\sqrt{2}}}$$

and

$$\Gamma\left(\frac{1}{8}\right)\Gamma\left(\frac{5}{8}\right) = 2^{\frac{1}{4}}\sqrt{2\pi}\,\Gamma\left(\frac{1}{4}\right)$$

we can rewrite (4.48) in the form

$$\Delta(\sqrt{2}i) = \frac{(\sqrt{2}-1)^6}{2^{21}}\frac{\Gamma\left(\frac{1}{8}\right)^{24}}{\Gamma\left(\frac{1}{4}\right)^{12}}.$$

Theorem 4.11. *For* $0 \neq \mu \in \mathbb{Z}[\rho]$, $\rho = \exp(\frac{2\pi i}{3})$, *let* $d(\mu) = \sum_{(\lambda)|(\mu)} 1$. *Then*

$$6 \sum_{0\neq\mu\in\mathbb{Z}[\rho]} d(\mu)\, K_0\left(\frac{4\pi|\mu|}{\sqrt{3}}\right) = 6\log\Gamma\left(\frac{1}{3}\right) + \left(1-\frac{3\sqrt{3}}{2\pi}\right)\log 3 - \gamma - 3\log 2\pi. \tag{4.49}$$

Proof. We draw from (4.21) $\lim_{s\to 0}\hat{E}_{\mathcal{O}}(j,s) = -6\log 3$. An evaluation of this limit by means of (3.11) requires the special value

$$g(\mathbb{Z}[\rho]) = (2\pi)^{-12}|\Delta(\rho)| \tag{4.50}$$

where $\rho = \exp(\frac{2\pi i}{3})$. Since $g_2(\rho) = 0$, we have $\Delta(\rho) = -27(g_3(\rho))^2$. Now Matter (1900) proved that

$$\sum_{\substack{m,n\in\mathbb{Z}\\(m,n)\neq(0,0)}} (m+n\rho)^{-6} = \frac{2^6}{6!}\cdot\frac{3^2}{2^2\cdot 7}\omega^6$$

with the elliptic integral

$$\omega = 2\int_0^1 \frac{dx}{\sqrt{1-x^6}} = 2\int_1^\infty \frac{dt}{\sqrt{4t^3-4}}.$$

This result is also given by Narishkina (1924). Evaluating the integral in terms of the gamma-function we find

$$\omega = \frac{\Gamma(\frac{1}{3})^3}{2\,\pi\,\sqrt[3]{2}}$$

and hence

$$\Delta(\rho) = -2^4 \cdot 3^3\,\omega^{12} = -3^3\,(2\pi)^{-12}\,\Gamma\left(\frac{1}{3}\right)^{36}. \tag{4.51}$$

This result can be traced back to Legendre (1811); see Whittaker and Watson (1927), pages 525–526. Formula (4.51) was also given by Herglotz (1922). Alternatively, the value $g(\mathbb{Z}[\rho])$ can be obtained from the Chowla–Selberg formula (see Chowla, Selberg (1967), page 110, or Weil (1976), page 92). The result now follows from (3.11). □

Theorem 4.12. *For* $0 \neq \mu \in \mathcal{O} := \mathbb{Z}[\frac{1}{2}(1+\sqrt{-7})]$ *let* $d(\mu) := \sum_{(\lambda)|(\mu)} 1$. *Then*

$$\begin{aligned} 2\sum_{0\neq\mu\in\mathcal{O}} d(\mu)\,K_0\left(\frac{4\pi|\mu|}{\sqrt{7}}\right) &= 2\log\left(\Gamma\left(\frac{1}{7}\right)\Gamma\left(\frac{2}{7}\right)\Gamma\left(\frac{4}{7}\right)\right) \\ &\quad - \frac{\sqrt{7}}{2\pi}\log 7 - \gamma - 3\log 2\pi. \end{aligned} \tag{4.52}$$

Proof. For $K = \mathbb{Q}(\sqrt{-7})$ we obtain from (4.22) $\lim_{s\to 0}\hat{E}_0(j,s) = -2\log 7$. Evaluating $|\Delta((1+\sqrt{-7})/2)|$ by means of the Chowla–Selberg formula (Chowla, Selberg (1967), page 110 or Weil (1976), page 92) we find

$$\left|\Delta\left(\frac{1+\sqrt{-7}}{2}\right)\right| = \left(\frac{2\pi}{7}\right)^6 \left(\frac{\Gamma(\frac{1}{7})\Gamma(\frac{2}{7})\Gamma(\frac{4}{7})}{\Gamma(\frac{3}{7})\Gamma(\frac{5}{7}))\Gamma(\frac{6}{7})}\right)^6,$$

and the multiplication formula for the gamma-function implies

$$\left|\Delta\left(\frac{1+\sqrt{-7}}{2}\right)\right| = 7^{-3}(2\pi)^{-12}\left(\Gamma\left(\frac{1}{7}\right)\Gamma\left(\frac{2}{7}\right)\Gamma\left(\frac{4}{7}\right)\right)^{12}. \tag{4.53}$$

Hence (3.11) yields the assertion. □

8.5 Applications to Zeta Functions and Asymptotics of Divisor Sums

We apply the Fourier expansion (2.21) to the study of zeta functions and divisor sums. The simple idea is as follows. Let the notations be as in Theorem 2.11 and put

$$T = \begin{pmatrix} 0 & 1 \\ -1 & 0 \end{pmatrix}, \qquad B_1 := TB. \tag{5.1}$$

B is assumed to be quasi-integral. Hence the associated modules $\mathfrak{u}_1 := \mathfrak{u}_{B_1}$, $\mathfrak{v}_1 := \mathfrak{v}_{B_1}$ satisfy the relations

$$\mathfrak{u}_1 = \mathfrak{v} = \mathfrak{u}^{-1}, \qquad \mathfrak{v}_1 = \mathfrak{u} = \mathfrak{v}^{-1}, \tag{5.2}$$

and B_1 is quasi-integral since B is. Define

$$P_1 := TP = z_1 + r_1 j, \tag{5.3}$$

$$z_1 = -\frac{\bar{z}}{|z|^2 + r^2}, \qquad r_1 = \frac{r}{|z|^2 + r^2}. \tag{5.4}$$

We have from the definitions

$$\hat{E}_{\mathfrak{m}}(B^{-1}P, s) = \hat{E}_{\mathfrak{m}}(B_1^{-1}P_1, s). \tag{5.5}$$

We now introduce the corresponding Fourier expansions (2.21) on both sides of (5.5). Letting $z = 0$ and replacing r by $N\mathfrak{u}^{-1}r$, we obtain for all $r > 0$

$$\begin{aligned} &\zeta(\mathfrak{m}, \mathfrak{u}, 1+s)\, r^{1+s} + \frac{2\pi}{\sqrt{|d_K|}\, s}\, \zeta(\mathfrak{m}, \mathfrak{u}^{-1}, s)\, r^{1-s} \\ &+ \frac{2^{2+s}\pi^{1+s}}{|d_K|^{\frac{s+1}{2}}\,\Gamma(1+s)} \sum_{0 \neq \omega \in \mathfrak{u}^2} |\omega|^s\, \sigma_{-s}(\mathfrak{m}, \mathfrak{u}, \omega)\, r\, K_s\left(\frac{4\pi|\omega|r}{\sqrt{|d_K|}N\mathfrak{u}}\right) \\ &= \zeta(\mathfrak{m}, \mathfrak{u}^{-1}, 1+s)\, r^{-1-s} + \frac{2\pi}{\sqrt{|d_K|}\, s}\, \zeta(\mathfrak{m}, \mathfrak{u}, s)\, r^{-1+s} \\ &+ \frac{2^{2+s}\pi^{1+s}}{|d_K|^{\frac{s+1}{2}}\,\Gamma(1+s)} \sum_{0 \neq \mu \in \mathfrak{u}^{-2}} |\mu|^s \sigma_{-s}(\mathfrak{m}, \mathfrak{u}^{-1}, \mu)\, r^{-1} K_s\left(\frac{4\pi|\mu|N\mathfrak{u}}{r\sqrt{|d_K|}}\right). \end{aligned} \tag{5.6}$$

For imaginary quadratic fields this relation between the zeta functions of ideal classes generalizes the result of Terras (1977), Theorem 2. In particular, we have for $\mathfrak{u} = \mathfrak{u}^{-1}$:

$$\zeta(\mathfrak{m},\mathfrak{u},1+s)\,(r^{1+s}-r^{-1-s})+\frac{2\pi}{\sqrt{|d_K|}s}\,\zeta(\mathfrak{m},\mathfrak{u},s)\,(r^{1-s}-r^{-1+s})$$

$$(5.7)\qquad =\frac{2^{2+s}\pi^{1+s}}{|d_K|^{\frac{s+1}{2}}\Gamma(1+s)}\sum_{0\neq\omega\in\mathcal{O}}|\omega|^s\sigma_{-s}(\mathfrak{m},\mathfrak{u},\omega)$$

$$\cdot\left(r^{-1}\,K_s\left(\frac{4\pi|\omega|}{r\sqrt{|d_K|}}\right)-r\,K_s\left(\frac{4\pi|\omega|r}{\sqrt{|d_K|}}\right)\right).$$

Theorem 5.1. *Let*

$$(5.8)\qquad Z(\mathfrak{m},\mathfrak{u},s):=\left(\frac{2\pi}{\sqrt{|d_K|}}\right)^{-s}\,\Gamma(s)\,\zeta(\mathfrak{m},\mathfrak{u},s),$$

$$(5.9)\qquad M_s(z):=K_s(z)+zK_s'(z)=\frac{d}{dz}zK_s(z),$$

and suppose that $\mathfrak{u}=\mathfrak{u}^{-1}$. *Then*

$$(5.10)\qquad \begin{aligned}&(1+s)\,Z(\mathfrak{m},\mathfrak{u},1+s)+(1-s)\,Z(\mathfrak{m},\mathfrak{u},s)\\&=-2\sum_{0\neq\omega\in\mathcal{O}}|\omega|^s\,\sigma_{-s}(\mathfrak{m},\mathfrak{u},\omega)\,M_s\left(\frac{4\pi|\omega|}{\sqrt{|d_K|}}\right).\end{aligned}$$

Proof. Apply the differential operator $r\frac{d}{dr}$ to both sides in (5.7) and then put $r=1$. □

The function $Z(\mathfrak{m},\mathfrak{u},s)$ satisfies the functional equation (3.7). Hence we have on the left-hand side of (5.10)

$$(5.11)\qquad \begin{aligned}&(1+s)\,Z(\mathfrak{m},\mathfrak{u},1+s)+(1-s)\,Z(\mathfrak{m},\mathfrak{u},s)\\&=(1+s)\,Z(\mathfrak{u},\mathfrak{m},-s)+(1-s)\,Z(\mathfrak{m},\mathfrak{u},s).\end{aligned}$$

That is, the left-hand side of (5.10) is invariant with respect to $s\mapsto -s$ provided that $\mathfrak{u}=\mathfrak{u}^{-1}$, $\mathfrak{m}=\mathfrak{m}^{-1}$. The same is obviously true for the right-hand side in view of (2.20) because $M_s=M_{-s}$.

For $s=n+1/2$ with $n\geq 0$ an integer, the Bessel function K_s is an elementary function (see Magnus, Oberhettinger, Soni (1966), page 72). The simplest case is $s=1/2$:

$$(5.12)\qquad K_{\frac{1}{2}}(z)=\sqrt{\frac{\pi}{2z}}\,e^{-z}.$$

Hence we find from (5.7) for $\mathfrak{u}=\mathfrak{u}^{-1}$:

$$\begin{aligned}
&\zeta\left(\mathfrak{m},\mathfrak{u},\frac{3}{2}\right)\left(r^{\frac{3}{2}}-r^{-\frac{3}{2}}\right)+\frac{4\pi}{\sqrt{|d_K|}}\zeta\left(\mathfrak{m},\mathfrak{u},\frac{1}{2}\right)\left(r^{\frac{1}{2}}-r^{-\frac{1}{2}}\right)\\
&=\frac{4\pi}{\sqrt{|d_K|}}\sum_{0\neq\omega\in\mathcal{O}}\sigma_{-\frac{1}{2}}(\mathfrak{m},\mathfrak{u},\omega)\left(r^{-\frac{1}{2}}e^{-\frac{4\pi|\omega|}{r\sqrt{|d_K|}}}-r^{\frac{1}{2}}e^{-\frac{4\pi|\omega|r}{\sqrt{|d_K|}}}\right).
\end{aligned}\tag{5.13}$$

We differentiate with respect to r and put $r = 1$. This yields the following corollary.

Corollary 5.2. *If* $\mathfrak{u} = \mathfrak{u}^{-1}$, *then*

$$\begin{aligned}
&\zeta\left(\mathfrak{m},\mathfrak{u},\frac{1}{2}\right)+\frac{3\sqrt{|d_K|}}{4\pi}\zeta\left(\mathfrak{m},\mathfrak{u},\frac{3}{2}\right)\\
&=\sum_{0\neq\omega\in\mathcal{O}}\sigma_{-\frac{1}{2}}(\mathfrak{m},\mathfrak{u},\omega)\left(\frac{8\pi|\omega|}{\sqrt{|d_K|}}-1\right)e^{-\frac{4\pi|\omega|}{\sqrt{|d_K|}}}.
\end{aligned}\tag{5.14}$$

This result generalizes a formula by A. Terras (1977), page 49, (2.1). There seems to be a minor error in this work, because one has to define M_s by (5.9) for the complex conjugates in the Corollary on page 48.

Combining Theorem 3.2 with (5.5) and choosing $z = 0$, we obtain the following result.

Theorem 5.3. *If the notation is as in Theorem 3.2, then*

$$\begin{aligned}
&r\,N\mathfrak{u}\left(\log(rN\mathfrak{u})+\gamma-\log 2\pi-\log|d_K|^{\frac{1}{2}}\right)\\
&-\frac{r\,N\mathfrak{u}}{12}\log(g(\mathfrak{m}\mathfrak{u}^{-1})g(\mathfrak{m}^{-1}\mathfrak{u}^{-1}))\\
&+r\,N\mathfrak{u}\sum_{0\neq\omega\in\mathfrak{u}^2}\sigma_0(\mathfrak{m},\mathfrak{u},\omega)\,K_0\left(\frac{4\pi|\omega|r}{\sqrt{|d_K|}}\right)\\
&=(r\,N\mathfrak{u})^{-1}\left(\log(rN\mathfrak{u})^{-1}+\gamma-\log 2\pi-\log|d_K|^{\frac{1}{2}}\right)\\
&-\frac{(r\,N\mathfrak{u})^{-1}}{12}\log(g(\mathfrak{m}\mathfrak{u})g(\mathfrak{m}^{-1}\mathfrak{u}))\\
&+(r\,N\mathfrak{u})^{-1}\sum_{0\neq\omega\in\mathfrak{u}^{-2}}\sigma_0(\mathfrak{m},\mathfrak{u}^{-1},\omega)K_0\left(\frac{4\pi|\omega|}{r\sqrt{|d_K|}}\right).
\end{aligned}\tag{5.15}$$

Corollary 5.4. *If the notation is as in Theorem 5.3, then*

$$\begin{aligned}
&\frac{\log(g(\mathfrak{m}\mathfrak{u}^{-1})g(\mathfrak{m}^{-1}\mathfrak{u}^{-1}))}{12}-\sum_{0\neq\omega\in\mathfrak{u}^2}\sigma_0(\mathfrak{m},\mathfrak{u},\omega)\,K_0\left(\frac{4\pi|\omega|}{N\mathfrak{u}\sqrt{|d_K|}}\right)\\
&=\frac{\log(g(\mathfrak{m}\mathfrak{u})g(\mathfrak{m}^{-1}\mathfrak{u}))}{12}-\sum_{0\neq\omega\in\mathfrak{u}^{-2}}\sigma_0(\mathfrak{m},\mathfrak{u}^{-1},\omega)K_0\left(\frac{4\pi|\omega|N\mathfrak{u}}{\sqrt{|d_K|}}\right).
\end{aligned}$$

Proof. Put $r = (N\mathfrak{u})^{-1}$ in (5.15). □

Corollary 5.5. *If the notation is as in Theorem 5.3 and $\mathfrak{u}^2 = \mathcal{O}$, then*

$$(r+r^{-1})\log r + \left(\gamma - \log 2\pi - \log|d_K|^{\frac{1}{2}} - \frac{\log(g(\mathfrak{m}\mathfrak{u})g(\mathfrak{m}^{-1}\mathfrak{u}))}{12}\right)(r-r^{-1})$$
$$= -\sum_{0\neq\omega\in\mathcal{O}} \sigma_0(\mathfrak{m},\mathfrak{u},\omega)\left(r\,K_0\left(\frac{4\pi|\omega|r}{\sqrt{|d_K|}}\right) - r^{-1}\,K_0\left(\frac{4\pi|\omega|}{r\sqrt{|d_K|}}\right)\right)$$

for all $r > 0$.

Proof. Put $N\mathfrak{u} = 1$ in (5.15). □

Corollary 5.6. *If the notation is as in Theorem 5.3, then*

$$(r+r^{-1})\log r + \left(\gamma - \log 2\pi - \log|d_K|^{\frac{1}{2}} - \frac{\log(g(\mathfrak{m})g(\mathfrak{m}^{-1}))}{12}\right)(r-r^{-1})$$
$$= -\sum_{0\neq\omega\in\mathcal{O}} \sigma_0(\mathfrak{m},\mathcal{O},\omega)\left(r\,K_0\left(\frac{4\pi|\omega|r}{\sqrt{|d_K|}}\right) - r^{-1}\,K_0\left(\frac{4\pi|\omega|}{r\sqrt{|d_K|}}\right)\right).$$

Proof. Put $\mathfrak{u} = \mathcal{O}$ in the formula of Corollary 5.5. □

We start again from (5.5) and focus our attention on the asymptotic behaviour. Choose $z = 0$ in (5.5) and let $r \to 0$. Then $r_1 = \frac{1}{r} \to \infty$ and the asymptotic behaviour of the right-hand side in (5.5) is obvious from the Fourier expansion since the infinite series converges exponentially to zero. This implies the following result.

Theorem 5.7. *For all $\mathfrak{m},\mathfrak{u} \in \mathcal{M}$*

$$\begin{aligned}
&\sum_{0\neq\omega\in\mathfrak{u}^2} |\omega|^s\,\sigma_{-s}(\mathfrak{m},\mathfrak{u},\omega)\,K_s\left(\frac{4\pi|\omega|r}{\sqrt{|d_K|}}\right)\\
(5.16)\qquad &= \frac{|d_K|^{\frac{s+1}{2}}\Gamma(1+s)}{2^{2+s}\pi^{1+s}}\,(N\mathfrak{u}^{-2-s}\,\zeta(\mathfrak{m},\mathfrak{u}^{-1},1+s)\,r^{-2-s}\\
&+\frac{2\pi}{\sqrt{|d_K|}\,s}N\mathfrak{u}^{-2+s}\,\zeta(\mathfrak{m},\mathfrak{u},s)\,r^{-2+s}) + O(r^{-s}) + O(r^s)
\end{aligned}$$

for $r \to +0$, provided that $s \neq 0$, $s \neq 1$. For $s = 0$ see Theorem 5.3. There is also an analogous result for $s = 1$.

The proof follows from the above remarks combined with Lemma 2.9, (3).

Corollary 5.8. *For all $\mathfrak{m},\mathfrak{u} \in \mathcal{M}$ the following asymptotic relation holds:*

$$\sum_{0 \neq \omega \in \mathfrak{u}^2} \sigma_{-\frac{1}{2}}(\mathfrak{m}, \mathfrak{u}, \omega)\, e^{-|\omega| r} = \frac{4\pi}{\sqrt{|d_K|}} N\mathfrak{u}^{-\frac{5}{2}} \zeta\left(\mathfrak{m}, \mathfrak{u}^{-1}, \frac{3}{2}\right) r^{-2} + O(r^{-1})$$

as $r \to +0$.

Proof. We put (5.12) into (5.16) and obtain the result. □

An application of Karamata's Tauberian Theorem to the formula of Corollary 5.8 yields the following result

Corollary 5.9. *For all* $\mathfrak{m}, \mathfrak{u} \in \mathcal{M}$ *the following asymptotic relation holds:*

$$(5.17) \quad \sum_{\substack{0 \neq \omega \in \mathfrak{u}^2 \\ |\omega| \leq x}} \sigma_{-\frac{1}{2}}(\mathfrak{m}, \mathfrak{u}, \omega) \sim \frac{2\pi}{\sqrt{|d_K|}} N\mathfrak{u}^{-\frac{5}{2}} \zeta\left(\mathfrak{m}, \mathfrak{u}^{-1}, \frac{3}{2}\right) x^2 \quad \text{as } x \to \infty.$$

Choosing $\mathfrak{u} = \mathfrak{m} = \mathcal{O}$ in (5.17), we obtain the next corollary.

Corollary 5.10. *For* $x \to \infty$,

$$(5.18) \qquad \sum_{\substack{0 \neq \omega \in \mathcal{O} \\ |\omega| \leq x}} \sum_{(\lambda) | (\omega)} \frac{1}{|\lambda|} \sim \frac{2\pi}{\sqrt{|d_K|}} \zeta\left(\mathcal{O}^{\#}, \frac{3}{2}\right) x^2.$$

If we take $K = \mathbb{Q}(i)$, then (5.18) yields:

$$(5.19) \qquad \sum_{\substack{0 \neq \omega \in \mathbb{Z}[i] \\ |\omega| \leq x}} \sum_{(\lambda) | (\omega)} \frac{1}{|\lambda|} \sim \pi \zeta\left(\frac{3}{2}\right) L\left(\frac{3}{2}\right) x^2 \qquad \text{for } x \to \infty,$$

and similar results hold for all imaginary quadratic fields of class number one. Relation (5.19) should be compared with the asymptotics for the circle problem:

$$\sum_{\substack{0 \neq \omega \in \mathbb{Z}[i] \\ |\omega| \leq x}} 1 \sim \pi x^2 \qquad \text{for } x \to \infty.$$

8.6 Non-Vanishing of L-Functions

The Fourier expansion of the Eisenstein series for $\mathbf{PSL}(2, \mathcal{O})$ enables us to give a rather unusual proof of Dicichlet's theorem on the non-vanishing of certain L-functions on the line $\operatorname{Re} s = 1$. We understand that this proof is a special instance of the much more general work of Jacquet, Shalika (1976). At the same time we give a quite unorthodox proof of the non-vanishing of the L-functions in the half-plane $\operatorname{Re} s > 1$ that makes no use of the Euler product expansion or the Dirichlet series expansion of the inverse L-function.

Definition 6.1. For a character $\chi : \mathcal{M}/K^* \to \mathbb{C}^*$ let

$$F_\chi(P,s) := \sum_{\mathfrak{m}\in\mathcal{M}/K^*} \chi(\mathfrak{m})\, \hat{E}_{\mathfrak{m}^{-1}}(P,s). \tag{6.1}$$

Proposition 6.2. *Let $\chi : \mathcal{M}/K^* \to \mathbb{C}^*$ be a character.*

(1) Suppose that $\chi \neq \chi_0$ where $\chi_0 = 1$ is the trivial character. Then for fixed $P \in \mathbb{H}$ the function $F_\chi(P,s)$ is holomorphic in the whole s-plane.
(2) The function $F_{\chi_0}(P,s)$ has a pole only at $s = 1$, this pole is simple with

$$\operatorname*{Res}_{s=1} F_{\chi_0}(P,s) = \frac{4\pi^2}{|d_K|}\, h. \tag{6.2}$$

Proof. By Theorem 3.1 the function $F_\chi(P,s)$ is holomorphic everywhere except possibly at $s = 1$. In $s = 1$ there is at most a simple pole with residue

$$\frac{4\pi^2}{|d_K|} \sum_{\mathfrak{m}\in\mathcal{M}/K^*} \chi(\mathfrak{m}).$$

Hence the singularity at $s = 1$ is removable for $\chi \neq \chi_0$, whereas for $\chi = \chi_0$ the residue is given by (6.2). □

Proposition 6.3. *Suppose that $s \in \mathbb{C}$ and $\chi : \mathcal{M}/K^* \to \mathbb{C}^*$ is a character.*

(1) If $-s \notin \mathbb{N}$, then $F_\chi(\cdot, s)$ is not identically zero.
(2) $F_\chi(\cdot, -n) = 0$ for all $n \in \mathbb{N}$, $n \geq 2$.
(3) $F_\chi(\cdot, -1) = 0$ for $\chi \neq \chi_0$, whereas $F_{\chi_0}(\cdot, -1) = -h$.

Proof. By means of (1.5), (1.11), (1.15) and (2.21) we find

$$\begin{aligned} F_\chi(B^{-1}P,s) &= |\mathcal{O}^*|\, N\mathfrak{u}^{1+s}\, \chi^{-1}(\mathfrak{u})\, L(1+s,\chi)\, r^{1+s} \\ &+ \frac{2\pi\,|\mathcal{O}^*|}{\sqrt{|d_K|}}\, N\mathfrak{u}^{1-s}\, \chi(\mathfrak{u})\, \frac{1}{s}\, L(s,\chi)\, r^{1-s} \\ &+ \frac{2^{2+s}\pi^{1+s} N\mathfrak{u}}{|d_K|^{\frac{s+1}{2}}\, \Gamma(1+s)} \sum_{0\neq\omega\in\mathfrak{u}^2} |\omega|^s \sum_{\mathfrak{m}\in\mathcal{M}/K^*} \chi(\mathfrak{m})\, \sigma_{-s}(\mathfrak{m}^{-1},\mathfrak{u},\omega) \\ &\cdot r\, K_s\!\left(\frac{4\pi\,|\omega|\, r}{\sqrt{|d_K|}}\right) e^{2\pi i \left\langle \frac{2\bar{\omega}}{\sqrt{d_K}}, z\right\rangle}. \end{aligned} \tag{6.3}$$

Choose $B = I, \mathfrak{u} = \mathcal{O}$, and Lemma 6.4 ahead shows that the Fourier coefficient of $F_\chi(\cdot, s)$ at $\omega = 1$ is different from zero provided that $-s \notin \mathbb{N}$. This proves (1), and (2), (3) are obvious from Theorem 3.1. □

Lemma 6.4. *Let* $\mathfrak{m}$, $\mathfrak{n} \in \mathcal{M}$ *be such that* $\mathfrak{n}^2 = \omega\mathcal{O}$ $(\omega \in K^*)$ *is principal. Then*

$$\sigma_s(\mathfrak{m}, \mathfrak{n}, \omega) = 0 \tag{6.4}$$

if $\mathfrak{m}$ *and* $\mathfrak{n}$ *do not belong to the same class. If* $\mathfrak{m}$ *and* $\mathfrak{n}$ *belong to the same class, then*

$$\sigma_s(\mathfrak{m}, \mathfrak{n}, \omega) = |\mathcal{O}^*| \, N\mathfrak{n}^s. \tag{6.5}$$

Proof. Assume that the sum

$$\sigma_s(\mathfrak{m}, \mathfrak{n}, \omega) = N\mathfrak{m}^s \sum_{\substack{\lambda \in \mathfrak{m}\mathfrak{n} \\ \omega \in \lambda\mathfrak{m}^{-1}\mathfrak{n}}} N\lambda^s \tag{6.6}$$

is not empty. Then $\omega = \lambda\mu$ for some $\lambda \in \mathfrak{m}\mathfrak{n}$ and $\mu \in \mathfrak{m}^{-1}\mathfrak{n}$ and hence $\mathfrak{n}^2 = \omega\mathcal{O} \subset \mu\mathfrak{m}\mathfrak{n} \subset \mathfrak{n}^2$. This implies that $\mathfrak{m}$ and $\mathfrak{n}$ are in the same class, and (6.4) is proved.

On the other hand, if $\mathfrak{m} = t\mathfrak{n}$ for some $t \in K^*$ and $u \in \mathcal{O}^*$ is arbitrary, then

$$\lambda := ut\omega \in \mathfrak{m}\mathfrak{n}, \quad \mu := t^{-1}u^{-1} \in \mathfrak{m}^{-1}\mathfrak{n} \tag{6.7}$$

satisfy $\lambda\mu = \omega$. In this way we find $|\mathcal{O}^*|$ terms for the sum (6.6) that give all the same contribution $N\mathfrak{n}^s$.

Conversely, if $\lambda \in \mathfrak{m}\mathfrak{n}$ is an arbitrary element such that $\omega \in \lambda\mathfrak{m}^{-1}\mathfrak{n}$, then $\omega = \lambda\mu$, where $\lambda = at\omega, \mu = bt^{-1}$ with certain $a, b \in \mathcal{O}$. Hence $ab = 1$ and λ, μ are of the form (6.7) with some unit $u \in \mathcal{O}^*$. □

Theorem 6.5. *Suppose that* $\chi : \mathcal{M}/K^* \to \mathbb{C}^*$ *is a character,* $t \in \mathbb{R}$ *and* $t \neq 0$ *if* $\chi = \chi_0$. *Then* $L(1 + it, \chi) \neq 0$.

Proof. Assume to the contrary that $L(1 + it, \chi) = 0$. Then for all B the coefficient of r^{1+it} in (6.3) is equal to zero. (Note that $L(s, \chi)$ is holomorphic at 1 for $\chi \neq \chi_0$.) The functional equation of the L-series implies that the coefficient of r^{1-it} in (6.3) is also equal to zero for all B, that is $F_\chi(\cdot, it)$ is a cusp function. Therefore $F_\chi(\cdot, it)$ is orthogonal to the space spanned by all Eisenstein series. But $F_\chi(\cdot, it)$ is an element of the latter space. Hence $F_\chi(\cdot, it) = 0$. This contradicts Proposition 6.3, (1). □

Theorem 6.6. *Suppose that* $\chi : \mathcal{M}/K^* \to \mathbb{C}^*$ *is a character. Then* $L(s, \chi) \neq 0$ *for* $\operatorname{Re} s > 1$.

Proof. Assume to the contrary, that $L(1+s_0,\chi)=0$, where $\operatorname{Re} s_0>0$. Then the coefficient of r^{1+s_0} in (6.3) vanishes for all B. Hence $F_\chi(\cdot,s_0)$ is an eigenfunction of $-\Delta$ with eigenvalue $1-s_0^2$ which is square integrable with respect to the hyperbolic volume form over a fundamental domain of $\mathbf{PSL}(2,\mathcal{O})$. Since $-\Delta$ is a self-adjoint positive operator with a natural domain containing $F_\chi(\cdot,s_0)$, we conclude that $1-s_0^2\geq 0$. Because $\operatorname{Re} s>0$, we infer that necessarily $s_0\in]0,1]$ and $L(1+s_0,\chi)=0$. By Proposition 6.2.2 $F_\chi(\cdot,s_0)$ can be reduced to a cusp form by subtraction of a suitable linear combination of residues of Eisenstein series at s_0. Hence by Theorem 3.1 $F_\chi(\cdot,s_0)$ is itself a cusp form if $s_0\neq 1$. But we just saw in the proof of Theorem 6.5 that $F_\chi(\cdot,s_0)$ cannot be a cusp form. Hence $s_0=1$. Now, if $\chi\neq\chi_0$, then $F_\chi(\cdot,1)$ is an eigenfunction of $-\Delta$ with eigenvalue 0. That is $F_\chi(\cdot,1)$ is a constant function. This contradicts Lemma 6.4 and (6.3). Hence $s_0=1$ and $\chi=\chi_0$. But then trivially $L(2,\chi_0)>0$, and the assertion is proved. □

8.7 Meromorphic Continuation by Integral Representation

We give a second proof of the meromorphic continuation of $\hat{E}_{\mathfrak{m}}(P,s)$ into the whole s-plane. Our approach follows the line of thought of Riemann's second proof of the meromorphic continuation of the Riemann zeta function by means of the functional equation of the theta-function. The same method was employed by Hecke in his famous work on the determination of Dirichlet series by their functional equation.

For $\mathfrak{m}\in\mathcal{M}$, $P=z+rj\in\mathbb{H}$, $\operatorname{Re} s>1$ we obtain from Euler's integral

$$\hat{E}_{\mathfrak{m}}(P,s)=\left(\frac{2\pi N\mathfrak{m}}{\sqrt{|d_K|}}\right)^{s+1}\Gamma(s+1)^{-1}\int_0^\infty\sum_{\substack{c,\,d\in\mathfrak{m}\\(c,d)\neq(0,0)}}e^{-\frac{2\pi}{r\sqrt{|d_K|}}\|cP+d\|^2t}\,t^{s+1}\,\frac{dt}{t}$$

$$=\left(\frac{2\pi N\mathfrak{m}}{\sqrt{|d_K|}}\right)^{s+1}\Gamma(s+1)^{-1}\int_0^\infty\left(\Theta_{\mathfrak{m}}(P,t)-1\right)t^{s+1}\,\frac{dt}{t}$$

with the theta function

$$\Theta_{\mathfrak{m}}(P,t)=\sum_{c,\,d\in\mathfrak{m}}e^{-\frac{2\pi}{r\sqrt{|d_K|}}\|cP+d\|^2t}\qquad(t>0).\tag{7.1}$$

We write (7.1) as a theta-series of a positive definite quadratic form in four variables. Introducing a $\mathbb{Z}$-basis $\{\alpha_1,\alpha_2\}$ of $\mathfrak{m}$ we have $c=u_1\alpha_1+u_2\alpha_2$, $d=u_3\alpha_1+u_4\alpha_2$ with some $u_1,\ldots,u_4\in\mathbb{Z}$ and we find

$$\frac{2}{r\sqrt{|d_K|}}\|cP+d\|^2=B[u]:=u^tBu$$

where $u = (u_1, \ldots, u_4)^t \in \mathbb{Z}^4$,

$$B = \frac{1}{r\sqrt{|d_K|}} \begin{pmatrix} \|P\|^2 S & T \\ T^t & S \end{pmatrix}$$

with

$$S = \begin{pmatrix} 2|\alpha_1|^2 & \alpha_1\bar{\alpha}_2 + \bar{\alpha}_1\alpha_2 \\ \alpha_1\bar{\alpha}_2 + \bar{\alpha}_1\alpha_2 & 2|\alpha_2|^2 \end{pmatrix},$$

$$T = \begin{pmatrix} |\alpha_1|^2(z+\bar{z}) & \alpha_1\bar{\alpha}_2 z + \bar{\alpha}_1\alpha_2\bar{z} \\ \bar{\alpha}_1\alpha_2 z + \alpha_1\bar{\alpha}_2\bar{z} & |\alpha_2|^2(z+\bar{z}) \end{pmatrix}.$$

This yields $\Theta_{\mathfrak{m}}(P,t) = \Theta_B(t)$ with

$$\Theta_B(t) = \sum_{u\in\mathbb{Z}^4} e^{-\pi B[u]t} \qquad (t>0).$$

We now cast the theta-transformation formula

$$\Theta_B(t) = (\det B)^{-\frac{1}{2}}\, t^{-2}\, \Theta_{B^{-1}}\left(\frac{1}{t}\right)$$

in terms of modules. Normalizing our $\mathbb{Z}$-basis $\{\alpha_1, \alpha_2\}$ of $\mathfrak{m}$ by the requirement

$$\det\begin{pmatrix} \alpha_1 & \alpha_2 \\ \overline{\alpha_1} & \overline{\alpha_2} \end{pmatrix} = \sqrt{d_K}\, N\mathfrak{m}$$

we find after some straightforward computations $\Theta_{B^{-1}}(t) = \Theta_{\mathfrak{m}^{-1}}(P,t)$ and

$$\Theta_{\mathfrak{m}}(P,t) = N\mathfrak{m}^{-2}\, t^{-2}\, \Theta_{\mathfrak{m}^{-1}}\left(P, \frac{1}{t}\right) \qquad (t>0). \tag{7.2}$$

This implies for $\operatorname{Re} s > 1$:

$$\begin{aligned}
&\left(\frac{2\pi}{\sqrt{|d_K|}}\right)^{-(1+s)} \Gamma(1+s)\, \hat{E}_{\mathfrak{m}}(P,s) \\
&= N\mathfrak{m}^{1+s}\left(\int_0^1 \Theta_{\mathfrak{m}}(P,t)\, t^{s+1}\frac{dt}{t} - \frac{1}{1+s} + \int_1^\infty (\Theta_{\mathfrak{m}}(P,t)-1)\, t^{1+s}\frac{dt}{t}\right) \\
&= N\mathfrak{m}^{s}\left(N\mathfrak{m}^{-1}\int_0^1 \Theta_{\mathfrak{m}^{-1}}(P,\frac{1}{t})\, t^{s-1}\frac{dt}{t} - \frac{N\mathfrak{m}}{1+s}\right. \\
&\left. + N\mathfrak{m}\int_1^\infty (\Theta_{\mathfrak{m}}(P,t)-1)\, t^{1+s}\frac{dt}{t}\right) \\
&= N\mathfrak{m}^{s}\left(N\mathfrak{m}\int_1^\infty (\Theta_{\mathfrak{m}}(P,t)-1)\, t^{1+s}\frac{dt}{t}\right. \\
&\left. + N\mathfrak{m}^{-1}\int_1^\infty (\Theta_{\mathfrak{m}^{-1}}(P,t)-1)\, t^{1-s}\frac{dt}{t} - \frac{N\mathfrak{m}}{1+s} - \frac{N\mathfrak{m}^{-1}}{1-s}\right).
\end{aligned}$$

The right-hand side is a meromorphic function of $s \in \mathbb{C}$ and the functional equation (3.1) is obvious. Moreover, it follows that $\hat{E}_{\mathfrak{m}}(P,s)$ has a simple pole at $s=1$ with residue (3.2), and that $\hat{E}_{\mathfrak{m}}(P,\cdot)$ has a removable singularity at $s=-1$ with $\hat{E}_{\mathfrak{m}}(P,-1)=-1$. $\hat{E}_{\mathfrak{m}}(P,\cdot)$ is holomorphic in $\mathbb{C}\setminus\{1\}$ and has zeroes at $-n$ where $n \in \mathbb{N}$, $n \geq 2$. Hence Theorem 3.1 is proved anew.

Our computations have the following integral representation as a consequence.

Theorem 7.1. *For* $\mathfrak{m} \in \mathcal{M}$, $P \in \mathbb{H}$, $s \in \mathbb{C}$, $s \neq 1$, *then*

$$\begin{aligned} &\left(\frac{2\pi}{\sqrt{|d_K|}}\right)^{-(1+s)} \Gamma(1+s)\,\hat{E}_{\mathfrak{m}}(P,s) \\ &= N\mathfrak{m}^s \left(N\mathfrak{m} \int_1^\infty (\Theta_{\mathfrak{m}}(P,t)-1)\,t^{1+s}\,\frac{dt}{t} \right. \\ &\left. + N\mathfrak{m}^{-1} \int_1^\infty (\Theta_{\mathfrak{m}^{-1}}(P,t)-1)\,t^{1-s}\,\frac{dt}{t} - \frac{N\mathfrak{m}}{1+s} - \frac{N\mathfrak{m}^{-1}}{1-s} \right). \end{aligned} \tag{7.3}$$

The case $\mathfrak{m} = \mathcal{O}$ is the simplest one, we have:

$$\begin{aligned} &\left(\frac{2\pi}{\sqrt{|d_K|}}\right)^{-(1+s)} \Gamma(1+s)\,\hat{E}_{\mathcal{O}}(P,s) \\ &= \int_1^\infty (\Theta_{\mathcal{O}}(P,t)-1)\,(t^{1+s}+t^{1-s})\,\frac{dt}{t} - \frac{2}{1-s^2}. \end{aligned} \tag{7.4}$$

8.8 Computation of the Volume

We give in this section two simple proofs of the result

$$\operatorname{vol}(\mathbf{PSL}(2,\mathcal{O})) = v(\mathbf{PSL}(2,\mathcal{O})\backslash\mathbb{H}) = \frac{|d_K|^{\frac{3}{2}}}{4\pi^2}\zeta_K(2) \tag{8.1}$$

which was already stated in Theorem 7.1.1, (3).

First Proof. It will turn out that (8.1) follows almost trivially from the Fourier expansion of the Eisenstein series when these are put into the Maaß-Selberg relations (Theorem 3.3.6). We apply formula 3.3.13 to $f := \hat{E}_{\mathfrak{m}}(\cdot,s)$, $g := \hat{E}_{\mathfrak{m}}(\cdot,s')$ where $\mathfrak{m} \in \mathcal{M}$ is fixed and $s, s' > 1, s \neq s'$. The cusps $\eta_1,\dots,\eta_h$ of $\mathcal{F}$ are indexed by a representative system $\mathfrak{u}_1,\dots,\mathfrak{u}_h$ of $\mathcal{M}/K^*$, where $B_1,\dots,B_h \in \mathbf{PSL}(2,K)$ are quasi-integral matrices such that $\eta_j = B_j^{-1}\infty$ and $\mathfrak{u}_j$ corresponds to B_j $(j=1,\dots,h)$. We choose $Y>0$ sufficiently large and let $Y_j = (N\mathfrak{u}_j)^{-1}Y$ $(j=1,\dots,h)$ and denote this choice of $Y_1, \dots, Y_h$ by a superscript Y. We put (2.12), the easily established fact $[\mathbf{PSL}(2,\mathcal{O})_{\eta_j} : \mathbf{PSL}(2,\mathcal{O})'_{\eta_j}] = |\mathcal{O}^*|/2$ and (2.21) into (3.3.13) and obtain

$$
\begin{aligned}
&\int_{\mathcal{F}} \hat{E}^Y_{\mathfrak{m}}(P,s)\,\hat{E}^Y_{\mathfrak{m}}(P,s')\,dv(P) \\
&= \frac{\sqrt{|d_K|}}{|\mathcal{O}^*|}\sum_{j=1}^{h}\Big(\frac{1}{s+s'}\,\zeta(\mathfrak{m},\mathfrak{u}_j,1+s)\,\zeta(\mathfrak{m},\mathfrak{u}_j,1+s')\,Y^{s+s'} \\
&+ \frac{1}{s-s'}\frac{2\pi}{\sqrt{|d_K|}s'}\,\zeta(\mathfrak{m},\mathfrak{u}_j,1+s)\,\zeta(\mathfrak{m},\mathfrak{u}_j^{-1},s')\,Y^{s-s'} \\
&- \frac{1}{s-s'}\frac{2\pi}{\sqrt{|d_K|}s}\,\zeta(\mathfrak{m},\mathfrak{u}_j^{-1},s)\,\zeta(\mathfrak{m},\mathfrak{u}_j,1+s')\,Y^{-s+s'} \\
&- \frac{1}{s+s'}\frac{4\pi^2}{\sqrt{|d_K|}ss'}\,\zeta(\mathfrak{m},\mathfrak{u}_j^{-1},s)\,\zeta(\mathfrak{m},\mathfrak{u}_j^{-1},s')\,Y^{-s-s'}\Big).
\end{aligned} \tag{8.2}
$$

Here $\mathcal{F}$ is a fundamental domain for $\mathbf{PSL}(2,\mathcal{O})$ as chosen in Section 2.3. Here we put $s = 1+2t$, $s' = 1+t$, $t > 0$, multiply this equation by $2t^2$ and let $t \to +0$. Taking the limit on the right-hand side, we find by (1.5), (3.9)

$$
\begin{aligned}
&\lim_{t\to+0} 2t^2 \int_{\mathcal{F}} \hat{E}^Y_{\mathfrak{m}}(P,1+2t)\,\hat{E}^Y_{\mathfrak{m}}(P,1+t)\,dv(P) \\
&= \frac{\sqrt{|d_K|}}{|\mathcal{O}^*|}\sum_{j=1}^{h}\frac{2\pi}{\sqrt{|d_K|}}\Big(\operatorname*{Res}_{s=1}\zeta(\mathfrak{m},\mathfrak{u}_j^{-1},s)\Big)\,\zeta(\mathfrak{m},\mathfrak{u}_j,2) + O(Y^{-2}) \\
&= \frac{4\pi^2}{\sqrt{|d_K|}}\,\zeta_K(2) + O(Y^{-2}).
\end{aligned} \tag{8.3}
$$

Consider now the left-hand side of (8.3). Let $\mathcal{F}_Y$ be the compact part of the fundamental domain with the cusps cut off at heights $Y_1, \dots, Y_h, Y_j = (N\mathfrak{u}_j)^{-1}Y$. Then the left-hand side is equal to the corresponding integral taken over $\mathcal{F}_Y$ up to an error term $O(e^{-\epsilon Y})$ $(\epsilon > 0)$. On the compact part $\mathcal{F}_Y$ the integrand converges uniformly to $(4\pi^2/\sqrt{|d_K|})^2$, see (3.2). Hence we find

$$
v(\mathcal{F}_Y) = \frac{|d_K|^{\frac{3}{2}}}{4\pi^2}\,\zeta_K(2) + O(Y^{-2}),
$$

and letting $Y \to \infty$, we have (8.1). □

We remark that the method of the first proof works for all discrete subgroups of finite covolume and gives expressions for the residues of the zeroth Fourier coefficients of $E_A(P,s)$ at $s = 1$ in terms of $v(\mathcal{F})$.

Second Proof. Let $\mathfrak{m} \in \mathcal{M}$ and let $\mathcal{F}$, $\eta_1, \dots, \eta_h$, $\mathfrak{u}_1, \dots, \mathfrak{u}_h$, $B_1, \dots, B_h$ be as in the first proof and let $\Lambda_1, \dots, \Lambda_h$ be the lattices associated with $\eta_1, \dots, \eta_h$, $B_1, \dots, B_h$. It is an immediate consequence of (2.21) that $\hat{E}_{\mathfrak{m}}(\cdot, s)$

is v-integrable over $\mathcal{F}$ for $|\operatorname{Re} s| < 1$. Moreover we know that the integration of $\hat{E}_{\mathfrak{m}}(\cdot, it)$ with respect to the real parameter t yields an eigenpacket of $-\Delta$, see Section 6.3. As proved in Section 6.2 this eigenpacket is orthogonal to the constant eigenfunction 1 of $-\Delta$. This means

$$\int_{\mathcal{F}} \int_0^u \hat{E}_{\mathfrak{m}}(P, it)\, dt\, dv(P) = 0 \quad (u \in \mathbb{R}).$$

Since this holds for all $u \in \mathbb{R}$, we conclude by analytic continuation

$$\int_{\mathcal{F}} \hat{E}_{\mathfrak{m}}(P, s)\, dv(P) = 0 \qquad \text{for } |\operatorname{Re} s| < 1. \tag{8.4}$$

Choose $\mathfrak{m} = \mathcal{O}$ in (8.4) and put (7.4) with $s = 0$ into (8.4). This gives

$$v(\mathcal{F}) = \int_{\mathcal{F}} \int_1^\infty (\theta_{\mathcal{O}}(P, t) - 1)\, dt\, dv(P). \tag{8.5}$$

Letting $A := 2\pi/\sqrt{|d_K|}$ we now evaluate $v(\mathcal{F})$ as follows:

$$\begin{aligned}
v(\mathcal{F}) &= \int_1^\infty \int_{\mathcal{F}} \sum_{c,d \in \mathcal{O}}{}' e^{-\frac{A}{r}\|cP+d\|^2 t}\, dv(P)\, dt \\
&= \frac{2}{|\mathcal{O}^*|} \sum_{j=1}^h \int_1^\infty \sum_{\lambda \in \mathfrak{u}_j^{-1}} \sum_{\left(\begin{smallmatrix} * & * \\ c & d \end{smallmatrix}\right) \in B_j \Gamma'_{\eta_j} B_j^{-1} \backslash B_j \Gamma} \int_{\mathcal{F}} e^{-\frac{A|\lambda|^2}{r}\|cP+d\|^2 t} dv(P)\, dt \\
&= \frac{2}{|\mathcal{O}^*|} \sum_{j=1}^h \sum_{\lambda \in \mathfrak{u}_j^{-1}} \int_1^\infty \int_{\Lambda_j \times]0,\infty[} e^{-\frac{A|\lambda|^2 t}{r}} \frac{dx\, dy\, dr}{r^3}\, dt \\
&= \frac{2}{|\mathcal{O}^*|} \sum_{j=1}^h |\Lambda_j| \sum_{\lambda \in \mathfrak{u}_j^{-1}} \int_1^\infty (A|\lambda|^2 t)^{-2} dt \\
&= \frac{\sqrt{|d_K|}}{|\mathcal{O}^*| A^2} \sum_{j=1}^h N\mathfrak{u}_j^{-2} \sum_{\lambda \in \mathfrak{u}_j^{-1}} N\lambda^{-2} = \frac{|d_K|^{\frac{3}{2}}}{4\pi^2 |\mathcal{O}^*|} \sum_{j=1}^h \zeta(\mathcal{O}, \mathfrak{u}_j, 2) = \frac{|d_K|^{\frac{3}{2}}}{4\pi^2} \zeta_K(2).
\end{aligned}$$

This finishes our second proof. □

8.9 Weyl's Asymptotic Law

In this section we shall prove Weyl's asymptotic law for the distribution of eigenvalues of the Laplace operator on the space $L^2(\mathbf{PSL}(2, \mathcal{O}) \backslash \mathbb{H})$. To formulate the result we put as in Section 6.5

$$A(\mathbf{PSL}(2, \mathcal{O}), T) = |\{\, n \in \mathfrak{D}(\mathbf{PSL}(2, \mathcal{O})) \;\mid\; \lambda_n \le T \,\}| \tag{9.1}$$

where the counting is done with multiplicities.

Theorem 9.1. *The counting function (9.1) satisfies the asymptotic relation*

$$A(\mathbf{PSL}(2,\mathcal{O}),T) \sim \frac{\mathrm{vol}(\mathbf{PSL}(2,\mathcal{O}))}{6\pi^2} T^{\frac{3}{2}} = \frac{|d_K|^{\frac{3}{2}}}{24\pi^4} T^{\frac{3}{2}} \tag{9.2}$$

as $T \to \infty$.

The proof is obtained from formula (3.30), Theorem 6.5.4 together with the following

Lemma 9.2. *Let L be a number field and $\zeta_L(s)$ its zeta function. Then there is a (positive) $\kappa \in \mathbb{R}$ so that*

$$\frac{\zeta_L'}{\zeta_L}(it) = O\left((\log t)^{\kappa}\right) \tag{9.3}$$

as $t \to \infty$.

Proof. First of all we use the functional equation of ζ_L and find that (9.3) is implied by

$$\frac{\zeta_L'}{\zeta_L}(1+it) = O\left((\log t)^{\kappa'}\right). \tag{9.4}$$

Doing this shift some expressions involving the Γ-function and its derivative occur, these are treated by standard estimates. The asymptotic relation (9.4) is proved as in Titchmarch (1951), page 44, where the case of the Riemann zeta function is discussed, see also Davenport (1980). □

It is relatively easy to extend the asymptotic relation to the case of congruence groups. To do this it is only necessary to compute the determinant of the scattering matrix for full congruence subgroups of $\mathbf{PSL}(2,\mathcal{O})$. The result is almost the same as in Theorem 3.10, the zeta function of the Hilbert class field has only to be modified suitably at finitely many Euler factors. For non-congruence subgroups of $\mathbf{PSL}(2,\mathcal{O})$ nothing is known so far.

Chapter 9. Integral Binary Hermitian Forms

Here we include some classical results from the theory of binary hermitian forms which originate from Hermite (1854). We discuss the reduction theory of binary hermitian forms as described for example in Bianchi (1892). Eventually our considerations lead to Humbert's computation of the covolume of $\mathbf{SL}(2, \mathcal{O})$ where $\mathcal{O}$ is the ring of integers in an imaginary quadratic number field. The work of Humbert on hermitian forms is contained in his papers (1915), (1919a)–(1919e). It contains an interesting error, we correct it in Section 9.6. We also develop a theory of representation numbers of binary hermitian forms which is analogous to the theory of binary quadratic forms as in Landau (1927).

9.1 Upper Half-Space and Binary Hermitian Forms

Here we shall report on some generalities from linear algebra. For complex matrices A we define

$$(1.1) \qquad A^* = \bar{A}^t.$$

The matrix $\bar{A}$ is obtained from A by applying complex conjugation to all entries. The matrix A^t is the transpose of A. An $n \times n$ matrix A with complex entries is called hermitian if $A^* = A$. R being a subring of $\mathbb{C}$ with $R = \bar{R}$, we write $\mathcal{H}(R)$ for the set of hermitian 2×2 matrices with entries in R, i.e.

$$(1.2) \qquad \mathcal{H}(R) = \{ \ A \in M_2(R) \ \mid \ A^* = A \ \}.$$

A 2×2 matrix A is in $\mathcal{H}(R)$ if it is of the form $A = \begin{pmatrix} a & b \\ \bar{b} & d \end{pmatrix}$ with $a, d \in R \cap \mathbb{R}$ and $b \in R$. Every $f \in \mathcal{H}(R)$ defines a binary hermitian form with coefficients in R. If $f = \begin{pmatrix} a & b \\ \bar{b} & d \end{pmatrix}$ then the associated binary hermitian form is the semi-quadratic map $f : \mathbb{C} \times \mathbb{C} \to \mathbb{R}$ defined by

$$(1.3) \qquad f(u, v) = (u, v) \begin{pmatrix} a & b \\ \bar{b} & d \end{pmatrix} (u, v)^* = au\bar{u} + bu\bar{v} + \bar{b}\bar{u}v + dv\bar{v}.$$

We shall often call an element $f \in \mathcal{H}(R)$ a binary hermitian form with coefficients in R. The discriminant $\Delta(f)$ of $f \in \mathcal{H}(R)$ is defined as

$$\Delta(f) = \det(f). \tag{1.4}$$

The group $\mathbf{GL}(2, R)$ acts on $\mathcal{H}(R)$ by the formula

$$\sigma(f) = \sigma f \sigma^* \tag{1.5}$$

for $\sigma \in \mathbf{GL}(2, R)$ and $f \in \mathcal{H}(R)$. If $\sigma = \begin{pmatrix} \alpha & \beta \\ \gamma & \delta \end{pmatrix} \in \mathbf{GL}(2, R)$ we have

$$\sigma f = \begin{pmatrix} (\alpha,\beta)f(\alpha,\beta)^* & (\alpha,\beta)f(\gamma,\delta)^* \\ (\gamma,\delta)f(\alpha,\beta)^* & (\gamma,\delta)f(\gamma,\delta)^* \end{pmatrix}. \tag{1.6}$$

Note that $\Delta(\sigma(f)) = \Delta(f) \cdot |\det \sigma|^2$ for every $\sigma \in \mathbf{GL}(2, R)$ and $f \in \mathcal{H}(R)$.

Two elements $f, g \in \mathcal{H}(R)$ are called $\mathbf{GL}(2, R)$-equivalent if $g = \sigma f$ for some $\sigma \in \mathbf{GL}(2, R)$; $\mathbf{SL}(2, R)$-equivalence is defined analogously.

A binary hermitian form $f \in \mathcal{H}(R)$ is positive definite if $f(u, v) > 0$ for all $(u, v) \in \mathbb{C} \times \mathbb{C} \setminus \{(0, 0)\}$. f is called negative definite if $-f$ is positive definite. f is called indefinite if $\Delta(f) < 0$.

We define

$$\mathcal{H}^+(R) = \{\ f \in \mathcal{H}(R) \ \mid \ f \text{ is positive definite } \}, \tag{1.7}$$

$$\mathcal{H}^-(R) = \{\ f \in \mathcal{H}(R) \ \mid \ f \text{ is indefinite } \}. \tag{1.8}$$

Clearly the group $\mathbf{GL}(2, R)$ leaves the $\mathcal{H}^{\pm}(R)$ invariant. It is easy to see that $f = \begin{pmatrix} a & b \\ \bar{b} & d \end{pmatrix} \in \mathcal{H}^+(R)$ if and only if $a > 0$ and $\Delta(f) > 0$. The group $\mathbb{R}^{>0}$ acts on $\mathcal{H}^+(\mathbb{C})$ by scalar multiplication. Similarly $\mathbb{R}^*$ acts on $\mathcal{H}^-(\mathbb{C})$. We define

$$\tilde{\mathcal{H}}^+(\mathbb{C}) := \mathcal{H}^+(\mathbb{C})/\mathbb{R}^{>0}, \qquad \tilde{\mathcal{H}}^-(\mathbb{C}) := \mathcal{H}^-(\mathbb{C})/\mathbb{R}^*. \tag{1.9}$$

For $f \in \mathcal{H}^{\pm}(\mathbb{C})$, $[f]$ stands for the class of f in $\tilde{\mathcal{H}}^{\pm}(\mathbb{C})$. The action (1.5) of $\mathbf{GL}(2, \mathbb{C})$ on $\mathcal{H}^{\pm}(\mathbb{C})$ clearly induces an action of $\mathbf{GL}(2, \mathbb{C})$ on $\tilde{\mathcal{H}}^{\pm}(\mathbb{C})$. The centre of $\mathbf{SL}(2, \mathbb{C})$ acts trivially on $\mathcal{H}(\mathbb{C})$, so we get an induced action of $\mathbf{PSL}(2, \mathbb{C})$ on $\mathcal{H}(\mathbb{C})$ and $\tilde{\mathcal{H}}^{\pm}(\mathbb{C})$.

Definition 1.1. The map $\Phi : \mathcal{H}^+(\mathbb{C}) \to \mathbb{H}$ is defined as

$$\Phi : f = \begin{pmatrix} a & b \\ \bar{b} & d \end{pmatrix} \to \frac{b}{d} + \frac{\sqrt{\Delta(f)}}{d} \cdot j.$$

Clearly Φ induces a map $\Phi : \tilde{\mathcal{H}}^+(\mathbb{C}) \to \mathbb{H}$. Φ is the analogue of the identification of the set of equivalence classes of binary positive definite quadratic forms with points of the upper half-plane.

Proposition 1.2. *The map* $\Phi : \tilde{\mathcal{H}}^+(\mathbb{C}) \to \mathbb{H}$ *is a* $\mathbf{PSL}(2,\mathbb{C})$*-equivariant bijection.* Φ *satisfies* $\Phi(\sigma(f)) = \sigma \cdot \Phi(f)$ *for every* $\sigma \in \mathbf{PSL}(2,\mathbb{C})$ *and* $f \in \tilde{\mathcal{H}}^+(\mathbb{C})$.

Proof. Clearly Φ is bijective. The equivariance of Φ is checked by a computation using the generators $\begin{pmatrix} 0 & 1 \\ -1 & 0 \end{pmatrix}, \begin{pmatrix} 1 & \beta \\ 0 & 1 \end{pmatrix}$ for $\mathbf{PSL}(2,\mathbb{C})$ with Proposition 1.1.2 and (1.6). □

For our treatment of indefinite hermitian forms we need the following constructions.

Definition 1.3. For a binary hermitian form $f = \begin{pmatrix} a & b \\ \bar{b} & d \end{pmatrix} \in \mathcal{H}^-(\mathbb{C})$ we define

$$\Psi(f) = \{\ z + rj \in \mathbb{H} \ \mid \ a + \bar{b}z + b\bar{z} + dz\bar{z} + r^2 d = 0\ \} \tag{1.10}$$

and $\mathbf{G} = \{\ \Psi(f) \ \mid \ f \in \mathcal{H}^-(\mathbb{C})\ \}$.

$\mathbf{G}$ is the set of geodesic planes in $\mathbb{H}$. If $f = \begin{pmatrix} a & b \\ \bar{b} & d \end{pmatrix}$ with $d \neq 0$ then $\Psi(f)$ is the following geodesic hemisphere

$$\Psi(f) = \{\ z + rj \ \mid \ |dz + b|^2 + |d|^2 r^2 = -\Delta(f)\ \}.$$

If $d = 0$ then $\Psi(f)$ is a vertical plane. The group $\mathbf{PSL}(2,\mathbb{C})$ acts on $\mathbf{G}$ by its induced action on subsets of $\mathbb{H}$.

Clearly Ψ induces a map $\Psi : \tilde{\mathcal{H}}^-(\mathbb{C}) \to \mathbf{G}$. The following is proved similarly to Proposition 1.2.

Proposition 1.4. *The map* $\Psi : \tilde{\mathcal{H}}^-(\mathbb{C}) \to \mathbf{G}$ *is a* $\mathbf{PSL}(2,\mathbb{C})$*-equivariant bijection.* Ψ *satisfies* $\Psi(\sigma(f)) = \sigma \cdot \Psi(f)$ *for every* $\sigma \in \mathbf{PSL}(2,\mathbb{C})$ *and* $f \in \tilde{\mathcal{H}}^-(\mathbb{C})$.

Definition 1.5. Let $R \subset \mathbb{C}$ be a subring with $\bar{R} = R$, and let $f \in \mathcal{H}(R)$. The group

$$E(f,R) = \{\ \sigma \in \mathbf{SL}(2,R) \ \mid \ \sigma(f) = f\ \}$$

is called the group of R-automorphs of f.

The automorphs of definite and indefinite hermitian forms differ drastically. The following easily established fact describes the complex automorphs of binary hermitian forms up to $\mathbf{GL}(2,\mathbb{C})$ conjugacy.

Proposition 1.6. *We have*

(1) $E\left(\begin{pmatrix} 1 & 0 \\ 0 & 1 \end{pmatrix}, \mathbb{C}\right) = \mathbf{SU}(2) = \left\{\begin{pmatrix} a & b \\ -\bar{b} & \bar{a} \end{pmatrix} \in \mathbf{SL}(2, \mathbb{C})\right\}$,

(2) $E\left(\begin{pmatrix} 0 & i \\ -i & 0 \end{pmatrix}, \mathbb{C}\right) = \mathbf{SL}(2, \mathbb{R})$,

(3) $E\left(\begin{pmatrix} 1 & 0 \\ 0 & 0 \end{pmatrix}, \mathbb{C}\right) = \left\{\begin{pmatrix} a & b \\ 0 & \bar{a} \end{pmatrix} \in \mathbf{SL}(2, \mathbb{C})\right\}$.

If $R \subset \mathbb{C}$ is a discrete ring with $\bar{R} = R$ and $f \in \mathcal{H}^+(R)$ is a positive definite hermitian form then the group $E(f, R)$ is finite. It is, for example, easy to see that

$$E\left(\begin{pmatrix} 1 & 0 \\ 0 & 1 \end{pmatrix}, \mathbb{Z}[i]\right) = \left\{\begin{matrix} \pm\begin{pmatrix} 1 & 0 \\ 0 & 1 \end{pmatrix}, \pm\begin{pmatrix} i & 0 \\ 0 & -i \end{pmatrix}, \\ \pm\begin{pmatrix} 0 & 1 \\ -1 & 0 \end{pmatrix}, \pm\begin{pmatrix} 0 & i \\ i & 0 \end{pmatrix} \end{matrix}\right\} \tag{1.11}$$

is the binary dihedral group of order 8.

For $f \in \mathcal{H}(R) \setminus (\mathcal{H}^+(R) \cup -\mathcal{H}^+(R))$, $f \neq 0$, the $E(f, R)$ are up to conjugacy discrete subgroups of $\mathbf{SL}(2, \mathbb{R})$ or the Borel subgroup of $\mathbf{SL}(2, \mathbb{C})$ according to $\Delta(f)$ being negative or zero.

9.2 Reduction Theory

Let $K = \mathbb{Q}(\sqrt{D}) \subset \mathbb{C}$ be an imaginary quadratic number field of discriminant $D < 0$. As in Chapter 7, we write $\mathcal{O}$ for the ring of integers in K. Here we study binary hermitian forms with coefficients $\mathcal{O}$. That is, we look at the $\mathbf{SL}(2, \mathcal{O})$ orbits in $\mathcal{H}(\mathcal{O})$. This theory was initiated by Hermite and further developed by Bianchi (1892) and Humbert (1915).

Definition 2.1. Let $\Delta \in \mathbb{Z}$ be an integer. We define

$$\mathcal{H}(\mathcal{O}, \Delta) = \{\, f \in \mathcal{H}(\mathcal{O}) \mid \Delta(f) = \Delta \,\},$$
$$\mathcal{H}^{\pm}(\mathcal{O}, \Delta) = \{\, f \in \mathcal{H}^{\pm}(\mathcal{O}) \mid \Delta(f) = \Delta \,\}.$$

Note that the sets $\mathcal{H}(\mathcal{O}, \Delta)$ and $\mathcal{H}^{\pm}(\mathcal{O}, \Delta)$ are invariant under $\mathbf{SL}(2, \mathcal{O})$. The main theorem of reduction theory is:

Theorem 2.2. *For any $\Delta \in \mathbb{Z}$ with $\Delta \neq 0$ the sets $\mathcal{H}(\mathcal{O}, \Delta)$ and $\mathcal{H}^{\pm}(\mathcal{O}, \Delta)$ split into finitely many $\mathbf{SL}(2, \mathcal{O})$ orbits.*

The following definition will be important in the proof of Theorem 2.2.

Definition 2.3. A positive definite hermitian form $f \in \mathcal{H}^+(\mathcal{O})$ is called reduced if $\Phi(f) \in \mathcal{F}_K$. A negative definite form f is called reduced if $-f$ is reduced. An indefinite form $f \in \mathcal{H}^-(\mathcal{O})$ is called reduced if $\Psi(f) \cap \mathcal{F}_K \neq \emptyset$.

Here $\mathcal{F}_K$ is the fundamental domain for $\mathbf{PSL}(2, \mathcal{O})$ acting on $\mathbb{H}$ described in Definition 7.3.1. We also use the further notation introduced there.

Theorem 2.4. *Let $\Delta \in \mathbb{Z}$ with $\Delta \neq 0$, then $\mathcal{H}(\mathcal{O}, \Delta)$ contains only finitely many reduced forms.*

Proof of Theorem 2.2. Clearly every $f \in \mathcal{H}(\mathcal{O}, \Delta)$ is $\mathbf{PSL}(2, \mathcal{O})$-equivalent to a reduced form. So Theorem 2.4 implies Theorem 2.2. □

Proof of Theorem 2.4. We distinguish two cases: (1): $\Delta > 0$: Here it is enough to show that $\mathcal{H}^+(\mathcal{O}, \Delta)$ contains only finitely many reduced binary hermitian forms. Let $f = \begin{pmatrix} \alpha & \beta \\ \bar{\beta} & \delta \end{pmatrix} \in \mathcal{H}^+(\mathcal{O}, \Delta)$ be a reduced form, that is

$$\Phi(f) = \frac{\beta}{\delta} + \frac{\sqrt{\Delta}}{\delta} \cdot j \in \mathcal{F}_K$$

with $\Delta = \alpha\delta - \beta\bar{\beta}$. Since F_K is compact, there is a $\kappa \in]0, \infty[$ such that $|z| < \kappa$ for all $z + rj \in \mathcal{F}_K$. Hence $|\frac{\beta}{\delta}| < \kappa$. This shows that it is enough to find a bound for $|\delta|$. As in Chapter 7 we write $\mathcal{S}_K \cap F_K = \{s_1, \dots, s_n\}$ for the set of singular points in F_K. By Lemma 7.3.6 there are an $\epsilon \in]0, \infty[$ and open neighbourhoods U_i of s_i such that $r^2 > \epsilon|z - s_i|$ for every $z + rj \in \mathcal{F}_K$ with $z \in U_i$. Put

$$M = F_K \setminus \left(\bigcup_{i=1}^{n} U_i \right).$$

M is a compact set. By making ϵ possibly a little smaller we may assume that every $z + rj \in \mathcal{F}_K$ with $z \in M$ satisfies $r > \epsilon$. If $\beta/\delta \in M$ then $|\delta| < \sqrt{\Delta}/\epsilon$. If $\beta/\delta \notin M$ then there is a $i \in \{1, \dots n\}$ with $\beta/\delta \in U_i$. Hence

$$\frac{\Delta}{\delta^2} > \epsilon \left| \frac{\beta}{\delta} - s_i \right|.$$

Writing $s_i = a/b$ with $a, b \in \mathcal{O}$ we get

$$\frac{|b|\Delta}{\delta} > \epsilon|\beta b - \delta a|.$$

The right-hand side is either zero or $\geq \epsilon$. The latter implies a bound for δ. Hence we have to prove that the set

$$\left\{ \begin{pmatrix} \alpha & \beta \\ \bar{\beta} & \delta \end{pmatrix} \in \mathcal{H}^+(\Delta) \;\middle|\; \frac{\beta}{\delta} \in \mathcal{S}_K \cap F_K \right\}$$

is finite. But this is obvious.

(2): $\Delta < 0$: Let $f = \begin{pmatrix} \alpha & \beta \\ \bar{\beta} & \delta \end{pmatrix} \in \mathcal{H}(\mathcal{O}, \Delta)$ be a reduced form. Then $\Psi(f) = S(\delta, \beta; -\Delta)$, see Definition 7.3.7. Leaving out finitely many f we may by Theorem 7.3.8 assume that f is not admissible. This means

$$\left| s + \frac{\beta}{\delta} \right| < \frac{\sqrt{|\Delta|}}{|\delta|} \tag{2.1}$$

for some singular point $s \in \mathcal{S}_K$. Since $\Psi(f) \cap \mathcal{F}_K \neq \emptyset$, only finitely many singular points s are suitable for inequality (2.1). Hence we may assume that (2.1) is satisfied for a fixed singular point s. Writing $s = a/b$ with $a, b \in \mathcal{O}$ we deduce

$$|a\delta + \beta b| < \sqrt{|\Delta|} \cdot |b|. \tag{2.2}$$

The set $T = \{ z \in \mathcal{O} \mid |z| < \sqrt{|\Delta|} \cdot |b| \}$ is finite. Hence we may assume that $a\delta + \beta b = \theta$ with some fixed $\theta \in T$. We put this equation into $\alpha\delta - \beta\bar{\beta} = \Delta$ and get

$$\delta(\alpha b\bar{b} + a\bar{\theta} + \bar{a}\theta - a\bar{a}\delta) = \Delta b\bar{b} + \theta\bar{\theta}. \tag{2.3}$$

The right-hand side is not zero because of $\theta \in T$. We deduce a bound for δ since δ is a divisor of the fixed right-hand side of (2.3).

Since the midpoint $-\beta/\delta$ of $\Psi(f)$ can only be of distance $\sqrt{|\delta|} \cdot \delta^{-1}$ from F_K we deduce a bound for $|\beta|$. This finishes the proof. □

Note that $\mathcal{H}(\mathcal{O}, 0)$ contains infinitely many $\mathbf{SL}(2, \mathcal{O})$ orbits. The binary hermitian forms $\begin{pmatrix} 0 & 0 \\ 0 & d \end{pmatrix}$, $d \in \mathbb{Z}$ are pairwise non-equivalent.

Definition 2.5. For any $\Delta \in \mathbb{Z}$ with $\Delta \neq 0$ we define

$$\tilde{\mathcal{H}}(\mathcal{O}, \Delta) = \mathbf{SL}(2, \mathcal{O}) \backslash \mathcal{H}(\mathcal{O}, \Delta), \qquad \tilde{\mathcal{H}}^+(\mathcal{O}, \Delta) = \mathbf{SL}(2, \mathcal{O}) \backslash \mathcal{H}^+(\mathcal{O}, \Delta),$$

$$h(\mathcal{O}, \Delta) = |\tilde{\mathcal{H}}(\mathcal{O}, \Delta)|, \qquad h^+(\mathcal{O}, \Delta) = |\tilde{\mathcal{H}}^+(\mathcal{O}, \Delta)|.$$

$h(\mathcal{O}, \Delta)$ is called the class number of binary hermitian forms of discriminant Δ and $h^+(\mathcal{O}, \Delta)$ is called the class number of positive definite binary hermitian forms of discriminant Δ.

Note that for $\Delta > 0$ we have $h(\mathcal{O}, \Delta) = 2 \cdot h^+(\mathcal{O}, \Delta)$. Given $\mathcal{O}$ and Δ it is always effectively possible to compute $h(\mathcal{O}, \Delta)$. This can be done by going through the proof of Theorem 2.4. We discuss now the example $\mathcal{O} = \mathbb{Z}[i]$.

Proposition 2.6. *Assume that* $\mathcal{O} = \mathbb{Z}[i]$ *and* $f = \begin{pmatrix} a & b \\ \bar{b} & d \end{pmatrix} \in \mathcal{H}(\mathcal{O})$ *with* $\Delta(f) \neq 0$. *Then the following hold.*

(1) If f is positive definite, then f is reduced if and only if $2\Delta(f)^2 \geq d^2$, $a \geq d$, $0 \leq \operatorname{Im} b \leq d/2$, $0 \leq |\operatorname{Re} b| \leq d/2$.

(2) If f is indefinite, reduced and $d \neq 0$, then

$$2\Delta(f)^2 \geq d^2, \qquad |b| \leq \frac{\sqrt{-\Delta(f)}}{\sqrt{2}} - \frac{\Delta(f)}{|d|};$$

if $d = 0$, then $|a| \leq \sqrt{2} \cdot |b|$.

Proof. Proposition 2.6 is easily deduced from the description of the fundamental domain $\mathcal{F}_{\mathbb{Q}(i)}$ given in Chapter 7. □

Using Proposition 2.6 it is easy to compute the following table of representatives of the $\mathbf{PSL}(2, \mathcal{O})$-classes of positive definite binary hermitian forms of discriminant Δ for $\Delta = 1, ..., 7$. The lower index gives the order of the group of automorphs.

$$\Delta = \mathbf{1}: \quad \begin{pmatrix} 1 & 0 \\ 0 & 1 \end{pmatrix}_8.$$

$$\Delta = \mathbf{2}: \quad \begin{pmatrix} 1 & 0 \\ 0 & 2 \end{pmatrix}_4, \quad \begin{pmatrix} 2 & 1+i \\ 1-i & 2 \end{pmatrix}_{24}.$$

$$\Delta = \mathbf{3}: \quad \begin{pmatrix} 1 & 0 \\ 0 & 3 \end{pmatrix}_4, \quad \begin{pmatrix} 2 & 1 \\ 1 & 2 \end{pmatrix}_{12}, \quad \begin{pmatrix} 2 & i \\ -i & 2 \end{pmatrix}_{12}.$$

$$\Delta = \mathbf{4}: \quad \begin{pmatrix} 1 & 0 \\ 0 & 4 \end{pmatrix}_4, \quad \begin{pmatrix} 2 & 0 \\ 0 & 2 \end{pmatrix}_8, \quad \begin{pmatrix} 2 & 1+i \\ 1-i & 3 \end{pmatrix}_4.$$

$$\Delta = \mathbf{5}: \quad \begin{pmatrix} 1 & 0 \\ 0 & 5 \end{pmatrix}_4, \quad \begin{pmatrix} 2 & 1 \\ 1 & 3 \end{pmatrix}_4, \quad \begin{pmatrix} 2 & i \\ -i & 3 \end{pmatrix}_4.$$

$$\Delta = \mathbf{6}: \quad \begin{pmatrix} 1 & 0 \\ 0 & 6 \end{pmatrix}_4, \quad \begin{pmatrix} 2 & 0 \\ 0 & 3 \end{pmatrix}_4, \quad \begin{pmatrix} 2 & 1+i \\ 1-i & 4 \end{pmatrix}_4.$$

$$\Delta = \mathbf{7}: \quad \begin{pmatrix} 1 & 0 \\ 0 & 7 \end{pmatrix}_4, \quad \begin{pmatrix} 2 & 1 \\ 1 & 4 \end{pmatrix}_4, \quad \begin{pmatrix} 2 & i \\ -i & 4 \end{pmatrix}_4, \quad \begin{pmatrix} 3 & 1+i \\ 1-i & 3 \end{pmatrix}_2.$$

9.3 Representation Numbers of Binary Hermitian Forms

We shall study now the representation numbers of binary hermitian forms with coefficients in the ring of integers $\mathcal{O}$ of an imaginary quadratic number field $K = \mathbb{Q}(\sqrt{d_K})$, d_K being always the discriminant of K. The theory we shall develop now is analogous to the theory of representation numbers of binary quadratic forms, see Landau (1918). Some complications arise from the fact that $\mathcal{O}$ need not have class number 1.

We keep our notation from Section 7.2. That is we for example write $\mathcal{M}$ for the group of fractional ideals of $\mathcal{O}$. Every $\mathfrak{m} \in \mathcal{M}$ is a free $\mathbb{Z}$-module of rank 2. For every $\mathfrak{m} \in \mathcal{M}$ there are integral ideals $\mathfrak{a}, \mathfrak{b} \subset \mathcal{O}$ such that $\mathfrak{m} = \mathfrak{a} \cdot \mathfrak{b}^{-1}$. The norm of $\mathfrak{m}$ is then defined as

$$N\mathfrak{m} = \frac{N\mathfrak{a}}{N\mathfrak{b}} = \frac{[\mathcal{O} : \mathfrak{a}]}{[\mathcal{O} : \mathfrak{b}]}.$$

By $\overline{\mathfrak{m}}$ we denote the complex conjugate of $\mathfrak{m} \in \mathcal{M}$. We have $\mathfrak{m} \cdot \overline{\mathfrak{m}} = < N\mathfrak{m} >$. If $B \subset A$ are abelian groups and $a, b \in A$ we write $a \equiv b \bmod B$ if and only if $a - b \in B$. $[a]$ stands for the class of a in A/B. For a binary hermitian form $f \in \mathcal{H}(\mathcal{O})$ we abbreviate Definition 1.5 to $E(f) := E(f, \mathcal{O})$.

Definition 3.1. For $f \in \mathcal{H}(\mathcal{O}), \mathfrak{m} \in \mathcal{M}$ and $k \in \mathbb{Z}$ we define

$$P(f, \mathfrak{m}, k) = \{ \ (u, v) \in \mathfrak{m} \times \mathfrak{m} \ \ | \ \ (N\mathfrak{m})^{-1} f(u, v) = k, \ < u, v >= \mathfrak{m} \ \}.$$

An element from $P(f, \mathfrak{m}, k)$ is called an $\mathfrak{m}$-primitive representation of k by f.

The above terminology is similar to that of Humbert (1919b), (1919c) and Landau (1918). Note that the numbers $(N\mathfrak{m})^{-1} f(u, v)$ are integers for $(u, v) \in \mathfrak{m} \times \mathfrak{m}$.

The group of automorphs $E(f)$ acts linearly on $P(f, \mathfrak{m}, k)$. The following will be of technical importance in describing the sets $P(f, \mathfrak{m}, k)$.

Definition 3.2. For $\mathfrak{m} \in \mathcal{M}$ and $k, \Delta \in \mathbb{Z}$ we define

$$\mathcal{R}(\mathfrak{m}, k, \Delta) = \{ \ [h] \in \mathfrak{m}^2 / k\mathfrak{m}^2 \ \ | \ \ h\bar{h} \equiv (N\mathfrak{m})^2 \cdot \Delta \mod (N\mathfrak{m})^2 k \cdot \mathbb{Z} \ \}.$$

Note that the condition in the definition of $\mathcal{R}(\mathfrak{m}, k, \Delta)$ is well defined.

Proposition 3.3. *Let $f \in \mathcal{H}(\mathcal{O})$, $\mathfrak{m} \in \mathcal{M}$, $k \in \mathbb{Z}$ and $(u, v) \in P(f, \mathfrak{m}, k)$.*

(1) There are $r, s \in \overline{\mathfrak{m}}$ such that

$$(a) \quad \det \begin{pmatrix} u & v \\ r & s \end{pmatrix} = N\mathfrak{m}.$$

Defining $h \in K$ by

$$(b) \quad \begin{pmatrix} u & v \\ r & s \end{pmatrix}(f) = \begin{pmatrix} * & h \\ \bar{h} & * \end{pmatrix},$$

we have $h \in \mathfrak{m}^2$ and $[h] \in \mathcal{R}(\mathfrak{m}, k, -\Delta(f))$.

(2) If $r', s' \in \bar{\mathfrak{m}}$, $h' \in \mathfrak{m}^2$ is a second triple satisfying (a) and (b), then $h \equiv h'$ mod $k \cdot \bar{\mathfrak{m}}^2$.

Proof. (1): The existence of r, s follows from $\mathfrak{m} \cdot \overline{\mathfrak{m}} =< N\mathfrak{m} >$. An explicit computation shows that

$$\begin{pmatrix} u & v \\ r & s \end{pmatrix}(f) = \begin{pmatrix} kN\mathfrak{m} & h \\ \bar{h} & tN\mathfrak{m} \end{pmatrix}$$

with $h \in \mathfrak{m}^2$ and $t \in \mathbb{Z}$. The discriminant of this hermitian form is $(N\mathfrak{m})^2 \cdot \Delta(f)$, hence

$$h\bar{h} - (N\mathfrak{m})^2 \cdot k \cdot t = -(N\mathfrak{m})^2 \cdot \Delta(f).$$

(2): From the hypotheses we have

$$\begin{pmatrix} u & v \\ r' & s' \end{pmatrix} \begin{pmatrix} u & v \\ r & s \end{pmatrix}^{-1} = \begin{pmatrix} 1 & 0 \\ (N\mathfrak{m})^{-1} \cdot \lambda & 1 \end{pmatrix}$$

with $\lambda \in \overline{\mathfrak{m}}^2$. We deduce $h' = h + \bar{\lambda} k$. □

Definition 3.4. For $f \in \mathcal{H}(\mathcal{O})$, $\mathfrak{m} \in \mathcal{M}$ and $k \in \mathbb{Z}$ we define a map

$$\varphi_f : P(f, \mathfrak{m}, k) \to \mathcal{R}(\mathfrak{m}, k, -\Delta(f))$$

by choosing for $(u, v) \in P(f, \mathfrak{m}, k)$ some $r, s \in \overline{\mathfrak{m}}$ and $h \in \mathfrak{m}^2$ with

$$\det \begin{pmatrix} u & v \\ r & s \end{pmatrix} = N\mathfrak{m} \quad \text{and} \quad \begin{pmatrix} u & v \\ r & s \end{pmatrix} \cdot f = \begin{pmatrix} * & h \\ \bar{h} & * \end{pmatrix}$$

and putting $\varphi_f((u, v)) = [h]$.

By Proposition 3.3 the procedure for the definition of φ_f makes sense and does not depend on the choice of r, $s \in \overline{\mathfrak{m}}$.

Proposition 3.5. *Let $f \in \mathcal{H}(\mathcal{O})$, $\mathfrak{m} \in \mathcal{M}$, $k \in \mathbb{Z}$ and $(u, v) \in P(f, \mathfrak{m}, k)$. Then the following hold.*

(1) For every $\sigma \in E(f)$ we have $\varphi_f((u, v) \cdot \sigma) = \varphi_f((u, v))$.

(2) For every $\sigma \in \mathbf{SL}(2, \mathcal{O})$ we have $\varphi_{\sigma f}((u, v)\sigma^{-1}) = \varphi_f((u, v))$.

The proof results from a straightforward consideration using part (2) of Proposition 3.3. The map φ_f induces now by part (1) of the last proposition a map

$$\varphi_f : P(f, \mathfrak{m}, k)/E(f) \to \mathcal{R}(\mathfrak{m}, k, -\Delta(f)).$$

Definition 3.6. For $\Delta \in \mathbb{Z}$ put $\mathcal{H}(\mathcal{O}, \Delta) = \{ f \in \mathcal{H}(\mathcal{O}) \mid \Delta(f) = \Delta \}$ (see Definition 2.1). Fix a set of representatives $f_i \in \mathcal{H}(\mathcal{O}, \Delta)$, $i \in I_\Delta$ for the classes in $\tilde{\mathcal{H}}(\mathcal{O}, \Delta) = \mathbf{SL}(2, \mathcal{O})\backslash\mathcal{H}(\mathcal{O}, \Delta)$. For $\Delta \in \mathbb{Z}$ we define

$$\varphi_\Delta : \bigcup_{i \in I_\Delta} P(f_i, \mathfrak{m}, k)/E(f_i) \to \mathcal{R}(\mathfrak{m}, k, -\Delta)$$

by composition of the maps φ_{f_i}.

The main theorem of the theory of representations of binary hermitian forms reads as follows.

Theorem 3.7. *For every $\Delta, k \in \mathbb{Z}$ the map φ_Δ is bijective.*

Proof. We shall construct a map

$$\psi_\Delta : \mathcal{R}(\mathfrak{m}, k, -\Delta) \to \bigcup_{i \in I_\Delta} P(f_i, \mathfrak{m}, k)/E(f_i)$$

which will be proved to be inverse to φ_Δ. For $[h] \in \mathcal{R}(\mathfrak{m}, k, -\Delta)$ we choose a $t \in \mathbb{Z}$ with

$$h\bar{h} - (N\mathfrak{m})^2 \cdot k \cdot t = -(N\mathfrak{m})^2 \cdot \Delta.$$

Define

$$g_1 = \begin{pmatrix} kN\mathfrak{m} & h \\ \bar{h} & tN\mathfrak{m} \end{pmatrix}, \qquad \eta_1 = (1, 0).$$

We have $g_1(\eta_1) = (N\mathfrak{m}) \cdot k$ and $\Delta(g_1) = (N\mathfrak{m})^2 \cdot \Delta$. Choose $u', v' \in \mathfrak{m}^{-1}$ and $r', s' \in \overline{\mathfrak{m}}^{-1}$ such that $< r', s' >= \overline{\mathfrak{m}}^{-1}$ and $u's' - r'v' = (N\mathfrak{m})^{-1}$ and define $\sigma_1 = \begin{pmatrix} u' & r' \\ v' & s' \end{pmatrix}$. Define furthermore $g_2 = \sigma_1 g_1$, $\eta_2 = \eta_1 \cdot \sigma_1^{-1}$. Obviously $g_2 \in \mathcal{H}(\mathcal{O})$ and $\eta_2 \in \mathfrak{m} \times \mathfrak{m}$. Trivially we have $g_2(\eta_2) = (N\mathfrak{m}) \cdot k$ and $\Delta(g_2) = \Delta$. It is also clear that the components of η_2 generate the module $\mathfrak{m}$. Hence $\eta_2 \in P(g_2, \mathfrak{m}, k)$. We choose an $\ell \in I_\Delta$ and $\sigma \in \mathbf{SL}(2, \mathcal{O})$ such that $\sigma g_2 = f_\ell$. We put $\eta = \eta_2 \sigma^{-1}$. Clearly we have $\eta \in P(f_\ell, \mathfrak{m}, k)$. We define the class of η in

$$\bigcup_{i \in I_\Delta} P(f_i, \mathfrak{m}, k)/E(f_i) \tag{3.1}$$

to be $\psi_\Delta([h])$. By Proposition 3.5 part (2) it is clear that $\varphi_\Delta \circ \psi_\Delta([h]) = [h]$. We need to prove that $\psi_\Delta \circ \varphi_\Delta(\lambda) = \lambda$ for any element λ of the set (3.1). We take $\ell_0 \in I_\Delta$ and $\eta_0 = (u_0, v_0)$ in the class of λ. Choose $h \in \mathfrak{m}^2$ satisfying $\varphi_\Delta(\lambda) = [h]$. We also consider $\ell \in I_\Delta$ and the elements f_ℓ, $\eta = (u, v)$ constructed above in the definition of $\psi_\Delta([h])$. Consider matrices

$$\sigma_3 = \begin{pmatrix} u_0 & v_0 \\ r_0 & s_0 \end{pmatrix}, \qquad \sigma_4 = \begin{pmatrix} u & v \\ r & s \end{pmatrix}$$

satisfying (1), (2) of Proposition 3.3 with f replaced by f_{ℓ_0} or f_ℓ. There is an $x \in \overline{\mathfrak{m}}^2$ such that

$$\sigma_3 f_{\ell_0} = \begin{pmatrix} 1 & 0 \\ N\mathfrak{m}^{-1} \cdot x & 1 \end{pmatrix}, \qquad \sigma_4 f_\ell = \begin{pmatrix} kN\mathfrak{m} & h \\ \bar{h} & tN\mathfrak{m} \end{pmatrix}.$$

Clearly

$$\sigma_3^{-1} \cdot \begin{pmatrix} 1 & 0 \\ N\mathfrak{m}^{-1}x & 1 \end{pmatrix} \cdot \sigma_4 \in \mathbf{SL}(2, \mathcal{O}).$$

This implies $\ell_0 = \ell$ and

$$\sigma_3^{-1} \cdot \begin{pmatrix} 1 & 0 \\ N\mathfrak{m}^{-1}x & 1 \end{pmatrix} \cdot \sigma_4 \in E(f_{\ell_0}).$$

This finishes the proof. □

An easy consequence of Theorem 3.7 is:

Theorem 3.8. *Let* $f \in \mathcal{H}(\mathcal{O}), \mathfrak{m} \in \mathcal{M}$ *and* $k \in \mathbb{Z}\backslash\{0\}$. *Then the set*

$$\{\ (u,v) \in \mathfrak{m} \times \mathfrak{m} \quad | \quad f(u,v) = k \cdot N\mathfrak{m}\ \}/E(f)$$

is finite.

Proof. If $(u,v) \in \mathfrak{m} \times \mathfrak{m}$ satisfies $f(u,v) = k \cdot N\mathfrak{m}$, then we have

$$k \cdot (N\mathfrak{m}) \cdot \mathcal{O} \subset < u, v > \cdot \overline{< u, v >} \subset \mathfrak{m} \cdot \overline{\mathfrak{m}}.$$

Since $k \neq 0$ there are only finitely many possibilities for $< u, v >$. By Theorem 3.7 there are only finitely many possibilities for u, v up to operation by $E(f)$. Note that $\mathfrak{m}^2/k\mathfrak{m}^2$ is finite since $k \neq 0$. □

For later use we have to compute the cardinalities of the sets $\mathcal{R}(\mathfrak{m}, k, \Delta)$.

Definition 3.9. For $\mathfrak{m} \in M, k, \Delta \in \mathbb{Z}$ with $k \neq 0$ we define

$$\begin{aligned} r(\mathfrak{m}, k, \Delta) &= |\mathcal{R}(\mathfrak{m}, k, \Delta)| \\ &= |\{\ [h] \in \mathfrak{m}^2/k\mathfrak{m}^2 \quad | \quad h\bar{h} \equiv (N\mathfrak{m})^2 \cdot \Delta \mod (N\mathfrak{m})^2 k\mathbb{Z}\ \}|. \end{aligned}$$

We need the following elementary lemma.

Lemma 3.10. *Let* $\mathfrak{m}, \mathfrak{a}$ *be elements of* $\mathcal{M}$ *so that* $\mathfrak{a}$ *is an ideal of* $\mathcal{O}$. *Then there is* $\lambda \in K\backslash\{0\}$ *such that* $\lambda\mathfrak{m} \subset \mathcal{O}$, *and* $< \mathfrak{a}, \lambda\mathfrak{m} >= \mathcal{O}$.

Proof. Let

$$\mathfrak{a} = \prod_{\mathcal{P}} \mathcal{P}^{a_\mathcal{P}}, \qquad \mathfrak{m} = \prod_{\mathcal{P}} \mathcal{P}^{b_\mathcal{P}}$$

be the prime ideal decompositions of $\mathfrak{a}, \mathfrak{m}$. Choose elements $x_\mathcal{P} \in \mathcal{O}$ satisfying $x_\mathcal{P} \in \mathcal{P}^{b_\mathcal{P}} \backslash \mathcal{P}^{b_\mathcal{P}+1}$ in case $b_\mathcal{P} > 0$ and $a_\mathcal{P} > 0$. By the Chinese remainder theorem we may solve the finite system of congruences $\lambda_1 \equiv x_\mathcal{P} \bmod \mathcal{P}^{b_\mathcal{P}+1}$ by an element $\lambda_1 \in \mathcal{O}$. We write

$$\lambda_1^{-1}\mathfrak{m} = \prod_{\mathcal{P}} \mathcal{P}^{c_\mathcal{P}}$$

for the prime ideal decomposition of $\lambda_1^{-1}\mathfrak{m}$. We have $c_\mathcal{P} \leq 0$ if $a_\mathcal{P} > 0$. Choose elements $y_\mathcal{P} \in \mathcal{O}$ satisfying $y_\mathcal{P} \in \mathcal{P}^{-c_\mathcal{P}} \backslash \mathcal{P}^{-c_\mathcal{P}+1}$ if $c_\mathcal{P} < 0$ or $a_\mathcal{P} > 0$. Solve the finite system of congruences $\lambda_2 \equiv y_\mathcal{P} \bmod \mathcal{P}^{-c_\mathcal{P}+1}$. Then the element $\lambda = \lambda_1^{-1}\lambda_2$ has the required properties. □

Of course, our proof works for any algebraic number field. By the Chebotarev density theorem, for any $\mathfrak{m} \in \mathcal{M}$ there are infinitely many prime ideals $\mathcal{P}$ such that $\lambda\mathfrak{m} = \mathcal{P}$ for an element $\lambda \in K\backslash\{0\}$. Using this deep fact our Lemma 3.10 becomes trivial.

Our first result concerning the $r(\mathfrak{m}, k, \Delta)$ is:

Proposition 3.11. *For every $k, \Delta \in \mathbb{Z}$ with $k \neq 0$ we have the following:*
(1) $r(\mathfrak{m}, k, \Delta) = r(\lambda\mathfrak{m}, k, \Delta)$ for all $\mathfrak{m} \in \mathcal{M}$, $\lambda \in K$, $\lambda \neq 0$,
(2) $r(\mathfrak{m}, k, \Delta) = r(\mathcal{O}, k, \Delta)$ for all $\mathfrak{m} \in \mathcal{M}$,
(3) if $k_1, k_2 \in \mathbb{Z}$ are coprime, then $r(\mathfrak{m}, k_1 \cdot k_2, \Delta) = r(\mathfrak{m}, k_1, \Delta) \cdot r(\mathfrak{m}, k_2, \Delta)$ for all $\mathfrak{m} \in \mathcal{M}$.

Proof. (1): Multiplication by λ^2 induces a bijection between $R(\mathfrak{m}, k, \Delta)$ and $R(\lambda\mathfrak{m}, k, \Delta)$.

(2): By (1) and Lemma 3.10 we may assume that $\mathfrak{m}$ is an ideal satisfying $\mathfrak{m} + k\mathcal{O} = \mathcal{O}$. We choose elements $a_0 \in \mathfrak{m}$ and $t_0 \in \mathcal{O}$ with $a_0 - kt_0 = 1$. Putting $a = a_0 \cdot N\mathfrak{m}$ we find

$$a = N\mathfrak{m} + k \cdot N\mathfrak{m} \cdot t_0. \tag{3.2}$$

The element a is contained in $\mathfrak{m}^2$. Next we look at the $\mathcal{O}/k\mathcal{O}$ linear map

$$g:\ \mathcal{O}/k\mathcal{O} \to \mathfrak{m}^2/k\mathfrak{m}^2, \qquad g:\ [\lambda] \mapsto [a \cdot \lambda].$$

If $g[\lambda_1] = g([\lambda_2])$ we have $a(\lambda_1 - \lambda_2) = kb$ with $b \in \mathfrak{m}^2$. Any prime ideal dividing $k\mathcal{O}$ has to divide $\lambda_1 - \lambda_2$ since it cannot divide $N\mathfrak{m}$ and a by assumption and (3.2). Hence g is injective. The finite sets $\mathcal{O}/k\mathcal{O}$ and $\mathfrak{m}^2/k\mathfrak{m}^2$ have the same orders, this implies that g is bijective. It is then easy to check that g induces a bijection $g : R(\mathcal{O}, k, \Delta) \longrightarrow R(\mathfrak{m}, k, \Delta)$.

(3): follows easily from the Chinese Remainder Theorem in case $\mathfrak{m} = \mathcal{O}$. For the general case use (2) of this Proposition. □

Next we compute explicitly the orders $r(\mathcal{O}, p^n, \Delta)$, where p is a prime number. To do this we introduce the character $\chi_K : \mathbb{Z} \to \mathbb{C}$ of the quadratic field K. χ_K is a completely multiplicative function satisfying $\chi_K(0) = 0$ and for every prime p

$$\chi_K(p) = \begin{cases} 0 & \text{if} \quad p \mid d_K, \\ \left(\dfrac{d_K}{p}\right) & \text{if} \quad p \nmid d_K. \end{cases} \tag{3.3}$$

Here $(\frac{d_K}{p})$ stands for the quadratic residue symbol if $p > 2$, whereas

$$\chi_K(2) = \left(\frac{d_K}{2}\right) = \begin{cases} +1 & \text{if} \quad d_K \equiv 1 \bmod 8, \\ -1 & \text{if} \quad d_K \equiv 5 \bmod 8. \end{cases}$$

If $\chi_K(p) = -1$ then $p\mathcal{O}$ is a prime ideal of $\mathcal{O}$. If $\chi_K(p) = 1$ then we have $p \cdot \mathcal{O} = \mathcal{P} \cdot \overline{\mathcal{P}}$ for a prime ideal $\mathcal{P} \subset \mathcal{O}$ satisfying $\mathcal{P} \neq \overline{\mathcal{P}}$. In case $\chi_K(p) = 0$ we have $p\mathcal{O} = \mathcal{P}^2$ for a prime ideal $\mathcal{P} \subset \mathcal{O}$. p is said to be inertial, split or ramified in these cases. For a general discussion see Hasse (1964).

To formulate our results it is advantageous to introduce the following formal power series.

Definition 3.12. Let p be a prime number and $\Delta \in \mathbb{Z}$. We define

$$Q_p(\Delta, X) = \sum_{n=0}^{\infty} r(\mathcal{O}, p^n, \Delta) \cdot X^n \in \mathbb{Z}[[X]].$$

It follows from general p-adic integration methods that $Q_p(\Delta, X)$ is a rational function. We compute it here explicitly. If p is a prime number and $\Delta \in \mathbb{Z}$ we write $p^t \| \Delta$ for $t \in \{0, 1, 2, \ldots, \infty\}$ if $\Delta \neq 0$ and p^t is the exact power of p dividing Δ or if $t = \infty$ and $\Delta = 0$.

Proposition 3.13. *Let p be a prime number which is split in K, that is $\chi_K(p) = 1$. Assume that $\Delta \in \mathbb{Z}$ and $t \in \mathbb{N} \cup \{0\}$ satisfy $p^t \| \Delta$. Then*

$$Q_p(\Delta, X) = \frac{1 - X}{1 - pX} \cdot \left(\sum_{\nu=0}^{t} (pX)^\nu \right) = \frac{(1 - X)(1 - (pX)^{t+1})}{(1 - pX)^2}. \tag{3.4}$$

Proof. Here we have $p\mathcal{O} = \mathcal{P} \cdot \overline{\mathcal{P}}$ with two distinct prime ideals $\mathcal{P}, \overline{\mathcal{P}}$. By the Chinese Remainder Theorem we have for any $n \geq 1$ a ring isomorphism

$$\mathcal{O}/p^n\mathcal{O} \to \mathcal{O}/\mathcal{P}^n \times \mathcal{O}/\overline{\mathcal{P}}^n \to \mathbb{Z}/p^n\mathbb{Z} \times \mathbb{Z}/p^n\mathbb{Z}, \qquad [\lambda] \mapsto ([\lambda], [\lambda]).$$

We infer that

$$r(\mathcal{O}, p^n, \Delta) = |\{ (x,y) \in \mathbb{Z}/p^n\mathbb{Z} \times \mathbb{Z}/p^n\mathbb{Z} \;\mid\; x \cdot y \equiv -\Delta \mod p^n\mathbb{Z} \}|.$$

If $p^t \| \Delta$ it is clear that $r(\mathcal{O}, p^n, \Delta) = r(\mathcal{O}, p^n, p^t)$. It is also immediate that $r(\mathcal{O}, p^n, 1) = p^{n-1}(p-1) = \varphi(p^n)$ where φ is Euler's φ-function. From this formula (3.4) follows for $t = 0$. Looking at decompositions $\epsilon p^\ell \cdot \eta p^m = p^{\ell+m}$ we get for $t < n$

$$r(\mathcal{O}, p^n, p^t) = \sum_{\mu=0}^{t} p^\mu \cdot r(\mathcal{O}, p^{n-\mu}, 1) = (t+1)p^{n-1}(p-1)$$

and for $t \geq n \geq 1$

$$r(\mathcal{O}, p^n, p^t) = \sum_{\mu=0}^{n} p^\mu r(\mathcal{O}, p^{n-\mu}, 1) + p^n \cdot = (n+1)p^n - np^{n-1}.$$

From these formula (3.4) follows. □

Proposition 3.14. *Let p be a prime number which is inertial in K, that is $\chi_K(p) = -1$. Assume that $\Delta \in \mathbb{Z}$ and $t \in \mathbb{N} \cup \{0\}$ satisfy $p^t \| \Delta$. Then*

$$(3.5)\quad Q_p(\Delta, X) = \begin{cases} \dfrac{1+X}{1-pX} & \text{if } t = 0, \\ (1+X)\displaystyle\sum_{\mu=0}^{t'-1} p^{2\mu} X^{2\mu} & \text{if } t = 2t'-1, \\ (1+X)\displaystyle\sum_{\mu=0}^{t'-1} p^{2\mu} X^{2\mu} + \dfrac{1+X}{1-pX} p^{2t'} X^{2t'} & \text{if } t = 2t' > 0. \end{cases}$$

This is equivalent to

$$Q_p(\Delta, X) = \frac{(1+X)(1-(-pX)^{t+1})}{1-(pX)^2}.$$

Proof. In this case $p\mathcal{O}$ is a prime ideal in $\mathcal{O}$ and $\mathcal{O}/p\mathcal{O} \cong \mathbb{F}_{p^2}$, where $\mathbb{F}_{p^2}$ is the field with p^2 elements. We shall first of all show that for every $n \in \mathbb{N}$ the map

$$(3.6)\qquad (\mathcal{O}/p^n\mathcal{O})^* \to (\mathbb{Z}/p^n\mathbb{Z})^*, \qquad [\lambda] \mapsto [\lambda\bar{\lambda}]$$

is surjective. We use induction on n. The case $n = 1$ we leave as an elementary exercise. For $\epsilon \in \mathbb{Z}$ and $p \nmid \epsilon$ we choose an $a \in \mathcal{O}$ such that $a\bar{a} - \epsilon = p^n u$ with $u \in \mathbb{Z}$. The equation $(a + \lambda p^n)(\bar{a} + \bar{\lambda} p^n) \equiv \epsilon \mod p^{n+1}\mathbb{Z}$ is equivalent to $u + \lambda\bar{a} + \bar{\lambda} a \equiv 0 \mod p\mathbb{Z}$. The trace form being non-degenerate for finite fields we find a solution λ of this congruence.

The surjectivity of the norm map (3.6) implies that $r(\mathcal{O}, p^n, \Delta) = r(\mathcal{O}, p^n, p^t)$ if $p^t \| \Delta$ and $r(\mathcal{O}, p^n, 1) = p^{n-1}(p+1)$. From these formulas (3.5) follows for $t = 0$. It is also clear that

$$(3.7)\quad |\{ b \in \mathcal{O}/p^n\mathcal{O} \mid b\bar{b} \equiv 0 \mod p^n\mathbb{Z} \}| = \begin{cases} p^n & \text{if } n \equiv 0 \mod 2, \\ p^{n-1} & \text{if } n \equiv 1 \mod 2. \end{cases}$$

From this formula (3.5) follows for t odd since then $r(\mathcal{O}, p^n, p^t) = 0$ for $n > t$.

Next we see that $r(\mathcal{O}, p^n, p^{2t'}) = p^{2t'} \cdot r(\mathcal{O}, p^{n-2t'}, 1)$ for all $n \geq t+1$. From formula (3.7) we infer that

$$\begin{aligned} Q_p(p^{2t'}, X) &= (1+X)\left(\sum_{\mu=0}^{t'-1} p^{2\mu} X^{2\mu}\right) \\ &\quad + p^{2t'} X^{2t'} + p^{2t'} \cdot \sum_{n=2t'+1}^{\infty} r(\mathcal{O}, p^{n-2t'}, 1) X^n \\ &= (1+X)\left(\sum_{\mu=0}^{t'-1} p^{2\mu} X^{2\mu}\right) + p^{2t'} X^{2t'} \frac{1+X}{1-pX}. \end{aligned}$$

So the proposition is proved. □

Proposition 3.15. *Let p be a prime number with $p \neq 2$ which is ramified in K, that is $\chi_K(p) = 0$. Assume that $\Delta \in \mathbb{Z}$ and $t \in \mathbb{N} \cup \{0\}$ satisfy $p^t \| \Delta$. Then the following hold.*

(1) In case $t = 0$ we have

$$(3.8)\qquad Q_p(\Delta, X) = \begin{cases} 1 & \text{if } \left(\frac{\Delta}{p}\right) = -1, \\ \dfrac{1+pX}{1-pX} & \text{if } \left(\frac{\Delta}{p}\right) = 1. \end{cases}$$

(2) In case $t > 0$ but $t \neq \infty$ define $D_0 = D/p$ and $\Delta_0 = \Delta/p^t$. Then

$$(3.9)\qquad Q_p(\Delta, X) = \begin{cases} \dfrac{1-(pX)^{t+1}}{1-pX} & \text{if } \left(\frac{-D_0}{p}\right)^t \left(\frac{\Delta_0}{p}\right) = -1 \\ \dfrac{1+(pX)^{t+1}}{1-pX} & \text{if } \left(\frac{-D_0}{p}\right)^t \left(\frac{\Delta_0}{p}\right) = 1. \end{cases}$$

(3) In case $\Delta = 0$ we have $t = \infty$ and $Q_p(0, X) = (1-pX)^{-1}$.

Proof. In this case we have $p\mathcal{O} = \mathcal{P}^2$ for a prime ideal $\mathcal{P}$. Assume that $\Delta \neq 0$, $t = 0$. For every $n \in \mathbb{N}$ we define

$$V(n) = \{ x \in (\mathbb{Z}/p^n\mathbb{Z})^* \mid x \text{ is a square modulo } p \}.$$

$V(n)$ is a subgroup of index 2 in $(\mathbb{Z}/p^n\mathbb{Z})^*$. The ring $\mathcal{O}$ has the $\mathbb{Z}$-basis 1, ω where $\omega = (d_K + \sqrt{d_K})/2$. For an element $a + b\omega \in \mathcal{O}$ with $a, b \in \mathbb{Z}$ we have

$$(a + b\omega)\overline{(a + b\omega)} = a^2 + a\,b\,d_K + \frac{d_K(d_K - 1)}{4}\,b^2.$$

Hence the image of the map

$$(\mathcal{O}/p^n\mathcal{O})^* \to (\mathbb{Z}/p^n\mathbb{Z})^*, \qquad [\lambda] \mapsto [\lambda \cdot \bar{\lambda}]$$

is contained in $V(n)$. An easy approximation argument like in Proposition 3.14 shows that this homomorphism is surjective. This shows $r(\mathcal{O}, p^n, \Delta) = 0$ if $(\frac{\Delta}{p}) = -1$ and furthermore $r(\mathcal{O}, p^n, \Delta) = r(\mathcal{O}, p^n, 1) = 2p^n$ if $(\frac{\Delta}{p}) = 1$. So we have

$$Q_p(1, X) = 1 + 2(1 + pX + p^2X^2 + \ldots) - 2 = -1 + \frac{2}{1 - pX} = \frac{1 + pX}{1 - pX}.$$

This establishes (1).

Next let $t > 0$ and put $\pi = [\sqrt{d_K}] \in \mathcal{O}/p^n\mathcal{O}$. We have $\pi^2 \equiv D_0 p$ and $\pi\bar{\pi} \equiv -D_0 p \bmod p^n\mathbb{Z}$. π is a generator of the image of the ideal $\mathcal{P}$ in $\mathcal{O}/p^n\mathcal{O}$. Every element $\alpha \in \mathcal{O}/p^n\mathcal{O}$ can be written as $\alpha = \epsilon\pi^\ell$ with a unique $\ell \in \{0, 1, \ldots, 2n\}$ and an $\epsilon \in (\mathcal{O}/p^n\mathcal{O})^*$ which is unique modulo $\mathcal{P}^{2n-\ell}$ if $\ell \neq 2n$. We have

$$\alpha\bar{\alpha} = \epsilon\bar{\epsilon} \cdot \pi^\ell \cdot \bar{\pi}^\ell = \epsilon\bar{\epsilon} \cdot (-D_0)^\ell \cdot p^\ell.$$

From this formula we infer that $r(\mathcal{O}, p^n, \Delta) = p^n$ if $t \geq n$. This proves (3). For $n > t$ we have

$$r(\mathcal{O}, p^n, \Delta) = |\{\ \epsilon \in (\mathcal{O}/\mathcal{P}^{2n-t})^* \ \mid\ \epsilon\bar{\epsilon}(-D_0)^t \equiv \Delta_0 \mod p^{n-t}\mathbb{Z}\ \}|.$$

This establishes the first case of the formula in (2) since if $n > t$ and $\left(\frac{-D_0}{p}\right)^t \left(\frac{\Delta_0}{p}\right) = -1$ then $r(\mathcal{O}, p^n, \Delta) = 0$ by formula (1). In the further case $\left(\frac{-D_0}{p}\right)^t \left(\frac{\Delta_0}{p}\right) = 1$ we put $k = n - t$ and get: $r(\mathcal{O}, p^n \Delta) = p^t \cdot r(\mathcal{O}, p^k, 1)$. Hence by case (1)

$$\begin{aligned} Q_p(\Delta, X) &= \frac{1 - (pX)^{t+1}}{1 - pX} + p^t \sum_{k=1}^{\infty} r(\mathcal{O}, p^k, 1) X^{t+k} \\ &= \frac{1 - (pX)^{t+1}}{1 - pX} + p^t X^t \cdot \left(\frac{1 + pX}{1 - pX}\right) - p^t X^t = \frac{1 + (pX)^{t+1}}{1 - pX}. \end{aligned}$$

This finishes the proof of this proposition. □

The formulation of the next results is easier if we use the following extension of our definitions.

Definition 3.16. Let p be a prime number and $a, b \in \mathbb{Z}$ with $p \not| a$. For every $n \in \mathbb{N}$ there is an $a_n \in \mathbb{Z}$ such that $aa_n \equiv 1 \bmod p^n\mathbb{Z}$. We define

$$r\left(\mathcal{O}, p^n, \frac{b}{a}\right) := r(\mathcal{O}, p^n, a_n \cdot b), \qquad Q_p\left(\frac{b}{a}, X\right) = Q_p(a_n b, X).$$

Clearly this definition is independent of the choices of the a_n.

Proposition 3.17. *Let the prime number 2 be ramified in K, that is $\chi_K(2) = 0$ and $4|d_K$. Putting $D_1 = d_K/4$ we have $D_1 \equiv 2, 6, 3, 7 \bmod 8\mathbb{Z}$. Assume furthermore that $\Delta \in \mathbb{Z}$ and $t \in \mathbb{N} \cup \{0\}$ satisfy $2^t \| \Delta$. Then the following hold.*

(1) In case $t = 0$ we have

$$(3.10) \quad Q_2(\Delta, X) = \begin{cases} 1 + 2X + 4X^2 & \text{if } D_1 \equiv 2, \quad \Delta \equiv 3,5 \mod 8, \\ 1 + 2X + 4X^2 & \text{if } D_1 \equiv 6, \quad \Delta \equiv 5,7 \mod 8, \\ 1 + 2X & \text{if } D_1 \equiv 3,7 \quad \Delta \equiv 3,7 \mod 8, \\ \dfrac{1 + 8X^3}{1 - 2X} & \text{if } D_1 \equiv 2, \quad \Delta \equiv 1,7 \mod 8, \\ \dfrac{1 + 8X^3}{1 - 2X} & \text{if } D_1 \equiv 6, \quad \Delta \equiv 1,3 \mod 8, \\ \dfrac{1 + 4X^2}{1 - 2X} & \text{if } D_1 \equiv 3,7 \quad \Delta \equiv 1,5 \mod 8. \end{cases}$$

(2) In case $t > 0$ but $t \neq \infty$ we define

$$D_0 = \begin{cases} -\dfrac{D_1}{2} & \text{if } D_1 \equiv 2,6 \mod 8, \\ \dfrac{1 - D_1}{2} & \text{if } D_1 \equiv 3,7 \mod 8 \end{cases}$$

and $\Delta_0 = \Delta/2^t$. Then

$$(3.11) \qquad Q_2(\Delta, X) = \frac{1 - (2X)^t}{1 - 2X} + 2^t X^t Q_2(\Delta_0 \cdot D_0^{-t}, X).$$

(3) In case $\Delta = 0$ we have $t = \infty$ and $Q_2(0, X) = (1 - 2X)^{-1}$.

Proof. In this case we have $2 \cdot \mathcal{O} = \mathcal{P}^2$ for a prime ideal $\mathcal{P}$. We define as usually $\omega = (d_K + \sqrt{d_K})/2$. For $a, b \in \mathbb{Z}$ we have

$$(3.12) \qquad (a + b\omega)\overline{(a + b\omega)} = a^2 + 4D_1 ab + D_1(4D_2 - 1)b^2.$$

This formula readily implies that $r(\mathcal{O}, 2, \Delta) = 2$ if $\Delta \equiv 0 \bmod 2\mathbb{Z}$ and

$$r(\mathcal{O}, 4, \Delta) = \begin{cases} 8 & \text{if} \quad D_1 \equiv 3, \quad \Delta \equiv 1 \mod 4, \\ 0 & \text{if} \quad D_1 \equiv 3, \quad \Delta \equiv 3 \mod 4, \\ 4 & \text{if} \quad D_1 \equiv 2, \quad \Delta \equiv 1,3 \mod 4. \end{cases}$$

Next we introduce for $n \geq 3$ the subgroups

$$V(D,n) = \begin{cases} \{ x \in (\mathbb{Z}/2^n\mathbb{Z})^* \mid x \equiv 1,3 \mod 8 \} \text{ if } D_1 \equiv 6 \mod 8, \\ \{ x \in (\mathbb{Z}/2^n\mathbb{Z})^* \mid x \equiv 1,7 \mod 8 \} \text{ if } D_1 \equiv 2 \mod 8, \\ \{ x \in (\mathbb{Z}/2^n\mathbb{Z})^* \mid x \equiv 1,5 \mod 8 \} \text{ if } D_1 \equiv 3,7 \mod 8. \end{cases}$$

$V(D,n)$ has index 2 in $\mathbb{Z}/2^n\mathbb{Z}$. Formula (3.12) shows that the image of the homomorphism

$$\varphi_n : (\mathcal{O}/2^n\mathcal{O})^* \to (\mathbb{Z}/2^n\mathbb{Z})^*, \qquad \varphi_n : [\lambda] \mapsto [\lambda\bar{\lambda}]$$

is contained in $V(D,n)$. We shall show now by induction that φ_n is surjective onto $V(D,n)$. By direct inspection we see that this is true if $n = 3$. Having $\alpha \in \mathcal{O}$, and $\epsilon \in \mathbb{Z}$ with $\epsilon \equiv 1 \bmod 2\mathbb{Z}$ and $u \in \mathbb{Z}$ with $\alpha\bar{\alpha} - \epsilon = 2^n u$ for $n \geq 3$ we note that the congruence $(\alpha + \lambda 2^{n-1})\overline{(\alpha + \lambda 2^{n-1})} \equiv \epsilon \bmod 2^{n+1}\mathbb{Z}$ is equivalent to $2n + \lambda\bar{\alpha} + \alpha\bar{\lambda} \equiv 0 \bmod 4$. Since $\alpha\bar{\alpha} \equiv 1 \bmod 2$, there is a $\lambda \in \mathcal{O}$ solving this last congruence.

We now have $|(\mathcal{O}/2^n\mathcal{O})^*| = 2^{2n-1}$, and $|V(D,n)| = 2^{n-2}$. for $n \geq 3$. We infer

$$r(\mathcal{O}, 2^n, \Delta) = \begin{cases} 0 & \text{if } \Delta \notin V(D,n) \mod 2^n, \\ 2^{n+1} & \text{if } \Delta \in V(D,n) \mod 2^n. \end{cases}.$$

This establishes case (1) of our proposition. Next we define

$$\pi = \begin{cases} \sqrt{D_1} & \text{if } D_1 \equiv 2,6 \mod 8, \\ 1 + \sqrt{D_1} & \text{if } D_1 \equiv 3,7 \mod 8. \end{cases}$$

π satisfies $\pi^2 \equiv 2\,\epsilon \bmod 2^n\mathcal{O}$ with a unit ϵ in $\mathcal{O}/2^n\mathcal{O}$. Hence the image of π in $\mathcal{O}/2^n\mathcal{O}$ generates the maximal ideal of that ring. We also have $\pi \cdot \bar{\pi} = 2 \cdot D_0$ where D_0 is defined as in the statement of the proposition. Note that D_0 satisfies $D_0 \equiv 1 \bmod 2$. Every element $\alpha \in \mathcal{O}/2^n\mathcal{O}$ can be written as $\alpha \equiv \epsilon \cdot \pi^\ell$ with a unique $\ell \in \{0, 1, \ldots, 2n\}$ and a $\epsilon \in (\mathcal{O}/2^n\mathcal{O})^*$ which is unique modulo $\mathcal{P}^{2n-\ell}$ if $\ell \neq 2n$. We have $\alpha\bar{\alpha} \equiv \epsilon\bar{\epsilon} \cdot 2^\ell \cdot D_0^\ell$. From this formula we see that $r(\mathcal{O}, 2^n\Delta) = 2^n$ if $t \geq n$. This proves (3). For $n > t$ we further have

$$\begin{aligned} r(\mathcal{O}, 2^n, \Delta) &= |\{ \epsilon \in \mathcal{O}/\mathcal{P}^{2n-t} \mid \epsilon\bar{\epsilon}D_0^t \equiv \Delta_0 \mod 2^{n-t} \}| \\ &= 2^t \cdot r(\mathcal{O}, 2^n, D_0^{-t}\Delta_0). \end{aligned}$$

This finally proves the proposition. □

The formulas contained in Propositions 3.15, 3.17 can be given uniformly, see Elstrodt, Grunewald, Mennicke (1987a). By putting our computations together we get a formula which is partly due to Humbert (1919b), (1919c).

Theorem 3.18. *Let $\mathfrak{m} \in \mathcal{M}$ and Δ be an integer. Let $f_1, \ldots, f_h$ be representatives for the* **SL**$(2, \mathcal{O})$*-classes of binary hermitian forms of discriminant Δ. Then for any $k \in \mathbb{Z}$ which is prime to $2 \cdot \Delta$ the following formula holds.*

$$\left|\bigcup_{i=1}^{h} \{ (u,v) \in \mathfrak{m} \times \mathfrak{m} \mid f_i(u,v) = kN\mathfrak{m},\ <u,v>= \mathfrak{m} \}/E(f_i)\right| \tag{3.13}$$
$$= k \cdot \prod_{\substack{p|k \\ p\nmid d_K}} \left(1 - \left(\frac{d_K}{p}\right)p^{-1}\right) \cdot \prod_{\substack{p|k \\ p|d_K}} \left(1 + \left(\frac{-\Delta}{p}\right)\right).$$

Proof. The left-hand side of formula (3.13) is equal to $r(\mathfrak{m}, k, -\Delta)$ which in turn equals $r(\mathcal{O}, k, -\Delta)$. By introducing a prime ideal decomposition of k the theorem is deduced from Propositions 3.13, 3.14, 3.15, 3.17. □

Humbert (1919b), (1919c) considers only definite hermitian forms in the case $d_K \equiv 1 \bmod 4\mathbb{Z}$ or $d_K \equiv 0 \bmod 8\mathbb{Z}$. He also introduces the notion of properly primitive forms, that is forms $\begin{pmatrix} a & b \\ \bar{b} & d \end{pmatrix}$ where a, b, d do not have a common divisor in $\mathbb{Z}$ and a, d are not both even. To see that formula (3.13) is equivalent to Humbert's formula, note that a form which is not properly primitive cannot represent an odd integer under the above assumptions on d_K.

Using the Propositions of this section it is easy to deduce some classical results in representation numbers of definite quaternary forms with integer coefficients such as Jacobi's 4-squares theorem. This is more elegantly done using the zeta functions attached to binary hermitian forms, see Corollary 4.11. Finally we note for later use

Corollary 3.19. *Let* $\mathfrak{m} \in \mathcal{M}$ *and* $\Delta \in \mathbb{Z}$. *Then* $r(\mathfrak{m}, k, \Delta) = O(|k|^{1+\delta})$ *for every* $\delta > 0$ *as* $|k| \to \infty$.

Proof. For integers k which are prime to 2Δ the corollary is immediately clear from formula (3.13). The other cases also can be treated by a look into Propositions 3.13, 3.14, 3.15, 3.17. □

9.4 Zeta Functions for Binary Hermitian Forms

Here we study certain Dirichlet series associated to binary hermitian forms with coefficients in the ring of integers of some imaginary quadratic number field. This part again is in several ways a generalization of the papers (1919b), (1919c), (1919d) by Humbert. In our treatment Humbert's arguments using quadratic forms are replaced by ideal theoretic arguments. The Dirichlet series studied here are special cases of the zeta functions associated to prehomogeneous vector spaces, see Sato, Shintani (1974). Some of the results proved in this section hold in this wider context.

K is again an imaginary quadratic number field of discriminant $d_K < 0$ and $\mathcal{O}$ is its ring of integers. We shall use the notation of the previous parts concerning hermitian forms.

Definition 4.1. For $\mathfrak{m} \in \mathcal{M}, f \in \mathcal{H}(\mathcal{O}), s \in \mathbb{C}$ with $\operatorname{Re} s > 1$ define

$$Z_{\mathfrak{m}}(f,s) = N\mathfrak{m}^{s+1} \sum_{\substack{(\alpha,\beta)\in K\times K/E(f) \\ f(\alpha,\beta)>0 \\ <\alpha,\beta>=\mathfrak{m}}} \left(\frac{1}{f(\alpha,\beta)}\right)^{s+1} \tag{4.1}$$

and

$$\hat{Z}_{\mathfrak{m}}(f,s) = N\mathfrak{m}^{s+1} \sum_{\substack{(\alpha,\beta)\in \mathfrak{m}\times \mathfrak{m}/E(f) \\ f(\alpha,\beta)>0}} \left(\frac{1}{f(\alpha,\beta)}\right)^{s+1}. \tag{4.2}$$

The series $Z_{\mathfrak{m}}(f,s)$ and $\hat{Z}_{\mathfrak{m}}(f,s)$ are called zeta functions of f.

Note that the zeta functions of negative definite forms are 0. For any $\lambda \in K$ we have $f(\lambda\alpha,\lambda\beta) = N\lambda \cdot f(\alpha,\beta)$, hence both $Z_{\mathfrak{m}}(f,s)$ and $\hat{Z}_{\mathfrak{m}}(f,s)$ do only depend on the class of $\mathfrak{m}$ in the ideal class group $\mathcal{J}_K = \mathcal{M}/K^*$. We may write formally

$$\hat{Z}_{\mathfrak{m}}(f,s) = \sum_{k=1}^{\infty} \frac{a_k}{k^{s+1}} \tag{4.3}$$

where $a_k = |\{\ (\alpha,\beta) \in \mathfrak{m}\times\mathfrak{m}/E(f)\ \mid\ f(\alpha,\beta) = kN\mathfrak{m}\ \}|$. The number a_k is finite, as was proved in Theorem 3.8. Analogously we have

$$Z_{\mathfrak{m}}(f,s) = \sum_{k=1}^{\infty} \frac{b_k}{k^{s+1}} \tag{4.4}$$

where $b_k = |P(f,\mathfrak{m},k)/E(f)|$. This follows from Definitions 3.1, 4.1. The convergence of the above Dirichlet series is established by the following proposition.

Proposition 4.2. *Let $\mathfrak{m} \in \mathcal{M}$, $f \in \mathcal{H}(\mathcal{O})$, then both $Z_{\mathfrak{m}}(f,s)$ and $\hat{Z}_{\mathfrak{m}}(f,s)$ converge for* $\operatorname{Re} s > 1$.

Proof. Consider expression (4.4). We have $0 \le b_k \le r(\mathfrak{m},k,-\Delta(f))$. By Corollary 3.19 the right-hand side is $O(|k|^{1+\delta})$ for every $\delta > 0$ as $|k| \to \infty$, and the result follows for $Z_{\mathfrak{m}}(f,s)$. For the convergence $\hat{Z}_{\mathfrak{m}}(f,s)$ use Proposition 4.5. □

We shall later prove that the functions $Z_{\mathfrak{m}}(f,s)$ and $\hat{Z}_{\mathfrak{m}}(f,s)$ have meromorphic continuations to all of $\mathbb{C}$. The continuations are holomorphic up to a simple pole at $s = 0$. We shall also prove a functional equation. There is a certain dependence amongst the $Z_{\mathfrak{m}}(f,s)$ and $\hat{Z}_{\mathfrak{m}}(f,s)$. To express this dependence we introduce as in Definition 8.1.2 the following ζ-functions.

Definition 4.3. For $\mathfrak{m}$, $\mathfrak{n} \in \mathcal{M}$ let

$$\zeta(\mathfrak{m},\mathfrak{n},s) = (N\mathfrak{m}\mathfrak{n}^{-1})^s \cdot \sum_{\substack{\lambda\in\mathfrak{m}\mathfrak{n}^{-1}\\ \lambda\neq 0}} N\lambda^{-s} \tag{4.5}$$

and for $\mathfrak{m}^{\#} \in \mathcal{M}/K^*$ define

$$\zeta(\mathfrak{m}^{\#},s) = \sum_{\substack{\mathfrak{a}\in\mathfrak{m}^{\#}\\ \mathfrak{a}\subset\mathcal{O}}} N\mathfrak{a}^{-s}. \tag{4.6}$$

The Dirichlet series (4.6) is the ζ-function of the ideal class associated with $\mathfrak{m}^{\#}$, it has abscissa of convergence 1.

Lemma 4.4. *Suppose that* $\mathfrak{m},\mathfrak{n} \in \mathcal{M}, \operatorname{Re} s > 1$. *Then* $\zeta(\mathfrak{m},\mathfrak{n},s) = |\mathcal{O}^*| \cdot \zeta((\mathfrak{m}^{-1}\mathfrak{n})^{\#},s)$.

For a proof see Lemma 8.1.3. For the zeta functions of binary hermitian forms we get the following result:

Proposition 4.5. *For* $\mathfrak{m} \in \mathcal{M}, f \in \mathcal{H}(\mathcal{O}), \operatorname{Re} s > 1$, *we have*

$$|\mathcal{O}^*|\,\hat{Z}_{\mathfrak{m}}(f,s) = \sum_{\mathfrak{n}^{\#}\in\mathcal{J}_K} \zeta(\mathfrak{m},\mathfrak{n},s+1)\,Z_{\mathfrak{n}}(f,s). \tag{4.7}$$

Proof. Let $\mathfrak{n}$ run through a representative system V of $\mathcal{J}_K = \mathcal{M}/K^*$. Consider a pair (γ,δ) of generators of an arbitrary element $\mathfrak{n} \in V$ and an arbitrary $\lambda \in \mathfrak{m}\mathfrak{n}^{-1}$ and consider the map $(\lambda,(\gamma,\delta)) \to (\alpha,\beta) := (\lambda\gamma,\lambda\delta) \in \mathfrak{m}\times\mathfrak{m}\backslash\{(0,0)\}$. This map is surjective and every (α,β) has precisely $|\mathcal{O}^*|$ different inverse images. This yields the assertion. □

The formulas on representation numbers for binary hermitian forms elaborated in Propositions 3.13, 3.14, 3.15, 3.17 will imply nice results on the following Dirichlet series.

Definition 4.6. Let $\mathfrak{m} \in \mathcal{M}$, $\Delta \in \mathbb{Z}$ and $s \in \mathbb{C}$ with $\operatorname{Re} s > 1$. We define

$$Z_{\mathfrak{m}}(\Delta,s) = \sum_{[f]\in\tilde{\mathcal{H}}(\mathcal{O},\Delta)} Z_{\mathfrak{m}}(f,s), \qquad \hat{Z}_{\mathfrak{m}}(\Delta,s) = \sum_{[f]\in\tilde{\mathcal{H}}(\mathcal{O},\Delta)} \hat{Z}_{\mathfrak{m}}(f,s).$$

Here the summation is extended over all $\mathbf{SL}(2,\mathcal{O})$ classes of integral binary hermitian forms of discriminant Δ.

These Dirichlet series are studied by Humbert (1919b), (1919c) for positive definite binary hermitian forms. From Theorem 3.7 and the definition of $r(\mathfrak{m}, k, -\Delta)$ we have

$$Z_{\mathfrak{m}}(\Delta, s) = \sum_{k=1}^{\infty} \frac{r(\mathfrak{m}, k, -\Delta)}{k^{s+1}}. \tag{4.8}$$

This formula leads to

Theorem 4.7. *Let $\Delta \in \mathbb{Z}$. Then the Dirichlet series $Z_{\mathfrak{m}}(\Delta, s)$ is an Euler product, that is the formula*

$$Z_{\mathfrak{m}}(\Delta, s) = \prod_{p} Q_p(-\Delta, p^{-s-1})$$

holds for $\operatorname{Re} s > 1$. *Here the $Q_p(-\Delta, p^{-s-1})$ are the rational functions in p^{-s-1} defined in Definition 3.12.*

Proof. From (4.8) and Proposition 3.11 we obtain

$$\begin{aligned} Z_{\mathfrak{m}}(\Delta, s) &= \sum_{k=1}^{\infty} \frac{r(\mathfrak{m}, k, -\Delta}{k^{s}} = \prod_{p} \left(\sum_{n=0}^{\infty} \frac{r(\mathfrak{m}, p^n, -\Delta)}{p^{n(s+1)}} \right) \\ &= \prod_{p} Q_p(-\Delta, p^{-s-1}). \end{aligned}$$

The rationality is proved in the propositions following Definition 3.12. □

We use our computation of the $Q_p(-\Delta, p^{-s})$ to determine the $Z_{\mathfrak{m}}(\Delta, s)$, $\hat{Z}_{\mathfrak{m}}(\Delta, s)$ explicitly. Our first result is

Proposition 4.8. *For every $\mathfrak{m} \in \mathcal{M}$ and $s \in \mathbb{C}$ with* $\operatorname{Re} s > 1$ *we have*

$$Z_{\mathfrak{m}}(0, s) = \zeta_K(s) \cdot L(\chi_K, s+1)^{-1}, \qquad \hat{Z}_{\mathfrak{m}}(0, s) = \zeta_K(s) \cdot \zeta_{\mathbb{Q}}(s+1).$$

The character $\chi_K : \mathbb{Z} \to \mathbb{C}$ of K is defined in (3.3). The L-function attached to χ_K has the Euler product

$$L(\chi_K, s) = \prod_{p} (1 - \chi_K(p) p^{-s})^{-1}. \tag{4.9}$$

By the decomposition law in K (Hasse (1968)) we have $\zeta_K(s) = \zeta_{\mathbb{Q}}(s) \cdot L(\chi_K, s)$ where $\zeta_{\mathbb{Q}}(s)$ is the ζ-function of $\mathbb{Q}$.

Proof of Proposition 4.8. In Proposition 3.13–3.17 we have found

$$Q_p(0,X) = \begin{cases} \dfrac{1-X}{(1-pX)^2} & \text{if } \chi_K(p) = 1, \\ \dfrac{1+X}{1-p^2X^2} & \text{if } \chi_K(p) = -1, \\ \dfrac{1}{1-pX} & \text{if } \chi_K(p) = 0. \end{cases}$$

The first formula of the proposition follows from $\zeta_K(s) = \prod_{\mathcal{P}}(1 - N\mathcal{P}^{-s})^{-1}$, where the product ranges over all prime ideals of $\mathcal{O}$. For the second part we obtain from Proposition 4.5

$$\begin{aligned} |\mathcal{O}^*|\, \hat{Z}_{\mathfrak{m}}(\Delta, s) &= \sum_{\mathfrak{n}^\# \in \mathcal{J}_K} \zeta(\mathfrak{m}, \mathfrak{n}, s+1)\; Z_{\mathfrak{n}}(\Delta, s) \\ &= |\mathcal{O}^*| \sum_{\mathfrak{n}^\# \in \mathcal{J}_K} \zeta((\mathfrak{m}^{-1}\mathfrak{n})^\#, s)\; Z_{\mathfrak{n}}(\Delta, s) \\ &= |\mathcal{O}^*|\, \zeta_K(s+1)\; \zeta_K(s)\; L(\chi_D, s+1)^{-1} \\ &= |\mathcal{O}^*|\, \zeta_K(s)\; \zeta_{\mathbb{Q}}(s+1). \end{aligned}$$

□

For the formulation of our general result it is advantageous to make the following definition.

Definition 4.9. For a prime number p and $\Delta \in \mathbb{Z}$, $\Delta \neq 0$ we define

$$\tilde{Q}_p(X) = \frac{1 - \chi_D(p)X}{1 - pX} \tag{4.10}$$

and

$$\Theta(\Delta, s) = \prod_{p \mid \Delta d_K} Q_p(\Delta, p^{-s-1}) \cdot \tilde{Q}_p^{-1}(p^{-s-1}) \tag{4.11}$$

where the $Q_p(\Delta, x)$ are given in Propositions 3.13–3.17.

A quick glance at Proposition 3.13–3.17 shows that $Q_p(\Delta, x) \cdot \tilde{Q}_p^{-1}(x)$ is a polynomial in x and hence $\Theta(\Delta, s)$ is a holomorphic function of s. Precise formulas for the Euler factors in (4.11) are written out in Lemma 5.8. Our result on the $Z_{\mathfrak{m}}(\Delta, s)$, $\hat{Z}_{\mathfrak{m}}(\Delta, s)$ is:

Theorem 4.10. *Let $\mathfrak{m} \in \mathcal{M}$ and $\Delta \in \mathbb{Z}$ with $\Delta \neq 0$. Then for any $s \in \mathbb{C}$ with* $\operatorname{Re} s > 1$ *we have*

$$\begin{aligned} Z_{\mathfrak{m}}(\Delta, s) &= \Theta(-\Delta, s)\; \zeta_{\mathbb{Q}}(s)\; L(\chi_K, s+1)^{-1}, \\ \hat{Z}_{\mathfrak{m}}(\Delta, s) &= \Theta(-\Delta, s)\; \zeta_{\mathbb{Q}}(s)\; \zeta_{\mathbb{Q}}(s+1). \end{aligned}$$

Proof. This result follows exactly as Proposition 4.8 from the computation of the $Q_p(\Delta, x)$ in Propositions 3.13–3.17. □

As an application of all these formulas we get Jacobi's theorem on sums of 4 squares.

Corollary 4.11. *We have*

$$\sum_{\substack{x,y,u,v\in\mathbb{Z}\\ x^2+y^2+u^2+v^2\neq 0}} \frac{1}{(x^2+y^2+u^2+v^2)^{s+1}} = 8\cdot\left(1-\frac{1}{4^s}\right)\cdot\zeta_{\mathbb{Q}}(s)\cdot\zeta_{\mathbb{Q}}(s+1), \tag{4.12}$$

and for every $k \in \mathbb{Z}$ with $k \geq 1$

$$|\{\ (x,y,u,v) \in \mathbb{Z}^4 \mid x^2+y^2+u^2+v^2 = k\ \}| = 8 \sum_{\substack{4\nmid d\mid k\\ d>0}} d. \tag{4.13}$$

Proof. We consider the case $d_K = -4$ and $\Delta = 1$. As we have seen in the part on reduction theory there is only one $\mathbf{SL}(2, \mathbb{Z}[i])$-class of positive definite hermitian forms of discriminant 1 over $\mathbb{Z}[i]$. A representative is $f_1 = \begin{pmatrix} 1 & 0 \\ 0 & 1 \end{pmatrix}$. We also know that $|E(f_1)| = 8$. Clearly the left-hand side in (4.12) is $\hat{Z}_{\mathbb{Z}[i]}(1, s)$. By Proposition 3.17 we find $\Theta(-1, s) = 1 - 4^{-s}$. Statement (4.13) is only a trivial reformulation of (4.12). □

9.5 The Mass-Formula

K is again an imaginary quadratic number field of discriminant $d_K < 0$ and $\mathcal{O}$ its ring of integers. We shall use here our notation concerning binary hermitian forms which is developed in the previous sections. We shall prove the following Mass-formula for binary hermitian forms.

Theorem 5.1. *Let $\Delta \in \mathbb{Z}$ with $\Delta > 0$, then*

$$\sum_{[f]\in\tilde{\mathcal{H}}^+(\mathcal{O},\Delta)} \frac{1}{|E(f)|} = \frac{|d_K|\ \Delta}{24}\ \Theta(-\Delta, 1). \tag{5.1}$$

Here $\tilde{\mathcal{H}}^+(\mathcal{O}, \Delta)$ (see Definition 2.5) is the set of $\mathbf{SL}(2, \mathcal{O})$-equivalence classes of positive definite binary hermitian forms with coefficients in $\mathcal{O}$ and discriminant Δ and $E(f) = E(f, \mathcal{O})$ is the (finite) group of units of f. $\Theta(-\Delta, s)$ is the finite Dirichlet series defined in Definition 4.9. Formula (5.1) stands in the tradition of Siegel's Mass-formulas for definite quadratic forms, see Siegel (1935). The more general Mass-formula developed by Braun (1941) gives, applied to binary hermitian forms, a result which differs from (5.1), since there the summation on the right-hand side extends over the classes in a genus. Also the right-hand side is not as explicit as in (5.1). The paper (1919d) by Humbert contains (5.1) in certain special cases. A formula analogous to (5.1) but for binary quadratic forms is explained in Cassels (1978). The proof of Theorem 5.1 uses a technique due to Dirichlet, which we shall explain now.

Definition 5.2. A subset $X \subset \mathbb{R}^n$ is called a cone if $x \in X$, $r \in [0,\infty[$ implies $rx \in X$. A function $F : \mathbb{R}^n \to [0,\infty[$ is called homogeneous of degree m if $F(rx) = r^m F(x)$ for all $x \in \mathbb{R}^n$ and $r \in]0,\infty[$. Given a cone $X \subset \mathbb{R}^n$, a function F on $\mathbb{R}^n$ homogeneous of degree m and $r \in]0,\infty[$ we define $X_F(r) = \{ \, x \in X \mid F(x) \le r \, \}$.

We write μ for the Lebesgue measure on $\mathbb{R}^n$. If $\Lambda \subset \mathbb{R}^n$ is a full lattice, we write $|\Lambda| = \mu(\mathcal{P})$ for the volume of a fundamental parallelogram $\mathcal{P}$ of Λ. The Lemma of Dirichlet is

Lemma 5.3. *Let $X \subset \mathbb{R}^n$ be a cone, $F : \mathbb{R}^n \to [0,\infty[$ a function homogeneous of degree n and $\Lambda \subset \mathbb{R}^n$ a lattice. Assume that $X_F(1)$ is a bounded Jordan-measurable set. Let $V = \mu(X_F(1))$. Then the Dirichlet series*

$$\zeta_\Lambda(F,X,s) = \sum_{\substack{x \in \Lambda \cap X \\ F(x) \neq 0}} \left(\frac{1}{F(x)} \right)^{s+1} \tag{5.2}$$

converges for $\operatorname{Re} s > 0$ *and*

$$\lim_{s \to 0+0} s \cdot \zeta_\Lambda(F,X,s) = \frac{V}{|\Lambda|}. \tag{5.3}$$

For a proof see Borevich, Shafarevich (1966), Chapter V. We shall consider the lattices in $\mathbb{C} = \mathbb{R}^2$ given by fractional ideals $\mathfrak{m} \in \mathcal{M}$.

Lemma 5.4. *For $\mathfrak{m} \in \mathcal{M}$ we have $|\mathfrak{m}| = \frac{1}{2}\sqrt{|d_K|} \cdot N\mathfrak{m}$.*

Proof. Since $\{1, \omega = (d_K + \sqrt{d_K})/2\}$ is a $\mathbb{Z}$-basis of $\mathcal{O}$ we have $|\mathcal{O}| = \sqrt{|d_K|}/2$. The rest follows by an index computation. □

Proof of Theorem 5.1. Let $f \in \mathcal{H}^+(\Delta)$ be a positive definite hermitian form. Then the function

$$f_0 : \mathbb{C} \times \mathbb{C} \to [0,\infty[, \qquad f_0 : (x,y) \mapsto (f(x,y))^2$$

is homogeneous of degree 4. If $\mathfrak{m} \in \mathcal{M}$ then we have by definition

$$\hat{Z}_{\mathfrak{m}}(f, 2s+1) = N\mathfrak{m}^{2s+2} \sum_{\substack{(\alpha,\beta)\in \mathfrak{m}\times\mathfrak{m}/E(f) \\ f(\alpha,\beta)\neq 0}} \left(\frac{1}{f(\alpha,\beta)}\right)^{2s+2}$$
$$= \frac{N\mathfrak{m}^{2s+2}}{|E(f)|}\, \zeta_{\mathfrak{m}\times\mathfrak{m}}(f_0, X, s),$$

where X is the cone $X = \mathbb{C} \times \mathbb{C} = \mathbb{R}^4$. It is an elementary exercise in integration theory to prove $\mu(X_{f_0}(1)) = \pi^2/2\Delta$. Using formula (5.3) and Lemma 5.4 we get

$$\lim_{s\to 0+0} s \cdot \hat{Z}_{\mathfrak{m}}(f, 2s+1) = \frac{N\mathfrak{m}^2}{|E(f)|} \cdot \frac{\pi^2}{2\Delta} \cdot 4 \cdot \frac{1}{|d_K| N\mathfrak{m}^2} = \frac{2\pi^2}{|E(f)| \cdot \Delta \cdot |d_K|}.$$

From Theorem 4.10 we obtain

$$\lim_{s\to 0+0} s \cdot \hat{Z}_{\mathfrak{m}}(\Delta, 2s+1) = \lim_{s\to 0+0} s \cdot \Theta(-\Delta, 2s+1) \cdot \zeta_{\mathbb{Q}}(2s+1) \cdot \zeta_{\mathbb{Q}}(2s+2).$$

The functions $\Theta(-\Delta, 2s+1)$, $\zeta_{\mathbb{Q}}(2s+2)$ are holomorphic in $s = 0$. We obtain from standard facts about $\zeta_{\mathbb{Q}}$:

$$\lim_{s\to 0+0} s \cdot \hat{Z}_{\mathfrak{m}}(\Delta, 2s+1) = \frac{\pi^2}{12} \cdot \Theta(-\Delta, 1).$$

This proves our Theorem 5.1. □

The entity on the left-hand side of the Mass-formula will play a special role in the next paragraph.

Definition 5.5. Let $\Delta \in \mathbb{Z}$ with $\Delta > 0$, then

$$M(\Delta) := \sum_{[f]\in \tilde{\mathcal{H}}^+(\mathcal{O},\Delta)} \frac{1}{|E(f)|}$$

is called the mass of $\mathbf{SL}(2,\mathcal{O})$-classes of integral positive definite binary hermitian forms of discriminant Δ over the fixed imaginary quadratic number field K.

Our Theorem 5.1 now immediately implies growth results for masses and class numbers.

Corollary 5.6. *For fixed* d_K *and* $\Delta \in \mathbb{Z}$ *with* $\Delta > 0$ *and* $\Delta \to \infty$ *we have:*

$$M(\Delta) = O(\Delta \log\log \Delta), \qquad h^+(\Delta) = O(\Delta \log\log \Delta).$$

Proof. To prove the first formula note that the right-hand side of (5.1) is

$$O\left(\Delta\prod_{p|\Delta}(1-p^{-1})^{-1}\right)=O\left(\frac{\Delta^2}{\varphi(\Delta)}\right).$$

By Hardy and Wright (1960), Theorem 328 we conclude the result. For the second formula note that the numbers $|E(f)|$ are uniformly bounded for all positive definite binary hermitian forms f. □

We shall also need some finer results on the masses. To express these we define the following Dirichlet series.

Definition 5.7. Given the imaginary quadratic number field K, we define

$$L_M(s)=\sum_{\Delta=1}^{\infty} M(\Delta)\ \Delta^{-s}.$$

The Dirichlet series $L_M(s)$ converges for $s\in\mathbb{C}$ with $\operatorname{Re}s>2$ by Corollary 5.6. Before computing $L_M(s)$ in terms of the usual L-series we for convenience collect some formulas for $\Theta(\Delta,s)$.

Lemma 5.8. *For $\Delta\in\mathbb{Z}$ with $\Delta>0$ and a prime number p define as in (4.10)*

$$\tilde{Q}_p(X)=\frac{1-\chi_K(p)X}{1-pX}$$

and $R_p(\Delta,X)=Q_p(-\Delta,X)\cdot\tilde{Q}_p^{-1}(X)$. Then we have

$$R_p(\Delta,X)=\begin{cases}\dfrac{1-(\chi_K(p)pX)^{t+1}}{1-\chi_K(p)pX} & \text{if } p\nmid d_K,\ p^t\|\Delta,\\[2ex] 1+\left(\dfrac{-D_0^t\Delta_0}{p}\right)(pX)^{t+1} & \text{if } p|d_K,\ p\neq 2,\ p^t\|\Delta,\\[2ex] 1+\left(\dfrac{8}{\Delta_0 D_2^t}\right)(2X)^{t+3} & \text{if } \begin{cases}p=2,\ 4|d_K,\ 2^t\|\Delta,\\ D_1\equiv 2 \bmod 8,\end{cases}\\[2ex] 1-\left(\dfrac{-8}{\Delta_0 D_2^t}\right)(2X)^{t+3} & \text{if } \begin{cases}p=2,\ 4|d_K,\ 2^t\|\Delta,\\ D_1\equiv 6 \bmod 8,\end{cases}\\[2ex] 1-\left(\dfrac{-4}{\Delta_0 D_2^t}\right)(2X)^{t+2} & \text{if } \begin{cases}p=2,\ 4|d_K,\ 2^t\|\Delta,\\ D_1\equiv 3,\,7 \bmod 8,\end{cases}\end{cases}$$

where $D_0:=-d_K/p$, $\Delta_0:=p^{-t}\Delta$, and where for $4|d_K$ we define $D_1:=d_K/4$ and

$$D_2:=\begin{cases}-\dfrac{D_1}{2} & \text{if}\quad D_1\equiv 2 \bmod 4,\\[2ex] \dfrac{1-D_1}{2} & \text{if}\quad D_1\equiv 3 \bmod 4.\end{cases}$$

Then

$$\Theta(-\Delta, s) = \prod_{p | d_K \Delta} R_p(\Delta, p^{-1-s}).$$

Proof. Everything follows from the formulas in Section 3. □

We use the above formulas to get some information about the Dirichlet series $L_M(s)$.

Theorem 5.9. *Let K be an imaginary quadratic number field. The Dirichlet series $L_M(s)$ from Definition 5.7 has the following properties:*

(1) $L_M(s)$ converges absolutely for $\operatorname{Re} s > 2$ and extends meromorphically to $\mathbb{C}$.

(2) $L_M(s)$ can be written as

$$L_M(s) = \frac{|d_K|}{24}\, \zeta_{\mathbb{Q}}(s-1)\, L(\chi_K, s) + \Lambda(s) \tag{5.4}$$

where $\Lambda(s)$ is holomorphic for $\operatorname{Re} s > 1$.

Before giving the proof we report on the following interesting special case which will also elucidate the principle of the general proof.

Theorem 5.10. *Let q be a rational prime with $q \equiv 3 \bmod 4$, and $K = \mathbb{Q}(\sqrt{-q})$. Then $d_K = -q$ and*

$$L_M(s) = \frac{|d_K|}{24}\, \zeta_{\mathbb{Q}}(s-1)\, L(\chi_K, s) - \frac{1}{24}\, \zeta_{\mathbb{Q}}(s)\, L(\chi_K, s-1). \tag{5.5}$$

Formula (5.5) makes the claim of Theorem 5.9 clear in the special case $K = \mathbb{Q}(\sqrt{-q})$. In the case of a general imaginary quadratic number field K a similar expression for $L_M(s)$ in terms of Dirichlet L-series can be developed but it is not so nice. Therefore we content ourselves with the cruder Theorem 5.9.

Proof. By the mass formula we find for $s \in \mathbb{C}$ with $\operatorname{Re} s > 2$

$$L_M(s) = \sum_{\Delta=1}^{\infty} M(\Delta)\, \Delta^{-s} = \frac{|d_K|}{24} \sum_{\Delta=1}^{\infty} \Theta(-\Delta, 1)\, \Delta^{-s+1}.$$

For $n \in \mathbb{N}$ with $(n, q) = 1$ we define

$$b(n) := \prod_{p|n} \frac{1 - (\chi_K(p)p^{-1})^{t_p+1}}{1 - \chi_K(p)p^{-1}}$$

where t_p is defined by $p^{t_p} \| n$. By the formulas in Lemma 5.8 we find

$$L_M(s) = \frac{|d_K|}{24} \sum_{t=0}^{\infty} \sum_{\substack{n=1 \\ (n,q)=1}}^{\infty} \left(1 + \left(\frac{-n}{q}\right) q^{-t-1}\right) b(n)\, q^{-ts+t}\, n^{-s+1}.$$

Summing the geometric series we get

$$\begin{aligned} L_M(s) = {} & \frac{|d_K|}{24} \frac{1}{1-q^{-s+1}} \sum_{\substack{n=1 \\ (n,q)=1}}^{\infty} b(n) n^{-s+1} \\ & - \frac{1}{24} \frac{1}{1-q^{-s}} \sum_{\substack{n=1 \\ (n,q)=1}}^{\infty} \left(\frac{n}{q}\right) b(n) n^{-s+1}. \end{aligned}$$

Both the functions $b(n)$ and $(\frac{n}{q})b(n)$ are weakly multiplicative in n, that is, we have $b(n_1 n_2) = b(n_1)b(n_2)$ if n_1, n_2 are coprime. From here we get

$$\begin{aligned} L_M(s) = {} & \frac{|d_K|}{24} \frac{1}{1-q^{-s+1}} \prod_{\substack{p \\ p \neq q}} \left(\sum_{t=0}^{\infty} \frac{1-(\chi_K(p)p^{-1})^{t+1}}{1-\chi_K(p)p^{-1}} p^{-ts+t} \right) \\ & - \frac{1}{24} \frac{1}{1-q^{-s}} \prod_{\substack{p \\ p \neq q}} \left(\sum_{t=0}^{\infty} \left(\frac{p}{q}\right)^t \frac{1-(\chi_K(p)p^{-1})^{t+1}}{1-\chi_K(p)p^{-1}} p^{-ts+t} \right). \end{aligned}$$

Summing all geometric series and using the fact $\chi_K(p) = (\frac{-q}{p}) = (\frac{p}{q})$ which is obtained from quadratic reciprocity we obtain the result. □

Proof of Theorem 5.9. Going through the same procedure as in the proof of Theorem 5.10 we see that

$$L_M(s) - \frac{|d_K|}{24} \zeta_{\mathbb{Q}}(s-1) L(\chi_K, s)$$

can be expressed as a sum of functions constant $\cdot\, G(s) \cdot L(\chi_1, s)L(\chi_2, s-1)$ where $G(s)$ is a finite product of Euler factors $(1 \pm q^{-s})^{\pm 1}$ or $(1 \pm q^{-s+1})^{\pm 1}$ for primes $q | d_K$ and χ_1, χ_2 are two Dirichlet characters with χ_2 nontrivial. The result follows now from standard facts about Dirichlet L-series. □

We point out that the summation of $L_M(s)$ can be explicitly carried out and yields an expression for $L_M(s)$ as a sum of products of the type indicated in the preceding proof. It follows from work of Sato, Shintani (1974) that $L_M(s)$ even satisfies a functional equation. For the $Z_{\mathfrak{m}}(f, s)$, $\hat{Z}_{\mathfrak{m}}(f, s)$ no similar formulas exist. Only the mean values $Z_{\mathfrak{m}}(\Delta, s)$, $\hat{Z}_{\mathfrak{m}}(\Delta, s)$ admit the simple expressions given in Theorem 4.10. There are other mean values for which an explicit summation is possible. These are certain weighted sums over a set of representatives of the classes in a genus. These results are given in Elstrodt, Grunewald, Mennicke (1987a), Section 7.

For use in the next section we remark that Theorem 5.9 implies that for any imaginary quadratic number field K the function $L_M(s)$ has a pole of order 1 at $s = 2$ with residue

$$\frac{|d_K|}{24} L(\chi_K, 2) = \frac{|d_K|}{4\pi^2} \zeta_K(2). \tag{5.6}$$

9.6 Computation of the Covolume of PSL$(2, \mathcal{O})$ à la Humbert

Let K be an imaginary quadratic number field of discriminant $d_K < 0$ and $\mathcal{O}$ its ring of integers. We shall compute here the covolume of the discontinuous group $\mathbf{PSL}(2, \mathcal{O}) < \mathbf{PSL}(2, \mathbb{C})$ acting on hyperbolic space $\mathbb{H}$. That is, we prove

$$\operatorname{vol}(\mathbf{PSL}(2, \mathcal{O})) = v(\mathbf{PSL}(2, \mathcal{O})\backslash\mathbb{H}) = \frac{|d_K|^{3/2}}{4\pi^2}\ \zeta_K(2). \tag{6.1}$$

This result appears already in Theorem 7.1.1, a proof can be found in Section 8.8. We shall reproduce a proof due to Humbert (1919b)–(1919e) which makes heavy use of the theory of binary hermitian forms developed in this chapter. The argument of Humbert works only for d_K satisfying a congruence condition. More interestingly it contains a gap in using the Dirichlet argument (Lemma 5.3) directly. We shall show how to close both gaps. The arguments used to close the second gap are taken from Grunewald, Kühnlein (1996). Before going into the proof we need a certain amount of constructions.

Definition 6.1. We put $V = \mathbb{R} \times \mathbb{R} \times \mathbb{C}$ and

$$\Delta : \ V \to \mathbb{R} \qquad \Delta : \ (a, d, b) \mapsto ad - b\bar{b}.$$

We furthermore define $V^+ = \{\ (a, d, b) \in V \ \mid \ a, d > 0,\ \Delta((a, d, b)) > 0\ \}$ and

$$\Phi : \ V^+ \to \mathbb{H} \qquad \Phi : \ x = (a, d, b) \mapsto \frac{b}{d} + \frac{\sqrt{\Delta(x)}}{d} j.$$

These definitions are of course reminiscent of our discussion of the space of binary hermitian forms in Section 1 of this chapter. We shall define now the ingredients for the application of Dirichlet's principle.

Definition 6.2. Let $\mathcal{F}_K$ be the fundamental domain for $\mathbf{PSL}(2, \mathcal{O})$ constructed in Definition 7.3.1 We define $Y = \Phi^{-1}(\mathcal{F}_K) \cup \{0\}$. Let $G : \ V \to [0, \infty[$ be the map $G : \ x \mapsto (\Delta(x))^2$. We also define the following lattice in V: $\Lambda_K = \mathbb{Z} \times \mathbb{Z} \times \mathcal{O}$.

Y is a cone in the 4-dimensional real vector space V. The function G is homogeneous of degree 4. In this case the set $Y_G(1) = \{ y \in Y \mid G(y) \leq 1 \}$ is not bounded. Hence we cannot apply the expanding domains argument directly to the pair Y, Λ_K. We need the following variation of Dirichlet's lemma.

Lemma 6.3. *Let $X \subset \mathbb{R}^n$ be a cone, $F : \mathbb{R}^n \to [0, \infty[$ a function homogeneous of degree n and $\Lambda \subset \mathbb{R}^n$ a lattice. Assume that $X_F(1)$ is Lebesgue-measurable with finite measure $V = \mu(X_F(1))$. Assume that the boundary $\partial X_F(1)$ of $X_F(1)$ is the union of finitely many regular hypersurfaces with piecewise differentiable boundaries. For $r \in]0, \infty[$ let $N(r) := |X_F(1) \cap r^{-1}\Lambda|$. Assume further that*

$$V = \mu(X_F(1)) = \lim_{r \to \infty} |\Lambda| \frac{N(r)}{r^n}. \tag{6.2}$$

Then the Dirichlet series (5.2) converges for $\operatorname{Re} s > 0$ *and (5.3) holds.*

The proof proceeds excactly as in the classical case in Borevich-Shafarevich (1966), Chapter V, 1. The extra hypothesis (6.2) will not be automatically satisfied. There exist examples, see Kühnlein (1996), of Lebesgue-measurable cones X with piecewise smooth boundary and lattices L in dimension 2 where for an appropriate function G homogeneous of degree 2 the set $X_G(1) = \{x \in X \mid G(x) \leq 1\}$ is unbounded of finite measure and $L \cap X_G(1)$ is infinite. We hence know that (6.2) also will not hold in general.

Lemma 6.4. *Let $Y \subset \mathbb{R}^n$ be the cone and G the homogeneous function given in Definition 6.2. Put $N(r) := |Y_G(1) \cap r^{-1}\Lambda|$. Then*

$$\mu(Y_G(1)) = \lim_{r \to \infty} |\Lambda| \frac{N(r)}{r^4}. \tag{6.3}$$

Proof. We denote by $\tilde{S} := \{\tilde{s}_1, \ldots, \tilde{s}_n\}$ the set of singular points in F_K (see Section 7.3). The set $S = \tilde{S} \cup \{\infty\}$ contains the points which are added to $\mathcal{F}_K$ when we close $\mathcal{F}_K$ in $\mathbb{H} \cup \partial\mathbb{H} = \mathbb{H} \cup \mathbb{P}^1\mathbb{C}$. To make our proof uniform we put $\mathcal{F} := g\mathcal{F}_K$ for an appropriate element $g \in \mathbf{SL}(2, \mathcal{O})$ and assume that the corresponding set $S = \{s_1, ..., s_{n+1}\}$ for $\mathcal{F}$ is contained in $K \subset \mathbb{P}^1\mathbb{C}$. We redefine the cone to $Y = \Phi^{-1}(\mathcal{F}) \cup \{0\}$ and prove (6.3) for the new cone also obtaining the result for the cone corresponding to $\mathcal{F}_K$. There is now a bound $\kappa_0 > 0$ such that for any $\kappa < \kappa_0$ the horoballs

$$B_{i,\kappa} := \{(z, r) \in \mathbb{H} \mid |z - s_i|^2 + (r - \kappa/2)^2 < \kappa^2/4\}$$

are pairwise disjoint. The procedure for proving (6.3) now is the following.

Let $\mathcal{F}_\kappa := \mathcal{F} \backslash (B_{1,\kappa} \cup \ldots \cup B_{n+1,\kappa})$ be the complement in $\mathcal{F}$ of the horoballs and define

$$Y_\kappa := \Phi^{-1}(\mathcal{F}_\kappa) \subset Y \quad \text{and} \quad Y_{\kappa,1} = Y_\kappa \cap Y_G(1).$$

Then it is clear that

$$\bigcup_{\kappa>0} Y_{\kappa,1} = Y_G(1).$$

Later we will count the number $M(r,\kappa)$ of points of $r^{-1}\Lambda$ in $Y_G(1)\backslash Y_{\kappa,1}$ and will estimate it by

$$M(r,\kappa) \leq C(\kappa) \cdot r^4 \quad \text{for} \quad r \to \infty, \tag{6.4}$$

where the function $C(\kappa)$ tends to 0 with κ. This and the fact that $Y_{\kappa,1}$ is bounded will allow us to calculate the right-hand side of (6.3), as for every $\kappa \leq \kappa_0$ we have

$$\begin{aligned}\liminf_{r\to\infty} N(r)\frac{|\Lambda|}{r^4} &= \liminf_{r\to\infty} \big[N(r) - M(r,\kappa) + M(r,\kappa)\big]\frac{|\Lambda|}{r^4} \\ &\geq \liminf_{r\to\infty} \big[N(r) - M(r,\kappa)\big]\frac{|\Lambda|}{r^4} = \mu(Y_{\kappa,1})\end{aligned}$$

and

$$\begin{aligned}\limsup_{r\to\infty} N(r)\frac{|\Lambda|}{r^4} &= \limsup_{r\to\infty} \big[N(r) - M(r,\kappa) + M(r,\kappa)\big]\frac{|\Lambda|}{r^4} \\ \leq \limsup_{r\to\infty} \big[N(r) - M(r,\kappa)\big]\frac{|\Lambda|}{r^4} &+ C(\kappa)|\Lambda| = \mu(Y_{\kappa,1}) + C(\kappa)|\Lambda|.\end{aligned}$$

With $\kappa \searrow 0$ formula (6.3) is proven.

We finally prove (6.4). We will first of all consider only one singular point $s_i =: s \in K$. We denote the horoball $B_{i,\kappa}$ for our fixed i by B_κ alone. For any $\kappa < \kappa_0$ and growing r we want to estimate the number of lattice-points $(a,d,b) \in \Lambda$ with

(a′) $ad - b\bar{b} \leq r^2$: this is the condition $(a,d,b) \in r^{-1}\Lambda \cap Y_G(1)$.
(b′) $\Phi(a,d,b) \in \mathcal{F}$.
(c′) $\Phi(a,d,b) \in B_\kappa$.

As the conditions (b′) and (c′) are quite cumbersome to write explicitly for $s \in K$ we transform everything to $\infty \in \mathbb{P}^1 K$. The matrix $g := \begin{pmatrix} 0 & 1 \\ -1 & s \end{pmatrix} \in$ $\mathbf{SL}(2,K)$ transforms s to ∞. The horoball B_κ is transformed by g to the horoball at infinity $\{\ (z,t) \ \mid \ t > T\ \}$, with $T := \kappa^{-1}$. The subgroup Γ'_s in $\mathbf{PSL}(2,\mathcal{O})$ of elements having the only eigenvalue 1 or -1 fixing s is conjugated by g to the group

$$\Gamma^*_s := \left\{ \pm\begin{pmatrix} 1 & \alpha \\ 0 & 1 \end{pmatrix} \ \Big|\ \alpha \in \mathfrak{a} \right\}$$

for the $\mathcal{O}$-ideal $\mathfrak{a} = \{\ c \in \mathcal{O} \ \mid \ sc \in \mathcal{O},\ s^2 c \in \mathcal{O}\ \} \subset \mathcal{O} \subset \mathbb{C}$. We denote by $\mathcal{P}$ a fundamental parallelogram for $\mathfrak{a}$ in $\mathbb{C}$ for which

$$g\mathcal{F} \cap \{\ (z,t) \ \mid \ t > T\} \subseteq \mathcal{P} \times \mathbb{R}^{>T}.$$

Finally, $\Phi(a, d, b)$ is transformed by g to

$$\left(\frac{\overline{s}d - \overline{b}}{s\overline{s}d - \overline{s}b - s\overline{b} + a}, \frac{\sqrt{ad - b\overline{b}}}{s\overline{s}d - \overline{s}b - s\overline{b} + a} \right).$$

Hence the number $M(r, \kappa)$ is less than the number of $(a, d, b) \in \Lambda$ satisfying the three conditions

(a) $ad - b\overline{b} \leq r^2$.
(b) $\frac{\overline{s}d - \overline{b}}{s\overline{s}d - \overline{s}b - s\overline{b} + a} \in \mathcal{P}$.
(c) $ad - b\overline{b} > T^2(s\overline{s}d - \overline{s}b - s\overline{b} + a)^2$.

Let $h := s\overline{s}d - \overline{s}b - s\overline{b} + a$. When s is written as v/u with $v \in \mathcal{O}$ and $u \in \mathbb{Z}$, $u > 0$, then clearly $k := u^2 h \in \mathbb{Z}$. The inequalities (a) and (c) imply $h < r/T$. For fixed h condition (b) says $\overline{s}d - \overline{b} \in h\mathcal{P}$. The points $x = \overline{s}d - \overline{b}$ all lie in the fractional $\mathcal{O}$-ideal generated by 1 and $\overline{s}$. This ideal contains a and the number of points x in the intersection of $h\mathcal{P}$ with this ideal is bounded by $C_1 h^2$ for some constant C_1. We have to analyse how many (a, d, b) can lead to one specific x. For this purpose we fix $h \in u^{-2}\mathbb{Z}$ and $x \in < 1, s > \cap h\mathcal{P}$ and count the possible choices for d. When d is chosen we have

$$\overline{b} = \overline{s}d - x, \qquad b = sd - \overline{x}, \qquad a = h - s\overline{s}d + \overline{s}b + s\overline{b} = h - \overline{s}\overline{x} - sx + s\overline{s}d.$$

The discriminant $ad - b\overline{b}$ is then expressed by

$$ad - b\overline{b} = hd - (\overline{s}\overline{x} + sx)d + s\overline{s}d^2 - s\overline{s}d^2 - x\overline{x} + (\overline{s}\overline{x} + sx)d = hd - x\overline{x}$$

and therefore (a) and (c) lead to $h^2T^2 < hd - x\overline{x} \leq r^2$. Hence there are at most $(r^2 - h^2T^2)/h + 1$ possibilities for d, when h and x are fixed. This means that the number of solutions of (a), (b) and (c) with points in Λ is bounded by (remember $k = u^2 h \in \mathbb{Z}$)

$$C_1 u^{-2} \sum_{k=1}^{u^2 r/T} k^2 \frac{r^2 - k^2 u^{-2} T^2 + u^{-2} k}{k} \leq C_2 \sum_{k=1}^{u^2 r/T} r^2(k+1) \leq C_3 r^2 \left(\frac{r}{T}\right)^2$$

for $r \to \infty$. Summing up these bounds for the finite number of singular points of $\mathcal{F}$ we conclude by seeing the existence of a constant C_∞ for which $M(r, \kappa) \leq C_\infty \kappa^2 r^4$ holds. This is a rather sharp form of (6.4). □

We are now ready to to finish the proof of (6.1).

Lemma 6.5. *The set $Y_G(1)$ is Lebesgue measurable and*

$$\mu(Y_G(1)) = \frac{1}{2} v(\mathcal{F}_K) = \frac{1}{2} \mathrm{vol}(\mathbf{PSL}(2, \mathcal{O})),$$

where μ is Lebesgue measure and v is the hyperbolic volume.

Proof. Using the transformation formula applied to

$$(x, y, r, u) \mapsto (ur^2 + x^2 + y^2, u, ux, uy)$$

we have

$$\int_{Y_G(1)} d\mu = 2 \cdot \int_{\mathcal{F}_K} \int_0^{r^{-1}} u^3 r \;\; du\, dx\, dy\, dr.$$

By Fubini's Theorem we obtain

$$\int_{Y_G(1)} d\mu = \frac{1}{2} \int_{\mathcal{F}_K} \frac{dx\, dy\, dr}{r^3}.$$

The coordinates x, y are chosen to be the real and imaginary part of z for a point $z + rj \in \mathbb{H}$. The integral on the right hand side is finite by Theorem 7.1.1, part (3). Arguing back we obtain the result. □

Proposition 6.6. *Let K be an imaginary quadratic number field. For $\Delta \in \mathbb{N}$ define $M_{red}(\Delta) := |\{\ f \in \mathcal{H}^+(\mathcal{O}, \Delta) \ \ |\ \ f$ reduced $\}|$ and*

$$L_{red}(s) := \sum_{\Delta=1}^{\infty} M_{red}(\Delta)\ \Delta^{-s}.$$

Then the Dirichlet series $L_{red}(s)$ has abscissa of convergence 2 and satisfies

$$\lim_{s \to 2+0} (s-2) L_{red}(s) = \frac{2\,\mathrm{vol}(\mathbf{PSL}(2, \mathcal{O}))}{\sqrt{|d_K|}}.$$

Proof. The zeta function of the cone Y from Lemma 5.3 differs from L_{red} only by a shift in the argument, in fact we have $L_{red}(2s+2) = \zeta_{\Lambda_K}(G, Y, s)$. From this and Lemmas 6.3, 6.4, 6.5 the above follows. □

Proof of formula (6.1). We compare the Dirichlet series $L_M(s)$ (see Definition 5.7) and $L_{red}(s)$. We first of all put $Z := \Phi^{-1}(\partial \mathcal{F}_K)$. Z is again a cone in $\mathbb{R}^4$ satisfying the hypothesis of Lemma 6.4. Note that we have $\mu(Z_G(1)) = 0$. Defining

$$\tilde{M}_{red}(\Delta) := |\{\ f \in \mathcal{H}^+(\mathcal{O}, \Delta) \ \ |\ \ \Phi(f) \in \partial \mathcal{F}_K \ \}|$$

and

$$\tilde{L}_{red}(s) := \sum_{\Delta=1}^{\infty} \tilde{M}_{red}(\Delta)\ \Delta^{-s},$$

we see that $\lim_{s \to 2+0} (s-2)\tilde{L}_{red}(s) = 0$. Let $f \in \mathcal{H}^+(\mathcal{O}, \Delta)$ be a reduced form with $|E(f, \mathcal{O})| > 2$. Then $\Phi(f)$ lies in $\partial \mathcal{F}_K$. Since on the other hand $|E(f, \mathcal{O})| \geq 2$ for every $f \in \mathcal{H}^+(\mathcal{O}, \Delta)$, we find that

$$\lim_{s \to 2+0} (s-2) L_{red}(s) = 2 \lim_{s \to 2+0} (s-2) L_M(s).$$

Putting Proposition 5.9 (see also (5.6)) together with the computation in this paragraph we obtain the result. □

9.7 Notes and Remarks

Sections 9.1–9.5 essentially follow Elstrodt, Grunewald, Mennicke (1987a). We already commented on Humbert's works (which are not contained in the Collected Papers of this author) at the beginning of Section 9.6.

Chapter 10. Examples of Discontinuous Groups

This chapter contains various constructions for discontinuous groups of isometries of hyperbolic 3-space. Since we often use the terminology of Coxeter groups we report on this here. For the general theory of these groups see Bourbaki (1968).

A Coxeter matrix of size $n \in \mathbb{N}$ is an $n \times n$-matrix

$$M = (m(i,j))_{i,j=1,\ldots,n}$$

with entries from $\mathbb{Z} \cup \{\infty\}$ having the following properties:

(1) M is symmetric,
(2) $m(i,i) = 1$ for all $i = 1, \ldots, n$,
(3) $m(i,j) > 1$ for all $i \neq j$.

The Coxeter group associated with M is the abstract group W_M generated by $s_1, \ldots s_n$ and defined by the relations: $(s_i s_j)^{m(i,j)} = 1$. The size n of M is called the rank of the Coxeter group W_M.

An important fact in the theory of Coxeter groups is that the order of the element $s_i s_j$ in W_M is exactly $m(i,j)$. Hence the elements s_i are involutions, that is elements of order 2. The information contained in the Coxeter matrix is often encoded in a graph G_M. It has a vertex i for every s_i. The vertex i is connected by $m(i,j) - 2$ edges to the vertex j. The graph G_M is called Coxeter graph.

Coxeter groups come with respresentations into orthogonal groups. They are constructed as follows. Let V_M be an n-dimensional real vector space with basis $e_1, \ldots, e_n$. On V_M we consider the bilinear form B_M defined by

$$B_M(e_i, e_j) := 2\cos\left(\frac{\pi}{m(i,j)}\right).$$

We write $Q_M(v) := B_M(v,v)$ for the corresponding quadratic form. For $i = 1, \ldots, n$ we denote by σ_i the reflection corresponding to e_i, that is σ_i is the linear map

$$\sigma_i(v) = v - 2\,\frac{B_M(v, e_i)}{Q_M(e_i)} \cdot e_i.$$

It is easily checked that $\sigma_i \in \mathbf{O}_n(\mathbb{R}, Q_M)$. From the theory of Coxeter groups we infer that the map $s_i \mapsto \sigma_i$ extends to an injective group homomorphism

$$\Theta_M : W_M \to \mathbf{O}_n(\mathbb{R}, Q_M).$$

10.1 Groups of Quaternions

In this section we use quaternion algebras to construct groups acting discontinuously on hyperbolic 3-space. We start off by giving certain generalities, which can for example be found in Deuring (1935), or Vignéras (1980a). We let K be a field of characteristic different from 2 and $a, b \in K$ be two non-zero elements. The quaternion algebra

$$\mathcal{H} = \mathcal{H}(a,b) = \mathcal{H}(a,b;K) \tag{1.1}$$

is defined to be the 4-dimensional K-vector space with basis $1, i, j, k$ together with the multiplication induced by the conventions

$$i^2 = a, \qquad j^2 = b, \qquad ij = -ji = k. \tag{1.2}$$

$\mathcal{H}(a,b;K)$ is then an associative, central simple K-algebra. It is either a skew field or isomorphic to the 2×2-matrix algebra $M(2,K)$. It has already appeared before in Chapter 1, Section 3 as the Clifford algebra corresponding to a 2-dimensional quadratic form. We write

$$^- : \mathcal{H}(a,b;K) \to \mathcal{H}(a,b;K), \quad \overline{x_0 + x_1 i + x_2 j + x_3 k} = x_0 - x_1 i - x_2 j - x_3 k$$

for the conjugation and

$$N : \mathcal{H}(a,b;K) \to K, \quad N(x) = x\bar{x}; \qquad \mathrm{Tr} : \mathcal{H}(a,b;K) \to K, \quad \mathrm{Tr}(x) = x + \bar{x}$$

for the norm and trace on $\mathcal{H}(a,b;K)$. These maps satisfy

$$\overline{xy} = \bar{y}\bar{x}, \qquad N(xy) = N(x) \cdot N(y)$$

for all $x, y \in \mathcal{H}(a,b;K)$. We also have the formula

$$N(x_0 + x_1 i + x_2 j + x_3 k) = x_0^2 - a x_1^2 - b x_2^2 + ab x_3^2$$

for $x_0, x_1, x_2, x_3 \in K$. An element $\alpha \in \mathcal{H}(a,b;K)$ annihilates the polynomial $x^2 - \mathrm{Tr}(\alpha)x + N(\alpha)$.

If L is a field extension of K with $\sqrt{a}, \sqrt{b} \in L$ then the map

$$\Phi : \mathcal{H}(a,b;K) \to M(2,L)$$
$$\Phi : x_0 + x_1 i + x_2 j + x_3 k \mapsto \begin{pmatrix} x_0 + x_1\sqrt{a} & x_2\sqrt{b} + x_3\sqrt{ab} \\ x_2\sqrt{b} - x_3\sqrt{ab} & x_0 - x_1\sqrt{a} \end{pmatrix} \tag{1.3}$$

is an injective K-algebra homomorphism satisfying

$$N(x) = \det(\Phi(x)) \tag{1.4}$$

for all $x \in \mathcal{H}(a, b; K)$.

Another general result which is of importance in invariance proofs is the Theorem of Skolem and Noether, see Deuring (1935). It says that for any K-algebra isomorphism $\varphi : A \to B$ of two central simple subalgebras A, B of a central simple K-algebra C there is an invertible element $a \in C$ such that $\varphi(x) = axa^{-1}$ for all $x \in A$.

Of particular importance for us is the case when K is an algebraic number field. We again write $V(K)$ for the set of places of K. We put further $V_\infty(K)$ for the archimedian and $V_f(K)$ for the non-archimedian places of K. The elements of $V_f(K)$ are sometimes interpreted as classes of non-archimedian valuations of K, sometimes as prime ideals in the ring of integers of K. The archimedian places $V_\infty(K)$ can be decomposed into the real ones $V_{\mathbb{R}}(K)$ and into pairs of conjugate complex places. We choose a representative for each pair of conjugate places and put them into the set $V_{\mathbb{C}}(K)$. We then have

$$V_\infty(K) = V_{\mathbb{R}}(K) \cup V_{\mathbb{C}}(K) \cup \overline{V_{\mathbb{C}}(K)}.$$

If $v \in V$ we put K_v for the completion of K with respect to v. See Lang (1968) for a more detailed discussion of these concepts.

For each v such that $K_v \neq \mathbb{C}$ there is a quaternion skew field $\mathcal{H}(v)$ which is unique up to K_v-algebra isomorphism. In case $v \in V_{\mathbb{R}}(K)$ the skew field $\mathcal{H}(v)$ is just the usual skew field of Hamilton quaternions $\mathcal{H}(-1, -1; \mathbb{R})$. In case v is a non-archimedian valuation the skew field $\mathcal{H}(v)$ can also be explicitly described, see Vignéras (1980a), chapitre II.

If $\mathcal{H}$ is a quaternion algebra over K we write

$$\mathcal{R}(\mathcal{H}) = \{ \, v \in V(K) \quad | \quad \mathcal{H} \otimes_K K_v \cong \mathcal{H}(v) \, \},$$

with the same definitions for $\mathcal{R}(\mathcal{H})_\infty$ and $\mathcal{R}(\mathcal{H})_f$. The elements of $\mathcal{R}(\mathcal{H})$ are called the places where $\mathcal{H}$ ramifies. $\mathcal{R}(\mathcal{H})$ is always a finite set of even cardinality. For every finite subset $\mathcal{R} \subset V_f(K) \cup V_{\mathbb{R}}(K)$ of even cardinality there is a quaternion algebra $\mathcal{H}$ over K which is unique up to K-algebra isomorphism such that $\mathcal{R}(\mathcal{H}) = \mathcal{R}$.

Let now $\mathcal{O}$ be the ring of integers of K and $\mathcal{O}_0$ a subring of finite index in $\mathcal{O}$. (We always assume that all rings contain an identity element 1.) Such subrings are usually called orders of K. Let $\mathcal{H}$ be a quaternion algebra over K. An $\mathcal{O}_0$-order $\mathfrak{R}$ in $\mathcal{H}$ is a subring which contains a K-basis of $\mathcal{H}$, is also an $\mathcal{O}_0$-submodule of $\mathcal{H}$ and is finitely generated as $\mathcal{O}_0$-module. An order is an $\mathcal{O}$-order of $\mathcal{H}$. An order is called maximal if it is not properly contained in another order. Two orders $\mathfrak{R}_1, \mathfrak{R}_2$ of $\mathcal{H}$ are called equivalent iff there is an invertible element $x \in \mathcal{H}$ with $\mathfrak{R}_2 = x\mathfrak{R}_1 x^{-1}$. There are always only finitely many equivalence classes of maximal orders in $\mathcal{H}$. We write $T_{\mathcal{H}}$ for the number of equivalence classes of maximal orders in $\mathcal{H}$. Assume now that $\mathcal{H}$ is unramified at some $v \in V_\infty(K)$. This condition is valid if as in the case of relevance here $V_{\mathbb{C}}(K) \neq \emptyset$. Then there is a simple formula for $T_{\mathcal{H}}$ due to Eichler (1938), see also Vignéras (1980b). We have:

$$T_{\mathcal{H}} = [\mathcal{M}_K : \mathcal{M}_K^2 \cdot D_{\mathcal{H}} \cdot P_{\mathcal{H}}]. \tag{1.5}$$

Here $\mathcal{M}_K$ is the group of fractional ideals of K, see Chapter 7 for more details. The subgroups $D_{\mathcal{H}}$, $P_{\mathcal{H}}$ of $\mathcal{M}_K$ are generated by the prime ideals where $\mathcal{H}$ is ramified and those principal ideals which have a generator which is positive at each real place where $\mathcal{H}$ is ramified.

Since the diagonally embedded ring $\mathcal{O}$ is a discrete subring of

$$\prod_{v \in V_{\mathbb{R}}(K) \cup V_{\mathbb{C}}(K)} K_v$$

any order is evidently a discrete subring of

$$\prod_{v \in V_{\mathbb{R}}(K) \cup V_{\mathbb{C}}(K)} \mathcal{H} \otimes_K K_v.$$

Definition 1.1. Let $\mathcal{H}$ be a quaternion algebra over K and R a subring of $\mathcal{H}$ closed under conjugation. We put

$$R^1 := \{\, x \in R \mid N(x) = 1 \,\}$$

and call it the norm 1 group of R.

If R is an order then $\mathcal{O} \subset R$ and if $x \in R$ then $x + \bar{x} \in \mathcal{O}$, hence every order is closed under conjugation and the above definition applies. Note that for any of the skew fields $\mathcal{H}(v)$ over the complete fields K_v the group $\mathcal{H}(v)^1$ is compact in the topology induced from K_v This is clear for $K_v = \mathbb{R}$ and an easy exercise in the case of a non-archimedian field K_v.

Our construction of discrete subgroups of $\mathbf{SL}(2, \mathbb{C})$ is incorporated in the following:

Theorem 1.2. *Let K be an algebraic number field with exactly one pair of complex places, that is $|V_{\mathbb{C}}(K)| = 1$. Let $\mathcal{H}$ be a quaternion algebra over K which is ramified at all real places of K, that is $V_{\mathbb{R}}(K) \subset \mathcal{R}(\mathcal{H})$. For $v_0 \in V_{\mathbb{C}}(K)$ let $\varphi : \mathcal{H} \otimes_K K_{v_0} \to M(2, \mathbb{C})$ be a $K_{v_0} = \mathbb{C}$-algebra isomorphism. For an order $\mathfrak{R} \subset \mathcal{H}$, put*

$$\Gamma = \varphi(\mathfrak{R}^1).$$

The group Γ has the following properties:

(1) Γ is a discrete subgroup of $\mathbf{SL}(2, \mathbb{C})$.
(2) Γ has finite covolume.
(3) Γ is cocompact if and only if $\mathcal{H}$ is a skew field.
(4) Suppose that K does not contain roots of the polynomials $x^2 + 1$, $x^2 + 3$. If $\mathcal{H}$ stays a skew field when extended to $K(\sqrt{-1})$ and to any quadratic extension field of K containing a non-trivial p-th root of unity for an odd prime p, then the image of Γ in $\mathbf{PSL}(2, \mathbb{C})$ is torsion free.

The theorem of Skolem and Noether implies that the subgroups Γ defined by the choice of two distinct $\mathbb{C}$-algebra isomorphisms φ will be $\mathbf{GL}(2,\mathbb{C})$- and hence $\mathbf{SL}(2,\mathbb{C})$-conjugate. Having K as a subfield of $\mathbb{C}$ and $\mathcal{H}$ in the form $\mathcal{H}=\mathcal{H}(a,b;K)$ we may then just use our algebra isomorphism from (1.3). If we switch from the place v to its conjugate, then Γ is changed to $\bar{\Gamma}$, where $\bar{\Gamma}$ is the complex conjugate of Γ. Note that the images of Γ and $\bar{\Gamma}$ in Iso($\mathbb{H}$) are conjugate. Note also that the condition under (4) is empty for fixed K and p large enough.

Examples of fields K having $|V_{\mathbb{C}}(K)| = 1$ are the imaginary quadratic number fields, or for example the field $\mathbb{Q}(\sqrt{1-\sqrt{2}})$. More examples will appear in the next section as certain quadratic extensions of totally real fields.

The first situation to which our theorem applies to is provided by taking an imaginary quadratic number field K and the quaternion algebra $M(2,K)$ with the order $M(2,\mathcal{O})$. Here $\mathcal{O}$ is the ring of integers of K. We find $\Gamma = \mathbf{SL}(2,\mathcal{O})$ by choice of the identity as algebra isomorphism φ. This group is never cocompact.

Another example of relevance later is

$$\mathcal{H}_1 = \mathcal{H}\left(\sqrt{5},\sqrt{5};K\right) \tag{1.6}$$

with $K=\mathbb{Q}\left(\sqrt{-\sqrt{5}}\right)\subset\mathbb{C}$. This quaternion algebra is a skew field because K has the real embedding

$$\iota : K \to \mathbb{R}, \qquad \iota\left(\sqrt{-\sqrt{5}}\right) = \sqrt{\sqrt{5}}$$

and because under that embedding $\mathcal{H}$ is isomorphic to a subring of the Hamilton quaternions $\mathcal{H}(-1,-1,\mathbb{R})$.

Still another example is the quaternion algebra

$$\mathcal{H}_2 = \mathcal{H}(2,5;K) \tag{1.7}$$

with $K=\mathbb{Q}(i)$. This quaternion algebra will also appear in the next section. It will be clear from there that it is a skew field.

Still another example can be constructed by taking $K=\mathbb{Q}(\sqrt{-5})$ and by specifying the ramified places. We define $\mathcal{H}_3$ to be the quaternion algebra over K with $\mathcal{R}(\mathcal{H}_3)=\{\wp_1,\ \wp_2\}$ where $\wp_1,\ \wp_2$ are the ideals generated by 11 and $3+2\sqrt{-5}$ respectively. Both $\wp_1$ and $\wp_2$ are prime ideals in the ring of integers of K. Let L be one of the fields $K(\sqrt{-1}),\ K(\sqrt{-3})$ then $\wp_1$ decomposes into two distinct factors in the ring of integers of L. Taking $\wp$ to be one of these factors we find that $\mathcal{H}_3\otimes_K L_\wp \cong \mathcal{H}_3\otimes_K K_{\wp_1}$ and hence $\mathcal{H}_3\otimes_K L$ are skew fields. We infer that the images of the discrete groups constructed in Theorem 1.2 in $\mathbf{PSL}(2,\mathbb{C})$ are torsion free.

In all the cases $\mathcal{H}_1,\mathcal{H}_2,\mathcal{H}_3$ it is clear how one might choose orders for the construction of discrete groups.

We will not add any more detailed discussion of concrete examples to this section since the case of $\Gamma = \mathbf{SL}(2, \mathcal{O})$, $\mathcal{O}$ a ring of imaginary quadratic integers has been discussed before. A subclass of the discrete groups constructed here will appear again in the next section as unit groups of certain quadratic forms.

Proof of Theorem 1.2. (1): As noted before the order $\mathfrak{R}$ embeds as a discrete subring diagonally into the direct product

$$\mathfrak{R} \to \prod_{v \in V_{\mathbb{R}}(K) \cup V_{\mathbb{C}}(K)} \mathcal{H} \otimes_K K_v.$$

Hence $\mathfrak{R}^1$ embeds as a discrete subgroup

$$\mathfrak{R}^1 \to \prod_{v \in V_{\mathbb{R}}(K) \cup V_{\mathbb{C}}(K)} (\mathcal{H} \otimes_K K_v)^1.$$

By our assumption the groups $(\mathcal{H} \otimes_K K_v)^1$ are compact for $v \in V_{\mathbb{R}}(K)$. Hence for $V_{\mathbb{C}}(K) = \{v_0\}$ the projection

$$p: \prod_{v \in V_{\mathbb{R}}(K) \cup V_{\mathbb{C}}(K)} (\mathcal{H} \otimes_K K_v)^1 \to (\mathcal{H} \otimes_K K_{v_0})^1$$

has compact kernel. It is easy to see that then $p(\mathfrak{R}^1)$ is a discrete subgroup of $(\mathcal{H} \otimes_K K_{v_0})^1$. Identifying K_{v_0} with $\mathbb{C}$ we find that $p(\mathfrak{R}^1)$ is conjugate in $(\mathcal{H} \otimes_K K_{v_0})^1 = \mathbf{SL}(2, \mathbb{C})$ to Γ.

(2) and (3): If $\mathcal{H}$ is not a skew field, then evidently $V_{\mathbb{R}}(K) = \emptyset$ and K is an imaginary quadratic number field. The group Γ is then commensurable with $\mathbf{SL}(2, \mathcal{O})$ and hence has finite covolume and is not cocompact.

The cocompactness of Γ in case $\mathcal{H}$ is a skew field remains to be shown. We will show that there is a compact subset $M \subset \mathbf{SL}(2, \mathbb{C})$ so that

$$\mathbf{SL}(2, \mathbb{C}) = \bigcup_{\gamma \in \Gamma} \gamma M.$$

It is then clear that Γ acts on hyperbolic 3-space $\mathbb{H}$ with a compact fundamental domain. We put $r := 2 + r_1$ where r_1 is the number of real places of K. Note that

$$K \otimes_{\mathbb{Q}} \mathbb{R} \cong \prod_{v \in V_\infty} K_v \cong \mathbb{C} \times \mathbb{R}^{r_1} \cong \mathbb{R}^r, \tag{1.8}$$

$$\mathcal{H} \otimes_{\mathbb{Q}} \mathbb{R} \cong M(2, \mathbb{C}) \times \mathcal{H}(-1, -1, \mathbb{R})^{r_1} \cong \mathbb{R}^{4r} \tag{1.9}$$

as real vector spaces. As before we consider $K, \mathcal{H}$ being embedded in $K \otimes_{\mathbb{Q}} \mathbb{R}$, $\mathcal{H} \otimes_{\mathbb{Q}} \mathbb{R}$ respectively. The ring of integers $\mathcal{O}$ and the order $\mathfrak{R}$ are then lattices in the appropriate real vector spaces. By linearity we extend the conjugation map from $\mathcal{H}$ to $\mathcal{H} \otimes_{\mathbb{Q}} \mathbb{R}$. The corresponding norm map provides us with a map

$$N : \mathcal{H} \otimes_{\mathbb{Q}} \mathbb{R} \to K \otimes_{\mathbb{Q}} \mathbb{R}, \qquad N(x) := x\bar{x}.$$

Using the norm map we define

$$(\mathcal{H} \otimes_{\mathbb{Q}} \mathbb{R})^1 := \{ \, x \in \mathcal{H} \otimes_{\mathbb{Q}} \mathbb{R} \;\mid\; N(x) = 1 \, \}.$$

We also choose some euclidean structures $\| \cdot \|$ on $K \otimes_{\mathbb{Q}} \mathbb{R}$, $\mathcal{H} \otimes_{\mathbb{Q}} \mathbb{R}$ and define for $\delta \in \mathbb{R}$ with $\delta > 0$

$$M_\delta := \{ \, x \in \mathcal{H} \otimes_{\mathbb{Q}} \mathbb{R} \;\mid\; \|x\| \leq \delta \, \}.$$

Notice that $M_\delta \cap (\mathcal{H} \otimes_{\mathbb{Q}} \mathbb{R})^1$ is compact.

An element $g \in (\mathcal{H} \otimes_{\mathbb{Q}} \mathbb{R})^1$ acts on $\mathcal{H} \otimes_{\mathbb{Q}} \mathbb{R}$ by right multiplication giving a linear map of determinant 1. Hence the covolumes of the lattices $\mathfrak{R}$ and $\mathfrak{R}g$ are the same. By Minkowski's lattice point theorem, see Hardy and Wright (1960) Chapter XXIV, we find a positive constant δ so that for every $g \in (\mathcal{H} \otimes_{\mathbb{Q}} \mathbb{R})^1$ there is a non-zero $h \in \mathfrak{R}$ with $hg \in M_\delta$. Since $\mathcal{H}$ is a skew field such an h is invertible and satisfies $N(h) \neq 0$. Since the norm map is continuous with respect to our Euclidean structures the norms $N(h)$ belong to a finite subset of $\mathcal{O}$. Using the subsequent Lemma 1.4 we infer that there are non-zero elements $h_1, ..., h_\ell \in \mathfrak{R}$ so that every $g \in (\mathcal{H} \otimes_{\mathbb{Q}} \mathbb{R})^1$ can be expressed as $g = \gamma h_i^{-1} m$ with $\gamma \in \mathfrak{R}^1$, $i \in \{1, ..., \ell\}$ and some $m \in M_\delta$. Noticing that the sets $(h_i^{-1} M_\delta) \cap (\mathcal{H} \otimes_{\mathbb{Q}} \mathbb{R})^1$ are compact and projecting from $(\mathcal{H} \otimes_{\mathbb{Q}} \mathbb{R})^1$ to $M(2, \mathbb{C})$ our proof is complete.

(4): Let $\tilde{\Gamma}$ be the image of Γ in $\mathbf{PSL}(2, \mathbb{C})$. Let p be an odd prime number so that $\tilde{\Gamma}$ contains an element of $\tilde{g}$ of order exactly p. Then $\mathfrak{R}^1$ also contains an element g of the same order. Then $g \notin K$. If $p = 3$ this is ensured by our condition, if $p \geq 5$ then K would have more than 2 complex embeddings. The K-subalgebra of $\mathcal{H}$ generated by g is then a quadratic extension field of K which splits $\mathcal{H}$ and which contains a non-trivial p-th root of unity. The case $p = 2$ is handled similarly. □

The following two lemmas result in a statement which has been used in the proof of Theorem 1.2.

Lemma 1.3. *Let K be an algebraic number field with ring of integers $\mathcal{O}$. For $n \in \mathbb{N}$ and $m \in \mathcal{O}$ with $m \neq 0$ let*

$$M(n, \mathcal{O})^m := \{ \, x \in M(n, \mathcal{O}) \;\mid\; \det x = m \, \}.$$

Then the group $\mathbf{SL}(n, \mathcal{O})$ acts by multiplication from the right with finitely many orbits on $M(n, \mathcal{O})^m$.

Proof. It is easy to see that there are, up to right multiplication by elements of $\mathbf{SL}(n, \mathcal{O})$, only finitely many distinct classes of lower triangular matrices in

$M(n,\mathcal{O})^m$. There is furthermore a finite extension K_1 of K (the Hilbert class field is suitable) with ring of integers $\mathcal{O}_1$ so that every ideal of $\mathcal{O}$ becomes principal when extended to $\mathcal{O}_1$, see Lang (1968). Let now A be in $M(n,\mathcal{O})^m$. As in the proof of Lemma 7.2.1 we find a $\gamma \in \mathbf{SL}(n,\mathcal{O}_1)$ so that

$$A\gamma = \begin{pmatrix} \lambda & 0 \\ v & A_1 \end{pmatrix}.$$

Repeating this argument we see that there is a finite extension K_0 with ring of integers $\mathcal{O}_0$ so that every element of $M(n,\mathcal{O})^m$ is lower triangular up to right multiplication by $\mathbf{SL}(n,\mathcal{O}_0)$. Hence the elements of $M(n,\mathcal{O})^m$ fall into finitely many classes up to right multiplication by $\mathbf{SL}(n,\mathcal{O}_0)$. If for $A, B \in M(n,\mathcal{O})^m$ there is $\gamma \in \mathbf{SL}(n,\mathcal{O}_0)$ with $A = B\gamma$ then $\gamma \in \mathbf{SL}(n,K) \cap M(n,\mathcal{O}_0)$. This set is equal to $\mathbf{SL}(n,\mathcal{O})$. □

The following is the final statement used in the proof of Theorem 1.2. We only sketch the proof which uses some computations similar to those in the proof of Lemma 1.3.

Lemma 1.4. *Let K be an algebraic number field with ring of integers $\mathcal{O}$. Let $\mathcal{H}$ be a quaternion algebra over K and $\mathfrak{R} \subset \mathcal{H}$ an order. For $m \in \mathcal{O}$ with $m \neq 0$ let*

$$\mathfrak{R}^m := \{\, x \in \mathfrak{R} \mid Nx = m \,\}.$$

Then the group $\mathfrak{R}^1$ acts by multiplication from the right with finitely many orbits on $\mathfrak{R}^m$.

Proof. By tensoring with the ring of integers in a suitable extension field of K we may assume that $\mathfrak{R}$ is free as an $\mathcal{O}$-module. Taking an $\mathcal{O}$-basis of $\mathfrak{R}$ we get an $\mathcal{O}$-algebra representation $\Theta : \mathfrak{R} \to M(4,\mathcal{O})$. The map Θ satisfies $N(x)^2 = \det \Theta(x)$ for all $x \in \mathfrak{R}$. Hence we may deduce Lemma 1.4 from Lemma 1.3. □

Sometimes questions of $\mathbf{PSL}(2,\mathbb{C})$-conjugacy arise for the groups constructed in Theorem 1.2. A partial answer is given in the following Proposition.

Proposition 1.5. *Let K be an algebraic number field with exactly one pair of complex places. Let $\mathcal{H}_1$, $\mathcal{H}_2$ be two quaternion algebras over K which are ramified at all real places of K. For $v_0 \in V_{\mathbb{C}}(K)$ let $\varphi_{1,2} : \mathcal{H}_{1,2} \otimes_K K_{v_0} \to M(2,\mathbb{C})$ be two $K_{v_0} = \mathbb{C}$-algebra isomorphisms. For orders $\mathfrak{R}_1 \subset \mathcal{H}_1$, $\mathfrak{R}_2 \subset \mathcal{H}_2$ put $\Gamma_{1,2} = \varphi_{1,2}(\mathfrak{R}^1_{1,2})$. Then the following hold.*

(1) If Γ_1 and Γ_2 are conjugate in $\mathbf{SL}(2,\mathbb{C})$ then $\mathcal{H}_1$ and $\mathcal{H}_2$ are K-isomorphic.
(2) Suppose that $\mathcal{H} = \mathcal{H}_1 = \mathcal{H}_2$ and that $\mathfrak{R}_1$, $\mathfrak{R}_2$ are two maximal orders in $\mathcal{H}$, then the groups $\Gamma_1 = \varphi(\mathfrak{R}^1_1)$, $\Gamma_2 = \varphi(\mathfrak{R}^1_2) < \mathbf{SL}(2,\mathbb{C})$ are $\mathbf{SL}(2,\mathbb{C})$-conjugate if and only if $\mathfrak{R}_1$, $\mathfrak{R}_2$ are equivalent.

Proof. (1): We write $\mathcal{H}$ for any of the quaternion algebras $\mathcal{H}_1$, $\mathcal{H}_2$ and use in a similar notation φ, $\mathfrak{R}$. Our setup is so that K is identified with a subfield of $\mathbb{C}$ and $\varphi(\mathcal{H})$ is a K-linear subspace of $M(2,\mathbb{C}) = \mathcal{H} \otimes_K K_{v_0}$. The subspace $K \cdot \mathfrak{R}^1$ generated by $\mathfrak{R}^1$ is a non-commutative K-subalgebra of $\mathcal{H}$. Notice that we already know that $\varphi(\mathfrak{R}^1)$ is a cofinite group and hence cannot be commutative. We conclude that $K \cdot \mathfrak{R}^1 = \mathcal{H}$. Any conjugation from Γ_1 to Γ_2 extends to a K-isomorphism from $\mathcal{H}_1$ to $\mathcal{H}_2$.

(2): One direction of this statement is obvious. Assume now that the groups $\Gamma_1 = \varphi(\mathfrak{R}_1^1)$ and $\Gamma_2 = \varphi(\mathfrak{R}_2^1)$ are conjugate in $\mathbf{SL}(2,\mathbb{C})$. The K-subspace $\varphi(\mathcal{H})$ contains a $\mathbb{C}$-basis $\mathbf{B}$ of $M(2,\mathbb{C})$. Let $g \in \mathbf{SL}(2,\mathbb{C})$ be an element with $g\Gamma_1 g^{-1} = \Gamma_2$. We consider an enumeration $\Gamma_1 = \{\gamma_1, \gamma_2, ...\}$ and write down the infinite system of equations

$$x \cdot \gamma_i = g\gamma_i g^{-1} \cdot x \qquad\qquad i = 1, 2, ...$$

Considering the unknown x expressed in terms of the basis $\mathbf{B}$ this is a set of homogeneous linear equations with coefficients in K. Since it has a non-zero complex solution (namely g) it also has a non-zero solution in $\varphi(\mathcal{H})$. Hence we have found an $h \in \mathcal{H}$ with $h \neq 0$ and

$$(1.10) \qquad\qquad h\,\mathfrak{R}_1^1\,h^{-1} = \mathfrak{R}_2^1.$$

We shall now finish the proof by showing that this implies $h\mathfrak{R}_1 h^{-1} = \mathfrak{R}_2$. For every $v \in V_f(K)$ we define $\mathcal{O}_v$ to be the ring of integers in K_v. We let $\mathfrak{R}$ be any maximal order in $\mathcal{H}$. Then $\mathfrak{R}_v := \mathfrak{R} \otimes_{\mathcal{O}} \mathcal{O}_v$ is a maximal $\mathcal{O}_v$-order in $\mathcal{H} \otimes_K K_v$. Considering the diagonal embedding

$$\mathcal{H} \to \prod_{v \in V_f(K)} \mathcal{H} \otimes_K K_v$$

we have by general theory (see for example Vignéras (1980a))

$$\mathfrak{R} = \mathcal{H} \cap \prod_{v \in V_f(K)} \mathfrak{R}_v.$$

We infer from the strong approximation theorem that $\mathfrak{R}^1$ is dense in $\mathfrak{R}_v^1$ in the v-adic topology. Hence (1.10) implies that $h(\mathfrak{R}_1)_v^1 h^{-1} = (\mathfrak{R}_2)_v^1$ for every $v \in V_f(K)$. We shall now prove that this implies $h(\mathfrak{R}_1)_v h^{-1} = (\mathfrak{R}_2)_v$ for every $v \in V_f(K)$. If $\mathcal{H}$ is ramified at v there is nothing to prove because there is only one maximal order in $\mathcal{H} \otimes_K K_v$. In the other cases we claim that the $\mathcal{O}_v$-module $\mathcal{O}_v(\mathfrak{R}_v)^1$ generated by $(\mathfrak{R}_v)^1$ satisfies $\mathcal{O}_v(\mathfrak{R}_v)^1 = \mathfrak{R}_v$. To do this we choose a $\mathcal{O}_v$-algebra isomorphism

$$\psi : \mathfrak{R}_v \to M(2, \mathcal{O}_v),$$

see Vignéras (1980a). We have $\psi((\mathfrak{R}_v)^1) = \mathbf{SL}(2, \mathcal{O}_v)$. But $\mathcal{O}_v\mathbf{SL}(2,\mathcal{O}_v) = M(2,\mathcal{O}_v)$ is obvious. □

We add the following interesting formula for the covolume of the discrete groups constructed above. It contains, as special case, the volume formula for $\mathbf{PSL}(2, \mathcal{O})$ where $\mathcal{O}$ is the ring of integers in an imaginary quadratic number field from Chapter 7.

Theorem 1.6. *Let K be an algebraic number field with exactly one pair of complex places. Let d_K be the absolute discriminant and r_1 the number of real places of K. Let $\mathcal{H}$ be a quaternion algebra over K that is ramified at all real places of K. Let $\varphi : \mathcal{H} \otimes_K K_v \to M(2, \mathbb{C})$ be an algebra isomorphism for $v \in V_{\mathbb{C}}(K)$. Let $\mathfrak{R} \subset \mathcal{H}$ be a maximal order in $\mathcal{H}$, then the image $\tilde{\Gamma}$ in $\mathbf{PSL}(2, \mathbb{C})$ of the group $\Gamma = \varphi(\mathfrak{R}^1)$ satisfies*

$$\text{(1.11)} \qquad \operatorname{vol}(\tilde{\Gamma}) = \prod_{v \in \mathcal{R}(\mathcal{H})_f} (N(v) - 1) \cdot \frac{|d_K|^{\frac{3}{2}}}{2^{2r_1+2}\, \pi^{2r_1+2}} \cdot \zeta_K(2).$$

Here $\mathcal{R}(\mathcal{H})_f$ stands for the set of non-archimedean valuations at which $\mathcal{H}$ ramifies. For $v \in \mathcal{R}(\mathcal{H})_f$ the number $N(v)$ is the norm of the prime ideal corresponding to v.

Theorem 1.6 can be proved by adelic volume computations. A clear exposition is contained in Borel (1981). Note that the volume formula for $\mathbf{PSL}(2, \mathcal{O})$ from Theorem 7.1.1 is a special case.

Finally we report on some results of M. Eichler which were used by Vignéras (1980b) for the construction of non-conjugate but isospectral subgroups of $\mathbf{PSL}(2, \mathbb{C})$. For the rest of this section we fix an algebraic number field K so that $V_{\mathbb{C}}(K) \neq \emptyset$. This condition ensures that the strong approximation theorem (see Vignéras (1980a)) is applicable to the norm 1 group of any quaternion algebra over K. To the quaternion algebra $\mathcal{H}$ over K which is also fixed we may hence apply the theorems from Eichler (1938).

We write $\mathcal{L}_{\mathcal{H}}$ for the set of those K-subalgebras in $\mathcal{H}$ which are commutative and not equal to K. All elements of $\mathcal{L}_{\mathcal{H}}$ are quadratic extension fields of K. A quadratic extension field L of K is K-isomorphic to an element of $\mathcal{L}_{\mathcal{H}}$ if and only if $\mathcal{H} \otimes_K L \cong M(2, L)$ as L-algebras. Another (local) condition for the quadratic extension field L of K to be K-isomorphic to an element of $\mathcal{L}_{\mathcal{H}}$ is that $L \otimes_K K_v$ is a field for every $v \in \mathcal{R}(\mathcal{H})$. Given a quadratic extension field L of K we write $L_{\mathcal{H}}$ for the set of K-algebra homomorphisms from L to $\mathcal{H}$. If $L_{\mathcal{H}}$ is not empty the multiplicative group of $\mathcal{H}$ acts transitively by conjugation on $L_{\mathcal{H}}$.

We let $\mathcal{O}$ now be the ring of integers of K and $\mathfrak{R}$ a maximal order in $\mathcal{H}$. We write $\Omega_{\mathcal{H}}$ for the set of subrings of $\mathcal{H}$ which are an $\mathcal{O}$-order in one of the elements of $\mathcal{L}_{\mathcal{H}}$. If $L \in \mathcal{L}_{\mathcal{H}}$ then $\mathfrak{R} \cap L$ is an element of $\Omega_{\mathcal{H}}$. The set of the elements of $\Omega_{\mathcal{H}}$ so obtained we denote by $\Omega(\mathfrak{R})$. Under the condition that for every $L \in \mathcal{L}_{\mathcal{H}}$

$$\text{(1.12)} \qquad [\mathcal{M}_K : N_{L/K}(\mathcal{M}_L) \cdot D_{\mathcal{H}} \cdot P_{\mathcal{H}}] = 1$$

we find in Eichler (1938) or Vignéras (1980b) that

$$\Omega_{\mathcal{H}} = \Omega(\mathfrak{R}). \tag{1.13}$$

Here $N_{L/K}$ is the relative norm from L to K. More on condition (1.12) can be found in the proof of Corollary 1.8. Given a quadratic extension field L of K and an $\mathcal{O}$-order S of L we define

$$(L,S)_{\mathfrak{R}} := \{ \ \psi \in L_{\mathcal{H}} \ \mid \ \psi(S) = \mathfrak{R} \cap \psi(L) \ \}.$$

If t is an element from $\mathcal{O}$ we write $P_t(x)$ for the polynomial $P_t(x) := x^2 - tx + 1$. Assume now that $t \in \mathcal{O}$ is so that P_t is irreducible over K. If S is an $\mathcal{O}$-order in a quadratic extension field L of K which contains a root of P_t we define

$$M(\mathfrak{R},S,t) := \{ \ x \in \mathfrak{R}^1 \ \mid \ \operatorname{tr}(x) = t, \ \exists \psi \in (L,S)_{\mathfrak{R}} \text{ with } x \in \psi(S) \ \}.$$

The group $\mathfrak{R}^1$ of elements of norm 1 in $\mathfrak{R}$ acts by conjugation on $M(\mathfrak{R},S,t)$, we write $m(\mathfrak{R},S,t)$ for the number of orbits. In Eichler (1938) we find the formula

$$m(\mathfrak{R},S,t) = \frac{[N(\mathfrak{R}^*) : N_{L/K}(S^*)] \ h(S)}{h_{\mathcal{H}}(\mathcal{O})} \cdot \prod_{\wp \in V_f(K)} \left(1 - \left(\frac{S}{\wp}\right)\right). \tag{1.14}$$

Here S^* stands for the group of units of the order S and $h(S)$ for its class number. The number $h_{\mathcal{H}}(\mathcal{O})$ is the order of the group of fractional ideals $\mathcal{M}_K$ divided by the subgroup generated by those principal ideals which have a generator which is positive at all the real places of K where $\mathcal{H}$ ramifies. Eichler's symbol $(\frac{S}{\wp})$ is defined as

$$\left(\frac{S}{\wp}\right) = \begin{cases} 1 & \text{if } \wp \text{ decomposes into 2 distinct factors in } L, \\ 1 & \text{if } \wp \text{ divides the conductor ideal of } S, \\ 0 & \begin{cases} \text{if } \wp \text{ is ramified in } L \text{ and does} \\ \text{not divide the conductor ideal of } S, \end{cases} \\ -1 & \text{in all other cases.} \end{cases}$$

The conductor ideal of S is the largest ideal of the ring of integers of L contained in S. The norm of a unit from $\mathfrak{R}$ is contained in the group of units $\mathcal{O}^*$ of $\mathcal{O}$. The image $N(\mathfrak{R}^*)$ of the group of units of $\mathfrak{R}$ under the norm map consists of those units of $\mathcal{O}$ which are positive at all the real places of K where $\mathcal{H}$ ramifies. This implies that the right-hand side of (1.14) does not depend on the maximal order $\mathfrak{R}$.

For more explanation on all of the above see Eichler (1938). As the outcome we get that the length spectra of the groups associated to maximal orders are the same under a certain condition. We use here the terminology of Chapter 5, in particular Section 3.

Theorem 1.7. *Let K be an algebraic number field with exactly one pair of complex places. Let $\mathcal{H}$ be a quaternion skew field over K which is ramified* *at all real places of K. For $v_0 \in V_{\mathbb{C}}(K)$ let $\varphi : \mathcal{H} \otimes_K K_{v_0} \to M(2, \mathbb{C})$ be a $K_{v_0} = \mathbb{C}$-algebra isomorphism. Assume further that condition (1.12) is satisfied for quadratic extension fields L of K which are isomorphic to an element of $\mathcal{L}_{\mathcal{H}}$. Let $\mathfrak{R}_1$, $\mathfrak{R}_2$ be maximal orders of $\mathcal{H}$. Let $\tilde{\Gamma}_1$, $\tilde{\Gamma}_2$ be the images of $\Gamma_1 = \varphi(\mathfrak{R}_1^1)$, $\Gamma_2 = \varphi(\mathfrak{R}_2^1)$ in* $\mathbf{PSL}(2, \mathbb{C})$. *Then $\tilde{\Gamma}_1$, $\tilde{\Gamma}_2$ have the same length spectrum.*

Proof. Let $\mathfrak{R}$ be one of $\mathfrak{R}_1$, $\mathfrak{R}_2$ and Γ be one of Γ_1, Γ_2. We write $\{\Gamma\}$ for the set of Γ-conjugacy classes of elements of Γ so that $\varphi(\gamma)$ is hyperbolic or loxodromic. We further write $\{\gamma\}_\Gamma$ for the conjugacy class of a non-trivial element $\gamma \in \Gamma$.

We shall construct a bijection

$$\Theta : \{\Gamma_1\} \to \{\Gamma_2\}, \qquad \Theta(\{\gamma\}_{\Gamma_1}) = \{\Theta(\gamma)\}_{\Gamma_2}$$

which in the notation of Chapter 5 satisfies

$$a(\gamma) = a(\Theta(\gamma)),\ N(\gamma) = N(\Theta(\gamma)),\ N(\gamma_0) = N(\Theta(\gamma_0)),\ |\mathcal{E}(\gamma)| = |\mathcal{E}(\Theta(\gamma_0))|.$$

Here $a(\gamma)$ is the eigenvalue of γ with $|a(\gamma)| > 1$ and $N(\gamma) = |a(\gamma)|^2$. The element γ_0 generates $\mathcal{C}(\gamma, \Gamma)/\mathcal{E}(\gamma)$ where $\mathcal{C}(\gamma, \Gamma)$ is the centralizer of γ in Γ and $\mathcal{E}(\gamma)$ is the maximal finite subgroup in $\mathcal{C}(\gamma, \Gamma)$. Taking in account the discussion in Section 3 of Chapter 5 the theorem is then proved. Note that since $-1 \in \Gamma$ the distinction between Γ_1, Γ_2 and $\tilde{\Gamma}_1$, $\tilde{\Gamma}_2$ creates no problems.

We shall construct our bijection Θ for the groups $\mathfrak{R}^1$, $\mathfrak{R}^2$ and then transport it to Γ_1, Γ_2 via φ. We shall use notations analogous to the above. Let $g \in \mathfrak{R}^1$, $g \neq \pm 1$. Since $g \notin K$ the K-algebra $K(g)$ generated by g is a quadratic extension field of K. We let $L[g]$ be the splitting field of the polynomial $P_g(x) = x^2 - \mathrm{Tr}(g)x + 1$ and choose a K-algebra isomorphism $\psi : L[g] \to K(g)$. We define $S[g]$ to be the $\mathcal{O}$-order $\psi^{-1}(K(g) \cap \mathfrak{R})$. The order $S[g]$ does not depend on the isomorphism ψ or the representative g of the conjugacy class $\{g\}_{\mathfrak{R}^1}$. We define σ to be the non-trivial element of the Galois group of $L[g]$ over K. We have $\sigma(y) = \psi^{-1}(\overline{\psi(y)})$ for every $y \in L[g]$. Since any $\mathcal{O}$-order is σ-invariant

$$S[g]^1 := \{\ s \in S[g] \ \mid\ s\sigma(s) = 1\ \}$$

is a subgroup of the group of units in $S[g]$. We find

$$K(g) \cap \mathfrak{R}^1 = \mathcal{C}(g, \mathfrak{R}^1) = \psi(S[g]^1). \tag{1.15}$$

We let now $t \in \mathcal{O} \subset \mathbb{C}$ be an element so that $P_t(x) = x^2 - tx + 1$ is irreducible over K and has no zero in $\mathbb{C}$ of absolute value 1. We let further S be an $\mathcal{O}$-order in the splitting field of $P_t(x)$. Then $M(\mathfrak{R}, S, t)$ consists of those conjugacy classes $\{g\}_{\mathfrak{R}^1}$ in $\{\mathfrak{R}^1\}$ with $\mathrm{Tr}(g) = t$ and $S[g] = S$. The sets $M(\mathfrak{R}_1, S, t)$ and $M(\mathfrak{R}_2, S, t)$ have by (1.14) the same cardinalities. Hence we choose any bijection between them. Finally we put all these bijections together to obtain Θ. From (1.15) it is clear that Θ satisfies the required compatibility relations. □

Corollary 1.8. *Let n be a positive integer. Then there are torsion free co-compact subgroups $\tilde{\Gamma}_1, ..., \tilde{\Gamma}_n < \mathbf{Iso}^+(\mathbb{H})$ which are pairwise non-conjugate in $\mathbf{Iso}(\mathbb{H})$ and which are all isospectral.*

Proof. We let D be an odd, square-free negative integer and consider the imaginary quadratic number field $K = \mathbb{Q}(\sqrt{D})$. We assume that its ideal class group $\mathcal{J}_K$ satisfies

$$\left|\mathcal{J}_K / \mathcal{J}_K^2\right| = 2^\alpha > 4n.$$

This situation occurs if D has sufficiently many distinct prime factors, see Borevich, Shafarevich (1966). We write d_K for its discriminant, $\mathcal{O}$ for its ring of integers and σ for the non-trivial element of its Galois group. We assume that d_K is odd and not divisible by 3.

We choose $\wp_1$ to be a prime ideal of $\mathcal{O}$ with $\sigma(\wp_1) \neq \wp_1$. We also choose a rational prime p so that

$$p \equiv 11 \text{ mod } 12, \qquad p \equiv \text{a non-square} \mod d_K.$$

The existence of p follows from Dirichlet's theorem on primes in arithmetic progressions. The congruences imply that $\wp_2 := p\mathcal{O}$ is a prime ideal in $\mathcal{O}$.

We put $\mathcal{R} := \{ \wp_1, \wp_2 \}$ and let $\mathcal{H}$ be the quaternion algebra with $\mathcal{R}(\mathcal{H}) = \mathcal{R}$. By (1.5) it contains at least n equivalence classes of maximal orders, let $\mathfrak{R}_1, ..., \mathfrak{R}_n$ be representatives for them. We put $\Gamma_\ell = \varphi(\mathfrak{R}_\ell^1)$ for $\ell = 1, ..., n$ and write $\tilde{\Gamma}_\ell$ for the images in $\mathbf{PSL}(2, \mathbb{C})$.

First of all we have to check that the $\tilde{\Gamma}_\ell$ are torsion free. We do this by looking at Theorem 1.2, statement (4). The only quadratic extension fields of K containing non-trivial roots of unity are $L_1 = K(\sqrt{-1})$ and $L_2 = K(\sqrt{-3})$. The prime $\wp_2$ is inertial in both fields, hence $\mathcal{H} \otimes_K L_1$ and $\mathcal{H} \otimes_K L_2$ are skew fields.

Secondly we have to prove that the $\tilde{\Gamma}_\ell$ are pairwise non-conjugate in $\mathbf{Iso}(\mathbb{H})$. We use Proposition 1.5. Notice that the prime ideal $\wp_1$ prevents a K-isomorphism between $\mathcal{H}$ and $\sigma(\mathcal{H})$.

Thirdly we have to check that the $\tilde{\Gamma}_\ell$ are pairwise isospectral. In order to apply Theorem 1.5 we have to check condition (1.12) for all subfields of $\mathcal{H}$ which are quadratic extensions of K. Let L be any quadratic extension of K. For any non-zero ideal $\mathfrak{m}$ of $\mathcal{O}$ we have a surjective homomorphism

$$(1.16) \qquad \rho_\mathfrak{m} : \mathcal{M}_K^\mathfrak{m} / N_{L/K}(\mathcal{M}_L^\mathfrak{m}) \cdot P_K^\mathfrak{m} \to \mathcal{M}_K / N_{L/K}(\mathcal{M}_L) \cdot D_\mathcal{H} \cdot P_\mathcal{H}$$

induced by the inclusion $\mathcal{M}_K^\mathfrak{m} \subset \mathcal{M}_K$. Here $\mathcal{M}_K^\mathfrak{m}$ is the subgroup of $\mathcal{M}_K$ generated by the prime ideals prime to $\mathfrak{m}$ with a similar meaning for $\mathcal{M}_L^\mathfrak{m}$. The symbol $P_K^\mathfrak{m}$ stands for the group generated by those principal ideals which have a generator congruent to 1 modulo $\mathfrak{m}$. Notice that in our case we have

$P_{\mathcal{H}} = P_K$ where P_K is the group of all principal ideals. By class field theory the group on the left hand side of (1.16) has order 1 or 2 and the order is 2 if and only if $\mathfrak{m}$ is a divisor of the conductor ideal $\mathfrak{f} = \mathfrak{f}(L)$ of L. We further know that exactly the prime divisors of $\mathfrak{f}(L)$ are ramified in L and that the prime ideals of $\mathcal{O}$ which decompose in L are those which are contained in $N_{L/K}(\mathcal{M}_L^{\mathfrak{f}}) \cdot P_K^{\mathfrak{f}}$. For all of this see Lang (1968). Assume now that L is some field which does not satisfy (1.12). Then the group on the right-hand side of (1.16) has order 2 and $\rho_{\mathfrak{m}}$ is an isomorphism for every ideal $\mathfrak{m}$ of $\mathcal{O}$. It follows that L is unramified over K. Since $\wp_2$ is trivial in the right-hand side of (1.16) it is trivial in the left-hand side and hence it decomposes in L. Since $\mathcal{H}$ is ramified at $\wp_2$ such a field L cannot be K-isomorphic to a subfield of $\mathcal{H}$. □

To get more concrete examples take the skew field $\mathcal{H}_3$ constructed above. It contains two equivalence classes of maximal orders. Then our discussion shows that the groups corresponding to them are an example of a pair of isospectral but not $\mathbf{Iso}(\mathbb{H})$-conjugate cocompact subgroups of $\mathbf{Iso}^+(\mathbb{H})$. This is the original example that occurred in Vignéras (1980b).

10.2 Unit Groups of Quadratic Forms

Here we give a general construction of certain integral orthogonal groups acting discontinuously on the Kleinian models $\mathbb{K}(q)$ introduced in Chapter 1, Section 5.

Let K_0 be a field of characteristic not equal to 2. We consider quadratic forms in 4 variables

$$Q(x_1, \ldots, x_4) = \sum_{i,j=1}^{4} a_{ij} x_i x_j$$

with coefficients in K_0 satisfying $a_{ij} = a_{ji}$. Writing

$$x = \begin{pmatrix} x_1 \\ x_2 \\ x_3 \\ x_4 \end{pmatrix} \in K_0^4$$

we may express Q as

$$Q(x) = x^t S_Q x$$

with the symmetric 4×4-matrix $S_Q = (a_{ij})$ which has coefficients in K_0. If $g \in \mathbf{GL}(4, K_0)$ then the symmetric matrix corresponding to the equivalent form $Q \circ g$ is $S_{Q \circ g} = g^t S_Q g$.

The quadratic form Q is said to be non-degenerate if $\det S_Q \neq 0$. It is called K_0-isotropic if there is an $x \in K_0^4$ with $x \neq 0$ and $Q(x) = 0$, and it is called K_0-anisotropic otherwise.

Let K_0 now be a totally real algebraic number field, that is $V_{\mathbb{C}}(K_0) = \emptyset$ and $|V_{\mathbb{R}}(K_0)| = r_1$ say. Here we keep the notation about the sets of real and complex places introduced in the previous Section 10.1. We write

$$V_{\mathbb{R}}(K_0) = \{\iota_1, \ldots, \iota_{r_1}\} \tag{2.1}$$

for the set of real places of K_0 which are thought of as embeddings

$$\iota_\ell : K_0 \to \mathbb{R}.$$

Examples of totally real fields are $\mathbb{Q}$, real-quadratic extensions of $\mathbb{Q}$ and also $\mathbb{Q}(\zeta + \zeta^{-1})$ where ζ is an m-th root of unity.

Given the quadratic form Q over the totally real algebraic number field K_0 we may consider the following r_1 quadratic forms over $\mathbb{R}$:

$$Q^{[\iota]}(x) = x^t \, \iota(S_Q) \, x \qquad (x \in \mathbb{R}^4, \ \iota \in V_{\mathbb{R}}(K_0)). \tag{2.2}$$

The embeddings ι are applied entrywise to the symmetric matrix S_Q.

Definition 2.1. We say that the quadratic form Q satisfies the hyperbolic signature condition if there is a $\kappa \in V_{\mathbb{R}}(K_0)$ so that $Q^{[\kappa]}$ is of signature (1,3) and if the $Q^{[\iota]}$ for all other embeddings ι are (positive or negative) definite.

In the situation of the definition there is a $g \in \mathbf{GL}(4, \mathbb{R})$ so that

$$Q^{[\kappa]} \circ g(x) = x_1^2 - x_2^2 - x_3^2 - x_4^2.$$

Going back to the map Ψ in Definition 1.3.9 we may use one of these g to get a group homomorphism

$$\Psi_0 : \mathbf{SL}(2, \mathbb{C}) \to \mathbf{PO_4}(\mathbb{R}, Q^{[\kappa]}). \tag{2.3}$$

The homomorphisms Ψ_0 for the various g are equal up to conjugation in $\mathbf{PO_4}(\mathbb{R}, Q^{[\kappa]})$, we fix one of them. Ψ_0 has the center of $\mathbf{SL}(2, \mathbb{C})$ as kernel and finite cokernel.

Examples of quadratic forms satisfying the hyperbolic signature condition are

$$q_1(x) = x_1^2 - x_2^2 - x_3^2 - x_4^2 \tag{2.4}$$

with $K_0 = \mathbb{Q}$ and

$$Q_2(x) = x_1^2 - \sqrt{5}x_2^2 - \sqrt{5}x_3^2 - \sqrt{5}x_4^2 \tag{2.5}$$

with $K_0 = \mathbb{Q}(\sqrt{5})$.

Note that in case $r_1 > 1$ any quadratic form satisfying the hyperbolic signature condition is automatically K_0-anisotropic because it is anisotropic at at least one real place.

If Q satisfies the hyperbolic signature condition as in the above definition the Kleinian model of 3-dimensional hyperbolic space associated with $Q^{[\kappa]}$ can be considered. It is discussed in Chapter 1, Section 5. The definition is

$$\mathbb{K}(Q^{[\kappa]}) = \{ [x] \in \mathbb{P}^3\mathbb{R} \mid Q^{[\kappa]}(x) > 0 \}.$$

The group of isometries was identified with $\mathbf{PO_4}(\mathbb{R}, Q^{[\kappa]})$ acting linearly on $\mathbb{K}(Q^{[\kappa]})$. As in Proposition 1.5.3 we have associated to the map Ψ_0 of (2.3) an isometry

$$\psi_0 : \mathbb{H} \to \mathbb{K}(Q^{[\kappa]}) \tag{2.6}$$

so that ψ_0 is Ψ_0-equivariant.

The discrete orthogonal groups constructed with the help of Q are described in the following theorem.

Theorem 2.2. *Let K_0 be a totally real number field with ring of integers $\mathcal{O}_{K_0}$. Let Q be a quadratic form in 4 variables over K_0 satisfying the hyperbolic signature condition. Let κ be the real place of K_0 so that $Q^{[\kappa]}$ is of signature (1,3). Define*

$$\Gamma(\mathcal{O}_{K_0}, Q) := \kappa(\mathbf{PO}_4(\mathcal{O}_{K_0}, Q)) < \mathbf{PO}_4(\mathbb{R}, Q^{[\kappa]}).$$

Then the following hold:

(1) $\Gamma(\mathcal{O}_{K_0}, Q)$ is a discrete subgroup of $\mathbf{PO}_4(\mathbb{R}, Q^{[\kappa]})$.
(2) $\Gamma(\mathcal{O}_{K_0}, Q)$ always has finite covolume acting on $\mathbb{K}(Q^{[\kappa]})$.
(3) $\Gamma(\mathcal{O}_{K_0}, Q)$ is cocompact if and only if Q is K_0-anisotropic.

The integral groups of the above theorem are defined as:

$$\mathbf{O}_4(\mathcal{O}_{K_0}, Q) := \mathbf{O}_4(K_0, Q) \cap \mathbf{GL}(4, \mathcal{O}_{K_0}), \tag{2.7}$$

$$\mathbf{PO_4}(\mathcal{O}_{K_0}, Q) := \mathbf{O}_4(\mathcal{O}_{K_0}, Q)/\mathbf{Z}(\mathcal{O}_{K_0}), \tag{2.8}$$

where $\mathbf{Z}(\mathcal{O})$ is the group of $\mathcal{O}$-invertible scalars in $\mathbf{O}_4(\mathcal{O}, Q)$.

We shall prove the above theorem by comparing the groups $\Gamma(\mathcal{O}_{K_0}, Q)$ under the homomorphism Ψ_0 of (2.3) with the groups constructed in the previous section.

We start off with some further remarks on Clifford algebras. We use the terminology of Chapter 1, Section 3. Let K_0 be a field of characteristic $\neq 2$ and $a, b, c \in K_0$. We assume that

$$d := abc \notin K_0^2. \tag{2.9}$$

We take a four-dimensional vector space E over K_0 with basis e_1, e_2, e_3, e_4 and equip it with the quadratic form

$$Q_{a,b,c}(x_1e_1 + x_2e_2 + x_3e_3 + x_4e_4) = x_1^2 + ax_2^2 + bx_3^2 + cx_4^2. \tag{2.10}$$

We write $\mathcal{C}(K_0, Q_{a,b,c})$ for the Clifford algebra of $Q_{a,b,c}$. Since we vary the fields of coefficients in this section we include them in the symbol for the Clifford algebra, changing the notation of Section 1.3 slightly. The algebra of even elements $\mathcal{C}^+(K_0, Q_{a,b,c})$ is an 8-dimensional K_0-subalgebra of $\mathcal{C}(K_0, Q_{a,b,c})$. We put

$$f := e_1e_2e_3e_4, \qquad K := K_0(f). \tag{2.11}$$

The element f satisfies $f^2 = abc$ and K is equal to the center of $\mathcal{C}^+(K_0, Q_{a,b,c})$. By our assumption (2.9) the field K is a quadratic extension of K_0 and the K-algebra $\mathcal{C}^+(K_0, Q_{a,b,c})$ is 4-dimensional with basis

$$1, \qquad i := e_1e_2, \qquad j := e_1e_3, \qquad k := e_3e_2. \tag{2.12}$$

These elements satisfy the usual relations (1.2) of the standard basis of a quaternion algebra over K, namely

$$i^2 = -a, \qquad j^2 = -b, \qquad ij = -ji = k. \tag{2.13}$$

We get an obvious K-algebra isomorphism

$$\Theta : \mathcal{H}(-a, -b; K) \to \mathcal{C}^+(K_0, Q_{a,b,c}). \tag{2.14}$$

A first important property of our construction is given in the following lemma.

Lemma 2.3. *Let K_0 be any field of characteristic $\neq 2$. Let a, b, $c \in K_0$ be so that $abc \notin K_0^2$ and put $K := K_0(\sqrt{abc})$. Let $Q_{a,b,c}$ be the quadratic form in 4 variables over K_0 defined in (2.10). Then the following statements are equivalent:*

(1) The quadratic form $Q_{a,b,c}$ is K_0-anisotropic.
(2) The quaternion algebra $\mathcal{H}(-a, -b; K)$ is a skew field.

Proof. Assume first that $Q_{a,b,c}$ is isotropic over K_0. This is the case if and only if the equation

$$abx_1^2 + bx_2^2 + ax_3^2 + abcx_4^2 = 0 \tag{2.15}$$

has a non-trivial K_0-solution. Then the element $\alpha := x_4\sqrt{abc} + x_3i + x_2j + x_1k$ is a non-zero element of $\mathcal{H}(-a, -b; K)$ having norm 0. This α cannot be invertible in $\mathcal{H}(-a, -b; K)$.

To prove the reverse implication we assume that $\mathcal{H}(-a, -b; K) \cong M(2, K)$. Since $\mathcal{H}(-a, -b; K) \cong \mathcal{H}(-a, -b; K_0) \otimes_{K_0} K$ the K_0-quaternion algebra

$\mathcal{H}(-a,-b;K_0)$ has to contain an element $\alpha = x_1 + x_2 i + x_3 j + x_4 k$ with $\alpha^2 = abc$. This equation amounts to

$$x_1 = 0, \qquad -ax_2^2 - bx_3^2 - abx_4^2 = abc.$$

We see then that (2.15) has a non-zero solution. □

We proceed with our general considerations. We introduce the fields

$$L_0 := K_0(\sqrt{-a}, \sqrt{-b}), \qquad L := K(\sqrt{-a}, \sqrt{-b}) \tag{2.16}$$

We have $L = L_0(\sqrt{abc})$. We assume that L is a quadratic extension field of L_0, this amounts to

$$c \notin L_0^2. \tag{2.17}$$

We consider the L-algebra isomorphism

$$\Phi : \mathcal{H}(-a,-b;L) \to M(2,L)$$

given in (1.3). Over the field L_0 we have the linear isomorphism

$$\tau : E \otimes_{K_0} L_0 \to E \otimes_{K_0} L_0 \tag{2.18}$$

which on the basis $\{e_1, ..., e_4\}$ is given by the matrix

$$\tau = \begin{pmatrix} 1 & 0 & 0 & 0 \\ 0 & \sqrt{-a}^{-1} & 0 & 0 \\ 0 & 0 & \sqrt{-b}^{-1} & 0 \\ 0 & 0 & 0 & \sqrt{ab} \end{pmatrix}. \tag{2.19}$$

Notice that $Q_{a,b,c} \circ \tau = Q_{-1,-1,abc}$. These maps give rise to the following diagram of K_0-algebra homomorphisms.

$$\begin{array}{ccc} \mathcal{H}(-a,-b;K) & \xrightarrow{\Theta} & \mathcal{C}^+(K_0, Q_{a,b,c}) \\ \downarrow{\scriptstyle \Phi} & & \downarrow \\ M(2,L) & \xrightarrow{\Theta} & \mathcal{C}^+(L_0, Q_{a,b,c}) \\ \downarrow{\scriptstyle id} & & \downarrow{\scriptstyle \tau} \\ M(2,L) & \xrightarrow{\psi} & \mathcal{C}^+(L_0, Q_{-1,-1,abc}) \end{array} \tag{2.20}$$

The map ψ is defined in Proposition 1.3.5. The map Θ in the second row is defined analogously to the Θ of the first row. A quick check through the definitions shows that diagram (2.20) is commutative.

The K-algebra isomorphism Θ described in (2.14) gives rise to a group isomorphism

$$\Theta : \mathcal{H}(-a,-b;K)^1 \to \mathbf{Spin_4}(K_0, Q_{a,b,c}). \tag{2.21}$$

Note that the first condition in Definition 1.3.6 of the spin-group is empty in our case and that the main anti-involution * of $\mathcal{C}(K_0, Q_{a,b,c})$ induces the usual conjugation on $\mathcal{H}(-a,-b;K)$. We write

$$\Lambda : \mathbf{Spin_4}(K_0, Q_{a,b,c}) \to \mathbf{O}_4(K_0, Q_{a,b,c}) \tag{2.22}$$

for the homomorphism defined in (1.3.10). The following lemma helps to settle questions of commensurability.

Lemma 2.4. *Let K_0 be an algebraic number field with ring of integers $\mathcal{O}_{K_0}$. Let a, b, $c \in \mathcal{O}_{K_0}$ be so that $abc \notin K_0^2$ and put $K := K_0(\sqrt{abc})$. Assume also (2.17). The ring*

$$\mathcal{O}_0 := \mathcal{O}_{K_0}[\sqrt{abc}] = \{ \ \alpha + \beta\sqrt{abc} \in K \ \ | \ \ \alpha, \beta \in \mathcal{O}_{K_0} \ \} \tag{2.23}$$

is an order in K. Let

$$\mathfrak{R} := \{ \ \alpha_1 + \alpha_2 i + \alpha_3 j + \alpha_4 k \in \mathcal{H}(-a,-b;K) \ \ | \ \ \alpha_1, ..., \alpha_4 \in \mathcal{O}_0 \ \}. \tag{2.24}$$

Then $\mathfrak{R}$ is an $\mathcal{O}_0$-order in $\mathcal{H}(-a,-b;K)$. Define the congruence group modulo 2 of $\mathfrak{R}^1$ by

$$\mathfrak{R}^1[2] := \{ \ \gamma \in \mathfrak{R}^1 \ \ | \ \ \gamma - 1 \in 2 \cdot \mathfrak{R} \ \}. \tag{2.25}$$

Then $\mathfrak{R}^1[2]$ is a subgroup of finite index in $\mathfrak{R}^1$. Let $Q_{a,b,c}$ be the quadratic form in 4 variables over K_0 defined in (2.10). Then

$$\Lambda(\Theta(\mathfrak{R}^1[2])) < \mathbf{O}_4(\mathcal{O}_{K_0}, Q_{a,b,c}). \tag{2.26}$$

Proof. We only prove statement (2.26), everything else is easy. Define the fields L_0, L as above, put $d := abc$ and consider the following commutative diagram derived from (2.20).

$$\begin{array}{ccccc}
\mathcal{H}(-a,-b;K)^1 & \xrightarrow{\Theta} & \mathbf{Spin}_4(K_0, Q_{a,b,c}) & \xrightarrow{\Lambda} & \mathbf{O}_4(K_0, Q_{a,b,c}) \\
\downarrow & & \downarrow & & \downarrow \\
\mathcal{H}(-a,-b;L)^1 & \xrightarrow{\Theta} & \mathbf{Spin}_4(L_0, Q_{a,b,c}) & \xrightarrow{\Lambda} & \mathbf{O}_4(L_0, Q_{a,b,c}) \\
\downarrow \Phi & & \downarrow \tau & & \downarrow \tau \\
\mathbf{SL}(2, L) & \xrightarrow{\psi} & \mathbf{Spin}_4(L_0, Q_{-1,-1,d}) & \xrightarrow{\Lambda} & \mathbf{O}_4(L_0, Q_{-1,-1,d})
\end{array}$$

Notice that the composition $\Lambda \circ \psi = \Psi$ of the maps of the last line occurs in Proposition 1.3.11. There we found formulas expressing the map Ψ in our coordinates. Define now orders in L and $\mathcal{H}(-a,-b;L)$ similar to $\mathcal{O}_0$ and $\mathfrak{R}$. The statement analogous to (2.26) but concerning the last line is obvious from Proposition 1.3.11. Statement (2.26) then follows by intersection with the images of the downwards maps. □

We now put everything together and consider the case of relevance here.

Proposition 2.5. *Let K_0 be a totally real number field with ring of integers $\mathcal{O}_{K_0}$. Let Q be a quadratic form in 4 variables over K_0 satisfying the hyperbolic signature condition. Let κ be the real place of K_0 so that $Q^{[\kappa]}$ is of signature (1,3). Put $d := \det S_Q$ and $K := K_0(\sqrt{d})$. Then K is a quadratic extension of K_0 having exactly one pair of complex places v_0, $\bar{v}_0$, they are the extensions of κ. The K_0-algebra $\mathcal{C}^+(K_0, Q)$ is a quaternion algebra $\mathcal{H}(Q)$ over its center K. It is extended from a quaternion algebra over K_0. The K-algebra $\mathcal{H}(Q)$ is a skew field if and only if Q is K_0 anisotropic. Choose maps*

$$\varphi : \mathcal{H}(Q) \otimes_K K_{v_0} \to M(2, \mathbb{C}), \qquad \Psi_0 : \mathbf{SL}(2, \mathbb{C}) \to \mathbf{O}_4(\mathbb{R}, Q^{[\kappa]})$$

where φ is a $K_{v_0} = \mathbb{C}$-algebra isomorphism and Ψ_0 is as in (2.3). Let $\mathfrak{R}$ be an order in $\mathcal{H}(Q)$. Then the group $\Psi_0(\varphi(\mathfrak{R}^1))$ is $\mathbf{O}_4(\mathbb{R}, Q^{[\kappa]})$-conjugate to a group commensurable with $\kappa(\mathbf{O}_4(\mathcal{O}_{K_0}, Q^{[\kappa]}))$.

Proof. We start off with the totally real field K_0 and the quadratic form Q satisfying the hyperbolic signature condition. We find $\lambda \in K_0$, a, b, $c \in \mathcal{O}_{K_0}$ and $g \in \mathbf{GL}(4, K_0)$ such that $Q_{a,b,c} = \lambda Q \circ g$. It is easy to see that $Q_{a,b,c}$ also satisfies the hyperbolic signature condition. A straightforward argument using Proposition 2.8.5 shows that the proposition follows for Q once it is proved for $Q_{a,b,c}$.

Notice that all assumptions made in Lemmas 2.3, 2.4 are automatically satisfied here. Hence we apply them and obtain all statements of our proposition. Perhaps one final remark about the commensurability statement is in order. By an argument analogous to Theorem 1.2, (4) $\kappa(\mathbf{O}_4(\mathcal{O}_{K_0}, Q^{[\kappa]}))$ is a discrete subgroup of $\mathbf{O}_4(\mathbb{R}, Q^{[\kappa]})$. By Lemma 2.4 there is a subgroup of finite index in $\Psi_0(\varphi(\mathfrak{R}^1))$ which is $\mathbf{O}_4(\mathbb{R}, Q^{[\kappa]})$-conjugate to a subgroup which we call Γ of $\kappa(\mathbf{O}_4(\mathcal{O}_{K_0}, Q^{[\kappa]}))$. Since the isometry ψ_0 is Ψ_0-equivariant the group Γ has finite covolume and the index $[\kappa(\mathbf{O}_4(\mathcal{O}_{K_0}, Q^{[\kappa]})) : \Gamma]$ is finite. □

Proof of Theorem 2.2. This is now obvious from Proposition 2.5. □

Proposition 2.5 shows that the set of cofinite or cocompact discrete subgroups of $\mathbf{Iso}(\mathbb{H})$ constructed in Theorem 2.2 is up to conjugacy and commensurability a subset of the set of groups constructed by the use of quaternion algebras in the previous section. Let K be a number field with exactly one pair of complex places, K_0 a totally real subfield with $[K : K_0] = 2$ and $\mathcal{H}$ a quaternion algebra over K which is extended from K_0. Then it is easy to see by the constructions leading to Proposition 2.5 that any discrete subgroup of $\mathbf{Iso}(\mathbb{H})$ constructed in Theorem 1.2 can up to conjugacy and commensurability also be obtained as an orthogonal group.

It is usually quite difficult to describe the relative position of the groups $\Psi_0(\varphi(\mathfrak{R}^1))$, $\kappa(\mathbf{O}_4(\mathcal{O}_{K_0}, Q^{[\kappa]}))$ of Proposition 2.5. We shall now give a case where this is easy. We take $K_0 = \mathbb{Q}$, $m \in \mathbb{N}$ square-free and consider the rational quadratic form

$$Q_m(x_1, x_2, x_3, x_4) := -x_1^2 - mx_2^2 + x_3x_4. \tag{2.27}$$

This form is rationally equivalent to $-x_1^2 - mx_2^2 - x_3^2 + x_4^2$. Hence Proposition 1.3.10 provides us with an isomorphism

$$\Psi_m : \mathbf{PSL}(2, \mathbb{Q}(\sqrt{-m})) \to \mathbf{SO}_4^+(\mathbb{Q}, Q_m) \tag{2.28}$$

where $\mathbf{SO}_4^+(\mathbb{Q}, Q_m)$ is the kernel of the spinor norm homomorphism Σ introduced in (1.3.12). In coordinates the formula analogous to that of Proposition 1.3.11 is

$$\Psi_m(g) = \begin{pmatrix} \frac{a\bar{d}+\bar{a}d+b\bar{c}+\bar{b}c}{2} & \frac{(a\bar{d}-\bar{a}d+b\bar{c}-\bar{b}c)\sqrt{-m}}{2} & \frac{a\bar{c}+\bar{a}c}{2} & \frac{b\bar{d}+\bar{b}d}{2} \\ \frac{a\bar{d}-\bar{a}d+b\bar{c}-\bar{b}c}{2\sqrt{-m}} & \frac{a\bar{d}+\bar{a}d-b\bar{c}-\bar{b}c}{2} & \frac{a\bar{c}-\bar{a}c}{2\sqrt{-m}} & \frac{b\bar{d}-\bar{b}d}{2\sqrt{-m}} \\ a\bar{b}+\bar{a}b & (a\bar{b}-\bar{a}b)\sqrt{-m} & a\bar{a} & b\bar{b} \\ c\bar{d}+\bar{c}d & (c\bar{d}-\bar{c}d)\sqrt{-m} & c\bar{c} & d\bar{d} \end{pmatrix} \tag{2.29}$$

for $g = \begin{pmatrix} a & b \\ c & d \end{pmatrix} \in \mathbf{PSL}(2, \mathbb{Q}(\sqrt{-m}))$.

We define $\mathbf{SO}_4^+(\mathbb{Z}, Q_m) := \mathbf{SO}_4(\mathbb{Z}, Q_m) \cap \mathbf{SO}_4^+(\mathbb{Q}, Q_m)$ and also

$$\mathbf{PSO}_4^+(\mathbb{Z}, Q_m) := \pi(\mathbf{SO}_4^+(\mathbb{Z}, Q_m)) < \mathbf{PO}_4(\mathbb{Z}, Q_m) \tag{2.30}$$

where $\pi : \mathbf{O}_4(\mathbb{Z}, Q_m) \to \mathbf{PO}_4(\mathbb{Z}, Q_m)$ is the quotient map. Note that the restriction of π to $\mathbf{SO}_4^+(\mathbb{Z}, Q_m)$ is injective. This can be read off from Proposition 1.3.10 and the fact that the entry $a\bar{a}$ in (2.29) is always non-negative, see also Artin (1957), Chapter 5.

Proposition 2.6. *For $m \in \mathbb{N}$ square-free let Q_m be the quadratic form (2.27). Then $\mathcal{R}_m = \mathbb{Z}[\sqrt{-m}]$ is an $\mathbb{Z}$-order in $\mathbb{Q}(\sqrt{-m})$. The isomorphism (2.28) gives rise to an isomorphism*

$$\Psi_m : \mathbf{PSL}(2, \mathcal{R}_m) \to \mathbf{SO}_4^+(\mathbb{Z}, Q_m) \cong \mathbf{PSO}_4^+(\mathbb{Z}, Q_m).$$

Let $\mathcal{O}$ be the ring of integers in $\mathbb{Q}(\sqrt{-m})$ then

$$[\mathbf{PSL}(2, \mathcal{O}) : \mathbf{PSL}(2, \mathcal{R}_m)] = \begin{cases} 1 & m \equiv 1, 2 \bmod 4, \\ 6 & m \equiv 7 \bmod 8, \\ 10 & m \equiv 3 \bmod 8. \end{cases}$$

Proof. The isomorphism (2.28) being established we have to show for a given $g \in \mathbf{PSL}(2, \mathbb{Q}(\sqrt{-m}))$ that the statements $g \in \mathbf{PSL}(2, \mathcal{R}_m)$ and $\Psi_m(g) \in \mathbf{SO}_4(\mathbb{Z}, Q_m)$ are equivalent. A glance at (2.29) shows that $g \in \mathbf{PSL}(2, \mathcal{R}_m)$ implies $\Psi_m(g) \in \mathbf{SO}_4(\mathbb{Z}, Q_m)$.

To show the remaining direction take $g = \begin{pmatrix} a & b \\ c & d \end{pmatrix} \in \mathbf{PSL}(2, \mathbb{Q}(\sqrt{-m}))$ and assume $\Psi_m(g) \in \mathbf{SO}_4(\mathbb{Z}, Q_m)$. Define $T = < 1/2, \sqrt{-m}/2m >$ where the brackets stand for the $\mathcal{R}_m$-module generated by the elements indicated. Note that T coincides with the $\mathbb{Z}$-module generated by $1/2$ and $\sqrt{-m}/2m$. From (2.29) we get

$$\text{(2.31)} \qquad a\bar{a},\ b\bar{b},\ c\bar{c},\ d\bar{d},\ a\bar{c},\ b\bar{d} \in \mathcal{R}_m, \qquad a\bar{b},\ c\bar{d},\ a\bar{d},\ b\bar{c} \in T.$$

Let us first consider the case $m \equiv 1, 2$ modulo 4. Then $\mathcal{R}_m$ coincides with the full ring of integers in $\mathbb{Q}(\sqrt{-m})$ and the prime 2 is ramified in $\mathbb{Q}(\sqrt{-m})$. If $M \subset \mathbb{Q}(\sqrt{-m})$ is a fractional ideal of $\mathcal{R}_m$ we call a prime ideal $\wp$ a divisor of M if $\wp$ occurs in the prime ideal decomposition of M. We then write $\wp|_+M$ if the exponent of $\wp$ in the decomposition of M is positive and $\wp|_-M$ if it is negative. The prime ideal divisors of T are the prime ideal dividing $< 2 >$ and the $< p, \sqrt{-m} >$ where $p \in \mathbb{Z}$ is a prime dividing m. These prime ideals $\wp$ all satisfy $\overline{\wp} = \wp$. Assume now that a, b, c, d are all non-zero and take a prime $\wp$ with $\wp|_- < a >$ which is not a divisor of T. From (2.31) we infer $\overline{\wp}|_+ < a, b, c, d >$. This contradicts $ad - bc = 1$. If $\wp$ is a divisor of T with $\wp|_- < a >$ then $a\bar{a} \in \mathcal{R}_m$ implies $\overline{\wp}|_+ < a >$. This leads to a contradiction because $\overline{\wp} = \wp$. Hence we have proved $a \in \mathcal{R}_m$. Similarly we find b, c, $d \in \mathcal{R}_m$. The cases that one or two of the entries of g are zero is also treated by the same argument.

Assume now that $m \equiv 3$ modulo 4. Then $\mathcal{R}_m$ is no longer a Dedekind ring and we have to be careful using prime ideal decompositions. The arguments used above show that a, b, c, $d \in \mathcal{O}$ where $\mathcal{O}$ is the full ring of integers in $\mathbb{Q}(\sqrt{-m})$. From (2.29) we see that

$$\text{(2.32)} \qquad a\bar{c},\ b\bar{d},\ a\bar{d} + b\bar{c} \in \mathcal{R}_m.$$

The ring $\mathcal{O}/2\mathcal{O}$ is isomorphic to $\mathbb{F}_4$ (the field with 4 elements) if $m \equiv 3$ mod 8 and to $\mathbb{F}_2 \times \mathbb{F}_2$ if $m \equiv 7$ mod 8. The ring $\mathcal{R}_m$ is always mapped to the prime field $\mathbb{F}_2 \subset \mathcal{O}/2\mathcal{O}$ under the quotient map. An element $g \in \mathbf{PSL}(2, \mathcal{O})$ gives by reduction rise to an element $g \in \mathbf{PSL}(2, \mathcal{O}/2\mathcal{O})$. A simple case by case study (in the finite group $\mathbf{PSL}(2, \mathcal{O}/2\mathcal{O})$) shows that (2.32) implies that all entries of the reduced g lie in $\mathbb{F}_2 = \mathcal{R}_m/2\mathcal{O}$.

The first statement of the proposition is proved now. After the remarks just made the rest is an exercise. □

We shall discuss now some examples of cofinite discrete subgroups of $\mathbf{Iso}(\mathbb{H})$ constructed by the procedure outlined in Theorem 2.2. We shall describe a fundamental region, generators and relations for the groups. We shall further discuss the cofinite subgroups of $\mathbf{SL}(2, \mathbb{C})$ corresponding to them as described in Proposition 2.5.

We start off by setting some general conventions. We define

(2.33) $$(\mathbb{R}^4 \setminus \{0\})^+ := \left\{ \begin{pmatrix} x_1 \\ x_2 \\ x_3 \\ x_4 \end{pmatrix} \in \mathbb{R}^4 \setminus \{0\} \;\middle|\; \begin{array}{l} x_1 > 0 \text{ or } x_i > 0 \\ \text{if } x_1 = ... = x_{i-1} = 0 \end{array} \right\}.$$

Three-dimensional real projective space $\mathbb{P}^3\mathbb{R}$ is covered under the projection map by $(\mathbb{R}^4 \setminus \{0\})^+$. We shall denote points of $\mathbb{P}^3\mathbb{R}$ as previously by $[x]$ where $x \in \mathbb{R}^4 \setminus \{0\}$ with the additional convention that we always take the representative x in $(\mathbb{R}^4 \setminus \{0\})^+$.

We shall consider quadratic forms Q on $\mathbb{R}^4$ of signature (1,3). For elements of projective 3-space we shall often write

(2.34) $$[x]_a$$

with the understanding that $x \in (\mathbb{R}^4 \setminus \{0\})^+$ and $Q(x) = a$. The Kleinian model $\mathbb{K}(Q)$ of hyperbolic space corresponding to Q is discussed in Section 1.5. We have

$$\mathbb{K}(Q) = \{ \ [x] \in \mathbb{P}^3\mathbb{R} \ \ | \ \ Q(x) > 0 \ \}.$$

The boundary of the Kleinian model is

$$\partial\mathbb{K}(Q) = \{ \ [x] \in \mathbb{P}^3\mathbb{R} \ \ | \ \ Q(x) = 0 \ \}.$$

An element $[S] \in \mathbb{P}^3\mathbb{R}$ with $Q(S) < 0$ defines the hyperbolic plane

(2.35) $$\mathbb{P}_{[S]} := \{ \ [y] \in \mathbb{K}(Q) \ \ | \ \ Q(S, y) = 0 \ \}.$$

Here we write $Q(x, y)$ for the bilinear form associated with Q. The two connected components of $\mathbb{K}(Q) \setminus \mathbb{P}_{[S]}$ are

(2.36) $$\mathbb{P}^+_{[S]} := \{ \ [y] \in \mathbb{K}(Q) \mid Q(S, y) > 0 \ \},$$

(2.37) $$\mathbb{P}^-_{[S]} := \{ \ [y] \in \mathbb{K}(Q) \mid Q(S, y) < 0 \ \}.$$

Notice that this definition makes sense due to our sign convention under (2.33). The reflection $\sigma_{[S]}$ in the plane $\mathbb{P}_{[S]}$ is given by

$$\sigma_{[S]}(v) = v - 2 \cdot \frac{Q(v, S)}{Q(S)} \cdot S \qquad (v \in \mathbb{R}^4).$$

A tetrahedron $\mathcal{T} \subset \mathbb{K}(Q)$ is the hyperbolic convex hull of 4 non-coplanar points $P_1, ..., P_4 \in \mathbb{K}(Q) \cup \partial\mathbb{K}(Q)$. We write

(2.38) $$\mathcal{T} = \mathcal{T}(P_1, ..., P_4).$$

The points $P_1, ..., P_4$ are the vertices of $\mathcal{T}$. By solving some linear equations we find four planes $\mathbb{P}_{[S_1]}, ..., \mathbb{P}_{[S_4]}$ so that $P_j \in \mathbb{P}_{[S_i]}$ if $j \neq i$. If $P_j \in \partial\mathbb{K}(Q)$ then $P_j \in \mathbb{P}_{[S_i]}$ has to be interpreted as in (2.35). The $\mathbb{P}_{[S_1]}, ..., \mathbb{P}_{[S_4]}$ are called boundary planes of $\mathcal{T}$, in fact the boundary of $\mathcal{T}$ consists of the union of certain triangles in the $\mathbb{P}_{[S_i]}$. We choose $\epsilon_1, ..., \epsilon_4 \in \{\pm 1\}$ so that $P_i \in \mathbb{P}^{\epsilon_i}_{[S_i]}$. We find that

$$(2.39) \qquad \mathcal{T}(P_1, ..., P_4) = \overline{\bigcap_{i=1}^{4} \mathbb{P}^{\epsilon_i}_{[S_i]}}.$$

The bar in formula (2.39) stands for the closure in $\mathbb{K}(Q)$. Notice that formula (2.39) shows that $\mathcal{T}(P_1, ..., P_4)$ may be described as the image in $\mathbb{P}^3\mathbb{R}$ of a subset of $(\mathbb{R}^4 \setminus \{0\})^+$ given by four linear inequalities. See the tetrahedron $\mathcal{T}$ in the example 1 below. For $i, j \in \{1, ..., 4\}$ with $i \neq j$ we define $\alpha = \angle_{\mathcal{T}}(i, j)$ by

$$(2.40) \qquad \cos\alpha = \cos(\angle_{\mathcal{T}}(i,j)) := \epsilon_i \epsilon_j \frac{Q(S_1, S_2)}{\sqrt{Q(S_i)Q(S_j)}}.$$

The square root of the positive number $Q(S_i)Q(S_j)$ is taken to be a positive real number and α is measured between 0 and 180 degrees or between 0 and π. Notice that the sign convention under (2.33) fixes signs on the right-hand side of (2.40). By formula (1.5.13) $\angle_{\mathcal{T}}(i, j)$ is one of the two angles between $\mathbb{P}_{[S_i]}$ and $\mathbb{P}_{[S_j]}$. It is in fact the inner angle between the corresponding faces of the tetrahedron. This is an exercise in hyperbolic geometry which may be done for a special form Q and a specific plane $\mathbb{P}_{[S_i]}$. We skip the details. The tetrahedron $\mathcal{T}(P_1, ..., P_4)$ has six inner angles between faces, we collect them in the symbol

$$(2.41) \quad \angle(\mathcal{T}) = [\angle_{\mathcal{T}}(1,2), \angle_{\mathcal{T}}(1,3), \angle_{\mathcal{T}}(1,4), \angle_{\mathcal{T}}(2,3), \angle_{\mathcal{T}}(2,4), \angle_{\mathcal{T}}(3,4)].$$

In the following examples we employ the notation $\mathcal{P} = \mathcal{P}(P_1, \ldots P_n)$ to be the convex hull of the non-coplanar points $P_1, \ldots, P_n \in \mathbb{K}$.

Example 1 :

Quadratic form: $Q(x_1, x_2, x_3, x_4) = -x_1^2 - x_2^2 + x_3 x_4$.
Quaternion algebra: $K_0 = \mathbb{Q}$, $K = \mathbb{Q}(i)$, $\mathcal{H}(Q) = M(2, \mathbb{Q}(i))$.
Discrete group: $\mathbf{PO}_4(\mathbb{Z}, Q) < \mathbf{PO}_4(\mathbb{R}, Q)$ (cofinite but not cocompact).
Fundamental domain: $\mathcal{T} = \mathcal{T}(P_1, ..., P_4)$ where

$$P_1 = \begin{bmatrix}1\\1\\2\\2\end{bmatrix}_2, \quad P_2 = \begin{bmatrix}1\\0\\2\\2\end{bmatrix}_3, \quad P_3 = \begin{bmatrix}0\\0\\0\\1\end{bmatrix}_0, \quad P_4 = \begin{bmatrix}0\\0\\1\\1\end{bmatrix}_1.$$

$\mathcal{T}$ has the faces $\mathbb{P}_{[S_1]}, ..., \mathbb{P}_{[S_4]}$ where

$$S_1 = \begin{bmatrix}0\\1\\0\\0\end{bmatrix}_{-1}, \quad S_2 = \begin{bmatrix}1\\-1\\0\\0\end{bmatrix}_{-2}, \quad S_3 = \begin{bmatrix}1\\0\\0\\1\end{bmatrix}_{-1}, \quad S_4 = \begin{bmatrix}0\\0\\1\\-1\end{bmatrix}_{-1}.$$

$$\angle(\mathcal{T}) = [45°, 90°, 90°, 45°, 90°, 60°].$$

Generators: $\sigma_1, \sigma_2, \sigma_3, \sigma_4$ with

$$\sigma_1 = \sigma_{[S_1]} = \begin{pmatrix} 1 & 0 & 0 & 0 \\ 0 & -1 & 0 & 0 \\ 0 & 0 & 1 & 0 \\ 0 & 0 & 0 & 1 \end{pmatrix}, \qquad \sigma_2 = \sigma_{[S_2]} = \begin{pmatrix} 0 & 1 & 0 & 0 \\ 1 & 0 & 0 & 0 \\ 0 & 0 & 1 & 0 \\ 0 & 0 & 0 & 1 \end{pmatrix},$$

$$\sigma_3 = \sigma_{[S_3]} = \begin{pmatrix} -1 & 0 & 1 & 0 \\ 0 & 1 & 0 & 0 \\ 0 & 0 & 1 & 0 \\ -2 & 0 & 1 & 1 \end{pmatrix}, \qquad \sigma_4 = \sigma_{[S_4]} = \begin{pmatrix} 1 & 0 & 0 & 0 \\ 0 & 1 & 0 & 0 \\ 0 & 0 & 0 & 1 \\ 0 & 0 & 1 & 0 \end{pmatrix}.$$

Relations: $\sigma_1^2 = \sigma_2^2 = \sigma_3^2 = \sigma_4^2 = (\sigma_1\sigma_2)^4 = (\sigma_1\sigma_3)^2 = (\sigma_1\sigma_4)^2 = (\sigma_2\sigma_3)^4 = (\sigma_2\sigma_4)^2 = (\sigma_3\sigma_4)^3 = 1$.
Commensurability class in $\mathbf{SL}(2,\mathbb{C})$: $\mathbf{SL}(2,\mathbb{Z}[i])$.

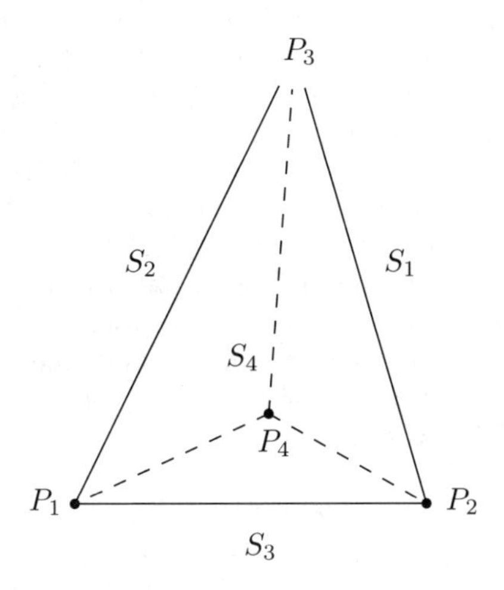

Fig. 2.1 A tetrahedron $\mathcal{T}$ with one vertex at infinity

Additional remarks: Let Γ be the subgroup of $\mathbf{PO}_4(\mathbb{Z}, Q)$ generated by $\sigma_1, \ldots, \sigma_4$. The group Γ acts discontinuously on $\mathbb{K}(Q)$ and has the tetrahedron $\mathcal{T}$ as a fundamental domain. A schematical picture of $\mathcal{T}$ is shown in figure 2.1. The hyperbolic volume of $\mathcal{T}$ is easily seen to be finite, in fact we have $v(\mathcal{T}) = 0.07633...$, see $\mathbf{CT}(1)$ in Section 4. From this and Proposition 2.5.6 we see that Γ is maximal discrete in $\mathbf{PO}_4(\mathbb{R}, Q)$, hence $\Gamma = \mathbf{PO}_4(\mathbb{Z}, Q)$.

We give now an alternative proof of this equality which is also applicable in the further examples of this section. By the method described in Definition 2.2.2 we find a fundamental domain $\mathcal{P} \subset \mathcal{T}$ for the now possibly bigger group $\mathbf{PO}_4(\mathbb{Z}, Q)$. The set $\mathcal{P}$ is a convex polyhedron and $\mathcal{T} = g_1\mathcal{P} \cup \ldots \cup g_m\mathcal{P}$ with

$g_1, \dots, g_n \in \mathbf{PO}_4(\mathbb{Z}, Q)$, $g_1 = 1$. The tetrahedron $\mathcal{T}$ can be described by the following inequalities

$$(2.42) \quad \mathcal{T} = \left\{ \begin{bmatrix} x_1 \\ x_2 \\ x_3 \\ x_4 \end{bmatrix} \in \mathbb{K}(Q) \;\middle|\; \begin{array}{c} x_4 > 0,\ -x_2 \leq 0,\ -x_1 + x_2 \leq 0, \\ -x_1 + \frac{x_3}{2} \geq 0,\ \frac{x_4}{2} - \frac{x_3}{2} \geq 0. \end{array} \right\}.$$

Note that this description is well defined. For $n \in \mathbb{N}$ we define

$$(2.43) \quad \mathbb{K}(\mathbb{Z}, Q)_n := \left\{ x \in \mathbb{K}(Q) \;\middle|\; \begin{array}{l} x \text{ has a representative} \\ x = [x_1, x_2, x_3, x_4]^t \text{ with } x_4 > 0, \\ x_1, x_2, x_3, x_4 \in \mathbb{Z},\ Q(x) = n \end{array} \right\}$$

and $\mathcal{T}(\mathbb{Z})_n = \mathcal{T} \cap \mathbb{K}(\mathbb{Z}, Q)$. The group $\mathbf{PO}_4(\mathbb{Z}, Q)$ leaves $\mathbb{K}(\mathbb{Z}, Q)_n$ invariant. Suppose that $x = [x_1, x_2, x_3, x_4]^t$ is in $\mathcal{T}(\mathbb{Z})_1$ with representative chosen as in (2.43), then $x_3 x_4 = 1 + x_1^2 + x_2^2$ and the inequalities in (2.42) imply $1 \geq 2x_1^2$ and furthermore $x_1 = x_2 = 0$, $x_3 = x_4 = 1$. We have shown $\mathcal{T}(\mathbb{Z})_1 = \{P_4\}$. This implies $P_4 \in g_i \mathcal{P}$ for all $i = 1, \dots, m$, we also have $P_3 \in g_i \mathcal{P}$ for all i. By a simple computation as above we find $\mathcal{T}(\mathbb{Z})_2 = \{P_1, P_1'\}$ with $P_1' = [0, 0, 1, 2]^t$. Since the distances of P_1, P_1' to P_4 are distinct we have $P_1 \in g_i \mathcal{P}$ for all i. This can only happen if $m = 1$ and $\Gamma = \mathbf{PO}_4(\mathbb{Z}, Q)$.

The presentation given above is computed by Theorem 2.7.5. The group $\mathbf{PO}_4(\mathbb{Z}, Q)$ is a Coxeter group of a very simple nature. It will appear again as $\mathbf{CT}(1)$ in Section 4. Under appropriate identifications $\mathbf{PO}_4(\mathbb{Z}, Q)$ gets mapped to the Bianchi extension of $\mathbf{PSL}(2, \mathbb{Z}[i])$ mentioned in Section 7.4. Let $\{3, 4, 4\}$ be the tesselating octahedron mentioned in Theorem 2.2.11. It is easy to see that this octahedron can be subdivided into 12 tetrahedra isometric to $\mathcal{T}$. The group of symmetries of the corresponding tesselation of hyperbolic space is $\mathbf{PO}_4(\mathbb{Z}, Q)$. □

Example 2 :

Quadratic form: $Q(x_1, x_2, x_3, x_4) = -x_1^2 - 2x_2^2 + x_3 x_4$.
Quaternion algebra: $K_0 = \mathbb{Q}$, $K = \mathbb{Q}(\sqrt{-2})$, $\mathcal{H}(Q) = M(2, \mathbb{Q}(\sqrt{-2}))$.
Discrete group: $\mathbf{PO}_4(\mathbb{Z}, Q) < \mathbf{PO}_4(\mathbb{R}, Q)$ (cofinite but not cocompact).
Fundamental domain: $\mathcal{P} = \mathcal{P}(P_1, \dots, P_5)$ where

$$P_1 = \begin{bmatrix} 0 \\ 1 \\ 2 \\ 2 \end{bmatrix}_2, \quad P_2 = \begin{bmatrix} 1 \\ 1 \\ 2 \\ 2 \end{bmatrix}_1, \quad P_3 = \begin{bmatrix} 1 \\ 0 \\ 2 \\ 2 \end{bmatrix}_3, \quad P_4 = \begin{bmatrix} 0 \\ 0 \\ 1 \\ 1 \end{bmatrix}_1, \quad P_5 = \begin{bmatrix} 0 \\ 0 \\ 0 \\ 1 \end{bmatrix}_0.$$

$\mathcal{P}$ has the faces $\mathbb{P}_{[S_1]}, \dots, \mathbb{P}_{[S_5]}$ where

$$S_1 = \begin{bmatrix} 0 \\ 1 \\ 0 \\ 0 \end{bmatrix}_{-2}, \quad S_2 = \begin{bmatrix} 1 \\ 0 \\ 0 \\ 0 \end{bmatrix}_{-1}, \quad S_3 = \begin{bmatrix} 0 \\ 0 \\ 1 \\ -1 \end{bmatrix}_{-1}, \quad S_4 = \begin{bmatrix} 1 \\ 0 \\ 0 \\ 1 \end{bmatrix}_{-1} \quad S_5 = \begin{bmatrix} 0 \\ 1 \\ 0 \\ 2 \end{bmatrix}_{-2}.$$

Generators: $\sigma_1 = \sigma_{[S_1]}, ..., \sigma_5 = \sigma_{[S_5]}$.
Relations: $\sigma_1^2 = ... = \sigma_5^2 = (\sigma_1\sigma_2)^2 = (\sigma_1\sigma_3)^2 = (\sigma_1\sigma_4)^2 = (\sigma_2\sigma_3)^2 = (\sigma_2\sigma_5)^2 = (\sigma_3\sigma_4)^3 = (\sigma_3\sigma_5)^4 = (\sigma_4\sigma_5)^2 = 1$.
Commensurability class in $\mathbf{SL}(2, \mathbb{C})$: $\mathbf{SL}(2, \mathbb{Z}[\sqrt{-2}])$.

Additional remarks: The method explained for Example 1 again shows that $\mathbf{PO}_4(\mathbb{Z}, Q)$ is generated by $\sigma_1, \ldots, \sigma_5$. The presentation is derived similarly.

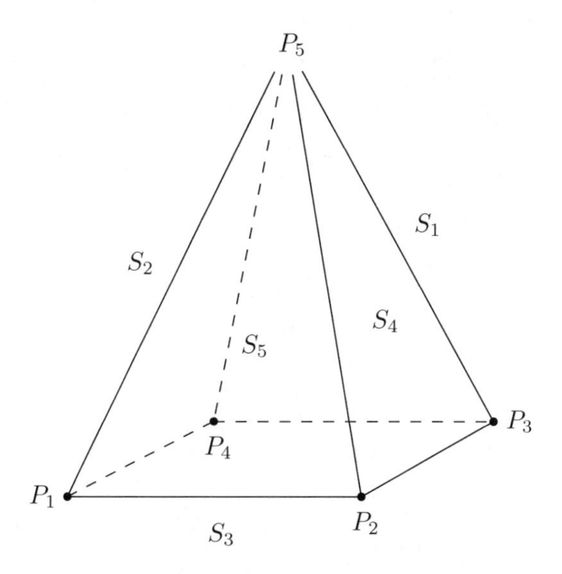

Fig. 2.2 The polyhedron $\mathcal{P}$ with one vertex at infinity

The polyhedron $\mathcal{P}$ is a chimney going to infinity with rectangular base, as indicated in figure 2.2. There is a semi-regular tesselation of hyperbolic space by cube octahedra with all vertices at infinity and right angles between adjacent faces. The group $\mathbf{PO}_4(\mathbb{Z}, Q)$ is its group of symmetries, see Grunewald, Gushoff, Mennicke (1982) for more details. Under appropriate identifications $\mathbf{PO}_4(\mathbb{Z}, Q)$ gets mapped to the Bianchi extension of $\mathbf{PSL}(2, \mathbb{Z}[\sqrt{-2}])$ mentioned in Section 7.4. □

Example 3 :

Quadratic form: $Q(x_1, x_2, x_3, x_4) = -x_1^2 - 3x_2^2 + x_3x_4$.
Quaternion algebra: $K_0 = \mathbb{Q}$, $K = \mathbb{Q}(\sqrt{-3})$, $\mathcal{H}(Q) = M(2, \mathbb{Q}(\sqrt{-3}))$.
Discrete group: $\mathbf{PO}_4(\mathbb{Z}, Q) < \mathbf{PO}_4(\mathbb{R}, Q)$ (cofinite but not cocompact).
Fundamental domain: $\mathcal{P} = \mathcal{P}(P_1, ..., P_5)$ where

$$P_1 = \begin{bmatrix} 0 \\ 1 \\ 2 \\ 2 \end{bmatrix}_1, \quad P_2 = \begin{bmatrix} 1 \\ 1 \\ 2 \\ 2 \end{bmatrix}_0, \quad P_3 = \begin{bmatrix} 1 \\ 0 \\ 2 \\ 2 \end{bmatrix}_3, \quad P_4 = \begin{bmatrix} 0 \\ 0 \\ 1 \\ 1 \end{bmatrix}_1, \quad P_5 = \begin{bmatrix} 0 \\ 0 \\ 0 \\ 1 \end{bmatrix}_0.$$

$\mathcal{P}$ has the faces $\mathbb{P}_{[S_1]}, \dots, \mathbb{P}_{[S_5]}$ where

$$S_1 = \begin{bmatrix} 0 \\ 1 \\ 0 \\ 0 \end{bmatrix}_{-3}, \quad S_2 = \begin{bmatrix} 1 \\ 0 \\ 0 \\ 0 \end{bmatrix}_{-1}, \quad S_3 = \begin{bmatrix} 0 \\ 0 \\ 1 \\ -1 \end{bmatrix}_{-1}, \quad S_4 = \begin{bmatrix} 1 \\ 0 \\ 0 \\ 1 \end{bmatrix}_{-1}, \quad S_5 = \begin{bmatrix} 0 \\ 1 \\ 0 \\ 3 \end{bmatrix}_{-3}.$$

Generators: $\sigma_1 = \sigma_{[S_1]}, \dots, \sigma_5 = \sigma_{[S_5]}$.
Relations: $\sigma_1^2 = \dots = \sigma_5^2 = (\sigma_1\sigma_2)^2 = (\sigma_1\sigma_3)^2 = (\sigma_1\sigma_4)^2 = (\sigma_2\sigma_3)^2 = (\sigma_2\sigma_5)^2 = (\sigma_3\sigma_4)^3 = (\sigma_3\sigma_5)^6 = (\sigma_4\sigma_5)^2 = 1$.
Commensurability class in $\mathbf{SL}(2,\mathbb{C})$: $\mathbf{SL}(2,\mathbb{Z}[(1+\sqrt{-3})/2])$.

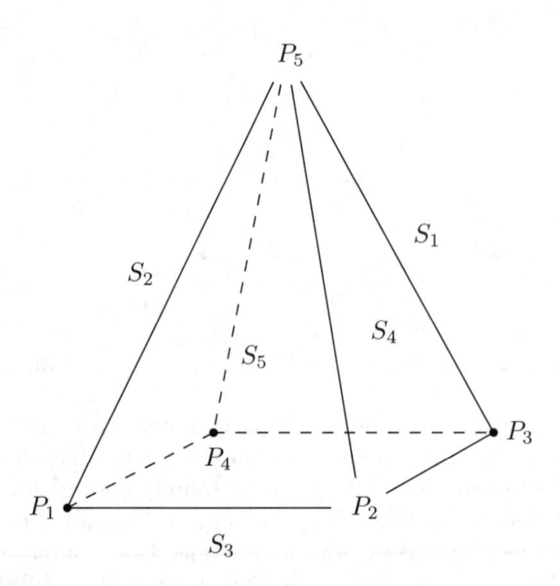

Fig. 2.3 The polyhedron $\mathcal{P}$ with two vertices at infinity

Additional remarks: The above is proved as under Example 1. The group $\mathbf{PO}_4(\mathbb{Z}, Q)$ can be seen to be maximal discrete in $\mathbf{PO}_4(\mathbb{R}, Q)$ but it is not the Bianchi extension of $\mathbf{PSL}(2,\mathbb{Z}[(1+\sqrt{-3})/2])$, see Example 6. A schematic picture of the fundamental region $\mathcal{P}$ is shown in figure 2.3. □

Example 4 :

Quadratic form: $Q(x_1,x_2,x_3,x_4) = -x_1^2 - 5x_2^2 + x_3x_4$.
Quaternion algebra: $K_0 = \mathbb{Q},\ K = \mathbb{Q}(\sqrt{-5}),\ \mathcal{H}(Q) = M(2,\mathbb{Q}(\sqrt{-5}))$.

Discrete group: $\mathbf{PO}_4(\mathbb{Z}, Q) < \mathbf{PO}_4(\mathbb{R}, Q)$ (cofinite but not cocompact).
Fundamental domain: $\mathcal{P} = \mathcal{P}(P_1, ..., P_7)$ where

$$P_1 = \begin{bmatrix} 5 \\ 3 \\ 10 \\ 10 \end{bmatrix}_{30}, \quad P_2 = \begin{bmatrix} 1 \\ 1 \\ 2 \\ 4 \end{bmatrix}_{2}, \quad P_3 = \begin{bmatrix} 0 \\ 2 \\ 5 \\ 5 \end{bmatrix}_{5}, \quad P_4 = \begin{bmatrix} 0 \\ 1 \\ 2 \\ 3 \end{bmatrix}_{1},$$

$$P_5 = \begin{bmatrix} 1 \\ 0 \\ 2 \\ 2 \end{bmatrix}_{3}, \quad P_6 = \begin{bmatrix} 0 \\ 0 \\ 1 \\ 1 \end{bmatrix}_{1}, \quad P_7 = \begin{bmatrix} 0 \\ 0 \\ 0 \\ 1 \end{bmatrix}_{0}.$$

$\mathcal{P}$ has the faces $\mathbb{P}_{[S_1]}, ..., \mathbb{P}_{[S_6]}$ where

$$S_1 = \begin{bmatrix} 0 \\ 1 \\ 0 \\ 0 \end{bmatrix}_{-5}, \quad S_2 = \begin{bmatrix} 1 \\ 0 \\ 0 \\ 0 \end{bmatrix}_{-1}, \quad S_3 = \begin{bmatrix} 0 \\ 0 \\ 1 \\ -1 \end{bmatrix}_{-1}, \quad S_4 = \begin{bmatrix} 1 \\ 0 \\ 0 \\ 1 \end{bmatrix}_{-1},$$

$$S_5 = \begin{bmatrix} 1 \\ 1 \\ 2 \\ 2 \end{bmatrix}_{-2}, \quad S_6 = \begin{bmatrix} 0 \\ 1 \\ 0 \\ 5 \end{bmatrix}_{-5}.$$

Generators: $\sigma_1 = \sigma_{[S_1]}, ..., \sigma_6 = \sigma_{[S_6]}$.
Relations: $\sigma_1^2 = ... = \sigma_6^2 = (\sigma_1\sigma_2)^2 = (\sigma_1\sigma_3)^2 = (\sigma_1\sigma_4)^2 = (\sigma_2\sigma_3)^2 = (\sigma_2\sigma_5)^4 = (\sigma_2\sigma_6)^2 = (\sigma_3\sigma_4)^3 = (\sigma_3\sigma_5)^2 = (\sigma_4\sigma_5)^2 = (\sigma_4\sigma_6)^2 = (\sigma_5\sigma_6)^2 = 1$.
Commensurability class in $\mathbf{SL}(2, \mathbb{C})$: $\mathbf{SL}(2, \mathbb{Z}[\sqrt{-5}])$.

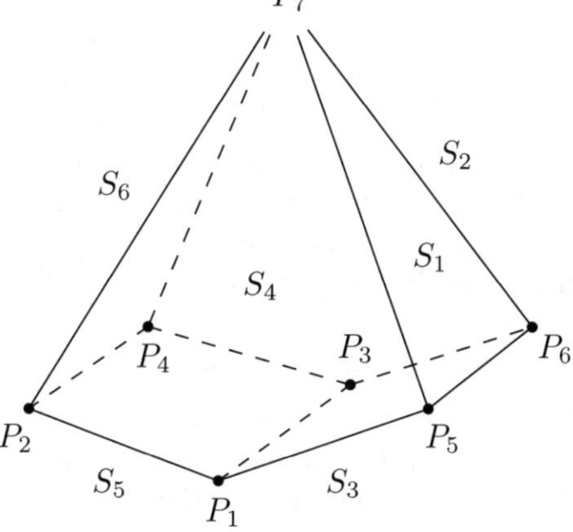

Fig. 2.4 The polyhedron $\mathcal{P}$ with one vertex at infinity

Additional remarks: To prove the above the same method can be used as in Example 1. Under appropriate identifications $\mathbf{PO}_4(\mathbb{Z}, Q)$ gets mapped to the Bianchi extension of $\mathbf{PSL}(2, \mathbb{Z}[\sqrt{-5}])$ mentioned in Section 7.4. A picture of the polyhedron P is shown in figure 2.4. □

Example 5 :

Quadratic form: $Q(x_1, x_2, x_3, x_4) = -x_1^2 - 6x_2^2 + x_3x_4$.
Quaternion algebra: $K_0 = \mathbb{Q}$, $K = \mathbb{Q}(\sqrt{-6})$, $\mathcal{H}(Q) = M(2, \mathbb{Q}(\sqrt{-6}))$.
Discrete group: $\mathbf{PO}_4(\mathbb{Z}, Q) < \mathbf{PO}_4(\mathbb{R}, Q)$ (cofinite but not cocompact).
Fundamental domain: $\mathcal{P} = \mathcal{P}(P_1, ..., P_7)$ where

$$P_1 = \begin{bmatrix} 3 \\ 2 \\ 6 \\ 6 \end{bmatrix}_3, \quad P_2 = \begin{bmatrix} 1 \\ 1 \\ 2 \\ 4 \end{bmatrix}_1, \quad P_3 = \begin{bmatrix} 0 \\ 1 \\ 3 \\ 3 \end{bmatrix}_3, \quad P_4 = \begin{bmatrix} 0 \\ 1 \\ 2 \\ 4 \end{bmatrix}_2,$$

$$P_5 = \begin{bmatrix} 1 \\ 0 \\ 2 \\ 2 \end{bmatrix}_3, \quad P_6 = \begin{bmatrix} 0 \\ 0 \\ 1 \\ 1 \end{bmatrix}_1, \quad P_7 = \begin{bmatrix} 0 \\ 0 \\ 0 \\ 1 \end{bmatrix}_0.$$

$\mathcal{P}$ has the faces $\mathbb{P}_{[S_1]}, ..., \mathbb{P}_{[S_6]}$ where

$$S_1 = \begin{bmatrix} 0 \\ 1 \\ 0 \\ 0 \end{bmatrix}_{-6}, \quad S_2 = \begin{bmatrix} 1 \\ 0 \\ 0 \\ 0 \end{bmatrix}_{-1}, \quad S_3 = \begin{bmatrix} 0 \\ 0 \\ 1 \\ -1 \end{bmatrix}_{-1}, \quad S_4 = \begin{bmatrix} 1 \\ 0 \\ 0 \\ 1 \end{bmatrix}_{-1},$$

$$S_5 = \begin{bmatrix} 0 \\ 1 \\ 2 \\ 2 \end{bmatrix}_{-2}, \quad S_6 = \begin{bmatrix} 0 \\ 1 \\ 0 \\ 6 \end{bmatrix}_{-6}.$$

Generators: $\sigma_1 = \sigma_{[S_1]}, ..., \sigma_6 = \sigma_{[S_6]}$.
Relations: $\sigma_1^2 = ... = \sigma_6^2 = (\sigma_1\sigma_2)^2 = (\sigma_1\sigma_3)^2 = (\sigma_1\sigma_4)^2 = (\sigma_2\sigma_3)^2 = (\sigma_2\sigma_5)^2 = (\sigma_2\sigma_6)^2 = (\sigma_3\sigma_4)^3 = (\sigma_3\sigma_5)^2 = (\sigma_4\sigma_5)^4 = (\sigma_4\sigma_6)^2 = (\sigma_5\sigma_6)^2 = 1$.
Commensurability class in $\mathbf{SL}(2, \mathbb{C})$: $\mathbf{SL}(2, \mathbb{Z}[\sqrt{-6}])$.

Additional remarks: To prove the above the same method as in Example 1 can be used. A picture of $\mathcal{P}$ can be seen in figure 2.4. Under appropriate identifications $\mathbf{PO}_4(\mathbb{Z}, Q)$ gets mapped to the Bianchi extension of the group $\mathbf{PSL}(2, \mathbb{Z}[\sqrt{-6}])$ mentioned in Section 7.4. □

Example 6 :

Quadratic form: $Q(x_1, x_2, x_3, x_4) = -x_1^2 - x_2^2 - x_3^2 + 3x_4^2$.
Quaternion algebra: $K_0 = \mathbb{Q}$, $K = \mathbb{Q}(\sqrt{-3})$, $\mathcal{H}(Q) = M(2, \mathbb{Q}(\sqrt{-3}))$.
Discrete group: $\mathbf{PO}_4(\mathbb{Z}, Q) < \mathbf{PO}_4(\mathbb{R}, Q)$ (cofinite but not cocompact).
Fundamental domain: $\mathcal{T} = \mathcal{T}(P_1, ..., P_4)$ where

$$P_1 = \begin{bmatrix} 0 \\ 1 \\ 1 \\ 1 \end{bmatrix}_1, \quad P_2 = \begin{bmatrix} 1 \\ 1 \\ 1 \\ 1 \end{bmatrix}_0, \quad P_3 = \begin{bmatrix} 0 \\ 0 \\ 1 \\ 1 \end{bmatrix}_2, \quad P_4 = \begin{bmatrix} 0 \\ 0 \\ 0 \\ 1 \end{bmatrix}_3.$$

$\mathcal{T}$ has the faces $\mathbb{P}_{[S_1]}, ..., \mathbb{P}_{[S_4]}$ where

$$S_1 = \begin{bmatrix} 1 \\ 0 \\ 0 \\ 0 \end{bmatrix}_{-1}, \quad S_2 = \begin{bmatrix} 1 \\ -1 \\ 0 \\ 0 \end{bmatrix}_{-2}, \quad S_3 = \begin{bmatrix} 0 \\ 0 \\ 3 \\ 1 \end{bmatrix}_{-6}, \quad S_4 = \begin{bmatrix} 0 \\ 1 \\ -1 \\ 0 \end{bmatrix}_{-2}.$$

$$\angle(\mathcal{T}) = [45°, 90°, 90°, 90°, 60°, 30°].$$

Generators: $\sigma_1 = \sigma_{[S1]}$, $\sigma_2 = \sigma_{[S2]}$, $\sigma_3 = \sigma_{[S_3]}$, $\sigma_4 = \sigma_{[S_4]}$.
Relations: $\sigma_1^2 = \sigma_2^2 = \sigma_3^2 = \sigma_4^2 = (\sigma_1\sigma_2)^4 = (\sigma_1\sigma_3)^2 = (\sigma_1\sigma_4)^2 = (\sigma_2\sigma_3)^2 = (\sigma_2\sigma_4)^3 = (\sigma_3\sigma_4)^6 = 1$.
Commensurability class in $\mathbf{SL}(2,\mathbb{C})$: $\mathbf{SL}(2,\mathbb{Z}[(1+\sqrt{-3})/2])$.

Additional remarks: Again the method of Example 1 can be used to prove the above. The group $\mathbf{PO}_4(\mathbb{Z}, Q)$ is a Coxeter group of a very simple nature. It will appear again as $\mathbf{CT}(7)$ in Section 4. Under appropriate identifications $\mathbf{PO}_4(\mathbb{Z}, Q)$ gets mapped to the Bianchi extension of $\mathbf{PSL}(2,\mathbb{Z}[(1+\sqrt{-3})/2])$ mentioned in Section 7.4. Let $\{3, 3, 6\}$ be the tesselating tetrahedron mentioned in Theorem 2.2.11. The group of symmetries of the corresponding tesselation of hyperbolic space is $\mathbf{PO}_4(\mathbb{Z}, Q)$. □

Example 7 :

Quadratic form: $Q(x_1, x_2, x_3, x_4) = -x_1^2 - x_2^2 - x_3^2 + \epsilon x_4^2$, $\epsilon = (1+\sqrt{5})/2$.
Quaternion algebra: $K_0 = \mathbb{Q}(\sqrt{5})$, $K = \mathbb{Q}(\sqrt{-\epsilon})$, $\mathcal{H}(Q) = \mathcal{H}(\epsilon, \epsilon; K)$.
Discrete group: $\mathbf{PO}_4(\mathbb{Z}[\epsilon], Q) < \mathbf{PO}_4(\mathbb{R}, Q)$ (cocompact).
Fundamental domain: $\mathcal{T} = \mathcal{T}(P_1, ..., P_4)$ *where*

$$P_1 = \begin{bmatrix} \epsilon \\ \epsilon \\ \epsilon \\ \epsilon^2 \end{bmatrix}_{2\epsilon}, \quad P_2 = \begin{bmatrix} 0 \\ 1 \\ 1 \\ \epsilon \end{bmatrix}_{2\epsilon-1}, \quad P_3 = \begin{bmatrix} 0 \\ 0 \\ 0 \\ \epsilon - 1 \end{bmatrix}_{\epsilon-1}, \quad P_4 = \begin{bmatrix} 0 \\ 0 \\ 1 \\ \epsilon \end{bmatrix}_{2\epsilon}.$$

$\mathcal{T}$ has the faces $\mathbb{P}_{[S_1]}, ..., \mathbb{P}_{[S_4]}$ where

$$S_1 = \begin{bmatrix} 1 \\ 0 \\ 0 \\ 0 \end{bmatrix}_{-1}, \quad S_2 = \begin{bmatrix} 1 \\ -1 \\ 0 \\ 0 \end{bmatrix}_{-2}, \quad S_3 = \begin{bmatrix} 0 \\ 0 \\ \epsilon \\ \epsilon - 1 \end{bmatrix}_{-2}, \quad S_4 = \begin{bmatrix} 0 \\ 1 \\ -1 \\ 0 \end{bmatrix}_{-2}.$$

$$\angle(\mathcal{T}) = [45°, 90°, 90°, 90°, 60°, 36°].$$

Generators: $\sigma_1 = \sigma_{[S_1]}$, $\sigma_2 = \sigma_{[S_2]}$, $\sigma_3 = \sigma_{[S_3]}$, $\sigma_4 = \sigma_{[S_4]}$.
Relations: $\sigma_1^2 = \sigma_2^2 = \sigma_3^2 = \sigma_4^2 = (\sigma_1\sigma_2)^4 = (\sigma_1\sigma_3)^2 = (\sigma_1\sigma_4)^2 = (\sigma_2\sigma_3)^2 = (\sigma_2\sigma_4)^3 = (\sigma_3\sigma_4)^5 = 1$.

Additional remarks: Again the method of Example 1 can be used to prove the above. The group $\mathbf{PO}_4(\mathbb{Z}[\epsilon], Q)$ is a Coxeter group of a very simple nature. It will appear again as $\mathbf{CT}(19)$ in Section 4. □

Examples 1–7 were computed together with many more by Mennicke around 1970. In our examples the unit groups of the quadratic forms turned out to be generated by reflections. This is no longer the case for slightly more complicated cases. Take $Q(x_1, x_2, x_3, x_4) = -x_1^2 - 10x_2^2 + x_3x_4$ or the form of the next example. Then $\mathbf{PO}_4(\mathbb{Z}, Q)$ does not have this property. There are now general results of Vinberg (1967), (1985), (1987) and Nikulin (1981) which say that there are only finitely many classes of integral quadratic forms whose unit groups can be generated by reflections.

Example 8 :

Consider the quadratic form

$$Q(x_1, x_2, x_3, x_4) = -2x_1^2 - 5x_2^2 - 10x_3^2 + x_4^2.$$

This form is $\mathbb{Q}$-anisotropic, hence by Theorem 2.2 the discontinuous group $\Gamma = \mathbf{PO}_4(\mathbb{Z}, Q) < \mathbf{PO}_4(\mathbb{R}, Q)$ is cocompact. We shall give now a description of a fundamental domain $\mathcal{P}$ for Γ. It is much more complicated than in the examples before.

The group Γ acts on the set $\mathbf{P}$ of geodesic planes S which have the property that $\sigma_{[S]} \in \Gamma$. The following are representatives for the four orbits

$$S_1 = \begin{bmatrix} 1 \\ 0 \\ 0 \\ 0 \end{bmatrix}_{-2}, \quad S_2 = \begin{bmatrix} 0 \\ 1 \\ 0 \\ 0 \end{bmatrix}_{-5}, \quad S_3 = \begin{bmatrix} 0 \\ 0 \\ 1 \\ 0 \end{bmatrix}_{-10}, \quad S_4 = \begin{bmatrix} 1 \\ 0 \\ 0 \\ 1 \end{bmatrix}_{-1}.$$

The stabilizer Γ_S in Γ of any of these planes acts cocompactly on the plane with a regular hexagon as fundamental domain. The angles at the vertices of the hexagon are all 90°. The group Γ_S is generated by the reflections in the edges of the hexagon. It is a Coxeter group with a graph as shown in figure 2.5.

Whenever two planes in $\mathbf{P}$ intersect they meet under an angle of 90°. In figure 2.6 a sequence of 6 hexagons on planes from $\mathbf{P}$ is shown which intersect consecutively and form a ring. Their representing vectors are S_2, S_3 and

$$S_5 = \begin{bmatrix} 0 \\ 2 \\ 1 \\ 5 \end{bmatrix}_{-5}, \quad S_6 = \begin{bmatrix} 0 \\ 2 \\ 3 \\ 10 \end{bmatrix}_{-10}, \quad S_7 = \begin{bmatrix} 0 \\ 0 \\ 1 \\ 3 \end{bmatrix}_{-1}, \quad S_8 = \begin{bmatrix} 0 \\ 1 \\ 0 \\ 2 \end{bmatrix}_{-1}.$$

This ring is extended by infinitely many similar rings to both sides. Together the hexagons on these rings form the boundary surface of a kind of rattlesnake. This rattlesnake extends along a geodesic line. A part of the configuration of the hexagons on the surface of the rattlesnake is shown in figure 2.7. The solid rattlesnake is a fundamental domain for the subgroup

N generated by the plane reflections in Γ. The group N is normal in Γ and has infinite index.

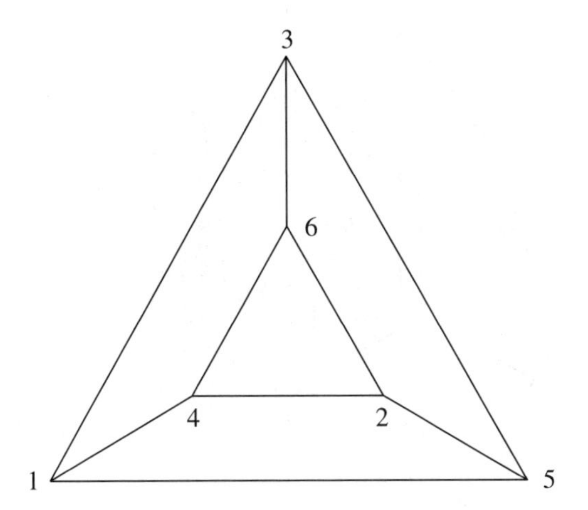

Fig. 2.5 The Coxeter graph of the group Γ_S, all edges carry the label ∞

A fundamental domain can be given as part of the rattlesnake as follows. The hexagons on the surface of the rattlesnake come in pairs. Examples are (S_1, S_9), (S_2, S_{10}), (S_4, S_5),... . Here the hexagons carry the same name as the planes containing them. Consider the centers of the two hexagons constituting the pair and the line connecting the two centers. Then there is a rotation of order 2 in Γ with this line as axis. The rotations map the rattlesnake to itself but reverse its orientation. The midpoints of the lines connecting the centers of the pairs lie on the central geodesic of the rattlesnake.

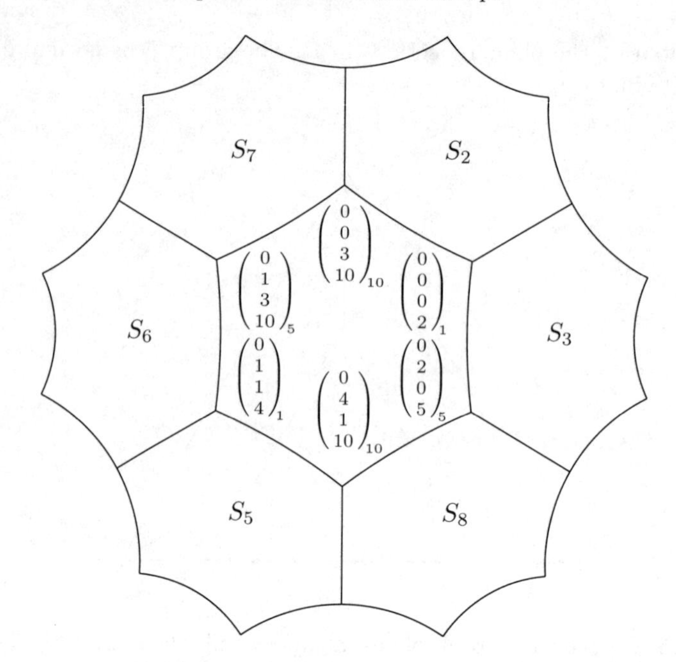

Fig. 2.6 A ring of hexagons

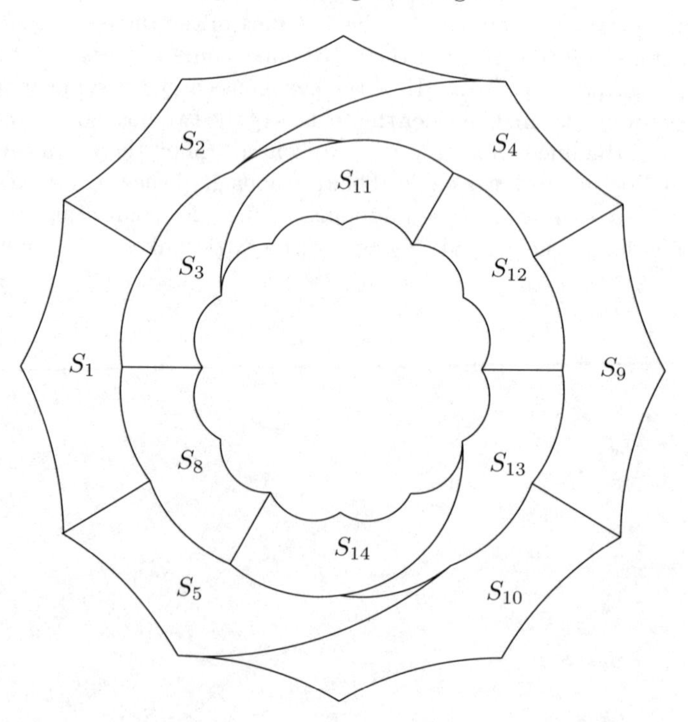

Fig. 2.7 A configuration of hexagons

The rotations for the pairs (S_1, S_9), (S_2, S_{10}) are given by

$$g_1 = \begin{pmatrix} 1 & 0 & 0 & 0 \\ 0 & -2 & -2 & 1 \\ 0 & -1 & -3 & 1 \\ 0 & -5 & -10 & 4 \end{pmatrix}, \qquad g_2 = \begin{pmatrix} -2 & 0 & -5 & 2 \\ 0 & 1 & 0 & 0 \\ -1 & 0 & -6 & 2 \\ -4 & 0 & -20 & 7 \end{pmatrix}.$$

The group generated by g_1, g_2 is an infinite dihedral group. Its central element $g_1 g_2$ acts as a translation along the axis of the rattlesnake. Any fundamental domain for the group generated by g_1, g_2 acting on the rattlesnake will be a fundamental domain for Γ.

Putting this geometric information together we find that Γ is generated by $\sigma_1, \ldots, \sigma_4$, g_1, g_2 where the notation $\sigma_i = \sigma_{[S_i]}$ is used. A presentation for Γ is given by the following system of relations.

$$\sigma_1^2 = \ldots = \sigma_4^2 = g_1^2 = g_2^2 = 1,$$

$$(\sigma_1\sigma_2)^2 = (\sigma_1\sigma_3)^2 = (\sigma_2\sigma_3)^2 = (\sigma_2\sigma_4)^2 = (\sigma_3\sigma_4)^2 = 1,$$

$$(\sigma_1\sigma_4^{g_2})^2 = (\sigma_3^{g_1}\sigma_4^{g_2})^2 = (\sigma_1^{g_2g_1}\sigma_3^{g_1g_2})^2 = (\sigma_1^{g_2g_1}\sigma_4^{g_1g_2})^2 = (\sigma_2^{g_1g_2}\sigma_4^{g_1})^2 = 1,$$

$$(\sigma_2^{g_1}\sigma_3^{g_1g_2})^2 = (g_1\sigma_1)^2 = (g_2\sigma_2)^2 = (g_1\sigma_3^{g_1g_2})^2 = (g_2\sigma_4^{g_1})^2 = 1.$$

Here we have used the notation $g^h := hgh^{-1}$ for group elements g, h. From the above presentation it is clear that the quotient group Γ/N is an infinite dihedral group generated by the images of g_1, g_2. It is also clear that Γ is isomorphic to the semidirect product of its subgroup generated by the plane reflections an infinite dihedral group which in fact is the symmetry group of the rattlesnake. These facts are special cases of a general theorem of Vinberg, see Vinberg, Shvartsman (1991), Chapter 5. The result of Vinberg says that a unit group of a quadratic form is isomorphic to the semidirect product of its reflection subgroup by the symmetry group of a fundamental polyhedron of the reflection subgroup.

10.3 Arithmetic and Non-Arithmetic Discrete Groups

The discrete subgroups $\Gamma < \mathbf{Iso}(\mathbb{H})$ constructed in the last two sections constitute, roughly speaking the class of arithmetic discrete groups. We shall give a precise definition below. First of all we have to recall some generalities concerning algebraic groups. The concepts are as in Borel (1991).

Let K be an algebraic number field with ring of integers $\mathcal{O}_K$. An affine algebraic group G defined over K is called a K-form of $\mathbf{SL}(2)$ if G and $\mathbf{SL}(2)$ are isomorphic as algebraic groups when extended to the algebraic closure of K. The K-forms of $\mathbf{SL}(2)$ are precisely the norm one groups of quaternion algebras over K. This can be read off from the table in Tits (1966). A subgroup $\Gamma < G(K)$ is called arithmetic if there is a faithful K-representation $\rho : G \to \mathbf{GL}(n)$ for some $n \in \mathbb{N}$ so that $\rho(\Gamma)$ is commensurable with $\mathbf{GL}(n, \mathcal{O}_K) \cap \rho(G(K))$. See Borel (1969) for a discussion of this concept. Examples are the norm 1 groups of $\mathcal{O}_K$-orders in quaternion algebras over K, see Section 1.

Definition 3.1. A subgroup $\Gamma < \mathbf{SL}(2, \mathbb{C})$ is called arithmetic if there are

(1) an algebraic number field K with exactly one pair of complex places $v_0, \bar{v}_0$,
(2) a K-form G of $\mathbf{SL}(2)$ so that $G(K \otimes_K K_v)$ is compact for every real place $v \in V_{\mathbb{R}}(K)$,
(3) an arithmetic subgroup $\Gamma_0 < G(K) < G(K \otimes_K K_{v_0}) \cong \mathbf{SL}(2, \mathbb{C})$

so that Γ is $\mathbf{SL}(2, \mathbb{C})$-conjugate to a group commensurable with Γ_0.

A subgroup $\Gamma < \mathbf{PSL}(2, \mathbb{C})$ is called arithmetic if its preimage in $\mathbf{SL}(2, \mathbb{C})$ has this property. A subgroup $\Gamma < \mathbf{Iso}(\mathbb{H})$ is called arithmetic if $\Gamma \cap \mathbf{Iso}^+(\mathbb{H}) < \mathbf{PSL}(2, \mathbb{C})$ is arithmetic.

There is a general definition of an arithmetic subgroup of a Lie group (see Mostow (1985)). The above treats the special case $\mathbf{Iso}(\mathbb{H})$. Note that an arithmetic subgroup $\Gamma < \mathbf{Iso}(\mathbb{H})$ is necessarily cofinite. All examples constructed in Theorem 1.2 are arithmetic groups.

By results of Borel (1981) and Thurston non-arithmetic cofinite subgroups $\Gamma < \mathbf{Iso}(\mathbb{H})$ exist in abundance. In fact the set of covolumes of the arithmetic groups is discrete in $\mathbb{R}_{>0}$ while the set of covolumes of all torsion free cofinite subgroups is of order type ω^ω, see Section 5 of Chapter 2 for some comments. Although it is thus known that so many non-arithmetic cofinite groups exist examples are difficult to locate. The first examples seem to be due to Makarov (1966). Later Vinberg (1967) gave a general procedure to decide non-arithmeticity for groups generated by reflections (and their higher dimensional analogs). Special cases will appear in the next section. In dimension 3 the method of Vinberg does only establish a version of non-arithmeticity weaker than the above. His argument does only exclude the groups to come from an arithmetic orthogonal group, see the remarks after Proposition 2.5 and also the appendix of Vinberg (1967). This gap was closed by Shvartsman (1969). The paper of Gromov, Piatetski-Shapiro (1986) gives a method to construct non-arithmetic cofinite and cocompact groups of isometries on hyperbolic spaces of arbitrary dimension.

Some non-arithmetic cofinite and cocompact groups will appear in the next section. We shall use the following necessary criterion for a group to be arithmetic. For groups which have a sufficiently concrete description these criteria seem to be applicable.

Proposition 3.2. *Let $\Gamma < \mathbf{Iso}(\mathbb{H})$ be an arithmetic group. Then there is a subgroup $\Gamma_0 < \Gamma$ of finite index so that $\Gamma_0 < \mathbf{PSL}(2, \mathbb{C}) = \mathbf{Iso}^+(\mathbb{H})$ and a number field K with exactly one pair of complex places $v_0, \bar{v}_0$ so that $\mathrm{tr}(g)$ is an integer in $v_0(K)$ for all $g \in \Gamma_0$. If Γ is not cocompact then K can be chosen to be imaginary quadratic.*

Proof. Let $\tilde{\Gamma}$ be the preimage of $\Gamma \cap \mathbf{PSL}(2,\mathbb{C}) = \mathbf{Iso}^+(\mathbb{H})$ in $\mathbf{SL}(2,\mathbb{C})$. From our definition and the classification of Tits (1966) we infer the existence of a number field K with exactly one pair of complex places v_0, $\bar{v}_0$ and a quaternion algebra $\mathcal{H}(a,b;K)$ and an $\mathcal{O}_K$-order $\mathcal{R}$ so that $\Phi(\mathcal{R}^1)$ is commensurable with a $\mathbf{SL}(2,\mathbb{C})$-conjugate of $\tilde{\Gamma}$. Here any isomorphism of algebraic groups $\Phi : \mathcal{H}(v_0(a), v_0(b), \mathbb{C})^1 \to \mathbf{SL}(2,\mathbb{C})$, for example that of (1.3) is suitable. With some straightforward arguments the result follows from formula (1.3) and Theorem 1.2. □

Definition 3.3. Let q be the quadratic form $q(x_1,x_2,x_3,x_4) = -x_1^2 - x_2^2 - x_3^2 + x_4^2$. A subgroup $\Gamma < \mathbf{PO}_4(\mathbb{R},q)$ is called arithmetic if there is an arithmetic subgroup $\Gamma_0 < \mathbf{PSL}(2,\mathbb{C})$ so that Γ is $\mathbf{PO}_4(\mathbb{R},q)$-conjugate to a group commensurable with $\Psi(\Gamma_0)$. Here $\Psi : \mathbf{SL}(2,\mathbb{C}) \to \mathbf{PSO}_4(\mathbb{R},q) < \mathbf{PO}_4(\mathbb{R},q)$ is the homomorphism constructed in Proposition 1.3.10. Let Q be a real quadratic form of signature (1,3). A subgroup $\Gamma < \mathbf{PO}_4(\mathbb{R},Q)$ is called arithmetic if $T\Gamma T^{-1} < \mathbf{PO}_4(\mathbb{R},q)$ is arithmetic for some (or all) $T \in \mathbf{GL}(4,\mathbb{R})$ with $Q = q \circ T^{-1}$.

It is easy to see that any other choice of a homomorphism of real algebraic groups $\Psi_0 : \mathbf{SL}(2,\mathbb{C}) \to \mathbf{PSO}_4(\mathbb{R},q)$ giving an isomorphism between $\mathbf{SL}(2,\mathbb{C})$ and $\mathbf{PSO}_4^+(\mathbb{R},q)$ leads to the same class of arithmetic groups.

Proposition 3.4. *Let Q be a real quadratic form of signature (1,3). Let $\Gamma < \mathbf{PO}_4(\mathbb{R},Q)$ be an arithmetic group. Then there is a subgroup $\Gamma_0 < \Gamma$ of finite index so that $\Gamma_0 < \mathbf{PSO}_4(\mathbb{R},Q)$ and a number field $K \subset \mathbb{C}$ with at most one pair of complex places so that $\mathrm{tr}(g)$ is an integer in K for all $g \in \Gamma_0$. If Γ is not cocompact then K can be chosen to be $\mathbb{Q}$.*

Proof. It is enough to prove the statement for the quadratic form q from Definition 3.3. Define $\tilde{\Gamma}$ to be the preimage of $\Gamma \cap \mathrm{Im}(\Psi)$ in $\mathbf{SL}(2,\mathbb{C})$. The group $\tilde{\Gamma}$ being arithmetic we find from Proposition 3.2 a subgroup of finite index $\Gamma_0 < \tilde{\Gamma}$ and a number field K_1 with exactly one pair of complex places v_0, $\bar{v}_0$ so that $\mathrm{tr}(g)$ is an integer in $v_0(K) \subset \mathbb{C}$ for all $g \in \Gamma_0$. A glance at the formulas in Proposition 1.3.11 shows that

$$\mathrm{tr}(\Psi(g)) = \pm|\mathrm{tr}(g)|^2.$$

It is an exercise to show that $|\alpha|^2$ is an algebraic integer (possibly lying in a proper subfield of $\mathbb{Q}(\alpha,\bar{\alpha})$) whenever $\alpha \in \mathbb{C}$ has this property. □

10.4 The Tetrahedral Groups

We shall describe now certain very simple discontinuous groups generated by reflections. They will turn out to be groups generated by the reflections in the faces of the tesselating tetrahedra. These are those tetrahedra satisfying the conditions of Theorem 2.2.8. A tetrahedron is the hyperbolic convex hull of 4 non-coplanar points from $\mathbb{K} \cup \partial\mathbb{K}$. Our groups are generalizations of the hyperbolic triangle groups, see Beardon (1983). An important difference is that there are infinitely many conjugacy classes of triangle groups whereas in our situation there are only 32 possibilities. For more general results on groups which can be generated by reflections see Vinberg, Shvartsman (1991).

We use the approach of Lanner (1950) which is also described in the exercises of Bourbaki (1968). We start off with a Coxeter matrix M of size 4. For the terminology see the introduction to this chapter. We assume that the quadratic form Q_M associated to the bilinear form B_M has signature (1,3) on $V_M = \mathbb{R}^4$. Then the rank 4 Coxeter group W_M has a representation

$$\Theta_M : W_M \to \mathbf{O}_4(\mathbb{R}, Q_M) \to \mathbf{PO}_4(\mathbb{R}, Q_M) = \mathbf{Iso}(\mathbb{K}(Q_M)),$$

with $\Theta_M(s_i) = \sigma_i$ and hence acts on the model $\mathbb{K}(Q_M)$ of hyperbolic space. The 4 elements

$$S_1 = \begin{bmatrix} 1 \\ 0 \\ 0 \\ 0 \end{bmatrix}, \quad S_2 = \begin{bmatrix} 0 \\ 1 \\ 0 \\ 0 \end{bmatrix}, \quad S_3 = \begin{bmatrix} 0 \\ 0 \\ 1 \\ 0 \end{bmatrix}, \quad S_4 = \begin{bmatrix} 0 \\ 0 \\ 0 \\ 1 \end{bmatrix} \in \mathbb{P}^3\mathbb{R}$$

satisfy $Q_M(S_i) = -2 < 0$, hence they represent geodesic planes in $\mathbb{K}(Q_M)$. For this terminology see Section 1.5. These planes are also denoted by S_i. The planes S_2, S_3, S_4 intersect in a point $P_1 \in \mathbb{K}(Q_M) \cup \partial\mathbb{K}(Q_M)$. The coordinates of P_1 are found by solving the following homogenous rank 3 system of linear equations:

$$x\, B_M\, S_2 = 0, \quad x\, B_M\, S_3 = 0, \quad x\, B_M\, S_4 = 0.$$

Define P_2, P_3, P_4 analogously. The points $P_1, \ldots, P_4$ are the vertices of a tetrahedron $\mathcal{T}_M$ which is bounded by $S_1, \ldots, S_4$. By formula (2.40) for the hyperbolic angle we see that S_i meets S_j under the inner angle $\pi/m(i,j)$.

Definition 4.1. A Coxeter matrix of size 4 is called hyperbolic if the corresponding quadratic form Q_M has signature (1,3) and if the group $\mathbf{CT}(M)$ generated by the reflections in the faces of the hyperbolic tetrahedron $\mathcal{T}_M$ acts discontinuously on $\mathbb{K}(Q_M)$. The group $\mathbf{CT}(M) \cong W_M$ is called a tetrahedral group.

Note that the condition of discontinuity is the same as to say that $\mathcal{T}_M$ tesselates hyperbolic space, see Theorem 2.2.10. If M is a hyperbolic Coxeter matrix then $\mathcal{T}_M$ is a fundamental domain for $\mathbf{CT}(M)$, hence the groups $\mathbf{CT}(M)$ are always cofinite.

Theorem 4.2. *There are exactly 32 hyperbolic Coxeter matrices. They are those listed in the table at the end of this section.*

There is a table of the tetrahedral groups in Vinberg, Shvartsman (1991) which omits a group.

Proof. Let M be a hyperbolic Coxeter matrix and P_i be one of the vertices of $\mathcal{T}_M$. Put $\Gamma_i =< \sigma_j \mid j \neq i >$. Then Γ_i is a Coxeter group corresponding to a Coxeter sub-matrix M_i of M of size 3, see Bourbaki (1968). Suppose now $P_i \in \mathbb{K}(Q_M)$. From the discontinuity and Theorem 2.2.10 we obtain an injective representation $\Gamma_i \to \mathbf{O}(3)$. Since Γ_i acts discontinuously (through this homomorphism) on S^2 it has to be a spherical triangle group. Using the abbreviation $(m(1,2), m(1,3), m(2,3))$ for M_i we have (up to permutation) the following possibilities:

$$(4.1) \qquad M_i = (2,3,3),\ (2,3,4),\ (2,3,5),\ (2,2,k) \qquad (k \geq 2),$$

see Magnus (1974). If P_i is in the boundary $\partial\mathbb{K}(Q_M)$, we similarly see that Γ_i is a Euclidean triangle group. The possibilities are:

$$(4.2) \qquad M_i = (2,3,6),\ (2,4,4),\ (3,3,3).$$

Suppose now that M has an entry $m(i,j) > 6$. From (4.1), (4.2) we see that all other off-diagonal entries have to be 2. Hence we may assume $m(3,4) > 6$ and $m(i,j) = 2$ in all other off-diagonal cases. The corresponding matrix

$$B_M = \begin{pmatrix} -2 & 0 & 0 & 0 \\ 0 & -2 & 0 & 0 \\ 0 & 0 & -2 & 2\xi \\ 0 & 0 & 2\xi & -2 \end{pmatrix} \qquad \left(\xi = \cos\left(\frac{\pi}{m(3,4)}\right)\right)$$

does not have signature (1,3), hence $m(i,j) \leq 6$ for all i, j.

Having proved that there are only finitely many matrices M to consider, it is now best to make a case distinction considering the number of entries of M which are equal to 2. Suppose for example that this number is 3. Then the Coxeter graph of M has two possibilities, one is

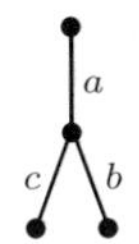

with multiplicities satisfying $3 \leq a,\, b,\, c \leq 6$. Using the symmetry of the graphs we may assume, without loss of generality, $3 \leq a \leq b \leq c \leq 6$.

Conditions (4.1), (4.2) imply that there are at most six possible triplets for (a, b, c). The signature condition is

$$\cos^2\left(\frac{\pi}{a}\right) + \cos^2\left(\frac{\pi}{b}\right) + \cos^2\left(\frac{\pi}{c}\right) > 1.$$

This leaves us with four triplets

$$(a, b, c) = (3, 3, 5),\ (3, 3, 6),\ (3, 4, 4),\ (4, 4, 4).$$

These can be found in the table below. In all other cases the arguments are of a similar nature. We skip them. □

We give now the complete table of the tetrahedral groups, with further information. We encode the Coxeter matrix into the Coxeter graph. The labels give the multiplicity of the edges. If an edge carries no label then the multiplicity is 3. The entry vol. gives the hyperbolic volume of the tetrahedron $\mathcal{T}_M$. It is computed by evaluation of the appropriate integral using Riemann sums. The number c counts the vertices of $\mathcal{T}_M$ which are in the boundary of hyperbolic space. The tetrahedron $\mathcal{T}_M$ is compact if and only if $c = 0$. The further entry gives information about the corresponding subgroup in $\mathbf{PSL}(2, \mathbb{C})$ or respectively on the fields of definition of the quadratic forms defining the arithmetic groups. The tesselation groups, that is the group generated by the reflections in the faces of the tetrahedra are denoted by the same symbols as the corresponding Coxeter graphs. We use the notation $\zeta = (1 + \sqrt{-3})/2$. We say that two subgroups $\Gamma_1 < \mathbf{PO}_4(\mathbb{R}, Q_1)$ and $\Gamma_2 < \mathbf{PO}_4(\mathbb{R}, Q_2)$ (Q_1, Q_2 two quadratic forms of signature (1,3)) are conjugate if there is a $T \in \mathbf{GL}(4, \mathbb{R})$ with $Q_1 = Q_2 \circ T^{-1}$ and $\Gamma_1 = T\Gamma_2 T^{-1}$.

Table of Tetrahedral Groups

CT(1): 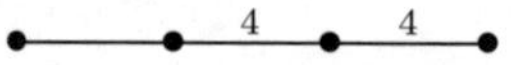

vol. : 0.07633..., $c = 1$, $\mathbf{PSL}(2, \mathbb{Z}[i])$

The tetrahedron $\mathcal{T}_M$ has the same configuration of inner angles as that of Example 1 in Section 2. We conclude from Proposition 2.6 that the subgroup of $\mathbf{PSL}(2, \mathbb{C})$ corresponding to $\mathbf{CT}(1)$ is $\mathbf{PSL}(2, \mathbb{Z}[i])$. Note that this means that $\mathbf{PSL}(2, \mathbb{Z}[i])$ is abstractly isomorphic to a subgroup of index 4 in $\mathbf{CT}(1)$.

CT(2): 

vol. : 0.15266..., $c = 1$, $\mathbf{PSL}(2, \mathbb{Z}[i])$

By a plane passing through the vertex in the boundary this tetrahedron can be divided into two tetrahedra which are both isometric to the tetrahedron of $\mathbf{CT}(1)$, hence $\mathbf{CT}(2)$ is a subgroup of index 2 in $\mathbf{CT}(1)$.

CT(3): 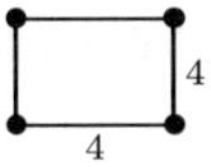 vol. : 0.30532..., $c = 1$, $\mathbf{PSL}(2, \mathbb{Z}[i])$

By a plane passing through the vertex in the boundary this tetrahedron can be divided into two tetrahedra which are both isometric to the tetrahedron of **CT**(2), hence **CT**(3) is conjugate to a subgroup of index 2 in **CT**(2).

CT(4): 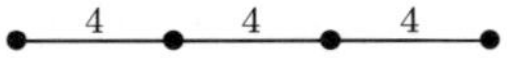 vol. : 0.22899..., $c = 2$, $\mathbf{PSL}(2, \mathbb{Z}[i])$

This tetrahedron can be subdivided into 3 tetrahedra which are isometric to the tetrahedron of **CT**(1). This implies that **CT**(4) is conjugate to a subgroup of index 3 in **CT**(1).

CT(5):  vol. : 0.45798..., $c = 3$, $\mathbf{PSL}(2, \mathbb{Z}[i])$

This tetrahedron can be subdivided into 2 tetrahedra which are isometric to the tetrahedron of **CT**(4). This implies that **CT**(5) is conjugate to a subgroup of index 2 in **CT**(4).

CT(6): 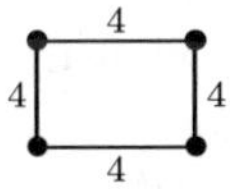 vol. : 0.91596..., $c = 4$, $\mathbf{PSL}(2, \mathbb{Z}[i])$

This tetrahedron can be subdivided into 2 tetrahedra which are isometric to the tetrahedron of **CT**(5). This implies that **CT**(6) is conjugate to a subgroup of index 2 in **CT**(5).

CT(7):  vol. : 0.10572..., $c = 1$, $\mathbf{PSL}(2, \mathbb{Z}[\zeta])$

The tetrahedron $\mathcal{T}_M$ has the same configuration of inner angles as that of Example 6 in Section 2. We conclude from Proposition 2.6 that the subgroup of $\mathbf{PSL}(2, \mathbb{C})$ corresponding to **CT**(7) is $\mathbf{PSL}(2, \mathbb{Z}[\zeta])$.

CT(8):  vol. : 0.08457..., $c = 1$, $\mathbf{PSL}(2, \mathbb{Z}[\zeta])$

Let B_M be the matrix corresponding to the bilinear form, then

$$T^t\, B_M\, T = \begin{pmatrix} -2 & 0 & 0 & 0 \\ 0 & -6 & 0 & 0 \\ 0 & 0 & 0 & 1 \\ 0 & 0 & 1 & 0 \end{pmatrix} =: B, \qquad T = \begin{pmatrix} 1 & -\frac{1}{2} & -\frac{1}{2} & 0 \\ 0 & \frac{1}{2} & -\frac{1}{2} & 0 \\ 0 & 0 & 0 & 1 \\ 0 & 0 & 1 & -1 \end{pmatrix}^{-1} .$$

Let Q_B be the quadratic form corresponding to B and define $\tilde{\sigma}_i := T\,\sigma_i\,T^{-1} \in$ $\mathbf{PO}_4(\mathbb{R}, Q_B)$, then

$$\tilde{\sigma}_1 = \begin{pmatrix} -1 & 0 & 0 & 0 \\ 0 & 1 & 0 & 0 \\ 0 & 0 & 1 & 0 \\ 0 & 0 & 0 & 1 \end{pmatrix}, \qquad \tilde{\sigma}_2 = \begin{pmatrix} \frac{1}{2} & \frac{3}{2} & 0 & 0 \\ \frac{1}{2} & -\frac{1}{2} & 0 & 0 \\ 0 & 0 & 1 & 0 \\ 0 & 0 & 0 & 1 \end{pmatrix},$$

$$\tilde{\sigma}_3 = \begin{pmatrix} \frac{1}{2} & -\frac{3}{2} & -\frac{1}{2} & 0 \\ -\frac{1}{2} & -\frac{1}{2} & -\frac{1}{2} & 0 \\ 0 & 0 & 1 & 0 \\ 1 & 3 & 1 & 1 \end{pmatrix}, \qquad \tilde{\sigma}_4 = \begin{pmatrix} 1 & 0 & 0 & 0 \\ 0 & 1 & 0 & 0 \\ 0 & 0 & 0 & 1 \\ 0 & 0 & 1 & 0 \end{pmatrix}.$$

It is easy to check that the integral matrices $\tilde{\sigma}_1, \tilde{\sigma}_4, \tilde{\sigma}_2\tilde{\sigma}_3\tilde{\sigma}_2, (\tilde{\sigma}_3\tilde{\sigma}_1\tilde{\sigma}_2)^2$ generate a subgroup of index 10 in $\mathbf{CT}(8)$. Hence $\mathbf{CT}(8)$ is conjugate to a group commensurable with $\mathbf{PO}_4(\mathbb{Z}, Q)$ where Q is the quadratic form of example 3 in Section 2.

$\mathbf{CT}(9)$:  vol. : 0.21144..., $c = 1$, $\mathbf{PSL}(2, \mathbb{Z}[\zeta])$

Let B_M be the matrix corresponding to the bilinear form, then

$$T^t\,B_M\,T = \begin{pmatrix} -2 & 0 & 0 & 0 \\ 0 & -6 & 0 & 0 \\ 0 & 0 & 0 & 1 \\ 0 & 0 & 1 & 0 \end{pmatrix} =: B, \qquad T = \begin{pmatrix} 1 & -\frac{1}{2} & -\frac{1}{2} & 0 \\ 0 & \frac{1}{2} & -\frac{1}{2} & 0 \\ 0 & 0 & 0 & \sqrt{2} \\ 0 & 0 & 1 & -\frac{\sqrt{2}}{2} \end{pmatrix}^{-1}.$$

Let Q_B be the quadratic form corresponding to B and define $\tilde{\sigma}_i := T\,\sigma_i\,T^{-1} \in$ $\mathbf{PO}_4(\mathbb{R}, Q_B)$, then

$$\tilde{\sigma}_1 = \begin{pmatrix} -1 & 0 & 0 & 0 \\ 0 & 1 & 0 & 0 \\ 0 & 0 & 1 & 0 \\ 0 & 0 & 0 & 1 \end{pmatrix}, \qquad \tilde{\sigma}_2 = \begin{pmatrix} \frac{1}{2} & \frac{3}{2} & 0 & 0 \\ \frac{1}{2} & -\frac{1}{2} & 0 & 0 \\ 0 & 0 & 1 & 0 \\ 0 & 0 & 0 & 1 \end{pmatrix},$$

$$\tilde{\sigma}_3 = \begin{pmatrix} \frac{1}{2} & -\frac{3}{2} & -\frac{1}{2} & 0 \\ -\frac{1}{2} & -\frac{1}{2} & -\frac{1}{2} & 0 \\ 0 & 0 & 1 & 0 \\ 1 & 3 & 1 & 1 \end{pmatrix}, \qquad \tilde{\sigma}_4 = \begin{pmatrix} 1 & 0 & 0 & 0 \\ 0 & 1 & 0 & 0 \\ 0 & 0 & 0 & 2 \\ 0 & 0 & \frac{1}{2} & 0 \end{pmatrix}.$$

It is easy to check that the integral matrices

$$\tilde{\sigma}_1, \tilde{\sigma}_2\tilde{\sigma}_3\tilde{\sigma}_2, (\tilde{\sigma}_2\tilde{\sigma}_1\tilde{\sigma}_2)^2, \tilde{\sigma}_4\tilde{\sigma}_2\tilde{\sigma}_3\tilde{\sigma}_2\tilde{\sigma}_4, \tilde{\sigma}_4(\tilde{\sigma}_2\tilde{\sigma}_1\tilde{\sigma}_3)^2\tilde{\sigma}_4, (\tilde{\sigma}_1\tilde{\sigma}_2\tilde{\sigma}_3\tilde{\sigma}_4)^2$$

generate a subgroup of index 12 in $\mathbf{CT}(9)$. Hence $\mathbf{CT}(9)$ is conjugate to a group commensurable with $\mathbf{PO}_4(\mathbb{Z}, Q)$ where Q is the quadratic form of example 3 in Section 2.

This tetrahedron can be subdivided into 2 tetrahedra which are isometric to the tetrahedron of $\mathbf{CT}(7)$. This implies that $\mathbf{CT}(9)$ is conjugate to a subgroup of index 2 in $\mathbf{CT}(7)$.

$\mathbf{CT}(10)$: •——•——•—6—• vol. : 0.04228..., $c = 1$, $\mathbf{PSL}(2, \mathbb{Z}[\zeta])$

The tetrahedron $\mathbf{CT}(8)$ can be subdivided into 2 tetrahedra which are isometric to the tetrahedron of $\mathbf{CT}(10)$. This implies that $\mathbf{CT}(8)$ is conjugate to a subgroup of index 2 in $\mathbf{CT}(10)$.

$\mathbf{CT}(11)$: vol. : 0.21144..., $c = 2$, $\mathbf{PSL}(2, \mathbb{Z}[\zeta])$

This tetrahedron can be subdivided into 2 tetrahedra which are isometric to the tetrahedron of $\mathbf{CT}(7)$. This implies that $\mathbf{CT}(11)$ is conjugate to a subgroup of index 2 in $\mathbf{CT}(7)$.

$\mathbf{CT}(12)$: vol. : 1.01494..., $c = 4$, $\mathbf{PSL}(2, \mathbb{Z}[\zeta])$

This tetrahedron can be divided into 12 tetrahedra which are isometric to the tetrahedron of $\mathbf{CT}(8)$, hence $\mathbf{CT}(12)$ is conjugate to a subgroup of index 12 in $\mathbf{CT}(8)$. Moreover this tetrahedron can be subdivided into 24 isometric copies of the tetrahedron of $\mathbf{CT}(10)$ hence $\mathbf{CT}(12)$ is conjugate to a subgroup of index 24 in $\mathbf{CT}(10)$.

This tetrahedron is completely regular with all vertices at infinity. Any two bounding planes meet under the angle $\pi/3$.

$\mathbf{CT}(13)$: vol. : 0.50747..., $c = 3$, $\mathbf{PSL}(2, \mathbb{Z}[\zeta])$

The tetrahedron of $\mathbf{CT}(12)$ can be divided into two tetrahedra which are both isometric to the tetrahedron of $\mathbf{CT}(13)$, hence $\mathbf{CT}(12)$ is conjugate to a subgroup of index 2 in $\mathbf{CT}(13)$.

$\mathbf{CT}(14)$: •—6—•——•—6—• vol. : 0.25373..., $c = 2$, $\mathbf{PSL}(2, \mathbb{Z}[\zeta])$

By a plane passing through one of the vertices in the boundary the tetrahedron of $\mathbf{CT}(13)$ can be divided into two tetrahedra which are both isometric to the tetrahedron of $\mathbf{CT}(14)$, hence $\mathbf{CT}(13)$ is conjugate to a subgroup of index 2 in $\mathbf{CT}(14)$.

$\mathbf{CT}(15)$: vol. : 0.16915..., $c = 2$, $\mathbf{PSL}(2, \mathbb{Z}[\zeta])$

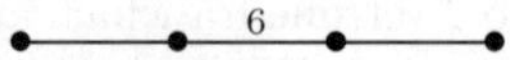

The tetrahedron of $\mathbf{CT}(12)$ can be divided into 6 tetrahedra which are all isometric to the tetrahedron of $\mathbf{CT}(15)$, hence $\mathbf{CT}(12)$ is conjugate to a subgroup of index 6 in $\mathbf{CT}(15)$. On the other hand $\mathbf{CT}(15)$ is up to conjugacy contained (with index 4) in $\mathbf{CT}(10)$.

$\mathbf{CT}(16)$: vol. : 0.42289..., $c = 2$, $\mathbf{PSL}(2, \mathbb{Z}[\zeta])$

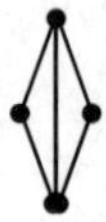

The matrix B_M has rational integral entries in this case and the matrices of the reflections $\sigma_1, \ldots, \sigma_4$ similarly. It is easy to find a matrix $T \in \mathbf{GL}(4, \mathbb{Z})$ so that the quadratic form Q_B corresponding to $B = T^t B_M T$ is $Q_B(x_1, x_2, x_3, x_4) = 2(-x_1^2 - 3x_2^2 + x_3 x_4)$. Hence $\mathbf{CT}(14)$ is conjugate to a group commensurable with $\mathbf{PO}_4(\mathbb{Z}, Q)$ where Q is the quadratic form of example 3 in Section 2.

This tetrahedron can be subdivided into 2 tetrahedra which are isometric to the tetrahedron of $\mathbf{CT}(9)$. This implies that $\mathbf{CT}(16)$ is conjugate to a subgroup of index 2 in $\mathbf{CT}(9)$.

$\mathbf{CT}(17)$: vol. : 0.84578..., $c = 4$, $\mathbf{PSL}(2, \mathbb{Z}[\zeta])$

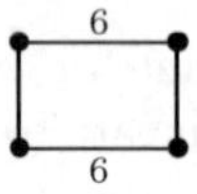

Let B_M be the matrix corresponding to the bilinear form, then

$$T^t B_M T = \begin{pmatrix} -2 & 1 & 0 & 3 \\ 1 & -2 & 3 & 0 \\ 0 & 3 & -6 & 3 \\ 3 & 0 & 3 & -6 \end{pmatrix} =: B, \qquad T = \begin{pmatrix} 1 & 0 & 0 & 0 \\ 0 & 1 & 0 & 0 \\ 0 & 0 & \sqrt{3} & 0 \\ 0 & 0 & 0 & \sqrt{3} \end{pmatrix}.$$

Let Q_B be the quadratic form corresponding to B and define $\tilde{\sigma}_i := T\,\sigma_i\,T^{-1} \in \mathbf{PO}_4(\mathbb{R}, Q_B)$, then the $\tilde{\sigma}_i$ are easily seen to be in $\mathbf{PO}_4(\mathbb{Z}, Q_B)$. Since $\det B = -108$ it follows from Proposition 2.6 that the subgroup of $\mathbf{PSL}(2, \mathbb{C})$ corresponding to $\mathbf{CT}(17)$ is commensurable to $\mathbf{PSL}(2, \mathbb{Z}[\zeta])$.

$\mathbf{CT}(18)$: vol. : 0.22222..., $c = 0$, ar. over $\mathbb{Q}$

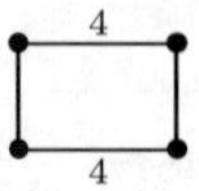

Let B_M be the matrix corresponding to the bilinear form, then

$$T^t\, B_M\, T = \begin{pmatrix} -2 & 1 & 0 & 1 \\ 1 & -2 & 2 & 0 \\ 0 & 2 & -4 & 1 \\ 1 & 0 & 1 & -1 \end{pmatrix} =: B \qquad T = \begin{pmatrix} 1 & 0 & 0 & 0 \\ 0 & 1 & 0 & 0 \\ 0 & 0 & \sqrt{2} & 0 \\ 0 & 0 & 0 & \frac{\sqrt{2}}{2} \end{pmatrix}.$$

Let Q_B be the quadratic form corresponding to B and define $\tilde{\sigma}_i := T\,\sigma_i\,T^{-1} \in$ $\mathbf{PO}_4(\mathbb{R}, Q_B)$, then

$$\tilde{\sigma}_1 = \begin{pmatrix} -1 & 1 & 0 & 1 \\ 0 & 1 & 0 & 0 \\ 0 & 0 & 1 & 0 \\ 0 & 0 & 0 & -1 \end{pmatrix}, \qquad \tilde{\sigma}_2 = \begin{pmatrix} 1 & 0 & 0 & 0 \\ 1 & -1 & 2 & 0 \\ 0 & 0 & 1 & 0 \\ 0 & 0 & 0 & 1 \end{pmatrix},$$

$$\tilde{\sigma}_3 = \begin{pmatrix} 1 & 0 & 0 & 0 \\ 0 & 1 & 0 & 0 \\ 0 & 1 & -1 & \frac{1}{2} \\ 0 & 0 & 0 & 1 \end{pmatrix}, \qquad \tilde{\sigma}_4 = \begin{pmatrix} 1 & 0 & 0 & 0 \\ 0 & 1 & 0 & 0 \\ 0 & 0 & 1 & 0 \\ 2 & 0 & 2 & -1 \end{pmatrix}.$$

The quadratic form Q_B is easily seen to be anisotropic over $\mathbb{Q}$. In fact it is integrally equivalent to $Q(x_1, x_2, x_3, x_4) = -x_1^2 - x_2^2 - x_3^3 + 7x_4^2$. We have $\tilde{\sigma}_1, \tilde{\sigma}_2, \tilde{\sigma}_4$, $\tilde{\sigma}_3\tilde{\sigma}_2\tilde{\sigma}_3, \tilde{\sigma}_3\tilde{\sigma}_4\tilde{\sigma}_3\tilde{\sigma}_4 \in \mathbf{PO}_4(\mathbb{Z}, Q_B)$. These elements generate a subgroup of index 3 in $\mathbf{CT}(18)$. By Theorem 2.2 $\mathbf{CT}(18)$ is conjugate to a group commensurable with $\mathbf{PO}_4(\mathbb{Z}, Q_B)$.

CT(19): •—4—•——•—5—• vol. : 0.03588..., $c = 0$, ar. over $\mathbb{Q}(\sqrt{5})$

The tetrahedron $\mathcal{T}_M$ has the same configuration of inner angles as that of Example 7 in Section 2. We conclude that $\mathbf{CT}(19)$ is conjugate to the unit group described there.

CT(20): [diagram: 5] vol. : 0.07177..., $c = 0$, ar. over $\mathbb{Q}(\sqrt{5})$

By a plane passing through one of the vertices this tetrahedron can be divided into two tetrahedra which are both isometric to the tetrahedron of $\mathbf{CT}(19)$, hence $\mathbf{CT}(20)$ is conjugate to a subgroup of index 2 in $\mathbf{CT}(19)$.

CT(21): [diagram: 5, 5] vol. : 0.50213..., $c = 0$, ar. over $\mathbb{Q}(\sqrt{5})$

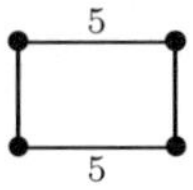

Put $\epsilon = (1 + \sqrt{5})/2$, then the matrix of the bilinear form B_M is

$$B_M = \begin{pmatrix} -2 & 1 & 0 & \epsilon \\ 1 & -2 & \epsilon & 0 \\ 0 & \epsilon & -2 & 1 \\ \epsilon & 0 & 1 & -2 \end{pmatrix}, \qquad \det B_M = -7\epsilon + 1.$$

The corresponding quadratic form Q_M is defined over $K_0 = \mathbb{Q}(\sqrt{5})$, and it is definite for one infinite place and indefinite for the other infinite place. So it satisfies the conditions of Theorem 2.2. It follows that $\mathbf{CT}(21)$ is a subgroup of finite index in $\mathbf{PO}_4(\mathcal{O}_{K_0}, Q_M)$.

$\mathbf{CT}(22)$: 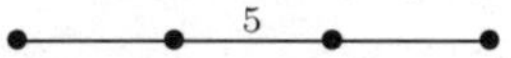

vol. : 0.03905..., $c = 0$, ar. over $\mathbb{Q}(\sqrt{5})$

The matrix B_M corresponding to the bilinear form has coefficients in $K_0 = \mathbb{Q}(\sqrt{5})$ and satisfies the conditions of Theorem 2.2. It follows that $\mathbf{CT}(22)$ is a subgroup of finite index in $\mathbf{PO}_4(\mathcal{O}_{K_0}, Q_M)$.

$\mathbf{CT}(23)$: 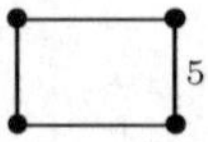

vol. : 0.205289..., $c = 0$, ar. over $\mathbb{Q}(\sqrt{5})$

The matrix B_M corresponding to the bilinear form has coefficients in $K_0 = \mathbb{Q}(\sqrt{5})$ and satisfies the conditions of Theorem 2.2. It follows that $\mathbf{CT}(23)$ is a subgroup of finite index in $\mathbf{PO}_4(\mathcal{O}_{K_0}, Q_M)$.

$\mathbf{CT}(24)$: 

vol. : 0.09332..., $c = 0$, ar. over $\mathbb{Q}(\sqrt{5})$

The matrix B_M corresponding to the bilinear form has coefficients in $K_0 = \mathbb{Q}(\sqrt{5})$ and satisfies the conditions of Theorem 2.2. It follows that $\mathbf{CT}(24)$ is a subgroup of finite index in $\mathbf{PO}_4(\mathcal{O}_{K_0}, Q_M)$.

$\mathbf{CT}(25)$: 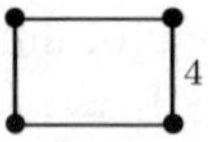

vol. : 0.08577..., $c = 0$, ar. over $\mathbb{Q}(\sqrt{2})$

The matrix B_M corresponding to the bilinear form has coefficients in $K_0 = \mathbb{Q}(\sqrt{2})$ and satisfies the conditions of Theorem 2.2. It follows that $\mathbf{CT}(25)$ is a subgroup of finite index in $\mathbf{PO}_4(\mathcal{O}_{K_0}, Q_M)$.

$\mathbf{CT}(26)$: 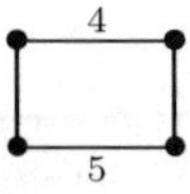

vol. : 0.35865..., $c = 0$, non-ar.

The matrix giving the bilinear form is

$$B_M = \begin{pmatrix} -2 & 1 & 0 & \epsilon \\ 1 & -2 & \sqrt{2} & 0 \\ 0 & \sqrt{2} & -2 & 1 \\ \epsilon & 0 & 1 & -2 \end{pmatrix}, \qquad \left(\epsilon = \frac{1+\sqrt{5}}{2}\right).$$

Put $\alpha = \sqrt{2} \cdot \epsilon$ then

$$T^t\, B_M\, T = \begin{pmatrix} -2 & 1 & 0 & \alpha \\ 1 & -2 & 2 & 0 \\ 0 & 2 & -4 & 2 \\ \alpha & 0 & 2 & -4 \end{pmatrix} =: B, \qquad T = \begin{pmatrix} 1 & 0 & 0 & 0 \\ 0 & 1 & 0 & 0 \\ 0 & 0 & \sqrt{2} & 0 \\ 0 & 0 & 0 & \sqrt{2} \end{pmatrix}.$$

Let Q_B be the quadratic form corresponding to B and define $\tilde{\sigma}_i := T\,\sigma_i\,T^{-1} \in \mathbf{PO}_4(\mathbb{R}, Q_B)$, then

$$\tilde{\sigma}_1 = \begin{pmatrix} -1 & 1 & 0 & 1 \\ 0 & 1 & 0 & 0 \\ 0 & 0 & 1 & 0 \\ 0 & 0 & 0 & 1 \end{pmatrix}, \qquad \tilde{\sigma}_2 = \begin{pmatrix} 1 & -1 & 2 & 0 \\ 0 & 0 & 1 & 0 \\ 0 & 0 & 1 & 0 \\ 0 & 0 & 0 & 1 \end{pmatrix},$$

$$\tilde{\sigma}_3 = \begin{pmatrix} 1 & 0 & 0 & 0 \\ 0 & 1 & 0 & 0 \\ 0 & 1 & -1 & 1 \\ 0 & 0 & 0 & 1 \end{pmatrix}, \qquad \tilde{\sigma}_4 = \begin{pmatrix} 1 & 0 & 0 & 0 \\ 0 & 1 & 0 & 0 \\ 0 & 0 & 1 & 0 \\ \frac{\alpha}{2} & 0 & 1 & -1 \end{pmatrix}.$$

Consider the element

$$X = \tilde{\sigma}_1\tilde{\sigma}_2\tilde{\sigma}_3\tilde{\sigma}_4 = \begin{pmatrix} \frac{3\alpha}{2} & 1 & 1 & -3 \\ 1+\alpha & 1 & 0 & -2 \\ \frac{\alpha}{2} & 1 & 0 & -1 \\ \frac{\alpha}{2} & 0 & 1 & -1 \end{pmatrix}.$$

Assume that $\mathbf{CT}(26)$ is arithmetic. By Proposition 3.4 it follows that $\mathrm{tr}(X^n)$ is for some $n \in \mathbb{N}$ an integer in the field $K = \mathbb{Q}(\alpha) = \mathbb{Q}(\sqrt{2}, \sqrt{5})$. The ring of integers $\mathcal{O}_K$ has the $\mathbb{Z}$-module basis $1, \sqrt{2}, \epsilon, \alpha$. Put $\mathfrak{a} := \sqrt{2}\mathcal{O}_K$, then $\mathfrak{a}$ is an prime ideal of index 4 in $\mathcal{O}_K$. From $\mathrm{tr}(X^n) \in \mathcal{O}_K$ we get $\mathrm{tr}((\sqrt{2}X)^n) \in \sqrt{2}^n\mathcal{O}_K \subset \mathfrak{a}$. By direct inspection we see $\mathrm{tr}((\sqrt{2}X)^n) \equiv \epsilon^n$ modulo $\mathfrak{a}$. This shows that $\mathbf{CT}(26)$ is not arithmetic.

CT(27): 5 vol. : 0.34300..., $c = 1$, non-ar.

Let $\sigma_1, \ldots, \sigma_4$ be the reflections corresponding to $S_1, \ldots, S_4$ and put $g = (\sigma_3\sigma_1\sigma_2)^2\sigma_4 \in \mathbf{CT}(27)$. A computation shows that the characteristic polynomial of g is

$$\det(g - xI) = (x-1)\,(x+1)\,(x^2 + (3\epsilon^2 + 2)x + 1)$$

where $\epsilon = (1+\sqrt{5})/2$. Hence g has four distinct eigenvalues which are $1, -1, \lambda_1, \lambda_2$ where $\lambda_1 = -9.751...$ and $\lambda_2 = -0.102....$ Write $\tau(\lambda_1)$, $\tau(\lambda_2)$ for the zeroes of $x^2 + \tau(3\epsilon^2+2)x + 1$ where $\tau : \mathbb{Q}(\sqrt{5}) \to \mathbb{Q}(\sqrt{5})$ is the non-trivial automorphism. We have $\tau(\lambda_1) = -2{,}787...$, $\tau(\lambda_2) = -0.358....$ If $\mathrm{tr}(g^{2n}) = \lambda_1^{2n} + \lambda_2^{2n} + 2$ was a rational number for some $n \in \mathbb{N}$ then $\lambda_1^{2n} + \lambda_2^{2n} = \tau(\lambda_1)^{2n} + \tau(\lambda_2)^{2n}$ would have to hold. This is impossible as is seen from the above numerical values. We conclude from Proposition 3.4 that $\mathbf{CT}(27)$ is non-arithmetic.

CT(28): 5 6 vol. : 0.17150..., $c = 1$, non-ar.

By a plane passing through the vertex in the boundary the tetrahedron corresponding to **CT**(27) can be divided into two tetrahedra which are both isometric to the tetrahedron of **CT**(28), hence **CT**(27) is a subgroup of index 2 in **CT**(28). This shows that **CT**(28) is non-arithmetic.

CT(29): [diagram: square with edge labels 5 (top), 6 (bottom)] vol. : 0.67298..., $c = 2$, non-ar.

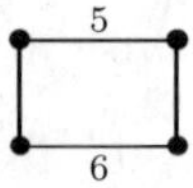

Let $\sigma_1, \ldots, \sigma_4$ be the reflections corresponding to $S_1, \ldots, S_4$ and put $g = \sigma_1\sigma_2\sigma_3\sigma_4 \in$ **CT**(29). A computation shows that the characteristic polynomial of g is

$$\det(g - xI) = x^4 - (\alpha + \epsilon + 2)x^3 - (2 - \epsilon)x^2 - (\alpha + \epsilon + 2)x + 1$$

where $\epsilon = (1 + \sqrt{5})/2$ and $\alpha = \sqrt{2}\epsilon$. An analysis of the zeroes of this polynomial similar to **CT**(27) shows that for no $n \in \mathbb{N}$ the number $\mathrm{tr}(g)^n$ is rational.

CT(30): [diagram: square with edge labels 4 (top), 6 (bottom)] vol. : 0.52584..., $c = 2$, non-ar.

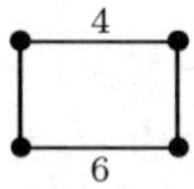

Let $\sigma_1, \ldots, \sigma_4$ be the reflections corresponding to $S_1, \ldots, S_4$ and put $g = \sigma_1\sigma_2\sigma_3\sigma_4 \in$ **CT**(30). A computation shows that the characteristic polynomial of g is

$$\det(g - xI) = x^4 - (3 + \sqrt{6})x^3 - x^2 - (3 + \sqrt{6})x + 1.$$

An analysis of the zeroes of this polynomial similar to **CT**(27) shows that for no $n \in \mathbb{N}$ the number $\mathrm{tr}(g)^n$ is rational.

CT(31): [diagram: square with edge labels 4 (top), 4 (right), 4 (bottom)] vol. : 0.55628..., $c = 2$, non-ar.

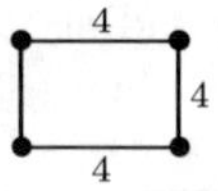

Let $\sigma_1, \ldots, \sigma_4$ be the reflections corresponding to $S_1, \ldots, S_4$ and put $g = \sigma_1\sigma_2\sigma_3\sigma_4 \in$ **CT**(31). A computation shows that the characteristic polynomial of g is

$$\det(g - xI) = x^4 - (3 + 2\sqrt{2})x^3 - 2x^2 - (3 + 2\sqrt{2})x + 1.$$

An analysis of the zeroes of this polynomial similar to **CT**(27) shows that for no $n \in \mathbb{N}$ the number $\mathrm{tr}(g)^n$ is rational.

CT(32): [diagram: square with edge label 6 (right)] vol. : 0.36410..., $c = 2$, non-ar.

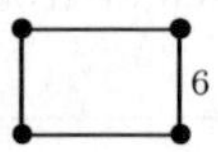

Let $\sigma_1, \dots, \sigma_4$ be the reflections corresponding to $S_1, \dots, S_4$ and put $g = \sigma_1\sigma_2\sigma_3\sigma_4 \in \mathbf{CT}(32)$. A computation shows that the characteristic polynomial of g is

$$\det(g - xI) = x^4 - (2+\sqrt{3})x^3 - 2x^2 - (2+\sqrt{3})x + 1.$$

An analysis of the zeroes of this polynomial similar to $\mathbf{CT}(27)$ shows that for no $n \in \mathbb{N}$ the number $\mathrm{tr}(g)^n$ is rational.

A table like the above is contained in Vinberg, Shvartsman (1991), there the group $\mathbf{CT}(11)$ is omitted. The inclusion relations between tetrahedral groups are analyzed in Maclachlan (1996).

We shall explain now the connection of the eight tesselating ideal polyhedra mentioned in Theorem 2.2.11 to the tetrahedra of this section.

Consider the graph

$\mathbf{CT}(10)$: •——•——•—6—•

and number its vertices by $1-4$ from left to right. Number the faces S_1, S_2, S_3, S_4 of the corresponding tetrahedron $\mathcal{T}$ accordingly. The vertices P_1, P_2, P_3, P_4 are numbered so that $P_i \notin S_i$. As before we denote the discontinuous group generated by the reflections in the faces of $\mathcal{T}$ also by $\mathbf{CT}(10)$. The stabilizer $\Sigma := \mathbf{CT}(10)_{P_4}$ of the vertex P_4 is generated by the reflections in the faces S_1, S_2, S_3. It follows that Σ is isomorphic to the Coxeter group corresponding to the graph

which is a finite group of order 24 known as the extended tetrahedral group $\mathbf{T}$. It is isomorphic to the symmetric group $\mathbf{S}_4$. See Coxeter, Moser (1965) for information on the extended tetrahedral, octahedral and icosahedral groups. We now form the polyhedron

$$\mathcal{P} := \bigcup_{\gamma \in \Sigma} \gamma\mathcal{T}.$$

It is easy to see that $\mathcal{P}$ is isometric to the tetrahedron $\{3,3,6\}$ which has already appeared under $\mathbf{CT}(12)$ in our table. Using our notation for the tesselation groups of polyhedra we get isomorphisms

$$\Gamma_{\{3,3,6\}} \cong \mathbf{CT}(12), \qquad \mathbf{CT}(12) \rtimes \mathbf{T} \cong \mathbf{CT}(10).$$

Using the same numbering and notation as in the first example and starting with

$\mathbf{CT}(7)$: •—4—•——•—6—•

we find the stabilizer Σ to be isomorphic to the Coxeter group corresponding to the graph

which is the extended octahedral group **O**. The group **O** is finite of order 48. The corresponding polyhedron $\mathcal{P}$ is isometric to the ideal cube $\{4,3,6\}$. As above we get an isomorphism

$$\Gamma_{\{4,3,6\}} \rtimes \mathbf{O} \cong \mathbf{CT}(7).$$

Again using the same numbering and notation as in the first example and starting with

$$\mathbf{CT}(28):$$

5 6

we find the stabilizer Σ to be isomorphic to the Coxeter group corresponding to the graph

5

which is the extended icosahedral group **I**. The group **I** is finite of order 120. The corresponding polyhedron $\mathcal{P}$ is isometric to the ideal dodecahedron $\{5,3,6\}$. As above we get an isomorphism

$$\Gamma_{\{5,3,6\}} \rtimes \mathbf{I} \cong \mathbf{CT}(28).$$

For the remaining five ideal polyhedra we only give the final isomorphisms between their tesselation groups extended by a suitable extended polyhedral group and one of our tetrahedral groups. The geometric situation is always that of the first example.

$$\Gamma_{\{3,4,4\}} \rtimes \mathbf{O} \cong \mathbf{CT}(1), \quad \Gamma_{\{4,3,5\}} \rtimes \mathbf{O} \cong \mathbf{CT}(19), \quad \Gamma_{\{5,3,5\}} \rtimes \mathbf{I} \cong \mathbf{CT}(24),$$
$$\Gamma_{\{5,3,4\}} \rtimes \mathbf{I} \cong \mathbf{CT}(19), \quad \Gamma_{\{3,5,3\}} \rtimes \mathbf{I} \cong \mathbf{CT}(22).$$

The connection of the tesselation groups of the ideal polyhedra in hyperbolic space to various arithmetic groups apparent from the above considerations have a long history. The first instances where discovered by Bianchi (1892), (1893) and Gieseking (1912).

The tetrahedral groups satisfy certain inclusion relations as indicated in the table. Hence some of these groups cannot be maximal as discrete subgroups of $\mathbf{Iso}(\mathbb{H})$. Another method to construct discrete groups containing one of the tetrahedral groups comes from the following observation.

Lemma 4.3. *Let $\Gamma < \mathbf{Iso}(\mathbb{H})$ be a cofinite subgroup and φ an automorphism of Γ of finite order. Assume that φ^n is an inner automorphism only in case $\varphi^n = \mathrm{id}_\Gamma$. Then there is a discrete subgroup $\Gamma_1 < \mathbf{Iso}(\mathbb{H})$ which contains Γ as a normal subgroup of finite index so that the group Γ_1 is isomorphic to the semidirect product $\Gamma \rtimes <\varphi>$.*

Proof. By application of Theorem 2.8.1 we find a $g \in \mathbf{Iso}(\mathbb{H})$ so that $gxg^{-1} = \varphi(x)$ for all $x \in \Gamma$. Since the centralizer of Γ in $\mathbf{Iso}(\mathbb{H})$ is trivial (Proposition 2.8.8) it follows that g has the same order as φ. It also follows that g^n can only be contained in Γ if $g^n = 1$. The group Γ_1 generated by Γ and g has the required properties. □

Lemma 4.3 can be applied to the tetrahedral groups in the following way. Let G be one of the 32 graphs listed in our table. Assume that G has a non-trivial graph-automorphism φ. This map induces in an obvious way a group automorphism of the corresponding Coxeter group which satisfies the requirements of Lemma 4.3. In this way we find proper discrete extension groups of **CT**(14), **CT**(17), **CT**(22), **CT**(29), **CT**(31), **CT**(32) which are not tetrahedral groups.

Finally for this chapter we add some remarks which will show that some of the groups in the above list are maximal discrete in **Iso**($\mathbb{K}$). To do this we need the following result.

Lemma 4.4. *Let $\Gamma <$ **Iso**($\mathbb{H}$) be a discrete subgroup which is generated by reflections. Let $M = \{ S \mid \sigma_{[S]} \in \Gamma \}$ be the set of geodesic planes so that the corresponding reflection is in Γ. Let*

$$L = \mathbb{H} \setminus \bigcup_{S \in M} S$$

and let Comp_Γ *be the set of connected components of L. Then Γ acts simply transitively on* Comp_Γ. *Let $C \in \mathrm{Comp}_\Gamma$ be any element then its closure in $\mathbb{H}$ is a fundamental domain for Γ.*

We leave the simple proof as an exercise, see Koszul (1965) for a much more general result.

Proposition 4.5. *The following of the 32 tetrahedral groups are maximal discrete in* **Iso**($\mathbb{K}$): **CT**(1), **CT**(7), **CT**(10), **CT**(19), **CT**(28).

Proof. Consider first of all the groups **CT**(1), **CT**(7), **CT**(10), **CT**(28). They all have one class of cusps. An application of Theorem 2.5.1 to the values of the covolumes given in the table gives the bounds 1, 2, 2, 4 for the indices of possible discrete overgroups. This already finishes the case of **CT**(1).

Let now Γ be one of **CT**(7), **CT**(10) and Γ_1 a discrete group properly containing Γ. By our bound on the index Γ is normal in Γ_1 and Γ_1 acts on the set of components Comp_Γ introduced in Lemma 4.4. The open kernel of the tetrahedron corresponding to Γ is one of the elements of Comp_Γ. By Lemma 4.4 an element of $\Gamma_1 \setminus \Gamma$ induces a symmetry of this tetrahedron. A glance at the graphs corresponding to **CT**(7), **CT**(10) shows that their tetrahedra have no non-trivial symmetries. This finishes these cases.

Take now $\Gamma =$ **CT**(28) and Γ_1 a discrete group properly containing Γ. Let $\mathcal{T}$ be the tetrahedron corresponding to Γ and P be its vertex at infinity. Let Γ_2 be the subgroup generated by the reflections in Γ_1. Clearly $\Gamma < \Gamma_2$. In case $\Gamma = \Gamma_2$ the group Γ is normal in Γ_1 and we are finished by the same argument as in the two cases before. Assume now $\Gamma \neq \Gamma_2$. An element $C \in \mathrm{Comp}_{\Gamma_2}$ is either contained in $\mathcal{T}$ or disjoint from $\mathcal{T}$. Hence there are $C_1, \ldots, C_n \in \mathrm{Comp}_{\Gamma_2}$ which are contained in $\mathcal{T}$ so that $\mathcal{T}$ is the union of the closures of $\mathcal{T}$. The number n is equal to the index of Γ in Γ_2. Any of the C_i has to contain P in its closure (in $\mathbb{H} \cup \mathbb{P}^1\mathbb{C}$) since they are all isometric and P is the unique point at infinity in $\mathcal{T}$. If $n > 1$ then there is a geodesic plane S passing through P so that the reflection in S is in Γ_2 but not in Γ. Hence the stabilizer of P in Γ_2 is a euclidean reflection group which properly contains the reflection group $(2, 3, 6)$ which is the stabilizer of P in Γ. A glance at the list of euclidean reflection groups shows that there is no such group.

The group $\mathbf{CT}(19)$ is treated similarly to the case of $\mathbf{CT}(28)$ with spherical reflection groups replacing the euclidean ones. □

10.5 Notes and Remarks

Explicit constructions for cofinite or cocompact subgroups of the group of isometries $\mathbf{Iso}(\mathbb{H}^n)$ of n-dimensional hyperbolic space have a long history, particularly in dimension $n = 2$. In this dimension there was an intensive study of the subgroups of finite index in $\mathbf{PSL}(2, \mathbb{Z})$ which occurred as invariance groups of certain elliptic modular functions, see Fricke (1914) and Klein, Fricke (1897) for this aspect. Klein, Fricke (1897) also treated many of the groups $\mathbf{PO}_3(\mathbb{Z}, Q)$ where Q is a rational quadratic form in 3 variables of signature (1,2). These are analogues of the groups which have occurred in our Section 10.2.

For arbitrary dimension n there are as obvious generalizations the groups $\mathbf{PO}_{n+1}(\mathbb{Z}, Q)$ where Q is a rational quadratic form in $n + 1$ variables of signature $(1, n)$. These are all discontinuous and cofinite. In dimensions $n \geq 5$ they are never cocompact. Such groups are obtained by using quadratic forms over totally real number fields, as we did in Section 10.2.

In dimension $n \neq 3$ the unit groups of quadratic forms constitute roughly speaking the class of arithmetic discontinuous subgroups of $\mathbf{Iso}(\mathbb{H}^n)$. The question of existence of further discontinuous groups comes to mind. In dimension 2 a whole series of examples can be constructed by considering the subgroup of $\mathbf{Iso}(\mathbb{H}^2)$ generated by the 3 reflections in the faces of tesselating triangles, see Beardon (1983). The two-dimensional case also is highly exceptional because of the uniformization theory of Riemann surfaces.

The method of constructing discontinuous non-arithmetic cofinite or cocompact subgroups of $\mathbf{Iso}(\mathbb{H}^n)$ by taking the group generated by the reflections in the faces of certain polyhedra was also successful for $n = 3$ (Section 10.4) and in some further dimensions. Finally Vinberg (1984) proved that for $n \geq 29$ no cocompact groups exist which can be generated by reflections. Prokhorov (1987) proved the same result in the cofinite case for $n \geq 995$.

In dimension $n = 3$ there are efforts to consider groups more general than groups generated by reflections. One possible concept are quasi-reflection groups. They have a fundamental domain which is a polyhedron bounded by two types of faces. The first type consists of reflective faces, this means that the corresponding hyperbolic reflection belongs to the discrete group. The faces of the second type have a central symmetry which is a point reflection. All faces of the polyhedron have boundaries which are intersections with planes of one of the above types. The discrete group is then generated by plane reflections and point reflections. An example of a quasi-reflection group is $\mathbf{PO}_4(\mathbb{Z}, Q)$ where Q is the quadratic form $Q = -2x^2 - 13y^2 - 26z^2 + u^2$.

The paper of Gromov, Piatetski-Shapiro (1986) gives a method to construct non-arithmetic cofinite and cocompact groups of isometries on hyperbolic spaces of arbitrary dimension.

Given a cofinite subgroup $\Gamma < \mathbf{Iso}(\mathbb{H})$ usually a whole series of cofinite groups can be constructed by Dehn surgery. See Thurston (1978) for an explanation of this method.

References

Abramowitz, M., Stegun, I. (1970): Handbook of Mathematical Functions. Dover Publ., New York

Adams, C.C. (1987): The noncompact hyperbolic 3-manifold of minimal volume. Proc. Am. Math. Soc. **100**, 601–606

Adams, C.C. (1988): Volumes of N-cusped hyperbolic 3-manifolds. J. London Math. Soc. **38**, 555–565

Agmon, S. (1965): Lectures on elliptic boundary value problems. D. van Nostrand Company, Inc., Princeton, N.J. Toronto New York London

Ahlfors, L.V. (1964): Eine Bemerkung über Fuchssche Gruppen. Math. Z. **84**, 244–245

Ahlfors, L.V. (1967): Hyperbolic motions. Nagoya Math. J. **29**, 163–166

Ahlfors, L.V. (1979): Complex analysis. Third edition. McGraw-Hill Book Company, New York London

Ahlfors, L.V. (1980): Ergodic properties of groups of Möbius transformations. (Analytic Functions, ed. by J. Lawrynowicz), Lect. Notes Math. **798** Springer, Berlin Heidelberg New York

Alekseevskij, D.V., Vinberg, E.B., Solodovnikov, A.S. (1991): Geometry of spaces of negative curvature. Encyclopaedia of Math. Sciences **29**, 1–139

Alonso, J.M., Brady, T., Cooper, D., Ferlini, V., Lustig, M., Mihalik, M., Shapiro, M., Short, H. (1991): Notes on word hyperbolic groups. In: Group theory from a geometric viewpoint. Word Scientific, Singapore

Andrianov, A. N. (1974): Euler products corresponding to Siegel modular forms of genus 2. Russian Math. Surveys **29**(3), 45–116 [= Uspekhi Mat. Nauk **29**(2) (1974), 43–110]

Andrianov, A.N., Fomenko, O.M. (1971): Distribution of the norms of the hyperbolic elements of the modular group and the class number of indefinite binary quadratic forms. Sov. Math. Dokl. **12**, 217–219 [= Dokl. Akad. Nauk SSSR **196** (1971), 743–745]

Apanasov, B.N. (1975a): A universal property of Kleinian groups in the hyperbolic metric. Sov. Math. Dokl. **16**, 1418–1421 [= Dokl. Akad. Nauk SSSR **222**(1) (1975)]

Apanasov, B.N. (1975b): On an analytic method in the theory of Kleinian groups on a multidimensional Euclidean space. Sov. Math. Dokl. **16**, 553–557 [= Dokl. Akad. Nauk SSSR **222**(1) (1975), 11–14]

Apanasov, B.N. (1976): Kleinian groups in space. Sib. Math. J. **16**, 679–684 [= Sib. Mat. Zh. **16**(5) (1975), 891–898]

Apanasov, B.N. (1979a): Entire automorphic functions on $\mathbb{R}^n$. Sib. Math. J. **19**, 518–528 [= Sib. Mat. Zh. **19**(4) (1978), 735–748]

Apanasov, B.N. (1979b): On Mostow's rigidity theorem. Sov. Math. Dokl. **19**(6), 1408–1412 [= Dokl. Akad. Nauk SSSR **243**(4) (1978), 829–832]

Apostol, T.M. (1976): Modular functions and Dirichlet series in number theory. Graduate texts in Math. 41, Springer, Berlin Heidelberg New York

Armstrong, M.A. (1965): On the fundamental group of an orbit space. Proc. Camb. Phil. Soc. **61**, 639–646

Armstrong, M.A. (1968): The fundamental group of a discontinuous group. Proc. Camb. Phil. Soc. **64**, 299–301

Artin E. (1957): Geometric Algebra. Interscience Publishers, New York

Asai, T. (1970): On a certain function analogous to $\log|\eta(z)|$. Nagoya Math. J. **40**, 193–211

Avakumović, V.G. (1956): Über die Eigenfunktionen auf geschlossenen Riemannschen Mannigfaltigkeiten. Math. Z. **65**, 327–344

Balslev, E., Venkov, A. (1997): The Weyl law for subgroups of the modular group. University of Aarhus, Dept. of Mathematics, Preprint Series No. 02

Bartels, H.J. (1982): Nichteuklidische Gitterpunktprobleme und Gleichverteilung in linearen algebraischen Gruppen. Comment. Math. Helv. **57**, 158–172

Bass, H. (1964): K-Theory and Stable Algebra. Publ. Math. IHES **22**, 489–544

Bauer, P. (1991): The Selberg trace formula for imaginary quadratic number fields of arbitrary class number. Bonner Mathematische Schriften, Bonn

Bauer, P. (1993): The Selberg Trace Formula for imaginary quadratic number fields. Japanese Journal of Mathematics **19**, 149–189

Beardon, A.F. (1977): The geometry of discrete groups. (Discrete Groups and Automorphic Functions, ed. by W.J. Harvey) Academic Press, London New York

Beardon, A.F. (1983): The geometry of discrete groups. Graduate texts in Mathematics 91, Springer, Berlin Heidelberg New York

Beardon, A.F., Nicholls, P.J. (1972): On classical series associated with Kleinian groups. J. London Math. Soc. **5**, 645–655

Bérard, P.H. (1976): Sur la fonction spectrale du Laplacien d'une variété riemannienne compacte sans points conjugués. C.R. Acad. Sci. Paris, Sér. A **283**, 45–48

Bérard, P.H. (1977): On the wave equation on a compact Riemannian manifold without conjugate points. Math. Z. **155**, 249–276

Bérard-Bergery, L. (1971): Sur les longueurs des géodésiques périodiques et le spectre des formes d'espace hyperbolique compactes. Séminaire de géométrie riemannienne 1970-1971 (M. Berger): Variétés à courbure négative, 84–126

Bérard-Bergery, L. (1973): Laplacien et géodésiques fermées sur les formes d'espace hyperbolique compactes. Séminaire Bourbaki, 24e année, 1971/72, exposé no 406. Lect. Notes Math. **317**, Springer, Berlin Heidelberg New York

Bérard-Bergery, L. (1981): La courbure scalaire des variétés riemanniennes. Sémin. Bourbaki, 32e année, Vol. 1979/80, Exp. 556, Lect. Notes Math. **842**, Springer, Berlin Heidelberg New York

Berger, M., Gauduchon, P., Mazet, E. (1971): Le spectre d'une variété riemannienne. Lect. Notes Math. **194**. Springer, Berlin Heidelberg New York

Best, L.A. (1971): On torsion free discrete subgroups of $\mathbf{PSL}(2,\mathbb{C})$ with compact orbit space. Canad. J. Math. **23**, 451–460

Bianchi, L. (1888): Sulle superficie Fuchsiane. Lincei Rend. **42**(4), 161–178

Bianchi, L. (1892): Sui gruppi di sostituzioni lineari con coefficienti appartenenti a corpi quadratici immaginari. Math. Ann. **40**, 332–412

Bianchi, L. (1893): Sopra alcune classi di gruppi di sostituzioni lineari a coefficienti complessi. Math. Ann. **43**, 101–112

Bianchi, L. (1952–1959): Opere. 11 Vols. Edizioni Cremonese, Roma

Boas, R.P. (1954): Entire functions. Academic Press, New York London

Bochner, S. (1932): Vorlesungen über Fouriersche Integrale. Akademische Verlagsgesellschaft, Leipzig; Reprint: Chelsea Publ. Comp., New York

Borel, A. (1960): Density properties for certain subgroups of semi-simple groups without compact components. Ann. Math. **72**, 179–188

Borel, A. (1963): Compact Clifford-Klein forms of symmetric spaces. Topology **2**, 111–122

Borel, A. (1969): Introduction aux groupes arithmétiques. Hermann, Paris

Borel, A. (1971): Sous-groupes discrets de groupes semi-simples (d'après D. A. Kajdan et G. A. Margulis). Séminaire Bourbaki, Exp. 358 (1968/69), Lect. Notes Math. **179**, Springer, Berlin Heidelberg New York

Borel, A. (1981): Commensurability classes and volumes of hyperbolic 3-manifolds. Ann. Sc. Norm. Super. Pisa, Cl. Sci., IV. Ser., **8**(1), 1–33

Borel, A. (1991): Linear algebraic groups. Graduate Texts in Math. **126**, Springer, Berlin Heidelberg New York

Borel, A., Casselman, W. (1979): Automorphic forms, Representations and L-functions I, II. Proceedings of Symposia in Pure Mathematics **33**, Amer. Math. Soc. Providence, R.I.

Borevich, Z.I., Shafarevich, I.R. (1966): Number theory. Academic Press, New York San Francisco London [Russian edition: Teoriya Čisel, Moscow (1964)]

Bourbaki, N. (1964a): Algèbre commutative. Éléments de Mathématiques, Hermann, Paris

Bourbaki, N. (1964b): Intégration. Éléments de Mathématiques, Hermann, Paris

Bourbaki, N. (1968): Groupes et algèbres de Lie. Éléments de Mathématiques, Hermann, Paris

Bowditch, B.H. (1991): Notes on Gromov's hyperbolicity criterion, In: Group theory from a geometric viewpoint. Word Scientific, Singapore

Braun, H. (1941): Zur Theorie der hermitischen Formen. Abh. Math. Semin. Univ. Hamburg **14**, 61–150

Brooks, R., Matelski, J.P. (1981): The dynamics of 2-generator subgroups of $\mathbf{PSL}(2, \mathbb{C})$. Ann. Math. Studies **97**, 65–71

Browder, F. E. (1959): Functional analysis and partial differential equations I. Math. Ann. **138**, 55–79

Browder, F. E. (1961): On the spectral theory of elliptic differential operators I. Math. Ann. **142**, 22–130

Browder, F. E. (1962): Functional analysis and partial differential equations II. Math. Ann. **145**, 81–226

Brüdern, J. (1995): Einführung in die analytische Zahlentheorie. Springer-Verlag, Berlin Heidelberg New York

Buser, P. (1976): Untersuchungen über den ersten Eigenwert des Laplace-Operators auf kompakten Flächen. Dissertation, Basel

Buser, P. (1977): Riemannsche Flächen mit Eigenwerten in $(0, \frac{1}{4})$. Comment. Math. Helv. **52**, 25–34

Buser, P. (1978a): Riemannsche Flächen mit großer Kragenweite. Comment. Math. Helv. **53**, 395–407

Buser, P. (1978b): The collar theorem and examples. Manuscr. Math. **25**, 349–357

Buser, P. (1978c): Über eine Ungleichung von Cheeger. Math. Z. **158**, 245–252

Buser, P. (1978d): Eine untere Schranke für λ_1 auf Mannigfaltigkeiten mit fast negativer Krümmung. Arch. Math. **30**, 528–531

Buser, P. (1978e): Cubic graphs and the first eigenvalue of a Riemann surface. Math. Z. **162**, 87–99

Buser, P. (1979a): Beispiele für λ_1 auf kompakten Mannigfaltigkeiten. Math. Z. **165**, 107–133.

Buser, P. (1979b): Dichtepunkte im Spektrum Riemannscher Flächen. Comment. Math. Helv. **54**, 431–439

Buser, P. (1979c): Über den ersten Eigenwert des Laplace-Operators auf kompakten Flächen. Comment. Math. Helv. **54**, 477–493
Buser, P. (1980a): Riemannsche Flächen und Längenspektrum vom trigonometrischen Standpunkt aus. Habilitationsschrift. Bonn
Buser, P. (1980b): On Cheeger's inequality $\lambda_1 \geq h^2/4$. Proc. Symp. Pure Math. **36**, 29–77
Buser, P. (1992): Geometry and spectra of Riemannian surfaces. Progress in Mathematics **106**, Birkhäuser, Boston Basel Berlin
Cassels, J.W.S. (1978): Rational Quadratic Forms. Academic Press, London New York San Francisco
Cao, C. (1995): Some trace inequalities for discrete groups of Möbius transformations. Proc. Am. Math. Soc. **123**, 3807–3815
Chavel, I. (1984): Eigenvalues in Riemannian Geometry. Academic Press, New York
Chen, S.S. (1979): Poincaré series and negatively curved manifolds. J. Reine Angew. Math. **305**, 77–81
Chen, S.S. (1981): Entropy of geodesic flow and exponent of convergence of some Dirichlet series. Math. Ann. **255**, 97–103
Chen, S.S., Greenberg, L. (1974): Hyperbolic spaces. (Contributions to analysis. A collection of papers dedicated to Lipman Bers), Academic Press, New York
Chernoff, P.R. (1973): Essential self-adjointness of powers of generators of hyperbolic equations. J. Funct. Anal. **12**, 401–414
Chevalley, C.C. (1946): Theory of Lie groups. Princeton University Press, Princeton
Chevalley, C.C. (1954): The algebraic theory of spinors. Columbia University Press, New York
Chinburg, T, Friedman, E. (1986): The smallest arithmetic hyperbolic three-orbifold. Invent. Math. **86**, 507–527
Chowla, T., Selberg, A. (1967): On Epstein's zeta-function. J. Reine Angew. Math. **227**, 86–110
Clozel, L. (1990): Motifs et Formes Automorphes: Applications du Principe de Fonctorialité. Proceedings of the 1988 Summer Conference on Automorphic Forms, Shimura Varieties and L-functions 77–159, Academic Press
Cogdell, J., Li, J.S., Piatetski-Shapiro, I., Sarnak, P. (1991): Poincaré Series for $\mathbf{SO}(n,1)$. Acta Math. **167**, 229–285
Cohen, P., Sarnak, P. (1979): Manuscript on discrete groups acting on hyperbolic spaces. Princeton
Cohn, H. (1980): Advanced number theory. Dover Publications, New York
Conder, M.D.E., Martin G.J. (1993): Cusps, Triangle Groups and Hyperbolic 3-Manifolds. J. Austral. Math. Soc. **55**, 149–182
Coxeter, H.S.M. (1956): Regular honeycombs in hyperbolic space. Proc. Internat. Congress Math., Amsterdam 1954, Vol. III, 155–169. Noordhoff N.V., Groningen and North-Holland Publishing Co., Amsterdam
Coxeter, H.S.M. (1956): The functions of Schläfli and Lobachevski. Quarterly J. Math. **6**, 13–29
Coxeter, H.S.M., Whitrow, G., J. (1950): World-structure and non-euclidean honey combs. Proc. Roy. Soc. A, **201**, 417–437
Coxeter, H.S.M., Moser, W.O.J. (1965): Generators and Relations for Discrete Groups. Springer-Verlag, Berlin Göttingen Heidelberg New York
Cremona, J. (1984): Hyperbolic tesselations, modular symbols and elliptic curves over complex quadratic fields. Compositio Math. **51**, 275–323
Davenport, H. (1980): Multiplicative number theory. Second edition, Springer Verlag, Berlin Heidelberg New York
Dedekind, R. (1931): Gesammelte mathematische Werke II. Braunschweig
Delsarte, J. (1942): Sur le gitter fuchsien. C.R. Acad. Sci., Paris **214**, 147–149

Delsarte, J. (1971): Le gitter fuchsien. Oeuvres de Jean Delsarte, Tome II, pp. 829-945. Éditions du Centre National de la Recherche Scientifique, Paris
Deshouillers, J.M., Iwaniec, H., Phillips, R., Sarnak, P. (1985): Maass cusp forms. Proc. Natl. Acad. Sci. USA **82**, 3533–3534
Deuring, M. (1935): Algebren. Ergebnisse der Mathematik, Vol. **4**, Springer, Berlin Heidelberg New York
Dickson, L. E. (1927): Algebren und ihre Zahlentheorie. Zürich [= Collected mathematical papers. IV, New York 1975, 7–320]
Dickson, L. E. (1952): History of the theory of numbers. Carnegie Institution of Washington Publ. (Reprinted New York 1952)
Dieudonné, J. (1960): Foundations of Modern Analysis. Academic Press, NewYork London
Dieudonné, J. (1971): La Géométrie des Groupes classiques. Ergebnisse der Mathematik, Vol. **5**, Springer, Berlin Heidelberg New York
Dintzl, E. (1909): Über die Zahlen im Körper $k(\sqrt{-2})$, welche den Bernoullischen Zahlen analog sind. Sitzungsber. Kaiserl. Akad. Wiss. Wien, Math.-Naturw. Kl. **118**, 173–200
Dintzl, E. (1914): Über die Entwicklungskoeffizienten der elliptischen Funktionen, insbesondere im Falle singulärer Moduln. Monatsh. Math. Phys. **25**, 125–151
Donnelly, H. (1982): On the cuspidal spectrum for finite volume symmetric spaces. J. Diff. Geom. **17**, 239–253
Doob, J.L. (1953): Stochastic processes. J. Wiley & Sons, Inc., New York
Duistermaat, J.J., Kolk, J.A.C., Varadarajan. V.S. (1979): Spectra of compact locally symmetric manifolds of negative curvature. Invent. Math. **52**, 27–93; Erratum, ibid. **54** (1979), 101
Dunford, N., Schwartz, J.T. (1958): Linear operators I, II. Interscience Publishers, New York
Earle, C.J. (1969): Some remarks on Poincaré series. Compos. Math. **21**, 167–176
Edwards, R.E. (1965): Functional analysis: Theory and applications. Holt, Rinehart and Winston, Inc., New York
Efrat, I., Sarnak, P. (1985): The determinant of the Eisenstein matrix and Hilbert class fields. Trans. Am. Math. Soc., **290**, 815–824
Eichler, M. (1938): Allgemeine Kongruenzklasseneinteilung der Ideale einfacher Algebren über algebraischen Zahlkörpern und ihre L-Reihen. J. Reine Angew. Math. **179**, 227–251
Eichler, M. (1952): Quadratische Formen und orthogonale Gruppen. Springer, Berlin Göttingen Heidelberg
Elstrodt, J. (1973a): Die Resolvente zum Eigenwertproblem der automorphen Formen in der hyperbolischen Ebene I. Math. Ann. **203**, 295–330
Elstrodt, J. (1973b): Die Resolvente zum Eigenwertproblem der automorphen Formen in der hyperbolischen Ebene II. Math. Z. **132**, 99–134
Elstrodt, J. (1973c): Einige Eigenschaften der Poincaréschen Reihen der Dimension −2. Manuscr. Math. **10**, 197–202
Elstrodt, J. (1974): Die Resolvente zum Eigenwertproblem der automorphen Formen in der hyperbolischen Ebene III. Math. Ann. **208**, 99–132
Elstrodt, J. (1981): Die Selbergsche Spurformel für kompakte Riemannsche Flächen. Jahresberichte Dtsch. Math. Verein. **83**, 45–77
Elstrodt, J. (1985): Note on the Selberg trace formula for the Picard group. Abh. Math. Sem. Univ. Hamburg **55**, 207–209
Elstrodt, J., Grunewald, F., Mennicke, J. (1982a): On the group $\mathbf{PSL}(2, \mathbb{Z}[i])$. Journées Arithmétiques 1980, LMS Lecture Notes **56**, 255–283
Elstrodt, J., Grunewald, F., Mennicke, J. (1982b): $\mathbf{PSL}(2)$ over imaginary quadratic integers. Journées Arithmétiques 1981, Astérisque **94**, 43–60

Elstrodt, J., Grunewald, F., Mennicke, J. (1983): Discontinuous groups on three-dimensional hyperbolic space: analytical theory and arithmetic applications. Russian Math. Surveys **38**(1), 137–168 [= Uspekhi Mat. Nauk **38**(1), (1983), 119–147]

Elstrodt, J., Grunewald, F., Mennicke, J. (1985): Eisenstein series on three-dimensional hyperbolic space and imaginary quadratic number fields. J. Reine Angew. Math. **360**, 160–213

Elstrodt, J., Grunewald, F., Mennicke, J. (1986): Eisenstein series for imaginary quadratic number fields. (The Selberg trace formula and related topics), Contemp. Math. **53**, 97–117

Elstrodt, J., Grunewald, F., Mennicke, J. (1987a): Zeta-functions of binary hermitian forms and special values of Eisenstein series. Math. Ann. **277**, 369–390

Elstrodt, J., Grunewald, F., Mennicke, J. (1987b): Vahlen's group of Clifford matrices and spin-groups. Math. Z. **196**, 369–390

Elstrodt, J., Grunewald, F., Mennicke, J. (1988): Arithmetic applications of the hyperbolic lattice point theorem. Proc. London Math. Soc. **57**, 239–283

Elstrodt, J., Grunewald, F., Mennicke, J. (1989): Some remarks on discrete subgroups of $\mathbf{SL}_2(\mathbb{C})$. J. Soviet Math. **46**, 1760–1788

Elstrodt, J., Grunewald, F., Mennicke, J. (1990): Kloosterman sums for Clifford algebras and a lower bound for the positive eigenvalues of the Laplacian for congruence subgroups acting on hyperbolic spaces. Invent. math. **101**, 641–685

Elstrodt, J., Roelcke, W. (1974): Über das wesentliche Spektrum zum Eigenwertproblem der automorphen Formen. Manuscr. Math. **11**, 391–406

Epstein, P. (1903): Zur Theorie allgemeiner Zetafunktionen I. Math. Ann. **56**, 614–644

Epstein, P. (1907): Zur Theorie allgemeiner Zetafunktionen II. Math. Ann. **63**, 205–216

Epstein, D., Petronio, C. (1994): An exposition of Poincaré's polyhedron theorem. L'Enseignement Math. **40**, 113–170

Epstein, B.A., Holt, D.F., Levy, S.V.F., Paterson, M.S., Thurston, W.P. (1992): Word processing in groups. Jones and Barlett Publishers

Erdélyi, A., Magnus, W., Oberhettinger, F., Tricomi, F. (1953): Higher transcendental functions Vol. 1–3. Bateman Manuscript Project, McGraw-Hill, New York Toronto London

Eymard, P. (1977): Le noyau de Poisson et la théorie des groupes. Symposia Mathematica **22**, 107–132, Academic Press, New York London

Faddeev, L. (1967): Expansion in eigenfunctions of the Laplace operator on the fundamental domain of a discrete group on the Lobachevskii plane. Trans. Moscow Math. Soc. **17**, 357–386 [= Tr. Mosk. Mat. O.-va **17**, 323–350 (1967) (Russian)]

Fay, J.D. (1977): Fourier coefficients of the resolvent for a Fuchsian group. J. Reine Angew. Math. **293/294**, 143–203

Fine, B. (1976): Fuchsian subgroups of the Picard group. Can. J. Math. **28**, 481–485

Fine, B. (1980): Congruence subgroups of the Picard group. Can. J. Math. **32**, 1474–1481

Fine, B. (1989): Algebraic theory of the Bianchi groups. Marcel Dekker, Inc., New York Basel

Fischer, J. (1987): An approach to the Selberg trace formula via the Selberg zeta-function. Lect. Notes Math. **1253**, Springer-Verlag: Berlin Heidelberg New York

Flöge, D. (1983): Zur Struktur der $\mathbf{PSL}_2$ über einigen imaginärquadratischen Zahlringen. Math. Zeit. **183**, 255–279

Ford, L.R. (1951): Automorphic functions. Second edition. Chelsea Publishing Company, New York

Fricke, R. (1914): Automorphe Funktionen mit Einschluß der elliptischen Modulfunktionen. (Encyklopädie der Mathematischen Wissenschaften, Vol. II B4, 349–470), B.G. Teubner Verlag, Leipzig

Fricke, R., Klein, F. (1897): Vorlesungen über die Theorie der automorphen Functionen I, II. B.G. Teubner, Leipzig. (Reprinted by Johnson Reprint Corp., New York, N.Y. and B.G. Teubner Verlagsgesellschaft, Stuttgart, 1965)

Fricker, F. (1968): Ein Gitterpunktproblem im dreidimensionalen hyperbolischen Raum. Comment. Math. Helv. **43**, 402–416

Fricker, F. (1971): Eine Beziehung zwischen der hyperbolischen Geometrie und der Zahlentheorie. Math. Ann. **191**, 293–312

Fricker, F. (1974): Die Geschichte des Kreisproblems. Mitt. Math. Semin. Gießen **111**, 1–34

Fricker, F. (1982): Einführung in die Gitterpunktlehre. Birkhäuser Verlag, Basel Boston Stuttgart

Frobenius, F. G. (1896): Über vertauschbare Matrizen. Sitzungsber. Königl. Preuß. Akad. Wiss. Berlin, 601–614 (Gesammelte Abhandlungen 2, 705–718, Springer, Berlin Heidelberg New York (1968))

Fueter, R. (1927): Über automorphe Funktionen in bezug auf Gruppen, die in der Ebene uneigentlich diskontinuierlich sind. J. Reine Angew. Math. **157**, 66–78

Fueter, R. (1931): Über automorphe Funktionen der Picardschen Gruppe, I. Comment. Math. Helv. **3**, 42–68

Fueter, R. (1936): Die Theorie der regulären Funktionen einer Quaternionenvariablen. C.R. Congrès International des Mathématiciens, Oslo, Tome I, 75–91

Gangolli, R. (1968): Asymptotic behaviour of spectra of compact quotients of certain symmetric spaces. Acta Math. **121**, 151–192

Gangolli, R. (1977a): The length spectra of some compact manifolds of negative curvature. J. Differ. Geom. **12**, 403–424

Gangolli, R. (1977b): Zeta functions of Selberg's type for compact space forms of symmetric spaces of rank one. Ill. J. Math. **21**, 1–42

Gangolli, R., Warner, G. (1975): On Selberg's trace formula. J. Math. Soc. Japan **27**, 328–343

Gangolli, R., Warner, G. (1980): Zeta functions of Selberg's type for some noncompact quotients of symmetric spaces of rank one. Nagoya Math. J. **78**, 1–44

Garland, H., Raghunathan, M. S. (1970): Fundamental domains for lattices in rank one semisimple Lie groups. Ann. Math. **92**, 279–326

Gehring, F. W., Martin, G. J. (1989): Stability and rigidity in Joergensen's inequality. Complex Variables **12**, 277–282

Gehring, F. W., Martin, G. J. (1991): Inequalities for Möbius transformations and discrete groups. J. Reine Angew. Math. **418**, 31–76

Gehring, F. W., Martin, G. J. (1993): 6-Torsion and hyperbolic volume. Proc. Am. Math. Soc. **117**, 727–735

Gelbart, S., S. (1975): Automorphic forms on adèle groups. Annals of Mathematics Studies **83**, Princeton Univ. Press, Princeton

Gelfand, I. M., Graev, M. I., Pyatetskii-Shapiro I. I. (1969): Representation Theory and Automorphic Functions. W. B. Saunders Company, Philadelphia London Toronto

Ghys, E., de la Harpe, P. (1990): Sur les groupes hyperboliques d'après Mikhael Gromov. Progress in Mathematics **83** Birkhäuser, Basel Boston Berlin

Gieseking, H. (1912): Analytische Untersuchungen über topologische Gruppen. Dissertation Münster

Godement, R. (1958): Série de Poincaré et Spitzenformen. Séminaire H. Cartan, 10e année (1957/58), Exposé 10. Secrétariat mathématique, École Normale Supérieure, Paris

Got, Th. (1933): Propriétés générales des groupes discontinus. Mémorial des Sciences Mathématiques **60**, Gauthier-Villars, Paris

Got, Th. (1934): Domaines fondamentaux des groupes fuchsiens et automorphes. Mémorial des Sciences Mathématiques **68**, Gauthier-Villars, Paris

Gradshteyn, I.S., Ryzhik, I.M. (1994): Table of Integrals, Series and Products. (Fifth Edition). Academic Press, London San Diego

Greenberg, L. (1966): Fundamental polyhedra for Kleinian groups. Ann. Math. **84**, 433–441

Greenberg, L. (1977): Finiteness theorems for Fuchsian and Kleinian groups. (Discrete Groups and Automorphic Functions, ed. by W.J. Harvey, 199–257) London New York

Gromov, M. (1978): Manifolds of negative curvature. J. Differ. Geom. **13**, 223–230

Gromov, M. (1987): Hyperbolic groups. In: S.M.Gersten (editor), Essays in group theory, MSRI Publ. **8**, 75–263

Gromov, M., Piatetski-Shapiro, I. (1986): Non-Arithmetic Groups in Lobachevsky Spaces. Publ. Math. IHES **66**, 93–103

Grunewald, F., Gushoff, A.-C., Mennicke, J. (1982): Komplex-quadratische Zahlkörper kleiner Diskriminante und Pflasterungen des dreidimensionalen hyperbolischen Raumes. Geometriae Dedicata **12**, 227–237

Grunewald, F., Helling, H., Mennicke, J. (1978): $\mathbf{SL}(2, \mathcal{O})$ over complex quadratic number fields. Algebra i Logika **17**, 512–580 (= Algebra and Logic **17** (1978), 332–382)

Grunewald, F., Hirsch, U. (1995): Link complements arising from arithmetic group actions. International Journal of Math. **6**, 337–370

Grunewald, F., Huntebrinker, W. (1996): A numerical Study of Eigenvalues of the Laplacian for Polyhedra with one Cusp. Experimental Mathematics, to appear

Grunewald, F., Kühnlein, S. (1996): On the proof of Humbert's volume formula. To appear

Grunewald, F., Mennicke, J. (1978): $\mathbf{SL}_2(\mathcal{O})$ and elliptic curves. Manuscript Bielefeld

Grunewald, F., Mennicke, J. (1980): Some 3-manifolds arising from $\mathbf{PSL}_2(\mathbb{Z}[i])$. Arch. Math. **35**, 275–291

Grunewald, F., Schwermer, J. (1981a): Free nonabelian quotients of $\mathbf{SL}_2$ over orders of imaginary quadratic numberfields. J. Algebra **69**, 298–304

Grunewald, F., Schwermer, J. (1981b): A nonvanishing result for the cuspidal cohomology of $\mathbf{SL}(2)$ over imaginary quadratic integers. Math. Ann. **258**, 183–200

Grunewald, F., Schwermer, J. (1981c): Arithmetic quotients of hyperbolic 3-space, cusp forms and link complements. Duke Math. J. **48**, 351–358

Grunewald, F., Schwermer, J. (1993): Subgroups of Bianchi groups and arithmetic quotients of hyperbolic 3-space. Trans. Am. Math. Soc. **335**, 47–78

Grunewald, F., Schwermer, J. (1996): On the concept of level for subgroups of $\mathbf{SL}_2$ over arithmetic rings. To appear

Gunning, R.C. (1962): Lectures on modular forms. Princeton University Press, Princeton, N.J.

Günther, P. (1979): Problème de réseaux dans les espaces hyperboliques. C.R. Acad. Sci., Paris, Sér. A. **288**, 49–52

Günther, P. (1980): Gitterpunktprobleme in symmetrischen Riemannschen Räumen vom Rang 1. Math. Nachr. **94**, 5–27

Hahn, H. (1912): Über die Integrale des Herrn Hellinger und die Orthogonalinvarianten der quadratischen Formen von unendlich vielen Veränderlichen. Monatsh. Math. Phys. **23**, 161–224

Harder, G. (1975a): On the cohomology of discrete arithmetically defined groups. Proc. of the Int. Colloq. on Discrete Subgroups of Lie groups and Applications to Moduli, 129–160, Oxford University Press

Harder, G. (1975b): On the cohomology of $\mathbf{SL}_2(\mathcal{O})$. Lie groups and their representations. Proc. of the summer school on group reps. 139-150, Hilger, London

Harder, G. (1979): Period integrals of cohomology classes which are represented by Eisenstein series. Automorphic forms, Representation theory and Arithmetic. Tata Institute of Fundamental Research, Bombay, 41–115

Harder, G. (1987): Eisenstein cohomology of arithmetic groups, the case $\mathbf{GL}_2$. Invent. Math. **89**, 37–118

Hardy, G.H., Wright, E.M. (1971): An Introduction to the Theory of Numbers. Oxford at the Clarendon Press, Oxford

Harris, M., Soudry, D., Taylor, R. (1993): l-adic representations associated to modular forms over imaginary quadratic fields I. Invent. Math. **112**, 377–411

Hartmann, D. (1974): Dreidimensionales Analogon zu den Heckeschen Substitutionsgruppen $G(\lambda)$. Diplomarbeit, Heidelberg

Harvey, W.J. (1977): Spaces of Discrete Groups. Discrete Groups and Automorphic Functions, ed. by W.J. Harvey, Academic Press, London San Francisco New York

Hasse, H. (1926): Bericht über neuere Untersuchungen und Probleme aus der Theorie der algebraischen Zahlkörper. Teil I. Jahresber. Dtsch. Math. Ver. **35**, 1–55; Teil Ia. Ibid. **36**, 233–311; Teil II. Ibid. Ergänzungsband VI; Berichtigungen ibid. **42**, 85–86

Hasse, H. (1964): Vorlesungen über Zahlentheorie. Grundlehren der math. Wissenschaften **19**, Springer, Berlin Heidelberg New York

Hatcher, A. (1983): Hyperbolic structures of arithmetic type on some link complements. J. London Math. Soc. **27**, 345–355

Heath-Brown, D.R., Patterson, S.J. (1979): The distribution of Kummer sums at prime arguments. J. Reine Angew. Math. **310**, 111–130

Hecke, E. (1923): Vorlesungen über die Theorie der algebraischen Zahlen. Leipzig (Reprinted New York (1970)), (English edition: Springer, Berlin Heidelberg New York (1983))

Hecke, E. (1927): Zur Theorie der elliptischen Modulfunktionen. Math. Ann. **97**, 210-242

Hejhal, D.A. (1975): The Selberg trace formula for congruence subgroups. Bull. Amer. Math. Soc. **81**, 752–755

Hejhal, D.A. (1976a): The Selberg trace formula for $\mathbf{PSL}(2,\mathbb{R})$ I. Lect. Notes Math. **548**. Springer, Berlin Heidelberg New York

Hejhal, D.A. (1976b): The Selberg trace formula and the Riemann zeta function. Duke Math. J. **43**, 441–482

Hejhal, D.A. (1978): Sur certaines séries de Dirichlet dont les pôles sont sur les lignes critiques. C.R. Acad. Sci., Paris, Sér. A **287**, 383–385

Hejhal, D.A. (1983): The Selberg trace formula for $\mathbf{PSL}(2,\mathbb{R})$ II. Lect. Notes Math. **1001**, Springer, Berlin Heidelberg New York

Hejhal, D.A., Rackner, B. (1992): On the topography of Maaß wave forms for $\mathbf{PSL}(2,\mathbb{Z})$: experiments and heuristics. Experimental Mathematics **1**, 275–305

Helgason, S. (1962): Differential geometry and symmetric spaces. Academic Press, New York London

Helling, H. (1966): Bestimmung der Kommensurabilitätsklasse der Hilbertschen Modulgruppe. Math. Z. **92**, 269–280

Helling, H., Mennicke, J., Vinberg, E.B. (1995): On some generalised triangle groups and 3-dimensional orbifolds. Trans. Moscow Math. Soc. **56**, 1–24

Helling, H., Kim, A.C., Mennicke, J. (1995): Some honeycombs in hyperbolic 3-space. Commun. Algebra **23**, 5169–5206

Hellinger, E. (1907): Die Orthogonalinvarianten quadratischer Formen von unendlich vielen Veränderlichen. Dissertation, Göttingen

Hellinger, E. (1909): Neue Begründung der Theorie quadratischer Formen von unendlich vielen Veränderlichen. J. Reine Angew. Math. **136**, 210–271

Hellwig, G. (1960): Partielle Differentialgleichungen. B.G. Teubner Verlagsgesellschaft, Stuttgart

Hellwig, G. (1964): Differentialoperatoren der mathematischen Physik, eine Einführung. Springer, Berlin Göttingen Heidelberg (English translation: Differential operators of mathematical physics. An introduction. Addison-Wesley Publishing Comp., Reading, Mass., (1967))

Herglotz, G. (1922): Über die Entwicklungskoeffizienten der Weierstraßschen $\wp$-Funktion. Berichte der Sächsischen Akad. Wiss. Leipzig. Math.-phys. Kl. **74**, 269–289 (= Gesammelte Schriften, Göttingen 1979, 436–456)

Hermite, Ch. (1854): Sur la théorie des formes quadratiques. J. Reine Angew. Math. **47**, 313–342

Hersonsky, S. (1993): Covolume estimates for discrete groups of hyperbolic isometries having parabolic elements. Illinois Math. J. **194**, 467–475

Hetrodt, G. (1994): Spectral theory on three-dimensional hyperbolic space and class numbers of biquadratic fields. C. R. Acad. Sci., Paris, Sér. I **319**,921–926

Hetrodt, G. (1996): Über einen Zusammenhang zwischen der Spektraltheorie automorpher Funktionen auf dem oberen Halbraum und den Klassenzahlen biquadratischer Körper. Schriftenreihe Math. Inst. Univ. Münster, 3. Ser., H. **17**, 1–80

Hilbert, D., Cohn-Vossen, S. (1932): Anschauliche Geometrie. Springer, Berlin (Reprinted by Wissenschaftliche Buchgesellschaft, Darmstadt, 1973), (English transl.: Geometry and the imagination. Chelsea Publ. Comp., New York, N.Y., 1952)

Hilden, H.M., Lozano, M.T., Montesinos-Amilibia, J.M. (1992a): On the Borromean orbifolds: Geometry and arithmetic. Topology '90, Contrib. Res. Semester Low Dimensional Topol., Columbus/OH (USA) 1990, Ohio State Univ. Math. Res. Inst. Publ. **1**, 133–167

Hilden, H.M., Lozano, M.T., Montesinos-Amilibia, J.M. (1992b): The arithmeticity of the figure eight knot orbifolds. Topology '90, Contrib. Res. Semester Low Dimensional Topol., Columbus/OH (USA) 1990, Ohio State Univ. Math. Res. Inst. Publ. **1**, 169–183

Hilden, H.M., Lozano, M.T., Montesinos-Amilibia, J.M. (1992c): A characterization of arithmetic subgroups of $\mathbf{SL}(2,\mathbb{R})$ and $\mathbf{SL}(2,\mathbb{C})$. Math. Nachr. **159**, 245–270

Hörmander, L. (1963): Linear partial differential operators. Springer-Verlag, Berlin Heidelberg New York

Hörmander, L. (1968): The spectral function of an elliptic operator. Acta Math. **121**, 193–218

Huber, H. (1956): Über eine neue Klasse automorpher Funktionen und ein Gitterpunktproblem in der hyperbolischen Ebene. Comment. Math. Helv. **30**, 20–62

Huber, H. (1959): Zur analytischen Theorie hyperbolischer Raumformen und Bewegungsgruppen I. Math. Ann. **138**, 1–26

Huber, H. (1961): Zur analytischen Theorie hyperbolischer Raumformen und Bewegungsgruppen II. Math. Ann. **142**, 385–398; Nachtrag zu II. Math. Ann. **143** (1961), 463–464

Huber, H. (1974): Über den ersten Eigenwert des Laplace-Operators auf kompakten Riemannschen Flächen. Comment. Math. Helv. **49**, 251–259

Huber, H. (1975): Über den ersten Eigenwert des Laplace-Operators auf kompakten Mannigfaltigkeiten konstanter negativer Krümmung. Arch. Math. **26**, 178–182

Huber, H. (1976): Über die Eigenwerte des Laplace-Operators auf kompakten Riemannschen Flächen I. Comment. Math. Helv. **51**, 215–231

Huber, H. (1977): Über die Darstellungen der Automorphismengruppe einer Riemannschen Fläche in den Eigenräumen des Laplace-Operators. Comment. Math. Helv. **52**, 177–184

Huber, H. (1978): Über die Eigenwerte des Laplace-Operators auf kompakten Riemannschen Flächen II. Comment. Math. Helv. **53**, 458–469

Huber, H. (1980a): On the spectrum of the Laplace operator on compact Riemann surfaces. Proc. Symp. Pure Math. **36**, 181–184

Huber, H. (1980b): Über die Dimension der Eigenräume des Laplace-Operators auf Riemannschen Flächen. Comment. Math. Helv. **55**, 390–397

Humbert, G. (1915): Sur la réduction des formes d'Hermite dans un corps quadratique imaginaire. C. R. Acad. Sci., Paris **161**, 189–196

Humbert, G. (1919a): Sur la transformation du domaine fondamental d'un groupe automorphe. C.R. Acad. Sci., Paris **169**, 205–211

Humbert, G. (1919b): Sur les représentations propres d'un entier par les formes positives d'Hermite dans un corps quadratique imaginaire. C. R. Acad. Sci., Paris **169**, 309–315

Humbert, G. (1919c): Sur les représentations d'un entier par les formes positives d'Hermite dans un corps quadratique imaginaire. C. R. Acad. Sci. Paris **169**, 360–365

Humbert, G. (1919d): Sur la mesure des classes d'Hermite de discriminant donné dans un corps quadratique imaginaire. C. R. Acad. Sci. Paris **169**, 407–414

Humbert, G. (1919e): Sur la mesure des classes d'Hermite de discriminant donné dans un corps quadratique imaginaire, et sur certaines volumes non euclidiens. C. R. Acad. Sci. Paris **169**, 448–454

Huntebrinker, W. (1991): Numerische Bestimmung von Eigenwerten des Laplace-Operators auf hyperbolischen Räumen mit adaptiven Finite-Element-Methoden. Bonner Mathematische Schriften **225**

Huntebrinker, W. (1994): Numerische Bestimmung von Eigenwerten des Laplace-Operators auf dreidimensionalen hyperbolischen Räumen mit Finite-Element-Methoden. Dissertation, Düsseldorf

Huppert, B. (1967): Endliche Gruppen I. Grundlehren der math. Wissenschaften **134**, Springer, Berlin Heidelberg New York

Hurwitz, A. (1895): Die unimodularen Substitutionen in einem algebraischen Zahlkörper. Nachr. k. Ges. Wiss. Göttingen, math.-phys. Kl. 332–356 (= Mathematische Werke **2**, Basel 1963, 244–268)

Hurwitz, A. (1899): Über die Entwicklungskoeffizienten der lemniskatischen Funktionen. Math. Ann. **51**, 196–226 (= Mathematische Werke **2**, Basel 1963, 342–373)

Huxley, M. (1984): Scattering matrices for congruence subgroups. In: Modular Forms, ed. by R. A. Rankin, Ellis Horwood Ltd., 141–156

Jacobson, N. (1980): Basic Algebra II. W.H. Freeman Comp., San Francisco

Jacquet, H., Shalika, J.A. (1976): A non-vanishing theorem for zeta functions of $\mathbf{GL}_n$. Invent. Math. **38**, 1–16

Jörgens, K., Rellich, F. (1976): Eigenwerttheorie gewöhnlicher Differentialgleichungen. Springer, Berlin Heidelberg New York

Jørgensen, T. (1976): On discrete groups of Möbius transformations. Amer. J. Math. **98**, 739–749

Jørgensen, T. (1978): Compact 3-manifolds of constant negative curvature fibering over the circle. Ann. Math. **106**, 61–72

Jørgensen, T., Marden A. (1988): Kleinian Groups with Quotients of Finite Volume. Ann. Acad. Sci. Fennicae **13**, 363–369

Karamata, J. (1931a): Neuer Beweis und Verallgemeinerung einiger Tauberian-Sätze. Math. Z. **33**, 294–299

Karamata, J. (1931b): Neuer Beweis und Verallgemeinerung der Tauberschen Sätze, welche die Laplacesche und Stieltjessche Transformation betreffen. J. Reine Angew. Math. **164**, 27–39

Kato, T. (1976): Perturbation Theory for Linear Operators. Grundlehren der math. Wissenschaften **132**, Springer, Berlin Heidelberg New York

Kazdhan, D. A., Margulis, G. A. (1968): A proof of Selberg's conjecture. Math. Sbornik N. S. **75**, 163–168

Kellerhals, R. (1991): The dilogarithm and volumes of hyperbolic polytopes. In: Structural properties of polylogarithms, ed. by L. Levin, Mathematical Surveys and Monographs **37**, 301–336, Amer. Math. Society, Providence, R.I.

Kellerhals, R. (1992): On the volumes of hyperbolic 5-orthoschemes and the trilogarithm. Comment. Math. Helv. **67**, 648–663

Kellerhals, R. (1995): Volumina von hyperbolischen Raumformen. Universität Bonn

Klein, F. (1872): Über Liniengeometrie und metrische Geometrie. Math. Ann. **5**, 256–277

Klein, F. (1884): Über die Transformation der allgemeinen Gleichung des zweiten Grades zwischen Linien-Coordinaten auf eine canonische Form. Math. Ann. **23**, 539–578

Klein, F. (1893): Vergleichende Betrachtungen über neuere geometrische Forschungen. Math. Ann. **43**, 63–100

Klein, F. (1921): Gesammelte mathematische Abhandlungen. 3 Bde. Springer, Berlin (Reprinted by Springer, Berlin Heidelberg New York 1973)

Klein, F. (1926): Vorlesungen über höhere Geometrie. 3. Aufl. Springer, Berlin (Reprinted by Springer, Berlin Heidelberg New York 1968, and Chelsea Publishing Company, New York, N.Y., 1957)

Klein, F. (1968): Vorlesungen über nicht-euklidische Geometrie. Springer, Berlin Heidelberg New York, 1968

Klein, F., Fricke, R. (1890): Vorlesungen über die Theorie der elliptischen Modulfunctionen. 2 Vols. B.G. Teubner Verlag, Leipzig (Reprinted by Johnson Reprint Corporation, New York, N.Y., and B.G. Teubner Verlagsgesellschaft, Stuttgart, 1965)

Klingholz, M. (1995): Kleine Eigenwerte auf hyperbolischen Räumen. Manuskript, Bonn

Kneser, M. (1956a): Klassenzahlen indefiniter quadratischer Formen in drei oder mehr Veränderlichen. Arch. Math. **7**, 323–332

Kneser, M. (1956b): Klassenzahlen definiter quadratischer Formen. Arch. Math. **8**, 241–250

Kneser, M. (1958): Klassenzahlen quadratischer Formen. Jahresber. Dtsch. Math.-Verein. **61**, 76–88

Kneser, M. (1974): Quadratische Formen. Vorlesungsausarbeitung, Mathematisches Institut der Universität Göttingen, Göttingen

Kolk, J. (1977a): The Selberg trace formula and asymptotic behaviour of spectra. Dissertation, Rijksuniversiteit Utrecht

Kolk, J. (1977b): Formule de Poisson et distribution asymptotique du spectre simultané d'opérateurs différentiels. C.R. Acad. Sci., Paris, Sér. A **284**, 1045–1048

Komatsu, H. (1962): A proof of Kotake and Narasimhan's theorem. Proc. Jap. Acad. **38**, 615–618

Koosis, P. (1992): The logarithmic integral I, II. Cambridge studies in advanced mathematics **21**, Cambridge University Press, Cambridge

Koszul, J., L. (1965): Lectures on groups of transformations. Tata Institute of Fundamental Research, Bombay,

Krushkal, S.L., Apanasov, B.N., Gusevskii, N.A. (1981): Kleinian groups and uniformization in examples and problems. Translations of Mathematical Monographs **62**, Amer. Math. Soc., Providence, R.I. (1986), (Russian edition: Nauka Sibirsk. Otdel., Novosibirsk)

Kubota, T. (1968a): Über diskontinuierliche Gruppen Picardschen Typus und zugehörige Eisensteinsche Reihen. Nagoya Math. J. **32**, 259–271

Kubota, T. (1968b): On a special kind of Dirichlet series. J. Math. Soc. Japan **20**, 193–207

Kubota, T. (1969): On automorphic functions and the reciprocity law in a number field. Lectures in Math. **2**, Kyoto University, Tokyo

Kubota, T. (1973): Elementary theory of Eisenstein series. J. Wiley & Sons, New York

Kühnlein, S. (1996): On a measure theoretic aspect of diophantine approximation. To appear in J. Reine Angew. Math.

Kuznetsov, N.V. (1978a): An arithmetical form of Selberg's trace formula and the distribution of norms of primitive hyperbolic classes of the modular group. Preprint. (Russ.) Far-eastern Science Centre, Academy of Sciences of the USSR, Habarovsk 1978. 44 pp. (Zentralblatt für Mathematik 381 (1979), review 10 022)

Kuznetsov, N.V. (1978b): Asymptotic formulas for the eigenvalues of the Laplacian on a fundamental domain of the modular group. Preprint. (Russ.) Far-eastern Science Centre, Academy of Sciences of the USSR, Habarovsk, 1978. 42 pp. (Zentralblatt für Mathematik 381 (1979), review 10 023).

Kuznetsov, N.V. (1978c): The distribution of norms of primitive hyperbolic classes of the modular group and asymptotic formulas for the eigenvalues of the Laplace-Beltrami operator on a fundamental region of the modular group. Soviet Math., Dokl. **19**, 1053–1056 (= Dokl. Akad Nauk SSSR **242** (1978), 40–43

Landau, E. (1918): Einführung in die elementare und analytische Theorie der algebraischen Zahlen und der Ideale. B.G. Teubner, Leipzig Berlin (Reprinted by Chelsea Publ. Comp., New York, N.Y.)

Landau, E. (1927): Vorlesungen über Zahlentheorie, I-III. Hirzel, Leipzig (Reprinted by Chelsea Publ. Comp., New York, N.Y. (1950))

Lang, S. (1968): Algebraic Number Theory. Addison Wesley, Reading Mass.

Lang, S. (1973): Elliptic functions. Reading, Mass.

Lang, S. (1975): $\mathbf{SL}(2,\mathbb{R})$. Addison-Wesley Publishing Company, 2nd printing: Springer, Berlin (1985)

Langlands, R. (1976): On the functional equations satisfied by Eisenstein series. Lect. Notes Math. 544, Springer, Berlin Heidelberg New York

Lanner, F. (1950): Hyperbolische Coxeter Gruppen. Commun. Sém. Math. Univ. Lund, **11**

Lax, P.D., Phillips, R. (1974): A scattering theory for automorphic functions. Séminaire Goulaouic-Schwartz 1973-1974, exposé no 23. École Polytechnique, Paris

Lax, P.D., Phillips, R. (1976): Scattering theory for automorphic functions. Annals of Mathematics Studies **87**, Princeton Univ. Press, Princeton

Lax, P.D., Phillips, R. (1980): Scattering theory for automorphic functions. Bull. Am. Math. Soc., New Ser. **2**, 261–295

Lax, P.D., Phillips, R. (1981): The asymptotic distribution of lattice points in Euclidean and non-Euclidean spaces. Functional analysis and approximation. Proc. Conf. Oberwolfach 1980. ISNM, Int. Ser. Numer. Math. **60**, 373–383

Lax, P.D., Phillips, R. (1982): The asymptotic distribution of lattice points in Euclidean and non-Euclidean spaces. J. Funct. Anal. **46**, 280–350

Lehner, J. (1964): Discontinuous groups and automorphic functions. American Mathematical Society, Providence, R.I.

Leutbecher, A. (1967): Über Spitzen diskontinuierlicher Gruppen von lineargebrochenen Transformationen. Math. Z. **100**, 183–200

Levitan, B.M. (1987): Asymptotic formulae for the number of lattice points in Euclidean and Lobachevski spaces. Russian Math. Surveys **42:3**, 13–42 (=Uspekhi Mat. Nauk **42:3**, (1987), 13–38)

Levin B.Y. (1996): Lectures on entire functions. Translations of Mathematical Monographs **150**, Am. Math. Soc., Providence, Rhode Island

Löbell, F. (1931): Beispiele geschlossener dreidimensionaler Clifford-Kleinscher Räume negativer Krümmung. Ber. Verh. Sächs. Akad. Wiss. **83**, 167–174

Löbell, F. (1955): Zur Konstruktion geschlossener Clifford-Kleinscher Räume negativer Krümmung. Sitzungsber. Bayer. Akad. Wiss., Math.-Naturwiss. Kl., München

Lubotzky, A. (1982): Free quotients and the congruence kernel of $\mathbf{SL}_2$. J. Algebra **77**, 411–418

Luo, W., Rudnick, Z., Sarnak, P. (1995): On Selberg's eigenvalue conjecture. Geometric and Functional Analysis **5**, 387–401

Lyndon, R., Schupp, P. (1977): Combinatorial group theory. Springer, Berlin Heidelberg New York

Lysionok, I.G. (1989): On some algorithmic properties of hyperbolic groups. Math. USSR Izvestia **35**, No. 4, 145–163

Maaß, H. (1948): Über die Erweiterungsfähigkeit der Hilbertschen Modulgruppe. Math. Z. **51**, 255–261

Maaß, H. (1949a): Über eine neue Art von nichtanalytischen automorphen Funktionen und die Bestimmung Dirichletscher Reihen durch Funktionalgleichungen. Math. Ann. **121**, 141–183

Maaß, H. (1949b): Automorphe Funktionen von mehreren Veränderlichen und Dirichletsche Reihen. Abh. Math. Semin. Univ. Hamb. **16**, 72–100

Maaß, H. (1953): Die Differentialgleichungen in der Theorie der elliptischen Modulfunktionen. Math. Ann. **125**, 235–263

Maaß, H. (1964): Lectures on modular functions of one complex variable. Tata Institute of Fundamental Research, Bombay

Maaß, H. (1971): Siegel's modular forms and Dirichlet series. Lect. Notes Math. **216**, Springer, Berlin Heidelberg New York

Macbeath, A.M. (1983): Commensurability of cocompact three-dimensional hyperbolic groups. Duke Math. J. **50**, 1245–1253 (Erratum: Duke Math. J. **56**, (1988), 219)

Maclachlan, C. (1996): Triangle subgroups of hyperbolic tetrahedral groups. Pacific J. Math. **176**, 195–203

Magnus, W. (1974): Noneuclidean tesselations and their groups. Academic Press, New York London

Magnus, W. (1975): Two generator subgroups of $\mathbf{PSL}(2, \mathbb{C})$. Nachr. Akad. Wiss. Gött., II. Math.-Phys. Kl. **7**, 81–94

Magnus, W., Karrass, A., Solitar, D. (1966): Combinatorial Group Theory. Interscience Publishers, New York London Sydney

Magnus, W., Oberhettinger, F., Soni, R.P. (1966): Formulas and theorems for the special functions of mathematical physics. Third edition. Springer, Berlin Heidelberg New York

Makarov, V.S. (1966): On a certain class of discrete Lobachevsky space groups with infinite fundamental domain of finite measure. Dokl. Akad. Nauk USSSR **167**, 30–33

Marden, A. (1974): Universal properties of Fuchsian groups in the Poincaré metric. (Discontinuous Groups and Riemann Surfaces), Proc. 1973 Conf. Univ. Maryland, 315–339

Marden, A. (1975): The geometry of finitely generated Kleinian groups. Ann. Math. **99**, 383–462 (Correction by T.W. Tucker, Ibid. **102** (1975), 565–566)

Marden, A. (1977): Geometrically finite Kleinian groups and their deformation spaces. (Discrete Groups and Automorphic Functions, ed. by W.J. Harvey), 259–293, Academic Press, London New York

Margulis, G. (1969): Applications of ergodic theory to the investigation of manifolds of negative curvature. Funct. Anal. Appl. **3**, 335–336, (transl. from Funkts. Anal. Prilozh.)

Margulis, G. (1974): Discrete groups of isometries of manifolds of nonpositive curvature. Proc. Int. Congress Math., Vancouver, Vol. 2, 21–34

Masani, P. (1968): Orthogonally scattered measures. Adv. Math. **2**, 61–117

Masani, P. (1970): Quasi-isometric measures and their applications. Bull. Am. Math. Soc. **76**, 427–528

Masani, P. (1972): Remarks on eigenpackets of self-adjoint operators. (Colloquia Math. Soc. J. Bolyai: 5. Hilbert space operators. Tihany, Hungary, Ed. by B. Sz.-Nagy) 415–441. North-Holland Publ. Comp., Amsterdam London

Maskit, B. (1965): On Klein's combination theorem. Trans. Amer. Math. Soc. **120**, 499–509

Maskit, B. (1968): On Klein's combination theorem II. Trans. Amer. Math. Soc. **131**, 32–39

Maskit, B. (1971): On Poincaré's theorem for fundamental polygons. Adv. Math. **7**, 219–230

Matter, K. (1900): Die den Bernoullischen Zahlen analogen Zahlen im Körper der dritten Einheitswurzeln. Vierteljahrsschrift d. Naturforsch. Gesellschaft in Zürich **45**, 238–269

McKean, H.P. (1972): Selberg's trace formula as applied to a compact Riemann surface. Commun. Pure Appl. Math. **25**, 225–246, Correction: ibid. **27** (1974), 134

Mendoza, E.R. (1980): Cohomology of PGL_2 over imaginary quadratic integers. Bonner Mathematische Schriften **128**, Mathematisches Institut der Universität Bonn, Bonn

Mennicke, J. (1967): On Ihara's modular group. Invent. Math. **4**, 202–228

Mennicke, J. (1980): Pflasterung des dreidimensionalen hyperbolischen Raumes. Math.-Phys. Semesterber., Neue Folge **27**, 55–68

Mennicke, J. (1984): Discontinuous groups. In: Groups. Proc. Conf. on combinatorial group theory, Kyongju/Korea. Lect. Notes Math. **1098**, 75–80, Springer Verlag, Berlin Heidelberg New York

Mennicke, J. (1989): On Fibonacci groups and some other groups. In: Group Theory. Proc. Conf. Pusan/Korea. Lect. Notes Math. **1398**, 117–123, Springer Verlag, Berlin Heidelberg New York

Meyerhoff, R. (1985): The cusped hyperbolic 3-orbifold of minimum volume. Bull. Am. Math. Soc. **13**, 154–156

Meyerhoff, R. (1987): A lower bound for the volume of hyperbolic 3-manifolds. Can. J. Math. **39**, 1038–1056

Meyerhoff, R. (1988a): A lower bound for the volume of hyperbolic 3-orbifolds. Duke Math. J. **57**, 185–203

Meyerhoff, R. (1988b): Sphere packing and volume in hyperbolic 3-space. Comment. Math. Helv. **61**, 271–278

Meyerhoff, R. (1992): Geometric invariants of 3-manifolds. Math. Intelligencer **14**, 37–53

Milnor, J. (1982): Hyperbolic geometry: The first 150 years. Bull. Am. Math. Soc., New Ser. **6**, 9–24

Milnor, J. (1982): Computation of volume. (in Thurston (1980))

Moreno, C. J. (1983): The Chowla-Selberg formula. J. Number Theory **17**, 226–245
Morokuma, T. (1976): On discontinuous groups acting on a real hyperbolic space I. Proc. Japan Acad., Ser. A, **52**, 359–362. II. Ibid. **53** (1977), 11–14
Mostow, G.D. (1973): Strong rigidity of locally symmetric spaces. Annals of Mathematics Studies, Princeton University Press, Princeton
Mostow, G.D. (1985): Discrete subgroups in Lie groups. Astérisque, 289–309
Müller, W. (1980a): Spectral theory for Riemannian manifolds with cusps and a related trace formula. Math. Nachr. **111**, 197–288
Müller, W. (1980b): Manifolds with cusps of rank one. Lect. Notes Math. **1244**, Springer, Berlin Heidelberg New York
Munkholm, H.J. (1980): Simplices of maximal volume in hyperbolic space, Gromov's norm, and Gromov's proof of Mostow's rigidity theorem (follow. Thurston). Lect. Notes Math. **788**, 109–124
Narishkina, E. A. (1924): On numbers analogous to the Bernoulli numbers in quadratic number fields of class number one and negative discriminant I, II. Bulletin de l'Académie des Sciences de Russie **19**, 145–176, 297–314 (Russian), (See also Proc. Int. Math. Congr. Toronto 1924. I, Toronto 1928, 299–307, reprinted Nendeln, Liechtenstein, 1967).
Neumann, P.M. (1973): The SQ-universality of some finitely presented groups. J. Austral. Math. Soc. **16**, 1–6
Neumann, W., Zagier, D. (1985): Volumes of hyperbolic 3-manifolds. Topology **24**, 307–332
Neunhöffer, H. (1968): Zur Spektralzerlegung von $L^2(G/\Gamma)$: Abspaltung des kontinuierlichen Teils des Spektrums für $G = \mathbf{SL}(2, \mathbb{C}), \Gamma = \mathbf{SL}(2, \mathcal{G})$, mit $\mathcal{G}$ = Hauptordnung im Gaußchen Zahlkörper. Diplomarbeit, Heidelberg
Neunhöffer, H. (1973): Über die analytische Fortsetzung von Poincaréreihen. Sitzungsber. Heidelb. Akad. Wiss., Math.-Naturwiss. Kl., Jahrgang 1973, 2. Abh., 33–90
Neunhöffer, H. (1978): Über Kronecker-Produkte irreduzibler Darstellungen von $\mathbf{SL}(2, \mathbb{R})$. Sitzungsber. Heidelb. Akad. Wiss., Math.-Naturwiss. Kl., Jahrgang 1978, 3 Abh., 167–260
Nicholls, P.: (1974): On the distribution of orbits in a Kleinian group. Proc. London Math. Soc., III. Ser., **29**, 193–215
Niebur, D. (1973): A class of nonanalytic automorphic functions. Nagoya Math. J. 52, 133–145
Nikulin, V.V. (1981): On arithmetic groups generated by reflections in Lobachevsky spaces. Math. USSR, Izv. **16**, 573–601, (= Izv. Akad. Nauk SSSR, Ser. Math. **44**, 637–669 (1980))
Oberhettinger, F. (1957): Tabellen zur Fourier–Transformation. Springer, Berlin Heidelberg New York
Parnovski, L. B. (1992): The Selberg trace formula and Selberg zeta-function for cocompact discrete subgroups of $\mathbf{SO}_+(1, n)$. Funct. Anal. Appl. **26**, 196–202
Parthasarathy, K.R. (1967): Probability measures on metric spaces. Academic Press, New York London
Patterson, S.J. (1975a): A lattice-point problem in hyperbolic space. Mathematika **22**, 81–88
Patterson, S.J. (1975b): The Laplacian operator on a Riemann surface. Compos. Math. **31**, 83–107. II. Ibid. **32** (1976), 71–112. III. Ibid. **33** (1976), 227–259
Patterson, S.J. (1976a): The limit set of a Fuchsian group. Acta Math. **136**, 241–273
Patterson, S.J. (1976b): Spectral theory and Fuchsian groups. Math. Proc. Camb. Philos. Soc. **81**, 59–75
Patterson, S.J. (1977): A cubic analogue of the theta series. J. Reine Angew. Math. **296**, 125–161. II. Ibid. **296** (1977), 217–220

Patterson, S.J. (1978a): On Dirichlet series associated with cubic Gauss sums. J. Reine Angew. Math. **303/304**, 102–125
Patterson, S.J. (1978b): On the distribution of Kummer sums. J. Reine Angew. Math. **303/304**, 126–143
Pepin, P. (1890): Sur quelques formes quadratiques quaternaires. J. Math. Pures Appl. **6**, 5–67
Petersson, H. (1938a): Zur analytischen Theorie der Grenzkreisgruppen, Teil II. Math. Ann. **115**, 175–204
Petersson, H. (1938b): Über die eindeutige Bestimmung und die Erweiterungsfähigkeit von gewissen Grenzkreisgruppen. Abh. Math. Sem. Univ. Hamburg **12**, 180–199
Petersson, H. (1982): Modulfunktionen und quadratische Formen. Springer, Berlin Heidelberg New York
Petrowskij, I.G. (1953): Vorlesungen über Integralgleichungen. Physica Verlag, Würzburg
Phillips, R.S., Sarnak, P. (1985a): On cusp forms for cofinite subgroups of the group $\mathbf{PSL}(2,\mathbb{R})$. Inv. Math. **80**, 339–364
Phillips, R.S., Sarnak, P. (1985b): The Weyl Theorem and deformation of discrete groups. Commun. Pure Appl. Math. **38**, 853–866
Phillips, R.S., Sarnak, P. (1991): The spectrum of Fermat curves. Geom. and Funct. Anal. **1**, 80–146
Phillips, R.S., Sarnak, P. (1992): Automorphic spectrum and Fermi's golden rule. J. Anal. Math. **59**, 179–187
Phillips, R.S., Sarnak, P. (1994): Cusp forms for character varieties. Geom. and Funct. Anal. **4**, 93–118
Picard, E. (1978): Oeuvres. 4 Vols. Éditions du Centre National de la Recherche Scientifique, Paris
Poincaré, H. (1882): Théorie des groupes fuchsiens. Acta Math. **1**, 1–62
Poincaré, H. (1883): Mémoire sur les groupes kleinéens. Acta Math. **3**, 49–92 (= Oeuvres de Henri Poincaré, Tome II, 258–299)
Poincaré, H. (1916): Oeuvres. 11 Vols. Gauthier-Villars, Paris
Postnikov, A.G. (1980): Tauberian theory and its applications. Proc. Steklov Inst. Math. Issue **2**
Prasad, G. (1973): Strong rigidity of $\mathbb{Q}$-rank 1 lattices. Invent. Math. **21**, 255–286
Prokhorov, M.N. (1987): The absence of discrete reflection groups with noncompact fundamental polyhedron of finite volume in Lobachevsky space of large dimension. Math. USSR Izvestiya **28**, 401–411
Ramanujan, S. (1927): Collected papers. Cambridge (Reprinted New York 1962)
Randol, B. (1974): Small eigenvalues of the Laplace operator on compact Riemann surfaces. Bull. Am. Math. Soc. **80**, 996–1000
Randol, B. (1975): On the analytic continuation of the Minakshisundaram–Pleijel zeta function for compact Riemann surfaces. Trans. Am. Math. Soc. **201**, 241–246
Randol, B. (1977): On the asymptotic distribution of closed geodesics on compact Riemann surfaces. Trans. Am. Math. Soc. **233**, 241–247
Randol, B. (1978): The Riemann hypothesis for Selberg's zeta-function and the asymptotic behavior of eigenvalues of the Laplace operator. Trans. Am. Math. Society **236**, 209–223
Rao, K.V. Rajeswara (1969): Fuchsian groups of convergence type and Poincaré series of dimension −2. J. Math. Mech. **18**, 629–644

Reed, M., Simon, B. (1972): Methods of modern mathematical physics. Vol. I: Functional analysis, Vol. II: Fourier analysis, self-adjointness, Vol. III: Scattering theory, Vol. IV: Analysis of operators. Academic Press, New York San Francisco London

Reid, A.W. (1990): A note on trace fields of Kleinian groups. Bull. London Math. Soc. **22**, 349–352

Reid, A.W. (1991): Arithmeticity of knot complements. J. London Math. Soc. **43**, 171–184

Reid, A.W. (1992): Isospectrality and commensurability of arithmetic hyperbolic 2- and 3-manifolds. Duke Math. J. **65**, 215–228

Rellich, F. (1951): Spectral theory of a second-order ordinary differential operator. Lectures delivered 1950-1951. Institute for Mathematics and Mechanics. New York University, New York

Rellich, F. (1952): Eigenwerttheorie partieller Differentialgleichungen I, II. Vorlesungen, gehalten an der Universität Göttingen. Mathematisches Institut der Universität Göttingen, Göttingen

Riggenbach, H. (1975): Freie Homotopieklassen und das Eigenwertspektrum des Laplace-Operators bei hyperbolischen Raumformen. Dissertation, Universität Basel

Riley, R. (1975): A quadratic parabolic group. Math. Proc. Camb. Phil. Soc. **77**, 281–288

Roelcke, W. (1956a): Über die Wellengleichung bei Grenzkreisgruppen erster Art. Sitzungsber. Heidelb. Akad. Wiss., Math.-Naturwiss. Kl. 4. Abh.

Roelcke, W. (1956b): Analytische Fortsetzung der Eisensteinreihen zu den parabolischen Spitzen von Grenzkreisgruppen erster Art. Math. Ann. **132**, 121–129

Roelcke, W. (1960): Über den Laplace-Operator auf Riemannschen Mannigfaltigkeiten mit diskontinuierlichen Gruppen. Math. Nachr. **21**, 131–149

Roelcke, W. (1961): Über diskontinuierliche Gruppen in lokal-kompakten Räumen. Math. Z. **75**, 36–52

Roelcke, W. (1966): Das Eigenwertproblem der automorphen Formen in der hyperbolischen Ebene, I. Math. Ann. **167**, 292–337

Roelcke, W. (1967): Das Eigenwertproblem der automorphen Formen in der hyperbolischen Ebene, II. Math. Ann. **168**, 261–324

Rohlfs, J. (1978): Über maximale arithmetisch definierte Gruppen. Math. Ann. **234**, 239–252

Rohlfs, J. (1985): On the cuspidal cohomology of the Bianchi modular groups. Math. Z. **188**, 253–269

Rubel, A., Taylor, B.A. (1968): A Fourier series method for meromorphic and entire functions. Bull. Soc. math. France, 53–96

Rudin, W. (1973): Functional Analysis. McGraw-Hill Book Company, New York London New Dehli

Rudin, W. (1974): Real and complex analysis. Second edition. McGraw-Hill Book Company, New York London. Third edition (1987)

Ruzmanov, O. (1990): Reflection subgroups in Bianchi groups. Russian Math. Surveys **45**, 227–228

Sarnak, P. (1980): Prime geodesic theorems. Dissertation, Stanford University

Sarnak, P. (1983): The arithmetic and geometry of some hyperbolic three manifolds. Acta Math. **151**, 253–295

Sarnak, P. (1986): On cusp forms. Contemp. Math. **53**, 393–407

Sarnak, P. (1990): On cusp forms II. Festschrift in honour of I. I. Piatetski-Shapiro. Israel mathematical conference proceedings **2**, 237–250

Satake, I. (1966): Spherical functions and Ramanujan conjecture. Proc. Symp. Pure Math. **9**, 258–254

Sato, M., Shintani, T. (1974): On zeta functions associated with prehomogeneous vector spaces. Ann. Math. **100**, 131–170

Schneider, J. (1985): Diskrete Untergruppen von $\mathbf{SL}(2,\mathbb{C})$ und ihre Operation auf dem drei-dimensionalen hyperbolischen Raum. Diplomarbeit, Bonn

Schoen, R. (1982): A lower bound for the first eigenvalue of a negatively curved manifold. J. Differ. Geom. **17**, 233–238

Schoeneberg, B. (1968): Über das unendliche Produkt $\prod_{k=1}^{\infty}(1-x^k)$. Mitt. Math. Ges. Hamburg **9**, 4–11

Schueth, D. (1993): Isospectral, non-isometric Riemannian manifolds. Suppl. Rend. Circ. Mat. Palermo, II. Ser. **37**, 207–231

Schwermer, J. (1980): A note on link complements and arithmetic groups. Math. Ann. **249**, 107–110

Schwermer, J., Vogtmann, K. (1983): The integral homology of $\mathbf{SL}_2$ and $\mathbf{PSL}_2$ of euclidean imaginary quadratic integers. Comment. Math. Helv. **58**, 573–598

Seifert, H., Weber, C. (1933): Die beiden Dodekaederräume. Math. Z. **37**, 237–253

Selberg, A. (1954): Harmonic analysis. Hand-written lecture notes, Göttingen (published in (1989a), Vol. 1)

Selberg, A. (1956): Harmonic analysis and discontinuous groups in weakly symmetric Riemannian spaces with applications to Dirichlet series. J. Indian Math. Soc. **20**, 47–87

Selberg, A. (1960): On discontinuous groups in higher-dimensional spaces. Contributions to function theory (International Colloquium on Function Theory), Tata Institute of Fundamental Research, Bombay, 147–164

Selberg, A. (1962): Discontinuous groups and harmonic analysis. Proceedings of the International Congress of Mathematicians, Stockholm, 177–189

Selberg, A. (1965): On the estimation of Fourier coefficients of modular forms. Proc. Symp. Pure Math. **8**, 1–15. Amer. Math. Soc., Providence, R.I.

Selberg, A. (1989a): Collected papers. 2 Vols. Springer, Berlin Heidelberg New York

Selberg, A. (1989b): Harmonic analysis. Introduction to the Göttingen lecture notes. (Collected Papers, Vol. 1), Springer, Berlin Heidelberg New York

Selberg, A. (1990): Remarks on the distribution of poles of Eisenstein series. Festschrift in honor of I.I. Piatetski-Shapiro, Weizmann Science Press, Israel **2**, 251–278, (Collected Papers, Vol. 2)

Serre, J.P. (1970): Le problème de groupes de congruence pour $\mathbf{SL}_2$. Ann. Math. **92**, 489–657

Serre, J.P. (1977): Modular forms of weight one and Galois representations. Algebraic Number Fields (ed. by A. Fröhlich), Acad. Press, 193–268

Serre, J.P. (1987): Sur les représentations modulaires de degrée 2 de $\mathrm{Gal}(\bar{\mathbb{Q}}/\mathbb{Q})$. Duke Math. J. **54**, 179–230

Shaikheev, M. (1987): Reflection subgroups in Bianchi groups. Problems in Group Theory and Homological Aspects, Jaroslawl, 127–134, (Engl. transl. in Sel. Math. Sov. **9**, (1990))

Shvartsman, O.V. (1969): On discrete arithmetic subgroups of complex Lie groups. Math. USSR Sbornik **6**, 501–503

Shvartsman, O.V. (1987): Reflection subgroups in Bianchi groups. Problems in Group Theory and Homological Aspects, Jaroslawl, 134–139 (Engl. transl. in Sel. Math. Sov. **9**, (1990))

Shimizu, H. (1963): On discontinuous groups operating on the product of the upper half planes. Ann. Math., II. Ser. **77**, 33–71

Shimura, G. (1971): Introduction to the arithmetic theory of automorphic functions. Iwanami Shoten, Tokyo and Princeton University Press, Princeton

Siegel, C.L. (1935): Über die analytische Theorie quadratischer Formen. Ann. Math. **36**, 527–606

Siegel, C.L. (1945): Some remarks on discontinuous groups. Ann. Math. **46**, 708–718

Siegel, C.L. (1955): Lectures on quadratic forms. Tata Institute of Fundamental Research, Bombay

Siegel, C.L. (1963): Lectures on the analytical theory of quadratic forms. The Institute for Advanced Study and Princeton University. Third revised edition. Buchhandlung R. Peppmüller, Göttingen

Siegel, C.L. (1966a): Gesammelte Abhandlungen. 4 Vols. Springer, Berlin Heidelberg New York

Siegel, C.L. (1966b): Vorlesungen über ausgewählte Kapitel der Funktionentheorie, Teil I-III. Lecture Notes, Mathematisches Institut der Universität Göttingen (= Topics in complex function theory. Wiley-Interscience, New York, Vol. I 1969, Vol. II 1971, Vol. III 1973)

Smithies, F. (1941): The Fredholm theory of Integral equations. Duke Math. J. **8**, 107–130

Spivak, M. (1965): Calculus on manifolds. W.A. Benjamin, Inc., Reading, Mass.

Stone, M.H. (1932): Linear transformations in Hilbert space and their applications to analysis. Amer. Math. Soc., New York

Stramm, K. (1994): Kleine Eigenwerte des Laplace-Operators zu Kongruenzuntergruppen. Schriftenreihe des Math. Instituts der Universität Münster **11**, Münster

Strichartz, R.S. (1983): Analysis of the Laplacian on a complex Riemannian manifold. J. Funct. Anal. **52**, 48–79

Sturm, J., Shinnar, M. (1974): The maximal inscribed ball of a Fuchsian group. (Discontin. Groups and Riemann Surfaces, Proc. 1973 Conf. Univ. Maryland) 439–443

Subia, N. (1975): Formule de Selberg et formes d'espaces hyperboliques compactes. (Séminaire Nancy-Strasbourg, 1973-75). Lect. Notes Math. **497**, 674–700 Springer, Berlin Heidelberg New York

Sudberry, A. (1979): Quaternionic analysis. Math. Proc. Camb. Philos. Soc. **85**, 199–224

Sullivan, D. (1978): On the ergodic theory at infinity of an arbitrary discrete group of hyperbolic motions. (Proceedings of the Stony Brook Conference on Riemann Surfaces and Kleinian Groups) Ann. Math. Stud. **97**, 465–496

Sullivan, D. (1979): The density at infinity of a discrete group of hyperbolic motions. Publ. Math., Inst. Hautes Etud. Sci. **50**, 171–202

Sullivan, D. (1986): Related aspects of positivity: λ-potential theory on manifolds, lowest eigenstates, Hausdorff geometry, renormalized Markoff processes... In: Aspects of Mathematics and its Applications, ed. by J.A. Barroso. Elsevier Science Publishers B.V., Amsterdam

Swan, R.G. (1971): Generators and relations for certain special linear groups. Adv. Math. **6**, 1–77

Szmidt, J. (1983): The Selberg trace formula for the Picard group $\mathbf{SL}(2, \mathbb{Z}[i])$. Acta Arith. **42**, 391–424

Tan, D. (1989): On two generator discrete groups of Möbius transformations. Proc. Am. Math. Soc. **106**, 763–770

Tanigawa, Y. (1977): Selberg trace formula for Picard groups. In: Int. Symp. Algebraic Number Theory, Kyoto, Japan Society for the Promotion of Science, 229–242

Taylor, R. (1994): l-adic representations associated to modular forms over imaginary quadratic fields. Invent. Math. **116**, 619–643

Terras, A. (1977): The Fourier expansion of Epstein's zeta function over an algebraic number field and its consequences for algebraic number theory. Acta Arith. **32**, 37–53

Terras, A. (1985): Harmonic Analysis on Symmetric Spaces and Applications I, II. Springer, New York, Berlin, Heidelberg, Tokyo

Thurnheer, P. (1979): Über den Restterm einer hyperbolischen Gitterpunktfunktion. Dissertation, ETH Zürich

Thurnheer, P. (1980): Le terme de reste dans un problème de réseau hyperbolique. C.R. Acad. Sci., Paris, Sér. A **290**, 581–583

Thurston, W. (1978): Geometry and topology of 3-manifolds. Mimeographed lecture notes. Princeton University, Princeton

Thurston, W. (1982): Three dimensional manifolds, Kleinian groups, and hyperbolic geometry. Bull. Am. Math. Soc. **6**, 357–381

Thurston, W. (1997): Three-Dimensional Geometry and Topology. Princeton University Press, Princeton, New Jersey

Titchmarsh, E.C. (1939): The theory of functions. Second edition. Oxford University Press, Oxford-London

Titchmarsh, E.C. (1951): The theory of the Riemann zeta-function. Oxford University Press, Oxford-London

Tits, J. (1966): Classification of algebraic semisimple groups. Proc. Symp. Pure Math. **9**, 32–62

Tsuji, M. (1959): Potential theory in modern function theory. Maruzen, Tokyo (Reprinted by Chelsea Publ. Comp., New York, 1975)

Unell, P.M. (1980): Self-adjointness of certain second order differential operators on Riemannian manifolds. J. Math. Anal. Appl. **73**, 351–365

Venkov, A.B. (1971): Expansion in automorphic eigenfunctions of the Laplace operator and the Selberg trace formula in the space $\mathbf{SO}_0(n,1)/\mathbf{SO}(n)$. Sov. Math., Dokl. **12**, 1363–1366 (= Dokl. Akad. Nauk SSSR **200** (1971), 266–269)

Venkov, A.B. (1973): Expansion in automorphic eigenfunctions of the Laplace-Beltrami operator in classical symmetric spaces of rank one, and the Selberg trace formula. Proc. Steklov Inst. Math. **125**, 6–55

Venkov, A.B. (1977a): On an asymptotic formula connected with the number of eigenvalues corresponding to odd eigenfunctions of the Laplace-Beltrami operator on a fundamental region of the modular group $\mathbf{PSL}(2,\mathbb{Z})$. Sov. Math., Dokl. **18**, 524–526 (= Dokl. Akad. Nauk SSSR **233** (1977), 1021–1023)

Venkov, A.B. (1977b): On the space of cusp forms for certain Fuchsian groups generated by reflections. Sov. Math., Dokl. **18**, 1214–1217 (= Dokl. Akad. Nauk SSSR **236** (1977), 525–527)

Venkov, A.B. (1978a): Selberg's trace formula for the Hecke operator generated by an involution, and the eigenvalues of the Laplace-Beltrami operator on the fundamental domain of the modular group $\mathbf{PSL}(2,\mathbb{Z})$. Math. USSR Izvestija **12**, 448–462 (= Izvestija Akad. Nauk SSSR, Ser. mat. **42** (1978) 484–499)

Venkov, A.B. (1978b): On the space of cusp functions for a Fuchsian group of the first kind with nontrivial commensurator. Sov. Math., Dokl. **19**, 343–347 (= Dokl. Akad. Nauk SSSR **239** (1978), 511-514)

Venkov, A.B. (1978c): Selberg's trace formula and non-euclidean vibrations of an infinite membrane. Sov. Math. Dokl. **19**, 708–712 (= Dokl. Akad. Nauk SSSR **240** (1978), 1021–1024)

Venkov, A.B. (1979a): The Selberg trace formula for $\mathbf{SL}(3,\mathbb{Z})$. J. Soviet Math. **12**, 384–424 (= Zap naucn. Sem. Leningrad. Otd. mat. Inst. Steklova **63** (1976), 8–66)

Venkov, A.B. (1979b): The Artin-Takagi formula for Selberg's zeta-function and the Roelcke conjecture. Sov. Math. Dokl. **20**, 745–748 (= Dokl. Akad. Nauk SSSR **247** (1979), 540–543)

Venkov, A.B. (1979c): Spectral theory of automorphic functions, the Selberg zeta-function and some problems of analytic number theory and mathematical physics. Russ. math. Surveys **34**, 79–153 (=Uspekhi Mat. Nauk **34**, No. 3 (1979), 69–135)

Venkov, A.B. (1979d): On the remainder term in the Weyl-Selberg asymptotic formula. (Russ.) Zapiski naučn. Sem. Leningrad. Otd. mat. Inst. Steklova **91**, 5–24 (= J. Soviet Math. **17** (1981), 2083–2097)

Venkov, A.B. (1980): Zeros of ζ- and L- functions of imaginary quadratic fields and eigenvalues of $\mathbf{PSL}(2,\mathbb{Z})$-automorphic Laplacian. Sov. Math., Dokl. **21**, 109–112 (= Dokl. Akad. Nauk SSSR **250** (1980), 528–531)

Venkov, A.B. (1981): Spectral theory of automorphic functions. Proc. Steklov Inst. Math. 1982, Issue 4, 163 pp. (=Trudy Mat. Inst. Steklova, Tom **153** (1981))

Venkov, A.B. (1990a): On essentially cuspidal noncongruence subgroups of the group $\mathbf{PSL}(2,\mathbb{Z})$. J. Funct. Anal. **92**, 1–7

Venkov, A.B. (1990b): Spectral theory of automorphic functions and its applications. Dordrecht Boston London: Kluwer Academic Publishers

Venkov, A.B., Kalinin, V.L., Faddeev., L.D. (1977): A nonarithmetic derivation of the Selberg trace formula. J. Soviet Math. **8**, 171–199 (= Zapiski naučn. Sem. Leningrad. Otd. mat. Inst. Steklova **37** (1973), 5-42)

Venkov, A.B., Skringanov, M.M. (1979): Weyl's formula in the spectral theory of automorphic functions. Funct. Anal. Appl. **13**, 54–55 (= Funkts. Anal. Prilozh. **13** (1979), 67–68)

Venkov, A.B., Vinogradov, A.I. (1978): The asymptotic distribution of the norms of hyperbolic classes and spectral characteristics of cusp forms of weight zero for a Fuchsian group. Sov. Math., Dokl. **19**, 1545–1548 (= Dokl. Akad. Nauk SSSR **243** (1978), 1373–1376)

de Verdière, Y.C. (1977): Quasi-modes sur les variétés Riemanniennes. Invent. Math. **43**, 15–77

de Verdière, Y.C. (1981): Une nouvelle démonstration du prolongement méromorphe des séries d'Eisenstein. C. R. Acad. Sci., Paris, Sér. A **293**, 361–363

de Verdière, Y.C. (1982): Pseudo-Laplaciens I. Ann. Inst. Fourier **32**, 275–286

de Verdière, Y.C. (1983): Pseudo-Laplaciens II. Ann. Inst. Fourier **33**, 87–113

Vignéras, M.-F. (1978): Exemples de sous-groupes discrets non conjugués de $\mathbf{PSL}(2,\mathbb{R})$ qui ont même fonction zêta de Selberg. C.R. Acad. Sci., Paris, Sér. A **287**, 47–49

Vignéras, M.-F. (1979): L'équation fonctionnelle de la fonction zêta de Selberg du groupe modulaire $\mathbf{PSL}(2,\mathbb{Z})$. Astérisque **61**, 235–249

Vignéras, M.-F. (1980a): Arithmétique des algèbres de quaternions. Lect. Notes Math. **800**. Springer, Berlin Heidelberg New York

Vignéras, M.-F. (1980b): Variétés riemanniennes isospectrales et non isométriqes. Ann. Math., II. Ser. **112**, 21–32

Vignéras, M.-F. (1985): Représentations Galoisiennes paires. Glasgow Math. J. **27**, 223–237

Vinberg, E.B. (1967): Discrete groups generated by reflections in Lobačevskii spaces. Math. USSR, Sb. **1**, 429–444 (= Mat. Sb. Nov. Ser. **72** (1967), 471–488)

Vinberg, E.B. (1984): The absence of crystallographic groups of reflections in Lobachevsky spaces of large dimensions. Trans. Moscow Math. Soc. **47**

Vinberg, E.B. (1985): Hyperbolic reflection groups. Russian Math. Surveys **40**, 31–75

Vinberg, E.B.. (1987): Reflection subgroups in Bianchi groups. Problems in Group Theory and Homological Aspects, Jaroslawl, 121–126 (Engl. transl. in Sel. Math. Sov. **9**, (1990))

Vinberg, E.B., Shvartsman, O.V. (1991): Discrete groups of motions of spaces of constant curvature. Encyclopaedia of Math. Sciences **29**, 139–249

Vishik, S.M. (1976): Analogs of Selberg's ζ-function. Funct. Anal. Appl. **9**, 256–257 (= Funkts. Anal. Prilozh. **9** (1975), 85–86)

Vogtmann, K. (1985): Rational homology of Bianchi groups. Math. Ann. **272**, 399–419

van der Waerden, B. (1935): Gruppen von linearen Transformationen. Ergebnisse der Mathematik und ihrer Grenzgebiete **4**, Springer Verlag, Berlin. Reprint: Chelsea Publishing Company (1948)

van der Waerden, B. (1968): Studien zur Theorie der quadratischen Formen. (Herausgegeben von B.L. van der Waerden und H. Gross). Basel, Birkhäuser

Wallach, N. R. (1973): Harmonic analysis on homogeneous spaces. Dekker, New York

Wallach, N. R. (1976): On the Selberg trace formula in the case of compact quotient. Bull. Am. Math. Soc. **82**, 171–195

Wang, H. C. (1969): On discrete nilpotent subgroups of Lie groups. J. Differ. Geom. **3**, 481–492

Wang, H. C. (1972): Topics in totally discontinuous groups. (Symmetric spaces, ed. by Boothby-Weiss), New York

Warner, F.W. (1970): Foundations of Differentiable Manifolds and Lie Groups. Scott, Foresman, Glenview London

Watson, G. N. (1966): The Theory of Bessel Functions. Cambridge at the University Press, Cambridge

Weidmann, J.: (1979): Lineare Operatoren in Hilbert-Räumen. B.G. Teubner Verlag, Stuttgart (= Linear operators on Hilbert spaces. Springer, Berlin Heidelberg New York, (1979))

Weil, A. (1961): Adèles and algebraic groups. (Notes by M. Demazure and T. Ono.) The Institute of Advanced Study, Princeton

Weil, A. (1967): Basic Number Theory. Grundl. Math. Wiss. **144**, Springer, Berlin Heidelberg New York

Weil, A. (1968): Sur une formule classique. J. Math. Soc. Japan **20**, 400–402

Weil, A. (1976): Elliptic functions according to Eisenstein and Kronecker. Springer, Berlin Heidelberg New York

Whittaker E. T., Watson G. N. (1927): A course of modern analysis. Fourth edition, Cambridge

Widder, D.V. (1941): The Laplace Transform. Princeton Univ. Press, Princeton N.J.

Wielenberg, N. J. (1977): Discrete Moebius groups: Fundamental polyhedra and convergence. Amer. J. Math. **99**, 861–877

Wielenberg, N. J. (1981): Hyperbolic 3-manifolds which share a fundamental polyhedron. (Riemann surfaces and related topics, Proc. 1978 Stony Brook Conf.), Ann. Math. Stud. **97**, 505–517

Wiener, N. (1923): Differential space. J. Math. and Phys. **2**, 131–174

Wiener, N. (1933): The Fourier integral and certain of its applications. Cambridge University Press, Cambridge

Wolf, J.A. (1977): Spaces of constant curvature. Fourth edition, Publish or Perish, Inc., Berkeley

Woodruff, W.M. (1967): Singular points of the fundamental domains for the groups of Bianchi. Ph.D. Thesis, University of Arizona

Wolpert, S. (1979): The length spectra as moduli for compact Riemann surfaces. Ann. Math., II Ser. **109**, 323–351

Wolpert, S. (1992): Spectral limits for hyperbolic surfaces, I, II. Inv. Math. **108**, 67–89, 91–129

Wolpert, S. (1994): Disappearance of cusp forms in special families. Ann. Math., II. Ser. **139**, 239–291

Zagier, D. (1981a): Zetafunktionen und quadratische Körper. Springer, Berlin Heidelberg New York

Zagier, D. (1981b): The Rankin–Selberg method for automorphic functions which are not of rapid decay. J. Fac. Sci. Univ. Tokyo I A **28**, 415–437

Zimmermann, B. (1995): Cyclic branched coverings of hyperbolic links. Topology Appl. **65**, 287–294

Zimmert, R. (1973): Zur $\mathbf{SL}_2$ der ganzen Zahlen eines imaginärquadratischen Zahlkörpers. Invent. Math. **19**, 73–82

Zucker, I. J. (1977): The evaluation in terms of Γ-functions of the periods of elliptic curves admitting complex multiplication. Math. Proc. Camb. Philos. Soc. **82**, 111–118

Subject Index

Print and Binding: Kösel GmbH & Co., Kempten